Exponential and Logarithmic Functions

- $e = \lim\limits_{n \to \infty} \left(1 + \dfrac{1}{n}\right)^n \approx 2.718$.

- **The exponential function** is $f(x) = e^x$.

- $y = \log_a x$ if and only if $x = a^y$.

- The **natural logarithm** of x is
  $\ln x = \log_e x$.

- $a^{\log_a x} = x;\; e^{\ln x} = x$.
  $\log_a (a^x) = x;\; \ln e^x = x$.

- $\log_a a = 1;\; \ln e = 1$.
  $\log_a 1 = 0;\; \ln 1 = 0$.

- $\log_a xy = \log_a x + \log_a y$.

- $\log_a \dfrac{x}{y} = \log_a x - \log_a y$.

- $\log_a \dfrac{1}{x} = -\log_a x$.

- $\log_a x^k = k \log_a x$.

- $\log_a x = \dfrac{\ln x}{\ln a}$.

- $\dfrac{d}{dx} [e^x] = e^x$.

- $\dfrac{d}{dx} [e^{g(x)}] = e^{g(x)} g'(x)$.

- $\dfrac{d}{dx} [e^u] = e^u \dfrac{du}{dx}$.

- $\dfrac{d}{dx} [\ln x] = \dfrac{1}{x}$ (assuming $x > 0$).

- $\dfrac{d}{dx} [\ln g(x)] = \dfrac{1}{g(x)} g'(x)$
  (assuming $g(x) > 0$).

- If $Q = Q_0 e^{kt}$, then $\dfrac{dQ}{dt} = kQ$.

Antiderivatives

- $F(x)$ is an antiderivative of $f(x)$ if
  $F'(x) = f(x)$ for all x in the domain of f.

- If $F(x)$ is any antiderivative of $f(x)$,
  then every antiderivative of $f(x)$ is
  given by
  $$\int f(x)\, dx = F(x) + C,$$
  where C is an arbitrary constant.

- Integration by substitution:
  $$\int g(f(x)) f'(x)\, dx = \int g(u)\, du.$$

APPLIED
CALCULUS

APPLIED
CALCULUS

BERNARD KOLMAN
Drexel University

CHARLES G. DENLINGER
Millersville University of Pennsylvania

Harcourt Brace Jovanovich, Publishers
and its subsidiary, Academic Press

San Diego   New York   Chicago   Austin   Washington, D.C.
London   Sydney   Tokyo   Toronto

To the memory of my father
B. K.

To Gloria, Joe, and Alison
C. G. D.

Cover: *Los Angeles* by Vasarely. Private Collection.

Copyright © 1989 by Harcourt Brace Jovanovich, Inc.

ISBN: 0-15-502903-7

Library of Congress Catalog Card Number: 87-83022

Printed in the United States of America

|||||||||||||||||||||||||||||||||||||||||||||||||||||||||||||PREFACE

Applied Calculus presents an elementary introduction to calculus for students majoring in business, economics, and the management, life, and social sciences. Calculus plays an important role in these areas; it is the mathematics of change and we, of course, live in a constantly changing world. The goal of this course is to equip students of these disciplines with the powerful analytical tools of calculus. This book is an expansion of the authors' *Calculus for the Management, Life, and Social Sciences*, Second Edition. It contains the following additional material: Section 8.6, Double Integrals; Chapter 10, Differential Equations; Chapter 11, Calculus in Probability Theory; and Chapter 12, Taylor Polynomials and Infinite Series. Both authors and publisher have been gratified by the wide acceptance of the first two editions of the brief version of this book. The new material has been written in keeping with our main objective: *to write a textbook that will help the instructor teach the material and that the student will find readable.*

|||||||||||| Features

Review of Algebra Appendix A contains a brief review of the necessary algebra material. This appendix can be used, in whole or in part, as the need arises.

The Diagnostic Test, which appears at the beginning of Appendix A, can be used as a class test or as a self-test. Each question is marked with the section in Appendix A that discusses the material covered by the question. This will allow the student to pinpoint weak areas in his or her background and to take appropriate remedial action.

Presentation The approach used in this book is informal, intuitive, and geometric. The amount of theoretical material covered has been minimized. The key ideas are carefully explained and each is fully illustrated with worked-out examples. We have not hesitated to omit some of the difficult proofs in certain discussions, but we have described the overall situation. Figures are used extensively to illustrate the ideas being presented. Whenever possible, procedures are described in an easy-to-follow, step-by-step manner.

Exercises Each exercise set contains routine exercises as well as plenty of applied problems drawn from business, economics, the social sciences, psychology, medicine, the life sciences, and ecology. The exercises within each section have been carefully graded, progressing naturally from easier to more difficult, and arranged in an order consistent with the textual exposition. Answers to all odd-numbered exercises and to all Chapter Test questions appear in the back of the book. Answers to all even-numbered exercises appear in the Instructor's Manual, which also includes a Test Bank (two tests per chapter, with solutions).

End-of-chapter Material Every chapter contains a summary of key ideas for review, a set of review exercises, and a chapter test.

Application A great many realistic and relevant applications have been included, both in the illustrative examples and in the exercises, from the areas of business, economics, the social sciences, psychology, medicine, the life sciences, and ecology. An entire section is devoted to applications of the derivative in business and economics. The wide range of applications can be seen by scanning the list on pp. vii–viii. The authors would like to call particular attention to the rich array of applications of differential equations given in Chapter 10.

Course Length There is more than enough material in this book for a full-year course. Some of the applications can also be used as individual student projects. A brief edition, entitled *Calculus for the Management, Life, and Social Sciences*, Second Edition, is available for a one-semester or one- or two-quarter course. It omits Chapters 10, 11, and 12—Differential Equations, Calculus in Probability Theory Taylor Polynomials and Infinite Series—and Section 8.6, Double Integrals.

|||||||||||| Acknowledgments

We would like to thank the following reviewers, whose suggestions, comments, and criticisms greatly improved the manuscript: Dennis DeTurck, University of Pennsylvania; Garret Etgen, University of Houston; Jerry W. Ferry, University of Arkansas; Robert L. Higgins, QUANTICS, Inc., and Drexel University; Stanley Lukawecki, Clemson University; James Snow, Whatcom Community College; Carol Achs, Mesa Community College; Kenneth J. Brown, University of Wisconsin-Stevens Point; Leon Gerber, St. John's University; Charles A. Loch, Duquesne University; Laurence Maher, North Texas State University; Roger J. Mergener, Moraine Valley Community College; Mark A. Miller, Mohawk Valley Community College; Roger B. Nelson, Ball State University; Robert G. Russell, West Valley College; and Susan Friedman, Bernard Baruch College of the City University of New York. We also thank Rajni Singal and Crystal Meck Evans for solving, and John Edenhofner, Jr., for checking, the exercises.

Finally, we wish to thank Michael Johnson, Thomas Thompson, Nancy Evans, and Robert Shelburne of Harcourt Brace Jovanovich for their support, encouragement, creative imagination, and unfailing cooperation during the conception, design, production, and marketing phases of this book. To all the above mentioned goes a sincere expression of gratitude.

B. K.
C. G. D.

IIIIIIIIIIII **Ancillaries**

Study Guide   A Students' Study Guide is available, containing key ideas for review of each section and completely detailed solutions of more than one-third of the odd-numbered exercises.

Instructor's   An Instructor's Manual contains the answers to all even-numbered exercises, as
Manual   well as a test bank containing two tests per chapter, with solutions.

SELECTED APPLICATIONS

Ecology

Economics

Energy

Medicine

Psychology

Bursts of knowledge: 102
Decrease in reaction time due to
    weightlessness: 435
Limits in learning: 569, 581
Performance and amount of training: 98,
    117, 256, 283, 326, 389
Probability of receiving a Ph.D.: 651

Rate of learning and amount of training:
    136, 150, 168, 217, 262, 283, 302
Reaction time: 650, 665
Speech understanding by dolphins: 192
Stimulus-response relationship: 562
Time to get out of a maze: 41, 648
Time to solve a puzzle: 503
Weber-Fecher law: 562

Social Science

Affirmative action: 595
Animal population with harvesting: 563
Arms race model: 622, 626, 631
Average population: 122
Latest glacial age: 310
Population growth: 48, 182, 283, 291,
    304, 306, 312, 328, 343, 555
Population growth with migration: 562,
    589, 595, 629

Radiocarbon dating: 310, 312
Spread of a fad: 584
Spread of a rumor: 582
"War-woods" model: 584
Work-incentive programs and prison
    expenses: 254
Work-incentive programs, remedial
    education, and public assistance: 502

Miscellaneous

Air conditioning: 563
Atmospheric pressure: 283, 291, 314
Autocatalytic reaction: 584, 630
Average air temperature: 402, 530
Average and instantaneous velocity: 110,
    118, 135, 143, 182
Average height of a tent: 498
Average number of units produced: 498
Average salt content in a lake: 503
Average wait in a dentist's office: 650
Bimolecular chemical reaction: 564, 630
Crop yield: 464, 481
Distance traveled by a bouncing ball:
    739, 775
Distance traveled by an object: 3, 264,
    328, 355, 383, 390, 537
Earthquake intensity: 292
Error estimate: 484, 486
Exponential decay: 307–309
Fine arts grant: 739
Fourth class mail: 248
Grades on a final examination: 687
Length of a telephone call: 651

Life expectancy of equipment: 665
Lifetime of a lightbulb: 687, 690
Light absorption: 283
Mixture problems: 556, 590, 595, 629,
    631
Newton's law of temperature change: 582
Parasites and hosts: 559
Predator-prey interaction: 520, 528, 557,
    563, 625
Production of gold: 390
Quality control: 679
Radioactive decay: 309, 312, 315, 339
Rate of change of distance: 183
Rate of change of temperature: 530,
    582, 583
Rate of reaction: 564
Romeo and Juliet: 626, 627
Skydiving: 571
Symbiotic relations: 630, 631
Temperature of a cold drink: 632
Temperature of a roast: 582
Terminal velocity: 570, 583, 630
Time to pay a bill: 691
Unrestricted growth and decay: 552

||||||||||||||||||||||||||||||||||||||||||||||||||CONTENTS

Appendix B: Tables

Answers to Odd-numbered Exercises and Chapter Tests

APPLIED
CALCULUS

FUNCTIONS

Note   Chapter 0 and Section 1.1 of the First Edition appear as Appendix A in this edition.

This chapter deals with functions and their applications. We thereby study one of the central ideas in mathematics. We shall start by carefully examining the notion of "function" and the idea of the "graph" of a function. We consider some practical examples. We then focus on "linear functions"—functions whose graphs are straight lines. After developing some important applications, we finish the chapter by discussing intersections of graphs. All of this is precalculus material, but very important to an understanding of the subject and its applications.

1.1 ‖‖‖‖‖‖‖‖ FUNCTIONS

Many applications in the physical sciences, the biological sciences, economics, business, and the social sciences often focus on the manner in which one quantity depends upon another. For example, suppose that a manufacturer's total revenue R (in dollars) received from selling x units of a product is given by the equation

$$R = 200x + \frac{x^2}{10}. \tag{1}$$

This equation describes R as a function of x. As we vary x, we obtain different values of R. If $x = 10$, then

$$R = 200(10) + \frac{(10)^2}{10} = 2000 + \frac{100}{10} = 2010.$$

But if $x = 20$, then

$$R = 200(20) + \frac{(20)^2}{10} = 4040.$$

The quantities R and x in Equation (1) are called **variables.** As soon as we specify a value of x, Equation (1) gives R a unique value. Consequently, we call x the **independent variable** and R the **dependent variable.**

In general, a **function** is a rule, or formula, that determines a unique value of one variable once the value of another variable has been specified. We shall often denote the independent variable by x and the dependent variable by y, although other letters may also be used.

||||||||||| Functional Notation

Suppose f is a function relating x and y; that is, f is a rule for associating a unique value of y with each given value of x. We denote the value of y by $f(x)$, and we write

$$y = f(x)$$

(Read "y equals f of x"). Thus, if $y = 5$ when $x = 2$, we merely write

$$5 = f(2).$$

Writing Equation (1) in functional notation,

$$R(x) = 200x + \frac{x^2}{10},$$

we have $R(10) = 2010$ and $R(20) = 4040$.

EXAMPLE 1 Suppose f is given by the formula

$$f(x) = x + 5.$$

Then f associates the number $x + 5$ with the given number x. Thus,

$$f(0) = 0 + 5 = 5, \qquad f(-1) = -1 + 5 = 4, \qquad f(2) = 2 + 5 = 7.$$

The function f associates the unique value $y = 5$ with $x = 0$; $y = 4$ with $x = -1$; and $y = 7$ with $x = 2$.

EXAMPLE 2 Let the function f be defined by

$$f(t) = t^2 - 1.$$

Find
(a) $f(2)$ (b) $f(-2)$ (c) $f(a)$ (d) $f(a + 1)$ (e) $f(a + h)$.

Solution
(a) $f(2) = 2^2 - 1 = 4 - 1 = 3$
(b) $f(-2) = (-2)^2 - 1 = 4 - 1 = 3$
(c) $f(a) = a^2 - 1$
(d) $f(a + 1) = (a + 1)^2 - 1 = a^2 + 2a + 1 - 1 = a^2 + 2a$
(e) $f(a + h) = (a + h)^2 - 1 = a^2 + 2ah + h^2 - 1.$

A function need not always be denoted by the letter f; any letter, including g, h, F, R, etc., will do. Similarly, we may represent the dependent variable by F, C, r, s, t, u, v, or any other letter. For example, the Celsius, or centigrade, temperature C and the Fahrenheit temperature F are related by

$$F = \tfrac{9}{5}C + 32. \qquad (2)$$

Here C is the independent variable and F is the dependent variable.

If Equation (2) is solved for C, we then express C as a function of F (see Exercise 5), with F becoming the independent variable and C the dependent variable. ∎

EXAMPLE 3
(Motion)
Suppose that the distance $s(t)$ (in feet) traveled in t seconds by an object starting from rest along a straight line is given by

$$s(t) = t + 2\sqrt{t}.$$

Find the distance traveled after (a) 9 seconds and (b) 36 seconds.

Solution (a) For $t = 9$, we have

$$s(9) = 9 + 2\sqrt{9} = 9 + 2 \cdot 3 = 15 \text{ feet}.$$

(b) For $t = 36$, we have

$$s(36) = 36 + 2\sqrt{36} = 36 + 2 \cdot 6 = 48 \text{ feet}.$$ ∎

The Domain of a Function

The values of the independent variable at which a function is defined and yields a real number form a set of real numbers called the **domain** of the function. For the function $s(t) = t + 2\sqrt{t}$ of Example 3, the independent variable t must be nonnegative since we are required to take its square root. Thus, the domain of the function $s(t)$ consists of all real numbers $t \geq 0$; that is, the interval $[0, \infty)$.

EXAMPLE 4 Let the function g be defined by

$$g(x) = \frac{1}{x - 2}. \qquad (3)$$

Find
  (a) $g(4)$ (b) $g(-3)$ (c) $g(0)$ (d) $g(2)$.

Also, find the domain of g.

Solution (a) $g(4) = \dfrac{1}{4 - 2} = \dfrac{1}{2}$

(b) $g(-3) = \dfrac{1}{-3 - 2} = -\dfrac{1}{5}$

(c) $g(0) = \dfrac{1}{0-2} = -\dfrac{1}{2}$

(d) $g(2)$ cannot be computed because $g(2) = 1/(2-2) = 1/0$, which is not a real number because we cannot divide by zero. Thus, g is not defined at $x = 2$.

Since 2 is the only value of x that does not yield a real number $g(x)$ in Equation (3), the domain of g consists of all real numbers except 2. ∎

EXAMPLE 5   Find the domain of the following functions:

(a) $f(x) = \dfrac{1}{x^2 - 9}$

(b) $g(u) = \sqrt{u - 5}$

Solution   (a) Since $f(3)$ and $f(-3)$ are not defined, the domain of f consists of all real numbers except 3 and -3.

(b) Since a negative number has no real square root, we must have $u - 5 \geq 0$. Thus, the domain of g consists of all real numbers that are greater than or equal to 5. ∎

A function f can also be viewed as an input/output machine that accepts as *input* any number x in its domain and produces as *output* the unique value $f(x)$ (see Figure 1).

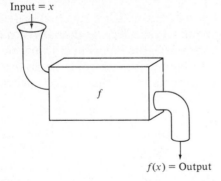

Input = x

f

$f(x) =$ Output

Figure 1

The rule, or equation, for f is "in the machine." For example, if f is defined by

$$f(x) = x^2,$$

then f is the function that associates to each number its square. The squaring "machine" is shown in Figure 2. If we feed the machine a value of x, the internal mechanism calculates its square, x^2, and produces the output $f(x) = x^2$.

Figure 2

Squaring machine

$f(x) = x^2$

Observe that a function must assign a unique number to each value of its domain. A function may assign the same number to two *different* values in the domain, however. For example, the function $f(x) = x^2$ assigns the number 4 to both 2 and -2, since $f(2) = 2^2 = 4$ and $f(-2) = (-2)^2 = 4$. Using the input/output analogy, an input to a function cannot produce two different outputs, but two different inputs may produce the same output.

Our definition of a function distinguishes between the function f and the *value* $f(x)$ of the function at the number x in its domain. But it is common practice, for convenience, to refer to $f(x)$ also as the function, and we shall do so at places in this book.

Graphing Functions

> The **graph** of a function f is the graph of the equation $y = f(x)$; that is, the graph of f consists of all points (x, y) whose coordinates satisfy the equation $y = f(x)$.
>
> Alternatively, the graph of f consists of all points $(x, f(x))$ such that x belongs to the domain of f.

Students who need a review of the coordinate system and graphing are advised to see Section A.8 (Appendix A).

EXAMPLE 6   Sketch the graph of the function

$$f(x) = 2x - 3.$$

Solution   We make a small table, plot the corresponding points, and connect them with a line or curve. The resulting graph is shown in Figure 3.

TABLE 1 $(y = 2x - 3)$

x	y
-2	-7
-1	-5
0	-3
1	-1
2	1
3	3
4	5

Figure 3

A function of the form

$$f(x) = ax + b,$$

such as in Example 6 (where $a = 2$, and $b = -3$), is a **first degree function** of x. As we shall prove in Section 1.2, the graph of a first degree function of x is always a straight line. For this reason such a function is called a **linear function.**

EXAMPLE 7   Sketch the graph of the function

$$g(x) = 2 - x^2.$$

Solution   We proceed as in Example 6. The points obtained in Table 2 are plotted and connected by a curve, as shown in Figure 5 on the following page.   ■

EXAMPLE 8   Sketch the graph of the function $f(x) = 3$.

Solution   The graph is shown in Figure 4. This function is called a **constant function** because it assigns the same value, 3, to each value of x.

Figure 4

TABLE 2 $(y = 2 - x^2)$

x	y
0	2
± 1	1
± 2	-2
± 3	-7
± 4	-14

$g(x) = 2 - x^2$

Figure 5

Warning Not every curve in the xy-plane is the graph of a function $y = f(x)$. For example, consider the curve in Figure 6 and suppose that it is the graph of $y = f(x)$ for a function f. Consider the value $x = a$. To find the corresponding value of y, we draw a vertical line through $x = a$. The y-coordinate of the point at which this line intersects the graph gives the value of y corresponding to $x = a$. In Figure 6 we see that the line intersects the graph at the points (a, b) and (a, c). Since these points lie on the graph, their coordinates satisfy the equation $y = f(x)$. Thus,

$$b = f(a) \quad \text{and} \quad c = f(a).$$

This result is impossible, however, if f is a function, because a function cannot associate two different y values to one value of x. Hence, the curve shown in Figure 6 is not the graph of *any* function f of x.

Figure 6

The preceding discussion leads to the following test for determining whether a given curve is the graph of a function.

The Vertical Line Test
If any vertical line intersects the curve at more than one point, then the curve is not the graph of any function of x; that is, we cannot express y as a *function* of x. If no vertical line intersects the curve at more than one point, then the curve is the graph of some function $y = f(x)$.

EXAMPLE 9 Which of the curves in Figure 7 are the graphs of functions of x?

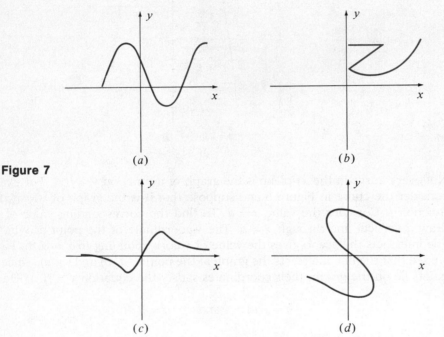

(a)

(b)

Figure 7

(c)

(d)

Solution Applying the vertical line test, we find that curves (a) and (c) are graphs of functions.   ■

1.1 EXERCISE SET ||||||||||||||||||||||||||||||||||||||||||||||||||||||||||||||||||||||||||||||||||

1. Consider the function f defined by
$$f(x) = -5x + 7.$$
Find
(a) $f(0)$ (f) $f(a + h)$
(b) $f(2)$ (g) $f(-x)$
(c) $f(-3)$ (h) $-f(x)$
(d) $f(a)$ (i) $x + f(x)$
(e) $f(a + 1)$

2. Consider the function g defined by
$$g(x) = x^2 + 2x.$$
Find
(a) $g(0)$ (f) $g(3 + h)$
(b) $g(3)$ (g) $g(-x)$
(c) $g(-3)$ (h) $-g(x)$
(d) $g(a)$ (i) $g(x^2)$
(e) $g(a - 2)$

3. Consider the function F defined by

$$F(x) = \frac{x^2 + 1}{3x - 1}.$$

Find

(a) $F(1)$ (f) $F(a - 2)$
(b) $F(-2)$ (g) $F(-x)$
(c) $F(4)$ (h) $F(x^2)$

(d) $F(0)$ (i) $\dfrac{1}{F(x)}$

(e) $F(a)$

4. Consider the function r defined by

$$r(t) = \frac{t - 2}{t^2 + 2t - 3}.$$

Find

(a) $r(-1)$ (f) $r(-x)$
(b) $r(2)$ (g) $-r(x)$
(c) $r(0)$ (h) $r(a + 1)$
(d) $r(3)$ (i) $r(t^2)$
(e) $r(a)$

5. The formula relating Fahrenheit temperature F to Celsius temperature C is given by

$$F = \tfrac{9}{5}C + 32.$$

(a) Write C as a function of F.
(b) Using the function in (a), find the Celsius temperatures for the following Fahrenheit temperatures.
 (i)   4°F
 (ii) 0°F
 (iii) −10°F
 (iv) 32°F (water freezes at this temperature)
 (v)   98.6°F (normal body temperature)
 (vi) 212°F (water boils at this temperature)

6. An ounce of hamburger yields 7 grams of protein, while an ounce of soybeans yields 10 grams of protein. Suppose a person wanted to get an entire daily requirement of 70 grams of protein from a combination of hamburger and soybeans alone using H ounces of hamburger and S ounces of soybeans. Then

$$7H + 10S = 70.$$

(a) Express H as a function of S.
(b) Find H when $S = 0$, $S = 1$, $S = 5$, $S = 7$ (ounces).
(c) Express S as a function of H.
(d) Find S when $H = 0$, $H = 2$, $H = 5$, $H = 10$ (ounces).

In Exercises 7–16, specify the domain of each function.

7. $f(x) = 2x^2 + x - 3$

8. $g(t) = \dfrac{1}{t - 2}$

9. $h(x) = \sqrt{x - 1}$

10. $f(u) = \dfrac{1}{u^2 - 2u - 3}$

11. $f(x) = \dfrac{x - 2}{x + 1}$

12. $g(x) = \dfrac{x - 3}{x^2 - 2x - 8}$

13. $h(t) = \sqrt{t^2 + 4}$

14. $g(u) = \dfrac{u - 2}{\sqrt{u^2 + 4}}$

15. $f(x) = \sqrt{x^2 - 4}$

16. $g(x) = \sqrt{3x - 6}$

In Exercises 17–24, sketch the graph of the given function.

17. $f(x) = 3x - 4$

18. $f(x) = -2x + 4$

19. $f(x) = 2x^2 + 3$

20. $f(x) = x^2 - 4$

21. $f(x) = x^2 - 4x$

22. $f(x) = x^2 + 2x + 1$

23. $f(x) = \dfrac{1}{x}$

24. $f(x) = \dfrac{1}{x^2}$

In Exercises 25–30, determine which of the given curves are the graphs of a function of x.

25.

26.

27.

28.

29.

30.

1.2 |||||||| MORE ABOUT FUNCTIONS

Piecewise-Defined Functions

Thus far, our functions have all been specified by a single formula or equation. In many applications, however, a function may be defined by several formulas, each formula applying over a different portion of the domain. Such functions are said to be defined "piecewise."

EXAMPLE 1
(The Absolute Value Function)

Sketch the graph of the function

$$f(x) = |x|.$$

Solution The graph of f is the graph of the equation $y = |x|$, which may also be written

$$y = \begin{cases} x & \text{if } x \geq 0 \\ -x & \text{if } x < 0. \end{cases}$$

We sketch the graph in two stages. If $x \geq 0$, we have Table 1, whose graph is the half line that bisects the first quadrant of Figure 6. If $x < 0$, we have Table 2, whose graph is the half line that bisects the second quadrant of Figure 1. Thus, the graph of f consists of the two half-lines meeting at the origin, shown in Figure 1.

TABLE 1

x	0	1	2
$f(x) = x$	0	1	2

TABLE 2

x	0	-1	-2
$f(x) = -x$	0	1	2

Figure 1

$f(x) = |x|$

EXAMPLE 2   Suppose that the annual dues of a union are as follows:

Employee's Annual Salary	Annual Dues
Less than $8000	$60
$8000 or more, but less than $15,000	$60 + 1% of the salary in excess of $8000
$15,000 or more	$130 + 2% of the salary in excess of $15,000

The relation between the annual dues d and the salary s can be expressed as

$$d(s) = \begin{cases} 60 & \text{if } s < 8000 \\ 60 + 0.01(s - 8000) & \text{if } 8000 \leq s < 15{,}000 \\ 130 + 0.02(s - 15{,}000) & \text{if } s \geq 15{,}000. \end{cases}$$

Thus, d is described by three equations. Of course, the value of s determines the appropriate equation for d. For example,

$$d(7000) = 60$$
$$d(10{,}000) = 60 + 0.01(10{,}000 - 8000) = 80$$
$$d(18{,}000) = 130 + 0.02(18{,}000 - 15{,}000) = 190.$$

Similarly, we obtain the values in Table 3.

TABLE 3

s	7000	10,000	12,000	15,000	18,000	20,000
$d(s)$	60	80	100	130	190	230

The graph of d is shown in Figure 2. Note that the units on the horizontal and vertical axes are of different lengths.

Figure 2

$$d(s) = \begin{cases} 60 & \text{if } s < 8000 \\ 60 + 0.01\,(s - 8000) & \text{if } 8000 \leq s < 15{,}000 \\ 130 + 0.02\,(s - 15{,}000) & \text{if } s \geq 15{,}000 \end{cases}$$

Salary s (in dollars)

EXAMPLE 3   Sketch the graph of the function defined by

$$f(x) = \begin{cases} x & \text{if } 0 \leq x \leq 1 \\ x + 2 & \text{if } 1 < x \leq 4. \end{cases}$$

Solution   We construct Table 4 and then draw the graph shown in Figure 3. An open circle marks the point $(1, 3)$ to indicate that it does not belong to the graph. A solid circle at point $(1, 1)$ indicates that it does belong to the graph. Notice that this graph has a jump, or gap at $x = 1$.

TABLE 4

x	0	$\frac{1}{2}$	1	$\frac{3}{2}$	2	3	4
$f(x)$	0	$\frac{1}{2}$	1	$\frac{7}{2}$	4	5	6

Figure 3

$$f(x) = \begin{cases} x & \text{if } 0 \le x \le 1 \\ x + 2 & \text{if } 1 < x \le 4 \end{cases}$$

■

||||||||||||| **Composite Functions**

If f and g are two given functions, we can define a new function $f \circ g$ by

$$(f \circ g)(x) = f(g(x))$$

which is called the **composite of f and g**. The domain of the function $f \circ g$ consists of all numbers x in the domain of g such that the number $g(x)$ is in the domain of f. To find $f(g(x))$, we merely substitute $g(x)$ for every occurrence of the variable x in the rule for f.

EXAMPLE 4   Let $f(x) = x^3$ and $g(x) = 1 + x^2$.
  (a) Find $f(g(3))$.
  (b) Find $f(g(x))$, or $(f \circ g)(x)$.

Solution   (a) We first compute

$$g(3) = 1 + 3^2 = 10.$$

Now we substitute $g(3) = 10$ for every occurrence of x in $f(x)$, and obtain

$$f(g(3)) = f(10) = (10)^3 = 1000.$$

(b) Substituting $g(x) = 1 + x^2$ for every occurrence of x in $f(x)$, we obtain

$$f(g(x)) = f(1 + x^2) = (1 + x^2)^3.$$

Equivalently we can write

$$(f \circ g)(x) = (1 + x^2)^3.$$

Note that we can now solve (a) using the result of (b),

$$(f \circ g)(3) = (1 + 3^2)^3 = 10^3 = 1000.$$

■

EXAMPLE 5 Let $f(x)$ and $g(x)$ be the same as in Example 4.
(a) Find $g(f(3))$.
(b) Find $g(f(x))$, or $(g \circ f)(x)$.

Solution (a) We first compute

$$f(3) = 3^3 = 27.$$

Now we substitute $f(3) = 27$ for every occurrence of x in $g(x)$ and obtain

$$g(f(3)) = 1 + (27)^2 = 730.$$

(b) Substituting $f(x) = x^3$ for every occurrence of x in $g(x)$, we obtain

$$g(f(x)) = 1 + (x^3)^2 = 1 + x^6,$$

or $$(g \circ f)(x) = 1 + x^6. \qquad \blacksquare$$

In general, $f(g(x))$ and $g(f(x))$ are different functions; that is,

$$\boxed{\text{in general, although not always,} \\ f \circ g \neq g \circ f.}$$

This is demonstrated in Examples 4 and 5, where

$$(f \circ g)(x) = (1 + x^2)^3,$$

while

$$(g \circ f)(x) = 1 + x^6.$$

These are not the same; in particular, $(f \circ g)(3) = 1000$, while $(g \circ f)(3) = 730$.

EXAMPLE 6 If $f(x) = \dfrac{1}{x}$ and $g(x) = 3x + 2$, then $(f \circ g)(x) = \dfrac{1}{3x + 2}$, while $(g \circ f)(x) = \dfrac{3}{x} + 2$.

The purpose of introducing the notion of composite functions will be clear later, especially in Section 3.5. It will often be useful to express a complicated function as a composite of two or more simpler functions.

EXAMPLE 7 Write each of the following functions as a composite of two simpler functions.

(a) $h(x) = (1 + x^2)^{100}$ (b) $h(x) = \sqrt{x + x^2}$ (c) $h(x) = \left(\dfrac{x^2}{x^2 - 5}\right)^{1/3}$

Solution   (a) We see that $h(x) = u^{100}$, where $u = 1 + x^2$. Thus, we let $f(u) = u^{100}$ and $g(x) = 1 + x^2$. Then

$$h(x) = f(1 + x^2) = f(g(x)) = (f \circ g)(x).$$

(b) We see that $h(x) = \sqrt{u}$, where $u = x + x^2$. Thus, we let $f(u) = \sqrt{u}$ and $g(x) = x + x^2$. Then

$$h(x) = f(x + x^2) = f(g(x)) = (f \circ g)(x).$$

(c) Here, $h(x) = u^{1/3}$, where $u = \dfrac{x^2}{x^2 - 5}$. We let $f(u) = u^{1/3}$ and $g(x) = \dfrac{x^2}{x^2 - 5}$. Then

$$h(x) = f\left(\frac{x^2}{x^2 - 5}\right) = f(g(x)) = (f \circ g)(x). \qquad \blacksquare$$

Reading and Interpreting Graphs of Functions

Recall that a function f is a rule that assigns to each input x in the domain of f a unique output $f(x)$. In the examples we have seen, the rule has been given by an equation. Frequently, however, we may be given only the *graph* of a function, and from this graph we may need to answer questions about the function itself.

EXAMPLE 8   Suppose the dollar value of a commodity, t days after some initial time, is given by the function $v = f(t)$ whose graph is shown in Figure 4.

Figure 4

(a) What is the domain of f?

(b) What is the value of the commodity at the beginning of the 10-day period? At the end?

(c) During the second day, is the value of the commodity rising or falling? What about the sixth day? The eighth day?

(d) If you owned the commodity on the third day, would you earn more by selling then or by waiting until later?

(e) When is the best time to sell this commodity?

(f) When is the best time to purchase this commodity?

Solution (a) Figure 4 shows the graph of f over the interval $0 \leq t \leq 10$. Thus the domain of f must include that interval. We cannot be sure, however, whether the domain includes more than this interval. In the absence of any additional information we can only assume from Figure 4 that the domain of f is [0, 10].

(b) At the beginning, the value is $v = f(0) = \$3000$ (see Figure 4). At the end, the value is $v = f(10) = \$6000$.

(c) During the second day, $1 < t \leq 2$, the graph "rises" as t increases from left to right. Thus, in the second day the value of the commodity is rising. During the sixth day, $5 < t \leq 6$, the graph "falls" as t increases, so the value of the commodity is falling. During the eighth day, the value is rising.

(d) After the third day, the value of the commodity will come down, but by the eighth day it will be back up again to rise even higher. Thus, it would not be wise to sell on the third day; you can earn more by waiting, provided you wait till after the eighth day.

(e) According to Figure 4, the best time to sell is at the end of the ninth day, for that is when the value is greatest ($7000).

(f) According to Figure 4, the best time to purchase the commodity is at the end of the sixth day, when the value is the lowest (about $2000). ∎

Mathematical Models

Problems in business, economics, and the physical and social sciences require step-wise solutions. The process used is often called **mathematical modeling.** A model of a situation is an idealized representation of the real-life problem under study. The model may involve simply a change in scaling as in the case of the hobbyist's HO railroad or the architect's display of a planned shopping complex.

An engineer often uses analogue models in which electrical properties take the place of mechanical properties because it is easier to work with them than with the real mechanical ones. For example, resetting a dial changes the mass of an object. In reality we would change the mass of the object by cutting off a part of it.

A **mathematical model** makes a statement of the problem in *mathematical* terms. *Functions, equations*, and *graphs* play key roles in the construction of the mathematical model.

Steps in the Modeling Process

Step 1. **Problem definition and formulation.** During this phase the problem is carefully defined and the desired objective in trying to solve the problem is established.

Step 2. **Model construction.** Once the problem has been carefully defined, an adequate mathematical model is sought. Variables are introduced to represent pertinent quantities, and functions or equations are used to express relationships that exist between the quantities. Graphs and additional data may also be used.

Step 3. **Model solution.** Solving the mathematical model may be an easy, moderate, or impossible task. If the model admits no exact solution, it may be appropriate to try an approximate solution or it may be necessary to reexamine and revise the model. Below we shall examine several models in business and economics. Later we shall introduce others.

Step 4. **Model evaluation.** Once we have the solution to the model, we must evaluate it to determine whether it is realistic and workable.

Step 5. **Implementation of the solution.** At this point we implement the solution to the problem. Often the story does not end here, however. Since humans are always striving to improve their current situation, we may soon find that a better solution is desirable, and so the cycle starts all over again.

||||||||||||| **Functions in Business and Economics**

Many useful and important functions occur in business and economics. We shall now present three such functions, which are of primary concern to a manufacturer.

One of the main problems faced by a manufacturer is that of determining the **level of production;** that is, how many units x of the product should be manufactured during a fixed time period, such as a day, week, month, and so on. In some businesses the variable x can take on only integer values; for example, an automobile manufacturer cannot manufacture 720.4 cars. Often, however, the variable x can take on any real number. For example, it is certainly possible to manufacture 7.87 million barrels of oil or 2.6 tons of steel. During a given time period, the manufacturer is concerned with the following three dependent variables:

> $C(x)$ = total cost of producing x units of the product;
> C is the **cost function.**
>
> $R(x)$ = total revenue received from selling x units of the product;
> R is the **revenue function.**
>
> $P(x)$ = total profit derived from selling x units of the product;
> P is the **profit function.**

The total profit is the difference between total revenue and total cost, so that if all the units that are manufactured are sold, then

$$P(x) = R(x) - C(x). \tag{1}$$

Let us examine the cost function C closely. Even if no items were produced (the level of production is kept at zero), certain costs, such as insurance, rent, and administrative salaries, would be incurred. Costs of this type, which are incurred when no items are produced, are called **fixed costs** and are denoted by $F(x)$. The costs that vary with the level of production are called **variable costs** and are denoted by $V(x)$. Variable costs include the cost of materials, labor, and transportation of the manufactured product. Hence, we may write

$$C(x) = F(x) + V(x). \tag{2}$$
$$\underset{\text{fixed}}{} \underset{\text{costs}}{} \underset{\text{variable}}{} \underset{\text{costs}}{}$$

EXAMPLE 9   Consider a manufacturer whose fixed cost $F(x)$ is $4000 and whose variable cost (in dollars) is given by $V(x) = x^2 + 4x$ when the level of production is x units per day. The cost function C (in dollars) is

$$C(x) = 4000 + x^2 + 4x.$$

If $x = 50$ units per day, then the total daily cost is

$$C(50) = 4000 + V(50) = 4000 + 2700$$
$$= \$6700.$$

EXAMPLE 10   Consider a manufacturer of animal feed whose fixed cost is $2000 per week. It costs $0.50 to produce 1 kilogram of feed. Write the cost function.

Solution   The total cost of producing x kilograms of feed is given by

$$C(x) = 2000 + 0.50x.$$

In Figure 5 we have plotted the fixed-cost and the variable-cost functions as dashed lines. The solid line represents total cost.

x	$F(x)$	$V(x)$	$C(x)$
0	2000	0	2000
2000	2000	1000	3000
10,000	2000	5000	7000
14,000	2000	7000	9000
16,000	2000	8000	10,000

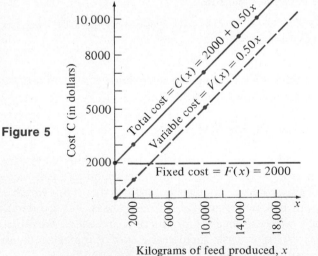

Figure 5

EXAMPLE 11   Suppose the feed manufacturer of Example 10 sells the feed for $0.75 per kilogram. Write the revenue and profit functions. Assume that all feed produced is sold.

Solution   The revenue received for x kilograms of feed is given by

$$R(x) = 0.75x$$

and the profit is [Equation (1)]

$$P(x) = R(x) - C(x)$$
$$= 0.75x - (2000 + 0.50x)$$
$$= 0.25x - 2000.$$ ∎

|||||||||||| **Cost Function Curves**

The total cost $C(x)$ of manufacturing x units of a product does not always increase proportionately with x. The cost per unit decreases as a result of mass production and the discounted cost of raw materials bought in large quantities. As x continues to increase, however, the cost per unit will increase after a certain point, due to payment of overtime wages and other inefficiencies intended to boost production. A more representative $C(x)$ is shown in Figure 6.

Figure 6

1.2 EXERCISE SET |||||||||||||||||||||||||||||||||||||||||||||||||||||||||||||||||||||||||||||||||||||||||

In Exercise 1–10, sketch the graph of the given function.

1. $f(x) = |x| + 1$

2. $f(x) = |x - 2|$

3. $f(x) = x + |x|$

4. $f(x) = 2x - |x|$

5. $f(x) = x|x|$

6. $f(x) = \dfrac{x}{|x|}$

7. $f(x) = \begin{cases} 1 & \text{if } x < 0 \\ x + 1 & \text{if } x \ge 0 \end{cases}$

8. $f(x) = \begin{cases} x & \text{if } 0 \le x \le 1 \\ 2x - 1 & \text{if } 1 < x \le 3 \end{cases}$

9. $f(x) = \begin{cases} x + 2 & \text{if } x < 0 \\ x - 2 & \text{if } x \ge 0 \end{cases}$

10. $f(x) = \begin{cases} x^2 & \text{if } -1 \le x \le 1 \\ 1 & \text{otherwise} \end{cases}$

11. If $f(x) = 2x - 1$ and $g(x) = x^3$, find
    (a) $f(g(3))$ (d) $(g \circ f)(x)$
    (b) $g(f(3))$ (e) $(f \circ f)(x)$
    (c) $(f \circ g)(x)$ (f) $(g \circ g)(x)$

12. If $f(x) = \sqrt{x^3}$ and $g(x) = (x^2 + 1)/x$, find
    (a) $f(g(2))$ (d) $(g \circ f)(x)$
    (b) $g(f(2))$ (e) $(f \circ f)(x)$
    (c) $(f \circ g)(x)$ (f) $(g \circ g)(x)$

13. If $f(x) = \dfrac{3}{x-1}$ and $g(x) = x^2 + 2$, find

   (a) $(f \circ g)(2)$ (d) $(g \circ f)(x)$
   (b) $(g \circ f)(2)$ (e) $(f \circ f)(1)$
   (c) $(f \circ g)(x)$ (f) $(g \circ g)(2)$

14. If $f(x) = \dfrac{1}{x}$ and $g(x) = -x$, find

   (a) $(f \circ g)(5)$ (d) $(g \circ f)(x)$
   (b) $(g \circ f)(5)$ (e) $(f \circ f)(x)$
   (c) $(f \circ g)(x)$ (f) $(g \circ g)(x)$

   Compare your answers to (c) and (d), and explain. Likewise, for your answers to (e) and (f).

In Exercises 15–20, write the given function as a composite of two simpler functions.

15. $h(x) = (5x - 3)^8$

16. $h(x) = \sqrt{3x^9 + 5x}$

17. $h(x) = 3\sqrt{x + 2} + 11$

18. $h(x) = (9 - 8x^3)^{20}$

19. $h(x) = \left(\dfrac{3x - 5}{x + 4}\right)^{1/3}$

20. $h(x) = \dfrac{1}{x^3 - 4x}$

21. Figure 7 shows the graph of the dollar value $v = f(t)$ of a commodity, t days after some initial time.
   (a) What is the domain of f, as shown in the figure?
   (b) What is the value of the commodity at the beginning of the period? At the end?
   (c) During the second day, is the value of the commodity rising or falling? What about the fourth day? The seventh day?
   (d) If you owned the commodity on the sixth day, would you earn more by selling then or by waiting until later?
   (e) When is the best time to sell the commodity?
   (f) When is the best time to purchase the commodity?

Figure 7

22. Answer the questions in Exercise 21, using Figure 8.

Figure 8

23. **(Postage Function)** The cost C of mailing a first-class letter is $0.25 for the first ounce or fraction thereof plus $0.20 for each additional ounce or fraction thereof, up to 12 ounces.
   (a) Write C as a function of the weight w (in ounces) for $0 < w \le 3$.
   (b) Sketch the graph of the function C for $0 < w \le 3$.
   (c) What is the cost of mailing a letter that weighs 2.25 ounces?

24. **(Business)** A salesman receives a monthly base salary of $1000 plus a commission of $100 for every $2000 of sales he makes that month in excess of $3000. Let x denote the salesman's sales for a month, and let S denote his corresponding total earnings for that month.
   (a) Express S as a function of x, for $0 \le x \le 9000$.

(b) Sketch the graph of the function S, for $0 \le x \le 9000$.

(c) How much would the salesman earn in a month in which his sales were $7800?

25. **(Business)** A manufacturer of fertilizer finds that the fixed cost of producing x tons is $F(x) = \$2438$ and the variable cost (in dollars) is $V(x) = 4x^2 - 2x$.
(a) Write the cost function.
(b) What is the cost of producing 50 units?

26. **(Business)** A producer of photographic developer finds that the fixed cost is $550 per week; the cost to produce 1 liter of the developer is $0.40. The manufacturer sells the product at $0.50 per liter.
(a) Write the weekly cost, revenue, and profit functions.
(b) Determine the cost, revenue, and profit if the level of production is 10,000 liters per week.

27. **(Business)** A tour operator has established the following pricing schedule for charter flights to Rome. The operator charges a fixed rental of $300,000 for the flight plus a tour fare for each person on the flight. For a group of no more than 100 people, the round trip fare per person is $300. For a group of more than 100 but less than 150 people, the fare decreases by $2 for each person in excess of 100 people. For a group of 150 or more people, the round-trip fare is $200 per person. Write the tour operator's total revenue as a function of the number of people x in the group.

28. **(Business)** A manufacturer of calculators can make a calculator for $15. Market research indicates that at a selling price of x dollars per calculator the manufacturer will sell $400 - x$ calculators daily.
(a) Determine the manufacturer's daily cost function of x.
(b) Determine the manufacturer's daily revenue function of x.
(c) Determine the daily profit function of x.
(d) If each calculator sells for $25, what is the daily profit?

29. **(Investment)** Suppose that x dollars are invested at 7% interest compounded annually. Express the amount A (in dollars) in the account at the end of one year as a function of x.

30. **(Investment)** Suppose that $1000 is invested at r% interest, compounded annually. Express the amount A in the account at the end of one year as a function of r.

1.3 |||||||| THE STRAIGHT LINE AND LINEAR FUNCTIONS

In this section we shall study the straight line and discuss both its equation and graph. There are many applications of straight lines. In the next section we shall examine a few of these. Others will come up later. Students who studied straight lines in high school may use this section as a review before going on to the applications in the next section.

|||||||||||| **Slope**

Consider a straight line L that is not parallel to the y-axis (see Figure 1a) and let $P_1(x_1, y_1)$ and $P_2(x_2, y_2)$ be any two distinct points on L. The graph shows the increments (or changes) $x_2 - x_1$ and $y_2 - y_1$ in the x- and y-coordinates, respectively, from P_1 to P_2. The increment $y_2 - y_1$ can be positive, negative, or zero, but the increment $x_2 - x_1$ can be only negative or positive since P_1 and P_2 are distinct.

Figure 1

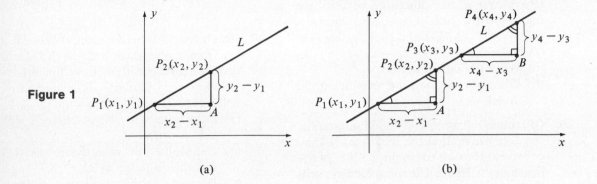

(a)   (b)

If we had chosen a different pair of points, say, $P_3(x_3, y_3)$ and $P_4(x_4, y_4)$ (see Figure 1b), we would have obtained different increments in the x- and y-coordinates. But triangles $P_1 A P_2$ and $P_3 B P_4$ are similar,* so corresponding sides are proportional, which means

$$\frac{y_2 - y_1}{x_2 - x_1} = \frac{y_4 - y_3}{x_4 - x_3}.$$

This observation leads us to the following definition, which we shall need in our study of the straight line.

> The **slope** of a nonvertical line L is the ratio
>
> $$m = \frac{y_2 - y_1}{x_2 - x_1}$$
>
> for any two distinct points $P_1(x_1, y_1)$ and $P_2(x_2, y_2)$ on L.

EXAMPLE 1   Find the slope of the line that passes through the points (2, 2) and (4, 8) on the graph in Figure 2.

---

* Corresponding angles of parallel lines are congruent, and right angles are congruent. Thus, the two triangles have congruent angles, as marked. Therefore, the triangles are similar.

Figure 2

Solution   We may choose either point as $P_1(x_1, y_1)$ and the other as $P_2(x_2, y_2)$. Let $P_1(x_1, y_1) = (2, 2)$ and $P_2(x_2, y_2) = (4, 8)$. Using the definition of slope, we have

$$m = \frac{y_2 - y_1}{x_2 - x_1} = \frac{8 - 2}{4 - 2} = \frac{6}{2} = 3.$$

If we had made the opposite choice, namely, $P_1(x_1, y_1) = (4, 8)$ and $P_2(x_2, y_2) = (2, 2)$, then we would have obtained the same value for m:

$$m = \frac{y_2 - y_1}{x_2 - x_1} = \frac{2 - 8}{2 - 4} = \frac{-6}{-2} = 3. \qquad \blacksquare$$

EXAMPLE 2   Find the slope of the line passing through the points (2, 8) and (5, 2) in Figure 3.

Figure 3

Solution   Let $P_1(x_1, y_1) = (2, 8)$ and $P_2(x_2, y_2) = (5, 2)$. Using the definition of slope, we have

$$m = \frac{y_2 - y_1}{x_2 - x_1} = \frac{2 - 8}{5 - 2} = \frac{-6}{3} = -2.$$ ∎

|||||||||||| Interpretation of Slope

From the definition of the slope of L, we see that

$$y_2 - y_1 = m(x_2 - x_1). \tag{1}$$

As x changes from x_1 to x_2, y changes from y_1 to y_2, and Equation (1) says that the change in y is proportional to the change in x. The slope m is the constant of proportionality; that is, the change in y is always m times the change in x. Thus, m measures the rate at which y changes with x along the line L. In Example 1, the line L has slope $m = 3$, which means that each time x increases by one unit, y **increases** by three units. In Example 2, the line L has slope $m = -2$, which means that each time x increases by one unit, y increases by -2 units; that is, y **decreases** by two units.

Figure 4 shows several lines with positive and negative slopes. If m is positive, the line **rises** from left to right; that is, as x increases, y also increases. If m is negative, the line **falls** from left to right; that is, as x increases, y decreases.

Figure 4

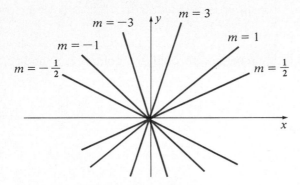

|||||||||||| Equations of Straight Lines

An equation of a given straight line is an equation in x and y whose graph in the xy-coordinate system is that line. That is, every point whose coordinates (x, y) satisfy the equation must be on the line, and conversely every point on the line has coordinates (x, y) that satisfy the equation.

We shall now determine an equation of a line L that is not parallel to the y-axis. Suppose L has slope m and intersects the y-axis at the point $(0, b)$. The number b is called the **y-intercept** of L. If $P_2(x, y)$ is any other point on L, we can find

Figure 5

the slope m of L by using the points $P_1(0, b)$ and $P_2(x, y)$, shown in Figure 5, to obtain

$$m = \frac{y - b}{x - 0} = \frac{y - b}{x}. \tag{2}$$

From Equation (2) we obtain the **slope-intercept** form of L.

$$y = mx + b \tag{3}$$

The straight line with y-intercept b and slope m consists of all points (x, y) satisfying Equation (3).

EXAMPLE 3   Find the slope-intercept form of the line having slope 2 and intersecting the y-axis at the point $(0, 3)$.

Solution   Given that $m = 2$ and $b = 3$ and substituting in Equation (3) we have

$$y = 2x + 3.$$ ■

EXAMPLE 4   Consider the equation

$$y + 3x + 5 = 0$$

of a line L. Find the slope and y-intercept of L.

Solution   We first solve the given equation for y, obtaining

$$y = -3x - 5.$$

Hence,

$$m = -3 \quad \text{and} \quad b = -5.$$ ■

The following principle is fundamental to the concept of slope:

> If two lines have the same slope, then they are parallel; conversely, if two lines are parallel, then they have the same slope (Figure 6).

Figure 6

(If L and L' are parallel, then $m = m'$; and, conversely.)

EXAMPLE 5 Find an equation of the line that is parallel to the line $y = -2x + 5$ and has a y-intercept of 4.

Solution The desired line L has slope $m = -2$ because it is parallel to $y = -2x + 5$. Hence, from Equation (3), its equation is

$$y = -2x + 4.$$ ∎

There is another principle that is fundamental to the concept of slope:

> Through any fixed point P_1 there is only one line L that has a given slope m.

Thus, suppose we know the slope m of L and a fixed point $P_1(x_1, y_1)$ on L. If $P(x, y)$ is an arbitrary point on L, then by definition of slope,

$$m = \frac{y - y_1}{x - x_1},$$

which gives the **point-slope** form of L,

$$y - y_1 = m(x - x_1). \tag{4}$$

The straight line L that has slope m and passes through the point $P_1(x_1, y_1)$ consists of all points P whose coordinates (x, y) satisfy Equation (4) (Figure 7).

Figure 7

Equation (4) is an extremely useful formula for determining the equation of a straight line. We shall use it often.

EXAMPLE 6 Find an equation of the line that passes through the point $P_1(-3, 4)$ and has slope $m = 2$.

Solution From Equation (4), the point-slope form of a line, with $x_1 = -3$, $y_1 = 4$, and $m = 2$, we have

$$y - 4 = 2(x + 3),$$

or, upon simplifying (verify),

$$y = 2x + 10. \qquad \blacksquare$$

EXAMPLE 7 Find an equation of the line determined by the points $(4, 3)$ and $(2, 8)$.

Solution Let $P_1(x_1, y_1) = (4, 3)$ and $P_2(x_2, y_2) = (2, 8)$. Then the slope of the line is given by

$$m = \frac{y_2 - y_1}{x_2 - x_1} = \frac{8 - 3}{2 - 4} = -\frac{5}{2}.$$

With P_1 as the point, we use the point-slope form in Equation (4) to obtain

$$y - 3 = -\tfrac{5}{2}(x - 4),$$

or, upon simplifying (verify),

$$y = -\tfrac{5}{2}x + 13. \qquad \blacksquare$$

Note In Example 7, we could have used P_2 instead of P_1 as the point to obtain the same equation of the line (verify).

|||||||||||||| **Vertical Lines**

Thus far, we have considered lines that are not parallel to the y-axis. Consider now a line L parallel to the y-axis (Figure 8).

Figure 8

Since every point on L has the same x-coordinate, if L intersects the x-axis at $x = a$, then L is described by the equation

$$x = a.$$

For example, the equation of a line parallel to the y-axis and two units to the right of it is

$$x = 2.$$

If we attempt to calculate the slope of a vertical line L, we choose two points $P_1(a, y_1)$ and $P_2(a, y_2)$ on L. The slope formula applied to L would yield

$$\frac{y_2 - y_1}{a - a},$$

which is impossible because we cannot divide by zero. That is why our slope formula applies only to nonvertical lines. Thus,

> A line parallel to the y-axis is said to have *no slope.*

|||||||||||||| **Horizontal Lines**

We now turn to lines that are parallel to the x-axis. Suppose that L is a line parallel to the x-axis and that it intersects the y-axis at $(0, b)$, as shown in Figure 9.

To calculate the slope of L we choose two distinct points $P_1(x_1, b)$ and $P_2(x_2, b)$ on L. Then

$$m = \frac{b - b}{x_2 - x_1} = \frac{0}{x_2 - x_1} = 0.$$

Figure 9

Hence,

> The slope of a line parallel to the x-axis is zero.

Using the slope-intercept form in Equation (3), with $m = 0$, we can write an equation of L:

$$y = 0 \cdot x + b,$$

or

$$y = b.$$

Hence, the equation of a line parallel to the x-axis and five units below it is

$$y = -5.$$

||||||||||||||| **The General Equation of a Straight Line**

THEOREM

The equation of a straight line can be written as

$$Ax + By = C, \tag{5}$$

where A and B are not both zero. Conversely, the graph of Equation (5) (A and B not both zero) is a straight line.

Proof   If the given line is parallel to the y-axis, then its equation is $x = a$, and we can write it as $1 \cdot x + 0 \cdot y = a$, which is in the form of Equation (5). If the given line is not parallel to the y-axis, then its slope-intercept form is $y = mx + b$, and we can write it as $-m \cdot x + 1 \cdot y = b$, which is also in the form of Equation (5).

To prove the converse, first suppose that $B \neq 0$. Then we can solve (5) for y, obtaining

$$y = -\frac{A}{B} x + \frac{C}{B}. \tag{6}$$

If we let

$$m = -\frac{A}{B} \quad \text{and} \quad b = \frac{C}{B},$$

then (6) is the slope-intercept form of a line. If $B = 0$ and, consequently, $A \neq 0$, then (5) becomes

$$Ax = C,$$

or

$$x = \frac{C}{A}.$$

which is the equation of a line parallel to the y-axis.

An equation that can be written in the form of Equation (5) is called a **linear equation** in x and y.

EXAMPLE 8   The following are linear equations in x and y:

$$3y = 4, \qquad 3x + 2y = 7, \qquad y = 2x - 4, \qquad x = 3y + 7, \qquad -\tfrac{1}{2}x = 7.$$

Moreover, the graph of each of these equations is a straight line.

|||||||||||| **Linear Functions**

It is clear from the vertical line test that a straight line that is not parallel to the y-axis is the graph of a function f of x. Such a function, called a **linear function,** is given by

$$f(x) = mx + b.$$

Of course, a straight line parallel to the y-axis cannot be the graph of a function of x.

|||||||||||| **Intersections of Lines**

If L_1 and L_2 are given lines, we can have the three following possibilities (Figure 10):
  1. The lines are two parallel lines, [Figure 10(a)].
  2. The lines are identical [Figure 10(b)].
  3. The lines intersect at only one point [Figure 10(c)].

Figure 10

(a)   (b)   (c)

If the first possibility occurs, then the slopes of L_1 and L_2 are equal. Moreover, L_1 and L_2 do not intersect.

If the second situation occurs, then the slopes and y-intercepts of L_1 and L_2 are equal. Moreover, L_1 and L_2 intersect at infinitely many points.

Given a pair of linear equations in x and y whose graphs are straight lines, we determine their point of intersection by solving the equations simultaneously. If we obtain no solution, then (1) holds. If we obtain infinitely many solutions, then (2) holds. If we obtain only one solution, then (3) holds.

EXAMPLE 9   Find the point of intersection of the following pairs of lines:
(a) $x + 3y = 7$
     $2x + 6y = 10$
(b) $x + 3y = 7$
     $-2x - 6y = -14$
(c) $x + 3y = 7$
     $x - 2y = 2$

Solution   (a) The system has no solution; that is, there is no pair of real numbers x and y such that $x + 3y = 7$ and $2x + 6y = 10$, for if

$$x + 3y = 7,$$

then

$$2x + 6y = 2(x + 3y) = 2 \cdot 7 = 14.$$

In this case, the lines are parallel.

(b) Since the second equation is -2 times the first, there are infinitely many solutions and the lines are identical.

(c) The only solution is $x = 4$, $y = 1$, so the lines intersect at the point $(4, 1)$. For a review of methods used to obtain the solution, the reader may consult the material on "linear equations in two unknowns" in Section A.3 (Appendix A).

■

1.3 EXERCISE SET ||||||||||||||||||||||||||||||||||||||||||||||||||||||||||||||||||||||||||

1. Find the slopes of the lines in Figure 11.

(a)

(b)

(c)

(d)

Figure 11

2. Find the slope of the line containing the two given points:
   (a) $(-1, 2)$ and $(-5, 4)$
   (b) $(-3, 5)$ and $(-3, -1)$
   (c) $(-1, -3)$ and $(4, 8)$
   (d) $(2, 3)$ and $(-7, 3)$

3. Find the slope-intercept form of the line with given slope m and intersecting the y-axis at the indicated point P. Sketch the line.
   (a) $m = -2$, $P(0, -2)$
   (b) $m = \frac{3}{2}$, $P(0, 2)$

4. Find an equation of the line that is parallel to the line $y = \frac{3}{2}x + 5$ and has a y-intercept of -2.

5. Find the slope of the given line:
   (a) $y = 3x + 2$ (c) $x = \frac{2}{3}y + 2$
   (b) $y = 3$

6. Find the slope of the given line:
   (a) $y = -\frac{2}{3}x - 4$ (c) $x = -5$
   (b) $3x + 4y = 5$

7. For each part, state whether the given line *rises* from left to right or *falls* from left to right.
   (a) $y = 2x + 3$ (c) $y = \frac{3}{4}x - 2$
   (b) $y = -\frac{3}{2}x + 5$ (d) $y = -\frac{4}{5}x - 6$

8. For each given line, state whether y increases or y decreases as x increases.
   (a) $y = 3x + 2$ (c) $y = \frac{4}{3}x - 3$
   (b) $y = -\frac{3}{2}x + 5$ (d) $y = -\frac{2}{5}x - 3$

9. Consider the line whose equation is $y = 2x + 5$. In each part, find the increase in y as x increases by the given amount. (Hint: use slope.)
   (a) x increases by one unit
   (b) x increases by three units

10. Consider the line whose equation is $y = -3x - 4$. In each part, find the decrease in y as x increases by the given amount. (Hint: use slope.)

(a) x increases by one unit

(b) x increases by two units

11. Let L be the line determined by P_1 and P_2 and let L' be the line determined by P_3 and P_4. Determine whether L and L' are parallel and sketch both L and L'.
    (a) $P_1(1, -1)$, $P_2(3, 4)$, $P_3(2, 3)$, $P_4(-1, 8)$
    (b) $P_1(2, 1)$, $P_2(4, 4)$, $P_3(0, -2)$, $P_4(-2, -5)$
    (c) $P_1(4, 2)$, $P_2(6, -1)$, $P_3(4, 5)$, $P_4(1, 8)$

12. In Exercises 11(a) and (c), change the x-coordinate of P_4 so that as a result of the change, L and L' become parallel.

13. Find the point-slope form of the straight line having the given slope m and passing through the given point P. Sketch the line.
    (a) $m = 2$, $P(3, 4)$
    (b) $m = -3$, $P(-4, 1)$
    (c) $m = 0$, $P(3, 2)$

14. Find the point-slope form of the line having the slope m and passing through the point P. Sketch the line.
    (a) $m = \frac{1}{2}$, $P(2, -1)$
    (b) $m = -3$, $P(-1, 2)$
    (c) $m = -\frac{2}{3}$, $P(0, 0)$

15. Find the point-slope form of the line determined by the given points.
    (a) $(-1, 2)$ and $(3, 5)$
    (b) $(-3, -4)$ and $(0, 0)$

16. Find the slope-intercept form of the line determined by the given points.
    (a) $(-2, -3)$ and $(3, 4)$
    (b) $(-1, 3)$ and $(2, -5)$

17. Find an equation of the line that is parallel to the line $x + 2y = 3$ and passes through the point $(1, -2)$.

18. Find an equation of the line that is parallel to the line $-3x + 4y = 5$ and passes through the point $(-2, 3)$.

19. Find an equation for each of the following lines.

(a)

(b)

(c)

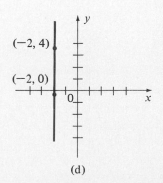

(d)

20. Find an equation for each of the following lines.

(a)

(b)

(c)

(d)

21. Express each of the following equations in the form of Equation (5), and state the values of A, B, and C.
    (a) $y = 2x - 3$ (c) $y = 3$
    (b) $y - 2 = \frac{3}{2}(x - 4)$ (d) $x = \frac{3}{4}y - 1$

22. Which of the following are linear equations in x and y?
    (a) $2x + 3y = 4$ (c) $2x^2 + y = 7$
    (b) $xy = 2$ (d) $2(x - \frac{3}{2}) + 5y = 4$

23. For each part, determine whether the given lines (i) are parallel, (ii) are identical, or (iii) intersect at only one point.
    (a) $x + y = 6$
           $2x + 3y = 15$
    (b) $x - 2y = 7$
           $3x - 6y = 14$
    (c) $x + 3y = 1$
           $-2x - 6y = -2$
    (d) $x - 3y = -5$
           $2x + 3y = -1$

24. Follow the directions in Exercise 23 for the following lines:
    (a) $x + 4y = 5$
           $2x + 8y = -2$
    (b) $x + y = 5$
           $2x - y = 4$
    (c) $x - 3y = 2$
           $-3x + 9y = -6$
    (d) $x + 3y = -8$
           $2x - 3y = 11$

1.4 |||||||| APPLICATIONS OF THE STRAIGHT LINE

In this section we shall consider a number of applications of the straight line and linear functions.

|||||||||||| **Simple Interest**

When we borrow money from a lender, we must pay a fee called **interest.** The type of interest that we shall discuss here is called **simple interest,** which is a fixed percentage of the amount borrowed over a specified period of time. Thus, borrowing $100 at the simple interest rate of 6% per year means that we must pay the lender 6% of the amount borrowed for each year that the money is held.

In borrowing transactions the amount of money borrowed is called the **principal** and is denoted by P. We use the letter r to denote the interest **rate** per year, expressed as a decimal. Thus, if the interest rate is 6%, then

$$r = \frac{6}{100} = 0.06.$$

The interest that must be paid to the lender for each year that the money is held is

$$Pr.$$

Thus, if we hold the money t years, the total interest due is

$$I = Prt. \tag{1}$$

Since the amount, S, repaid after t years is the principal plus the interest, we have

$$S = P + I, \tag{2}$$

or

$$S = P + Prt. \tag{3}$$

In Equation (3), P and r are known constants, so this is a linear equation in S and t; the slope is Pr. We can also write Equation (3) as

$$S(t) = P(1 + rt) \tag{4}$$

by factoring out P, and using function notation to show that S is a function of t.

EXAMPLE 1 Suppose that you borrow $6000 at the simple interest rate of 7%.
    (a) How much money do you owe after t years?
    (b) Sketch the graph of the function $S(t)$ found in (a).
    (c) How much money do you owe after 10 years?
    (d) How much money do you owe after 8 months?

Solution   (a) We have

$$r = \frac{7}{100} = 0.07 \quad \text{and} \quad P = 6000.$$

From Equation (3),

$$S(t) = 6000 + 6000(0.07)t,$$

or

$$S(t) = 6000 + 420t \text{ (dollars)}. \tag{5}$$

(b) The graph of Equation (5) is a straight line, with s-intercept 6000 and slope 420, as shown in Figure 1.

Figure 1

(c) After 10 years, $t = 10$, so we want

$$S(10) = 6000 + 420(10)$$
$$= \$10,200.$$

(d) After 8 months, $t = \frac{8}{12} = \frac{2}{3}$, so we want

$$S(\tfrac{2}{3}) = 6000 + 420(\tfrac{2}{3}) = \$6,280. \qquad ■$$

||||||||||||| **Markup and Discount**

Markup   Retail merchants earn their profit by purchasing goods from suppliers (producers or wholesalers) and selling them at a somewhat higher price. The difference between the merchant's selling price and his or her cost is called the **markup**. If the merchant's markup is 30%, then the selling price of an item costing the merchant c dollars is

$$S = c + .3c,$$

or

$$S = (1 + .3)c.$$

In functional notation,

$$S(c) = (1 + r)c \text{ dollars}. \tag{6}$$

Discount A merchant often attempts to stimulate sales of an item by placing it for sale at a "discount". If the list price (normal selling price) is L dollars, and the merchant allows a 30% discount, then the selling price is

$$S = L - .3L,$$

or

$$S = (1 - .3)L.$$

In functional notation,

$$S(L) = .7L.$$

In general, when an item whose list price is L dollars is sold at a discount r (percent, changed to a decimal), the selling price is

$$S(L) = (1 - r)L \text{ dollars.} \tag{7}$$

Note For a fixed markup r, Equation (6) shows that the selling price is a linear function of the cost c. Similarly, Equation (7) shows that for a fixed discount rate r, the selling price is a linear function of the list price L.

EXAMPLE 2 A bicycle retailer has a uniform markup of 40% on all items he sells. During several weeks when his sales are slumping, he advertises a special sale of "40% off, on everything in the store."
   (a) Find the list price of an item that costs the retailer c dollars as a function of c.
   (b) Find the special sale price of an item whose list price is L, as a function of L.
   (c) Find the list price, and the special sale price, of bicycle tires that cost the retailer $5 each.
   (d) Find the special sale price of an item that cost the retailer c dollars each, as a function of c.

Solution (a) Using formula (6), the list price of an item costing the retailer c dollars is

$$L(c) = (1 + .4)c = 1.4c \text{ dollars.}$$

(b) Using formula (7), the special sale price of an item whose list price is L dollars is

$$S(L) = (1 - .4)L = .6L \text{ dollars.}$$

(c) When the retailer's cost is $c = \$5$, the retailer's list price will be

$$L(5) = 1.4(5) = \$7.00,$$

and the special sale price will be

$$L(7) = .6(7) = \$4.20.$$

Observe that a 40% markup followed by a 40% discount will not cause the retailer to break even. (Why?)

(d) For an item that cost the retailer c dollars, the special sale price will be

$$
\begin{aligned}
S(L) &= S(1.4c) &&\text{[from part (a)]} \\
&= .6(1.4c) &&\text{[from part (b)]} \\
&= .84c \text{ dollars.}
\end{aligned}
$$

Thus, the special sale price is only 84% of the retailer's cost.   ∎

|||||||||||| Linear Depreciation

A great many things that we buy lose all or part of their value as time goes on. This decrease in value is called the **depreciation** of the property; we also say that the property **depreciates.** For example, if you buy a car for $6000 and later sell it for $2000, then the depreciation is

$$
\$6000 - \$2000 = \$4000.
$$

For a business, certain assets such as buildings and equipment are continually depreciating, due to normal wear and tear as well as obsolescence. These assets have limited life spans, after which they must be replaced. Such annual depreciation (loss of value) must be considered as an "expense" (loss), and the business must recover this loss at approximately the same rate so that it can afford to purchase replacements as they are needed. The Internal Revenue Service (IRS) allows a business to claim its depreciation as a "business expense," so that in reporting its annual earnings the business may subtract its **annual depreciation.** But first the annual depreciation must be calculated.

There are several methods of calculating the annual depreciation that meet IRS requirements, each having different advantages. The simplest is the method of **linear** or **straight-line depreciation**, in which the decrease in value over a stated period of time is a **fixed percentage of the original value.** For example, if a piece of machinery depreciates linearly at the rate of 5% per year, then the decrease in value for each year that the equipment is owned is 5% of its original value. After twenty years of ownership, the equipment would depreciate to zero.

Suppose that the **original cost** of an item is V_0 dollars, and that the item is expected to have a **useful life** of n years, after which it has a **salvage** (or resale) **value** of V_n dollars. Then the net cost (or **depreciable value**) of the item is

$$
C = V_0 - V_n. \tag{8}
$$

In the straight-line depreciation method, the **annual depreciation** is

$$
D = \frac{C}{n} = \frac{V_0 - V_n}{n}. \tag{9}
$$

Sometimes, as in the above illustration, this annual depreciation is expressed as a percentage of the net cost.

After t years of ownership, the total loss of value (accumulated depreciation) of the item is

$$Dt, \tag{10}$$

and the **book value** of the item in the accounting records of the company is

$$V(t) = V_0 - Dt, \tag{11}$$

which is a linear function of t.

EXAMPLE 3   Suppose that a piece of equipment that cost \$60,000 depreciates linearly at the rate of 4% of its original value per year, with no salvage value.
(a) Find the useful life of the equipment.
(b) Find the annual depreciation.
(c) Find the accumulated depreciation after t years; after 12 years.
(d) Find the book value after t years; after 12 years.

Solution   (a) At the rate of 4% of the original value per year it will take $\frac{100}{4} = 25$ years to depreciate to \$0 in value. Thus the expected useful life is 25 years.

(b) From (9) the annual depreciation is $D = \dfrac{60,000}{25} = \$2400$.

(c) From (10) the accumulated depreciation after t years is

$$Dt = 2400t \text{ dollars.}$$

When $t = 12$, this yields $2400 \times 12 = \$28,800$.
(d) From (11) the book value after t years is

$$V(t) = 60,000 - 2400t \text{ dollars.}$$

When $t = 12$, this yields

$$V(12) = 60,000 - 28,800$$
$$= \$31,200.$$

The graph of the linear function $V(t)$ is shown in Figure 2.

Figure 2

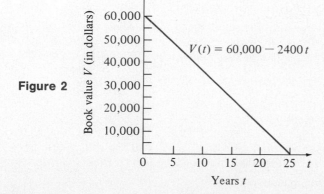

|||||||||||| **Break-Even Analysis**

A manufacturer can tell, by comparing the total cost and total revenue functions, whether the firm is operating at a profit or at a loss. If the revenue exceeds the cost, the firm is operating at a profit; if the cost exceeds the revenue, the firm is operating at a loss. The level of production x for which $R(x) = C(x)$, so that the profit is zero, is called the **break-even point.** At the break-even point, the manufacturer neither makes money nor loses money.

EXAMPLE 4   Determine the break-even point for the animal feed manufacturer of Examples 10 and 11 in Section 1.2.

Solution   The cost and revenue functions for the feed manufacturer were determined in Section 1.2 to be

$$C(x) = 2000 + 0.50x,$$

and

$$R(x) = 0.75x.$$

Thus, we are interested in finding the value of x for which

$$C(x) = R(x),$$

or

$$2000 + 0.50x = 0.75x.$$

Solving for x, we obtain

$$x = \frac{2000}{0.25} = 8000.$$

That is, producing 8000 kilograms of feed will yield the manufacturer zero profit. The total revenue, which equals the total cost at the break-even point, is $6000. The situation is shown graphically in Figure 3.

Figure 3

The break-even point is the x-coordinate of the point where the straight lines representing the functions $R(x)$ and $C(x)$ intersect. If the level of production is greater than 8000 kilograms, then $R(x) > C(x)$ and the manufacturer makes a profit. If the level of production is less than 8000 kilograms, then $R(x) < C(x)$ and the manufacturer incurs a loss. These conclusions can also be reached as follows. If we form the profit function, we obtain

$$P(x) = R(x) - C(x)$$
$$= 0.75x - (2000 + 0.50x)$$
$$= 0.25x - 2000.$$

The graph of this function, shown in Figure 4, reveals that $P(8000) = 0$, whereas $P(x)$ is positive (profit) for $x > 8000$ and $P(x)$ is negative (loss) for $x < 8000$.

Figure 4

Line of Best Fit; Least Squares Method

In almost every area of the sciences, engineering, business, economics, and the social sciences, a primary task of problem-solving is the gathering and analysis of data. During this stage, the analyst will usually measure a value of y for a given value of x and then plot the point (x, y) on graph paper. The resulting graph may yield an equation relating x and y that will predict new values of y for given values of x.

EXAMPLE 5
(Psychology)
In an experiment designed to determine the extent of a person's natural orientation, a subject is placed in a special room and kept there for a certain length of time. He or she must then find the way through a maze. The researcher records the time it takes subjects to complete this task and obtains the following data:

Time in room (hours)	1	2	3	4	5	6	7	8
Time to go through maze (minutes)	0.6	2.3	2.4	1.8	3.3	3.6	3.1	4.0

The points in this table are plotted in Figure 5.

Figure 5

The relationship between x and y in this experiment is **probabilistic** in nature. That is, there is an element of chance involved. If the experiment is repeated, there is no reason to believe that exactly the same data would result. Corresponding to the same value of x we would expect slightly different values of y. (If there is no chance involved—that is, if we can expect the same results in every subsequent repetition of the experiment—then we say that the relationship between x and y is **deterministic.**)

In Figure 5 we observe that y tends to increase as x increases, and the data seem to be scattered about a straight line. We might imagine that y is approximately a linear function of x,

$$y = mx + b. \tag{12}$$

We cannot graph this imagined function (12) by merely joining the plotted points in Figure 5. Indeed, joining these points would not produce a straight line at all. Instead, we shall try to find the straight line that "best fits" the given data points. The technique we shall use in finding such a line is called the **method of least squares,** which we shall now outline.

Suppose that we are given r data points, $(x_1, y_1), (x_2, y_2), \ldots, (x_r, y_r)$, with $x_1 < x_2 < \cdots < x_r$. We want to find values of b and m so that the line

$$y = mx + b \tag{13}$$

is the one that "best fits" these data points. Our method interprets "best fits" to mean that the sum of squares of the signed vertical deviations of the points (x_i, y_i) from the line $y = mx + b$ is as small as possible.

Figure 6 shows the five data points $(x_1, y_1), (x_2, y_2), (x_3, y_3), (x_4, y_4)$, and (x_5, y_5). Let the signed vertical deviation from the point (x_i, y_i) to the (not yet known) line of best fit be denoted by d_i. Then the line of best fit in this figure will be the one for which the expression

$$E = d_1^2 + d_2^2 + d_3^2 + d_4^2 + d_5^2$$

Figure 6

is as small as possible. In the general case where there are r data points, we want the expression

$$E = d_1^2 + d_2^2 + \cdots + d_r^2 \tag{14}$$

to be as small as possible.

To find the line of best fit, Equation (13), we must determine m and b. The y-coordinate of the point on the line of best fit whose x-coordinate is x_i is

$$mx_i + b$$

by Equation (13). Then

$$d_i = y_i - mx_i - b.$$

Thus, the expression in (14), which we are trying to make as small as possible, is

$$E = (y_1 - mx_1 - b)^2 + (y_2 - mx_2 - b)^2 + \cdots + (y_r - mx_r - b)^2. \tag{15}$$

In Section 8.3 we shall establish the following result:

The slope m and the y-intercept b of the line of best fit can be found by solving the following system of two linear equations in two unknowns (m and b):

$$rb + (x_1 + x_2 + \cdots + x_r)m = y_1 + y_2 + \cdots + y_r$$
$$(x_1 + x_2 + \cdots + x_r)b + (x_1^2 + x_2^2 + \cdots + x_r^2)m = x_1y_1 + x_2y_2 + \cdots + x_ry_r \tag{16}$$

For a review of the methods of solving a system of two linear equations in two unknowns, see Section A.3 (Appendix A).

EXAMPLE 6   Find the line of best fit for the data (1, 2), (2, 5), (3, 3), and (4, 6).

Solution   We have $r = 4$. To solve the pair of linear equations in (16) for b and m, we first form the coefficients. Specifically,

$$x_1 + x_2 + x_3 + x_4 = 1 + 2 + 3 + 4 = 10$$
$$x_1^2 + x_2^2 + x_3^2 + x_4^2 = 1^2 + 2^2 + 3^2 + 4^2 = 30$$
$$y_1 + y_2 + y_3 + y_4 = 2 + 5 + 3 + 6 = 16$$
$$x_1y_1 + x_2y_2 + x_3y_3 + x_4y_4 = 2 + 10 + 9 + 24 = 45.$$

Equations (16) now become

$$4b + 10m = 16$$
$$10b + 30m = 45.$$

Multiplying both sides of the first equation by 3, our equations become

$$12b + 30m = 48$$
$$10b + 30m = 45.$$

Subtracting the second equation from the first, we obtain

$$2b = 3, \quad \text{or} \quad b = \tfrac{3}{2}.$$

Then, since $4b + 10m = 16$, we have

$$4(\tfrac{3}{2}) + 10m = 16$$
$$6 + 10m = 16$$
$$10m = 10$$
$$m = 1.$$

The solution obtained is thus

$$b = 1.5 \quad \text{and} \quad m = 1.$$

Figure 7

Hence, the line of best fit is

$$y = x + 1.5,$$

which is shown in Figure 7 along with the four given data points. ∎

EXAMPLE 7 (a) Find the equation of the line of best fit for the data in Example 5
(b) Use the equation obtained in part (a) to *predict* how long it will take the subject to go through the maze after 24 hours in the room.

Solution (a) Here $r = 8$. To obtain Equations (16), we arrange our computations as follows:

x_i (time in room)	x_i^2	y_i (time to go through maze)	$x_i y_i$
1	1	0.6	0.6
2	4	2.3	4.6
3	9	2.4	7.2
4	16	1.8	7.2
5	25	3.3	16.5
6	36	3.6	21.6
7	49	3.1	21.7
8	64	4.0	32.0
Totals: 36	204	21.1	111.4

Equations (16) now become

$$8b + 36m = 21.1$$
$$36b + 204m = 111.4.$$

Multiplying both sides of the first equation by $\frac{9}{2}$, our equations become

$$36b + 162m = 94.95$$
$$36b + 204m = 111.4.$$

Subtracting, we obtain $-42m = -16.45$, or $m = 0.392$. Substituting this value of m into the first equation, we obtain

$$8b + 36(0.392) = 21.1$$
$$8b = 21.1 - 14.1 = 7$$
$$b = \tfrac{7}{8} = 0.875.$$

The solution, obtained to three decimals, is

$$b = 0.875 \qquad \text{and} \qquad m = 0.392.$$

Then the line of best fit, shown in Figure 8, is

$$y = 0.392x + 0.875. \tag{17}$$

Figure 8

Number of hours in room

(b) If $x = 24$, then substituting in (17) we obtain (verify)

$$y = 10.283.$$

Thus, we predict that it would take the subject 10.283 minutes to go through the maze. ∎

1.4 EXERCISE SET ||||||||||||||||||||||||||||||||||||||||||||||||||||||||||||||||||||||||||||

1. **(Business)** Suppose that you borrowed $8000 from a finance company at the simple interest rate of 12% per year.
   (a) How much money do you owe after t years?
   (b) Sketch the graph of the equation obtained in (a).
   (c) How much money do you owe after 6 years?
   (d) How much money do you owe after 9 months?

2. **(Business)** Suppose that a business firm borrowed $10,000 at the simple interest rate of 10% per year.
   (a) How much money does the firm owe after t years?
   (b) Sketch the graph of the equation obtained in (a).
   (c) How much money does the firm owe after 8 years?
   (d) How much money does the firm owe after 6 months?

3. Suppose that you borrowed $12,000 for 2 years at the simple interest rate of $100r\%$ per year. If you owe $13,440, find r.

4. Susan borrowed $800 from her uncle at simple interest. Four years later she repaid her uncle in one lump sum of $1104. What annual rate of interest was she paying?

5. **(Business)** A sporting goods store has a 25% markup on all warm-up suits.
   (a) What is the selling price of a warm-up suit that costs the store $38?
   (b) What did the store pay for a warm-up suit that it sells for $54.95?
   (c) Members of the local high school wrestling team are entitled to a 10% discount at the store. How much must they pay for a warm-up suit that costs the store $65?

6. **(Business)** A jeweler has a markup of 45% on all gold bracelets, but is running a special sale of 30% off.

(a) What is the store's regular price for a gold bracelet that costs them $85?

(b) What is the special sale price of this bracelet?

(c) Express the store's regular selling price P as a function of store's cost x.

(d) Express the special sale price S as a function of the store's cost x. Sketch the graph of this function.

7. (**Business**) During a special sale a merchant sells an item at 20% off the regular selling price, which was marked up 30% above her cost, x. Write the special sale price as a function of x, and sketch the graph of this function. What is the merchant's percentage profit (or loss)?

8. (**Business**) A bookstore decides to stimulate sales of a new novel by a local author by running a special sale. The usual selling price is 30% above the store's cost. What is the maximum percentage discount that the store can allow without losing money?

9. (**Medical Economics**) A Computerized Axial Tomographic Scanner (a piece of medical equipment that shows abnormalities in soft tissue) costs $400,000. Suppose it depreciates linearly at the rate of 10% of its original value per year.

(a) What is the value V after t years?

(b) Sketch the graph of the equation obtained in (a).

(c) What is the value after 5 years?

(d) After how many years will the equipment have depreciated to zero value?

10. (**Business**) A delivery service purchases a new truck for $90,000. It is expected to have a useful life of 8 years, after which it will have a salvage value of $5000. Assuming linear depreciation,

(a) Find the annual depreciation, in dollars.

(b) Find the accumulated depreciation, after t years.

(c) Find the book value $V(t)$ of the truck, after t years.

(d) Sketch the graph of the function $V(t)$.

11. (**Business**) Suppose that a car costing $8000 depreciates linearly at a rate of r% of its original cost per year. After 2 years the value of the car is $4800. Find r.

12. (**Business**) A factory purchases an assembly line machine for $300,000, which depreciates linearly. After 7 years its value is $132,000.

(a) Find the annual depreciation in dollars and as a percent of the original cost.

(b) Find the book value $V(t)$ of the machine t years after it is purchased.

(c) Sketch the graph of the function $V(t)$ found in (b).

13. (**Break-Even Analysis**) The annual cost and revenue functions of a steel producer (in millions of dollars) are given by

$$C(x) = 20 + 0.4x$$
$$R(x) = 0.8x,$$

where x is in millions of tons.

(a) Sketch the cost and revenue functions.

(b) Find the break-even point.

(c) Find the total revenue at the break-even point.

14. (**Break-Even analysis**) A manufacturer of electronic components finds that in making x units of a product weekly it has a cost of $2 per unit, plus a fixed cost of $1800. Each unit sells for $5.

(a) Find the cost and revenue functions.

(b) Sketch these functions.

(c) Find the break-even point.

(d) Find the total weekly revenue at the break-even point.

15. (**Profit**) A small manufacturer of a new solar energy device finds that the annual cost and revenue functions (in dollars) are given by

$$C(x) = 24,000 + 55x$$
$$R(x) = 95x,$$

where x is the number of units manufactured and sold. Sketch the profit function.

If 560 units were sold during the year, did the manufacturer make money (how much?) or lose money (how much?)?

16. Sketch the profit function for the steel producer of Exercise 13.

17. **(Population Growth)** Suppose that in a certain state there are 3.4 million people who are over 65 years old. Assume that the number N (in millions) of people who are over 65 years old is increasing at the constant rate of 8% per year. Let t denote time in years and let $t = 0$ designate the present time.
    (a) Find an equation relating N and t.
    (b) Compute the number of people who will be over 65 years old 15 years from now.

18. For each job, a computer center charges $1.50 plus $0.04 per second of time used. Write an equation giving the cost C for a job lasting x seconds.

19. **(Biology)** The amount of solid food an animal consumes daily depends upon its weight. Suppose that an animal weighing 10 kilograms requires 1.5 kilograms of food each day, and an animal weighing 40 kilograms requires 4.5 kilograms of food each day. Assume that the amount A of food required daily and the weight w are related by a linear equation.
    (a) Find A in terms of w.

(b) How much food will be required each day by an animal weighing 60 kilograms?

In Exercises 20–23, find the line of best fit for the given data points.

20. (2, 2), (3, 3), (4, 4), (5, 6)

21. (3, 2), (4, 3), (5, 5), (6, 7), (7, 9)

22. (2, 1), (3, 2), (4, 4), (5, 5), (6, 8), (7, 10)

23. (3, 2), (4, 4), (5, 6), (6, 7), (7, 8), (8, 12)

24. **(Business)** A plastics manufacturer obtains the data shown below in Table 1. Represent the years 1981, 1982, . . . , 1986 as 0, 1, 2, 3, 4, 5, respectively, and let x denote the year. Let y denote the annual sales (in millions of dollars).
    (a) Find the line of best fit relating x and y.
    (b) Use the equation obtained in (a) to predict the annual sales for the year 1994.

25. **(Ecology)** A large new industrial plant has opened near a stream. The municipality's water inspection facility has obtained the data shown in Table 2 for the amount of lead in the stream:
    (a) Find the line of best fit relating the amount y of lead (in parts per million) and the number x of weeks the plant has been in operation.

Table 1

Year	1981	1982	1983	1984	1985	1986
Annual sales (millions of dollars)	1.2	2.4	3.4	4.1	5.0	6.2

Table 2

Weeks plant has been open	5	10	15	20	30	52
Amount of lead (parts per million)	0.0040	0.0060	0.0080	0.0090	0.0150	0.020

(b) Use the equation obtained in (a) to predict the amount of lead in the stream 2 years (104 weeks) after the opening of the plant.

26. **(Medicine)** The Peabody Picture Vocabulary Test is used to evaluate the comprehension of receptive vocabulary by a patient receiving language therapy after a stroke. For a typical patient undergoing a specific treatment, the following data was accumulated:

Number of weeks of therapy	5	10	20	30
Raw score on test	75	80	85	95

(a) Find the line of best fit relating the raw score y and the number x of weeks of therapy.
(b) Use the equation obtained in (a) to predict the raw score after 52 weeks of therapy.

1.5 ||||||| QUADRATIC FUNCTIONS; PARABOLAS

Functions that can be written in the form

$$p(x) = a_0 + a_1 x + a_2 x^2 + \ldots + a_n x^n, \tag{1}$$

where the coefficients $a_0, a_1, a_2, \ldots, a_n$ are constants, are called **polynomial functions.** Examples of polynomial functions are

$$p(x) = 7 - 6x + 4x^2 + 10x^3 - 5x^4,$$
$$q(x) = x^{10} - 7x^7 + 4x^5 - x + 12,$$
$$f(x) = x^8.$$

Polynomials are studied in elementary algebra and are the simplest functions studied in calculus. A brief review of the algebra of polynomials may be found in Section A.5 (Appendix A).

In Equation (1) the number n is called the **degree** of $p(x)$, assuming that $a_n \neq 0$. A polynomial of degree 0 is called a **constant function,** and a polynomial of degree 1 is called a **linear function.** In this section we are concerned with polynomials of degree 2, which are called **quadratic functions.** The general form of a quadratic function is

$$f(x) = ax^2 + bx + c, \qquad (a \neq 0) \tag{2}$$

where a, b, and c are constants.
   Quadratic equations

$$ax^2 + bx + c = 0 \tag{3}$$

are studied in detail in elementary algebra. The reader will need to be able to use the two principal methods of solving them: by factoring and by the **quadratic formula**

$$x = \frac{-b \pm \sqrt{b^2 - 4ac}}{2a}. \tag{4}$$

These methods, along with other considerations such as using the discriminant

$$D = b^2 - 4ac,$$ (5)

are reviewed in Section A.6 (Appendix A).

|||||||||||| **Graphing**

The graph of a quadratic function is called a **parabola.** All parabolas are similar in shape; the shape of parabolas in general may be seen by inspecting just one of them.

EXAMPLE 1   Sketch the graph of the parabola $y = x^2$.

Solution   We make a table of values for this function.

x	0	1	-1	2	-2	3	-3	4	-4
y	0	1	1	4	4	9	9	16	16

We plot the points determined by this table and join them with a curve. The resulting curve is shown in Figure 1.

Figure 1

In general, the graph of any quadratic function

$$f(x) = ax^2 + bx + c \tag{6}$$

is a parabola with the following geometric characteristics.

(1) The parabola has an **axis of symmetry** that is vertical (parallel to the y-axis). If the plane were folded along this axis (see Figure 2), the two halves of the graph would coincide exactly.

(2) The equation of the axis of symmetry is

$$x = -\frac{b}{2a}. \tag{7}$$

(3) The point where the parabola intersects its axis of symmetry is called the **vertex** of the parabola.

Figure 2

(a) $a > 0$ (b) $a < 0$

(4) The parabola (6) **opens upward** (Figure 2a) if $a > 0$, and **opens downward** (Figure 2b) if $a < 0$.

(5) If $a > 0$ the vertex is the **lowest point** of the parabola (6). If $a < 0$ the vertex is the **highest point** of the parabola.

(6) The **x-intercepts** of the parabola (6) are the solutions of the quadratic equation $ax^2 + bx + c = 0$.

(7) The **discriminant** $D = b^2 - 4ac$ gives us the following information:

   (a) If $D > 0$, the parabola has two x-intercepts (Figure 3a).

   (b) If $D = 0$, the parabola is tangent to the x-axis at its vertex, and thus has only one x-intercept (Figure 3b).

   (c) If $D < 0$, the parabola has no x-intercept (Figure 3c).

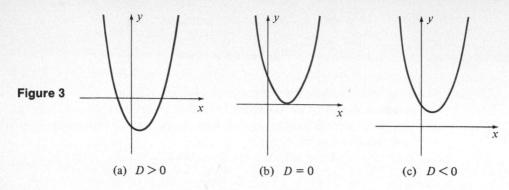

Figure 3

(a) $D > 0$ (b) $D = 0$ (c) $D < 0$ ■

EXAMPLE 2   Sketch the graph of the function

$$f(x) = -3 + 4x - x^2.$$

Solution   In this example $a = -1$, so the parabola opens downward. The axis of symmetry is the line

$$x = -\frac{b}{2a} = -\frac{4}{2(-1)} = 2.$$

Thus, the vertex has x-coordinate $x = 2$ and y-coordinate

$$y = f(2) = -3 + 4(2) - 2^2 = 1.$$

The point $(2, 1)$ is the vertex, and will be the highest point on the parabola. The x-intercepts are found:

$$-3 + 4x - x^2 = 0$$

or

$$x^2 - 4x + 3 = 0$$
$$(x - 1)(x - 3) = 0$$
$$x = 1 \quad \text{or} \quad x = 3.$$

We make a table of values including a few points on both sides of the vertex.

x	2	1	3	0	4	-1	5
y	1	0	0	-3	-3	-8	-8

vertex

The graph is shown in Figure 4.

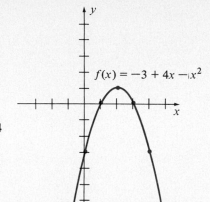

$f(x) = -3 + 4x - x^2$

Figure 4

■

EXAMPLE 3   Sketch the graph of the function

$$f(x) = x^2 + 2x + 5.$$

Solution   In this example $a = 1$, so the parabola opens upward. The axis of symmetry is the line

$$x = -\frac{b}{2a} = -\frac{2}{2} = -1.$$

The vertex has x-coordinate $x = -1$, and y-coordinate

$$y = f(-1) = 1 - 2 + 5 = 4.$$

Thus, the vertex is the point $(-1, 4)$ and is the lowest point on the graph.
When we attempt to find the x-intercepts, we first note that the discriminant is

$$D = b^2 - 4ac = 4 - 4(1)(5) < 0.$$

Thus, this parabola has no x-intercepts.
We make a table of values, including a few points on both sides of the vertex.

x	-1	-2	0	-3	1	-4	2
y	4	5	5	8	8	13	13

↑
vertex

The graph is shown in Figure 5.

Figure 5

$$f(x) = x^2 + 2x + 5$$

||||||||||||| **Applications**

We now provide a few examples showing how quadratic functions arise in applications. Further applications will be seen in the exercises. More substantial applications will be encountered as we progress through the course.

EXAMPLE 4   An 8-inch by 10-inch picture is to be surrounded by a uniform border that is x inches wide. Find the total area of the border as a function of x.

Solution   The area of the picture alone is

$$A_1 = 8 \cdot 10 = 40 \text{ square inches.}$$

Figure 6

The combined area of the picture plus frame (see Figure 6) is

$$A_2 = (8 + 2x) \cdot (10 + 2x)$$
$$= 80 + 36x + 4x^2 \text{ square inches.}$$

Therefore, the area of the frame is

$$A = A_2 - A_1$$
$$= (80 + 36x + 4x^2) - 80$$
$$= 36x + 4x^2 \text{ square inches.} \qquad ■$$

EXAMPLE 5
(Optimizing a
Quadratic
Function)

A farmer wishes to fence off a small portion of a large field into two adjacent rectangular garden plots, as shown in Figure 7.

Figure 7

He has a total of 600 feet of fencing available, which he wants to use up entirely.
(a) Write the total enclosed area as a function of x.
(b) Determine the dimensions x and y that will make the total enclosed area as large as possible.

Solution   (a) The total enclosed area is

$$A = xy. \qquad (8)$$

Since the total amount of fencing available is 600 feet, we want to have

$$3x + 2y = 600.$$

Thus,

$$2y = 600 - 3x$$
$$y = 300 - \tfrac{3}{2}x. \qquad (9)$$

Therefore, substituting (9) into (8), we find the total area as a function of x:

$$A(x) = x(300 - \tfrac{3}{2}x), \quad \text{or} \quad A(x) = 300x - \tfrac{3}{2}x^2.$$

(b) Notice that $A(x)$ is a quadratic function, and that the coefficient of x^2 is negative. Thus, the graph of $A(x)$ is a parabola opening downward, and has its highest point at its vertex, where $x = -\dfrac{b}{2a}$. For this function, $a = -\tfrac{3}{2}$ and $b = 300$. Thus, at the vertex,

$$x = -\frac{300}{-3} = 100.$$

Therefore, the area A has its largest value when

$$x = 100 \text{ ft.}$$

and, using Equation (9),

$$y = 300 - (\tfrac{3}{2})(100)$$
$$y = 150 \text{ ft.} \qquad \blacksquare$$

|||||||||||||| **Profit vs. Revenue**

A business or industry sells a certain commodity at a selling price of x dollars (or cents) per unit. Over one time period (1 day, 1 week, 1 month, and so on) the following functions are relevant:

$R(x) = $ **revenue** over one time period, when the selling price is x dollars (or cents) per unit.

$C(x) = $ **cost** of obtaining or producing the commodity sold during one time period, when the selling price is x.

$P(x) = $ **profit** over one time period, when the selling price is x.

In our model, in each time period the business obtains or produces only as many units as it can sell during that time period. Thus, these three functions satisfy the equation

$$P(x) = R(x) - C(x). \qquad (10)$$

EXAMPLE 6   Rex's Drugstore obtains regular size AA batteries at a wholesale cost of 30 cents each. Experience has shown Rex that if the store sells the batteries at x cents each then on the average they will sell $80 - x$ batteries daily.
(a) Find formulas for $R(x)$, $C(x)$, and $P(x)$.
(b) Find the daily profit when $x = 40$ cents; when $x = 60$ cents.
(c) Find the most profitable selling price.

Solution   (a) Since the store will sell $80 - x$ batteries daily at x cents each, the daily revenue will be

$$R(x) = x(80 - x) \text{ cents.}$$

Since the store will need $80 - x$ batteries at 30 cents, the daily cost will be

$$C(x) = 30(80 - x) \text{ cents.}$$

Therefore, from Equation (10), the daily profit from the sale of the batteries will be

$$
\begin{aligned}
P(x) &= x(80 - x) - 30(80 - x) \\
&= (x - 30)(80 - x) \\
&= -(x - 30)(x - 80).
\end{aligned}
\tag{11}
$$

Thus,

$$P(x) = -x^2 + 110x - 2400. \tag{12}$$

(b) From (11),

$$P(40) = (40 - 30)(80 - 40) = 10(40) = 400 \text{ cents} = \$4.00.$$

and

$$P(60) = (60 - 30)(80 - 60) = 30(20) = 600 \text{ cents} = \$6.00.$$

(c) The profit function $P(x)$ is a quadratic function, with $a = -1 < 0$. Thus, its graph is a parabola opening downward. The vertex is at the point where

$$x = -\frac{b}{2a} = -\frac{110}{2(-1)} = 55.$$

This is the x-coordinate of the highest point on the parabola (see Figure 8).

Figure 8

Therefore, $P(x)$ is greatest when $x = 55$. That is, the most profitable selling price for the batteries at Rex's Drugstore is 55 cents.

In fact, when $x = 55$ the daily profit from selling the batteries will be

$$P(55) = (55 - 30)(80 - 55) = 25(25) = 625 \text{ cents}$$
$$= \$6.25.$$

If the selling price is set any higher or lower than 55 cents each, the daily profit will be lower.   ∎

1.5 EXERCISE SET ||||||||||||||||||||||||||||||||||||||||||||||||||||||||||||||||||||||||||||||||||||||||||

In Exercises 1–8, sketch the graph of the given quadratic function. In each exercise, find the vertex and the axis of symmetry.

1. $f(x) = x^2 - x$

2. $f(x) = 2x - x^2$

3. $f(x) = -x^2 + x + 6$

4. $f(x) = x^2 + x - 2$

5. $g(x) = -x^2 + 4x + 3$

6. $g(x) = 3 + 2x - x^2$

7. $g(x) = 2x^2 - 8x + 9$

8. $g(x) = -3x^2 + 4x - 2$

In Exercises 9–16, use the discriminant to determine how many x-intercepts the associated graph has.

9. $f(x) = x^2 - 6x + 9$

10. $f(x) = x^2 - 5$

11. $f(x) = x^2 + x + 1$

12. $f(x) = 3 + x^2$

13. $f(x) = 2x^2 + 3x - 3$

14. $f(x) = 4x^2 - 12x + 9$

15. $f(x) = (x + 5)^2$

16. $f(x) = 3x^2 - 2x + 1$

17. In preparing a border for framing a picture, the center of a 12-inch by 16-inch rectangular mat is removed, leaving a border x inches wide (see Figure 9).

(a) Express the area of the piece removed, as a function of x.

(b) Express the area of the border as a function of x.

Figure 9

Figure 10

18. A 3-feet by 7-feet doorway is surrounded by a border x feet wide (see Figure 10). Find the area of the border as a function of x.

19. A gardener has 100 feet of fencing available to be entirely used in fencing off a small rectangular plot, of length l and width w.
    (a) Find the area enclosed as a function of the length l.
    (b) Determine the length and width that will make this area as large as possible.

20. Before purchasing paint to cover the walls and ceiling of a rectangular room, we must find the total area to be covered. The room is 8 feet high, and the floor is a rectangle whose length is 3 feet shorter than twice the width, w. Find the total area of the walls and ceiling as a function of w.

21. The sum of two numbers is 100. Let x denote one of the numbers.
    (a) Express the product of the numbers as a function of x.
    (b) Show that this product has a largest possible value. Find it, and find the two numbers for which this product is largest.

22. One number is 60 more than twice a second number. Let x denote the second number.
    (a) Express the product of the two numbers as a function of x.
    (b) Show that this product has a smallest value. Find it, and find the two numbers for which this product is smallest.

23. **(Profit vs. Revenue)** Yums bakery makes their special pastry at a cost of 50 cents each. They estimate that if they sell them at x cents each they will sell $600 - 5x$ of these pastries daily.
    (a) Express their daily profit from these pastries as a function of x.
    (b) Find the daily profit when $x = 70$ cents.

(c) Find the most profitable selling price x.

24. **(Profit vs. Revenue)** It costs a clothing store $20 per pair to stock Brand X blue jeans. The store finds that it sells $100 - 2x$ pairs monthly when the selling price is x per pair.
    (a) Express the monthly profit as a function of x.
    (b) Find the monthly profit when $x = 25$; when $x = 40$.
    (c) Find the most profitable selling price.

25. **(Profit vs. Revenue)** Organizers of a benefit concert determine that if they sell all 2500 seats for the coming performance at $8 each, they will exactly break even (that is, show no profit). They estimate that for each additional $1 charged per ticket, 100 fewer people will come. Suppose they set the price per ticket at $(8 + x)$ dollars. Find
    (a) the total cost of the concert.
    (b) $R(x) =$ anticipated revenue, as a function of x.
    (c) $P(x) =$ anticipated profit, as a function of x.
    (d) the most profitable price per ticket. The anticipated attendance, and the profit, at that price.

26. **(Profit vs. Revenue)** A campus store normally sells 80 school blazers each semester, at a profit of $20 per blazer. The store management estimates that for each $1 they lower the selling price (that is, $1 less profit) they will sell 5 additional blazers per semester. Suppose they lower the price x. Find
    (a) $P(x) =$ profit per semester, as a function of x.
    (b) $P(0) =$ the normal profit per semester.
    (c) the profit per semester if the profit per blazer is lowered to $19.
    (d) the most profitable value of x. The profit per blazer, and the profit per semester, at this value of x.

1.6 ||||||| INTERSECTIONS OF GRAPHS

In Example 9 of Section 1.3 we found the point of intersection of two straight lines, and in Example 4 of Section 1.4 we saw that the graphs of the revenue and cost functions intersect at the break-even point. Sometimes it is necessary to find the domain value(s) at which two given functions have the same y-value. Equivalently, we are interested in finding the point(s) at which the graphs of the given functions intersect. In Figure 1, the graphs of f and g intersect at the points $P_1(a, f(a))$ $(= P_1(a, g(a)))$ and $P_2(b, f(b))$ $(= P_2(b, g(b)))$.

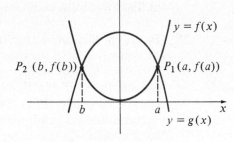

Figure 1

EXAMPLE 1   Find the point(s) of intersection of the graphs of the functions $f(x) = x^2 + 1$ and $g(x) = 2x$.

Solution   Since a point $P(x, y)$ at which the graphs intersect must lie on both curves, we have to solve the equations

$$y = x^2 + 1$$
$$y = 2x \tag{1}$$

simultaneously. Equating the expressions for y, we have

$$x^2 + 1 = 2x$$

or

$$x^2 - 2x + 1 = 0$$
$$(x - 1)^2 = 0.$$

Hence,

$$x = 1.$$

From (1), we find $y = 2$ when $x = 1$. Thus, the only point of intersection of the graphs is $(1, 2)$. The graphs are shown in Figure 2.

Figure 2

EXAMPLE 2   Find the point(s) of intersection of the graphs of the functions $f(x) = x^3$ and $g(x) = x$.

Solution   Equating the values of the functions, we set

$$x^3 = x$$

or

$$x^3 - x = 0$$
$$x(x^2 - 1) = 0$$
$$x(x - 1)(x + 1) = 0.$$

Hence,

$$x = 0, \qquad x = 1, \quad \text{or} \quad x = -1.$$

Substituting these values of x into the formula for $f(x)$ or $g(x)$ we find that $y = 0$ when $x = 0$, $y = 1$ when $x = 1$, and $y = -1$ when $x = -1$. The points of intersection are then

$$(0, 0), \qquad (1, 1), \qquad \text{and} \qquad (-1, -1).$$

The graphs are shown in Figure 3.

Figure 3

EXAMPLE 3
**(The Method
of Substitution)**

Find the points of intersection of the curves $x^2 + y^2 = 34$ and $y = x + 2$.

Solution

We must algebraically solve the equations

$$x^2 + y^2 = 34 \qquad\qquad (2)$$
$$y = x + 2 \qquad\qquad (3)$$

simultaneously; that is, we must find all (x, y) that satisfy both equations at the same time. We **substitute** the expression for y given by Equation (3) into Equation (2):

$$x^2 + (x + 2)^2 = 34$$
$$x^2 + x^2 + 4x + 4 = 34$$
$$2x^2 + 4x - 30 = 0$$
$$x^2 + 2x - 15 = 0$$
$$(x + 5)(x - 3) = 0.$$

Thus, either $x = -5$ or $x = 3$. We then substitute each of these x-values into the first degree equation (3) to obtain the corresponding y-value.

When $x = -5$ we have $y = -5 + 2 = -3$, from (3), and when $x = 3$ we have $y = 3 + 2 = 5$. Therefore, the curves intersect at the two points $(-5, -3)$ and $(3, 5)$.

The two curves and their points of intersection are shown in Figure 4.

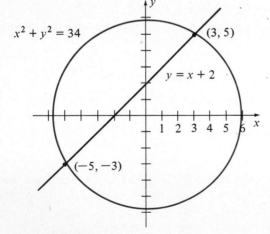

Figure 4

EXAMPLE 4
**(Break-Even
Analysis)**

In maintaining a production level of x units a day, a manufacturer has fixed daily costs of \$5000 and variable costs of $80x - x^2$ dollars per day. The manufacturer sells all units produced at \$30 each.

(a) Find the daily cost function $C(x)$ and revenue function $R(x)$.
(b) Find the break-even point [where $R(x) = C(x)$].

Solution   (a) Adding the fixed and variable costs, we obtain the total daily cost function

$$C(x) = 5000 + 80x - x^2 \text{ (dollars).}$$

Since all x units produced daily will be sold at \$30, the daily revenue function is

$$R(x) = 30x \text{ (dollars).}$$

(b) To find the break-even point, we solve $R(x) = C(x)$ algebraically:

$$5000 + 80x - x^2 = 30x$$
$$x^2 - 50x - 5000 = 0$$
$$(x - 100)(x + 50) = 0.$$

Either $x = 100$ or $x = -50$. Since the break-even point cannot be negative (we cannot produce a negative number of units per day), we discard $x = -50$ but keep $x = 100$. The break-even point is thus at 100 units per day.

Figure 5 shows the graphs of $C(x)$ and $R(x)$, and the break-even point.

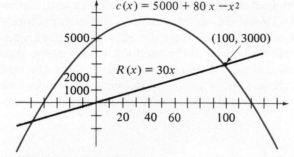

Figure 5

■

||||||||||||| **Supply and Demand**

Intersections of graphs are especially useful in economics and business applications. For example, a manufacturer is free to set any price p (in dollars) for each unit of the product. Of course, if the price is too high not enough people will buy the product; if the price is too low, so many people will rush to buy the product that the producer will not be able to satisfy the demand. Thus, in setting the price, the manufacturer must take into consideration the demand for the product.

Let $p = S(x)$ be the unit price at which a manufacturer is willing to supply x units of the product; S is called the **supply** function. Generally, as p increases, x increases; that is, the manufacturer is willing to supply more of the product as the unit price p increases. Let $p = D(x)$ be the unit price at which a consumer is willing to buy x units of the product; D is called the **demand** function. Generally, as p increases, x decreases; that is, consumers are willing to buy fewer units of the product as the unit price rises.

The general graph of a pair of supply and demand functions for the same product is shown in Figure 6. The point of intersection is denoted by (x_E, p_E). The price p_E at which supply $S(x)$ and demand $D(x)$ are equal is called the **equilibrium price.** At this price, every unit that is supplied is purchased. Thus, there is neither a surplus nor a shortage. The quantity x_E is the amount supplied at the equilibrium price.

Figure 6

If we are in an economic system in which there is pure competition, then the **law of supply and demand** states that the selling price of a product will be its equilibrium price. In other words, if the selling price is higher than the equilibrium price, then the reduced consumer demand would leave the manufacturer with an unsold surplus. This situation would force the manufacturer to reduce the selling price. If the selling price is below the equilibrium price, then the increased demand would cause a shortage of the product. In this event, the manufacturer, would raise the selling price. Of course, in actual practice, the market-place does not follow the principle of pure competition because manufacturers check each others' selling prices, governments try to influence selling prices, and many other factors intervene.

EXAMPLE 5 Suppose that the supply and demand functions for a product are given (in dollars) by

$$p = S(x) = x^2 + 3x$$

and

$$p = D(x) = 12 - x.$$

Find the equilibrium price and the number of units sold at this price.

Solution Setting $S(x) = D(x)$, we have

$$x^2 + 3x = 12 - x$$

or

$$x^2 + 4x - 12 = 0$$
$$(x + 6)(x - 2) = 0$$
$$x = -6, \qquad x = 2.$$

Since the quantity of units sold cannot be negative, we reject the solution $x = -6$. Thus, the number of units sold at the equilibrium price is $x_E = 2$ units. The equilibrium price is

$$p_E = S(2) = 2^2 + 3 \cdot 2 = 10.$$

Hence, the equilibrium price is $10 per unit and 2 units are sold at this price. The graphs of this supply and demand function are shown in Figure 7.

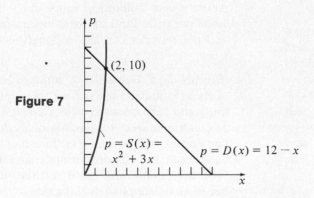

Figure 7

■

1.6 EXERCISE SET ||||||||||||||||||||||||||||||||||||||||||||||||||||||||||||||||||||||||||||||||||||||||||||||

In Exercises 1–18, find the point(s) of intersection of the graphs of the given functions.

1. $f(x) = x^2$, $g(x) = 2x$

2. $f(x) = x^2 - 4$, $g(x) = 3x$

3. $F(x) = x^2 + 3$, $g(x) = 4x$

4. $f(x) = x^2 + 4x$, $g(x) = x + 10$

5. $f(x) = 3 - x$, $g(x) = x^2 + x$

6. $f(x) = x^3 + 2x$, $g(x) = 3x^2 + 2x$

7. $f(x) = x^2$, $g(x) = -x^2 + 8$

8. $f(x) = x^2 + 4$, $g(x) = -x^2 + 22$

9. $f(x) = x^2$, $g(x) = \dfrac{1}{x}$

10. $f(x) = x$, $g(x) = \dfrac{4}{x}$

11. $x^2 + y^2 = 25$
    $3x - 4y = 0$

12. $3x^2 + y^2 = 12$
    $3x + y = 6$

13. $f(x) = \dfrac{2}{x}$, $g(x) = x - 1$

14. $f(x) = \dfrac{1}{x}$, $g(x) = \dfrac{5x + 2}{3}$

15. $x^2 = 9y$
      $x^2 - y^2 = 8$

16. $y^2 = 6x$
      $x^2 - 2y^2 = -27$

17. $x^2 + y^2 = 10$
      $9x^2 + y^2 = 18$

18. $x^2 + 4y^2 = 65$
     $x^2 - 4y^2 = 33$

19. Prove algebraically that the graphs of the functions $f(x) = x^2 - x$ and $g(x) = x - 2$ do not intersect.

20. Prove algebraically that the graphs of the curves $x^2 + y^2 = 1$ and $x + y = 3$ do not intersect.

21. **(Break-Even Analysis)** When producing x units daily, a manufacturer has fixed costs of $1000 and variable costs of $\sqrt{x+5}$ (thousands of dollars) daily. The manufacturer sells all units produced at $1000 dollars each.
    (a) Find the daily cost and revenue functions.
    (b) Find the break-even point.

22. **(Break-Even Analysis)** It costs a craftsman $12 each, in addition to a $100 fixed daily cost, to make his special bracelet. As an incentive to induce the craftsman to produce more of these bracelets, the distributor offers to pay him $(27 + x)$ each if he makes x bracelets a day.
    (a) Find the craftsman's daily cost and revenue functions, assuming he makes x bracelets a day.
    (b) Find the break-even point.

23. **(Supply and Demand)** Suppose a manufacturer of ball-point pens finds that the supply and demand functions for a new model are given by

$$p = S(x) = x^2 + 2x$$
$$p = D(x) = 24 - x^2$$

where x is in millions of units and p is in cents per unit. Find the equilibrium price and the number of units supplied at this price.

24. **(Supply and Demand)** A manufacturer of hand calculators is willing to supply x thousand calculators at $x^2 + x$ dollars per calculator. Market research indicates that consumers will purchase x thousand calculators at $18 - 2x$ dollars per calculator. Find the equilibrium price and the number of units supplied at this price.

KEY IDEAS FOR REVIEW ||||||||||||||||||||||||||||||||||||||||||||||||||||||||||||||||||||||||

- A function is a rule or formula that determines the unique value of one variable (the dependent variable) once the value of another variable (the independent variable) has been specified.

- A single equation is not the only way to define a function. Sometimes a function is defined by a table, chart, or by several equations, in a piecewise definition.

- The domain of a function consists of the set of all real numbers at which the function is defined and yields a real number.

- The graph of the function f is the graph of the equation $y = f(x)$.

- The vertical line test: if any vertical line cuts a curve at more than one point, then the curve is not the graph of any function of x.

- $(f \circ g)(x) = f(g(x))$.

- In business and economics problems, the cost, revenue, and profit functions occur frequently; moreover, $P(x) = R(x) - C(x)$. The "break-even point" occurs where $R(x) = C(x)$.

- The slope m of a nonvertical line is given by

$$m = \frac{(y_2 - y_1)}{(x_2 - x_1)}$$

where $P_1(x_1, y_1)$ and $P_2(x_2, y_2)$ are any two distinct points on the line.

- A vertical line has no slope.

- If the slope m is positive, then y increases as x increases (the line *rises* from left to right); if m is negative, then y decreases as x increases (the line *falls* from left to right).

- The slope-intercept form of a line is $y = mx + b$.

- If two nonvertical lines have the same slope, then they are parallel. Conversely, if two nonvertical lines are parallel, then they have the same slope.

- The point-slope form of a line that passes through the point $P(x_1, y_1)$ and has slope m is $y - y_1 = m(x - x_1)$.

- The slope of a line parallel to the x-axis is zero.

- The equation of a vertical line through (a, b) is $x = a$. The equation of a horizontal line through (a, b) is $y = b$.
- The equation of a straight line can be written as $Ax + By = C$, where A and B are not both zero. Conversely, the graph of the linear equation $Ax + By = C$ (A and B not both zero) is a straight line.
- Two lines are parallel or identical or intersect at only one point.
- When a principal P is invested for t years at simple interest rate r, the interest earned is $I = Prt$ and the value of the investment after t years is $S = P + I = P(1 + rt)$.
- If a merchant marks up an item that costs him C by the rate r, then the selling price is

$$S = C + rC = C(1 + r).$$

Similarly, if an item with list price L is discounted at the rate r, then the selling price is

$$S = L - rL = L(1 - r).$$

- If an item with net cost C is depreciated linearly over n years, then the annual depre-ciation is $D = \dfrac{C}{n}$. After t years the total accumulated depreciation is Dt, and the book value of the item is $V(t) = V(0) - Dt$.
- Given a set of n data points (x_1, y_1), (x_2, y_2), ..., (x_n, y_n), the line of best fit obtained by the method of least squares can be found by solving the system (16) of Section 1.4 for the slope m and the y-intercept b.
- The graph of a second-degree function $f(x) = ax^2 + bx + c$ is a parabola, opening upward if $a > 0$ and downward if $a < 0$. The vertex has x-coordinate $x = -\dfrac{b}{2a}$. The axis of symmetry is the vertical line through the vertex.
- The parabola $y = ax^2 + bx + c$ has either a highest point (if $a < 0$) or a lowest point (if $a > 0$). This is important when we are looking for the largest or smallest value of a quadratic function.
- The point of intersection of the graphs of two equations can be found algebraically by solving the two equations simultaneously.

REVIEW EXERCISES |||||||||||||||||||||||||||||||||||||||||||||||||||||||||||||||||||||||||||||||||||

1. If $f(x) = 3x^2 - 2$, find
   (a) $f(-1)$ (c) $f(2)$
   (b) $f(0)$ (d) $f(2x)$

2. If $f(u) = (u + 1)/(u^2 - 2)$, find
   (a) $f(-3)$ (c) $f(2)$
   (b) $f(0)$ (d) $f(x - 1)$

3. Give the domain of the function $f(x) = \sqrt{2x - 1}$.

4. Sketch the graph of the function
$$f(s) = \begin{cases} s + 1 & \text{if } s \geq 1 \\ s - 1 & \text{if } s < 1 \end{cases}$$

5. Which of the following is the graph of a function of x?

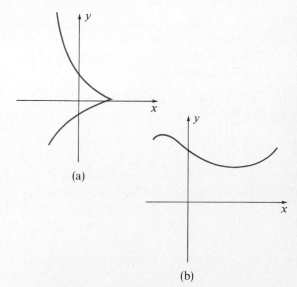

(a)

(b)

6. Express the function $h(x) = 3(x^2 + x - 7)^5$ as a composite of two simpler functions.

7. Find the slope of the line in the following figure.

8. For the line $2y + 3x - 2 = 0$, does y increase as x increases?

9. Are the lines $x + \frac{1}{2}y - 4 = 0$ and $y = 8 - 2x$ parallel?

10. Find an equation of the line parallel to the line $x + 2y + 3 = 0$ with a y intercept of -3.

11. Sketch the line having slope $-\frac{1}{2}$ and passing through the point $(-1, 3)$.

12. Find the point-slope form of the line having slope $m = 2$ and passing through the point $(2, 1)$.

13. Find the point-slope form of the line determined by the points $(3, 2)$ and $(-2, -3)$.

14. (a) Write the equation of the horizontal line through $(-3, 7)$.
    (b) Write the equation of the vertical line through $(-3, 7)$.

15. Find an equation of the line in the following figure.

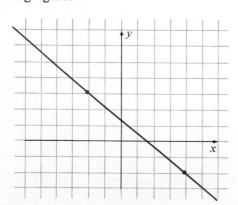

16. Write the equation $y + 3 = \frac{2}{3}(x - 2)$ in the form $Ax + By + C = 0$ and give the values of A, B, and C.

17. Which of the following are linear equations in x and y?
    (a) $3x^2 + y = 4$
    (b) $2(x + 1) + \frac{1}{2}(y - 4) = 5$

18. Consider the pair of lines $3x - 2y = 6$ and $-x + \frac{2}{3}y = -2$. Determine whether they intersect at only one point, are parallel, or are identical.

19. If you borrow $5000 at the simple interest rate of 7% per year, how much do you owe after t years? After 4 years? After 10 years? How long will it take your debt to grow to $10,000?

20. A car rental firm's daily charges are $14 plus $0.08 per mile.
    (a) Express the cost per day of renting a car as a function of the number of miles traveled.
    (b) What is the domain of this function?
    (c) How much would it cost to rent a car for a 100-mile day trip?

21. A record club offers the following sale. Buy 3 records at the regular price of $7.98 each and you can buy up to 7 more records at half price.
    (a) Express the total cost as a function of the number x of records bought, $0 \le x \le 10$.
    (b) How much will it cost to buy a total of 8 records?

22. A manufacturer of attaché cases finds that the fixed cost is $1200 per week and the cost to make one attaché case is $9. The manufacturer sells each case for $12.
    (a) Write the weekly cost, revenue, and profit functions.
    (b) Determine the weekly cost, revenue, and profit if the level of production is 2000 cases per week.

23. An office purchases a new desk-top computer system for $7300. It is expected to have a useful life of 9 years, after

which it will have a salvage value of $1000. Assuming linear depreciation,
(a) find the annual depreciation in dollars, and as a percent of the original cost.
(b) find the accumulated depreciation after t years.
(c) find the book value of the computer system after t years.
(d) sketch the graph of the function $V(t)$.

24. Find the point(s) of intersection of the graphs of the functions $f(x) = x^2$ and $g(x) = 3x + 10$.

25. Find the point(s) of intersection of the graphs of the curves $7x^2 - 2xy = 24$ and $5x - y = 9$.

26. Sketch the graph of the quadratic function $f(x) = 3x^2 + 10x - 8$. Find the x-intercepts, the vertex, and the axis of symmetry.

27. Use the discriminant to determine how many x-intercepts the parabola $y = 25x^2 - 30x + 9$ has.

28. A 20-inch length of wire is bent into the shape of a rectangle of width w. Write a formula expressing the area enclosed by the wire as a function of w.

29. A campus store obtains sweatshirts for $9 each. They know that if they sell the sweatshirts for $x each, they will sell $50 - 2x$ of them weekly.
(a) Express the anticipated weekly profit as function of x.
(b) Find the weekly profit when $x = 10$; when $x = 13$; when $x = 20$.
(c) Find the most profitable selling price.

30. Suppose that a manufacturer of toaster ovens finds that the supply and demand functions for a certain model are given by

$$p = S(x) = x^2 + 6x$$
$$p = D(x) = 39 - 4x$$

where x is in millions of units and p is in dollars per unit. Find the equilibrium price and the number of units supplied at this price.

31. Find the line which best fits the data points $(-3, 3)$, $(2, 5)$, $(4, 8)$ and $(5, 7)$ by the method of least squares.

CHAPTER TEST ||||||||||||||||||||||||||||||||||||||||||||||||||||||||||||||||||||||||||||||||||

1. For the function $f(x) = \dfrac{x - 5}{3x + 1}$, find $f(0)$, $f(3)$, and $f(2x)$.

2. Specify the domain of the function $f(x) = \sqrt{3x + 5}$.

3. Sketch the graph of the function $g(x) = \begin{cases} 2 + x, & \text{if } x < 0 \\ 2, & \text{if } x \geq 0. \end{cases}$

4. An ice-cream bar has a fixed cost of $160 per week associated with making ice cream, and an additional cost of $2.30 per gallon in making the ice cream. The ice cream is sold for $3.50 per gallon.
   (a) Write the cost, revenue and profit functions.
   (b) Determine the cost, revenue and profit for a week in which 400 gallons are sold.

5. Find the equation of the straight line passing through the points $(2, -1)$ and $(4, 2)$.

6. Find the equation of the straight line passing through $(-1, 3)$ and parallel to the line $2x + 3y = 4$.

7. Consider the line $y = -2x + 7$.
   (a) Does the line *rise* or *fall* from left to right?
   (b) Does y increase or decrease as x increases?
   (c) When x increases by 3 units, by how much does y increase or decrease?
   (d) Sketch the graph of this line.

8. A small manufacturer has weekly cost and revenue functions (in thousands of dollars)

$$C(x) = 30 + 0.3x$$
$$R(x) = 0.9x$$

   when it is producing x units of its product weekly.
   (a) Sketch the cost and revenue functions in the same coordinate system.
   (b) Find the break-even point.
   (c) Find the total weekly revenue at the break-even point.

9. A piece of equipment costing $50,000 when new depreciates linearly at the rate of 8% of its original value annually.
   (a) How much (in dollars) does it depreciate annually?
   (b) What is its book value V after t years?
   (c) Sketch the graph of the function $V = V(t)$ obtained in (b).
   (d) How long will it take for the value V to decline to $20,000?

10. Use the discriminant to determine how many x-intercepts the parabola $y = 3x^2 - 4x + 2$ has.

11. Sketch the graph of the function $f(x) = 2x^2 - 5x - 3$. Show the vertex.

12. A college snack bar makes hot dog sandwiches at a cost of 60¢ each. They determine that when the selling price is x¢ each, they sell $100 - \dfrac{x}{2}$ hot dogs daily.
   (a) Find formulas for $R(x)$, $C(x)$, and $P(x)$.
   (b) Find the most profitable selling price, and the daily profit from hot dogs at that selling price.

13. Find the point(s) of intersection of the graphs of the functions $f(x) = x^2 - x$ and $g(x) = x + 3$.

LIMITS, CONTINUITY, AND RATES OF CHANGE

In this chapter we shall actually begin the study of calculus, which was developed in the seventeenth century by Isaac Newton* and Gottfried Wilhelm von Leibniz.[†] One of the greatest developments in the history of science and mathematics, *calculus is the mathematics of change*. Since everything around us is changing, calculus applies to the solution of a great many pressing problems in almost every field of human endeavor. These include applications in the physical sciences, engineering, business and economics, management, medicine and the biological sciences, psychology, and the social sciences. Two basic geometric problems that calculus is instrumental in solving are:

1. Finding a tangent to a curve, and
2. Finding the area under a curve.

We shall see, in the next chapter, that problem (1) is closely related to the problem of calculating the rate at which a variable quantity changes in value. Calculus is generally divided into differential calculus, which solves problem (1), and integral calculus, which solves problem (2).

In the first section of this chapter we shall introduce the concept of a limit. This important notion, which lies at the very heart of calculus, underlies the development of both differential and integral calculus. It will be seen to have immediate practical value in Section 2.3, where we shall want to determine when the graph of a function is an unbroken curve. In Section 2.4 we shall apply limits to study the rate at which one quantity changes with respect to another quantity.

---

* Isaac Newton (1642–1727) was born in Wools Thorpe, Lincolnshire, England. He is considered one of the greatest names in the history of human thought because of his fundamental contributions in mathematics, physics, and astronomy. In a period of eighteen months (1665–67) he developed the calculus, discovered the theory of gravity, and made contributions to the theory of light and color (for example, he showed that white light is made up of many colors). His work in astronomy provided an explanation of the motion of planets in their orbits around the sun. He constructed a new type of telescope that used a reflecting mirror.

[†] Gottfried Wilhelm von Leibniz (1646–1716) was a German philosopher, mathematician, physicist, and historian. He was a genius who made major contributions to mathematics, logic, philosophy, mechanics, geology, law, and theology, all while pursuing a career as a civil servant. He was named a Baron in 1700. Leibniz invented differential and integral calculus later than, but independently of, Isaac Newton, and in his later years he became embroiled in a bitter dispute with friends of Newton over who invented calculus.

2.1 ‖‖‖‖‖‖ LIMITS

In many applied problems, it is necessary to describe the behavior of a function f when x is near, but different from, a. For example, consider a manufacturer of a single product whose profit $P(x)$ is a function of the number of units x that the firm makes and sells. It is important to know the level of production x at which $P(x)$ becomes negative; that is, the value $x = a$ at which the producer starts to lose money. It is just as important, however, to know how the profit $P(x)$ behaves when x is *near* the crucial value a. This information might help the manufacturer to develop a plan for avoiding losses.

EXAMPLE 1   Consider the function f defined by

$$f(x) = 2x + 1.$$

How do the values $f(x)$ behave as x gets close to (but remains different from) $x = 2$? As shown in Figure 1, we consider two different ways in which x can get close to, or approach, the value 2: either from the right of 2 or from the left of 2.

Figure 1

Table 1 shows the values of $f(x)$ as x approaches 2 from the right and Table 2 shows these values as x approaches 2 from the left. We have plotted the data in Tables 1 and 2 on the graph shown in Figure 2.

TABLE 1 x approaches 2 from the right

x	3	2.5	2.3	2.1	2.05	2.01	2.001	2.0001
$f(x) = 2x + 1$	7	6	5.6	5.2	5.1	5.02	5.002	5.0002

TABLE 2 x approaches 2 from the left

x	1	1.5	1.7	1.9	1.95	1.99	1.999	1.9999
$f(x) = 2x + 1$	3	4	4.4	4.8	4.9	4.98	4.998	4.9998

Figure 2

It is intuitively clear from Figure 2 that the values of $f(x) = 2x + 1$ get close to 5 as x approaches 2 from either side. The number 5 is called the **limit** of f as x approaches 2, and we write

$$\lim_{x \to 2} f(x) = 5 \quad \text{or} \quad \lim_{x \to 2} (2x + 1) = 5. \tag{1}$$

Here $x \to 2$ means "x approaches 2 (either from the right or from the left) but $x \neq 2$." Equation (1) is read, "the limit of $f(x)$ as x approaches, but remains different from 2, is 5."

It is important to note that the value of the limit of f as $x \to 2$ in Equation (1) was obtained *intuitively*. It is possible to give a mathematically more precise definition of limit and then to carefully show that the above limit is indeed 5, but the intuitive approach is adequate for our needs.

||||||||||||| **Intuitive Definition of Limit**

> A function f is said to **approach the limit** L, as x approaches a if the values of $f(x)$ get closer and closer to the unique real number L as x gets closer and closer to (but remains different from) a. We write this statement as
>
> $$\lim_{x \to a} f(x) = L.$$

Let us examine the notion of a limit a little more closely.

By definition, the statement

$$\lim_{x \to a} f(x) = L$$

means that it is possible to make the value of $f(x)$ as close to L as we want by taking x sufficiently close to (but different from) a. It makes no statement whatsoever about the value of $f(x)$ when x actually equals a.

The condition "$f(x)$ gets closer and closer to" specified in our definition must be taken somewhat loosely. This is true, in particular, if $f(x)$ already equals L even though x continues to get closer to a. Thus, for example,

$$\lim_{x \to 4} 3 = 3$$

even though the constant function $f(x) = 3$ doesn't get "closer" to 3 as x approaches 4. In general, for any constant function $f(x) = c$ and any number a,

$$\lim_{x \to a} c = c. \tag{2}$$

You may have noticed that in Example 1 we have $f(2) = 5$, so in *this case*, $\lim_{x \to 2} f(x)$ is simply $f(2)$. As we shall see in the following example, it is not always true that $\lim_{x \to a} f(x)$ is $f(a)$.

EXAMPLE 2   Consider the function f defined by

$$f(x) = \frac{x^2 - 9}{x - 3}.$$

First, observe that this function is not defined for $x = 3$ because we cannot divide by zero. Thus, we cannot possibly find $\lim_{x \to 3} f(x)$ merely by evaluating $f(3)$. The behavior of $f(x)$ as $x \to 3$ from the right and left is shown in Tables 3 and 4, respectively. Table 3 shows that the values of $f(x)$ approach 6 as x approaches 3 from the right. Table 4 shows that the values of $f(x)$ approach 6 as x approaches 3 from the left.

TABLE 3 $x \to 3$ from the right

x	4	3.5	3.1	3.01	3.001	3.0001
$f(x) = \dfrac{x^2 - 9}{x - 3}$	7	6.5	6.1	6.01	6.001	6.0001

TABLE 4 $x \to 3$ from the left

x	2	2.5	2.9	2.99	2.999	2.9999
$f(x) = \dfrac{x^2 - 9}{x - 3}$	5	5.5	5.9	5.99	5.999	5.9999

Consequently,

$$\lim_{x \to 3} f(x) = 6, \qquad \text{or} \qquad \lim_{x \to 3} \frac{x^2 - 9}{x - 3} = 6.$$

To see what is happening geometrically, we graph the function f. Observe that if $x \neq 3$, then

$$f(x) = \frac{x^2 - 9}{x - 3} = \frac{(x + 3)(x - 3)}{(x - 3)} = x + 3.$$

Thus, when $x \neq 3$, the graph of f is the graph of $y = x + 3$, a straight line. If $x = 3$, $f(x)$ is not defined. Hence, the graph of f, shown in Figure 3, is a straight line with a missing point or gap at $x = 3$. As usual, we have marked the point where $x = 3$ with an open circle to indicate that it is not on the line. As the graph shows, as x approaches 3, the values of $f(x)$ approach the number 6.

Figure 3

||||||||||||| **One-Sided Limits**

In the two previous examples the value approached by $f(x)$ was the same finite number, whether x approached a from the right or from the left. This need not be the case, as the following examples show.

EXAMPLE 3   Let f be defined by

$$f(x) = \frac{|x|}{x}.$$

We are interested in

$$\lim_{x \to 0} f(x).$$

First, observe that $f(0)$ is undefined. Moreover, we can write the given function as

$$f(x) = \begin{cases} \dfrac{x}{x} = 1, & \text{if } x > 0 \\[2ex] -\dfrac{x}{x} = -1, & \text{if } x < 0. \end{cases}$$

As $x \to 0$ from the right, then $f(x)$ is always 1, so $f(x)$ approaches 1. If $x \to 0$ from the left, then $f(x)$ is always -1, so $f(x)$ approaches -1. Hence, the values of $f(x)$ do not approach a single finite number as $x \to 0$. This behavior is evident in the graph of f, which is shown in Figure 4.

Figure 4 $f(x) = \dfrac{|x|}{x}$

If the values of $f(x)$ do not approach a unique real number as x approaches the value a, we say that

$$\lim_{x \to a} f(x)$$

does not exist.

Thus, for the function in Example 3, $\lim\limits_{x \to 0} f(x)$ does not exist; that is, $\lim\limits_{x \to 0} \dfrac{|x|}{x}$ does not exist.

Nevertheless, the function $f(x) = \dfrac{|x|}{x}$ has two **one-sided limits** as $x \to 0$. The **limit, as x approaches 0 from the right,** of the function $f(x) = \dfrac{|x|}{x}$ is $+1$, written as

$$\lim_{x \to 0^+} \frac{|x|}{x} = +1.$$

The **limit, as x approaches 0 from the left,** of the function $f(x) = \dfrac{|x|}{x}$ is -1; written as

$$\lim_{x \to 0^-} \frac{|x|}{x} = -1.$$

These two "limits" are called **one-sided limits.**

In general, a function $f(x)$ may have one or both one-sided limits at $x = a$, but they need not be the same. It may also happen that $f(x)$ has neither one-sided limit at $x = a$.

EXAMPLE 4   Consider the function f defined by

$$f(x) = \begin{cases} x + 2, & \text{if } x \le 3 \\ \frac{1}{3}x + 5, & \text{if } x > 3. \end{cases}$$

The graph of this function is shown in Figure 5.

Figure 5

$$f(x) = \begin{cases} x + 2 \text{ if } x \le 3 \\ \frac{1}{3}x + 5 \text{ if } x > 3 \end{cases}$$

This function has two different one-sided limits as x approaches 3. As $x \to 3$ from the left, $f(x) \to 5$. As $x \to 3$ from the right, $f(x) \to 6$. Thus,

$$\lim_{x \to 3^-} f(x) = 5 \qquad \text{and} \qquad \lim_{x \to 3^+} f(x) = 6. \qquad\blacksquare$$

It is important to recognize that a function $f(x)$ has a limit L as x approaches a if and only if both one-sided limits, as x approaches a from the left and as x approaches a from the right, exist and are equal to L. That is,

$$\lim_{x \to a} f(x) = L \qquad \text{if and only if}$$
$$\text{both} \qquad \lim_{x \to a^-} f(x) = L \quad \text{and} \quad \lim_{x \to a^+} f(x) = L.$$

Therefore, in Example 4 $\lim_{x \to 3} f(x)$ does not exist, even though $f(3)$ exists and both $\lim_{x \to 3^-} f(x)$ and $\lim_{x \to 3^+} f(x)$ exist.

One might be inclined to conjecture that when the rule for a function is given by several equations, the limit of the function does not exist at the values in the domain where the rule changes. This was the case in Example 4, where the rule for the function changed at $x = 3$ and $\lim_{x \to 3} f(x)$ did not exist. We shall see in Example 5, however, that a function defined by two or more equations may have a limit at a value in the domain where the rule changes.

EXAMPLE 5   Consider the function f defined by

$$f(x) = \begin{cases} -\frac{2}{3}x + 6, & \text{if } x \geq 3 \\ x + 1, & \text{if } x < 3. \end{cases}$$

The graph of this function is shown in Figure 6.

Figure 6

We can see that $\lim_{x \to 3^-} f(x) = 4$ and $\lim_{x \to 3^+} f(x) = 4$, and, therefore,

$$\lim_{x \to 3} f(x) = 4.$$ ∎

It is possible also that as $x \to a$, the values of $f(x)$ do not approach any real number. This phenomenon occurs in Example 6.

EXAMPLE 6   Consider the function f defined by

$$f(x) = \frac{1}{x^2}.$$

The graph is shown in Figure 7.

Figure 7

$$f(x) = \frac{1}{x^2}$$

To find $\lim_{x \to 0} f(x)$, we observe that as $x \to 0$ from the right, $f(x)$ gets larger and larger without bound, and as $x \to 0$ from the left, $f(x)$ also gets larger and larger without bound. This can also be seen from Tables 5 and 6.

TABLE 5 $x \to 0$ from the right

x	2	1	0.7	0.5	0.2	0.1	0.05	0.01	0.001
$f(x) = \dfrac{1}{x^2}$	0.25	1	2.04	4	25	100	400	1×10^4	1×10^6

TABLE 6 $x \to 0$ from the left

x	-2	-1	-0.7	-0.5	-0.2	-0.1	-0.05	-0.01	-0.001
$f(x) = \dfrac{1}{x^2}$	0.25	1	2.04	4	25	100	400	1×10^4	1×10^6

Since $f(x)$ approaches no real number as x approaches 0, we conclude that $\lim_{x \to 0} f(x)$ does not exist. Similarly, neither of the one-sided limits exists at 0.

||||||||||||| **Properties of Limits**

We now list seven important rules, or "properties of limits." These properties will make it easier to evaluate limits of functions.

Although Properties 1 through 7 are stated for ordinary limits, it should be understood that they are true for one-sided limits as well.

Property 1

> If f is a constant function defined by $f(x) = k$, then for any fixed a,
>
> $$\lim_{x \to a} f(x) = k,$$
>
> which can also be written
>
> $$\lim_{x \to a} k = k.$$

EXAMPLE 7

$$\lim_{x \to 2} 4 = 4, \qquad \lim_{x \to 0} 4 = 4, \qquad \lim_{x \to -2} 4 = 4,$$

because in each case we want to find the limit of the constant function f defined by $f(x) = 4$.

Property 2

> If f is the function defined by $f(x) = x$, then
>
> $$\lim_{x \to a} f(x) = \lim_{x \to a} x = a.$$

EXAMPLE 8

$$\lim_{x \to -3} x = -3, \qquad \lim_{x \to 0} x = 0, \qquad \lim_{x \to 4} x = 4.$$

Property 3 | If $\lim_{x \to a} f(x)$ and $\lim_{x \to a} g(x)$ both exist, then

$$\lim_{x \to a} [f(x) + g(x)] = \lim_{x \to a} f(x) + \lim_{x \to a} g(x),$$

$$\lim_{x \to a} [f(x) - g(x)] = \lim_{x \to a} f(x) - \lim_{x \to a} g(x).$$

EXAMPLE 9 (a) $\lim_{x \to 2} (x + 3) = \lim_{x \to 2} x + \lim_{x \to 2} 3 = 2 + 3 = 5,$

(b) $\lim_{x \to 2} (x - 3) = \lim_{x \to 2} x - \lim_{x \to 2} 3 = 2 - 3 = -1.$

Property 4 | If $\lim_{x \to a} f(x)$ and $\lim_{x \to a} g(x)$ both exist, then

$$\lim_{x \to a} [f(x)g(x)] = \lim_{x \to a} f(x) \cdot \lim_{x \to a} g(x).$$

EXAMPLE 10 (a) $\lim_{x \to 2} x(x + 3) = \lim_{x \to 2} x \cdot \lim_{x \to 2} (x + 3) = 2 \cdot 5 = 10$

(b) $\lim_{x \to 3} x^2 = \lim_{x \to 3} x \cdot \lim_{x \to 3} x = 3 \cdot 3 = 9$

(c) $\lim_{x \to 2} 4x = \lim_{x \to 2} 4 \cdot \lim_{x \to 2} x = 4 \cdot 2 = 8.$

Observe that by using Property 4 repeatedly, we can obtain the following property:

If $\lim_{x \to a} f(x)$ exists and n is a positive integer, then

$$\lim_{x \to a} [f(x)]^n = \left[\lim_{x \to a} f(x) \right]^n.$$

Using Properties 1 and 4, we obtain Property 5.

Property 5 | If f is a function, k is a constant, and $\lim_{x \to a} f(x)$ exists, then

$$\lim_{x \to a} k f(x) = k \lim_{x \to a} f(x).$$

EXAMPLE 11

$$\lim_{x \to 3} \tfrac{1}{2}x^2 = \tfrac{1}{2}\lim_{x \to 3} x^2 = \tfrac{1}{2}(3^2) = \tfrac{9}{2}.$$

|||||||||||| **Limits of Polynomials**

If

$$p(x) = c_0 x^n + c_1 x^{n-1} + \cdots + c_{n-1}x + c_n$$

is a polynomial, then extension of Properties 1–5 to more than two functions gives

$$\lim_{x \to a} p(x) = c_0 a^n + c_1 a^{n-1} + \cdots + c_{n-1}a + c_n = p(a).$$

Thus, if $p(x)$ is any polynomial and a is any fixed number,

$$\lim_{x \to a} p(x) = p(a).$$

In particular,

$$\lim_{x \to 2} (3x^2 - 5x + 4) = 3 \cdot 2^2 - 5 \cdot 2 + 4 = 6.$$

Property 6   If $\lim_{x \to a} f(x)$ exists and $\lim_{x \to a} g(x)$ exists and is nonzero, then

$$\lim_{x \to a} \frac{f(x)}{g(x)} = \frac{\lim_{x \to a} f(x)}{\lim_{x \to a} g(x)}.$$

EXAMPLE 12

$$\lim_{x \to 5} \frac{2x^2 - 4}{x - 3} = \frac{\lim_{x \to 5}(2x^2 - 4)}{\lim_{x \to 5}(x - 3)} = \frac{2 \cdot 5^2 - 4}{5 - 3} = \frac{46}{2} = 23.$$

EXAMPLE 13   Consider once more the function in Example 2,

$$f(x) = \frac{x^2 - 9}{x - 3},$$

where we showed that

$$\lim_{x \to 3} f(x) = 6.$$

This result cannot be obtained by using Property 6, since $\lim_{x \to 3} (x - 3) = 0$. However,

$$\lim_{x \to 3} \frac{x^2 - 9}{x - 3} = \lim_{x \to 3} \frac{(x + 3)(x - 3)}{(x - 3)} = \lim_{x \to 3} (x + 3) = 6.$$

Property 7

If n is a positive integer, then

$$\lim_{x \to a} \sqrt[n]{x} = \sqrt[n]{a} \tag{3}$$

under either of the following two conditions:
(a) when n is odd and a is any real number, or
(b) when n is even and $a > 0$.

When $a = 0$ and n is even, Equation (3) remains true if it is replaced by the one-sided limit statement

$$\lim_{x \to 0^+} \sqrt[n]{x} = 0. \tag{4}$$

The conditions in Property 7 merely serve to prevent us from indicating an even root of a negative number, which cannot exist in the real number system.

EXAMPLE 14

(a) $\lim_{x \to -64} \sqrt[3]{x} = \sqrt[3]{-64} = -4$

(b) $\lim_{x \to 81} \sqrt[4]{x} = \sqrt[4]{81} = 3.$

(c) $\lim_{x \to 0} \sqrt{x}$ does not exist, because $\lim_{x \to 0^-} \sqrt{x}$ does not exist.

(d) $\lim_{x \to 0^+} \sqrt{x} = 0.$

Extended Property 7

If $\lim_{x \to a} f(x) = L$ exists and n is a positive integer, then

$$\lim_{x \to a} \sqrt[n]{f(x)} = \sqrt[n]{\lim_{x \to a} f(x)} = \sqrt[n]{L} \tag{5}$$

under either of the following two conditions:
(a) when n is odd and L is any real number, or
(b) when n is even and $L > 0$.

When $L = 0$ and n is even, Equation (5) is valid only if $f(x) \geq 0$ for all x near a.

EXAMPLE 15 (a) $\lim_{x \to 4} \sqrt{3x + 5} = \sqrt{\lim_{x \to 4} (3x + 5)} = \sqrt{3 \cdot 4 + 5} = \sqrt{17}.$

(b) $\lim_{x \to 3} (x^2 + 2)^{2/3} = \lim_{x \to 3} \sqrt[3]{(x^2 + 2)^2} = \sqrt[3]{\lim_{x \to 3} [(x^2 + 2)^2]}$

$$= \sqrt[3]{\left[\lim_{x \to 3} (x^2 + 2)\right]^2}$$

$$= \sqrt[3]{(3^2 + 2)^2} = \sqrt[3]{11^2} = 11^{2/3}.$$

(c) $\lim_{x \to 2} \sqrt[4]{3x - 10}$ does not exist, because $L = \lim_{x \to 2} (3x - 10) = -4 < 0.$

EXAMPLE 16 In the next chapter we shall have occasion to compute limits of expressions involving x and h by treating x as fixed and letting $h \to 0$. For example,

(a) $\lim_{h \to 0} (2 + h) = \lim_{h \to 0} 2 + \lim_{h \to 0} h = 2 + 0 = 2.$

(b) $\lim_{h \to 0} (2x + 3h) = \lim_{h \to 0} 2x + \lim_{h \to 0} 3h$

$$= \lim_{h \to 0} 2x + 3 \lim_{h \to 0} h = 2x + 3 \cdot 0 = 2x.$$

(In this case, $2x$ does not involve h, so $2x$ acts as a constant in the calculation of $\lim_{h \to 0} 2x$.)

(c) $\lim_{h \to 0} [3x^2 + 2xh + 5h^2] = \lim_{h \to 0} 3x^2 + 2x \lim_{h \to 0} h + 5 \lim_{h \to 0} h^2$

$$= 3x^2 + 2x \cdot 0 + 5 \cdot 0$$

$$= 3x^2.$$

2.1 EXERCISE SET ||||||||||||||||||||||||||||||||||||||||||||||||||||||||||||||||||||||||||||||

1. (a) Find $\lim_{x \to 4} (2x - 3)$ by completing the following tables of values for $2x - 3$.

x	4.4	4.2	4.1	4.01	4.001
$2x - 3$					

x	3.6	3.8	3.9	3.99	3.999
$2x - 3$					

(b) Sketch the graph of $f(x) = 2x - 3$ and verify your conclusion.

2. (a) Find $\lim_{x \to 1} \dfrac{x^2 + x - 2}{x - 1}$ by completing the following tables.

x	2	1.5	1.1	1.01	1.001
$\dfrac{x^2 + x - 2}{x - 1}$					

x	0	.5	.9	.99	.999
$\dfrac{x^2 + x - 2}{x - 1}$					

(b) Sketch the graph of $f(x) = \dfrac{x^2 + x - 2}{x - 1}$

to verify your conclusion.

3. Let $f(x) = \dfrac{2x + 4}{|x + 2|}$.

(a) Investigate $\lim\limits_{x \to -2^-} f(x)$, $\lim\limits_{x \to -2^+} f(x)$ and $\lim\limits_{x \to -2} f(x)$ by completing the following tables.

x	-1	-1.5	-1.9	-1.99	-1.999
$\dfrac{2x + 4}{|x + 2|}$					

x	-3	-2.5	-2.1	-2.01	-2.001
$\dfrac{2x + 4}{|x + 2|}$					

(b) Sketch the graph of $f(x) = \dfrac{2x + 4}{|x + 2|}$ to verify your conclusions.

4. Let $f(x) = \begin{cases} 2x + 1, & \text{if } x \le 1 \\ x - 1, & \text{if } x > 1. \end{cases}$

(a) Investigate $\lim\limits_{x \to 1^-} f(x)$, $\lim\limits_{x \to 1^+} f(x)$, and $\lim\limits_{x \to 1} f(x)$ by completing the following tables.

x	0	.5	.9	.99	.999
$f(x)$					

x	2	1.5	1.1	1.01	1.001
$f(x)$					

(b) Sketch the graph of $f(x)$ to verify your conclusion.

5. Let $f(x) = \begin{cases} x, & \text{if } x \le 0 \\ x^2, & \text{if } x > 0. \end{cases}$

(a) Investigate $\lim\limits_{x \to 0^-} f(x)$, $\lim\limits_{x \to 0^+} f(x)$ and $\lim\limits_{x \to 0} f(x)$ by completing the following tables.

x	1	.5	.1	.01	.001
$f(x)$					

x	-1	$-.5$	$-.1$	$-.01$	$-.001$
$f(x)$					

(b) Sketch the graph of $f(x)$ to verify your conclusions.

6. (a) Investigate $\lim\limits_{x \to 2} 1/(x - 2)$ by completing the following tables of values for $1/(x - 2)$. What about the one-sided limits?

x	2.5	2.2	2.1	2.05	2.01	2.001	2.0001
$\dfrac{1}{x - 2}$							

x	1.5	1.7	1.9	1.95	1.99	1.999	1.9999
$\dfrac{1}{x - 2}$							

(b) Sketch the graph of $f(x) = 1/(x - 2)$ and verify your conclusion.

In Exercises 7–36, find the limit, if it exists, using the properties of limits.

7. $\lim\limits_{x \to 3} 2x$

8. $\lim\limits_{x \to -1} (2x + 3)$

9. $\lim\limits_{x \to -2} (3x + 4)$

10. $\lim\limits_{x \to 4} (x^2 - 1)$

11. $\lim\limits_{x \to 2} (3x^2 + 2x - 5)$

12. $\lim\limits_{x \to -1} (x^3 - 1)$

13. $\lim\limits_{x \to 4} \dfrac{2x + 5}{x^2 + 9}$

14. $\lim\limits_{x \to 1} \dfrac{3x^2 + 2x + 1}{4x^2 - 5x + 6}$

15. $\lim\limits_{x \to 2} \dfrac{x^2 - 16}{x - 4}$

16. $\lim\limits_{x \to -1} \dfrac{x^2 - 2x - 3}{x + 1}$

17. $\lim\limits_{x \to 0} \dfrac{x^2 - 2x}{x}$

18. $\lim\limits_{x \to -6} \dfrac{x^2 - 36}{x + 6}$

19. $\lim\limits_{x \to 4} \dfrac{x + 1}{x^2 - 3x - 4}$

20. $\lim\limits_{x \to 0} \dfrac{x}{3x^2 - 2x}$

21. $\lim\limits_{x \to 0} |x|$

22. $\lim\limits_{x \to 1^-} \sqrt{x - 1}$

23. $\lim\limits_{x \to 1^+} \sqrt{x - 1}$

24. $\lim\limits_{x \to 1} \sqrt{x - 1}$

25. $\lim\limits_{x \to 3^-} \dfrac{|x - 3|}{x - 3}$

26. $\lim\limits_{x \to 2^+} \dfrac{3x - 6}{|x - 2|}$

27. $\lim\limits_{x \to 0} \sqrt[4]{x}$

28. $\lim\limits_{x \to 3} \sqrt[3]{1 - x^2}$

29. $\lim\limits_{x \to 4^+} \sqrt[3]{5x - 21}$

30. $\lim\limits_{x \to 5} \sqrt{x^2 - 4x - 5}$

31. $\lim\limits_{x \to 0} \dfrac{2}{x}$

32. $\lim\limits_{x \to 1} \dfrac{3x}{x - 1}$

33. $\lim\limits_{h \to 0} \dfrac{3(x + h) - 3x}{h}$

34. $\lim\limits_{h \to 0} \dfrac{[a(x + h) + b] - [ax + b]}{h}$

35. $\lim\limits_{h \to 0} \dfrac{(x + h)^2 - x^2}{h}$

36. $\lim\limits_{h \to 0} \dfrac{(x + h)^3 - x^3}{h}$

37. Let f be defined by

$$f(x) = \begin{cases} 2x + 1 & \text{if } x \le 2 \\ -\frac{5}{2}x + 10 & \text{if } x > 2 \end{cases}$$

(a) Find $\lim\limits_{x \to 2} f(x)$.

(b) Sketch the graph of f.

38. Let f be defined by

$$f(x) = \begin{cases} x & \text{if } x \le 1 \\ 2x + 1 & \text{if } x > 1 \end{cases}$$

(a) Find $\lim\limits_{x \to 1} f(x)$.

(b) Sketch the graph of f.

In Exercises 39–42, find $\lim\limits_{x \to a^-} f(x)$, $\lim\limits_{x \to a^+} f(x)$, and $\lim\limits_{x \to a} f(x)$, if they exist, and sketch the graph of $f(x)$.

39. $f(x) = \dfrac{x^2 - 25}{x + 5}$, $a = -5$

40. $f(x) = \dfrac{1}{2x - 1}$, $a = \dfrac{1}{2}$

41. $f(x) = \dfrac{x - 4}{x^2 - 3x - 4}$, $a = -1$

42. $f(x) = \dfrac{x^2 + x - 2}{x + 2}$, $a = -2$

In Exercises 43–48, the graph of a function $f(x)$ is given and a value $x = a$ is specified. Investigate $\lim\limits_{x \to a^-} f(x)$, $\lim\limits_{x \to a^+} f(x)$ and $\lim\limits_{x \to a} f(x)$.

43.

44.

47.

45.

48.

46.

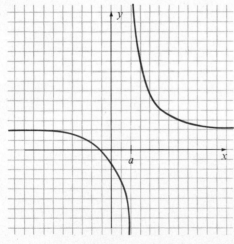

2.2 |||||||| **INFINITY IN LIMITS**

The Symbols $+\infty$ and $-\infty$

There is no largest real number; for every real number there is a larger real number. Geometrically, this means that the real number line extends indefinitely to the right without endpoint. Similarly, there is no largest negative real number;

the real number line extends indefinitely to the left without endpoint. Since the real number system has no upper bound (on the right) and no lower bound (on the left) we say that it is **unbounded.** We use the two symbols, $+\infty$ and $-\infty$, to denote this unboundedness. These symbols do not represent real numbers; they indicate only unboundedness and direction (see Figure 1).

Figure 1

As we shall see, these symbols for infinity are often quite helpful in the study of limits.

Infinity as a Limit

EXAMPLE 1   Let us investigate

$$\lim_{x \to 1} \frac{1}{x - 1}.$$

We cannot use Property 6 from Section 2.1 here, because the denominator has limit 0 as $x \to 1$. The graph of $f(x) = \dfrac{1}{x - 1}$ is shown in Figure 2.

Figure 2

Tables 1 and 2 show the behavior of $f(x)$ as $x \to 1$.

TABLE 1 $x \to 1$ from the right

x	2	1.5	1.25	1.01	1.001	1.0001
$f(x) = \dfrac{1}{x - 1}$	1	2	4	100	1000	10,000

TABLE 2 $x \to 1$ from the left

x	0	0.5	0.75	0.99	0.999	0.9999
$f(x) = \dfrac{1}{x-1}$	-1	-2	-4	-100	-1000	$-10{,}000$

From the graph and Table 1 it is evident that as $x \to 1$ from the right, $f(x)$ grows larger and larger positively without bound. We express that fact by saying that $f(x) \to +\infty$ as $x \to 1$ from the right, or

$$\lim_{x \to 1^+} \frac{1}{x-1} = +\infty. \tag{1}$$

Similarly, from the graph and Table 2 it is evident that as $x \to 1$ from the left, $f(x)$ grows larger and larger negatively without bound. We express that fact by saying that $f(x) \to -\infty$ as $x \to 1$ from the left, or

$$\lim_{x \to 1^-} \frac{1}{x-1} = -\infty. \tag{2}$$

It is important to remember that $+\infty$ *and* $-\infty$ *are not real numbers* and thus we do not say that the limits in Equations (1) and (2) exist. Limit statements (1) and (2) are valuable because they express the behavior of the function as x approaches 1, but in neither case does the limit "exist." Moreover, since the two one-sided limits do not agree, we cannot say that $\lim\limits_{x \to 1} \dfrac{1}{x-1}$ is $+\infty$ or that it is $-\infty$; it is neither. We can only say that

$$\lim_{x \to 1} \frac{1}{x-1} \textbf{ does not exist.}$$

If $f(x)$ grows larger and larger positively without bound, as x approaches a, then we write

$$\lim_{x \to a} f(x) = +\infty.$$

If $f(x)$ grows larger and larger negatively without bound, as x approaches a, then we write

$$\lim_{x \to a} f(x) = -\infty.$$

We have similar limit statements as $x \to a^+$ and as $x \to a^-$.

Property 1

Evaluating $\lim\limits_{x \to a} \dfrac{f(x)}{g(x)}$, when $\lim\limits_{x \to a} g(x) = 0$:

Suppose $\lim\limits_{x \to a} g(x) = 0$, $\lim\limits_{x \to a} f(x)$ exists but is not 0, and the quotient $\dfrac{f(x)}{g(x)}$ maintains the same sign as $x \to a$. Then

(a) If $\dfrac{f(x)}{g(x)} > 0$ as $x \to a$, then $\lim\limits_{x \to a} \dfrac{f(x)}{g(x)} = +\infty$.

(b) If $\dfrac{f(x)}{g(x)} < 0$ as $x \to a$, then $\lim\limits_{x \to a} \dfrac{f(x)}{g(x)} = -\infty$.

These conclusions are also valid if "$x \to a$" is everywhere replaced by "$x \to a^+$" or by "$x \to a^-$."

EXAMPLE 2   (a) $\lim\limits_{x \to 3^+} \dfrac{x + 1}{x - 3} = +\infty$ because

1. $\lim\limits_{x \to 3^+} (x - 3) = 0$,

2. $\lim\limits_{x \to 3^+} (x + 1) = 4 \neq 0$,

3. $\dfrac{x + 1}{x - 3} = \dfrac{+}{+} > 0$ as $x \to 3^+$ $(x > 3)$.

(b) $\lim\limits_{x \to 2^+} \dfrac{1}{2 - x} = -\infty$ because

1. $\lim\limits_{x \to 2^+} (2 - x) = 0$,

2. $\lim\limits_{x \to 2^+} 1 = 1 \neq 0$,

3. $\dfrac{1}{2 - x} = \dfrac{+}{-} < 0$ as $x \to 2^+$ $(x > 2)$.

(c) $\lim\limits_{x \to 1^-} \dfrac{x}{1 - x} = +\infty$ because

1. $\lim\limits_{x \to 1^-} (1 - x) = 0$,

2. $\lim\limits_{x \to 1^-} x = 1 \neq 0$,

3. $\dfrac{x}{1 - x} = \dfrac{+}{+} > 0$ as $x \to 1^-$ $(x < 1)$.

||||||||||||| **Vertical Asymptotes**

A vertical line $x = a$ is an **asymptote** of a curve $y = f(x)$ if

$$\text{either} \qquad \lim_{x \to a^-} f(x) = +\infty \text{ or } -\infty,$$

$$\text{or} \qquad \lim_{x \to a^+} f(x) = +\infty \text{ or } -\infty$$

or both.

In Example 1 the vertical line $x = 1$ is a vertical asymptote of the curve $y = \dfrac{1}{x - 1}$ (see Figure 2).

Property 1 is often useful in determining the vertical asymptotes of a curve $y = f(x)$.

EXAMPLE 3 Find all vertical asymptotes of the curve $y = \dfrac{x^2}{x^2 - 4}$.

Solution We write $f(x) = \dfrac{x^2}{(x - 2)(x + 2)}$. Since the denominator approaches 0 as $x \to 2$ or $x \to -2$, we anticipate that the vertical lines $x = -2$ and $x = 2$ may be asymptotes. Using Property 1 we verify that

$$\lim_{x \to 2^-} \frac{x^2}{(x - 2)(x + 2)} = -\infty, \qquad \text{and} \qquad \lim_{x \to 2^+} \frac{x^2}{(x - 2)(x + 2)} = +\infty.$$

Therefore, the line $x = 2$ is a vertical asymptote. Using Property 1 again we find that

$$\lim_{x \to -2^-} \frac{x^2}{(x - 2)(x + 2)} = +\infty, \qquad \text{and} \qquad \lim_{x \to -2^+} \frac{x^2}{(x - 2)(x + 2)} = -\infty.$$

Therefore, the line $x = -2$ is a vertical asymptote. The graph of this curve is shown in Figure 3. In addition to the two vertical asymptotes just found, the curve

Figure 3

has one horizontal asymptote. Horizontal asymptotes, and methods of finding them, will be discussed shortly.   ∎

The problem of finding the vertical asymptotes of a rational function is made simple by the following theorem.

THEOREM

A rational function

$$f(x) = \frac{p(x)}{q(x)}$$

(where $p(x)$ and $q(x)$ are polynomials) has vertical asymptote $x = a$ for any number a for which $q(a) = 0$ but $p(a) \neq 0$.

This theorem may be proved using the properties of limits and the definition of vertical asymptote.

EXAMPLE 4   Find all vertical asymptotes of the curve $y = \dfrac{x^2}{x^2 - 4}$ given in Example 3.

Solution   Writing $y = \dfrac{x^2}{(x-2)(x+2)}$, we see that the denominator is 0 when $x = 2$ or $x = -2$, and that the numerator is not 0 when x equals either of these numbers. Therefore, the vertical asymptotes are $x = 2$ and $x = -2$.   ∎

Limits at Infinity

In some applications we have to examine the behavior of the function f as x gets larger and larger without any bound.

EXAMPLE 5   Consider the function defined by

$$f(x) = 3 - \frac{1}{x}, \qquad \text{if } x \geq 1.$$

To determine the behavior of f as x increases without any bound, we calculate values of $f(x)$ for increasingly larger values of x shown in Table 3. The graph of f is shown in Figure 4.

TABLE 3

x	1	2	4	10	100	1000	10,000
$f(x) = 3 - \dfrac{1}{x}$	2	2.5	2.75	2.9	2.99	2.999	2.9999

Figure 4

From Table 3 and the graph of f, it seems intuitively clear that the value of $f(x) = 3 - (1/x)$ gets closer and closer to the number 3 as x gets larger and larger without bound. We express this fact by writing

$$\lim_{x \to +\infty} \left(3 - \frac{1}{x}\right) = 3.$$ ■

Here is our *intuitive definition* of limits at infinity.

(a) We write

$$\lim_{x \to +\infty} f(x) = L$$

(read "the limit of $f(x)$ as x goes to positive infinity equals L") if the values of $f(x)$ get closer and closer to the unique real number L as x gets larger and larger positively without bound.

(b) We write

$$\lim_{x \to -\infty} f(x) = L$$

(read "the limit of $f(x)$ as x goes to negative infinity equals L") if the values of $f(x)$ get closer and closer to the unique real number L as x gets larger and larger negatively without bound.

Keep in mind that in this definition the symbol ∞ does not represent any number; $x \to +\infty$ merely means that x is increasing without bound.

EXAMPLE 6
$$\lim_{x \to +\infty} \frac{1}{x} = 0 \quad \text{and} \quad \lim_{x \to -\infty} \frac{1}{x} = 0.$$

These limits may be seen from Tables 4 and 5 on the following page.

TABLE 4   As $x \to +\infty$

x	1	2	4	10	100	1000	10,000
$f(x) = \dfrac{1}{x}$	1	0.5	0.25	0.1	0.01	0.001	0.0001

TABLE 5   As $x \to -\infty$

x	-1	-2	-4	-10	-100	-1000	$-10,000$
$f(x) = \dfrac{1}{x}$	-1	-0.5	-0.25	-0.1	-0.01	-0.001	-0.0001

EXAMPLE 7   Find $\lim\limits_{x \to +\infty} 2x$ and $\lim\limits_{x \to -\infty} 2x$.

Solution   As x gets larger and larger positively without bound, so does $2x$. Therefore $\lim\limits_{x \to +\infty} 2x = +\infty$. As x gets larger and larger negatively without bound, so does $2x$. Therefore, $\lim\limits_{x \to -\infty} 2x = -\infty$. ■

EXAMPLE 8   Find

$$\lim_{x \to +\infty} \frac{5}{x^2 + 4} \quad \text{and} \quad \lim_{x \to -\infty} \frac{5}{x^2 - 4}.$$

Solution   As x gets larger and larger, positively or negatively without bound, the denominator $x^2 - 4$ gets larger and larger positively without bound, so

$$\lim_{x \to +\infty} \frac{5}{x^2 + 4} = 0 \quad \text{and} \quad \lim_{x \to -\infty} \frac{5}{x^2 - 4} = 0. \quad ■$$

Algebraic Properties 1–7 set forth in Section 2.1 remain valid when $\lim\limits_{x \to a}$ is replaced by $\lim\limits_{x \to +\infty}$ or $\lim\limits_{x \to -\infty}$, except that Property 7 must be replaced by Extended Property 7. Examples 6 and 8 suggest one more useful property.

Property 2

If k is a constant and n is a positive integer, then

$$\lim_{x \to +\infty} \frac{k}{x^n} = 0 \quad \text{and} \quad \lim_{x \to -\infty} \frac{k}{x^n} = 0.$$

EXAMPLE 9   Find

$$\lim_{x \to +\infty} \left(2 + \frac{5}{x^2} \right).$$

Solution
$$\lim_{x \to +\infty} \left(2 + \frac{5}{x^2} \right) = \lim_{x \to +\infty} 2 + \lim_{x \to +\infty} \frac{5}{x^2} = 2 + 0 = 2. \qquad \blacksquare$$

To find the limit of a quotient when both numerator and denominator have no limits as $x \to \pm\infty$, we cannot use Property 6 from Section 2.1, the quotient property of limits. In this case, we usually use some algebra to rewrite the quotient so that the limit properties become applicable. The following three examples illustrate this approach.

EXAMPLE 10   Find

$$\lim_{x \to +\infty} \frac{3x^2 - 4x - 5}{2x^2 - 3}.$$

Solution   In this case, the limits of the numerator and denominator do not exist. If we divide numerator and denominator by the highest power of x in the denominator, we have

$$\lim_{x \to +\infty} \frac{3x^2 - 4x - 5}{2x^2 - 3} = \lim_{x \to +\infty} \frac{\dfrac{3x^2}{x^2} - \dfrac{4x}{x^2} - \dfrac{5}{x^2}}{\dfrac{2x^2}{x^2} - \dfrac{3}{x^2}}$$

$$= \lim_{x \to +\infty} \frac{3 - \dfrac{4}{x} - \dfrac{5}{x^2}}{2 - \dfrac{3}{x^2}} = \frac{3 - 0 - 0}{2 - 0} = \frac{3}{2},$$

since $\lim_{x \to +\infty} (4/x)$, $\lim_{x \to +\infty} (5/x^2)$, and $\lim_{x \to +\infty} (3/x^2)$ are each zero by Property 2. $\blacksquare$

EXAMPLE 11   Find

$$\lim_{x \to -\infty} \frac{x^2 - 2x}{5x}.$$

Solution   In this case, the limits of the numerator and denominator do not exist. We write

$$\lim_{x \to -\infty} \frac{x^2 - 2x}{5x} = \lim_{x \to -\infty} \frac{x^2}{5x} - \lim_{x \to -\infty} \frac{2x}{5x}$$

$$= \frac{1}{5} \lim_{x \to -\infty} x - \lim_{x \to -\infty} \frac{2}{5}$$

$$= \frac{1}{5} \lim_{x \to -\infty} x - \frac{2}{5}$$

$$= -\infty. \qquad \blacksquare$$

EXAMPLE 12 Find

$$\lim_{x \to +\infty} \frac{3x + 2}{x}.$$

Solution In this case, the limits of the numerator and denominator do not exist. We write

$$\lim_{x \to +\infty} \frac{3x + 2}{x} = \lim_{x \to +\infty} \frac{3x}{x} + \lim_{x \to +\infty} \frac{2}{x}$$

$$= \lim_{x \to +\infty} 3 + \lim_{x \to +\infty} \frac{2}{x}$$

$$= 3 + 0 = 3. \qquad \blacksquare$$

EXAMPLE 13
(Business) Suppose that when the price of a certain product is p dollars per unit, the number $N(p)$ of units sold is given by

$$N(p) = \frac{50,000}{p^3} + \frac{2000}{p^2}.$$

Then $\lim_{p \to +\infty} N(p) = 0$, which means that as the price increases without bound, the number of units sold will approach zero.

||||||||||||| **Horizontal Asymptotes**

> A horizontal line $y = b$ is an **asymptote** of a curve $y = f(x)$ if
>
> $$\text{either} \quad \lim_{x \to +\infty} f(x) = b,$$
>
> $$\text{or} \quad \lim_{x \to -\infty} f(x) = b,$$
>
> or both.

EXAMPLE 14 Find all horizontal asymptotes for each of the following curves.

(a) $y = \dfrac{1}{x - 1}$

(b) $y = \dfrac{x^2}{x^2 - 4}$

(c) $y = 3 - \dfrac{1}{x}$, for $x \geq 1$.

Solution   (a) Since $\lim\limits_{x \to +\infty} \dfrac{1}{x-1} = 0$ and $\lim\limits_{x \to -\infty} \dfrac{1}{x-1} = 0$, the only horizontal asymptote

for the curve $y = \dfrac{1}{x-1}$ is the line $y = 0$, the x-axis. The graph of this curve was given in Figure 2.

(b) $\lim\limits_{x \to \pm\infty} \dfrac{x^2}{x^2 - 4} = \lim\limits_{x \to \pm\infty} \dfrac{\dfrac{x^2}{x^2}}{\dfrac{x^2}{x^2} - \dfrac{4}{x^2}} = \lim\limits_{x \to \pm\infty} \dfrac{1}{1 - \dfrac{4}{x^2}}$

$$= \frac{1}{1 - 0} = 1.$$

Therefore, the line $y = 1$ is the only horizontal asymptote for the curve $y = \dfrac{x^2}{x^2 - 4}$. The graph of this curve was given in Figure 3.

(c) As shown in Example 5, $\lim\limits_{x \to +\infty} \left(3 - \dfrac{1}{x}\right) = 3$. Therefore, the line $y = 3$ is the only horizontal asymptote. The curve is shown in Figure 4.   ∎

2.2 EXERCISE SET ||||||||||||||||||||||||||||||||||||||||||||||||||||||||||||||||||||||||||||||

In Exercises 1–36, investigate the given limit.

1. $\lim\limits_{x \to 0^-} \dfrac{1}{x}$

2. $\lim\limits_{x \to 0^+} \dfrac{1}{x}$

3. $\lim\limits_{x \to 0^+} \dfrac{-2}{x}$

4. $\lim\limits_{x \to 0^-} \dfrac{-4}{x}$

5. $\lim\limits_{x \to 2^+} \dfrac{x}{x-2}$

6. $\lim\limits_{x \to 2^-} \dfrac{x}{x-2}$

7. $\lim\limits_{x \to 3^-} \dfrac{1}{x^2 - 9}$

8. $\lim\limits_{x \to 3^+} \dfrac{1}{x^2 - 9}$

9. $\lim\limits_{x \to -3^-} \dfrac{x}{x+3}$

10. $\lim\limits_{x \to -1^+} \dfrac{x}{x+1}$

11. $\lim\limits_{x \to 1^+} \dfrac{x+1}{x^2 - 1}$

12. $\lim\limits_{x \to 3^-} \dfrac{x+3}{x^2 - 9}$

13. $\lim\limits_{x \to 2} \dfrac{x-2}{x^2 - x - 2}$

14. $\lim\limits_{x \to 2} \dfrac{x+1}{x^2 - x - 2}$

15. $\lim\limits_{x \to 4^-} \dfrac{x+1}{x^2 - 3x - 4}$

16. $\lim\limits_{x \to -1^+} \dfrac{x-4}{x^2 - 3x - 4}$

17. $\lim\limits_{x \to 5} \dfrac{x+5}{x-5}$

18. $\lim\limits_{x \to 5^-} \dfrac{x+5}{x-5}$

19. $\lim\limits_{x \to +\infty} \dfrac{1}{x+1}$

20. $\lim\limits_{x \to -\infty} \dfrac{1}{2x+1}$

21. $\lim\limits_{x \to -\infty} \left(5 + \dfrac{6}{x}\right)$

22. $\lim\limits_{x \to +\infty} \left(8 - \dfrac{7}{x+1}\right)$

23. $\lim\limits_{x \to +\infty} (3 + 4x)$

24. $\lim\limits_{x \to +\infty} (8 - 3x)$

25. $\lim\limits_{x \to +\infty} \dfrac{5x}{x+3}$

26. $\lim\limits_{x \to -\infty} \dfrac{7x+2}{3x-1}$

27. $\lim\limits_{x \to -\infty} \dfrac{2x^2}{x^2 + x - 10}$

28. $\lim\limits_{x \to +\infty} \dfrac{x}{x^2 + 5}$

29. $\lim\limits_{x \to -\infty} \dfrac{x^3}{x^2 + 7}$

30. $\lim\limits_{x \to -\infty} \dfrac{x + 4x^2}{1 + x^2}$

31. $\lim\limits_{x \to +\infty} \dfrac{2x^2 + 5x - 3}{5x^2 - 3x + 1}$

32. $\lim\limits_{x \to +\infty} \dfrac{2 - 5x}{3 + 2x}$

33. $\lim\limits_{x \to +\infty} \dfrac{2x}{x^2 - 3x}$

34. $\lim\limits_{x \to -\infty} \dfrac{5x^2 - 3}{x^2}$

35. $\lim\limits_{x \to -\infty} (3x^2 + 2)$

36. $\lim\limits_{x \to -\infty} \dfrac{4}{x^3 - 2}$

In Exercises 37–42, find all horizontal and vertical asymptotes for the given curve.

37. $y = \dfrac{x + 1}{x + 2}$

38. $y = \dfrac{3x + 5}{2x - 4}$

39. $y = \dfrac{x}{x^2 - x - 2}$

40. $y = \dfrac{2x^2}{x^2 - 9}$

41. $y = \dfrac{x + 1}{x^2 + 5x + 4}$

42. $y = 7 - \dfrac{3}{x}$

43. **(Ecology)** A small source of nuclear radiation sends its energy uniformly into three-dimensional space. The intensity I of radiation received at a distance r from the source is given by the formula

$$I = \frac{E}{4\pi r^2}$$

where E is the energy received by the surface of a sphere per second. Find $\lim\limits_{r \to +\infty} I$.

44. **(Biology)** Suppose that the growth of a bacteria culture is given by the formula

$$N(t) = \frac{80{,}000t^2 + 5000}{t^2 + 1000}$$

where $N(t)$ is the number of bacteria after t days. Find $\lim\limits_{t \to +\infty} N(t)$.

45. **(Educational Psychology)** Suppose that a typing instructor has determined the following graph for $N(t)$, the number of words typed per minute by an average student after t weeks of instruction.

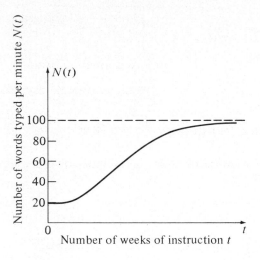

(a) Find $\lim\limits_{t \to 0} N(t)$. Interpret your answer.

(b) Find $\lim\limits_{t \to +\infty} N(t)$. Interpret your answer.

2.3 ||||||| CONTINUITY

In this section we shall discuss functions whose graphs form unbroken or continuous curves. Such functions are important in many applications.

Most of you should have an intuitive feeling that a continuous curve is one that has no gaps or jumps. That is, from an intuitive point of view, a continuous curve is one that can be drawn without having to lift the pencil from the paper (Figure 1).

Figure 1

To gain precision we shall think of continuity as a *local property* of functions. Thus, a function may be continuous at one point but not at another.

Having this intuitive notion of a continuous curve, we can turn back to Figure 3 of Section 2.1, and see that the curve is not continuous at $x = 3$ because there is a gap at $x = 3$ (f is not defined at $x = 3$). Look also at Figure 5 of Section 2.1, for which f is defined at $x = 3$, but since $\lim_{x \to 3} f(x)$ does not exist, the curve has a jump at $x = 3$ and is, therefore, not continuous at $x = 3$. For the curve shown in Figure 2, below, f is defined at $x = a$ and $\lim_{x \to a} f(x)$ exists, but since

$$\lim_{x \to a} f(x) \neq f(a)$$

there is a jump and the curve is not continuous at $x = a$.

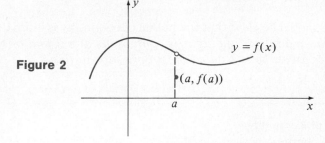

Figure 2

Based on the preceding discussion, we can state precise mathematical conditions on f under which the graph of f will not have a gap or jump at $x = a$.

Definition of Continuity

A function f is said to be **continuous** at $x = a$ if the following conditions hold:

1. $f(a)$ is defined.
2. $\lim_{x \to a} f(x)$ exists.
3. $\lim_{x \to a} f(x) = f(a)$.

If f is not continuous at $x = a$, we say that f is **discontinuous** at $x = a$, or, equivalently, that $x = a$ is a **point of discontinuity** for f. A function f is called **continuous on an interval** if f is continuous at every point of the interval; f is called **discontinuous on an interval** if there is at least one point of discontinuity in the interval. A function f is said to be **continuous everywhere** if it is continuous at every real number.

EXAMPLE 1   Every constant function is continuous everywhere. For, let $f(x) = k$ be a constant function. Then for any number a,

1. $f(a) = k$.
2. $\lim_{x \to a} f(x) = \lim_{x \to a} k = k$ exists.
3. $\lim_{x \to a} f(x) = f(a)$.

EXAMPLE 2   Let f be defined by $f(x) = x$. Where is f continuous?

Solution   Let a be any real number. Verifying conditions (1), (2), and (3) for the value $x = a$:

1. $f(a) = a$.
2. $\lim_{x \to a} f(x) = \lim_{x \to a} x = a$ exists.
3. $\lim_{x \to a} f(x) = f(a)$.

Hence, f is continuous at a. Therefore, f is continuous everywhere.   ∎

From our definition it should be clear that the notion of continuity is intimately tied to the notion of limit. Indeed, *a function* f *is continuous at* a *if and only if its limit as* x *approaches* a *can be evaluated by simply substituting* x = a *into the formula for* f(x).

Algebraic Properties of Continuity

Property 1 (Sums, Differences, Products)   If $f(x)$ and $g(x)$ are continuous at $x = a$, then $f(x) + g(x)$, $f(x) - g(x)$ and $f(x)g(x)$ are also continuous at $x = a$; that is, the sum, difference, and product of continuous functions are continuous.

Property 2 (Quotients)   If $f(x)$ and $g(x)$ are continuous at $x = a$, then $f(x)/g(x)$ is also continuous at $x = a$, unless $g(a) = 0$. If $g(a) = 0$, then $x = a$ is a point of discontinuity for $f(x)/g(x)$.

Property 3 (Roots)   The function $f(x) = \sqrt[n]{x}$ is
(a) continuous everywhere if n is odd.
(b) continuous on the interval $(0, \infty)$ if n is even.

Property 4 (Composites)   If $f(x)$ is continuous at $x = a$ and $g(y)$ is continuous at $y = f(a)$, then the composite function $g(f(x))$ is continuous at $x = a$.

||||||||||||| **Polynomials and Rational Functions**

It follows from Examples 1 and 2 and Property 1 that every monomial

$$f(x) = cx^n$$

where c is a constant, is continuous because it is a product of continuous functions. Hence,

Every polynomial is an everywhere continuous function.

(Why?)

Recall that any quotient of two polynomials is called a "rational expression" (see Section A.5, Appendix A). Since a rational expression is a function of x, we call it a **rational function.** Property 2 above assures us that

Every rational function is continuous except at
a point where the denominator has zero value.

That is, a rational function is continuous at each point of its domain.

EXAMPLE 3   Discuss the continuity of the functions

(a) $f(x) = x^2 + 3x + 3$

(b) $g(x) = \dfrac{x + 2}{x^2 + 4}$

(c) $h(x) = \sqrt{x + 5}$.

Solution   (a) $f(x)$ is continuous everywhere because it is a polynomial.

(b) $g(x)$ is continuous everywhere because it is a ratio of two polynomials and the denominator is never zero.

(c) The function $h(x)$ is a composite function, $h(x) = v(u(x))$, where $u(x) = x + 5$ and $v(y) = \sqrt{y}$. Now, $u(x)$ is continuous everywhere because it is a polynomial. By Property 3, $v(y)$ is continuous wherever $y > 0$. Therefore, by Property 4, the composite $v(u(x))$ is continuous when $u(x) > 0$; that is, when $x + 5 > 0$ or when $x > -5$. Therefore, $h(x)$ is continuous on the interval $(-5, +\infty)$.

∎

EXAMPLE 4   Discuss the continuity of the functions

(a) $f(x) = \dfrac{x}{x^2 - 4}$

(b) $g(x) = \begin{cases} \dfrac{x^2 - 4}{x - 2} & \text{if } x \neq 2 \\ 4 & \text{if } x = 2. \end{cases}$

Solution   (a) As a rational function, $f(x)$ is continuous everywhere, except at $x = -2$ and $x = 2$ because the denominator is zero at these points.

(b) $g(x)$ is clearly continuous if $x \neq 2$. When $x = 2$, $g(2)$ is defined, since $g(2) = 4$. Moreover, since

$$\lim_{x \to 2} g(x) = \lim_{x \to 2} \frac{(x+2)(x-2)}{(x-2)} = \lim_{x \to 2} (x+2) = 4 = g(2),$$

we conclude that $g(x)$ is also continuous at $x = 2$. Therefore, $g(x)$ is continuous everywhere.

Note that for all x, the rule for the function g can be simplified to $g(x) = x + 2$.   ∎

EXAMPLE 5   Where is the function $f(x) = |x|$ continuous?

Solution   Recall that $|x| = \begin{cases} x & \text{if } x \geq 0 \\ -x & \text{if } x \geq 0. \end{cases}$

Thus, the absolute value function is continuous on both $(0, +\infty)$ and $(-\infty, 0)$, since it is a polynomial on each of these intervals: x on $(0, +\infty)$ and $-x$ on $(-\infty, 0)$. But is f continuous at $x = 0$? We have

1. $f(0) = |0| = 0$. Thus, $f(0)$ exists.

2. (a) $\displaystyle \lim_{x \to 0^+} f(x) = \lim_{x \to 0^+} |x|$

   $\displaystyle = \lim_{x \to 0^+} x$ (when $x > 0$, $|x| = x$)

   $= 0$.

   (b) $\displaystyle \lim_{x \to 0^-} f(x) = \lim_{x \to 0^-} |x|$

   $\displaystyle = \lim_{x \to 0^-} (-x)$ (when $x < 0$, $|x| = -x$)

   $\displaystyle = -\lim_{x \to 0^-} x$ (Property 5 of limits)

   $= 0$.

3. From parts (1) and (2), $\displaystyle \lim_{x \to 0} f(x) = f(0)$.

Thus, f is continuous at $x = 0$. Therefore, f is continuous everywhere. The graph of the absolute value function should be familiar; it is shown in Example 1 of Section 1.2.   ∎

EXAMPLE 6
(Educational
Psychology)   In some learning situations, students learn the basic skills at a somewhat slow pace. Sometimes a student will experience a burst of knowledge leading to a jump in the learning curve. Such a learning curve is shown in Figure 3. This curve is discontinuous at time t_1, at which point a large jump in mastering the material has taken place.

Figure 3

EXAMPLE 7   A dog breeder recorded the number of dogs $N(t)$ in the kennel at time t and
(Biology)   obtained the curve shown in Figure 4. Every birth, death, and sale of a dog is
marked by a discontinuity in the curve.

Figure 4

2.3 EXERCISE SET |||||||||||||||||||||||||||||||||||||||||||||||||||||||||||||||||||||||||||||||||||

1. In each part, determine whether the function graphed is continuous everywhere. If
it is not, then where is it discontinuous?

(a)

(b)

(c)

(d)

2. In each part, determine whether the function graphed is continuous everywhere. If it is not, then where is it discontinuous?

(a)

(b)

(c)

(d)

3. In each part, determine whether the given function is continuous at $x = 6$. If it is not continuous there, state which of the three continuity conditions is (are) violated. Graph each function.

(a) $f(x) = \dfrac{x^2 - 36}{x - 6}$

(b) $f(x) = \begin{cases} \dfrac{x^2 - 36}{x - 6} & \text{if } x \neq 6 \\ 12 & \text{if } x = 6 \end{cases}$

(c) $f(x) = \begin{cases} \dfrac{x^2 - 36}{x - 6} & \text{if } x \neq 6 \\ 8 & \text{if } x = 6. \end{cases}$

4. In each part, the given function is discontinuous at the indicated point. State which of the three continuity conditions is (are) violated. Graph each function.

(a) $f(x) = \dfrac{x^2 + x - 2}{x - 1}$, at $x = 1$

(b) $f(x) = \begin{cases} \dfrac{x^2 + x - 2}{x - 1} & \text{if } x \neq 1 \\ 2 & \text{if } x = 1 \end{cases}$ at $x = 1$

(c) $f(x) = \begin{cases} x + 3 & \text{if } x \neq 3 \\ 1 & \text{if } x = 3 \end{cases}$ at $x = 3$.

In Exercises 5–12, determine whether the given function is continuous at the indicated point.

5. $f(x) = 2x^2 - 5x + 4$, $x = 3$

6. $f(x) = \dfrac{1}{2x - 1}$, $x = \dfrac{1}{2}$

7. $f(x) = \dfrac{(x - 2)(x - 3)}{(x - 3)}$, $x = 3$

8. $f(x) = \begin{cases} 2x & \text{if } x \leq 2 \\ 4 & \text{if } x > 2 \end{cases}$ $x = 2$

9. $f(x) = \begin{cases} 2x & \text{if } x \neq 2 \\ 5 & \text{if } x = 2 \end{cases}$ $x = 2$

10. $f(x) = \begin{cases} 2x & \text{if } x \leq 2 \\ 2x + 1 & \text{if } x > 2 \end{cases}$ $x = 2$

11. $f(x) = |x - 3|$, $x = 3$

12. $f(x) = \sqrt{x - 1}$, $x = 1$

In Exercises 13–22, determine where the function is continuous.

13. $f(x) = \dfrac{x^2 - 9}{x - 3}$

14. $f(x) = 5x^2 - 3x + 4$

15. $f(x) = |x + 1|$

16. $f(x) = \dfrac{2}{x - 4}$

17. $f(x) = \begin{cases} 3x & \text{if } x \le 1 \\ 2x + 1 & \text{if } x > 1 \end{cases}$

18. $f(x) = \begin{cases} \dfrac{x}{x - 3} & \text{if } x \ne 3 \\ 5 & \text{if } x = 3 \end{cases}$

19. $f(x) = \sqrt{x + 2}$

20. $f(x) = \sqrt{x^2 - 1}$

21. $f(x) = \dfrac{3}{(x - 2)(x + 3)}$

22. $f(x) = |3x - 6|$

23. **(The Postage Function)** Consider the postage function of Exercise 23, Section 1.2. For what values of w, if any, is the function discontinuous? Explain.

24. **(The Light Bulb Function)** Let $c(t)$ denote the electric current passing through a light bulb at time t, which draws 1 ampere when on. Suppose the light bulb is turned on at time $t = a$ and off at time $t = b$. Figure 5 shows the graph of the current function $c(t)$.

For what values of t is $c(t)$ discontinuous? Explain.

25. **(Business)** Suppose that the function $P(x)$ (in thousands of dollars) gives the daily profit of a manufacturer that produces x tons of a product per day. The daily capacity of the plant is 6 tons, and to produce more than 6 tons would require that employees work overtime, at a higher pay

scale. Thus, suppose that $P(x)$ is defined by

$$P(x) = \begin{cases} 3x - 4 & \text{if } 2 \le x \le 6 \\ 3x - 6 & \text{if } 6 < x \le 10. \end{cases}$$

(a) Is $P(x)$ continuous at $x = 6$?
(b) Sketch the graph of $P(x)$.
(c) Compare the values of $P(5.5)$ and $P(6.1)$. Explain.

26. **(Business)** A heating oil distributor has set the following price schedule:
$1.30 per gallon for 100 or fewer gallons;
$1.15 per gallon for more than 100 but no more than 150 gallons;
$1.00 per gallon for more than 150 but fewer than 200 gallons;
$0.85 per gallon for 200 or more gallons.

(a) Determine the function $C(x)$ that gives the cost (in dollars) of purchasing x gallons of heating oil.
(b) Sketch the graph of $C(x)$.
(c) Find all points, if any, where $C(x)$ is discontinuous.

27. **(Medicine)** A medical research experiment established that the mass $M(t)$ of a tumor as a function of the length of time t that a patient is exposed to radiation during treatment is given by

$$M(t) = \dfrac{t^2 - 5t + 6}{t - 3}$$

where $M(t)$ is in milligrams and t is in seconds. Due to a malfunction in the equipment being used, it is impossible to expose the patient for exactly 3 seconds of radiation therapy. What value should $M(3)$ be assigned so that $M(t)$ is a continuous function?

28. **(The Intermediate Value Property)** Explain why the following property of continuous functions holds. Suppose that the function f is continuous throughout the closed interval $[a, b]$ and that $f(a)$ and $f(b)$ have opposite signs. There is at least one value of x, say, $x = c$, $a < c < b$ for which $f(c) = 0$. Sketch a graph of a function that satisfies these conditions.

2.4 ||||||| AVERAGE AND INSTANTANEOUS RATES OF CHANGE

Most quantities in the real world are in a state of constant change. For example, the velocity of a car changes with time, the profits of a manufacturer change with the number of items sold, the number of people unemployed in the United States changes with time, the size of a tumor changes with the amount of radiation to which the patient is exposed, the number of items of a product bought by consumers changes with the price at which the item is offered, and so on. Often we are not as interested in measuring the change itself as in measuring the *rate* at which the change occurs. The *differential calculus* provides the basic tools for measuring rates of change.

In this section we shall put the notion of limit to work. We shall develop a close connection between two seemingly unrelated problems: drawing a tangent line to a curve at a given point on the curve and finding the rate at which one quantity changes with respect to another quantity. The fundamental tool enabling us to solve both of these problems is the notion of limit.

||||||||||||| **Average Velocity**

Suppose that a driver leaves home at 8 AM for a 530-mile trip. The record of distances driven appears in Table 1.

TABLE 1

Time	Distance driven from home (in miles)
8 AM	0
11 AM	150
3 PM	310
7 PM	530

To find the **average velocity** of the car between 8 AM and 11 AM, we divide the distance traveled by the time elapsed. Thus,

$$\text{Average velocity} = \frac{150}{3} = 50 \text{ miles per hour.}$$

Similarly, the average velocity between 11 AM and 3 PM is

$$\text{Average velocity} = \frac{310 - 150}{4} = \frac{160}{4} = 40 \text{ miles per hour,}$$

and the average velocity between 3 PM and 7 PM is

$$\text{Average velocity} = \frac{530 - 310}{4} = \frac{220}{4} = 55 \text{ miles per hour.}$$

In general, suppose that after t_1 hours the car is d_1 miles from home and after t_2 hours it is d_2 miles from home. In this case, the average velocity during the time period from t_1 to t_2 is

$$\text{Average velocity} = \frac{d_2 - d_1}{t_2 - t_1} \text{ miles per hour.} \tag{1}$$

Stated simply,

$$\text{Average velocity} = \frac{\text{total change in distance}}{\text{total change in time}}.$$

It is useful to look at the quotient in (1) from a geometric point of view. Suppose that we know the distance traveled as a function of the time elapsed, t:

$$d = f(t).$$

Imagine that the curve shown in Figure 1 is the graph of this function.

Figure 1

Looking at the graph, we see that the quotient in (1), which gives the average velocity over the time period from t_1 to t_2, is the *slope of the line joining the points* (t_1, d_1) *and* (t_2, d_2). Thus, we begin to see a relationship between slopes and velocities.

||||||||||||| **Average Rate of Change**

To generalize the notion of average velocity, we shall define the average rate of change of a function.

> Suppose that the quantity y is a function $y = f(x)$ of the quantity x. As x changes from x_1 to x_2, y changes from $y_1 = f(x_1)$ to $y_2 = f(x_2)$. The **average rate of change of y with respect to x** between x_1 and x_2 is defined as the ratio
> $$\frac{y_2 - y_1}{x_2 - x_1} = \frac{f(x_2) - f(x_1)}{x_2 - x_1} = \frac{\text{change in } y}{\text{change in } x}. \tag{2}$$

Figure 2, the graph of $y = f(x)$, shows that the average rate of change of y with respect to x between x_1 and x_2 is the *slope of the line joining* $P_1(x_1, y_1)$ *and* $P_2(x_2, y_2)$. This line determined by points P_1 and P_2 is called the **secant line.**

Figure 2

Important Note   According to the above definition, the average velocity of the car is the average rate of change in the distance traveled with respect to time. *Thus, velocity is a rate of change.*

**EXAMPLE 1
(Medicine)**   Suppose that the weight of a tumor prior to treatment is given by the function

$$W(t) = 5t^2,$$

where the weight W is in milligrams, and the time t is in weeks. After 5 weeks of growth, the weight is $W(5) = 5(5^2) = 125$ milligrams. After 10 weeks, the weight is $W(10) = 5(10)^2 = 500$ milligrams. The average rate of change in weight over the 5-week period is

$$\frac{W(10) - W(5)}{10 - 5} = \frac{500 - 125}{5} = 75 \text{ milligrams per week.}$$

Figure 3 is the graph of the weight function $W(t)$. The average rate of change is the slope of the secant line joining the points $(5, 125)$ and $(10, 500)$ on the graph of $W(t)$. The slope is *positive*, which means that the weight of the tumor has increased during the 5-week period.

Figure 3

EXAMPLE 2
(Business)

Suppose that a small manufacturer of a chemical product finds that the weekly profit P (in dollars) from the manufacture and sale of x kilograms of the product per week is given by

$$P(x) = -x^2 + 180x - 4500.$$

Thus, the profit received from making 100 kilograms per week is $P(100) = \$3500$, and the profit received from making 130 kilograms is $P(130) = \$2000$. The average rate of change in profit with respect to the number of kilograms manufactured and sold is

$$\text{Average rate of change} = \frac{P(130) - P(100)}{130 - 100} = \frac{2000 - 3500}{30} = -\frac{1500}{30}$$
$$= -\$50 \text{ per kilogram.}$$

That is to say, if the weekly production is increased from 100 to 130 kilograms, then the profit from the average kilogram produced goes down by $50.

    The negative sign tells us that profit decreases as production increases from 100 to 130 kilograms. One of the reasons that the profit function behaves this way could be the need to pay overtime rates as the level of production increases. As the graph of the profit function $P(x)$ shows (see Figure 4) the average rate of change of $P(x)$ with respect to x is the slope of the secant line joining the points (100, 3500) and (130, 2000).

Instantaneous Velocity

In many applications it is not enough to know the average rate of change of y with respect to x. We need another concept, called the **instantaneous rate of change.** The driver in the problem at the start of this section, for example, might want to know the instantaneous velocity at exactly 10:18 AM. The following example will point the way to a suitable definition of the instantaneous rate of change.

Figure 4

EXAMPLE 3
(Motion—
Free-Falling
Objects)

When an object, such as a stone or a piece of metal, is dropped from a given height, such as a building or a cliff, the distance that it has fallen (in feet) after t seconds is given approximately (neglecting air resistance) by the formula

$$f(t) = 16t^2$$

(Figure 5). We want to find the instantaneous velocity of the object exactly 2 seconds after it has been dropped.

Solution

We can obtain an approximate value of the instantaneous velocity when $t = 2$ seconds by computing the average velocity over a small time interval of length h between $t_1 = 2$ and $t_2 = 2 + h$. For example, if $h = 0.1$ second, we calculate the average velocity over the interval from $t_1 = 2$ to $t_2 = 2.1$ and obtain

$$\text{Average velocity} = \frac{f(2.1) - f(2)}{2.1 - 2} = \frac{16(2.1)^2 - 16(2)^2}{0.1}$$
$$= 65.6 \text{ feet per second.}$$

Hence, an *approximate* value for the instantaneous velocity when $t = 2$ is 65.6 feet per second. To improve this approximation, we can make the time interval's length h smaller and obtain the average velocity over this smaller time interval. As the size of the interval decreases, we should expect to get a better approximation

Distance fallen in 1 second.
$f(1) = 16$ feet

Distance fallen in 2 seconds.
$f(2) = 64$ feet

Figure 5

to the instantaneous velocity. The results of a number of such calculations are shown in Table 2.

TABLE 2   As $h \to 0^+$

Value of h (sec.)	Time Interval (sec.)	Average Velocity (ft. per sec.)
0.1	$t = 2$ to $t = 2.1$	65.6
0.01	$t = 2$ to $t = 2.01$	64.16
0.001	$t = 2$ to $t = 2.001$	64.016
0.0001	$t = 2$ to $t = 2.0001$	64.0016
0.00001	$t = 2$ to $t = 2.00001$	64.00016

We can also approximate the instantaneous velocity when $t = 2$ by taking small time intervals that immediately precede, rather than follow, time $t = 2$. The results of such calculations are shown in Table 3.

TABLE 3   As $h \to 0^-$

Value of h (sec.)	Time Interval (sec.)	Average Velocity (ft. per sec.)
-0.1	$t = 1.9$ to $t = 2$	62.4
-0.01	$t = 1.99$ to $t = 2$	63.84
-0.001	$t = 1.999$ to $t = 2$	63.984
-0.0001	$t = 1.9999$ to $t = 2$	63.9984
-0.00001	$t = 1.99999$ to $t = 2$	63.99984

Because the average velocities approach 64 feet per second, we conclude that the instantaneous velocity of the object when $t = 2$ is 64 feet per second.

We could arrive at the same conclusion without the tedious work of drawing up Tables 2 and 3, by the following analysis. For arbitrary real number h, the average velocity over the interval $t_1 = 2$ to $t_2 = 2 + h$ is

$$\text{Average velocity} = \frac{f(2 + h) - f(2)}{(2 + h) - 2} = \frac{16(2 + h)^2 - 16(2)^2}{h}$$

$$= \frac{16(4 + 4h + h^2) - 16(4)}{h}$$

$$= \frac{64h + 16h^2}{h}$$

$$= 64 + 16h, \quad \text{if } h \neq 0.$$

We now find the limit of this average velocity as the size of the interval gets smaller and smaller; that is, as $h \to 0$. We have

$$\lim_{h \to 0} (64 + 16h) = 64,$$

which confirms the earlier conclusion.   ■

Guided by this example, we can formulate the following definition.

If an object travels a linear distance of

$$d = f(t) \text{ feet}$$

in t seconds, then its instantaneous velocity at the precise time $t = t_0$ is

$$v(t_0) = \lim_{h \to 0} \frac{f(t_0 + h) - f(t_0)}{h} \text{ ft./sec.}, \qquad (3)$$

if this limit exists.

Of course, a given problem may involve units other than feet and seconds, resulting in a velocity of feet per hour, miles per hour, and so on.

EXAMPLE 4   Consider the falling object described in Example 3. Exactly 3 seconds after it is dropped, how far has it fallen and what is its velocity?

Solution   (a) When $t = 3$ sec., it has fallen $f(3)$ feet. Since $f(3) = 16(3)^2 = 16 \cdot 9 = 144$, it has fallen 144 feet.

(b) According to the definition above, the velocity at time $t = 3$ is

$$
\begin{aligned}
v(3) &= \lim_{h \to 0} \frac{f(3 + h) - f(3)}{h} \\
&= \lim_{h \to 0} \frac{16(3 + h)^2 - 16(3)^2}{h} \\
&= \lim_{h \to 0} \frac{16(9 + 6h + h^2) - 16(9)}{h} \\
&= \lim_{h \to 0} \frac{144 + 96h + 16h^2 - 144}{h} \\
&= \lim_{h \to 0} \frac{h(96 + 16h)}{h} \\
&= \lim_{h \to 0} (96 + 16h) \\
&= 96 \text{ ft./sec.}
\end{aligned}
$$

∎

‖‖‖‖‖‖‖‖‖‖ Instantaneous Rate of Change

Recall that velocity is a special case of the more general concept of *rate of change*. Suppose that we know that one quantity, y, is a function of another, x:

$$y = f(x).$$

As x changes from one specific value, x_0, to another, $x_0 + h$, the average rate of change in y with respect to x is calculated according to Equation (2):

$$\frac{f(x_0 + h) - f(x_0)}{h}.$$

Remember that h may be negative as well as positive.

It is often useful to take the limit of this average rate of change, as the interval size h gets smaller and smaller. We are thus led to the following definition.

Suppose that the quantity y is a function $y = f(x)$ of the quantity x. The **instantaneous rate of change of y with respect to x at the value $x = x_0$** is

$$\lim_{h \to 0} \frac{f(x_0 + h) - f(x_0)}{h}, \tag{4}$$

if this limit exists.

||||||||||||| **Tangent Lines to Curves**

We can obtain a useful geometric interpretation of the instantaneous rate of change as follows. The quotient

$$\frac{f(x_0 + h) - f(x_0)}{h}$$

represents the slope of the secant line joining the points $A(x_0, f(x_0))$ and $B(x_0 + h, f(x_0 + h))$ (Figure 6).

Figure 6

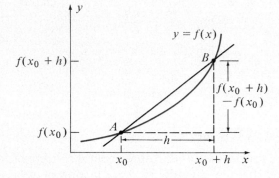

As h gets closer and closer to 0, the point B moves along the graph of f toward the point A, and the secant lines from A to B approach a line called the **tangent line** to the graph of f at the point A (Figure 7).

Figure 7

Thus, *the instantaneous rate of change at* x_0 *given by (4) is the slope of the tangent line to the graph of* f *at the point* $(x_0, f(x_0))$. The slope of this tangent line is called the **slope of the curve** $y = f(x)$ at $(x_0, f(x_0))$ or merely the **slope of** $y = f(x)$ **at** x_0.

EXAMPLE 5   Consider the manufacturer of Example 2 again. Find the instantaneous rate of change of profit $p(x) = -x^2 + 180x - 4500$ with respect to weekly production x when the level of production is

(a) $x_0 = 50$ kilograms   (b) $x_0 = 120$ kilograms   (c) $x_0 = 90$ kilograms.

Solution   Using (4) to obtain the instantaneous rate of change when the level of production is x_0, we write

Instantaneous rate of change at x_0

$$= \lim_{h \to 0} \frac{P(x_0 + h) - P(x_0)}{h}$$

$$= \lim_{h \to 0} \frac{[-(x_0 + h)^2 + 180(x_0 + h) - 4500] - [-x_0^2 + 180x_0 - 4500]}{h}$$

$$= \lim_{h \to 0} \frac{[-x_0^2 - 2x_0h - h^2 + 180x_0 + 180h - 4500] - [-x_0^2 + 180x_0 - 4500]}{h}$$

$$= \lim_{h \to 0} \frac{-2x_0h - h^2 + 180h}{h}$$

$$= \lim_{h \to 0} (-2x_0 - h + 180) = -2x_0 + 180.$$

Thus, we have obtained the following formula:

Instantaneous rate of change $= -2x_0 + 180$ dollars per kilogram.   (5)

(a) When $x_0 = 50$ kilograms, we use (5) to obtain the instantaneous rate of change and find

$$-2(50) + 180 = \$80 \text{ per kilogram.}$$

That is, when the level of production is exactly 50 kilograms per week, the profit is *increasing* at the rate of $80 per kilogram. As we shall prove in the next chapter, this means that the 51st kilogram produced will bring in about $80 profit.

Geometrically, this also means that the slope of the curve $y = -x^2 + 180x - 4500$, shown in Figure 8, at the point (50, 2000) is 80.

(b) When $x_0 = 120$ kilograms, the instantaneous rate of change according to (5) is

$$-2(120) + 180 = -\$60 \text{ per kilogram.}$$

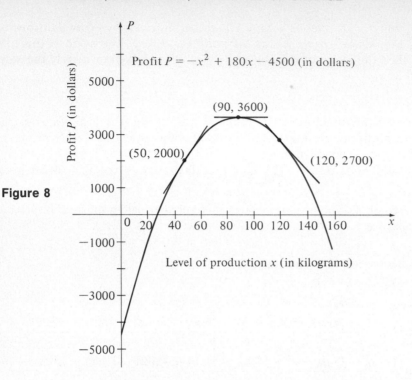

Figure 8

That is, when the level of production is exactly 120 kilograms per week, the profit is *decreasing* at the rate of $60 per kilogram. Thus, the slope of the curve $y = -x^2 + 180x - 4500$ (Figure 8) at the point $(120, 2700)$ is -60, and producing the 121st kilogram will result in a *loss* of about $60.

(c) When $x_0 = 90$ kilograms, the instantaneous rate of change is

$$-2(90) + 180 = \$0 \text{ per kilogram.}$$

That is, when the level of production is exactly 90 kilograms per week, the profit is neither increasing nor decreasing. The slope of the curve (Figure 8) $y = -x^2 + 180x - 4500$ at the point $(90, 3600)$ is zero, so the tangent line at this point is horizontal. This fact has considerable significance, which we shall explore in later sections.   ■

2.4 EXERCISE SET ||||||||||||||||||||||||||||||||||||||||||||||||||||||||||||||||||||||||||||||||||||||||||||||||

1. A projectile fired upward into the air from a height of 10 feet reaches a height of

$$h(t) = 10 + 80t - 16t^2$$

   t seconds after it is fired. Find the average velocity
   (a) during the first second.
   (b) during the first two seconds.

   (c) during the second second.
   (d) during the third second.
   Explain your answer to (d).

2. An object moving along the x-axis has x-coordinate

$$x(t) = 5t - t^2 \text{ (inches)}$$

   t seconds after it starts. Find the average

velocity
(a) during the first second.
(b) during the first two seconds.
(c) during the second and third seconds (from $t = 1$ to $t = 3$).
(d) during the fourth and fifth seconds (from $t = 3$ to $t = 5$).

In Exercises 3–8, find the average rate of change of y with respect to x between the values of x_1 and x_2.

3. $y = 3x$; $x_1 = 1$, $x_2 = 7$

4. $y = 4x - 2$; $x_1 = 3$, $x_2 = 6$

5. $y = 2x^2 + 3$; $x_1 = -2$, $x_2 = 5$

6. $y = \dfrac{3x^2 - 1}{x - 2}$; $x_1 = 3$, $x_2 = 5$

7. $y = 2$; $x_1 = -3$, $x_2 = 5$

8. $y = \dfrac{1}{x}$; $x_1 = 1$, $x_2 = 4$

9. Suppose $y = x^2 - 2x$.
   (a) Find the average rate of change of y with respect to x between $x_1 = 0$ and $x_2 = 3$.
   (b) Sketch the graph of the curve and draw the secant line whose slope is the average rate of change obtained in (a).
   (c) Find the average rate of change of y with respect to x between $x_1 = 1$ and $x_2 = 1 + h$.
   (d) Find the instantaneous rate of change of y with respect to x at $x_0 = 1$.
   (e) Find the slope of the tangent line to the curve $y = x^2 - 2x$ at $(1, -1)$.

10. Suppose $y = x^2 + 1$.
    (a) Find the average rate of change of y with respect to x between $x_1 = 0$ and $x_2 = 2$.
    (b) Find the average rate of change of y with respect to x between $x_1 = -1$ and $x_2 = -1 + h$.
    (c) Sketch the graph of the curve.
    (d) Draw the secant line whose slope is the average rate of change obtained in (c).

(e) Find the instantaneous rate of change of y with respect to x at $x_0 = -1$.
(f) Find the slope of the tangent line to the curve $y = x^2 + 1$ at $(-1, 2)$.

11. For the function $y = x^2 - 2x$ given in Exercise 9, find the instantaneous rate of change of y with respect to x at $x = x_0$.

12. For the function $y = x^2 + 1$ given in Exercise 10, find the instantaneous rate of change of y with respect to x at $x = x_0$.

13. For the function whose graph is shown in Figure 9, find the average rate of change of y with respect to x between the values of x_1 and x_2 given below.
    (a) $x_1 = 2$ and $x_2 = 5$
    (b) $x_1 = 3$ and $x_2 = 8$
    (c) $x_1 = 2$ and $x_2 = 10$.

Figure 9

14. For the function whose graph is shown in Figure 10, find the average rate of change of y with respect to x between the values of x_1 and x_2 given below.

Figure 10

(a) $x_1 = 2$, $x_2 = 4$
(b) $x_1 = 2$, $x_2 = 7$
(c) $x_1 = 7$, $x_2 = 10$.

In Exercises 15–18, find the instantaneous rate of change of y with respect to x at the given value of x_0.

15. $y = 3x$, $x_0 = 2$

16. $y = 2x - 5$, $x_0 = -2$

17. $y = x^2$, $x_0 = 1$

18. $y = 3x^2$, $x_0 = 4$

19. **(Medicine)** A hospital patient's chart shows the following data:

Time	6 AM	12 noon	3 PM	9 PM
Temperature (in °F)	99.2	100.4	101.3	100.3

(a) What is the average rate of change of temperature with respect to time between 6 AM and 12 noon?
(b) What is the average rate of change of temperature with respect to time between 3 PM and 9 PM?
(c) Explain why the instantaneous rate of change of temperature at 1:15 PM cannot be determined.

20. **(Economics)** The wholesale price index for commodities from 1980 to 1986 is given approximately in the following table.

Year	1980	1981	1982	1983	1984	1985	1986
Index	110	114	119	135	160	175	183

(a) Find the average rate of change in the wholesale price index between 1980 and 1984.
(b) Find the average rate of change in the wholesale price index between 1983 and 1985.

(c) Explain why the instantaneous rate of change in the wholesale price index at noon on 1 January 1984 cannot be found from this table.

21. **(Business)** A manufacturer of a catalytic agent finds the cost (in millions of dollars) of producing x million gallons of the product is given by the formula

$$C(x) = x^2 - 4x + 8.$$

(a) Find the average rate of change of cost when the level of production changes from $x_1 = \frac{1}{2}$ to $x_2 = 2$.
(b) Find the average rate of change of cost when the level of production changes from $x_1 = 3$ to $x_2 = 5$.
(c) Find the instantaneous rate of change of cost when $x_0 = 1$.
(d) Find the instantaneous rate of change of cost when $x_0 = 2$.
(e) Find the instantaneous rate of change of cost when $x_0 = 3$.

22. **(Ecology)** The cooling system of a nuclear power plant is leaking waste into a nearby stream. The amount A (in gallons) of nuclear waste that flowed into a stream during the first t days after the leak occurred is given by the formula

$$A = 3t^2 + 2t.$$

At what rate is the waste flowing into the stream when $t = 3$?

23. **(Motion)** Suppose that a car starts from rest and travels a distance d (in miles) in t hours, according to the formula

$$d = 2t + 6t^2.$$

(a) Find the average velocity between $t_1 = 0$ and $t_2 = 4$.
(b) Find the average velocity between $t_1 = 2$ and $t_2 = 5$.
(c) Find the instantaneous velocity when $t_0 = 3$.

KEY IDEAS FOR REVIEW ||||||||||||||||||||||||||||||||||||||||||||||||||||||||||||||||||||||||||||||||

- $\lim\limits_{x \to a} f(x) = L$ means that the values of $f(x)$ get close to the unique real number L as x gets close to (but remains different from) a.

- If the values of $f(x)$ do not approach a unique real number as x approaches the value a, we say that $\lim\limits_{x \to a} f(x)$ does not exist.

- The limit from the left, $\lim\limits_{x \to a^-} f(x)$, or the limit from the right, $\lim\limits_{x \to a^+} f(x)$, may exist even when $\lim\limits_{x \to a} f(x)$ does not exist.

- $\lim\limits_{x \to a} f(x)$ exists and equals L only when both one-sided limits exist and equal L.

- Algebraic properties of limits.

  1. $\lim\limits_{x \to a} k = k$.

  2. $\lim\limits_{x \to a} x = a$.

  3a. $\lim\limits_{x \to a} [f(x) + g(x)] = \lim\limits_{x \to a} f(x) + \lim\limits_{x \to a} g(x)$.

  3b. $\lim\limits_{x \to a} [f(x) - g(x)] = \lim\limits_{x \to a} f(x) - \lim\limits_{x \to a} g(x)$.

  4. $\lim\limits_{x \to a} [f(x) \cdot g(x)] = \lim\limits_{x \to a} f(x) \cdot \lim\limits_{x \to a} g(x)$.

  5. $\lim\limits_{x \to a} k\, f(x) = k \lim\limits_{x \to a} f(x)$.

  6. $\lim\limits_{x \to a} \dfrac{f(x)}{g(x)} = \dfrac{\lim\limits_{x \to a} f(x)}{\lim\limits_{x \to a} g(x)}$ (if $\lim\limits_{x \to a} g(x) \neq 0$).

  7. $\lim\limits_{x \to a} [f(x)]^{1/n} = \left[\lim\limits_{x \to a} f(x) \right]^{1/n}$
  (where we do not take an even root of a negative number).

- Properties 1–7 are also valid for one-sided limits.

- $+\infty$ and $-\infty$ are not numbers.

- If $f(x)$ grows larger and larger positively (or negatively) without bound as $x \to a$, then we write $\lim\limits_{x \to a} f(x) = +\infty$ (or $\lim\limits_{x \to a} f(x) = -\infty$).

- Suppose that $\lim\limits_{x \to a} g(x) = 0$ and $\lim\limits_{x \to a} f(x) = L \neq 0$.

  1. if $\dfrac{f(x)}{g(x)} > 0$ as $x \to a$, then $\lim\limits_{x \to a} \dfrac{f(x)}{g(x)} = +\infty$

  2. if $\dfrac{f(x)}{g(x)} < 0$ as $x \to a$, then $\lim\limits_{x \to a} \dfrac{f(x)}{g(x)} = -\infty$

- A line $x = a$ is a vertical asymptote of a curve $y = f(x)$ if either $\lim\limits_{x \to a^-} f(x) = \pm\infty$ or $\lim\limits_{x \to a^+} f(x) = \pm\infty$, or both.

- A rational function $\dfrac{p(x)}{q(x)}$ has vertical line $x = a$ as an asymptote if $q(a) = 0$ but $p(a) \neq 0$.

- $\lim\limits_{x \to +\infty} f(x) = L$ if the values of $f(x)$ get closer and closer to the unique real number L as x gets larger and larger positively without bound.

- $\lim\limits_{x \to -\infty} f(x) = L$ if the values of $f(x)$ get close to the unique real number L as x gets larger and larger negatively without bound.

- $\lim\limits_{x \to \infty} \dfrac{k}{x^n} = 0$ and $\lim\limits_{x \to -\infty} \dfrac{k}{x^n} = 0$, for any constant k and positive integer n.

- A line $y = b$ is a horizontal asymptote of a curve $y = f(x)$ if $\lim\limits_{x \to +\infty} f(x) = b$, or $\lim\limits_{x \to -\infty} f(x) = b$, or both.

- f is continuous at $x = a$ if and only if

  1. $f(a)$ is defined.

  2. $\lim\limits_{x \to a} f(x)$ exists.

  3. $\lim\limits_{x \to a} f(x) = f(a)$.

- If $f(x)$ is continuous at every number in an interval I, then we say that f is continuous on the interval I. If f is continuous on $(-\infty, +\infty)$ we say that f is continuous everywhere. If f is not continuous at $x = a$ we say that f is discontinuous there.

- Every polynomial is a continuous function everywhere.

- Every rational function is continuous everywhere except where the denominator has zero value.

- Average velocity $= \dfrac{\text{change in distance}}{\text{change in time}}$.

- If $y = f(x)$, then the average rate of change of y with respect to x between x_1 and x_2 is

$$\frac{y_2 - y_1}{x_2 - x_1} = \frac{f(x_2) - f(x_1)}{x_2 - x_1} = \frac{\text{change in } y}{\text{change in } x}.$$

- Velocity is a rate of change.

- If $y = f(x)$, then the instantaneous rate of change of y with respect to x at the value $x = x_0$ is

$$\lim_{h \to 0} \frac{f(x_0 + h) - f(x_0)}{h}.$$

- The instantaneous rate of change of f with respect to x at x_0 is the slope of the tangent line to the graph of f at the point $(x_0, f(x_0))$. The slope of this tangent line is called the slope of the curve $y = f(x)$ at x_0.

REVIEW EXERCISES ||||||||||||||||||||||||||||||||||||||||||||||||||||||||||||||||||||||||||||||

In Exercises 1–8, find the limit, if it exists.

1. $\lim\limits_{x \to 2} (2x^2 - 3x + 5)$

2. $\lim\limits_{x \to -4} \dfrac{x^2 + x - 12}{x + 4}$

3. $\lim\limits_{x \to -5^+} \dfrac{3}{x + 5}$

4. $\lim\limits_{x \to 2^-} \dfrac{x - 3}{x - 2}$

5. $\lim\limits_{x \to 8} \dfrac{64 - x^2}{8 - x}$

6. $\lim\limits_{x \to 0} \dfrac{2x^2}{x^2 - 4}$

7. $\lim\limits_{x \to +\infty} \dfrac{4 - 2x}{3 + 5x}$

8. $\lim\limits_{x \to -\infty} \dfrac{3}{x - 4}$

9. Let f be defined by

$$f(x) = \begin{cases} 2 - x & \text{if } x \le 1 \\ \frac{4}{3}x - \frac{1}{3} & \text{if } x > 1. \end{cases}$$

Investigate $\lim\limits_{x \to 1^-} f(x)$, $\lim\limits_{x \to 1^+} f(x)$ and $\lim\limits_{x \to 1} f(x)$.

10. Use the graph to find $\lim\limits_{x \to -2} f(x)$.

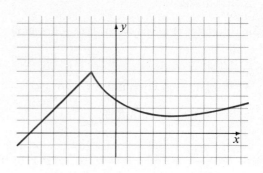

11. Use the graph to find $\lim\limits_{x \to \infty} f(x)$.

12. Determine whether the function graphed below is continuous at $x = 4$.

13. If

$$f(x) = \frac{x^2 - 3x - 10}{x + 2}, \qquad x \neq -2$$

how do we define $f(-2)$ so that f will be continuous everywhere?

In Exercises 14 and 15, determine whether the given function is continuous at $x = -2$. If it is not continuous there, state which of the three continuity conditions is (are) violated. Graph each function.

14. $f(x) = \begin{cases} \dfrac{x^2 - 4}{x + 2} & \text{if } x \neq -2 \\ 1 & \text{if } x = -2 \end{cases}$

15. $f(x) = \begin{cases} \dfrac{x^2 + 3x + 2}{x + 2} & \text{if } x \neq -2 \\ -1 & \text{if } x = -2 \end{cases}$

In Exercises 16–19, determine whether the functions are continuous at the point specified.

16. $f(x) = |x - 2|$, $x = 2$

17. $f(x) = \dfrac{x^2 - 2x - 3}{x - 3}$, $x = 3$

18. $f(x) = \begin{cases} 2 & \text{if } x \leq 1 \\ x + 1 & \text{if } x > 1 \end{cases}$ $x = 1$

19. $f(x) = \dfrac{1}{x + 1}$, $x = -1$

In Exercises 20–23, determine where the given function is continuous.

20. $f(x) = 3x^2 + 2x - 1$

21. $f(x) = \dfrac{2}{x^2 + 4}$

22. $f(x) = \sqrt{2x + 1}$

23. $f(x) = \dfrac{x - 1}{x}$

24. Suppose $y = 3x^2 - 4x$.
    (a) Find the average rate of change of y with respect to x between $x_1 = 4$ and $x_2 = 7$.
    (b) Find the slope of the tangent line at $(1, -1)$.

25. If $y = 2x^2 - 1$, find the average rate of change of y with respect to x between $x_1 = -2$, and $x_2 = 3$.

26. For the function graphed below, find the average rate of change of y with respect to x between $x_1 = 2$ and $x_2 = 5$.

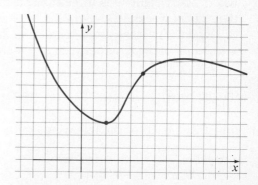

27. If $y = 2x^2 + 2$, find the instantaneous rate of change of y with respect to x at $x_0 = 3$.

28. (**Business**) The number $N(p)$ of calculators of a certain manufacturer that can be sold at a price of p dollars per unit is given by

$$N(p) = \frac{500}{p^2}$$

Find $\lim_{p \to 0} N(p)$ and interpret the result.

29. **(Business)** A manufacturer of X-ray developers finds that the profit received from the sale of x million liters of the product is given (in millions of dollars) by

$$P(x) = 20x^2 + 12x.$$

(a) Find the average rate of change of profit when the level of production changes from $x_1 = 20$ to $x_2 = 40$.

(b) Find the instantaneous rate of change of profit at $x_0 = 15$.

30. **(Population)** The population $P(t)$ (in tens of thousands) of a newly incorporated city t years after incorporation is given by

$$P(t) = 10t^2 + 2t.$$

(a) Find the average rate of change of the population between $t_1 = 3$ and $t_2 = 8$.

(b) Find the instantaneous rate of change of the population at $t_0 = 6$.

31. **(Medicine)** Suppose that the growth of a human fetus is given by the function

$$W(t) = 3t^2, \qquad 0 \le t \le 39$$

where the weight w is in grams and the time t is in weeks.

(a) Find the average rate of change in the weight of the fetus between $t_1 = 4$ and $t_2 = 10$.

(b) Find the instantaneous rate of change of the weight of the fetus at $t_0 = 20$.

32. **(Medicine)** A medical research team studying effects of choline on the memory of elderly patients determined the number $N(x)$ of words remembered 5 hours after reading a story as a function of the amount x (in milligrams) of choline administered. Assume that the team constructed the following graph.

Amount of choline x given to a patient (in milligrams)

(a) Find $\lim\limits_{x \to 10} N(x)$.

(b) Where is the function $N(x)$ discontinuous?

CHAPTER TEST ||||||||||||||||||||||||||||||||||||||||||||||||||||||||||||||||||||||||||||||||||||

1. Investigate each of the following limits:

   (a) $\lim\limits_{x \to 3} \dfrac{x+2}{x-1}$ (b) $\lim\limits_{x \to 1} \dfrac{x^2-1}{x-1}$

2. Investigate each of the following limits:

   (a) $\lim\limits_{x \to 5} \dfrac{|x-5|}{x-5}$ (b) $\lim\limits_{x \to 2^+} \dfrac{x}{2-x}$

3. Investigate each of the following limits:

   (a) $\lim\limits_{x \to +\infty} \dfrac{3x}{1-6x}$ (b) $\lim\limits_{x \to -\infty} \dfrac{2x}{x^2-3}$

4. Find all horizontal and vertical asymptotes for the curve $y = \dfrac{x^2}{x^2-4}$.

5. In each part determine whether the given function is continuous everywhere. If it is not, tell where it is discontinuous.

(a)

(b)

(c)

(d)

6. In each part determine whether the given function is continuous at $x = 2$.

(a) $f(x) = \dfrac{x^2 - x - 2}{x - 2}$

(b) $f(x) = \begin{cases} \dfrac{x^2 - x - 2}{x - 2} & \text{if } x \neq 2 \\ 3 & \text{if } x = 2 \end{cases}$

(c) $f(x) = \begin{cases} \dfrac{1}{x - 2} & \text{if } x \neq 2 \\ 0 & \text{if } x = 2. \end{cases}$

7. Where is the function $f(x) = \dfrac{1}{2x^2 - 7x - 4}$ continuous?

8. For the function whose graph is shown below, find the average rate of change of y with respect to x between

(a) $x_1 = 2$ and $x_2 = 4$

(b) $x_1 = 2$ and $x_2 = 7$.

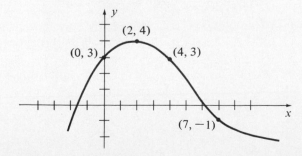

9. Consider the function $f(x) = x^2 - 3x$.
   (a) Find the average rate of change in y with respect to x, from $x_1 = 1$ to $x_2 = 4$.
   (b) Sketch the graph of the curve $y = x^2 - 3x$, and draw the secant line whose slope is the average rate found in (a).
   (c) Find the average rate of change in y with respect to x, from $x_1 = 1$ to $x_2 = 1 + h$. Simplify your answer.
   (d) Find the instantaneous rate of change of y with respect to x, at $x_0 = 1$.

10. For the curve $y = x^2 + 2$, find the slope of the tangent line at $(2, 6)$.

THE DERIVATIVE

3.1 |||||||| DEFINITION OF THE DERIVATIVE

One of the most important concepts in calculus is the concept of *derivative*. In Chapter 2 we laid the groundwork for this central idea by developing the notion of limit, and applying it to obtain the instantaneous rate of change of $y = f(x)$ at x_0 as the expression

$$\lim_{h \to 0} \frac{f(x_0 + h) - f(x_0)}{h},$$

if this limit exists. Upon this foundation we are now ready to build a systematic study of the derivative.

The word "derivative" itself is used in two different ways: as a limit and as a function.

The Derivative as a Limit

For a given function f, and a number x_0 in its domain, **the derivative of f at x_0** is the limit

$$f'(x_0) = \lim_{h \to 0} \frac{f(x_0 + h) - f(x_0)}{h}, \tag{1}$$

provided this limit exists.

Notice that we use a special symbol $f'(x_0)$, read "f prime of x_0," to denote this important limit.

When the limit (1) exists we say that the function f is **differentiable at x_0**; when the limit (1) does not exist we say that f is **not differentiable at x_0**. If f is differentiable at every number of an interval I, then we say that f is **differentiable on the interval I**. If f is differentiable on $(-\infty, +\infty)$ we say that f is **differentiable everywhere**.

> **The Derivative as a Function**
> For a given function f, the **derivative of f** is the function $f'(x)$ whose domain consists of all numbers where f is differentiable, and whose values $f'(x)$ are found by the limit formula (1).

Because of our interest in the derivative as a function of x, we follow the customary procedure of replacing the constant symbol x_0 in Equation (1) by the variable symbol x.

EXAMPLE 1   Let f be defined by $f(x) = x^2 - 4x$. Then the derivative of f is

$$f'(x) = \lim_{h \to 0} \frac{f(x + h) - f(x)}{h}$$

$$= \lim_{h \to 0} \frac{[(x + h)^2 - 4(x + h)] - [x^2 - 4x]}{h}$$

$$= \lim_{h \to 0} \frac{x^2 + 2xh + h^2 - 4x - 4h - x^2 + 4x}{h}$$

$$= \lim_{h \to 0} \frac{2xh + h^2 - 4h}{h}$$

$$= \lim_{h \to 0} \frac{h(2x + h - 4)}{h}$$

$$= \lim_{h \to 0} (2x + h - 4) = 2x - 4.$$

Thus,

$$f'(x) = 2x - 4. \tag{2}$$

Because this is valid *for all x*, f is differentiable everywhere. To find the value of the derivative $f'(x_0)$ at the specific value x_0, we merely substitute $x = x_0$ in Equation (2). Thus,

$$f'(-1) = 2(-1) - 4 = -6$$
$$f'(2) = 2(2) - 4 = 0$$
$$f'(4) = 2(4) - 4 = 4.$$

Tangent Lines

Geometrically, the derivative $f'(x_0)$ gives the slope of the tangent line to the graph of f at the point $(x_0, f(x_0))$.

EXAMPLE 2   Consider the function in Example 1: $f(x) = x^2 - 4x$. The slopes of the tangent lines to the curve $y = f(x) = x^2 - 4x$ at the values $x_0 = -1$, $x_0 = 2$, and $x_0 = 4$ are -6, 0, and 4, respectively, as we saw in Example 1. The tangent lines at the corresponding points are shown in Figure 1.

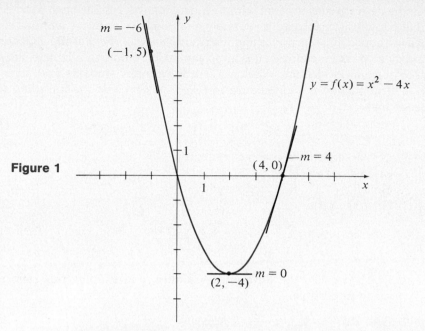

Figure 1

Observe that when the slope of the tangent line is zero, for instance at $x_0 = 2$ in Example 2, the tangent line is horizontal. ∎

Using the point-slope form of a straight line, we have:

The **equation of the tangent line** to the graph of f at the point $(x_0, f(x_0))$ is

$$y - f(x_0) = f'(x_0)(x - x_0). \tag{3}$$

EXAMPLE 3   For the function in Example 1, find the equation of the tangent line to the curve $y = x^2 - 4x$ at the points
(a) $(-1, 5)$ (b) $(2, -4)$ (c) $(4, 0)$.

Solution   Using the slopes obtained in Example 2 and substituting in Equation (3), we have the following equations of the tangent lines:
(a) Tangent line at $(-1, 5)$:

$$y - 5 = -6(x + 1) \qquad \text{or} \qquad y = -6x - 1.$$

(b) Tangent line at $(2, -4)$:

$$y + 4 = 0(x - 2) \qquad \text{or} \qquad y = -4.$$

(c) Tangent line at $(4, 0)$:

$$y - 0 = 4(x - 4) \qquad \text{or} \qquad y = 4x - 16. \qquad ∎$$

Since the derivative (1) is of such importance, many formulas have been developed that allow us to calculate it routinely and painlessly, almost without effort. We shall introduce these formulas and make ample use of them at the appropriate time and place. In the meantime, it is important to learn how to calculate a derivative directly from the defining equation (1). Since many students find it helpful to have a clear step-by-step procedure to follow, we present the following **four-step procedure.**

Four-Step Procedure for Finding $f'(x)$

Step 1. Calculate $f(x + h)$.

Step 2. Find $f(x + h) - f(x)$, and simplify.

Step 3. Form the ratio

$$\frac{f(x + h) - f(x)}{h},$$

which is called the **difference quotient,** and simplify this ratio as much as possible.

Step 4. Find $\lim\limits_{h \to 0} \dfrac{f(x + h) - f(x)}{h}$, if it exists.

EXAMPLE 4   Consider the function f defined by

$$f(x) = ax + b,$$

where a and b are constants. Find the derivative function, and give a geometric interpretation.

Solution   *Step 1.* $f(x + h) = a(x + h) + b.$

Step 2. $f(x + h) - f(x) = a(x + h) + b - (ax + b)$
$$= ax + ah + b - ax - b$$
$$= ah.$$

Step 3. $\dfrac{f(x + h) - f(x)}{h} = \dfrac{ah}{h} = a \qquad$ if $h \neq 0.$

Step 4. $\lim\limits_{h \to 0} \dfrac{f(x + h) - f(x)}{h} = \lim\limits_{h \to 0} a = a.$

Figure 2

Hence, $f'(x) = a$. For example, if $f(x) = 3x - 2$, then $f'(x) = 3$.

The graph of f, shown in Figure 2, is a straight line with slope a and y-intercept b.

It is clear from the graph of f that the tangent line to the graph of f at any point $(x_0, f(x_0))$ is the line itself. Thus, the slope of the tangent line should be a. Our inference is confirmed by the fact that we obtained $f'(x) = a$.   ■

EXAMPLE 5 It follows from Example 4 that if f is a constant function, say,

$$f(x) = 6$$

then, for all x,

$$f'(x) = 0.$$

EXAMPLE 6 Consider the function f defined by

$$f(x) = \frac{2}{x}.$$

Find $f'(x)$.

Solution *Step 1.* $f(x + h) = \dfrac{2}{x + h}$.

Step 2. $f(x + h) - f(x) = \dfrac{2}{x + h} - \dfrac{2}{x}$

$$= \frac{2x - 2(x + h)}{x(x + h)}$$

$$= -\frac{2h}{x(x + h)}.$$

Step 3. $\dfrac{f(x + h) - f(x)}{h} = -\dfrac{2h}{hx(x + h)}$

$$= -\frac{2}{x(x + h)} \qquad \text{if } h \neq 0.$$

Step 4. $\displaystyle\lim_{h \to 0} \frac{f(x + h) - f(x)}{h} = \lim_{h \to 0} -\frac{2}{x(x + h)}$

$$= -\frac{2}{x^2}. \qquad \text{Therefore, } f'(x) = \frac{-2}{x^2}.$$

Since $f'(x) = -\dfrac{2}{x^2}$, we see that $f'(x)$ is not defined for $x = 0$, but it is defined for every $x \neq 0$. Therefore, f is differentiable everywhere except at 0.

The graph of f (shown in Figure 3) also illustrates this fact as follows: at any point $(x_0, f(x_0))$ with $x_0 \neq 0$, we can draw a tangent line to the graph. In Figure 3, we have drawn such a tangent line at the point $(2, 1)$; its slope is $f'(2) = -\frac{1}{2}$. Because f is not defined at $x_0 = 0$, we cannot draw a tangent line to f at 0. Notice that f is also discontinuous at $x_0 = 0$.

Figure 3

$y = f(x) = \dfrac{2}{x}$

Slope $= -\dfrac{1}{2}$

$(2, 1)$

∎

Differentiability Versus Continuity

Differentiability is a stronger property than continuity. In fact, differentiability implies continuity, but the converse is not true. The following two statements correctly summarize this relationship.

> (a) If a function f is differentiable at $x = x_0$, then f must be continuous at $x = x_0$.
> (b) However, a function may be continuous at x_0 without being differentiable there.

Statement (a) is a significant theorem in the theory of calculus; it can be proved directly from the definition of derivative and the properties of limits. We shall illustrate (b) by providing Examples 7 and 8, below.

EXAMPLE 7   Consider the absolute value function f defined by

$$f(x) = |x|, \qquad \text{for all } x.$$

Investigate $f'(0)$.

Solution   We have

$$f'(0) = \lim_{h \to 0} \frac{f(0 + h) - f(0)}{h}$$

$$= \lim_{h \to 0} \frac{|0 + h| - |0|}{h}$$

$$= \lim_{h \to 0} \frac{|h|}{h}.$$

In Example 3 of Section 2.1, we saw that $\lim\limits_{h \to 0} \dfrac{|h|}{h}$ does not exist. In fact,

$$\lim_{h \to 0^+} \frac{|h|}{h} = 1, \qquad \text{but} \quad \lim_{h \to 0^-} \frac{|h|}{h} = -1.$$

Hence, $f'(0)$ does not exist so that *f is not differentiable at 0*, but it is continuous there. From the graph of f, shown in Figure 4, we see that we can draw a tangent line to the graph at any point $(x_0, f(x_0))$ where $x \neq 0$. For example, the tangent line to the graph at the point $(2, 2)$ is the extension of the right-hand branch of the curve itself. On the other hand, there is *no* tangent line to the graph at the point $(0, 0)$.

Figure 4

$$y = f(x) = |x|$$

∎

Roughly speaking, a continuous function is said to have a "sharp corner" at $x = a$ if its direction changes sharply at $x = a$. A function with a sharp corner at $x = a$ is not differentiable at a. Examples of such a function are the one in Figure 4 and those in Figure 5.

Figure 5

Another way in which a function can be continuous at a number $x = x_0$ without being differentiable there is for the graph of the function to have a vertical tangent line at $x = x_0$.

EXAMPLE 8   Consider the function f defined by

$$f(x) = \sqrt[3]{x}.$$

Investigate $f'(0)$.

Solution   We have

$$f'(0) = \lim_{h \to 0} \frac{f(0 + h) - f(0)}{h}$$

$$= \lim_{h \to 0} \frac{\sqrt[3]{(0 + h)} - \sqrt[3]{0}}{h}$$

$$= \lim_{h \to 0} \frac{h^{1/3} - 0}{h}$$

$$= \lim_{h \to 0} \frac{1}{\sqrt[3]{h^2}} = +\infty.$$

Since this limit does not exist, we conclude that f is not differentiable at 0. The graph of f is shown in Figure 6. In this case, f is continuous at $x = 0$, but the tangent line to the graph at the point (0, 0) is a vertical line (the y-axis), so it has no slope and $f'(0)$ does not exist.

Figure 6

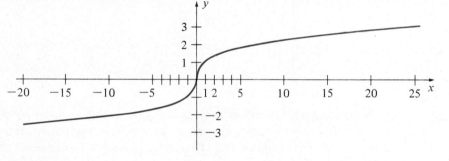

EXAMPLE 9   The function $f(x) = x^{2/3}$ combines the features of both functions found in Examples 7 and 8. Its graph (Figure 7) has both a sharp corner at $x = 0$ and a vertical tangent line there.

When we attempt to find $f'(0)$, we get

$$f'(0) = \lim_{h \to 0} \frac{f(0 + h) - f(0)}{h}$$

$$= \lim_{h \to 0} \frac{\sqrt[3]{(0 + h)^2} - \sqrt[3]{0^2}}{h}$$

$$= \lim_{h \to 0} \frac{h^{2/3} - 0}{h} = \lim_{h \to 0} \frac{1}{h^{1/3}}$$

$$= \lim_{h \to 0} \frac{1}{\sqrt[3]{h}}.$$

This limit does not exist, since

$$\lim_{h \to 0^+} \frac{1}{\sqrt[3]{h}} = +\infty \qquad \text{while} \qquad \lim_{h \to 0^-} \frac{1}{\sqrt[3]{h}} = -\infty.$$

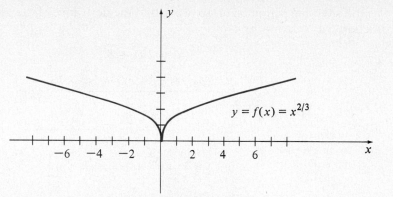

Figure 7

$$y = f(x) = x^{2/3}$$

We conclude this section by restating our definition of the derivative (1) in an alternate form that is widely used. This form is sometimes easier to use than Equation (1).

Alternate Definition of Derivative
For a given function f, and a number x_0 in its domain, **the derivative of f at x_0** is the limit

$$f'(x_0) = \lim_{x \to x_0} \frac{f(x) - f(x_0)}{x - x_0}, \tag{4}$$

if this limit exists.

Equation (4) is sometimes called the **symmetric definition** of derivative. It is easy to see that Equation (4) is equivalent to Equation (1) by making the substitution $x = x_0 + h$, so that $x - x_0 = h$, and $h \to 0$ if and only if $x \to x_0$.

EXAMPLE 10   Use the alternate definition (4) to find the derivative of $f(x) = 2x^2 - 5x - 4$ at $x_0 = 3$.

Solution   First, note that $f(x_0) = f(3) = 2 \cdot 9 - 5 \cdot 3 - 4 = -1$. Thus,

$$f'(3) = \lim_{x \to 3} \frac{f(x) - f(3)}{x - 3}$$

$$= \lim_{x \to 3} \frac{(2x^2 - 5x - 4) - (-1)}{x - 3}$$

$$= \lim_{x \to 3} \frac{2x^2 - 5x - 3}{x - 3}$$

$$= \lim_{x \to 3} \frac{(2x + 1)(x - 3)}{x - 3}$$

$$= \lim_{x \to 3} (2x + 1)$$

$$= 7.$$

Therefore, $f'(3) = 7$.

EXAMPLE 11   Use the alternate definition (4) to prove that the derivative of the general quadratic
function $f(x) = ax^2 + bx + c$ is

$$f'(x) = 2ax + b. \tag{5}$$

Solution

$$f'(x_0) = \lim_{x \to x_0} \frac{(ax^2 + bx + c) - (ax_0^2 + bx_0 + c)}{x - x_0}$$

$$= \lim_{x \to x_0} \frac{ax^2 - ax_0^2 + bx - bx_0 + c - c}{x - x_0}$$

$$= \lim_{x \to x_0} \left[a\frac{x^2 - x_0^2}{x - x_0} + b\frac{x - x_0}{x - x_0} + 0 \right]$$

$$= \lim_{x \to x_0} [a(x + x_0) + b]$$

$$= a(x_0 + x_0) + b$$

$$= 2ax_0 + b. \qquad \blacksquare$$

3.1 EXERCISE SET ||||||||||||||||||||||||||||||||||||||||||||||||||||||||||||||||||||||||||||||||||||

1. In each part, estimate $f'(x_0)$ from the graph.

(a)

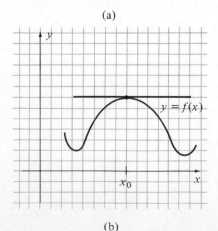

(b)

2. In each part, estimate $f'(x_0)$ from the graph.

(a)

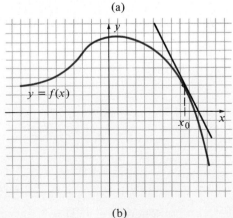

(b)

In Exercises 3–6, calculate the difference quotient $\dfrac{f(x + h) - f(x)}{h}$ for the given function $f(x)$.

3. $f(x) = 3x - 2$ 4. $f(x) = 4$

5. $f(x) = 3x^2 + 2$ 6. $f(x) = 1/x$

In Exercises 7–10, find the derivative of $f(x)$, using the four-step procedure.

7. $f(x) = 2x + 5$ 8. $f(x) = 4 - x^2$

9. $f(x) = 8$ 10. $f(x) = \dfrac{1}{x + 1}$

11. Suppose $f(x) = x^2 - 2x + 4$.
    (a) Find $f'(x)$ using the four-step procedure.
    (b) Calculate $f'(2)$.
    (c) Find the slope of the tangent line to the curve at the point where $x = 2$.
    (d) Find an equation of the tangent line to the curve at the point where $x = 2$.
    (e) Sketch the graph of f and draw the tangent line obtained in (d).

12. Suppose $f(x) = -x^2 + 3x + 2$.
    (a) Find $f'(x)$ using the four-step procedure.
    (b) Calculate $f'(3)$.
    (c) Find the slope of the tangent line to the curve at the point where $x = 3$.
    (d) Find an equation of the tangent line to the curve at the point where $x = 3$.
    (e) Sketch the graph of f and draw the tangent line obtained in (d).

13. Suppose $f(x) = 2x^2 - 3x + 2$. Use alternate definition (4) to compute
    $$f'(-2), \quad f'(0), \quad f'(3).$$

14. Suppose $f(x) = x^3$. Use alternate definition (4) to compute
    $$f'(3), \quad f'(0), \quad f'(4).$$

In Exercises 15–20, you are encouraged to make use of Equation (5).

15. Suppose $f(x) = 2x^2 - 3x + 2$.
    (a) Find the point on the graph of f where the slope is 5.

(b) Find the point on the graph of f where the tangent line is horizontal.
    (c) Find the point on the graph of f where the tangent line is parallel to the line $3x + y - 5 = 0$.

16. Suppose $f(x) = -x^2 + 3x + 2$.
    (a) Find the point on the graph of f where the slope is -3.
    (b) Find the point on the graph of f where the tangent line is horizontal.
    (c) Find the point on the graph of f where the tangent line is parallel to the line $5x - y + 1 = 0$.

17. **(Motion)** Recall from Example 3 in Section 2.4, that when an object is dropped from a given height, the distance that it has fallen (in feet) after t seconds is given approximately by the formula
    $$f(t) = 16t^2.$$
    (a) Find the instantaneous velocity t seconds after the object has been released.
    (b) Find the instantaneous velocity after 5 seconds.
    (c) How many seconds will it take for the object to reach an instantaneous velocity of 128 feet per second?

18. **(Advertising)** Suppose that x days after an advertising campaign for a certain product has ended, the number $N(x)$ of units sold per day is given approximately by $N(x) = -2x^2 + 20x + 200$. Find the rate of change of sales when $x = 6$. That is, find $N'(6)$.

19. **(Epidemic)** An epidemic of Victorian flu has broken out in a certain town. Public health officials have determined that the number $N(x)$ of cases reported x days after the start of the epidemic is given approximately by
    $$N(x) = 40x - x^2, \quad 0 \le x \le 40.$$
    What is the rate at which the epidemic is spreading after
    (a) 10 days   (b) 20 days   (c) 30 days?
    That is, find (a) $N'(10)$, (b) $N'(20)$, and (c) $N'(30)$.

20. **(Educational Psychology)** A typing school has determined that the number of words $N(x)$ typed by a typical student after x hours of training is given by

$$N(x) = 5 + 5x + x^2, \qquad 0 \le x \le 6.$$

Find the rate of learning after training for (a) 1 hour (b) 2 hours (c) 4 hours. That is, find (a) $N'(1)$, (b) $N'(2)$, and (c) $N'(4)$.

In Exercises 21–24, graph each function and find the values of x at which it fails to be differentiable.

21. $f(x) = \begin{cases} -3x + 7 & \text{if } x \ge 1 \\ 4x & \text{if } x < 1 \end{cases}$

22. $f(x) = 2\sqrt{x}$

23. $f(x) = \dfrac{1}{x - 1}$

24. $f(x) = |x - 1|$

25. Consider the function graphed in Figure 8. At which of the points is the function (a) continuous, but not differentiable? (b) neither continuous nor differentiable? Explain your answers.

Figure 8

26. Consider the function graphed in Figure 9. At which of the points is the function (a) continuous, but not differentiable? (b) neither continuous nor differentiable? Explain your answers.

Figure 9

3.2 ||||||| BASIC RULES OF DIFFERENTIATION

First, we acknowledge some alternative notation for the derivative. The symbol

$$D_x[f(x)], \qquad \text{or} \qquad D(f(x)),$$

is sometimes used instead of $f'(x)$. Recall from Section 3.1 that the formula for the derivative of $f(x) = ax^2 + bx + c$ is

$$f'(x) = 2ax + b. \tag{1}$$

Thus,

$$D_x[ax^2 + bx + c] = 2ax + b = D[ax^2 + bx + c],$$

and

$$D_x[x^2 - 4x] = 2x - 4 = D[x^2 - 4x].$$

The letter "D" tells us to differentiate the indicated function, and the subscript "x" tells us that x is the independent variable.

Sometimes we use an even simpler notation. If $y = f(x)$, we may write y' instead of $f'(x)$. For example,

$$\text{if} \quad y = 5x^2 - 7x - 11, \quad \text{then} \quad y' = 10x - 7.$$

Leibniz Notation

The most popular notation requires some background explanation. A special symbol, Δ, is often used to denote the **change** (or **increment**) in a variable. As a variable x changes value from x_0 to x_1, the amount of change is denoted $\mathbf{\Delta x}$ (read "delta x"); thus,

$$\Delta x = x_1 - x_0. \tag{2}$$

The symbol "Δx" represents a single number; it is *not* to be thought of as a product of Δ and x, since Δ by itself does not represent a number.

If $y = f(x)$, then as x changes from x_0 to x_1, y also changes, from $f(x_0)$ to $f(x_1)$. Thus, the change in y is

$$\Delta y = f(x_1) - f(x_0) \tag{3}$$

The average rate of change in y with respect to x is, according to Equation (2) of Section 2.4,

$$\frac{\text{change in } y}{\text{change in } x} = \frac{\Delta y}{\Delta x}. \tag{4}$$

Using the Δ-notation in the difference quotient that defines the derivative [Equation (1) of Section 3.1] we let $x_1 = x_0 + h$. Then

$$\Delta x = x_1 - x_0 = (x_0 + h) - x_0 = h, \tag{5}$$

and

$$\Delta y = f(x_1) - f(x_0) = f(x_0 + h) - f(x_0). \tag{6}$$

Thus,

$$\frac{\Delta y}{\Delta x} = \frac{f(x_0 + h) - f(x_0)}{h}. \tag{7}$$

Since $f'(x) = \lim\limits_{h \to 0} \dfrac{f(x_0 + h) - f(x_0)}{h}$ and $h = \Delta x$, we have

$$f'(x_0) = \lim_{\Delta x \to 0} \frac{\Delta y}{\Delta x}. \tag{8}$$

To Gottfried Leibniz (1646–1716), one of the two principal originators of the calculus, the fact that the derivative of a function is a **limit of a difference quotient** was so impressive that he felt compelled to use a notation that would serve as a constant reminder. Thus, instead of $f'(x)$, Leibniz chose to write $\dfrac{dy}{dx}$ (pronounced "dee y dee x"). Hence,

$$\frac{dy}{dx} = \lim_{\Delta x \to 0} \frac{\Delta y}{\Delta x}. \qquad (9)$$

For example,

$$\text{if} \quad y = 3x^2 + 7x - 10, \qquad \text{then} \quad \frac{dy}{dx} = 6x + 7.$$

Finally, the notation $\dfrac{d}{dx}$ is often used to mean D_x. Thus,

$$\frac{d}{dx}\,[f(x)] = D_x[f(x)]; \qquad \frac{d}{dx}\,[3x^2 + 7x - 10] = 6x + 7.$$

In summary,

$$\text{(a)} \quad \frac{dy}{dx} = \frac{d}{dx}\,(y) = D_x(y) = y'; \qquad (10)$$

$$\text{(b)} \quad \frac{d}{dx}\,[f(x)] = D_x[f(x)] = f'(x). \qquad (11)$$

All of these notations are in widespread use in the mathematical literature. To assist you in becoming familiar with them, we shall often express problems in alternate notation. One notation may have an advantage in one particular situation but not in another.

By now you have undoubtedly concluded that the procedure for finding the derivative $f'(x)$ by using the definition

$$f'(x) = \lim_{h \to 0} \frac{f(x + h) - f(x)}{h}$$

can be a rather tedious process for all but the simplest of functions. In this section, we shall discuss a number of differentiation techniques that will greatly simplify the calculation of derivatives.

The following rule was established in Example 5 of Section 3.1.

The Constant Rule

If $f(x) = k$ is a constant function, then $D_x[f(x)] = 0$. That is, the derivative of any constant function is the zero constant function.

From Example 4 of Section 3.1 we see that $D_x x = 1$, and from Example 11 of Section 3.1 we have $D_x x^2 = 2x$. In general, we can differentiate any positive integral power of x by the following rule.

The Power Rule

$$D_x[x^n] = nx^{n-1} \qquad \text{for any positive integer } n.$$

Proof   Let n denote a positive integer ≥ 2, and let $f(x) = x^n$. Using the alternate definition of derivative [Equation (4) of Section 3.1],

$$f'(x_0) = \lim_{x \to x_0} \frac{f(x) - f(x_0)}{x - x_0}$$

$$= \lim_{x \to x_0} \frac{x^n - x_0^n}{x - x_0}.$$

The numerator is factorable [see Factoring Special Forms in Section A.5 (Appendix A)].

$$f'(x_0) = \lim_{x \to x_0} \frac{(x - x_0)(x^{n-1} + x^{n-2}x_0 + x^{n-3}x_0^2 + \cdots + xx_0^{n-2} + x_0^{n-1})}{x - x_0}$$

$$= \lim_{x \to x_0} (x^{n-1} + x^{n-2}x_0 + x^{n-3}x_0^2 + \cdots + xx_0^{n-2} + x_0^{n-1})$$

$$= x_0^{n-1} + x_0^{n-2}x_0 + x_0^{n-3}x_0^2 + \cdots + x_0 x_0^{n-2} + x_0^{n-1}$$

$$= x_0^{n-1} + x_0^{n-1} + x_0^{n-1} + \cdots + x_0^{n-1} + x_0^{n-1} \qquad (n \text{ terms})$$

$$= nx_0^{n-1}.$$

Therefore, for all x, $f'(x) = nx^{n-1}$.

EXAMPLE 1

$$\frac{d}{dx}[x^5] = 5x^{5-1} = 5x^4$$

$$\frac{d}{dx}[x] = 1x^{1-1} = 1x^0 = 1$$

$$\frac{d}{dx}[x^{20}] = 20x^{19}.$$

The Constant Times a Function Rule
If k is a constant, then

$$\frac{d}{dx}[kf(x)] = k\frac{d}{dx}[f(x)].$$

That is, the derivative of a constant times a function is the constant times the derivative of the function.

Proof   By the definition of the derivative,

$$\frac{d}{dx}[kf(x)] = \lim_{h \to 0} \frac{kf(x+h) - kf(x)}{h}$$

$$= \lim_{h \to 0} k\frac{f(x+h) - f(x)}{h}$$

$$= k\lim_{h \to 0} \frac{f(x+h) - f(x)}{h} \qquad \text{(by Limit Property 5)}$$

$$= kf'(x).$$

EXAMPLE 2

$$\frac{d}{dx}[4x^5] = 4 \cdot \frac{d}{dx}[x^5] = 4 \cdot 5x^4 = 20x^4$$

$$\frac{d}{dx}[-x^3] = -1 \cdot \frac{d}{dx}[x^3] = -1 \cdot 3x^2 = -3x^2$$

$$\frac{d}{dx}[\pi x^2] = \pi \cdot \frac{d}{dx}[x^2] = 2\pi x$$

$$\frac{d}{dx}[-7x^8] = -7\frac{d}{dx}[x^8] = -7 \cdot 8x^7 = -56x^7.$$

The Sum and Difference Rule

$$\frac{d}{dx}[f(x) + g(x)] = \frac{d}{dx}[f(x)] + \frac{d}{dx}[g(x)]$$

$$\frac{d}{dx}[f(x) - g(x)] = \frac{d}{dx}[f(x)] - \frac{d}{dx}[g(x)]$$

That is, the derivative of a sum is the sum of the derivatives; the derivative of a difference is the difference of the derivatives.

Proof $\dfrac{d}{dx} [f(x) + g(x)] = \lim\limits_{h \to 0} \dfrac{[f(x + h) + g(x + h)] - [f(x) + g(x)]}{h}$

$= \lim\limits_{h \to 0} \dfrac{[f(x + h) - f(x)] + [g(x + h) - g(x)]}{h}$

$= \lim\limits_{h \to 0} \left[\dfrac{f(x + h) - f(x)}{h} + \dfrac{g(x + h) - g(x)}{h} \right]$

$= \lim\limits_{h \to 0} \dfrac{f(x + h) - f(x)}{h} + \lim\limits_{h \to 0} \dfrac{g(x + h) - g(x)}{h}$

(by Limit Property 3a)

$= f'(x) + g'(x).$

The proof for $\dfrac{d}{dx} [f(x) - g(x)]$ is entirely similar, and is left as an exercise (see Exercise 32).

||||||||||||| **Derivatives of Polynomials**

One important consequence of the above rules is that *every* polynomial function is differentiable everywhere. No further rules are needed for differentiating polynomials.

EXAMPLE 3   (a) $\dfrac{d}{dx} [x^2 - 4x] = \dfrac{d}{dx} [x^2] - \dfrac{d}{dx} [4x]$

$= 2x - 4$ (see Example 1 in Section 3.1)

(b) $\dfrac{d}{dx} [2 + x^3] = \dfrac{d}{dx} [2] + \dfrac{d}{dx} [x^3]$

$= 0 + 3x^2 = 3x^2$

(c) $\dfrac{d}{dx} [x^6 + x^4] = \dfrac{d}{dx} [x^6] + \dfrac{d}{dx} [x^4]$

$= 6x^5 + 4x^3$

(d) $\dfrac{d}{dx} [x^5 - 7x^4] = \dfrac{d}{dx} [x^5] - \dfrac{d}{dx} [7x^4]$

$= 5x^4 - 28x^3.$

EXAMPLE 4   Find where the slope of the tangent line to the curve $y = f(x) = x^3 - x^2$ has the value 1.

Solution   The slope of the tangent line at the point $(x, f(x))$ is

$$f'(x) = \dfrac{d}{dx} [x^3 - x^2] = 3x^2 - 2x.$$

Thus, we want to find all values of x for which

$$3x^2 - 2x = 1,$$

or

$$3x^2 - 2x - 1 = 0,$$

or

$$(3x + 1)(x - 1) = 0.$$

Thus,

$$3x + 1 = 0 \quad \text{or} \quad x - 1 = 0.$$

Hence, the curve has slope 1 when

$$x = -\tfrac{1}{3} \quad \text{or} \quad x = 1.$$

When $x = -\tfrac{1}{3}$,

$$y = (-\tfrac{1}{3})^3 - (-\tfrac{1}{3})^2 = -\tfrac{4}{27}.$$

When $x = 1$,

$$y = 1^3 - 1^2 = 0.$$

Thus, the slope is 1 at the points $(-\tfrac{1}{3}, -\tfrac{4}{27})$ and $(1, 0)$. ∎

EXAMPLE 5 Consider the function $f(x)$ defined by

$$f(x) = x^3 - 4x^2 + 4x.$$

Find all values of x for which
(a) $f'(x) = 4$ (b) $f'(x) = 0$

Solution We have

$$f'(x) = 3x^2 - 8x + 4.$$

(a) If $f'(x) = 4$, then

$$3x^2 - 8x + 4 = 4$$

or

$$3x^2 - 8x = 0$$
$$x(3x - 8) = 0$$
$$x = 0 \quad \text{or} \quad x = \tfrac{8}{3}.$$

Then $f'(x) = 4$ at $x = 0$ and at $x = \tfrac{8}{3}$.

(b) If $f'(x) = 0$, then

$$3x^2 - 8x + 4 = 0$$
$$(3x - 2)(x - 2) = 0$$
$$x = \tfrac{2}{3} \quad \text{or} \quad x = 2.$$

Then $f'(x) = 0$ at $x = \tfrac{2}{3}$ and at $x = 2$.   ■

3.2 EXERCISE SET ||||||||||||||||||||||||||||||||||||||||||||||||||||||||||||||||||||||||||||||||||||||

In Exercises 1–6, for the given function $f(x)$ and the given values of x_0 and x_1, find Δx and Δy.

1. $f(x) = 5x + 4$, $x_0 = 3$, $x_1 = 7$.

2. $f(x) = 3x^2 - 5x$, $x_0 = 1$, $x_1 = 3$.

3. $f(x) = \sqrt{x - 1}$, $x_0 = 10$, $x_1 = 17$.

4. $f(x) = x^3$, $x_0 = -1$, $x_1 = 2$.

5. $f(x) = 2x - x^2$, $x_0 = 2$, $x_1 = 5$.

6. $f(x) = \dfrac{1}{x + 3}$, $x_0 = 0$, $x_1 = 3$.

In Exercises 7–24, use the rules for differentiation to find the derivative of each function and simplify the answer.

7. $f(x) = 5$ 8. $f(x) = -\sqrt{2}$

9. $f(x) = x^7$ 10. $f(t) = t^{15}$

11. $f(t) = 6t^3$ 12. $f(x) = -5x^7$

13. $f(x) = 3x^4 - 2x + 2$

14. $f(u) = 4u^6 - 2u^4 + 6$

15. $g(x) = 3^5$ 16. $g(x) = 5^{20} + x$

17. $g(u) = 2u - 5$ 18. $g(u) = 18 - 7x$

19. $h(u) = 11u^5 - 7u^4 + 8^3 + 5$

20. $h(u) = 25u^3 - u^4 + 5u^6$

21. $h(x) = x^2(x + 1)$

22. $h(x) = (3x + 1)(2x - 1)$

23. $h(x) = (2x + 3)(x^2 - 1)$

24. $h(x) = x^3(2x^2 + x - 5)$

25. Suppose $f(x) = 2x^3 - 3x^2 + 2$. Compute
    (a) $f'(-2)$ (b) $f'(0)$ (c) $f'(3)$.

26. Find the slope of the tangent line, and the equation of the tangent line, to the graph of $f(x) = x^3 - 2x^2 + 3x - 5$ at the point where
    (a) $x = 0$ (b) $x = 1$ (c) $x = 2$.

27. Suppose $f(x) = x^3 + x^2 - 8x + 2$. Find all values of x for which
    (a) $f'(x) = -8$ (c) $f'(x) = 8$.
    (b) $f'(x) = 0$

28. Suppose f is defined by $f(x) = x^3 + 2x^2 + 3x + 1$. Find all points on the graph of f where the tangent line is parallel to the line $y = 3x + 2$.

29. **(Advertising)** The marketing research department of a sporting goods manufacturer has determined that when the company spends x thousand dollars on advertising, the number of units it sells is given by

    $$N(x) = -x^2 + 50x + 5000, \quad 0 \le x \le 50.$$

    (a) Find the instantaneous rate of change of the number of units sold with respect to the amount x spent on advertising.
    (b) Compute $N'(5)$ and $N'(30)$.
    (c) Interpret the results obtained in (b).

30. **(Motion)** Suppose that an object moves from rest along a straight line and that the distance s (in feet) it has traveled after t seconds is given by

    $$s = 6t^2 - t^3, \quad 0 \le t \le 6.$$

    (a) Find the instantaneous velocity at time t.
    (b) After how many seconds will the object come to rest?

31. Find the rate of change of the area A of a circle with respect to the radius r. (Recall that $A = \pi r^2$.)

32. Prove the *Difference Rule*

$$\frac{d}{dx}[f(x) - g(x)] = f'(x) - g'(x)$$

directly from the definition of derivative.

3.3 |||||||| FURTHER TECHNIQUES OF DIFFERENTIATION: PRODUCTS, QUOTIENTS, AND RATIONAL POWERS

Since the derivative of a sum of two functions is the sum of the derivatives, we might conjecture that the derivative of a product of two functions is the product of the derivatives. This conjecture is readily seen to be false by taking the product of

$$f(x) = x^3 \quad \text{and} \quad g(x) = x^2.$$

The derivative of the product of the two functions is

$$\frac{d}{dx}[f(x) \cdot g(x)] = \frac{d}{dx}[x^3 \cdot x^2] = \frac{d}{dx}[x^5] = 5x^4, \tag{1}$$

but the product of the derivatives of the two functions is

$$\frac{d}{dx}[f(x)] \cdot \frac{d}{dx}[g(x)] = \frac{d}{dx}[x^3] \cdot \frac{d}{dx}[x^2]$$

$$= (3x^2)(2x) = 6x^3. \tag{2}$$

We can easily see that results (1) and (2) are not the same.

The derivative of a product is given by the following rule.

The Product Rule

$$\frac{d}{dx}[f(x)g(x)] = f(x)\frac{d}{dx}[g(x)] + g(x)\frac{d}{dx}[f(x)]$$

That is, the derivative of a product of two functions is the first function times the derivative of the second plus the second function times the derivative of the first.

Proof

$$\frac{d}{dx}[f(x)g(x)] = \lim_{h \to 0} \frac{f(x + h)g(x + h) - f(x)g(x)}{h}.$$

To evaluate this limit we must employ two "tricks." First, to allow us to separate the difference quotient into the sum of two useful quotients, we add

$$0 = -f(x + h)g(x) + f(x + h)g(x)$$

to the numerator. Thus,

$$\frac{d}{dx}[f(x)g(x)] = \lim_{h \to 0} \frac{f(x+h)g(x+h) - f(x+h)g(x) + f(x+h)g(x) - f(x)g(x)}{h}$$

$$= \lim_{h \to 0} \frac{f(x+h)[g(x+h) - g(x)] + g(x)[f(x+h) - f(x)]}{h}$$

$$= \lim_{h \to 0} \left[f(x+h)\frac{g(x+h) - g(x)}{h} + g(x)\frac{f(x+h) - f(x)}{h} \right]$$

$$= \lim_{h \to 0} f(x+h) \cdot \lim_{h \to 0} \frac{g(x+h) - g(x)}{h} + g(x)\lim_{h \to 0} \frac{f(x+h) - f(x)}{h} \cdot$$

(by Limit Properties 3, 4, and 5)

$$= \left[\lim_{h \to 0} f(x+h) \right] \cdot \frac{d}{dx}g(x) + g(x)\frac{d}{dx}[f(x)]. \tag{3}$$

The second "trick" is to recall that a differentiable function must be continuous (Section 3.1). That is, if $\frac{d}{dx}f(x)$ exists, then $\lim_{\bar{x} \to x} f(\bar{x}) = f(x)$. Taking $\bar{x} = x + h$, and observing that $\bar{x} \to x$ as $h \to 0$, we see that $\lim_{h \to 0} f(x+h) = f(x)$ whenever $\frac{d}{dx}f(x)$ exists. Substituting this result into (3) above, we obtain

$$\frac{d}{dx}[f(x)g(x)] = f(x)\frac{d}{dx}[g(x)] + g(x)\frac{d}{dx}[f(x)].$$

EXAMPLE 1   Consider again $f(x) = x^3$, $g(x) = x^2$. By the product rule,

$$\frac{d}{dx}[x^3 \cdot x^2] = x^3 \frac{d}{dx}[x^2] + x^2 \frac{d}{dx}[x^3]$$

$$= x^3(2x) + x^2(3x^2)$$

$$= 2x^4 + 3x^4$$

$$= 5x^4,$$

as we obtained in (1) by finding $(d/dx)[x^5]$ directly.

EXAMPLE 2   If

$$h(x) = (x^3 + 4x^2 + 1)(x^2 - 2x),$$

then by the product rule we have

$$h'(x) = (x^3 + 4x^2 + 1)\frac{d}{dx}[x^2 - 2x] + (x^2 - 2x)\frac{d}{dx}[x^3 + 4x^2 + 1]$$

$$= (x^3 + 4x^2 + 1)(2x - 2) + (x^2 - 2x)(3x^2 + 8x)$$

$$= (2x^4 + 6x^3 - 8x^2 + 2x - 2) + (3x^4 + 2x^3 - 16x^2)$$

$$= 5x^4 + 8x^3 - 24x^2 + 2x - 2.$$

Of course, we could have solved this problem without the product rule by multiplying out the given factors of $h(x)$, collecting similar terms, and then differentiating. The method using the product rule is faster, however.

EXAMPLE 3
(Biology)

The population of a certain species is given by the formula

$$N(t) = (2t + 40)(200 - t) \qquad 0 \le t \le 200$$

where $N(t)$ is the number or organisms after t days.
  (a) Find the instantaneous rate of change of the population after 20 days.
  (b) When will the instantaneous rate of change of the population be 200 organisms per day?
  (c) When will the population stop growing?

Solution

The instantaneous rate of change of the population is given by the product rule as

$$N'(t) = \frac{d}{dt}[(2t + 40)(200 - t)]$$

$$= (2t + 40)\frac{d}{dt}[200 - t] + (200 - t)\frac{d}{dt}[2t + 40]$$

$$= (2t + 40)(-1) + (200 - t)(2)$$

$$= 360 - 4t.$$

(a) When $t = 20$,

$$N'(20) = 360 - 4(20) = 360 - 80 = 280 \text{ organisms per day.}$$

(b) We want to find a value for t so that

$$N'(t) = 360 - 4t = 200,$$

or

$$4t = 160$$
$$t = 40.$$

Thus, the instantaneous rate of change of the population will be 200 organisms per day after 40 days.
  (c) The population stops growing when the instantaneous rate of change of the population is zero. That is, when

$$N'(t) = 360 - 4t = 0,$$

or

$$t = 90.$$

Thus, the population stops growing after 90 days. ∎

|||||||||||| **Derivatives of Quotients**

Just as the derivative of a product is not the product of the derivatives, so the derivative of a quotient is not the quotient of the derivatives. The derivative of a quotient is given by the following rule.

The Quotient Rule

$$\frac{d}{dx}\left[\frac{f(x)}{g(x)}\right] = \frac{g(x)\dfrac{d}{dx}[f(x)] - f(x)\dfrac{d}{dx}[g(x)]}{[g(x)]^2}$$

That is, the derivative of a quotient is the denominator times the derivative of the numerator minus the numerator times the derivative of the denominator, all divided by the square of the denominator.

The proof of the quotient rule is a bit more complicated than that of the product rule. We omit it here, but you can find a guide to its proof in Exercises 37–39.

EXAMPLE 4   Suppose $x \neq 0$ and

$$y = \frac{5x^4 - 2x^5}{x^2}.$$

Find dy/dx.

Solution   Using the quotient rule, we have

$$\frac{dy}{dx} = \frac{x^2 \dfrac{d}{dx}[5x^4 - 2x^5] - (5x^4 - 2x^5)\dfrac{d}{dx}[x^2]}{(x^2)^2}$$

$$= \frac{x^2(20x^3 - 10x^4) - (5x^4 - 2x^5)(2x)}{x^4}$$

$$= \frac{20x^5 - 10x^6 - 10x^5 + 4x^6}{x^4}$$

$$= \frac{10x^5 - 6x^6}{x^4} = 10x - 6x^2, \qquad \text{if } x \neq 0.$$

Observe that dividing by x^2 at the beginning would have enabled us to write the given function as

$$y = 5x^2 - 2x^3.$$

Then

$$\frac{dy}{dx} = \frac{d}{dx}[5x^2 - 2x^3] = 10x - 6x^2. \qquad \blacksquare$$

Clearly, in this example, the latter approach is simpler than using the quotient rule. Thus, it is important to carefully examine the given function for alternate approaches. For many quotient problems, however, we *must* use the quotient rule.

EXAMPLE 5   If

$$y = \frac{x^3 + 2x^2}{x^2 + 1}$$

then, by the quotient rule, we have

$$\frac{dy}{dx} = \frac{(x^2 + 1)\frac{d}{dx}[x^3 + 2x^2] - (x^3 + 2x^2)\frac{d}{dx}[x^2 + 1]}{(x^2 + 1)^2}$$

$$= \frac{(x^2 + 1)(3x^2 + 4x) - (x^3 + 2x^2)(2x)}{(x^2 + 1)^2}$$

$$= \frac{x^4 + 3x^2 + 4x}{(x^2 + 1)^2}.$$

EXAMPLE 6   For the curve,

$$y = f(x) = \frac{x}{x + 2},$$

find an equation of the tangent line at the point $(1, \frac{1}{3})$.

Solution   The slope of the tangent line at the point $(x, f(x))$ is, by the quotient rule,

$$\frac{dy}{dx} = \frac{d}{dx}\left[\frac{x}{x + 2}\right]$$

$$= \frac{(x + 2)\frac{d}{dx}[x] - x\frac{d}{dx}[x + 2]}{(x + 2)^2}$$

$$= \frac{(x + 2) \cdot (1) - x \cdot (1)}{(x + 2)^2}$$

$$= \frac{2}{(x + 2)^2}.$$

When $x = 1$, the slope is

$$m = \frac{2}{(1 + 2)^2} = \frac{2}{9}.$$

Thus, the equation of the tangent line is

$$y - \tfrac{1}{3} = \tfrac{2}{9}(x - 1)$$

or

$$9y - 2x = 1.$$ ∎

The Power Rule for Negative and Fractional Exponents

We have stated and proved the power rule $\dfrac{d}{dx}[x^n] = nx^{n-1}$ for positive integer exponents only. The quotient rule allows us to extend this rule to include negative integer exponents as well. For any positive integer n and $x \neq 0$,

$$
\begin{aligned}
\frac{d}{dx}[x^{-n}] &= \frac{d}{dx}\left[\frac{1}{x^n}\right] \\
&= \frac{x^n \dfrac{d}{dx}[1] - 1\dfrac{d}{dx}[x^n]}{(x^n)^2} \qquad \text{by the quotient rule} \\
&= \frac{0 - nx^{n-1}}{x^{2n}} \\
&= -nx^{n-1-2n} \\
&= -nx^{-n-1}.
\end{aligned}
$$

Thus, the power rule is valid for negative integer exponents. Moreover,

$$\frac{d}{dx}[x^0] = \frac{d}{dx}[1] = 0 = 0x^{-1} = 0x^{0-1}, \qquad (\text{if } x \neq 0)$$

so that the power rule is valid when the exponent is 0. Finally, the power rule is true when the exponent is *any rational number*. Thus, when p and q are integers and $q \neq 0$,

$$\frac{d}{dx}[x^{p/q}] = \frac{p}{q}x^{p/q-1}.$$

The proof of this fact requires a method to be presented in Section 3.7.

In summary,

The Rational Power Rule

$$\frac{d}{dx}[x^r] = rx^{r-1}, \qquad \text{for any rational number.}$$

EXAMPLE 7 (a) $\dfrac{d}{dx}[x^{-3}] = -3x^{-3-1} = -3x^{-4}$.

(b) $\dfrac{d}{dx}[x^{1/2}] = \dfrac{1}{2}x^{(1/2)-1} = \dfrac{1}{2}x^{-(1/2)} = \dfrac{1}{2\sqrt{x}}$.

(c) $\dfrac{d}{dx}[x^{-1/3}] = -\dfrac{1}{3}x^{-(1/3)-1} = -\dfrac{1}{3}x^{-(4/3)}$.

(d) $\dfrac{d}{dx}\left[\dfrac{10}{x^5}\right] = \dfrac{d}{dx}[10x^{-5}] = 10\dfrac{d}{dx}[x^{-5}] = 10(-5x^{-5-1}) = -50x^{-6}$.

EXAMPLE 8 (a) If $y = (\sqrt{x} + x^3)(2x^3 - 3x^2)$, then by the product rule, we have

$$\frac{dy}{dx} = (\sqrt{x} + x^3)\frac{d}{dx}[2x^3 - 3x^2] + (2x^3 - 3x^2)\frac{d}{dx}[\sqrt{x} + x^3]$$
$$= (\sqrt{x} + x^3)(6x^2 - 6x) + (2x^3 - 3x^2)(\tfrac{1}{2}x^{-1/2} + 3x^2)$$
$$= 7x^2\sqrt{x} - \tfrac{15}{2}x\sqrt{x} + 12x^5 - 15x^4.$$

(b) If

$$y = \frac{3\sqrt{x}}{x^3 + 8}$$

then, by the quotient rule,

$$\frac{dy}{dx} = \frac{(x^3 + 8)\dfrac{d}{dx}[3\sqrt{x}] - 3\sqrt{x}\dfrac{d}{dx}[x^3 + 8]}{(x^3 + 8)^2}$$
$$= \frac{(x^3 + 8)\cdot 3(\tfrac{1}{2})x^{-1/2} - (3\sqrt{x})(3x^2)}{(x^3 + 8)^2}$$
$$= \frac{-\tfrac{15}{2}x^{5/2} + 12x^{-1/2}}{(x^3 + 8)^2}.$$

**EXAMPLE 9
(Educational
Psychology)**
A computer manufacturer gives each new assembly-line worker a 12-hour training course. After t hours of training, a worker can assemble y units each day, where

$$y = f(t) = 18\sqrt{t}, \qquad 0 \le t \le 12.$$

Find the rate at which a worker learns the assembly skill at the end of
(a) 4 hours   (b) 9 hours of training.

Solution The rate of learning is measured by the rate of change in the worker's daily output:

$$\frac{dy}{dt} = f'(t) = \frac{d}{dt}[18\sqrt{t}] = 18\frac{d}{dt}[t^{1/2}] = 18\left(\frac{1}{2}\right)t^{-1/2}$$
$$= \frac{9}{\sqrt{t}}.$$

(a) When $t = 4$, the rate of learning is $9/\sqrt{4} = 9/2$ additional units per day per hour of training.

(b) When $t = 9$, the rate of learning is $9/\sqrt{9} = 9/3 = 3$ additional units per day per hour of training.

Thus, although the *rate* of learning slows down as training increases, the worker's output increases with additional training. For example, the number of units produced after 4 hours of training is $y = f(4) = 18\sqrt{4} = 36$ units daily, but the number of units produced daily after 9 hours of training is $y = f(9) = 18\sqrt{9} = 54$ units. ∎

Summary of the Basic Rules for Differentiation

1. **The Constant Rule.** For any constant c, $\dfrac{d}{dx}[c] = 0$.

2. **The Power Rule.** For any rational number n, $\dfrac{d}{dx}[x^n] = nx^{n-1}$.

3. **The Constant Times a Function Rule.** For any constant k,

$$\frac{d}{dx}[kf(x)] = kf'(x).$$

4. **The Sum and Difference Rules.**

$$\frac{d}{dx}[f(x) + g(x)] = f'(x) + g'(x)$$

$$\frac{d}{dx}[f(x) - g(x)] = f'(x) - g'(x).$$

5. **The Product Rule.**

$$\frac{d}{dx}[f(x)g(x)] = f(x)g'(x) + g(x)f'(x).$$

6. **The Quotient Rule.**

$$\frac{d}{dx}\left[\frac{f(x)}{g(x)}\right] = \frac{g(x)f'(x) - f(x)g'(x)}{[g(x)]^2}.$$

These rules imply that:
  (a) Every polynomial function is differentiable everywhere.
  (b) Every rational function is differentiable everywhere, except where its denominator is 0.
  (c) The nth root function $\sqrt[n]{x}$ is differentiable
    (1) everywhere, if n is odd.
    (2) on $(0, +\infty)$, if n is even.

3.3 EXERCISE SET ||||||||||||||||||||||||||||||||||||||||||||||||||||||||||||||||||||||||||||||||||

In Exercises 1–26, use the rules for differentiation to find the derivative of each function, and simplify the answer.

1. $f(x) = (4x + 3)(5x - 1)$ (two ways)

2. $f(x) = (\frac{1}{2}x - 7)(6x + 2)$ (two ways)

3. $f(x) = x^4(3x^2 + 7)$ (two ways)

4. $f(x) = (x^2 - 3)(x^2 + 3)$ (two ways)

5. $g(x) = (x^2 - 3x)(x + 1)$ (two ways)

6. $f(x) = x^3(x^3 + 7x^2 - 8)$ (two ways)

7. $g(s) = (s^3 - 5)(s^2 + s + 3)$

8. $h(x) = (x^3 + 2x + 1)(x^2 - 4)$

9. $h(x) = \dfrac{3x + 2}{x - 1}$ 10. $h(u) = \dfrac{u + 3}{5 - 2u}$

11. $h(u) = \dfrac{u^2}{u^2 + 6}$ 12. $g(s) = \dfrac{1}{u^3 + 5u}$

13. $h(x) = \dfrac{x^2 - 4}{x^2 - 1}$ 14. $h(t) = \dfrac{3t^2 - 2t}{2t^2 + 1}$

15. $f(x) = 4x^{-2}$

16. $f(x) = 3x^{-4} + 85$

17. $f(x) = \dfrac{2}{x^2} - \dfrac{3}{x^4} + 5x^4 + 7$

18. $f(s) = \dfrac{4}{\sqrt[3]{s^2}} - 3s^5 + 2s + 9$

19. $g(x) = \sqrt[3]{x}$ 20. $g(x) = \sqrt[4]{x^7}$

21. $g(x) = 8x^{-3/2}$ 22. $g(t) = -10t^{-3/5}$

23. $h(t) = (12t^3 - 4t)(\sqrt[3]{t^2} + 5t)$

24. $h(x) = (3x^2 - \sqrt{x} + 2)(4x^3 - x + 3)$

25. $h(x) = \dfrac{3\sqrt{x} + x^2}{2 + 4\sqrt{x}}$

26. $h(x) = \dfrac{5x^{1/3} + 4x^2 + 2}{x^{1/3}}$ (two ways)

27. (a) If $f(x) = 3x^2 + 2\sqrt{x} + 3$, find $D_x[f(x)]$.

(b) If $h(t) = (t + 1)(t^2 - 3t)$, find $\dfrac{d}{dt}[h(t)]$.

(c) If $v = \dfrac{u^2 + 1}{u^2 - u}$, find $\dfrac{dv}{du}$.

28. (a) If $y = 2x^4 - 3x + 5\sqrt{x}$, find $\dfrac{dy}{dx}$.

(b) If $f(t) = \dfrac{3t + 1}{t^2 + t}$, find $D_t[f(t)]$.

(c) If $y = \dfrac{2t^2 + t}{t^4 - 1}$, find y'.

29. Find dy/dx for
(a) $y = (x^3 - 3x^2)(x^2 + x)$

(b) $y = \dfrac{2x^3}{2x - 3}$.

30. Find ds/dt if
(a) $s = (\sqrt{t} + 4t)(t^{3/2} - 2t)$

(b) $s = \dfrac{3t^2 + t}{2t^4 - 1}$.

31. Suppose $h(x) = (2x - 4)/(3x + 2)$. Compute
(a) $h'(-3)$ (b) $h'(0)$ (c) $h'(2)$.

32. Find the slope of the tangent line to the graph of $f(x) = (2x + 1)/(3x - 2)$ at the point where
(a) $x = -2$ (b) $x = 0$ (c) $x = 3$.

33. Find the slope of the tangent line to the graph of $f(x) = \dfrac{x}{x - 1}$ at the point where
(a) $x = 0$ (b) $x = -1$ (c) $x = 3$.

34. Suppose $f(x) = x^{3/2}$. Find all values of x for which
(a) $f'(x) = \frac{3}{2}$ (b) $f'(x) = 0$ (c) $f'(x) = 3$.

35. **(Ecology)** The amount A of sulfur dioxide sent into the atmosphere by a new petrochemical plant is given by

$$A = 4t^{3/2},$$

where A is in tons and t is the number of years that the plant has been in operation.

The sulfur dioxide is absorbed into the atmosphere at the constant rate of 48 tons per year. At what time will the rate at which sulfur dioxide is given off by the plant equal the rate at which it is absorbed into the atmosphere?

36. **(Medicine)** A bacteria colony starts out with a population of 10,000 bacteria. After t hours, the population is given by

$$P(t) = 10,000(1 + 0.4t^{3/2} + 0.2t^2).$$

(a) Find the rate of change of the population P with respect to time.
(b) How many bacteria will be present after 16 hours?

37. Show that the difference quotient for $\dfrac{1}{f(x)}$ is

$$\frac{-1}{f(x)f(x+h)} \cdot \frac{f(x+h) - f(x)}{h}.$$

38. Use the result of Exercise 37 to prove that if f is differentiable at x and $f(x) \neq 0$, then

$$\frac{d}{dx}\left[\frac{1}{f(x)}\right] = \frac{-f'(x)}{[f(x)]^2}.$$

39. Use the result of Exercise 38, along with the product rule, to prove the quotient rule.

3.4 ||||||| MARGINAL ANALYSIS IN BUSINESS AND ECONOMICS

In business and economics it is important to know how changes in level of production, demand, advertising, supply, price, and other such quantities affect cost, revenue, and profit. The use of derivatives in business and economics to study these relationships is called **marginal analysis.** Thus, marginal analysis deals with the rate at which one business or economic quantity varies with respect to another quantity.

In Section 1.2, we introduced three important functions: during a given time period, such as a day, a week, or a month,

$C(x)$ = total cost of producing x units of the product (the **cost** function);
$R(x)$ = total revenue received from selling x units of the product (the **revenue** function);
$P(x)$ = total profit derived from making and selling x units of the product (the **profit** function).

The derivatives of each of these functions have, respectively, the following names and notations:

$C'(x)$ = **marginal cost** (MC), or the rate of change of total cost per unit change in the level of production, when x items are made per time period.
$R'(x)$ = **marginal revenue** (MR), or the rate of change of total revenue per unit change in the level of sales, when x items are sold per time period.
$P'(x)$ = **marginal profit** (MP), or the rate of change of total profit per unit change in the level of production, when x items are made and sold per time period

Observe that for the cost function $C(x)$, x represents the number of units *produced*, whereas in the revenue function x represents the number of units *sold*. If all the items that are made are also sold, then x is the same in both functions, and

$$P(x) = R(x) - C(x). \qquad (1)$$

Differentiating both sides of Equation (1) we obtain

$$P'(x) = R'(x) - C'(x)$$

or

$$MP = MR - MC \qquad\qquad (2)$$

EXAMPLE 1   Consider a manufacturer whose weekly cost and revenue functions (in dollars) are

$$C(x) = 5000 + 300x - \frac{x^2}{10} \quad \text{and} \quad R(x) = 400x + \frac{x^2}{5}.$$

Find and interpret the marginal cost, marginal revenue, and marginal profit when the manufacturer is producing and selling 50 units per week.

Solution   The marginal cost and marginal revenue are given by

$$MC = C'(x) = 300 - \frac{2x}{10} = 300 - \frac{x}{5} \quad \text{and} \quad MR = R'(x) = 400 + \frac{2x}{5}.$$

The marginal profit is

$$MP = P'(x) = R'(x) - C'(x)$$
$$= \left(400 + \frac{2x}{5}\right) - \left(300 - \frac{x}{5}\right)$$
$$= 100 + \frac{3x}{5}.$$

When the level of production is 50 items per week ($x = 50$), then the total weekly cost is given by

$$C(50) = 5000 + 300(50) - \frac{(50)^2}{10} = \$19{,}750.$$

The marginal cost, when $x = 50$ items, is

$$C'(50) = 300 - \frac{50}{5} = \$290 \text{ per item},$$

which means that the weekly cost is increasing at the rate of $290 per item produced when the level of production is 50 items per week.

The total revenue received from the sale of 50 items is

$$R(50) = 400(50) + \frac{(50)^2}{5} = \$20{,}500,$$

and the marginal revenue, when $x = 50$ items, is

$$R'(50) = 400 + \frac{2(50)}{5} = \$420 \text{ per item,}$$

which means that the weekly revenue is increasing at the rate of $420 per item at the point when their weekly sales is 50 items.

From Equation (1), the total profit received from the production and sale of 50 items is

$$P(50) = R(50) - C(50) = 20,500 - 19,750 = \$750.$$

The marginal profit, when $x = 50$ items, is

$$P'(50) = 100 + \frac{3(50)}{5} = \$130 \text{ per item,}$$

which means that at this level of production and sales the profit is increasing at the rate of $130 per item. ∎

Estimating the Cost, Revenue, and Profit from Producing and Selling the Next Unit

The marginal cost $MC = C'(x)$ is the *rate of change* of the total cost $C(x)$, with respect to a unit change in the level of production x. Thus, increasing the level of production by one unit, from x to $x + 1$ units, will change the cost function by approximately $C'(x)$ dollars. The mathematical explanation is easily obtained. Recall that

$$C'(x) = \frac{dC}{dx} = \lim_{\Delta x \to 0} \frac{\Delta C}{\Delta x}.$$

For small values of Δx, $C'(x) \approx$ (is approximately equal to) $\dfrac{\Delta C}{\Delta x}$. Thus, when $\Delta x = 1$ (when the production level is increased by 1),

$$C'(x) \approx \Delta C.$$

But when $\Delta x = 1$, $\Delta C = C(x + 1) - C(x) =$ the cost of the $(x + 1)$st unit. Therefore, we have proved that

$$MC = C'(x) \approx \text{cost of producing the } (x + 1)\text{st unit.} \tag{3}$$

Similarly, we can show that

$$MR = R'(x) \approx \text{revenue from selling the } (x + 1)\text{st unit, and} \tag{4}$$

$$MP = P'(x) \approx \text{profit from producing and selling the } (x + 1)\text{st unit.} \tag{5}$$

EXAMPLE 2   For the manufacturer described in Example 1, use the marginal functions to estimate the cost, revenue, and profit from producing and selling the 51st unit. As a check, also calculate the *exact* cost, revenue, and profit for the 51st unit.

Solution   The marginal cost $C'(x)$ gives the approximate cost of producing the $(x + 1)$st unit. In Example 1 we calculated $C'(50) = 290$. Thus, we estimate the cost of producing the 51st unit to be approximately \$290. The exact cost of making item 51 is

$$C(51) - C(50) = 5000 + 300(51) - \frac{(51)^2}{10} - 19{,}750$$

$$= \$20{,}039.90 - \$19{,}750 = \$289.90.$$

Similarly, the marginal revenue $R'(x)$ gives the approximate revenue from selling the $(x + 1)$st unit. In Example 1 we calculated $R'(50) = 420$. Thus, we estimate the revenue from selling the 51st unit to be approximately \$420. The *exact* revenue would be

$$R(51) - R(50) = \left[400(51) + \frac{(51)^2}{5} \right] - 20{,}500$$

$$= \$20{,}400 + \$520.20 - \$20{,}500$$

$$= \$420.20.$$

Finally, the marginal profit $P'(x)$ gives the approximate profit from producing and selling the 51st unit. In Example 1 we found $P'(50) = 130$. Thus, the profit from the 51st unit is approximately \$130. The *exact* profit is

$$P(51) - P(50) = [R(51) - C(51)] - P(50)$$

$$= \$20{,}920.20 - \$20{,}039.90 - \$750$$

$$= \$130.30. \qquad \blacksquare$$

We must remember that the marginal functions, *MC*, *MR*, and *MP* are *functions* of the production and sales level x. Thus, for the manufacturer in Examples 1 and 2, when $x = 40$, we find (verify) that

$$P(40) = -\$520 \qquad \text{and} \qquad P'(40) = \$124 \text{ per item,}$$

which means that when the level of production is 40 items per week, the manufacturer loses \$520 weekly. At this level of production, however, the manufacturer will make a profit of approximately \$124 on item 41. The profit earned by the 41st item will reduce the weekly losses to approximately \$396.

If $x = 500$ in Examples 1 and 2, the marginal cost of production is $C'(500) = \$200$ per item, which is less than $C'(50) = 290$. Thus, as often happens, the marginal cost *decreases* as the level of production *increases*. The **economics of size** resulting from the more efficient use of labor and capital at higher levels of production account for this change. In some cases, however, there is a level of production at which further production increases will increase the marginal cost due to **diseconomics of size** resulting from payment of overtime wages to laborers, higher interest rates for borrowing additional capital, and so on.

Marginal analysis can be used to make important business decisions. The next two examples illustrate how.

EXAMPLE 3 A manufacturer of calculators finds that the cost (in dollars) of producing x calculators per week is given by

$$C(x) = 0.02x^2 + 50x + 4000.$$

The present level of production is 749 calculators per week and the manufacturer sells each calculator for $79.95. Because this output has not entirely met the demand, the manufacturer is considering expanding production.
  (a) Compute the marginal cost when the level of production is 749 calculators per week.
  (b) Should production be expanded?

Solution (a) We have

$$C'(x) = 0.04x + 50.$$

Then

$$C'(749) = (0.04)(749) + 50 = \$79.96 \text{ per calculator}$$

(b) Since the marginal cost is $79.96 per calculator, the cost of calculator 750 is approximately $79.96. [The actual cost is (verify)

$$C(750) - C(749) = \$52,750 - \$52,670.02 = \$79.98.]$$

Thus, the cost of making calculator 750 is greater than the revenue its sale would bring. It would *not* be financially advisable for the manufacturer to expand production. ∎

EXAMPLE 4 A radio manufacturer wants to determine the weekly level of production of a new model that will maximize sales profit. Based on past sales information and consumer surveys, if x is the number of radios that will be bought per week at a price of p dollars each, then

$$p = 60 - \frac{x}{100}. \tag{6}$$

Suppose that the total cost (in dollars) of producing x radios per week is given by

$$C(x) = 10x + 8000. \tag{7}$$

  (a) Find the profit function and the marginal profit function.
  (b) Find an interpret $P'(2000)$, $P'(2500)$, and $P'(3000)$.
  (c) What production level x will produce the greatest weekly profit for the manufacturer?

Solution   (a) The total revenue $R(x)$ received from selling x radios per week is given by

$$R(x) = (\text{number of radios sold}) \cdot (\text{price per radio})$$

$$= x\left(60 - \frac{x}{100}\right)$$

$$= 60x - \frac{x^2}{100}.$$

From Equation (1), the profit function is

$$P(x) = R(x) - C(x)$$

$$= 60x - \frac{x^2}{100} - (10x + 8000),$$

or

$$P(x) = -\frac{x^2}{100} + 50x - 8000. \tag{8}$$

Therefore, the marginal profit function is

$$P'(x) = -\frac{2x}{100} + 50,$$

or

$$P'(x) = 50 - \frac{x}{50}. \tag{9}$$

(b) When the levels of production are 2000, 2500, and 3000, Equation (9) gives

$$P'(2000) = \$10 \text{ per radio}$$
$$P'(2500) = \$0 \text{ per radio}$$
$$P'(3000) = -\$10 \text{ per radio}$$

This means that when the production level is 2000 radios per week, the rate of change of profit per unit change in the level of production is $10 per radio. In this case, at increased production levels, if all units can be sold, the profit will increase. In contrast, at the 2500 level of production, a slight increase in production will *not* result in additional profits, and at the 3000 level of production, we see that an increase in production will actually decrease profits. These results are obvious on the graph of $P(x)$, shown in Figure 1.

(c) Since $P(x)$ is a quadratic function, and the coefficient of its x^2 term is negative, the graph of $P(x)$ is a parabola, opening downward (Figure 1). Such a parabola has a highest point at its vertex, which in this case has x-value

$$x = \frac{-b}{2a} = \frac{-50}{2\left(-\dfrac{1}{100}\right)} = \frac{5000}{2} = 2500.$$

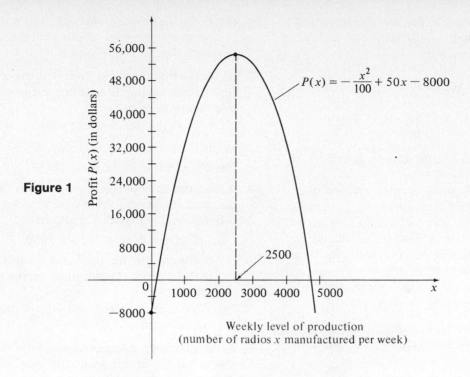

Figure 1

$$P(x) = -\frac{x^2}{100} + 50x - 8000$$

Profit $P(x)$ (in dollars)

Weekly level of production
(number of radios x manufactured per week)

Therefore, the maximum profit occurs when the level of production is 2500 radios per week. ∎

In the next chapter we shall discuss a more general procedure for finding the production level at which the profit will be greatest—one that is applicable to a wider class of functions. Once we know the optimal level of production, we can use Equation (6) to determine the selling price that will yield the maximum profit.

3.4 EXERCISE SET ||||||||||||||||||||||||||||||||||||||||||||||||||||||||||||||||||||||||||||||||

1. A manufacturer of motorcycles finds that the weekly cost and revenue functions for manufacturing x motorcycles (in dollars) are given by

$$C(x) = 2000 + 40x - \frac{x^2}{20} \quad \text{and}$$

$$R(x) = 80x + \frac{x^2}{10}.$$

(a) Find the marginal cost and marginal revenue functions.

(b) What is the rate of change of cost when the level of production is $x = 30$ motorcycles per week?

(c) What is the rate of change of revenue when the level of production is $x = 30$ motorcycles per week?

(d) Estimate the cost of manufacturing motorcycle 31.

(e) Estimate the revenue from selling motorcycle 31.

2. Repeat Exercise 1 for a manufacturer of electric toasters whose daily cost and revenue functions are

$$C(x) = 100 + 30x + \frac{x^2}{5} \quad \text{and}$$

$$R(x) = 50x + \frac{x^2}{20}.$$

3. Consider the manufacturer described in Exercise 1.
   (a) Find the profit function and the marginal profit function.
   (b) When the level of production is 40 units, what is the rate of change in profit relative to a one-unit increase in production?
   (c) Estimate the profit from manufacturing and selling the 41st unit.

4. For the manufacturer described in Exercise 2, follow instructions a, b, and c of Exercise 3.

5. Suppose the daily cost function and profit function for a manufacturer are

$$C(x) = 300 + 20x - \frac{x^2}{20}, \quad \text{and}$$

$$P(x) = \frac{x^2}{4} + 60x - 300.$$

   (a) Find the revenue function $R(x)$, and the marginal revenue function.
   (b) Estimate the cost, revenue, and profit for the 101st unit produced and sold, each day.

6. Suppose the daily revenue function and profit function for a manufacturer of battery-powered model cars are

$$R(x) = 40x + \frac{x^3}{90}, \quad \text{and}$$

$$P(x) = \frac{x^2}{60} + 20x - 200.$$

   (a) Find the cost function $C(x)$, and the marginal cost function.

   (b) Estimate the cost, revenue, and profit for the 21st model produced and sold, each day.

7. A camera manufacturer finds that the cost (in dollars) of manufacturing x cameras per month is given by

$$C(x) = 50 + 3x + 2x^{1/2}.$$

   What is the level of production when the marginal cost is $3.25 per camera?

8. A manufacturer of refrigerators finds that the cost (in dollars) of manufacturing x refrigerators per week is given by

$$C(x) = 0.03x^2 + 200x + 2000.$$

   (a) Compute the marginal cost when the level of production is 1000 refrigerators per week.
   (b) What is the approximate cost of making refrigerator 1001?

9. A manufacturer of tennis rackets finds that the profit (in cents) from the sale of x rackets is given by

$$P(x) = 200x - x^2.$$

   (a) Find the marginal profit when the level of production is $x = 40$.
   (b) Find the level of production above which profits are no longer increased by making more rackets.

10. A manufacturer of a custom-made film developer finds that the cost (in dollars) of making x gallons per month is given by

$$C(x) = 0.006x^2 + 0.4x + 500.$$

   (a) Compute the marginal cost when the level of production is 99 gallons per month.
   (b) Estimate the cost of increasing production to 100 gallons per month?
   (c) If each gallon of developer is sold for $1.50, should production be expanded beyond the present level of 99 gallons per month?
   (d) Find the level of production above which an expansion of production will no longer produce an increase in profits.

11. A calculator manufacturer knows that if each calculator is priced at p dollars, then the number of calculators sold per week is given by

$$x = 2000 - 20p.$$

The cost of manufacturing x calculators per week (in dollars) is given by

$$C(x) = 20x + 6000.$$

   (a) Find the total revenue $R(x)$.
   (b) Find the marginal cost.
   (c) Find the marginal profit.
   (d) Find $P'(200)$, $P'(800)$, and $P'(10,000)$, and interpret your results.
   (e) Find the most profitable level of production.

12. A coat manufacturer knows that at the selling price of p dollars per coat the relation between the number x of coats sold and the price is given by

$$p = 80 - \frac{x}{50}.$$

The cost of manufacturing x coats per month (in dollars) is given by

$$C(x) = 20x + 5000.$$

   (a) Find the total revenue $R(x)$.
   (b) Find the marginal cost.
   (c) Find the marginal profit.
   (d) Estimate the additional profit (or loss) that results from increasing production by one coat when the level of production is 1000 coats per month.
   (e) Estimate the additional profit (or loss) that results from increasing production by one coat when the level of production is 2500 coats per month.
   (f) Find the most profitable level of production.

3.5 |||||||| THE CHAIN RULE

Composite Functions

If we want to find the derivative of the function

$$f(x) = (1 + x^2)^3$$

we can expand the right side of the equation to get

$$f(x) = 1 + 3x^2 + 3x^4 + x^6.$$

Then

$$f'(x) = 6x + 12x^3 + 6x^5.$$

This approach is not practical if we want to find the derivative of

$$h(x) = (1 + x^2)^{100}.$$

We need another method for this type of problem. Another differentiation problem that we cannot yet handle, except by the definition of the derivative, is the calculation of the derivative of the function

$$k(x) = \sqrt{x + x^2}$$

Both of these functions, however, can be analyzed as *composite functions*, built out of simpler functions that we *can* differentiate.

Recall from Section 1.2 that if f and g are two given functions, we can define a new function $f \circ g$ by

$$(f \circ g)(x) = f(g(x)),$$

which is called the **composite of f and g**. The domain of the function $f \circ g$ consists of all numbers x in the domain of g such that the number $g(x)$ is in the domain of f. To find $f(g(x))$, we merely substitute $g(x)$ for every occurrence of the variable x in the rule for f.

We observed in Section 1.2 that, in general, the composite functions $f \circ g$ and $g \circ f$ are different.

The variables and functions involved in a composite function form a "chain." If $y = f(u)$ and $u = g(x)$, we can picture the variables and functions as in Figure 1, (a) or (b).

Figure 1

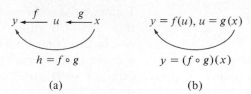

(a)   (b)

The chain rule will tell us how, by knowing the derivatives at each link of this chain $\left(\dfrac{dy}{du} \text{ and } \dfrac{du}{dx} \right)$, we can find the derivative $\dfrac{dy}{dx}$ of the composite function.

Suppose we know that $\dfrac{dy}{du} = 3$ and $\dfrac{du}{dx} = 2$. Since derivatives tell us the rate of change in one variable relative to another, this means that y is changing 3 times as fast as u, and u is changing 2 times as fast as x. How fast is y changing, relative to x? To most readers it will be intuitively clear that the correct answer is 6; y is changing 6 times as fast as x.

The formal statement of the general principle underlying this reasoning is one of the most important rules of differentiation, called the **chain rule.**

Chain Rule

If $y = f(u)$ and $u = g(x)$ are differentiable functions, so that y is a composite function of x,

$$y = h(x) = f(g(x)),$$

then the derivative $\dfrac{dy}{dx} = h'(x)$ is given by

$$h'(x) = \frac{d}{dx}[f(g(x))] = f'(g(x)) \cdot g'(x). \tag{1}$$

In other words, the derivative $h'(x)$ is obtained by first differentiating $f(x)$ and replacing every appearance of x in $f'(x)$ by $g(x)$ and then multiplying this expression by $g'(x)$.

Equation (1) can also be stated more explicitly as

$$\frac{dy}{dx} = \frac{dy}{du} \cdot \frac{du}{dx} \tag{2}$$

In this equation, after calculating dy/du, we may substitute $u = g(x)$ for every occurrence of u, although it will not always be necessary to do so.

Equation (2) says that dy/du, the rate at which y varies with respect to u, times du/dx, the rate at which u varies with respect to x, is dy/dx, the rate at which y varies with respect to x.

The proof of the chain rule is beyond the scope of this book. We shall not attempt it here.

To apply the chain rule, we must introduce an appropriate variable $u = g(x)$. Although there are no rules for selecting an appropriate u, Examples 1–3 show how to proceed.

EXAMPLE 1   If $y = (1 + x^2)^{100}$, find dy/dx.

Solution   If we let

$$u = 1 + x^2,$$

then

$$y = u^{100}. \quad \text{[See Example 7(a), Section 1.2.]}$$

We have

$$\frac{du}{dx} = \frac{d}{dx}[1 + x^2] = 2x, \quad \text{and}$$

$$\frac{dy}{du} = \frac{d}{du}[u^{100}] = 100u^{99} = 100(1 + x^2)^{99}, \quad \text{since } u = 1 + x^2.$$

Then, by Equation (2),

$$\frac{dy}{dx} = \frac{dy}{du} \cdot \frac{du}{dx} = 100(1 + x^2)^{99} \cdot 2x$$

$$= 200x(1 + x^2)^{99}. \qquad \blacksquare$$

EXAMPLE 2   If $y = \sqrt{x + x^2}$, find dy/dx.

Solution   If we let

$$u = x + x^2,$$

then

$$y = u^{1/2}. \quad \text{[See Example 7(b), Section 1.2]}$$

We have

$$\frac{du}{dx} = \frac{d}{dx}[x + x^2] = 1 + 2x, \quad \text{and}$$

$$\frac{dy}{du} = \frac{d}{du}[u^{1/2}] = \frac{1}{2}u^{-1/2} = \frac{1}{2}(x + x^2)^{-1/2}, \quad \text{since } u = x + x^2.$$

Then, by Equation (2),

$$\frac{dy}{dx} = \frac{dy}{du} \cdot \frac{du}{dx} = \frac{1}{2}(x + x^2)^{-1/2} \cdot (1 + 2x)$$

$$= \frac{1 + 2x}{2\sqrt{x + x^2}}. \quad \blacksquare$$

EXAMPLE 3 If $y = \left(\dfrac{x^2}{x^2 - 5}\right)^{1/3}$, find dy/dx.

Solution If we let

$$u = \frac{x^2}{x^2 - 5},$$

then

$$y = u^{1/3}. \quad \text{[See Example 7(c), Section 1.2.]}$$

We have

$$\frac{du}{dx} = \frac{d}{dx}\left[\frac{x^2}{x^2 - 5}\right]$$

$$= \frac{(x^2 - 5)\dfrac{d}{dx}[x^2] - x^2\dfrac{d}{dx}[x^2 - 5]}{(x^2 - 5)^2}$$

$$= \frac{(x^2 - 5)(2x) - x^2(2x)}{(x^2 - 5)^2} = -\frac{10x}{(x^2 - 5)^2}, \quad \text{and}$$

$$\frac{dy}{du} = \frac{d}{du}[u^{1/3}] = \frac{1}{3}u^{-2/3} = \frac{1}{3}\left(\frac{x^2}{x^2 - 5}\right)^{-2/3}, \quad \text{since } u = \frac{x^2}{x^2 - 5}.$$

Then, by Equation (2),

$$\frac{dy}{dx} = \frac{dy}{du} \cdot \frac{du}{dx} = \frac{1}{3}\left(\frac{x^2}{x^2 - 5}\right)^{-2/3} \cdot \left(-\frac{10x}{(x^2 - 5)^2}\right)$$

$$= -\frac{10x^{-1/3}}{3(x^2 - 5)^{4/3}}. \quad \blacksquare$$

Of course, the chain rule may be applied when letters other than x, y, and u are used. For example, if w is a function of v and v is a function of u, the chain rule becomes

$$\frac{dw}{du} = \frac{dw}{dv} \cdot \frac{dv}{du}.$$

Thus, if

$$w = v^2 + 4v \qquad \text{and} \qquad v = 4u^3,$$

then

$$\frac{dw}{dv} = \frac{d}{dv}[v^2 + 4v] = 2v + 4 = 2(4u^3) + 4 = 8u^3 + 4, \qquad \text{and}$$

$$\frac{dv}{du} = \frac{d}{du}[4u^3] = 12u^2,$$

from which we get

$$\frac{dw}{du} = \frac{dw}{dv} \cdot \frac{dv}{du} = (8u^3 + 4)(12u^2) = 96u^5 + 48u^2.$$

|||||||||||| **The Generalized Power Rule**

We often have to find the derivative of a function of the form $[g(x)]^r$, where r is a rational number; in fact, the functions in Examples 1, 2, and 3 are of this type. We shall now develop a general formula for the derivative of this type of function. Letting

$$y = [g(x)]^r \qquad \text{and} \qquad u = g(x),$$

we have

$$y = u^r.$$

By the chain rule, Equation (2),

$$\frac{dy}{dx} = \frac{d}{dx}[g(x)]^r = \frac{dy}{du} \cdot \frac{du}{dx}. \tag{3}$$

First,

$$\frac{dy}{du} = \frac{d}{du}[u^r] = ru^{r-1} = r[g(x)]^{r-1}, \tag{4}$$

and then

$$\frac{du}{dx} = g'(x). \tag{5}$$

Substituting the results of (4) and (5) into (3), we have $\dfrac{dy}{dx} = r[g(x)]^{r-1} \cdot g'(x)$. We state this result as a formal rule.

The Generalized Power Rule

If $g(x)$ is differentiable and r is a constant rational number, then

$$\frac{d}{dx}[g(x)]^r = r[g(x)]^{r-1} \cdot g'(x). \tag{6}$$

Alternatively, letting $u = g(x)$, Equation (6) becomes

$$\frac{d}{dx}[u^r] = ru^{r-1} \cdot \frac{du}{dx}. \tag{7}$$

EXAMPLE 4 By Equation (6),

$$\frac{d}{dx}[x^3 + 2x^2]^9 = 9(x^3 + 2x^2)^8 \frac{d}{dx}[x^3 + 2x^2]$$

$$= 9(x^3 + 2x^2)^8 \cdot (3x^2 + 4x).$$

Alternatively, by Equation (7) with $u = x^3 + 2x^2$,

$$\frac{d}{dx}[x^3 + 2x^2]^9 = \frac{d}{dx}u^9 = 9u^8 \cdot \frac{du}{dx}$$

$$= 9(x^3 + 2x^2)^8 \cdot (3x^2 + 4x).$$

EXAMPLE 5 If $y = \dfrac{2}{(x^2 + 5x - 3)^5}$, find dy/dx.

Solution 1 We *could* use the quotient rule;

$$\frac{dy}{dx} = \frac{(x^2 + 5x - 3)^5 \dfrac{d}{dx}[2] - 2 \dfrac{d}{dx}[(x^2 + 5x - 3)^5]}{[(x^2 + 5x - 3)^5]^2}$$

$$= \frac{(x^2 + 5x - 3)^5(0) - 2(5)(x^2 + 5x - 3)^4(2x + 5)}{(x^2 + 5x - 3)^{10}}$$

$$= \frac{-10(x^2 + 5x - 3)^4(2x + 5)}{(x^2 + 5x - 3)^{10}}$$

$$= \frac{-10(2x + 5)}{(x^2 + 5x - 3)^6}.$$

∎

Solution 2   It is better to first write

$$y = 2(x^2 + 5x - 3)^{-5}.$$

Then, by the generalized power rule (6),

$$\frac{dy}{dx} = 2(-5)(x^2 + 5x - 3)^{-6} \frac{d}{dx} [x^2 + 5x - 3]$$

$$= -10(x^2 + 5x - 3)^{-6}(2x + 5)$$

$$= \frac{-10(2x + 5)}{(x^2 + 5x - 3)^6}.$$ ∎

As Example 5 illustrates, the quotient rule should never be used when the generalized power rule is sufficient.

EXAMPLE 6   Find the derivative of $f(x) = \sqrt[3]{\dfrac{x^2 + 1}{4x - 1}}$.

Solution   Our function is $f(x) = \left(\dfrac{x^2 + 1}{4x - 1}\right)^{1/3}$. Using the generalized power rule,

$$f'(x) = \frac{1}{3} \left(\frac{x^2 + 1}{4x - 1}\right)^{-2/3} \frac{d}{dx}\left[\frac{x^2 + 1}{4x - 1}\right],$$

and using the quotient rule,

$$f'(x) = \frac{1}{3}\left(\frac{4x - 1}{x^2 + 1}\right)^{2/3} \cdot \left[\frac{(4x - 1)(2x) - (x^2 + 1)(4)}{(4x - 1)^2}\right]$$

$$= \frac{1}{3}\frac{(4x - 1)^{2/3}}{(x^2 + 1)^{2/3}}\left[\frac{8x^2 - 2x - 4x^2 - 4}{(4x - 1)^2}\right]$$

$$= \frac{4x^2 - 2x - 4}{3(x^2 + 1)^{2/3}(4x - 1)^{4/3}}.$$ ∎

3.5 EXERCISE SET ||||||||||||||||||||||||||||||||||||||||||||||||||||||||||||||||||||||||||||||||||||

In Exercises 1–4, write the given function of x as a function of u by introduction an appropriate variable $u = g(x)$.

1. $y = (3x + 2)^6$

2. $y = (x^3 + 2x^2 + 1)^5$

3. $y = (x^3 - 2x^2)^{1/3}$

4. $y = \left(\dfrac{x^2 + 2x}{x^3 - 1}\right)^{3/2}$

In Exercises 5–24, find dy/dx.

5. $y = (3x^2 + 1)^{20}$

6. $y = (3 - 2x^3)^{30}$

7. $y = \sqrt{4 - x}$

8. $y = \sqrt{2x^2 - x + 2}$

9. $y = (2 - 5x^2)^{-10}$

10. $y = \dfrac{1}{(3x^2 + 2x)^8}$

11. $y = x^2(2x + 3)^5$

12. $y = x^3\sqrt{7x + 1}$

13. $y = x^2\sqrt{4x - 2x^2}$

14. $y = x^{3/5}(2x^2 + 5)^4$

15. $y = (4x - 9)^{10}(5x + 2)^7$

16. $y = (x^2 + 1)^8(5x^3)^6$

17. $y = (5x^3 + 4x^2 - 2x + 4)^{1/5}$

18. $y = \dfrac{4}{(2x^3 + 4)^{1/2}}$

19. $y = \dfrac{1}{\sqrt[3]{x^2 + 4x - 3}}$

20. $y = x^2\sqrt[3]{x^2 + 10}$

21. $y = (2x^5 + 6)^{10}(3x^3 - 5)^{20}$

22. $y = (8x^3 + 7x)^5(5x^2 + 8)^{12}$

23. $y = \dfrac{x - 2}{x + 5}$

24. $y = \sqrt[3]{\dfrac{x^2 - 5}{x - 2}}$

25. Find an equation of the tangent line to the graph of the function $y = 4/\sqrt{x^2 + 2x + 1}$ at the point $(2, \frac{4}{3})$.

26. Find the equation of the line tangent to the graph of

$$y = \sqrt[3]{\frac{2x}{x + 1}} \text{ at the point } (1, 1).$$

27. **(Educational Psychology)** A training program for workers on a radio assembly line has revealed that the number $N(t)$ of radios assembled daily by a typical worker after t hours of training is given by $N(t) = \sqrt{2t^3 + 9}$. Find the rate of learning after 2 hours of training.

28. A nationwide weight reduction clinic analyzes its records and finds that of the 4000 clients who begin their program on a typical day, the number remaining in the program after t days is

$$r(t) = 4000\left(\frac{100}{100 + t}\right)^3.$$

Find and interpret $r'(1)$, $r'(30)$, and $r'(100)$.

3.6 ||||||| RELATED RATES PROBLEMS

Related rates problems are among the most fascinating applications of derivatives. Unlike problems from algebra, in these problems an equation relating two variables will be used to relate the *rates of change* of these variables. We will be interested in how the rate of change in one variable affects the rate of change of other variables.

In a **related rates** problem we know y as a function of x, $y = f(x)$, and we know that x is a function of another variable, say, t. Given dx/dt, the rate at which x varies with respect to t, what is dy/dt, the rate at which y varies with respect to t? The basic idea is the chain rule:

$$\frac{dy}{dt} = \frac{dy}{dx} \cdot \frac{dx}{dt}.$$

EXAMPLE 1 When a rock falls into a pond it creates a circular ripple, the radius of which is increasing at the rate of 3 feet per second. At what rate is the area of the ripple increasing when the radius is 15 feet?

Solution   If r is the radius (in feet) of the ripple, then its area A (in square feet) is given by

$$A = \pi r^2, \tag{1}$$

where $r = r(t)$ is an unknown function of t. We want to find dA/dt when $r = 15$. By the chain rule,

$$\frac{dA}{dt} = \frac{dA}{dr} \cdot \frac{dr}{dt}. \tag{2}$$

From Equation (1) we get $\dfrac{dA}{dr} = 2\pi r$, and we are given in the problem that $\dfrac{dr}{dt} = 3$. Substituting these into (2) we get

$$\frac{dA}{dt} = 6\pi r.$$

If $r = 15$, we have

$$\frac{dA}{dt} = 6\pi(15) = 90\pi \approx 282.6. \qquad (\pi \approx 3.14)$$

Thus, the area of the ripple is increasing at approximately 283 square feet per second.   ∎

In Example 1 there are two variables, A and r, both of which vary relative to a third variable, t. We are given the rate of change of one of them relative to t $\left(\dfrac{dr}{dt}\right)$ and are asked to find the rate of change of the other relative to t $\left(\dfrac{dA}{dt}\right)$. Our strategy is this: first we relate the variables to each other [Equation (1)]; then we differentiate both sides of this equation, using the chain rule, to relate their rate of change. The final solution is then obtained by substituting the appropriate values given in the statement of the problem.

EXAMPLE 2   A manufacturer of nuclear reactors finds that the profit (in dollars) received from
(Business)   making and selling x reactors is given by

$$P(x) = (x^2 + x)^2. \tag{3}$$

If the rate of production is kept at 5 units per month, what is the rate of change of profit when 40 units have been made?

Solution   First observe the role of t, the unmentioned third variable in this problem. The variable x is a function of t, in that x represents the number of reactors made up to time t. The problem tells us the rate of change of x, relative to t:

$$\frac{dx}{dt} = 5. \tag{4}$$

The variable P is also a function of time t, since P is a function of x and x is a function of t. We are interested in calculating dP/dt when $x = 40$. By the chain rule, we have

$$\frac{dP}{dt} = \frac{dP}{dx} \cdot \frac{dx}{dt}. \tag{5}$$

From Equation (3) we obtain

$$\frac{dP}{dx} = 2(x^2 + x)(2x + 1). \tag{6}$$

Substituting the results of (6) and (4) into Equation (5) we get

$$\frac{dP}{dt} = 2(x^2 + 1)(2x + 1) \cdot 5$$

$$= 10(x^2 + 1)(2x + 1).$$

Finally, when $x = 40$, we have

$$\frac{dP}{dt} = 10[(40)^2 + 40][2(40) + 1]$$

$$= \$1,328,400.$$

Thus, when 40 reactors are produced and sold, the profit is increasing at the rate of \$1,328,400 per month. ∎

EXAMPLE 3 Water is flowing into a conical tank at the rate of 3 cubic feet per minute. The tank has a radius of 2 feet at the top (Figure 1) and a depth of 4 feet. How fast is the water level rising
(a) when the water is 1 foot deep?
(b) when the water is 2 feet deep?
(c) when the water is 3 feet deep?

Figure 1

Solution   This is a related rates problem, and can be solved by the methods we have given. The variables to be used in solving this related rates problem are

$$t = \text{time, in minutes,}$$
$$h = \text{height of the water in the tank (in feet) at time } t,$$
$$r = \text{radius of the circular surface of the water (in feet) at time } t,$$

and

$$V = \text{volume of the water in the tank (cubic feet) at time } t.$$

The variables V, r, and h all vary with time, and are related by the formula for the volume of a cone:

$$V = \tfrac{1}{3}\pi r^2 h. \tag{7}$$

We are asked to find $\dfrac{dh}{dt}$ when h = 1, $h = 2$, and $h = 3$. The variables r and h are related. By similar triangles (see Figure 2)

$$\frac{r}{2} = \frac{h}{4},$$

so

$$r = \frac{2h}{4} = \frac{h}{2}. \tag{8}$$

Figure 2

By substituting (8) into (7) we can eliminate the variable r so that Equation (7) becomes

$$V = \tfrac{1}{3}\pi(\tfrac{1}{2}h)^2 h,$$

or

$$V = \frac{\pi}{12} h^3. \tag{9}$$

By the chain rule,

$$\frac{dV}{dt} = \frac{dV}{dh} \cdot \frac{dh}{dt}. \tag{10}$$

We know that $\dfrac{dV}{dt} = 3$, since we are given that water is flowing into the tank at the rate of 3 cubic feet per minute. From Equation (9) we get

$$\frac{dV}{dh} = \frac{3\pi}{12} h^2 = \frac{\pi}{4} h^2.$$

Substituting these facts into (10) we get

$$3 = \frac{\pi}{4} h^2 \frac{dh}{dt},$$

or

$$\frac{12}{\pi h^2} = \frac{dh}{dt}. \tag{11}$$

(a) When $h = 1$, Equation (11) yields

$$\frac{dh}{dt} = \frac{12}{\pi(1)} \approx 3.82 \text{ feet per minute} \approx 45.8 \text{ inches per minute.}$$

(b) When $h = 2$, Equation (11) yields

$$\frac{dh}{dt} = \frac{12}{\pi(4)} = \frac{3}{\pi} \approx 0.95 \text{ feet per minute} \approx 11.5 \text{ inches per minute.}$$

(c) When $h = 3$, Equation (11) yields

$$\frac{dh}{dt} = \frac{12}{\pi(9)} = \frac{4}{3\pi} \approx 0.424 \text{ feet per minute} \approx 5.1 \text{ inches per minute.}$$

This example shows mathematically that, although the water flows in at a constant rate, the water level does not rise at a constant rate. Indeed, when the water level is only 1 foot high, it is rising 9 times as fast as when it is 3 feet high. ∎

Suggested Procedure for Solving Related Rates Problems
1. Identify all relevant variables.
2. If the problem is geometric, sketch and label a helpful figure.
3. Indicate the variables whose rates of change are given, and the one (or more) whose rate of change you are asked to find.
4. Obtain one or more equations relating the relevant variables.
5. Differentiate both sides of this equation to obtain the desired derivative. (The chain rule will be used.)
6. Substitute the given constant values of the variables in the appropriate places.

It is important to realize that Step 6 must be carried out last. When variables are "frozen" at constant values, then their derivatives become zero. The variables must remain free to vary until after Step 5 has been completed. In Example 1 the last step was to set $r = 15$. In Example 2 the last step was to set $x = 40$, and in Example 3 the last step was to set $h = 1$, $h = 2$, and $h = 3$.

Before proceeding to the exercises, we point out the significance of the *sign of the derivative*. When $f'(x) > 0$ for all x in an interval surrounding x_0 (see Figure 3), the curve $y = f(x)$ is "rising" as it passes through $(x_0, f(x_0))$.

When $f'(x) < 0$ for all x in an interval surrounding x_1 (see Figure 3), the curve is "falling" as it passes through $(x_1, f(x_1))$. In the former case we say that the function f *is* **increasing** at x_0, and in the latter case we say that f is **decreasing** at x_1. The concepts of increasing and decreasing functions will be examined in greater detail in Section 4.1. Meanwhile, an understanding of the difference between positive and negative derivatives will be helpful in some related rates problems.

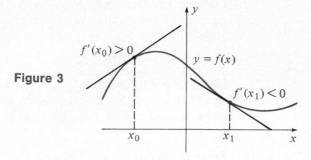

Figure 3

3.6 EXERCISE SET ||||||||||||||||||||||||||||||||||||||||||||||||||||||||||||||||||||||||||||||||||||||

1. **(Business)** The price y (in dollars) of a ton of wheat depends on the weekly supply x (in tons) and is given by the equation $y = 200 + 300/x$. How fast is the price of wheat changing when the weekly supply is 30 tons and is decreasing at the rate of 3 tons per week? (Note: a function has a *negative derivative* when it is decreasing.)

2. **(Business)** A manufacturer finds that the cost of producing x units weekly is

    $$C(x) = 4000 - 100x + 7x^2.$$

    If the production level is increasing steadily at the rate of 2 units per week, how fast are production costs rising when the production level reaches

    (a) 20 units per week?
    (b) 25 units per week?

3. When a pebble is dropped into still water it creates circular waves that travel outward from the point of impact. For a wave whose radius is increasing at the constant rate of 6 inches per second, how fast is its area increasing when its radius is (a) 1 foot? (b) 2 feet? (c) 3 feet?

4. Water is dripping slowly onto a metal countertop, creating a circular wet spot whose area is increasing at the constant rate of $\pi/2$ square inches per second. How fast is its radius increasing when the radius is (a) 1 inch? (b) 4 inches? (c) 25 inches?

5. Water is flowing into a conical tank at the rate of 2 cubic feet per minute. The tank has a diameter of 6 feet at the top, and a depth of 4 feet. How fast is the water level rising when the water is (a) 1 foot deep? (b) 2 feet deep? (c) 4 feet deep? (See Example 3.)

6. Sand is falling at the rate of 27 cubic feet per minute onto a conical pile whose base diameter is always twice its height (see Figure 4). How fast is the height of the pile growing when the height is exactly (a) 3 feet? (b) 6 feet? (c) 9 feet?

$$V = \frac{1}{3}\pi r^2 h$$

Figure 4

7. A spherical balloon is being inflated with helium at the rate of 12 cubic inches per second. How fast is the radius of the balloon expanding at the instant when
   (a) the diameter of the balloon is 6 inches?
   (b) the volume of the balloon is $9\pi/2$ cubic inches? (The volume of a sphere of radius r is $V = \frac{4}{3}\pi r^3$.)

8. A block of ice in the shape of a cube melts so that each side decreases at the rate of 2 centimeters per hour. How rapidly is the volume of the block decreasing when each side is 10 centimeters?

9. **(Ecology)** On a golden-eagle preserve the size of the golden-eagle population depends upon the size of the population of rodents, the golden eagle's primary source of food. Suppose that the number of eagles N is given by

$$N = 0.004x + 0.008x^2$$

where $x =$ the number of rodents on the preserve. If the number of rodents available is increasing at the rate of 200 per year, how fast is the golden-eagle population increasing when there are 500 rodents on the preserve?

10. **(Price versus Demand)** Suppose that the demand x for a product selling at price p (in dollars) is

$$x = 5000 - 75p.$$

When the selling price is $8 and decreasing at the rate of $0.40 per month, what is the instantaneous rate of change in demand?

11. **(Price versus Demand)** Suppose that the demand x for a product at selling price p (in dollars) satisfies the equation

$$p = 800 - 15x + \frac{x^2}{10}.$$

When the monthly demand is 60 units, and the price per unit is decreasing at the rate of $15 per month, what is the instantaneous rate of change in the demand?

3.7 |||||||| IMPLICIT DIFFERENTIATION

Consider the equation

$$y = x^2 - 4 \tag{1}$$

and the equation

$$xy - y = 6. \tag{2}$$

Equation (1) expresses y as a function of x, $y = f(x)$, where $f(x) = x^2 - 4$. Thus we say that (1) defines y **explicitly.** Equation (2) also determines a value of y

for each value of x, since we can solve (2) for y:

$$y = \frac{6}{x - 1},$$

obtaining y explicitly as a function of x. Equation (2) is said to define y **implicitly**, as a function of x. For some equations, however, it is very difficult or impossible to solve for y explicitly as a function of x. For example, in the equation

$$xy^5 + y^3 - y^2 + 2x + 5 = 0,$$

it is rather hopeless to try to express y explicitly as a function of x.

A number of applications in which y is defined implicitly as a function of x require finding dy/dx. The method for finding dy/dx without first obtaining y explicitly is called **implicit differentiation.** The following examples illustrate this approach.

Example 1   If $y^4 + 3y^2 - x^3 - 1 = 0$, find dy/dx.

Solution   We treat y as an unknown differentiable function of x, $y = f(x)$, which is implied by the given equation. Substituting $f(x)$ for y in the given equation we have

$$[f(x)]^4 + 3[f(x)]^2 - x^3 - 1 = 0. \tag{3}$$

Both sides of Equation (3) may be regarded as functions of x. Since Equation (3) says that these are the same function, they must have the same derivative. We thus proceed to differentiate both sides of (3). Remember that by the chain rule (Section 3.5),

$$\frac{d}{dx}[f(x)]^4 = 4[f(x)]^3 f'(x)$$

or

$$\frac{d}{dx}[y^4] = 4y^3 y',$$

and

$$\frac{d}{dx}[f(x)]^2 = 2f(x)f'(x)$$

or

$$\frac{d}{dx}[y^2] = 2yy'.$$

Thus, after differentiating both sides of Equation (3), we have

$$4[f(x)]^3 \cdot f'(x) + 6f(x) \cdot f'(x) - 3x^2 = 0$$

or

$$4y^3y' + 6yy' - 3x^2 = 0. \qquad (4)$$

Of course, in differentiating (3), we use

$$\frac{d}{dx}[x^3] = 3x^2, \qquad \frac{d}{dx}[-1] = 0, \quad \text{and} \quad \frac{d}{dx}[0] = 0.$$

We can now solve (4) for y':

$$(4y^3 + 6y)y' = 3x^2.$$

Thus,

$$\frac{dy}{dx} = y' = \frac{3x^2}{4y^3 + 6y}. \qquad \blacksquare$$

It is important to note that the derivative in the last example depends both on x and y, so we must know both coordinates to determine a value for the derivative at a point.

||||||||||||| **Notation**

Suppose that (a, b) is a point whose coordinates satisfy the equation that defines y implicitly as a function of x. When we write

$$\left.\frac{dy}{dx}\right|_{(a, b)}$$

we mean the derivative dy/dx evaluated at $x = a$, $y = b$. In Example 1 the coordinates of point $(3, 2)$ satisfy the equation that defines y implicitly. Thus, $(3, 2)$ lies on the curve, and at this point the slope of the tangent line is

$$\left.\frac{dy}{dx}\right|_{(3, 2)} = \frac{3 \cdot 3^2}{4 \cdot 2^3 + 6 \cdot 3} = \frac{27}{50}.$$

EXAMPLE 2 If $xy + x^2y^2 + 2x - 4 = 0$, find

$$\left.\frac{dy}{dx}\right|_{(1, 1)}.$$

Solution We must differentiate both sides of the given equation with respect to x and treat y as an unknown differentiable function of x. Since xy is a product, we must use the product rule to obtain

$$\frac{d}{dx}[xy] = x\frac{dy}{dx} + y\frac{d}{dx}[x] = xy' + y \cdot 1 = xy' + y.$$

Similarly, x^2y^2 is a product, so

$$\frac{d}{dx}[x^2y^2] = x^2\frac{d}{dx}[y^2] + y^2\frac{d}{dx}[x^2] = x^2 \cdot 2yy' + y^2 \cdot 2x = 2x^2yy' + 2xy^2.$$

Thus, when we differentiate both sides of the given equation, we obtain

$$xy' + y + 2x^2yy' + 2xy^2 + 2 = 0.$$

We can now solve for y':

$$(x + 2x^2y)y' = -2 - y - 2xy^2,$$

or

$$y' = \frac{dy}{dx} = \frac{-2 - y - 2xy^2}{x + 2x^2y}.$$

Hence,

$$\left.\frac{dy}{dx}\right|_{(1,\,1)} = -\frac{5}{3}. \qquad\blacksquare$$

||||||||||||| Implicit Differentiation in Related Rates Problems

Many related rates problems are most easily solved using implicit differentiation. The following example illustrates a typical situation.

EXAMPLE 3   A 13-foot ladder is leaning against a wall. As the bottom of the ladder is pulled away from the wall, the top of the ladder slides down the wall (see Figure 1).

If the bottom of the ladder is pulled away at the rate of 3 feet per second, how fast is the top of the ladder sliding down the wall when the bottom is 12 feet from the wall?

Figure 1

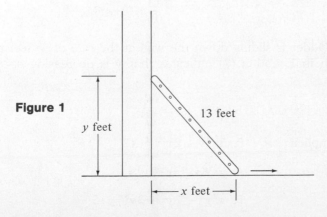

y feet

13 feet

x feet

Solution   Let x denote the distance (in feet) between the bottom of the ladder and the wall and let y denote the distance (in feet) between the top of the ladder and the ground. Observe that x and y are functions of time t (in seconds). We want to find dy/dt when $x = 12$.

By the Pythagorean theorem, we have

$$x^2 + y^2 = 13^2. \tag{5}$$

Equation (5) defines y as a function of x implicitly. Differentiating both sides of (5) with respect to t, we have

$$\frac{d}{dt}[x^2] + \frac{d}{dt}[y^2] = \frac{d}{dt}[13^2]$$

or

$$2x\frac{dx}{dt} + 2y\frac{dy}{dt} = 0.$$

Solving for dy/dt, we obtain

$$2y\frac{dy}{dt} = -2x\frac{dx}{dt}$$

or

$$\frac{dy}{dt} = -\frac{x}{y} \cdot \frac{dx}{dt}. \tag{6}$$

When $x = 12$, Equation (5) gives us

$$y^2 = 13^2 - 12^2 = 169 - 144 = 25 = 5^2$$

or

$$y = 5.$$

Substituting $x = 12$, $y = 5$, and $dx/dt = 3$ in (6), we obtain

$$\left.\frac{dy}{dt}\right|_{(12,\,5)} = -\frac{12}{5} \cdot 3 = -\frac{36}{5}. \tag{7}$$

Hence, the top of the ladder is sliding down the wall at the rate of $\frac{36}{5}$ feet per second [the minus sign in Equation (7) indicates that y is decreasing as t is increasing].   ∎

3.7 EXERCISE SET ||||||||||||||||||||||||||||||||||||||||||||||||||||||||||||||||||||||||||||

In Exercises 1–12, find dy/dx by implicit differentiation.

1. $x^2 + 3x + y^2 = 4$

2. $2x^2 + 4y - y^2 = 16$

3. $3x^2y + 2xy^3 = 12$

4. $x^2y + 4xy - y^3 = 6$

5. $x^2y^2 - 2x^2 + 3y^3 = 15$

6. $\dfrac{1}{x} + \dfrac{1}{y} = 4$

7. $(x + y)^2 + (x - y)^2 = 5$

8. $\sqrt{x} + \sqrt{y} = 25$

9. $xy + 3x = y^2 + 12$

10. $\dfrac{x}{y} + 3y = 5x^2 + 10$

11. $xy^2 = y^3 - x^2$

12. $\dfrac{x^3}{x + y} = 3y$

In Exercises 13–16, evaluate dy/dx at the given point.

13. $xy - y^3 = 2;\ (5, 2)$

14. $x^2y^2 - 2x^2 - y^3 = 4;\ (-1, -2)$

15. $y^3 - xy^2 = 2 - 4x;\ (1, -1)$

16. $y^2 + 3 = x^2y;\ (2, 1)$

17. Find an equation of the tangent line to the graph of the equation $x^2 + y^2 = 25$ at the point $(3, 4)$.

18. Find an equation of the tangent line to the graph of the equation $x^2y + y = 2$ at the point $(1, 1)$.

19. **(Business)** The weekly demand for coffee in a supermarket is given by the equation $px + 16p - 800 = 0$, where p is the price (in dollars) per pound, and x is the number of pounds of coffee demanded per week. If this week the price of coffee is $5.00 per pound and the price is increasing at the rate of $.06 per week, at what rate is the demand changing?

20. **(Business)** The wholesale price of apples is related to its daily supply (in thousands of crates) by the equation $px - 10p - 2x + 172 = 0$, where p is the price (in dollars) per crate, and x is the number of thousands of crates supplied per day. If today's supply is decreasing at the rate of 200 crates per day and there are 6000 crates available, at what rate is the price changing?

21. A 10-foot ladder is leaning against a wall, as in Example 3. If the bottom of the ladder is pulled away from the wall at the rate of 2 feet per second, how fast is the top of the ladder sliding down the wall at the instant it is (a) 8 feet above the ground? (b) 2 feet above the ground? (c) 1 foot above the ground?

22. A baseball "diamond" is a square, with a distance of 90 feet between the bases. A runner is running from first base to second at the constant rate of 20 feet per second. How fast is the distance between the runner (R) and the catcher (C) changing when the runner is halfway from first to second? (See Figure 2.)

Figure 2

23. As a boy is flying a kite the wind is blowing it horizontally, so that the kite is at a constant 50 feet above the ground but moving horizontally away from the boy at the rate of 4 feet per second. At what rate is the boy paying out the string when exactly 130 feet of string is out?

24. A car is 400 feet from an intersection, and approaching it from the south at the speed of 80 feet per second. Five seconds earlier, a truck crossed the intersection headed east, maintaining a speed of 60 feet per

second. At the present instant, how far apart are the car and the truck, and how fast is the distance changing? Are they getting closer, or farther apart, at this instant? (See Figure 3.)

Figure 3

25. **(Supply versus Price)** A study shows that when the market price is p (in dollars) per unit, the number x of units that manufacturers of a certain commodity will produce monthly satisfies the equation

$$400p = x^2 + 200x + 10,000.$$

If the current price is $100 per unit and increasing at the rate of $3 per month, find the current rate of change in the supply x.

26. **(Demand versus Price)** The demand equation for a certain commodity is

$$150p = x^2 - 10x,$$

where the demand for the commodity is x thousand units per year when the price is p (in dollars) per unit. If the unit price is currently $20 and increasing at the rate of $11 per year, find the current rate of change in demand.

27. Suppose that an airplane is being pulled into a hangar by a rope attached to a pulley that is 8 feet higher than the nose of the plane (see Figure 4). If the rope is being pulled at the rate of 3 feet per second, how fast is the plane coming into the hangar when the length of rope from the pulley to the nose of the plane is 17 feet?

Figure 4

KEY IDEAS FOR REVIEW ||||||||||||||||||||||||||||||||||||||||||||||||||||||||||||||||||||||||||

- The derivative of the function f at x_0 is defined by

$$f'(x_0) = \lim_{h \to 0} \frac{f(x_0 + h) - f(x_0)}{h}$$

if this limit exists.

- If $f'(x_0)$ exists, then we say that f is differentiable at x_0.
- $f'(x_0)$ is the slope of the tangent line to the curve $y = f(x)$ at the point where $x = x_0$.
- The equation of the tangent line to the graph of f at the point $(x_0, f(x_0))$ is

$$y - f(x_0) = f'(x_0)(x - x_0).$$

- If f is differentiable at x_0, then f must be continuous at x_0. However, we have seen examples of functions that are continuous at x_0 but not differentiable there.

- Alternate definition of derivative:

$$f'(x_0) = \lim_{x \to x_0} \frac{f(x) - f(x_0)}{x - x_0}.$$

- Delta notation: If $y = f(x)$, then $\Delta x = $ a change in x, while

$$\Delta y = \text{the corresponding change in } y$$
$$= f(x + \Delta x) - f(x).$$

Then

$$\frac{dy}{dx} = \lim_{\Delta x \to 0} \frac{\Delta y}{\Delta x}.$$

■ Alternative notation for derivatives:

$$\frac{dy}{dx} = \frac{d}{dx}(y) = D_x(y) = y';$$

$$\frac{d}{dx}[f(x)] = D_x[f(x)] = f'(x).$$

■ If f is a constant function, then $f'(x) = 0$.

■ For any rational number r, $\dfrac{d}{dx}[x^r] = rx^{r-1}$.

■ For any constant k, $\dfrac{d}{dx}[kf(x)] = k\dfrac{d}{dx}[f(x)]$.

■ $\dfrac{d}{dx}[f(x) + g(x)] = \dfrac{d}{dx}[f(x)] + \dfrac{d}{dx}[g(x)]$.

■ $\dfrac{d}{dx}[f(x) - g(x)] = \dfrac{d}{dx}[f(x)] - \dfrac{d}{dx}[g(x)]$.

■ $\dfrac{d}{dx}[f(x)g(x)] =$

$$f(x)\frac{d}{dx}[g(x)] + g(x)\frac{d}{dx}[f(x)].$$

■ $\dfrac{d}{dx}\left[\dfrac{f(x)}{g(x)}\right] = \dfrac{g(x)f'(x) - f(x)g'(x)}{[g(x)]^2}$.

■ When x items have been made, the marginal cost $C'(x)$ is the rate of change of the total cost per unit change in the level of production. Similar remarks are true for the marginal revenue $R'(x)$ and marginal profit $P'(x)$.

■ The cost of making item $(x + 1)$ is approximately $C'(x)$. Similar interpretations of $R'(x)$ and $P'(x)$ are valid.

■ The chain rule for composite functions: If $y = f(u)$ and $u = g(x)$, then

$$\frac{dy}{dx} = \frac{dy}{du} \cdot \frac{du}{dx}.$$

■ For any rational number r, $\dfrac{d}{dx}[g(x)]^r = r[g(x)]^{r-1}g'(x)$.

■ To solve related rates problems, first relate the variables by an equation, then differentiate both sides using the chain rule, and finally substitute in constants.

■ Given an equation that defines y implicitly as a function of x, dy/dx can be obtained from that equation by implicit differentiation.

■ Implicit differentiation is often useful in related rates problems.

REVIEW EXERCISES ||||||||||||||||||||||||||||||||||||||||||||||||||||||||||||||||||||||||||||||

1. Calculate $f'(x_0)$ from the graph below.

In Exercises 2–5, calculate $f'(x)$ using the definition of the derivative.

2. $f(x) = 3$ 3. $f(x) = 2 - 3x$

4. $f(x) = 2x^2 + 1$ 5. $f(x) = \dfrac{1}{x}$

In Exercises 6–11, find the derivative using appropriate rules.

6. $f(x) = -3$

7. $f(x) = 3x^3 - 2x^2 + 5x - 2$

8. $f(t) = 3t^{2/3} - \dfrac{2}{t^{1/3}}$

9. $f(x) = 2\sqrt{x} - \dfrac{1}{x}$

10. $h(x) = (2x + \sqrt{x})(3x^3 - x^2 + 1)$

11. $h(s) = \dfrac{3s + 2}{s^2 - s}$

In Exercises 12 and 13, find dy/dx.

12. $y = (3x^3 + 2x^2 - 1)(4x^2 - x^{2/3})$

13. $y = \dfrac{x - 1}{x^2 + 1}$

14. If $h(x) = (2x + 1)/(2 - x)$, find
    (a) $h'(-3)$ (b) $h'(0)$ (c) $h'(2)$

15. Suppose $f(x) = 2x^2 + x - 1$.
    (a) Find $f'(x)$.
    (b) Calculate $f'(3)$.
    (c) Find an equation for the tangent line to the curve $y = f(x)$ at the point where $x = 2$.
    (d) Sketch the graph of f and draw the tangent line obtained in (c).

16. If $f(x) = 3x^2 + 6x - 2$, find the point on the graph of f where the tangent line is parallel to the line $y - 18x + 2 = 0$.

17. If $f(x) = 2x^3 + 5x^2 - 4x$, find all points on the graph of f where the tangent line is horizontal.

18. Consider the function graphed below.

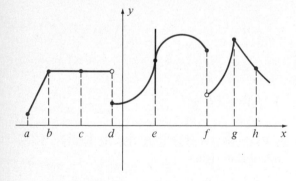

At what points is the function
(a) continuous but not differentiable?
(b) neither continuous nor differentiable? Explain your answer.

In Exercises 19–24, find dy/dx.

19. $y = (2x + 3)^{-8}$

20. $y = \left(\dfrac{x^3 - 2x^2}{x^2 + 1}\right)^8$

21. $y = \sqrt{x^2 + 3x}$

22. $y = \dfrac{3}{(3x^4 - x^3)^{1/3}}$

23. $y = (3x^4 - 2x^3 + x^2)^{3/5}$

24. $y = (3x^2 + 2x)^8(2x^3 - 3x^2)^6$

25. **(Business)** A computer manufacturer has determined that spending x million dollars on research yields a profit (in millions of dollars) given by

$$P(x) = 3x^2 + 2x + 80.$$

Find the rate of change of profit with respect to money invested in research when the amount spent on research is $2.5 million.

26. **(Social Science)** Suppose that the population of a new city is 80,000 people and after t years, the population is given by

$$P(t) = 80,000(2 - \tfrac{1}{4}\sqrt{t}).$$

(a) Find the rate of change of the population P with respect to time.
(b) What will be the population after 25 years?

27. **(Motion)** An object moving from rest along a straight line travels s feet in t seconds, where

$$s = 2t^4 - 80t^3, \qquad 0 \le t \le 40.$$

(a) Find the instantaneous velocity at time t.
(b) After how many seconds will the object come to rest?

28. Find the rate of change of the volume V of a sphere with respect to the radius r. (Recall that $V = \tfrac{4}{3}\pi r^3$.)

29. **(Medicine)** Suppose that the radius of a circular duodenal ulcer is increasing at the rate of 0.002 centimeters per month. How rapidly is the area increasing when the radius of the ulcer is 0.12 centimeters ($A = \pi r^2$)?

30. **(Economics)** Suppose that, to stabilize the dollar, the United States begins to sell gold and that the amount of gold (in millions of of ounces) sold after t weeks is given by

$$A = 2t^{4/3}.$$

The international money markets can absorb the gold at the maximum rate of 4 million ounces per week. How long will it be until the government starts selling more gold than the international money markets can absorb?

31. A manufacturer of pens finds that the cost (in cents) of manufacturing x pens per day is given by

$$C(x) = 0.05x^2 + 100x + 800.$$

(a) Compute the marginal cost when the level of production is 2000 pens per day.
(b) What is the approximate cost of making pen 2001?

32. A manufacturer finds that when producing x units per month its monthly costs and revenue functions are given by

$$C(x) = 7000 + 50x - \frac{x^2}{100}$$

and

$$R(x) = 150x + \frac{x^2}{10}.$$

(a) Find the marginal cost and the marginal revenue when the manufacturer is producing 500 units per month.
(b) Find the profit function and the marginal profit function.
(c) Estimate the cost, the revenue, and the profit from the 501st unit in a month.

33. A (spherical) ball of ice is melting at the rate of 4 cubic inches per minute. How fast is its radius changing when the radius is 3 inches?

34. A chemical is draining out of a conical container at the rate of 2 cubic centimeters per second. The container is 18 centimeters deep and has a radius of 9 centimeters at the top. How fast is the level of the chemical falling when it is 4 centimeters high?

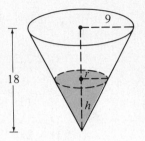

In Exercises 35 and 36, compute dy/dx by implicit differentiation.

35. $x^2 + 2xy - y^2 = 0$

36. $xy^3 - 2y^2 + x^3 + 2 = 0$

37. Find the equation of the tangent line to the graph of the equation $xy^2 + y = 3$ at the point $(2, 1)$.

38. A ball is dropped from a height of 104 feet. From a position 30 feet from the point directly beneath the ball, an observer watches the ball fall. When it has fallen exactly 64 feet, how fast is the distance z between the observer and the ball changing? (During the first t seconds the ball falls $16t^2$ feet.)

CHAPTER TEST ||||||||||||||||||||||||||||||||||||||||||||||||||||||||||||||||||||||||||||||||||||||||||

1. Given $f(x) = x - 2x^2$, find $f'(1)$ using the four-step procedure.

2. Given $f(x) = 3x^2 - 4$, use the alternate definition of the derivative (Equation [4], Section 3.1) to calculate $f'(x_0)$.

3. Where does the function $f(x) = |x + 3|$ fail to be differentiable? Explain briefly. Where is this function continuous?

In Exercises 4–9, find the derivative of the given function using the appropriate rules.

4. $f(x) = 7x^5 - 8x^2 - 3x + 4$

5. $f(x) = 49^3 + 2x$

6. $g(x) = \dfrac{1}{(3x - 11)^4}$

7. $g(x) = \dfrac{3x + 8}{x^2 - 1}$

8. $f(x) = 2x^{4/3} - \dfrac{5}{x^{2/3}}$

9. $f(x) = (3x^4 + 7x + 2)^8(x^3 - 4x^2)$

10. An object moving along an x-axis is located at

$$x = \frac{t}{4} - 2\sqrt{t}$$

t seconds after initial time $t = 0$.
   (a) Find the instantaneous velocity at time t.
   (b) After how many seconds will the object reach zero velocity?

11. Find the equation of the line tangent to the curve

$$x^2y^3 - 3y^2 = x^4 - 15$$

at the point (2, 1).

12. A cylinder of base radius r and altitude h is expanding in such a way that h is always twice the value of r. How fast is the volume changing when the radius is 10 inches and increasing at the rate of $\frac{1}{2}$ inch per second?

(vol. $= \pi r^2 h$)

APPLICATIONS OF THE DERIVATIVE

We have already seen some applications of the derivative. In Chapter 3 we used derivatives to determine rates of change and slopes of tangent lines. We saw these ideas at work in marginal analysis and in related rates problems. The applications to be studied in this chapter fall into four broad categories:

I. Analysis of increasing and decreasing functions.

II. Application of derivatives to optimization problems.

III. Use of derivatives in sketching graphs of functions.

IV. The differential, and linear approximation of functions.

As we have seen, many applied problems require analysis of a function and its graph. In this chapter we shall see that the derivative provides considerable information about the graph of a function. In fact, this may be viewed as the major thrust of the chapter.

Optimization problems arise in many different professions. A manufacturer, for example, wants to maximize profits and minimize costs; a trucking firm wants to minimize travel time for deliveries; a farmer with a certain amount of fencing material on hand to enclose a rectangular grazing field wants to know the dimensions that will maximize the area of the field. Using the derivative, we shall solve such basic optimization problems.

The final section in this chapter explains how to use the tangent line to the graph of a function to approximate the values of the function near the point of tangency.

4.1 |||||||| INCREASING AND DECREASING FUNCTIONS

In many applied problems it is important to know when a function is increasing or decreasing. For example, after making a certain capital investment in a business, we would want to know whether the profit function is increasing or decreasing; after administering a certain dosage of a drug a physician might want to know if the patient's temperature is increasing or decreasing. In this section we shall use the derivative to see when a function is increasing and when it is decreasing.

A function f is said to be **increasing on an interval** (a, b) if $f(x_1) < f(x_2)$ whenever x_1 and x_2 are values in (a, b) with $x_1 < x_2$. That is, as we move along the x-axis in the direction of increasing x-coordinate, we find that the respective y-coordinate also *increases* in value. The functions in Figure 1(a) and (b) are increasing on the interval (a, b)

Figure 1

(a)   (b)

Similarly, we say that f is **decreasing on an interval** (a, b) if $f(x_1) > f(x_2)$ whenever x_1 and x_2 are values in (a, b) with $x_1 < x_2$. That is, as we move along the x-axis in the direction of increasing x-coordinate, we find that the respective y-coordinate *decreases* in value. The functions in Figure 2(a) and (b) are decreasing on the interval (a, b).

Figure 2

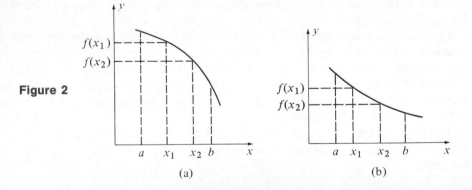

(a)   (b)

Of course, we can tell whether a function is increasing or decreasing on a given interval (a, b) by examining its graph. But sometimes we can obtain the same information more simply by examining the first derivative to see when it is positive and when it is negative. Recall from Section 1.3 that if f is a function whose graph is a straight line, then f is increasing when the slope is positive and decreasing when the slope is negative. An analogous result can be proved for functions whose graphs are not straight lines. The following theorem is a precise statement of this result.

THEOREM

Suppose the function f is differentiable on (a, b).
(a) If $f'(x) > 0$ for all values of x in (a, b), then f is an increasing function on (a, b).
(b) If $f'(x) < 0$ for all values of x in (a, b), then f is a decreasing function on (a, b).

Figure 3

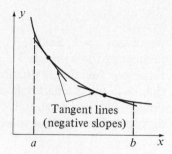

(a) Increasing function   (b) Decreasing function

Figure 3(a) shows a function that is increasing on (a, b) because $f'(x)$, the slope of the tangent line at $(x, f(x))$, is positive at every value x in (a, b). Similarly, the function in Figure 3(b) is decreasing on (a, b), because the tangent line at $(x, f(x))$ has a negative slope at every value x in (a, b).

We can summarize the above theorem as follows:

Behavior of $f(x)$ on (a, b)

Sign of $f'(x)$	Graph of f
$+$	f is increasing; the graph is rising
$-$	f is decreasing; the graph is falling

EXAMPLE 1

Consider the function f defined by

$$f(x) = x^2 - 2x + 4.$$

(a) Find the open interval(s) on which f is increasing and those on which f is decreasing.
(b) Find the values of x at which the graph has a horizontal tangent.
(c) Sketch the graph of f.

Solution

(a) We find the derivative of f and determine the open intervals where $f'(x)$ is positive and where it is negative. We have

$$f'(x) = 2x - 2 = 2(x - 1),$$

so

$$f'(x) > 0 \quad \text{if} \quad x - 1 > 0, \quad \text{or} \quad x > 1$$
$$f'(x) < 0 \quad \text{if} \quad x - 1 < 0, \quad \text{or} \quad x < 1.$$

Thus, f is increasing on the interval $(1, +\infty)$ and f is decreasing on the interval $(-\infty, 1)$.

(b) The graph has a horizontal tangent when its slope is zero (see Example 2 in Section 3.1). Setting

$$f'(x) = 2x - 2 = 0$$

we obtain $x = 1$. Since $f(1) = 3$, we conclude that the graph has a horizontal tangent at the point $(1, 3)$.

(c) Since our function is quadratic and the coefficient of x^2 is positive, we know that its graph is a parabola opening upward. The vertex is the point where the tangent line is horizontal, found in (b) to be $(1, 3)$. The graph is shown in Figure 4. Observe that the results of (a) are verified by this graph.

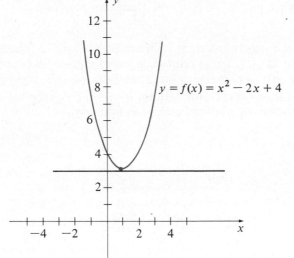

Figure 4

$$y = f(x) = x^2 - 2x + 4$$

■

EXAMPLE 2   Consider the function f defined by

$$f(x) = x^3 - 3x + 3.$$

(a) Find the open interval(s) on which f is increasing and those on which f is decreasing.

(b) Find the values of x at which the graph has a horizontal tangent.

(c) Sketch the graph of f.

Solution   (a) We find the derivative of f and determine the open intervals where $f'(x)$ is positive and where it is negative. We have

$$f'(x) = 3x^2 - 3 = 3(x^2 - 1) = 3(x + 1)(x - 1).$$

Thus, $f'(x) > 0$ when $(x + 1)$ and $(x - 1)$ have the same sign and $f'(x) < 0$ when $(x + 1)$ and $(x - 1)$ have opposite signs. Figure 5 shows how to determine the intervals on which f is increasing and those on which it is decreasing by analyzing the dependence on x of the sign of the "test quantities" $x + 1$, $x - 1$, and $f'(x)$. A plus sign designates values of x on the number line for which the test quantity is positive. A minus sign marks values of x for which the test quantity is negative.

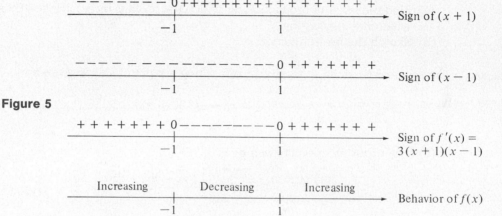

Figure 5

We see that $f'(x) > 0$ when $x < -1$ and when $x > 1$, and that $f'(x) < 0$ when $-1 < x < 1$. Thus, f is increasing on the intervals $(-\infty, -1)$ and $(1, +\infty)$ and decreasing on the interval $(-1, 1)$.

(b) The graph has a horizontal tangent when $f'(x) = 0$, which occurs when $x = -1$ and $x = 1$. We find that $f(-1) = 5$ and $f(1) = 1$. Thus, there are horizontal tangents at the points $(-1, 5)$ and $(1, 1)$.

(c) By plotting several points and using the information in (a) and (b), we can sketch the graph of f, shown in Figure 6.

Figure 6

EXAMPLE 3   A manufacturer of skis finds that the cost (in dollars) of making x pairs of skis per
(Business)   week is given by

$$C(x) = 60 + 40x.$$

The demand equation is known to be

$$p = 120 - x,$$

where p is the price (in dollars) at which they can sell x pairs of the skis per week.
  (a) Find the levels of production for which the profit is increasing and those for
      which it is decreasing.
  (b) Sketch the profit function.

Solution   The revenue (in dollars) received from selling x pairs of skis is given by

$$R(x) = xp = x(120 - x)$$
$$= 120x - x^2.$$

  (a) The profit function is given by

$$P(x) = R(x) - C(x) = (120x - x^2) - (60 + 40x)$$
$$= 80x - x^2 - 60.$$

Then

$$P'(x) = 80 - 2x = 2(40 - x).$$

Now $P'(x) > 0$ when $40 - x > 0$, or $40 > x$. Thus, if $0 \le x < 40$, the profit
is increasing and if $x > 40$, the profit is decreasing.
  (b) The profit function is sketched in Figure 7.

Figure 7

■

Note that in all of the preceding examples we determined the behavior of a function by analyzing the sign of its first derivative function at every value of x. In each case, we completely factored the derivative function before analyzing the sign.

4.1 EXERCISE SET ||||||||||||||||||||||||||||||||||||||||||||||||||||||||||||||||||||||||||||||||||||||||||||

1. Consider the function f graphed in Figure 8.

Figure 8

(a) Find all open intervals where f is increasing.

(b) Find all open intervals where f is decreasing.

2. Consider the function f graphed in Figure 9.

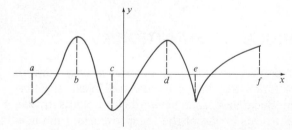

Figure 9

(a) Find all open intervals where f is increasing.

(b) Find all open intervals where f is decreasing

In Exercises 3–18, find where the given function is increasing and where it is decreasing. Also, for what values of x does the graph of f have a horizontal tangent?

3. $f(x) = 3x - 2$

4. $f(x) = -2x + 3$

5. $f(x) = x^2 - 4x + 2$

6. $f(x) = -3x^2 + 6x + 5$

7. $f(x) = 3x^2 + 3$

8. $f(x) = x^2 + 5x$

9. $f(x) = \frac{1}{3}x^3 - x$

10. $f(x) = 2x^3 - 3x^2 - 12x + 4$

11. $f(x) = x^3 - 3x^2 + 3x + 3$

12. $f(x) = x^3 + 6x^2 + 12x - 10$

13. $f(x) = \frac{1}{4}x^4 - \frac{1}{2}x^2$

14. $f(x) = x^4 + 2x^2$

15. $f(x) = \dfrac{1}{x}$

16. $f(x) = \dfrac{1}{x^2}$

17. $f(x) = \dfrac{x^2}{x^2 + 1}$

18. $f(x) = \dfrac{x - 1}{x + 1}$

19. Sketch the graph of the function in Exercise 5.

20. Sketch the graph of the function in Exercise 6.

21. Sketch the graph of the function in Exercise 9.

22. Sketch the graph of the function in Exercise 10.

23. Sketch the graph of the function in Exercise 13.

24. Sketch the graph of the function in Exercise 14.

25. **(Business)** A manufacturer of mopeds finds that the profit (in hundreds of thousands of dollars) received from making and selling x mopeds per week is given by

$$P(x) = 100x - 400 - 0.002x^2.$$

(a) For what levels of production is the profit increasing and for what levels is it decreasing?

(b) For what levels of production is the profit neither increasing nor decreasing?

26. **(Business)** A manufacturer of stereo turntables finds that the cost of making x turntables per week is

$$C(x) = 40 + 76x.$$

The demand equation is

$$p = 400 - 3x.$$

(a) Find the profit function. (See Example 3.)

(b) Find the levels of production for which the profit is increasing, and those for which it is decreasing.

(c) For what levels of production is the profit neither increasing nor decreasing?

27. **(Educational Psychology)** A psychologist who is training dolphins to understand human speech finds that the number $N(t)$ of words learned after t weeks of training is given by

$$N(t) = 80t - t^2.$$

For what value of t does $N(t)$ increase? Decrease?

28. **(Biology)** It has been found that after x hours, a dog that has received a certain drug shows a change in blood pressure (in millimeters of mercury) given by

$$P(x) = \frac{x^3}{3} - \frac{7}{2} x^2 + 10x + 40, \ 0 \le x \le 6.$$

(a) During 6 hours of observation, when will the blood pressure be increasing? When will it be decreasing?

(b) Sketch the graph of $P(x)$.

4.2 |||||||| RELATIVE EXTREME VALUES OF FUNCTIONS

This section is the first of several in this chapter that are concerned with maximum and minimum (extreme) values of functions. The principal tool used in determining these values is the derivative. Because so many everyday problems involve maximizing or minimizing a function, this is one of the most important applications of the derivative. Our first concern will be to locate the high and low points on the graph of a function.

Figure 1

||||||||||||| ## Relative Extrema

A function f is said to have a **relative maximum** (or **local maximum**) at $x = c$ if there is an open interval containing c such that for all x in this interval, $f(c) \geq f(x)$. Thus, a relative maximum occurs at a value of x where f has a "peak." The function graphed in Figure 1 has relative maxima (maxima is the plural of maximum) at $x = c_1$, c_2, and c_3.

A function f is said to have a **relative minimum** (or **local minimum**) at $x = c$ if there is an open interval containing c such that for all x in this interval, $f(c) \leq f(x)$. Thus, a relative minimum occurs at a value x where f has a "valley." Figure 2 is the graph of the function that has relative minima (minima is the plural of minimum) at $x = c_4$, c_5, and c_6.

Figure 2

If the function f has either a relative maximum or a relative minimum at $x = c$, we say that f has a **relative extremum** at $x = c$ (the plural of extremum is *extrema*). The value $f(c)$ is the relative extremum of the function. Given a function f, how do we find the values x at which f has relative extrema? Figures 3(a) and 3(b) focus on the four relative extrema from Figures 1 and 2 where the tangent line to the curve is horizontal and two where the derivative does not exist. The precise statement suggested by Figure 3 takes the form of the following theorem.

Figure 3

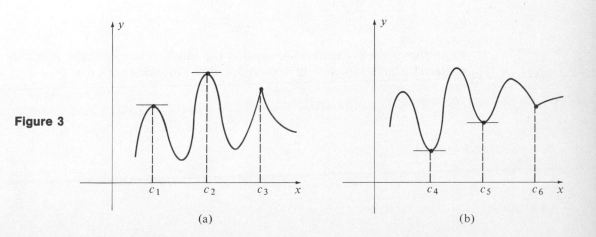

(a)   (b)

THEOREM | If a function f has a relative extremum at c, then

$$\text{either } f'(c) = 0 \text{ or } f'(c) \text{ does not exist.}$$

Thus, if f is a differentiable function, this theorem tells us that the only *possible* values of x at which f can have a relative extremum are those where $f'(x) = 0$.

EXAMPLE 1   Locate the relative extrema of the quadratic function $f(x) = ax^2 + bx + c$.

Solution   Suppose the function $f(x) = ax^2 + bx + c$ $(a \neq 0)$ has a relative extremum at x_0. Since a quadratic function is differentiable everywhere, the above theorem tells us that $f'(x_0) = 0$. Now

$$f'(x) = 2ax + b.$$

Thus, we must have $2ax_0 + b = 0$, or

$$x_0 = -\frac{b}{2a}.$$

Recall from Chapter 1 that $-\dfrac{b}{2a}$ is the x-coordinate of the vertex of the parabola. Therefore, the (only) relative extremum of a quadratic function occurs at its vertex. ∎

Let us now examine the two relative extrema in Figures 3(a) and 3(b) where $f'(c) = 0$ is *not* true. The tangent line at $(c_3, f(c_3))$ in Figure 3(a) is vertical. Consequently, $f'(c_3)$ does not exist. Similarly, the graph in Figure 3(b) has a sharp corner at $(c_6, f(c_6))$, so $f'(c_6)$ does not exist. Another example of a function having a relative extremum at c for which $f'(c)$ does not exist follows.

EXAMPLE 2   Consider the function f defined by

$$f(x) = |x|.$$

From the graph in Figure 4 we see that the function has a relative minimum at 0. Recall from Example 7 in Section 3.1, that $f'(0)$ does not exist.

Figure 4

$y = f(x) = |x|$

∎

The following terminology is useful in summarizing our observations regarding relative extrema.

If f is defined at c and either $f'(c) = 0$ or $f'(c)$ does not exist, then c is called a **critical number** of f.

Thus, if f has a relative extremum at c, then c is a critical number of f.

EXAMPLE 3   Find the critical numbers of the following functions:

(a) $f(x) = x^2 - 3x + 2$ (b) $f(x) = x^{2/3}$ (c) $f(x) = \dfrac{x^2 - 3x}{x + 1}$

Solution   (a) We have

$$f'(x) = 2x - 3.$$

To find the critical numbers we set

$$f'(x) = 0,$$

obtaining $x = \frac{3}{2}$. Since $f'(x)$ exists for all values of x, we conclude that $x = \frac{3}{2}$ is the only critical number of f.

(b) In Example 9 of Section 3.1 we showed that $f'(0)$ does not exist. For all $x \neq 0$, we have (verify)

$$f'(x) = \frac{2}{3} x^{-1/3} = \frac{2}{3\sqrt[3]{x}}.$$

Here $f'(x)$ is never zero, but $f'(0)$ does not exist. Hence, $x = 0$ is the only critical number of f.

(c) We have (verify)

$$f'(x) = \frac{x^2 + 2x - 3}{(x + 1)^2} = \frac{(x + 3)(x - 1)}{(x + 1)^2}.$$

To find the critical numbers, we set

$$f'(x) = 0,$$

obtaining $x = -3$ and $x = 1$. Observe that $f'(x)$ does not exist for $x = -1$. Because f is not defined at $x = -1$, we conclude that $x = -1$ is *not* a critical number of f. Thus, the critical numbers of f are $x = -3$ and $x = 1$.   ∎

Warning   Critical numbers of a function f are numbers in its domain where the function *may* (but may not) have a relative extremum. Read the theorem again! It does *not* say that if either $f'(c) = 0$ or $f'(c)$ does not exist then f must have a relative extremum at c. Indeed, there are many cases in which f has a critical member at c but does not have a relative extremum there. The next two examples illustrate this point, with Example 4 showing that $f'(x)$ can be zero at $x = c$ and f need not have a relative extremum at c.

EXAMPLE 4   Consider the function f defined by

$$f(x) = x^3,$$

whose graph is sketched in Figure 5. Since $f'(x) = 3x^2$, we see that $f'(0) = 0$. However, $f(x) < 0$ if $x < 0$ and $f(x) > 0$ if $x > 0$. Hence, f does not have a relative extremum at $x = 0$.

Figure 5

EXAMPLE 5   Let

$$f(x) = \begin{cases} x + 2 & \text{if } x \le 1 \\ 3x & \text{if } x > 1. \end{cases}$$

Figure 6

The graph of f is shown in Figure 6. Note that the graph has a sharp corner at the point $(1, 3)$. Thus, $f'(1)$ does not exist. Therefore, $x = 1$ is a critical number for f, but the function f has no relative extremum since it is always increasing.

|||||||||||| **A Test for Relative Extrema**

Since the critical numbers are only *candidates* for the values of x at which f may have relative extrema, we need a test for determining at which of the critical numbers f *does* have relative extrema. We can obtain this information by examining the behavior of the function in the vicinity of the critical number. Referring to Figure 7(a), we see that if f is increasing on an interval extending to the left of c and decreasing on an interval extending to the right of c, then f has a relative maximum at c. Figure 7(b) shows that if f is decreasing on an interval extending to the left of c and increasing on an interval extending to the right of c, then f has a relative minimum at c.

Figure 7

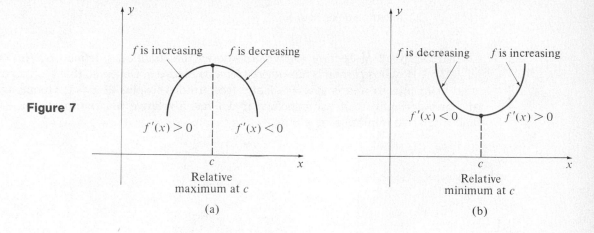

Recall from Section 4.1 that if $f'(x)$ is positive on (a, b), then f is increasing on (a, b) and if $f'(x)$ is negative on (a, b) then f is decreasing on (a, b) (see Figure 7). Thus, we can state the following important test for relative extrema.

THEOREM

The First Derivative Test

Suppose the function f is continuous at every x in (a, b) and $f'(x)$ exists at every x in (a, b), except possibly at $x = c$.

(a) If $f'(x) > 0$ on an interval extending to the left from c and $f'(x) < 0$ on an interval extending to the right from c, then f has a *relative maximum* at c.

(b) If $f'(x) < 0$ on an interval extending to the left from c and $f'(x) > 0$ on an interval extending to the right from c, then f has a *relative minimum* at c.

We can summarize the first derivative test as follows:

For a function f that is continuous at $x = c$:
If the sign of f' changes from plus to minus at c, then f has a relative maximum at c; if the sign of f' changes from minus to plus at c, then f has a relative minimum at c (see Figure 8).

Relative maximum at c

Figure 8

Relative minimum at c

In Example 2 of Section 4.1 we considered the function f defined by $f(x) = x^3 - 3x + 3$, whose graph is reproduced here as Figure 9. Observed that f' changes sign from plus to minus at $x = -1$ and from minus to plus at $x = 1$. Hence, by the first derivative test, we conclude that f has a relative maximum at $x = -1$ and a relative minimum at $x = 1$.

Figure 9

Notice that if f' does not change sign at c, then f does not have a relative extremum at c. For example, consider the function defined by $f(x) = x^3$, discussed in Example 4 (see Figure 5). Since $f'(x) = 3x^2$, f' does not change sign at the critical number $x = 0$; therefore f does not have a relative extremum at $x = 0$.

The first derivative test gives us the following method of finding the relative extrema of a function.

> **Finding the Relative Extrema of f(x)**
> *Step 1.* Find $f'(x)$.
> *Step 2.* Find the critical numbers of f.
> *Step 3.* Apply the first derivative test to each critical number.

EXAMPLE 6   Find the relative extrema of

$$f(x) = x^2 - 6x.$$

Solution   *Step 1.* $f'(x) = 2x - 6$. There are no values of x where the derivative fails to exist.

Step 2. Solving $f'(x) = 2x - 6 = 2(x - 3) = 0$, or $x - 3 = 0$, we see that $x = 3$ is the only critical number of f.

Step 3. If $x < 3$ then $x - 3 < 0$ so $f'(x) < 0$, and if $x > 3$, then $x - 3 > 0$ so $f'(x) > 0$. Hence, f has a relative minimum at $x = 3$ because the sign of f' changes from minus to plus at $x = 3$. The relative minimum is $f(3) = -9$. Figure 10 shows the graph of f.

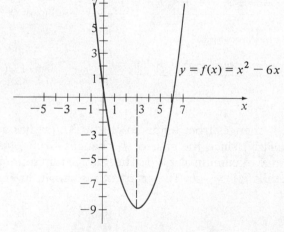

Figure 10

In problems in which the function f has more than one relative extremum, it becomes difficult to analyze the sign of f' because $f'(x)$ is nonlinear. We must then solve the nonlinear inequality

$$f'(x) > 0 \qquad (\text{or } f'(x) < 0).$$

In Section A.6 (Appendix A) and in Example 2 of Section 4.1 we have shown how to handle a nonlinear inequality.

We first factor the derivative f' completely and determine the sign of each factor. Knowing the sign of each factor for a value of x, we can determine the sign of the product for that value of x. This approach is used in the next example.

EXAMPLE 7   Find the relative extrema of

$$f(x) = x^3 - 6x^2 + 9x + 2.$$

Solution   We have

$$f'(x) = 3x^2 - 12x + 9 = 3(x^2 - 4x + 3).$$

Since $f(x)$ is a polynomial, there are no critical numbers of f for which $f'(x)$ does not exist. Hence, to find the critical numbers of f, we set

$$f'(x) = 3(x^2 - 4x + 3) = 0$$
$$= 3(x - 3)(x - 1) = 0,$$

so the critical numbers of f are $x = 3$ and $x = 1$. Figure 11 shows how to determine the sign of f' on each side of the critical numbers.

Figure 11

Since the sign of f' changes from minus to plus at 3, $f(x)$ has a relative minimum at $x = 3$. Similarly, since the sign of f' changes from plus to minus at $x = 1$, $f(x)$ has a relative maximum at $x = 1$. The relative minimum is $f(3) = 2$ and the relative maximum is $f(1) = -6$. The graph of f is shown in Figure 12.

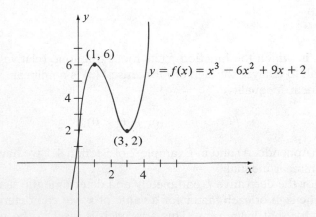

Figure 12

■

EXAMPLE 8   Find the relative extrema of

$$f(x) = x^{2/3}.$$

Solution   From Example 3, we know that $f'(x) = \frac{2}{3}x^{-1/3}$ and that $x = 0$ is the only critical number of f because $f'(0)$ does not exist. If $x < 0$, then $f'(x) < 0$, and if $x > 0$, then $f'(x) > 0$. Hence, f has a relative minimum at $x = 0$ because the sign of f' changes from minus to plus at $x = 0$. The relative minimum is $f(0) = 0$. The graph of f is shown in Figure 13. In Section 4.5 we shall develop some techniques that will enable us to sketch such curves.

Figure 13

EXAMPLE 9   Find the relative extrema of

$$f(x) = x + \frac{4}{x}.$$

Solution   We have

$$f'(x) = 1 - \frac{4}{x^2} = \frac{x^2 - 4}{x^2} = \frac{(x - 2)(x + 2)}{x^2},$$

so the critical numbers of f are $x = 2$ and $x = -2$. Note that $f'(0)$ does not exist, but $x = 0$ is not a critical number of f. (Why not?)

Figure 14 shows how to determine the sign of f' on each side of the critical numbers. Observe that the sign of x^2 is always positive. Since the sign of f'

Figure 14

changes from plus to minus at $x = -2$, $f(x)$ has a relative maximum at $x = -2$. Similarly, since the sign of f' changes from minus to plus at $x = 2$, $f(x)$ has a relative minimum at $x = 2$. The relative maximum is $f(-2) = -4$ and the relative minimum is $f(2) = 4$. We defer sketching the graph (see Exercise Set 4.6, number 16). ∎

4.2 EXERCISE SET |||||||||||||||||||||||||||||||||||||||||||||||||||||||||||||||||||||||||||||||||||||

Exercises 1 and 2 refer to Figure 15.

Figure 15

1. At what points does the function have a relative maximum?
2. At what points does the function have a relative minimum?

In Exercises 3–26, locate the critical numbers of the given functions and classify each according to whether it produces a relative maximum, relative minimum, or neither.

3. $f(x) = x^2 - 8x + 5$

4. $f(x) = 2x^2 + 12x - 5$

5. $f(x) = -x^2 + 6x + 3$

6. $f(x) = -\frac{1}{2}x^2 - 4x + 8$

7. $f(x) = -3x + 2$

8. $f(x) = -4x + 8$

9. $f(x) = 4x^2 + 4$

10. $f(x) = x^2 + 5x$

11. $f(x) = 2x^3 - 3x^2 - 12x + 6$

12. $f(x) = x^3 + 3x^2 - 9x + 5$

13. $f(x) = \dfrac{1}{x^2}$

14. $f(x) = 2x + \dfrac{1}{2x}$

15. $f(x) = 5x^4 - 4x^5$

16. $f(x) = (x - 3)^{2/3}$

17. $f(x) = -x^3 + 3x^2 - 3x + 2$

18. $f(x) = \frac{1}{3}x^3 - x^2 + 5$

19. $f(x) = x^{1/3}$

20. $f(x) = \dfrac{x^2}{x^2 + 1}$

21. $f(x) = \dfrac{1}{x^2 + 1}$

22. $f(x) = \dfrac{x}{x + 1}$

23. $f(x) = x - \frac{1}{3}x^3$

24. $f(x) = x^4 - 4x^3 + 8$

25. $f(x) = x^4 - 2x^3$

26. $f(x) = \frac{1}{4}x^4 - \frac{1}{2}x^2 + 2$

4.3 |||||||| ABSOLUTE EXTREME VALUES OF FUNCTIONS

Many practical problems require finding not merely a relative extremum of f, but rather the smallest or largest value of $f(x)$ for x in some closed interval. For example, a manager will want to choose a course of action that makes the profit maximum in the absolute sense, not just in the relative sense. A function f, defined on an interval $[a, b]$ is said to have an **absolute maximum** at $x = c$ (c is in the interval) if $f(c)$ is the largest value of $f(x)$ (that is, $f(c) \geq f(x)$) for all x in the interval. Similarly, a function f defined on an interval $[a, b]$ is said to have an **absolute minimum** at $x = c$ if $f(c)$ is the smallest value of $f(x)$ (that is, $f(c) \leq f(x)$) for all x in the interval. If f has either an absolute maximum or an absolute minimum at $x = c$, we say that f has an **absolute extremum** at $x = c$.

Figure 1

(a)

(b)

In Figure 1(a), the function f has an absolute maximum at $x = c$ and an absolute minimum at the endpoint $x = a$. Observe that f also has a relative maximum at $x = c$ and at $x = d$, and a relative minimum at $x = e$. Figure 1(b) shows a function f that has an absolute minimum at $x = 0$ and absolute maxima at the endpoints $x = a$ and $x = b$. Of course, f also has a relative minimum at $x = 0$.

If f is a function that is continuous on a closed interval $[a, b]$, then we can draw the graph of f without lifting the pencil from the paper. Therefore it is intuitively clear that there is a highest and a lowest point on the graph. The following theorem is a precise statement of this observation.

EXTREME VALUE THEOREM

If a function f is *continuous* on a closed interval $[a, b]$, then f has an absolute maximum and an absolute minimum on $[a, b]$.

The Extreme Value Theorem has several limitations that must be kept in mind. First of all, it is an *existence theorem*. It tells us that under certain conditions a function *has* an absolute maximum and an absolute minimum, but it does not tell us what these values are nor how to find them. Secondly, the theorem requires two conditions:

(1) that the interval of x-values over which we seek the maximum (or minimum) value of $f(x)$ be a *closed* interval, and

(2) that the function f be *continuous* over this interval.

We cannot drop either of these two conditions from the Extreme Value Theorem, as the following example shows.

EXAMPLE 1   (a) A function f that is continuous over an interval I, and that has no maximum or minimum value over the interval I.

Let $f(x) = x$ and $I = (0, 1)$. As we see in Figure 2, f has no maximum and no minimum over I.

Figure 2

This example shows that the condition that the interval be *closed* is a necessary part of the Extreme Value Theorem.

(b) A function g that has no maximum or minimum value over the closed interval I.

Let $g(x) = -\dfrac{1}{x}$ and $I = [-1, 1]$. As we see in Figure 3, g has no maximum and no minimum value over the closed interval I.

Figure 3

This example shows the role of the condition that the function be continuous, for g is not continuous at the value $x = 0$ in the interval $[-1, 1]$.

|||||||||||| **Finding Absolute Extrema**

Suppose that a function f has an absolute extremum at some number c in a closed interval $I = [a, b]$. Then one of the following two possibilities must occur:

1. c is in the open interval (a, b). By definition, this means that f has a *relative* extremum at c. By the theorem of Section 4.2, this means that c is a *critical number* of f.

2. c is not in the open interval (a, b). Since c must be in the closed interval $I = [a, b]$, this means that either $c = a$ or $c = b$.

Therefore, c *must be either a critical number of* f *or an endpoint of the interval* I = [a, b].

We can now summarize the procedure for finding the absolute extrema of a function that is continuous on $[a, b]$.

Finding the Absolute Extrema of a Continuous Function f(x) on [a, b]
Step 1. Find the critical numbers of f on $[a, b]$.
Step 2. Evaluate f at the critical numbers and at the endpoints a and b.
Step 3. The largest of the values obtained in step 2 is the absolute maximum of f and the smallest is the absolute minimum of f.

EXAMPLE 2   Find the absolute extrema of

$$f(x) = x^3 - 3x^2 - 24x + 2,$$

(a) on the interval $[-6, 5]$, and (b) on the interval $[0, 5]$.

Solution   (a) *Step 1.* Since f is a polynomial function that is differentiable everywhere, we find the critical numbers of f by setting $f'(x) = 0$. We have

$$f'(x) = 3x^2 - 6x - 24 = 0$$
$$3(x^2 - 2x - 8) = 0$$
$$3(x - 4)(x + 2) = 0,$$

so the critical numbers of f are $x = 4$ and $x = -2$.
Step 2. We evaluate f at the critical numbers and at the endpoints and obtain

$$f(4) = (4)^3 - 3(4)^2 - 24(4) + 2 = -78$$
$$f(-2) = (-2)^3 - 3(-2)^2 - 24(-2) + 2 = 30$$
$$f(-6) = (-6)^3 - 3(-6)^2 - 24(-6) + 2 = -178$$
$$f(5) = (5)^3 - 3(5)^2 - 24(5) + 2 = -68.$$

Step 3. For the interval $[-6, 5]$, the absolute maximum of f is 30, which occurs at the critical number $x = -2$. The absolute minimum of f is -178, which occurs at the endpoint $x = -6$.

(b) *Step 1.* As in part (a),

$$f'(x) = 3(x - 4)(x + 2),$$

so the critical numbers of f are $x = 4$ and $x = -2$. But only one of these critical numbers, $x = 4$, lies in the interval $[0, 5]$.

Step 2. Evaluating f at the critical number and at the endpoints of $[0, 5]$ we have

$$f(4) = -78 \quad \text{as in part (a)}$$
$$f(0) = 0^3 - 3(0)^2 - 24(0) + 2 = 2$$
$$f(5) = -68 \quad \text{as in part (a)}.$$

Step 3. For the interval $[0, 5]$, the absolute maximum of f is 2, which occurs at the endpoint $x = 0$. The absolute minimum of f is -78, which occurs at the critical number $x = 4$. ∎

In Section 4.7 we shall examine a wide variety of applied *optimization problems* in which we shall use the techniques of this chapter to maximize or minimize a function for a concrete application. We get a start here by considering the following simpler example, and by solving Exercises 15–20 in the exercise set at the end of this section.

EXAMPLE 3   Consider the radio manufacturer discussed in Example 4 of Section 3.4. The profit $P(x)$ (in dollars) received from producing and selling x radios per week is given by

$$P(x) = -\frac{x^2}{100} + 50x - 8000. \tag{1}$$

Suppose that the manufacturer's production facilities limit the output to 3000 radios per week. The equation

$$p = 60 - \frac{x}{100} \tag{2}$$

gives the relation between the number of radios x that will be bought per week at a price of p dollars each. Assuming that every radio manufactured is sold,

(a) What level of production will produce the maximum profit?
(b) What will be the maximum profit?
(c) What should be the selling price per radio at this level of production?

Solution   (a) and (b) We need to determine a value of x in the interval $[0, 3000]$ that will maximize $P(x)$. We set

$$P'(x) = -\frac{x}{50} + 50 = 0.$$

The only critical number of $P(x)$ is

$$x = 2500.$$

Evaluating $P(x)$ at the endpoints of the interval $[0, 3000]$ and at the critical number, we have

x	0	2500	3000
$P(x) = -\dfrac{x^2}{100} + 50x - 8000$	-8000	54,500	52,000

Hence, the maximum profit is \$54,500 when the level of production is 2500 radios per week.

(c) Using Equation (2), we find that

$$p = 60 - \frac{2500}{100} = 35.$$

Thus, the optimal selling price per radio is \$35. ∎

4.3 EXERCISE SET ||||||||||||||||||||||||||||||||||||||||||||||||||||||||||||||||||||||||||||||||||||||||

Exercises 1 and 2 refer to Figure 4.

Figure 4

1. At what points does the function have an absolute maximum for the given interval?
   (a) $[a, x_1]$ (c) $[x_3, x_6]$ (e) $[x_7, x_9]$
   (b) $[x_2, x_4]$ (d) $[x_5, x_7]$ (f) $[a, b]$

2. At what points does the function have an absolute minimum for the given interval?
   (a) $[a, x_1]$ (c) $[x_3, x_6]$ (e) $[x_8, b]$
   (b) $[x_1, x_3]$ (d) $[x_5, x_7]$ (f) $[a, b]$

In Exercises 3–14, locate the absolute maximum and absolute minimum for each function over the given interval(s).

3. $f(x) = x^2 + 4x - 5$, $[-6, 2]$

4. $f(x) = 6 + 3x - x^2$, $[-1, 2]$

5. $f(x) = x^3 - 3x + 2$,
   (a) $[0, 2]$ (b) $[-2, 2]$

6. $f(x) = x^3 - 6x^2 + 9x + 10$,
   (a) $[0, 4]$ (b) $[-1, 2]$

7. $f(x) = \frac{1}{3}x^3 - x^2 + 10$,
   (a) $[-2, 1]$ (b) $[1, 4]$

8. $f(x) = x^4 - 2x^2 + 3$,
   (a) $[-2, -1]$ (b) $[-1, 1]$

9. $f(x) = \frac{1}{x}$, $[1, 3]$

10. $f(x) = \frac{1}{2 - x}$, $[-2, 1]$

11. $f(x) = |x - 1|$, $[0, 4]$

12. $f(x) = 2x - 4\sqrt{x}$, $[0, 4]$

13. $f(x) = x^{2/3} + 2$, $[-8, 1]$

14. $f(x) = \frac{x}{x^2 + 1}$, $[-2, 2]$

15. **(Business)** An automobile dealer's weekly profit (in dollars) is given by $P(x) = 1000x - 25x^2$, where x is the number of cars sold. Because of capital and storage restrictions, the dealer cannot sell more than 30 cars per week. How many cars must be sold to yield a maximum profit?

16. **(Business)** A manufacturer of custom-made cars takes in a weekly profit $P(x)$ (in dollars) defined by

$$P(x) = -15,000x^2 + 120,000x - 5000,$$

where x is the number of cars made per week. If the production capacity is limited to 8 cars per week, how many cars should be made each week to maximize the profit?

17. **(Medicine)** An experimental drug is being tested on a bacteria colony. It is found that t days after treating the colony, the number $N(t)$ of bacteria per cubic centimeter is given by

$$N(t) = 20t^2 - 120t + 800, \qquad 0 \le t \le 7.$$

How many days after treatment is the number of bacteria per cubic centimeter at a minimum? What is the minimum number?

18. **(Epidemic)** Suppose that the number of people in a certain city who are sick t days after the outbreak of a flu epidemic is given by

$$P(t) = -t^2 + 120t + 20.$$

On what day will the maximum number of people be sick, and how many people will be sick on that day?

19. **(Business)** A book publisher finds that the total weekly cost $C(x)$ (in dollars) of publishing x copies of a book is given by

$$C(x) = 300 + 6x + 0.02x^2.$$

If each book sells for $12, what weekly level of production will maximize profit, assuming that each book produced is sold?

20. **(Business)** A manufacturer of tennis rackets finds that the total daily cost $C(x)$ (in dollars) of producing x rackets per day is given by

$$C(x) = 400 + 4x + 0.0001x^2.$$

Each racket can be sold at a price p (in dollars), given by

$$p = 10 - 0.0004x.$$

If all the rackets that are manufactured can be sold, find the daily level of production that will result in the maximum profit.

4.4 |||||||| **THE SECOND DERIVATIVE**

So far, we have used the sign of f' in this chapter to determine the open intervals over which f is increasing or decreasing. This information has enabled us to find the relative extrema of f. For example, if $f'(x) > 0$ on (a, b), then f is increasing on (a, b). Yet, this information alone does not tell us the precise shape of the graph of f, which could match either of the graphs in Figure 1. The curve in Figure 1a is said to be concave upward, and the curve in Figure 1b is said to be concave downward. We shall define these terms precisely below.

Figure 1

(a)   (b)

The mathematical tool that will enable us to distinguish between these two types of curves is the derivative of the derivative, which we call the second derivative. Because the derivative $f'(x)$ of $f(x)$ is a function, we can sometimes compute the derivative of $f'(x)$,

$$\frac{d}{dx}[f'(x)],$$

which is called the **second derivative** of $f(x)$ and is denoted by

$$f''(x) \qquad \text{or} \qquad \frac{d^2}{dx^2}[f(x)].$$

If $y = f(x)$, the second derivative is also denoted by y'' or

$$\frac{d^2y}{dx^2}.$$

EXAMPLE 1   If $f(x) = x^5 + \sqrt{x}$, then the first derivative of f is

$$f'(x) = \frac{d}{dx}[x^5 + \sqrt{x}] = 5x^4 + \tfrac{1}{2}x^{-1/2},$$

and the second derivative of f is

$$f''(x) = \frac{d}{dx}[f'(x)] = \frac{d}{dx}[5x^4 + \tfrac{1}{2}x^{-1/2}]$$
$$= 20x^3 + \tfrac{1}{2}(-\tfrac{1}{2})x^{-3/2}$$
$$= 20x^3 - \tfrac{1}{4}x^{-3/2}.$$

|||||||||||| ## The Second Derivative and Concavity

Suppose we know that the second derivative satisfies $f''(x) > 0$ for all x in the interval (a, b). From this we infer that the first derivative $f'(x)$ is increasing on (a, b). Geometrically, this relationship means that the *slope* of the tangent line to the graph of f is *increasing* as we move along the x-axis from a to b. Thus, the curve $y = f(x)$ "bends upward" over the interval (a, b) [see Figure 2(a)]. Such a curve is said to be **concave upward** over (a, b). Alternatively we can say that a curve is concave upward over (a, b) if *the curve lies above its tangent line at each* x *in* (a, b).

Figure 2

Concave upward

(a) Slopes increasing

Concave downward

(b) Slopes decreasing

Similarly, if $f''(x) < 0$ for each x in (a, b), then $f'(x)$ is decreasing on (a, b), so the *slope* of the tangent line to the graph of f is *decreasing* as we move along the x-axis from a to b. Thus, the curve $y = f(x)$ "bends downward" over the interval (a, b), as shown in Figure 2(b). Such a curve is said to be **concave downward** over (a, b). Also, we can say that a curve is concave downward over (a, b) if *the curve lies below its tangent line at each* x *in* (a, b).

We can formulate our observations as follows.

THEOREM

(a) If $f''(x) > 0$ for all x in (a, b), then the curve $y = f(x)$ is concave upward over (a, b).

(b) If $f''(x) < 0$ for all x in (a, b), then the curve $y = f(x)$ is concave downward over (a, b).

The following description is sometimes helpful: a curve that is concave upward can be said to "hold water" and a curve that is concave downward can be said to "spill water" (see Figure 3).

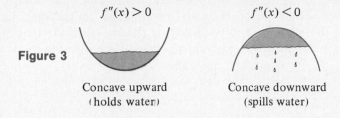

Figure 3

Concave upward
(holds water)

Concave downward
(spills water)

EXAMPLE 2 Let

$$f(x) = 2x^3 + 9x^2 + 12x + 2.$$

When is $y = f(x)$ concave upward? Concave downward?

Solution We have

$$f'(x) = 6x^2 + 18x + 12, \quad \text{and} \quad f''(x) = 12x + 18 = 6(2x + 3).$$

Then $f''(x) > 0$ if $2x + 3 > 0$ or $x > -\frac{3}{2}$, and $f''(x) < 0$ if $2x + 3 < 0$ or $x < -\frac{3}{2}$. Hence, the curve is concave upward if $x > -\frac{3}{2}$ and concave downward if $x < -\frac{3}{2}$ (see Figure 4).

Figure 4

$y = f(x) = 2x^3 + 9x^2 + 12x + 2$

$\left(-\frac{3}{2}, -\frac{5}{2}\right)$

■

|||||||||||| **Inflection Points**

A point $(c, f(c))$ on a curve $y = f(x)$ where the curve changes from concave upward to concave downward, or vice versa, is called an **inflection point** of the curve. Inflection points are among the most important points on the graph of a function. Knowledge of their presence or absence is a great help in sketching the graph.

EXAMPLE 3 For the curve shown in Figure 4, the point $(-\frac{3}{2}, -\frac{5}{2})$ is an inflection point, because the curve is concave downward when $x < -\frac{3}{2}$ and concave upward when $x > -\frac{3}{2}$. ∎

Suppose that $(c, f(c))$ is an inflection point of the curve $y = f(x)$. Can we say anything about the value $f''(c)$? Under quite general conditions, the answer is "yes." If the second derivative $f''(x)$ exists on an open interval containing c, then we may apply the first derivative test to the function $f'(x)$ to conclude that $f'(x)$ has a relative extreme value at $x = c$. Therefore, either $f''(c) = 0$ (Figure 5) or $f''(c)$ does not exist.

Figure 5

$f''(x) < 0$ inflection point

$f''(x) > 0$

$f''(x) = 0$ or does not exist

For the function $f(x) = 2x^3 + 9x^2 + 12x + 2$ of Examples 2 and 3, the point $(-\frac{3}{2}, -\frac{5}{2})$ was found to be an inflection point. Since $f''(x) = 6(2x + 3)$ in those examples, we have

$$f''(-\tfrac{3}{2}) = 0.$$ ∎

As the next example shows, a curve may also have an inflection point at $(c, f(c))$ where $f''(c)$ does not exist.

EXAMPLE 4 Find the inflection points of the graph of

$$f(x) = x^{1/3}.$$

Solution We have

$$f'(x) = \frac{1}{3} x^{-2/3}, \quad \text{and} \quad f''(x) = \frac{1}{3}\left(-\frac{2}{3}\right)x^{-5/3} = -\frac{2}{9} \cdot \frac{1}{x^{5/3}}.$$

If $x < 0$, then $f''(x) > 0$ (because $x^{5/3} < 0$), but if $x > 0$, then $f''(x) < 0$. Thus, the curve is concave upward if $x < 0$ and concave downward if $x > 0$. The point $(0, 0)$ is an inflection point, but $f''(0)$ does not exist. The curve is shown in Figure 6.

Figure 6

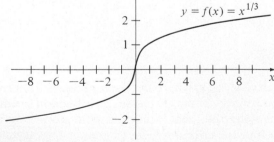

$y = f(x) = x^{1/3}$

∎

We can summarize our discussion of inflection points as follows.

> A point $(c, f(c))$ on a curve is a *candidate* for an inflection point of the curve if $f''(c) = 0$ *or* if $f''(c)$ does not exist. If $f''(x)$ changes sign at c, then $(c, f(c))$ is an inflection point.

Warning   Not every candidate for an inflection point actually turns out to be an inflection point. That is, we may have $f''(c) = 0$ or $f''(c)$ failing to exist even when $(c, f(c))$ is not an inflection point of the curve $y = f(x)$.

EXAMPLE 5   Consider the curve

$$f(x) = x^4.$$

Then $f'(x) = 4x^3$ and $f''(x) = 12x^2$. Thus, $f''(0) = 0$, but $(0, 0)$ is not an inflection point because $f''(x) > 0$ for all $x \neq 0$. Consequently, the curve does not change concavity at $x = 0$ and $(0, 0)$ is not an inflection point. Indeed, this curve has no inflection point. Its graph is shown in Figure 7.

Figure 7

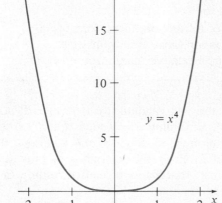

EXAMPLE 6
(Control of
an Epidemic)   An epidemic of Russian flu spreads throughout a community in the manner shown in Figure 8. At time $t = 0$, the number $N(t)$ of people infected is 0. The disease starts to spread slowly and then the *rate* of infection increases up to time t_1, when public health measures start to bring it under control. The curve is concave upward to the left of $(t_1, N(t_1))$ and concave downward to the right of $(t_1, N(t_1))$, so $(t_1, N(t_1))$ is an inflection point. At time t_1, the *rate* of the infection starts to decline, although the *number* of people infected continues to increase. We expect that the epidemic will soon peak and that after this time the number of people infected will begin to decline. In Figure 8 we see that the epidemic peaks at time

t_2. We can understand why public health officials would view the inflection point $(t_1, N(t_1))$ optimistically.

Figure 8

We shall conclude this section by discussing the use of the second derivative to determine whether a function has a relative extremum at a critical number. The precise statement of this test is given as follows.

THEOREM

The Second Derivative Test for Relative Extrema
Let c be a *critical number* of a function f such that $f'(c) = 0$.
  (a) If $f''(c) > 0$, then f has a relative minimum at c.
  (b) If $f''(c) < 0$, then f has a relative maximum at c.

A brief and intuitive discussion will validate (a). If $f''(c) > 0$, it can be shown that there is an interval (a, b) containing $x = c$ such that $f''(x) > 0$ for all x in (a, b). Therefore f' is increasing on (a, b). Since $f'(c) = 0$, $f'(x) < 0$ for $x < c$ and $f'(x) > 0$ for $x > c$. The sign of f' changes from minus to plus at c, so we conclude, by the first derivative test, that f has a relative minimum at c (see Figure 9).

Figure 9

A similar intuitive argument will validate (b). But what happens when $f''(c) = 0$? The above theorem does not cover this case.

If f''(c) = 0, *the second derivative test is not applicable.* The function may have a relative minimum at c or a relative maximum at c or neither type of extrema at c. Functions exhibiting these properties are shown in Figure 10. In each case, $f'(0) = 0$ and $f''(0) = 0$.

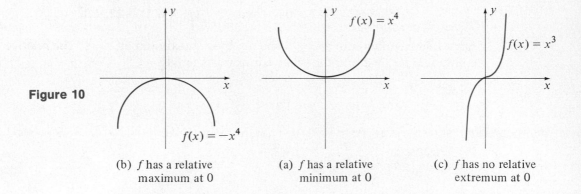

Figure 10

(b) f has a relative maximum at 0   (a) f has a relative minimum at 0   (c) f has no relative extremum at 0

The second derivative test is a convenient alternative to the first derivative test. Students often prefer it, finding it easier to use than the first derivative test. Once the critical numbers c of f, for which $f'(c) = 0$, have been found, it is the test to try first. Of course, if the conditions for its applicability are not satisfied or if the test provides no information or if the second derivative of f is too difficult to calculate, then we can always fall back on the first derivative test.

EXAMPLE 7   Use the second derivative test to find the relative extrema of

$$f(x) = x^2 - 6x.$$

(See Example 6, Section 4.2.)

Solution   We have

$$f'(x) = 2x - 6$$
$$f''(x) = 2.$$

The only critical number of f is $x = 3$. Because $f'(3) = 0$ and

$$f''(3) = 2 > 0,$$

f has a relative minimum at $x = 3$. The relative minimum is $f(3) = -9$.   ■

EXAMPLE 8   Use the second derivative test to find the relative extrema of

$$f(x) = x^3 - 6x^2 + 9x + 2.$$

(See Example 7, Section 4.2.)

Solution   We have

$$f'(x) = 3x^2 - 12x + 9$$
$$f''(x) = 6x - 12.$$

We already know that the critical numbers of f are $x = 3$ and $x = 1$. Because $f'(3) = 0$, $f'(1) = 0$, and

$$f''(3) = 6(3) - 12 > 0 \quad \text{and} \quad f''(1) = 6(1) - 12 < 0,$$

f has a relative minimum at $x = 3$ and a relative maximum at $x = 1$. The relative minimum is $f(3) = 2$ and the relative maximum is $f(1) = 6$.   ■

4.4 EXERCISE SET |||||||||||||||||||||||||||||||||||||||||||||||||||||||||||||||||||||||||||||||||||||

1. Consider the function graphed in Figure 11.

Figure 11

Figure 12

   (a) Over what open intervals is the curve concave upward?
   (b) Over what open intervals is the curve concave downward?
   (c) Find the inflection points of f over (x_1, x_7).

2. Consider the function graphed in Figure 12.
   (a) Over what open intervals is the curve concave upward?
   (b) Over what open intervals is the curve concave downward?
   (c) Find the inflection points of f over (x_1, x_6).

In Exercises 3–10, find the inflection points and determine where the given curves are concave upward and where they are concave downward.

3. $f(x) = x^2 + 3x - 8$

4. $f(x) = -x^2 - 2x + 5$

5. $f(x) = x^3 + 6x^2 - 15x + 8$

6. $f(x) = 2x^3 + 3x^2 - 36x + 10$

7. $f(x) = \frac{1}{3}x^3 - x^2$

8. $f(x) = 3x^4 - 4x^3 - 6x^2 + 6$

9. $f(x) = x^4 - 4x^3 + 6$

10. $f(x) = (x - 2)^5$

In Exercises 11–18, locate the relative extrema of the given functions by using the second derivative test.

11. $f(x) = x^2 - 6x + 5$

12. $f(x) = -2x^2 + 5x - 3$

13. $f(x) = 2x^3 - 3x^2 - 12x + 5$

14. $f(x) = 2x^3 - 21x^2 + 36x - 8$

15. $f(x) = 2x^3 + 4x^2 + 2x + 3$

16. $f(x) = x^3 - 3x + 2$

17. $f(x) = 3 - x^3$

18. $f(x) = x + \dfrac{1}{x}$

In Exercises 19–30, locate the relative extrema of the given functions by using either the first derivative test or the second derivative test.

19. $f(x) = x^2 + 6x - 3$

20. $f(x) = -x^2 + 10x - 5$

21. $f(x) = 2x^3 - 9x^2 + 12x + 3$

22. $f(x) = \frac{1}{3}x^3 - x - 4$

23. $f(x) = x^3 - 12x^2 + 36x$

24. $f(x) = (x - 2)^5$

25. $f(x) = (x - 3)^{2/3}$

26. $f(x) = (x - 1)^{1/3}$

27. $f(x) = \dfrac{1}{x^2 + 1}$

28. $f(x) = x^4 - 2x^3$

29. $f(x) = \frac{1}{4}x^4 - \frac{1}{2}x^2 + 2$

30. $f(x) = \dfrac{1}{x - 1}$

31. **(Educational Psychology)** The acquisition of a skill, such as the ability to learn a foreign language, starts rather quickly and then slows down. Using a standard test to measure performance, a psychologist finds that an average person's score $P(t)$ on the test after t weeks of study is given by

$$P(t) = 12t^2 - t^3, \qquad 0 \le t \le 8.$$

After how many weeks of study would the psychologist conclude that the *rate* of learning has started to decrease?

32. **(Economics)** Suppose that the cost $C(x)$ of manufacturing x units of a product is as shown in Figure 13. What does the point $(x_0, C(x_0))$ represent? Discuss the economic implications.

Figure 13

33. The curve shown in Figure 14 is an unemployment curve.

Figure 14

(a) What is happening in May?
(b) What is happening in August?

4.5 |||||||| **APPLICATIONS TO CURVE SKETCHING**

There are essentially two ways to gain an understanding of a particular function: *algebraically,* through the equation that represents the function, and *geometrically,* through the graph of the function. Without denying the importance of the equation, especially for computational purpose, we must emphasize that we often find the graph to be even more helpful in gaining an understanding of the function. Truly, a picture can be worth a thousand words. The graph displays features not easily seen from the formula: increasing and decreasing behavior, concavity, relative maxima and minima, inflection points, discontinuities, asymptotes, and so on. Because we aim to *understand* functions and their behavior, it is important that we be able to sketch their graphs with confidence and ease. In this section and the next we show how to use the tools of calculus developed so far to make a reasonable sketch of a curve $y = f(x)$.

We begin with a summary in Table 1 of the basic techniques for curve sketching that apply in general, whenever we graph a continuous function $y = f(x)$.

TABLE 1

Basic Procedure for Sketching a Continuous Curve $y = f(x)$
(First, check to make sure the function f is continuous everywhere.)

Step 1. Find $f'(x)$ and $f''(x)$.

Step 2. Locate the critical numbers of f. These are the values c in the *domain* of f for which $f'(c) = 0$ or $f'(c)$ does not exist.

Step 3. Use the first derivative test or the second derivative test to find out whether, at each critical number, the function has a relative maximum, relative minimum, or neither.

Step 4. Find the open intervals over which f is increasing and those over which f is decreasing:

If $f'(x) > 0$ for all x in (a, b), then f is increasing on (a, b).
If $f'(x) < 0$ for all x in (a, b), then f is decreasing on (a, b).

Step 5. Find the inflection points of the curve. A value d for which $f''(d) = 0$ or $f''(d)$ does not exist gives rise to a *candidate* for an inflection point. Test each of these candidates: if $f''(x)$ changes sign at d, then $(d, f(d))$ is an inflection point.

Step 6. Find the open intervals over which the curve is concave upward and those over which it is concave downward:

If $f''(x) > 0$ for all x in (a, b), then the curve is concave upward over (a, b).
If $f''(x) < 0$ for all x in (a, b), then the curve is concave downward over (a, b).

Step 7. Make a table of values for f that includes all critical numbers and inflection points, and just a few more well-chosen numbers.

Step 8. Use the above information to sketch the curve $y = f(x)$.

EXAMPLE 1   Sketch the graph of

$$f(x) = \tfrac{1}{3}x^3 - x^2 - 3x + 1.$$

Solution   *Step 1.* We have

$$f'(x) = x^2 - 2x - 3 = (x - 3)(x + 1), \tag{1}$$
$$f''(x) = 2x - 2 = 2(x - 1). \tag{2}$$

Step 2. The critical numbers of f are $x = 3$ and $x = -1$.

Step 3. Using the second derivative test, we find from (2) that

$$f''(3) = 2(3) - 2 = 4 > 0$$

and

$$f''(-1) = 2(-1) - 2 = -4 < 0,$$

so f has a relative minimum at $x = 3$ and a relative maximum at $x = -1$.

Step 4. We determine where f is increasing and where it is decreasing by the following analysis.

Thus, f is increasing on $(3, +\infty)$ and $(-\infty, -1)$ and decreasing on $(-1, 3)$. Of course, this analysis would also have told us that f has a relative minimum at $x = 3$ and a relative maximum at $x = -1$, by the first derivative test.

Step 5. By (2),

$$f''(1) = 0.$$

We check whether $(1, -\tfrac{8}{3})$ is an inflection point of the curve. From (2) we know that

$$f''(x) > 0 \qquad \text{if } x > 1, \qquad \text{and}$$
$$f''(x) < 0 \qquad \text{if } x < 1,$$

so $(1, -\frac{8}{3})$ is an inflection point of the curve since f'' changes from minus to plus at $x = 1$.

Step 6. From (2), the curve is concave upward if $x > 1$ and concave downward if $x < 1$.

Step 7.

x	$f(x)$
3	-8 (relative minimum)
-1	$\frac{8}{3}$ (relative maximum)
1	$-\frac{8}{3}$ (inflection point)
-3	-8
5	$\frac{8}{3}$

Step 8. The graph of f is sketched in Figure 1.

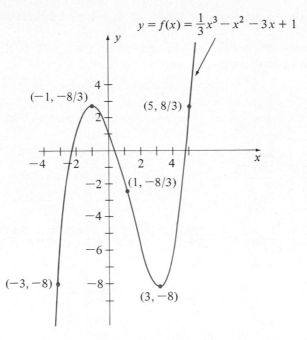

Figure 1

We have encountered the function $f(x) = x^{2/3}$ several times already: in Example 9 of Section 3.1, and in Examples 3 and 8 of Section 4.2. At this time we can finally establish the correct shape of its graph.

EXAMPLE 2   Sketch the graph of

$$f(x) = x^{2/3}.$$

Solution   We have

$$f'(x) = \frac{2}{3} x^{-1/3} = \frac{2}{3} \frac{1}{x^{1/3}},$$

$$f''(x) = \frac{2}{3}\left(-\frac{1}{3}\right) x^{-4/3} = -\frac{2}{9x^{4/3}}.$$

Hence, the only critical number of f is $x = 0$, because $f'(0)$ does not exist and $f(0)$ is defined.

Using the first derivative test, we have

$$f'(x) > 0 \quad \text{if } x > 0,$$
$$f'(x) < 0 \quad \text{if } x < 0.$$

Thus, f has a relative minimum at $x = 0$, and f is increasing if $x > 0$ and decreasing if $x < 0$. We previously made this observation in Example 8, Section 4.2.

Since $f''(0)$ does not exist, we check whether $(0, 0)$ is an inflection point. We have

$$f''(x) < 0 \quad \text{if } x > 0,$$
$$f''(x) < 0 \quad \text{if } x < 0,$$

so $(0, 0)$ is *not* an inflection point of the curve. We also conclude that the curve is concave downward if $x < 0$ as well as if $x > 0$. Figure 2 shows the graph of f.

Figure 2

$y = f(x) = x^{2/3}$

EXAMPLE 3 Sketch the graph of

$$f(x) = \frac{x^4}{4} + 2x^3 + \frac{9}{2}x^2 - 1.$$

Solution We have

$$f'(x) = x^3 + 6x^2 + 9x = x(x^2 + 6x + 9)$$
$$= x(x + 3)^2, \tag{3}$$

and

$$f''(x) = 3x^2 + 12x + 9$$
$$= 3(x^2 + 4x + 3)$$
$$= 3(x + 3)(x + 1). \tag{4}$$

From (3), the critical numbers of f are $x = 0$ and $x = -3$.

Since $f''(-3) = 0$, we cannot use the second derivative test to determine whether f has a relative extremum at $x = -3$. We could use it to test the critical number $x = 0$, because $f''(0) = 9$, but we shall examine both critical numbers by the first derivative test.

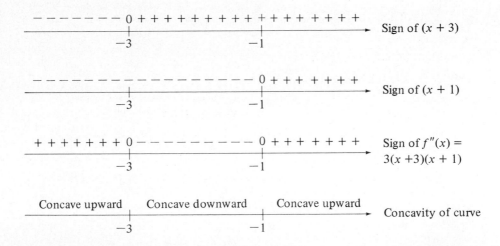

By the first derivative test, f does not have a relative extremum at $x = -3$, but it does have a relative minimum at $x = 0$. Also f is decreasing on $(-\infty, 0)$ and increasing on $(0, +\infty)$.

From (4), the candidates for inflection points are at points whose x-coordinates are $x = -3$ and $x = -1$. To see if $x = -3$ and $x = -1$ are the x-coordinates of inflection points of the curve, we proceed to analyze the sign of $f''(x) = 3(x + 3)(x + 1)$.

We now calculate $f(-3) = \frac{23}{4}$ and $f(-1) = \frac{7}{4}$.

Thus, $(-3, \frac{23}{4})$ and $(-1, \frac{7}{4})$ are inflection points of the curve. The curve is concave upward over $(-\infty, -3)$ and over $(-1, +\infty)$; it is concave downward over $(-3, -1)$. The resulting graph of f is sketched in Figure 3.

The procedure outlined in Table 1 is a *suggested* one. It is not always necessary to follow steps 1 through 8 exactly as given there. Sometimes we follow them very closely, as we did in Examples 1 and 3. Other times we may combine several steps together, or even omit one or more of the steps.

x	$f(x)$
0	-1 (relative minimum)
-3	$\frac{23}{4}$ (inflection point)
-1	$\frac{7}{4}$ (inflection point)
1	$\frac{23}{4}$
-4	7
2	37 (out of sight)
-5	$17\frac{3}{4}$ (out of sight)

$$y = f(x) = \frac{x^4}{4} + 2x^3 + \frac{9}{2}x^2 - 1$$

Figure 3   ■

4.5 EXERCISE SET ||||||||||||||||||||||||||||||||||||||||||||||||||||||||||||||||||||||||||||||||||

In Exercise 1–16, sketch the graph of the given function.

1. $f(x) = x^3 - 12x^2 + 36x$

2. $f(x) = x^3 - 3x^2 + 2$

3. $f(x) = x^3 - 3x + 3$

4. $f(x) = 2x^3 - 3x^2 - 12x + 18$

5. $f(x) = x^3 - x^2$

6. $f(x) = x^3 - 9x$

7. $f(x) = -x^3 + x^2 + 8x - 3$

8. $f(x) = (x - 2)^3$

9. $f(x) = x^4 - 2x^3$

10. $f(x) = \frac{1}{4}x^4 - \frac{1}{2}x^2$

11. $f(x) = x^4 + 4x^3$

12. $f(x) = 3x^4 - 4x^3 - 12x^2 + 2$

13. $f(x) = 3x^{4/3}$

14. $f(x) = 3x^{5/3}$

15. $f(x) = 2\sqrt{x}$

16. $f(x) = 6\sqrt[3]{x}$

4.6 |||||||| ADDITIONAL TECHNIQUES FOR CURVE SKETCHING

While the methods we have seen so far are always useful in obtaining the graph of a function $y = f(x)$, they do not always tell the whole story. We must occasionally supplement these methods with others. We present a few of these here.

|||||||||||||| ***x*-Intercepts and *y*-Intercepts**

In Step 7 of Section 4.5 you are encouraged to make a small table of values for $y = f(x)$. One of the easiest points to obtain is the **y-intercept,** the point where the curve intersects the y-axis. We merely substitute $x = 0$, calculate $f(0)$, and obtain the point $(0, f(0))$.

Less easy (perhaps even impossible) to find are the **x-intercepts,** if there are any. These are the points $(x, 0)$ where the curve $y = f(x)$ intersects the x-axis. To find them we set $y = 0$ and solve for x, if possible.

EXAMPLE 1   Find the y-intercept and the x-intercept of the curve $y = \dfrac{x^2 - 4}{x - 1}$.

Solution   To find the y-intercept we set $x = 0$ and solve:

$$y = \frac{0 - 4}{0 - 1} = 4.$$

Thus, the (only) y-intercept is $(0, 4)$. To find the x-intercepts we set $y = 0$ and solve:

$$\frac{x^2 - 4}{x - 1} = 0$$
$$x^2 - 4 = 0$$
$$(x - 2)(x + 2) = 0$$

Thus, the x-intercepts are $(2, 0)$ and $(-2, 0)$.   ∎

||||||||||||| Continuity

If the function f is continuous everywhere, then its graph is "all of one piece." If it has one point of discontinuity, in an interval on which f is otherwise continuous, then the graph splits into separate pieces at this point of discontinuity. Thus, we can use points of discontinuity to tell us the number of separate "continuous pieces" that comprise the graph.

EXAMPLE 2   Tell how many separate continuous pieces the graph of each function has.

(a) $f(x) = x^4 - 7x^3 + 15$ (c) $f(x) = \sqrt{x - 2}$ (e) $f(x) = \dfrac{x}{x^2 + 5}$.

(b) $f(x) = \dfrac{x}{x^2 - 4}$ (d) $f(x) = \dfrac{x + 3}{x^3 + x^2 - 2x}$

Solution   (a) Since the function $f(x) = x^4 - 7x^3 + 15$ is a polynomial, it is continuous everywhere. Thus, its graph consists of only one continuous piece.

(b) Since a rational function is continuous everywhere except where its denominator is 0, the function

$$f(x) = \frac{x}{x^2 - 4}$$

is continuous everywhere, except when $x^2 - 4 = 0$; that is, when $x = 2$ or $x = -2$. Thus, the graph of this function splits apart at two places, and so it must consist of three separate continuous pieces.

(c) By Properties 3 and 4 of Section 2.3, the function $f(x) = \sqrt{x - 2}$ is continuous everywhere on the interval $(2, +\infty)$. Moreover,

$$\lim_{x \to 2^+} \sqrt{x - 2} = 0 = f(2).$$

Hence, there are no points where the graph splits apart into separate pieces. Therefore, the graph of f is in one piece.

(d) We rewrite the function, factoring its denominator:

$$f(x) = \frac{x + 3}{x(x - 1)(x + 2)}.$$

Since f is discontinuous when $x = 0$, $x = 1$, and $x = -2$, the graph must be in four separate continuous pieces.

(e) The rational function $f(x) = \dfrac{x}{x^2 + 5}$ is continuous everywhere, because there are no real numbers x that make the denominator zero. Therefore, the graph is in one continuous piece. ∎

|||||||||||| Excluded Regions

It is often possible to tell the general location of a curve, even before a single point has been plotted. We can do so by systematically excluding regions of the plane that we are sure the curve cannot enter. A few examples will show how to do this.

EXAMPLE 3 By excluding regions of the plane that the given curve cannot enter, determine the general location of the curve.

(a) $y = \sqrt{x - 2}$ (b) $y = \dfrac{1}{x^2 + 1}$ (c) $y = \dfrac{1}{\sqrt{4 - x^2}}$ (d) $y = \dfrac{x}{x^2 - 9}$

Solution (a) Since the function $f(x) = \sqrt{x - 2}$ requires the square root of $x - 2$, we know that $x - 2$ must be ≥ 0, so we must have $x \geq 2$. Also, the radical always denotes the nonnegative square root, so $y = \sqrt{x - 2} \geq 0$. Therefore, the graph must lie to the right of (or on) the vertical line $x = 2$, and above (or on) the x-axis. In Figure 1 we have shaded the excluded regions. The curve must lie in the unshaded region.

Figure 1

(b) Observe that the values of the function

$$y = \frac{1}{x^2 + 1}$$

are always positive, and the denominator is always greater than or equal to the numerator. Thus, $0 < \frac{1}{x^2 + 1} \leq 1$. The curve must therefore lie in the unshaded region of Figure 2.

Figure 2

(c) The function

$$y = \frac{1}{\sqrt{4 - x^2}}$$

requires that $4 - x^2 > 0$, or $x^2 < 4$, which means $-2 < x < 2$. Also, $y > 0$. Therefore, the curve must lie in the unshaded region of Figure 3.

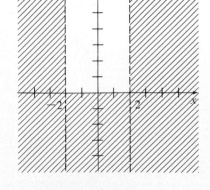

Figure 3

(d) For the function

$$y = \frac{x}{x^2 - 9},$$

we apply our usual analysis of signs.

From this analysis we see that
1. when $x < -3$, then $y < 0$.
2. when $-3 < x < 0$, then $y > 0$.
3. when $0 < x < 3$, then $y < 0$.
4. when $x > 3$, then $y > 0$.
Therefore, the curve must lie in the unshaded regions of Figure 4.

Figure 4

|||||||||||||| **Asymptotes**

As we have seen in Section 2.2,

(a) A line $x = a$ is a **vertical asymptote** of the curve $y = f(x)$ if either

$$\lim_{x \to a^-} f(x) = +\infty \text{ or } -\infty$$

or

$$\lim_{x \to a^+} f(x) = +\infty \text{ or } -\infty.$$

(b) A line $y = b$ is a **horizontal asymptote** of the curve $y = f(x)$ if either

$$\lim_{x \to +\infty} f(x) = b \qquad \text{or} \qquad \lim_{x \to -\infty} f(x) = b.$$

EXAMPLE 4   Find the horizontal and vertical asymptotes of the curve

$$y = \frac{2x^2}{x^2 + x - 6}.$$

Solution   (a) Vertical asymptotes. By rewriting the equation as

$$y = \frac{2x^2}{(x - 2)(x + 3)},$$

we can see that as $x \to 2$ or $x \to 3$ (one-sided), $y \to +\infty$ or $-\infty$. Therefore, the lines $x = 2$ and $x = -3$ are vertical asymptotes.

(b) Horizontal asymptotes. By evaluating the limit,

$$\lim_{x \to +\infty} \frac{2x^2}{x^2 + x - 6} = \lim_{x \to +\infty} \frac{2}{1 + \dfrac{1}{x} - \dfrac{6}{x^2}}$$

$$= \frac{2}{1 + 0 - 0}$$

$$= 2,$$

we see that the line $y = 2$ is a horizontal asymptote. We obtain the same result when we evaluate $\lim_{x \to -\infty}$, so there is no other horizontal asymptote.

A sketch of the curve $y = \dfrac{2x^2}{x^2 + x - 6}$ is shown in Figure 5.

Figure 5

■

|||||||||||| **Symmetry with Respect to the _y_-Axis**

A curve $y = f(x)$ is said to be **symmetric with respect to the _y_-axis** if, for every point (x, y) on the curve, its symmetric counterpart $(-x, y)$ also lies on the curve (see Figure 6).

Figure 6

This suggests a simple algebraic test.

If $f(x) = f(-x)$, for all x in the domain of f, then the graph of the function f is symmetric with respect to the _y_-axis.

EXAMPLE 5   Which of the functions of Example 3 have graphs that are symmetric with respect to the _y_-axis?

Solution   (a) For the function $f(x) = \sqrt{x - 2}$, we have $f(-x) = \sqrt{-x - 2}$. Since $f(-x) \neq f(x)$, the graph is _not_ symmetric with respect to the _y_-axis.

(b) For the function $f(x) = \dfrac{1}{x^2 + 1}$, we have

$$f(-x) = \frac{1}{(-x)^2 + 1} = \frac{1}{x^2 + 1} = f(x).$$

Since $f(-x) = f(x)$, the graph of this function *is* symmetric with respect to the y-axis.

(c) For the function $f(x) = \dfrac{1}{\sqrt{4 - x^2}}$, we have

$$f(-x) = \frac{1}{\sqrt{4 - (-x)^2}} = \frac{1}{\sqrt{4 - x^2}} = f(x).$$

Thus, the graph *is* symmetric with respect to the y-axis.

(d) When $f(x) = \dfrac{x}{x^2 - 9}$ we have

$$f(-x) = \frac{(-x)}{(-x)^2 - 9} = -\frac{x}{x^2 - 9} = -f(x).$$

Since $f(-x) \neq f(x)$, the graph is *not* symmetric with respect to the y-axis.

∎

EXAMPLE 6 Use methods presented in this section to sketch the graph of the function

$$f(x) = \frac{x}{x + 2}. \tag{1}$$

Solution First, we obtain the first and second derivatives,

$$f'(x) = \frac{(x + 2) - x}{(x + 2)^2} = \frac{2}{(x + 2)^2} \tag{2}$$
$$= 2(x + 2)^{-2},$$

$$f''(x) = -4(x + 2)^{-3} = \frac{-4}{(x + 2)^3}. \tag{3}$$

This rational function is continuous everywhere except at $x = -2$. Thus, the graph is in two separate continuous pieces.

From (2), the function is increasing on both intervals $(-\infty, -2)$ and $(-2, +\infty)$. Thus, the function has no relative extremum, by the first derivative test.

From (3), the curve is concave upward when $x + 2 < 0$; that is, for $x < -2$. The curve is concave downward for $x > -2$. But there is no inflection point because there is no point at all on the curve when $x = -2$.

The intercepts are easily obtained from (1). When $x = 0$, $y = \dfrac{0}{0 + 2} = 0$. When $y = 0$, we must have $0 = \dfrac{x}{x + 2}$, so $x = 0$. Thus $(0, 0)$ is the only x-intercept and the only y-intercept.

Excluded regions are found from the following sign analysis:

The excluded regions are shaded in Figure 7.

The graph is not symmetric with respect to the y-axis, since Equation (1) clearly shows that $f(-x) \neq f(x)$.

To find horizontal asymptotes we evaluate the limits:

$$\lim_{x \to +\infty} \frac{x}{x + 2} = \lim_{x \to \infty} \frac{1}{1 + \dfrac{2}{x}} = \frac{1}{1 + 0} = 1$$

$$\lim_{x \to -\infty} \frac{x}{x + 2} = \lim_{x \to -\infty} \frac{1}{1 + \dfrac{2}{x}} = \frac{1}{1 + 0} = 1.$$

Therefore, the line $y = 1$ is the only horizontal asymptote. For vertical asymptotes, we observe the behavior of $f(x)$ as $x \to -2$ from both sides separately. By Property 8 of Section 2.2,

$$\lim_{x \to -2^-} \frac{x}{x+2} = +\infty$$

$$\lim_{x \to -2^+} \frac{x}{x+2} = -\infty.$$

Thus, the line $x = -2$ is the (only) vertical asymptote.

Finally, we make a small table of values. The graph of f is shown in Figure 8.

x	y
0	0
1	$\frac{1}{3}$
2	$\frac{1}{2}$
-1	-1
-3	3
-4	2
-6	$\frac{3}{2}$

Figure 8   ■

EXAMPLE 7   Use the methods of this section to obtain the graph of the function

$$f(x) = \frac{x}{x^2 - 4}. \tag{4}$$

Solution   We rewrite (4) as

$$f(x) = \frac{x}{(x-2)(x+2)}. \tag{5}$$

Since this rational function is continuous at every number except $x = -2$ and $x = 2$, the graph of f is broken apart at these x-values into three continuous pieces. Using the quotient rule we obtain the derivative,

$$f'(x) = \frac{(x^2 - 4)(1) - x(2x)}{(x^2 - 4)^2}$$

$$= -\frac{x^2 + 4}{(x^2 - 4)^2}. \tag{6}$$

From (6) we see that $f'(x)$ is never 0. The derivative does not exist when $x = \pm 2$, but these numbers are not in the domain of f. Therefore, this function has no critical points.

From (6) we can also see that $f'(x) < 0$ whenever $f'(x)$ exist. Therefore, f is decreasing on each of its three pieces.

We now test for asymptotes. By inspection of (5) we see that the graph of f has vertical asymptotes $x = -2$ and $x = 2$. Also,

$$\lim_{x \to \pm\infty} f(x) = \lim_{x \to \pm\infty} \frac{x}{x^2 - 4} = \lim_{x \to \pm\infty} \frac{\dfrac{1}{x}}{1 - \dfrac{4}{x^2}}$$

$$= \frac{0}{1 - 0} = 0.$$

Therefore, the graph of f has horizontal asymptote $y = 0$.

From (6) we calculate the second derivative,

$$\begin{aligned}
f''(x) &= -\frac{-(x^2 - 4)(2x) - (x^2 + 4)2(x^2 - 4)(2x)}{(x^2 - 4)^4} \\
&= \frac{-2x(x^2 - 4)\left[(x^2 - 4) - 2(x^2 + 4)\right]}{(x^2 - 4)^4} \\
&= \frac{-2x(-x^2 - 12)}{(x^2 - 4)^3} \\
&= \frac{2x(x^2 + 12)}{(x - 2)^3(x + 2)^3}.
\end{aligned} \tag{7}$$

To determine the concavity of the graph of f we analyze the sign of $f''(x)$. Since $x^2 + 12$ is always positive, this factor is ignored in our analysis.

Thus, $f''(x) > 0$ when $-2 < x < 0$ or $x > 2$, and $f''(x) < 0$ when $x < -2$ or $0 < x < 2$. Therefore, the graph of f is concave upward on $(-2, 0)$ and $(2, \infty)$, and is concave downward on $(-\infty, -2)$ and $(0, 2)$.

The graph of f changes concavity at $x = -2$, 0 and 2. From (4), $y = 0$ when $x = 0$. However, there is no point on the graph corresponding to $x = -2$ or $x = 2$. Therefore, there is only one inflection point on this curve: $(0, 0)$.

We obtain excluded regions by applying sign analysis to Equation (5):

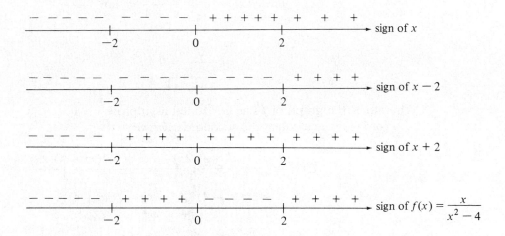

Thus, $f(x) > 0$ when $-2 < x < 0$ and when $x > 2$; $f(x) < 0$ when $x < -2$ and when $0 < x < 2$. In Figure 9 we have shaded the excluded regions and indicated the asymptotes.

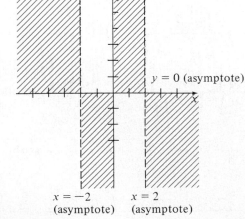

Figure 9

We make a brief table of values. The graph of f is sketched in Figure 10.

x	$f(x) = \dfrac{x}{x^2 - 4}$
0	0
1	$-\frac{1}{3}$
3	$\frac{3}{5}$
6	$\frac{3}{10}$
-1	$\frac{1}{3}$
-3	$-\frac{3}{5}$
-6	$-\frac{3}{10}$

Figure 10 ∎

4.6 EXERCISE SET ||||||||||||||||||||||||||||||||||||||||||||||||||||||||||||||||||||||||||||||||||

Use the techniques presented in Sections 4.5 and 4.6 to sketch the graph of the given function.

1. $f(x) = \dfrac{-2}{x}$

2. $f(x) = \dfrac{1}{x^2}$

3. $f(x) = \sqrt{x - 2}$
   [See Example 3(a)]

4. $f(x) = \sqrt{2x + 6}$

5. $f(x) = \dfrac{1}{x^2 + 1}$
   [See Example 3(b)]

6. $f(x) = \dfrac{1}{\sqrt{4 - x^2}}$
   [See Example 3(c)]

7. $f(x) = \dfrac{1}{x - 3}$

8. $f(x) = \dfrac{x}{x + 1}$

9. $f(x) = \dfrac{x}{x^2 - 9}$
   [See Example 3(d)]

10. $f(x) = \dfrac{x^2}{x^2 - 4}$

11. $f(x) = \dfrac{2x + 1}{1 - x}$

12. $f(x) = \dfrac{1}{x^2 - 1}$

13. $f(x) = \dfrac{1}{\sqrt{x}}$

14. $f(x) = \dfrac{1}{\sqrt[3]{x}}$

15. $f(x) = \dfrac{1}{\sqrt{x + 2}}$

16. $f(x) = x + \dfrac{4}{x}$

4.7 ||||||| APPLIED OPTIMIZATION PROBLEMS

Now we shall apply the techniques developed in this chapter to a wide variety of optimization problems. Most of these problems are presented in verbal terms. Before attempting to solve them, therefore, we must translate them into mathematical terms. The following suggestions will prove helpful.

Solving Verbal Optimization Problems

Step 1. Read the problem carefully.

Step 2. Choose variables to represent the situation under consideration. Draw a figure, if appropriate, to aid in the solution of the problem.

Step 3. Determine the variable that is to be optimized (minimized or maximized) and write an equation expressing this variable as a function of all the other variables in the problem. Thus, the variable to be optimized is the dependent variable.

Step 4. Using the information provided, eliminate all but one of the independent variables in the equation obtained in Step 3. Thus, the dependent variable to be optimized is written as a function f of one independent variable.

Step 5. Find the domain of values that the independent variable of Step 4 can take. The domain is often a closed interval $[a, b]$, so the extreme value theorem (see Section 4.3) applies when f is continuous. When the domain is not a closed interval, the extreme value theorem does not apply. The latter situation will be considered below.

Step 6. Find the maximum or minimum of the function obtained in Step 4 by using the methods explained earlier in this chapter.

In most problems, we can find the absolute maximum or absolute minimum of a function without having to sketch the graph of the function.

**EXAMPLE 1
(Oil Pricing
Policies)**

An oil-producing country sells 1 million barrels per day at $30 per barrel. The oil ministry is planning a price increase, but careful studies indicate that for each $.30 increase per barrel 5000 fewer barrels will be sold per day. How large an increase, if any, can be made to maximize daily revenue?

Solution

Let x denote the number of $.30 increases per barrel. If the country sold $5000x$ fewer barrels per day, the number of barrels sold per day would be

$$n = 1,000,000 - 5000x. \tag{1}$$

At a price increase in dollars of $0.3x$ per barrel, the new selling price would be $p = 30 + 0.3x$ dollars per barrel and the daily revenue $R(x)$ (in dollars) would be

$$R(x) = (\text{barrels sold per day}) \cdot (\text{price per barrel})$$
$$= np$$
$$= (1{,}000{,}000 - 5000x)(30 + 0.3x)$$
$$= 30{,}000{,}000 + 150{,}000x - 1500x^2. \tag{2}$$

Now $n \geq 0$, since the oil-producing country will not export a negative number of barrels of oil in a day. Thus, from (1),

$$1{,}000{,}000 - 5{,}000x \geq 0$$

or

$$1{,}000{,}000 \geq 5{,}000x$$
$$200 \geq x$$

Similarly $x \geq 0$, since the price per barrel is going to *increase* rather than decrease. Thus $0 \leq x \leq 200$.

Hence, our problem is to find the maximum value of $R(x)$ on the interval $[0, 200]$. The extreme value theorem (see Section 4.3) tells us that $R(x)$ does have a maximum value over this interval. We know that it must occur either at a critical number of R or at an endpoint of the interval.

We first find the critical numbers of $R(x)$:

$$\text{from (2),} \quad R'(x) = 150{,}000 - 3000x,$$
$$\text{so} \qquad x = 50$$

is the only critical number of $R(x)$ in the interval $[0, 200]$.

We then calculate the value of $R(x)$ at the critical number and at the endpoints.

x	$R(x) = 30{,}000{,}000 + 150{,}000x - 1500x^2$
50	33,750,000
0	30,000,000
200	0

Therefore, $R(x)$ has an absolute maximum for the interval $[0, 200]$ at $x = 50$.

This means that the oil ministry should make fifty \$.30 increases per barrel. That is, the price per barrel should be raised by \$15 to \$45 per barrel. ■

EXAMPLE 2   A rectangular box with no top is to be made from a piece of cardboard that measures 8 centimeters by 15 centimeters by cutting out and discarding a square from each corner and then folding up the sides along the perforation, as shown in Figure 1.

Figure 1

What size square will give a box of maximum volume? What is the maximum volume?

Solution   Let x denote the length (in centimeters) of a side of the square cut from each corner. The volume V of the box is the product of its length, width, and height. Thus,

$$V = (15 - 2x)(8 - 2x)x = 4x^3 - 46x^2 + 120x. \qquad (3)$$

Since x is the height of the box, $x \geq 0$. Also, $2x \leq 15$ and $2x \leq 8$ because $15 - 2x$ and $8x - 2x$ are the length and width of the box, respectively. These inequalities are both satisfied when $x \leq 4$. Thus, our problem is to find the maximum value of V over the interval $[0, 4]$. From (3) we have

$$\frac{dV}{dx} = 12x^2 - 92x + 120$$

$$= 4(3x^2 - 23x + 30)$$

$$= 4(3x - 5)(x - 6).$$

Hence, the critical numbers of V are $x = \frac{5}{3}$ and $x = 6$. Since $x = 6$ is not in the interval $[0, 4]$, we have to compute only the value of V at the endpoints of $[0, 4]$ and at the critical number $x = \frac{5}{3}$:

x	0	$\frac{5}{3}$	4
$V = (15 - 2x)(8 - 2x)x$	0	$\frac{2450}{27}$	0

Thus, when $x = \frac{5}{3}$ centimeters, we can make a box of maximum volume

$$V = \frac{2450}{27} \approx 90.7 \text{ cubic centimeters.} \qquad \blacksquare$$

EXAMPLE 3   A 20-inch-long piece of wire will be cut into two pieces. One piece will be bent into a circle and the other will be formed into a square. How should the wire be cut to maximize the sum of the areas (see Figure 2)?

Solution Let x denote the length (in inches) of wire used to make the circle (see Figure 2). Thus, the circumference of the circle will be x and the perimeter of the square will be $20 - x$.

For circle	For square
x	$20 - x$

Figure 2

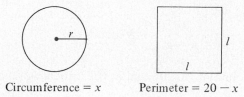

Circumference = x | Perimeter = $20 - x$

Since

$$\text{Circumference} = 2\pi \cdot \text{radius}$$

we find that the radius of the circle will be

$$r = \frac{x}{2\pi},$$

and since

$$\text{Area} = \pi r^2$$

we find that the area of the circle will be

$$A_c = \pi \left(\frac{x}{2\pi}\right)^2 = \frac{x^2}{4\pi}.$$

The length of a side of the square is one fourth of the perimeter, or

$$l = \frac{20 - x}{4},$$

and so its area will be

$$A_s = \left(\frac{20 - x}{4}\right)^2.$$

Hence, the total area A is given by

$$A = A_c + A_s = \frac{x^2}{4\pi} + \left(\frac{20 - x}{4}\right)^2. \tag{4}$$

Our problem is to maximize A over the interval $[0, 20]$. The chain rule applied to (4) gives us

$$\frac{dA}{dx} = \frac{2x}{4\pi} + 2\left(\frac{20 - x}{4}\right) \cdot \left(-\frac{1}{4}\right)$$

$$= \frac{x}{2\pi} - \frac{20 - x}{8}. \tag{5}$$

Setting $dA/dx = 0$, we find (verify) the (only) critical number of A:

$$x = \frac{20\pi}{4 + \pi} \approx 8.8 \text{ inches.}$$

Evaluating A at the endpoints of $[0, 20]$ and at the critical number, we obtain

x	0	8.8	20
A	25	14	31.8

Thus, $x = 20$ will give the maximum area. Consequently, we should use the entire wire for the circle and not make the square at all. ∎

Warning   Example 3 shows us that in solving optimization problems we must not blindly assume that the solution we are seeking occurs at a critical number of the function. As the example shows, the solution may very well occur at an endpoint. For Example 3 it turns out that the total area is *minimum*, rather than maximum at the critical number. If we had blindly taken the critical number $x = 8.8$, we would have chosen the *worst* possible number rather than the best possible number.

Another way of knowing that the critical number in Example 3 does not maximize the area is to use the second derivative test. From (5),

$$A'' = \frac{1}{2\pi} + \frac{1}{8} > 0.$$

Since $A'' > 0$ at the critical number, the function must have a relative *minimum* there.

|||||||||||||| **Inventory Control**

An important optimization problem in a merchandising business is **inventory control.** The business must keep enough inventory in stock to meet anticipated demand over a period of time (say, a year); however, it must not keep too much inventory because the resulting costs of storage, insurance, and interest (on the money borrowed to buy the inventory) would be excessive. Thus, the business has two alternate courses of action: place one order to fill all the anticipated demand or place a number of smaller orders throughout the time period under consideration. The second approach will keep down **holding costs** (which include interest, storage, and insurance) but will incur **reorder costs** (which include paperwork, shipping, and handling costs). The first approach will eliminate reorder costs but will entail holding costs. The number of units ordered each time is called the **lot size.** Thus, the business person's problem is to determine the lot size that will minimize **total annual inventory cost** = annual holding cost + annual reorder cost, or in symbols,

$$K = H + O. \tag{6}$$

We assume that the lot size x is the same for each order and that each order arrives as soon as the inventory on hand falls to zero (Figure 3). Then the **largest inventory** on hand at any one time is x, and assuming sales occur at a uniform rate, it is reasonable to say that the **average inventory** during the year is $x/2$ (Figure 3). Since both the annual holding cost H and the annual reorder cost O are functions of the lot size x, we may rewrite Equation (6) as

$$K(x) = H(x) + O(x). \tag{7}$$

The lot size x, which minimizes the total annual inventory cost $K(x)$, is called the **economic ordering quantity**, abbreviated **EOQ.**

Figure 3

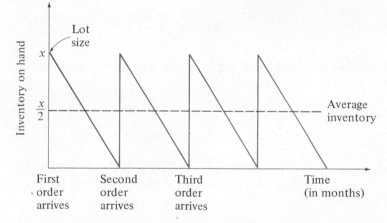

EXAMPLE 4 A camera dealer expects to sell 4000 cameras of a particular make during the year at a rather uniform rate. The annual holding cost per camera is $8 and each time the dealer places an order with the distributor there is a flat fee of $40 per order. How many times per year should the dealer reorder, and what lot size will minimize the total annual inventory cost?

Solution Let $x =$ the lot size. Then the average number of cameras on hand during the year is $x/2$. The annual holding cost $H(x)$ (in dollars), then, is given by

$$H(x) = \begin{pmatrix} \text{annual holding} \\ \text{cost per camera} \end{pmatrix} \cdot \begin{pmatrix} \text{average number} \\ \text{of cameras} \end{pmatrix}$$

$$= 8\left(\frac{x}{2}\right) = 4x. \tag{8}$$

To obtain the annual reorder cost $O(x)$, we proceed as follows. The number of orders is

$$\text{Number of orders} = \frac{(\text{number of cameras sold per year})}{\text{lot size}} = \frac{4000}{x}.$$

Thus, $O(x)$ (in dollars) is given by

$$O(x) = \left(\begin{array}{c}\text{cost}\\\text{per order}\end{array}\right)\left(\begin{array}{c}\text{number of}\\\text{orders}\end{array}\right) = 40\left(\frac{4000}{x}\right) = \frac{160{,}000}{x}. \tag{9}$$

Hence, from (8) and (9), the total annual inventory cost $K(x)$ is

$$K(x) = H(x) + O(x)$$

$$= 4x + \frac{160{,}000}{x}. \tag{10}$$

Since the dealer expects to sell 4000 cameras during the year, the lot size can be as small as 1 and as large as 4000. Thus, our problem is to find the minimum value of $K(x)$ over the interval [1, 4000]. From (10),

$$K'(x) = 4 - \frac{160{,}000}{x^2}.$$

The critical numbers for K are found by solving the equation

$$4x^2 = 160{,}000$$

or

$$x^2 = 40{,}000.$$

Thus, $x = 200$ or $x = -200$, but $x = 200$ is the only critical number of $K(x)$ in the interval [1, 4000]. Evaluating $K(x)$ at the critical number and at the endpoints, we find

x	1	200	4000
$K(x) = 4x + \dfrac{160{,}000}{x}$	160,004	1600	16,040

The minimum total annual inventory cost of \$1600 will occur when the EOQ is 200 cameras. Thus, the dealer should place

$$\frac{4000}{200} = 20 \text{ orders}$$

of 200 cameras each.   ■

Observe that in Example 4 the actual cost of the camera plays no role whatsoever.

|||||||||||| Optimizing Functions on More General Intervals

If a function f is continuous on a closed interval $[a, b]$, then the extreme value theorem (see Section 4.3) assures us that f has both an absolute maximum and an absolute minimum on $[a, b]$. If f is continuous on an interval for which the extreme value theorem does not apply, however, we have no assurance that f has absolute extrema on that interval. In this connection the following theorem is useful in optimization problems in which the function has only one critical number in the interval under consideration.

THEOREM

Suppose that f is a continuous function on an interval I, and that I contains *only one critical number c of f*.
(a) If f has a relative maximum at c, then $f(c)$ is the absolute maximum value of f on I.
(b) If f has a relative minimum at c, then $f(c)$ is the absolute minimum value of f on I.

EXAMPLE 5 A container manufacturer is designing a closed rectangular cargo box having a square base and a volume of 1024 cubic meters. Since the containers will be stacked on top of each other, the tops and bottoms must be stronger than the sides. Suppose that the material for the top and bottom costs $4 per square meter and the material for the sides costs $2 per square meter. Find the dimensions of the box that will minimize the total cost. What is the minimum cost?

Solution Let x denote the length (in meters) of a side of the base and let y denote the height (in meters) of the box (Figure 4).

Figure 4

The total area of the top and bottom will be

$$x^2 + x^2 = 2x^2$$

and the total area of the sides will be $4xy$. Hence, the total cost C (in dollars) will be

$$C = 4(2x^2) + 2(4xy)$$
$$= 8x^2 + 8xy. \tag{11}$$

The volume of the box is to be 1024 cubic meters, so we have

$$x^2y = 1024. \tag{12}$$

Solving Equation (12) for y and substituting in (11), we obtain

$$C = 8x^2 + \frac{8(1024)}{x} = 8x^2 + \frac{8192}{x}. \tag{13}$$

Thus, we have expressed C as a function of a single variable, x. Since x represents a length, $x > 0$. Thus, we must find the minimum value of C on the interval $(0, +\infty)$. Since this is not a closed interval, the extreme value theorem does not apply and we have no assurance that C has a minimum value on $(0, +\infty)$. However, the last theorem can be used.

To find the critical numbers of C, we differentiate (13) and obtain

$$\frac{dC}{dx} = 16x - \frac{8192}{x^2}.$$

The derivative dC/dx does not exist when $x = 0$, but C is not defined at $x = 0$, so 0 is not a critical number. Thus, we set $dC/dx = 0$ and obtain

$$16x - \frac{8192}{x^2} = 0,$$

or, multiplying by x^2,

$$16x^3 - 8192 = 0$$
$$x^3 = \frac{8192}{16} = 512,$$

or

$$x = \sqrt[3]{512} = 8 \text{ meters.}$$

Hence, $x = 8$ is the only critical number of C in the interval $(0, +\infty)$. To use the second derivative test we compute

$$\frac{d^2C}{dx^2} = 16 + \frac{2(8192)}{x^3},$$

which is positive when $x = 8$. Thus, by the second derivative test, C has a relative minimum at $x = 8$. Applying the above theorem, we conclude that the absolute minimum value of C on $(0, +\infty)$ occurs at $x = 8$. From Equation (12),

$$y = \frac{1024}{x^2} = \frac{1024}{64} = 16 \text{ meters.}$$

Hence, the dimensions of the minimum cost box are 8 meters by 8 meters by 16 meters. The minimum cost, obtained from Equation (13), is \$1536. ∎

The theorem used in the last example is useful also in maximizing or minimizing a function over a closed interval $[a, b]$. Thus, suppose f is continuous on $[a, b]$ and c is the only critical number of f in $[a, b]$. If f has a relative maximum (or minimum) at c, then $f(c)$ is the absolute maximum (or minimum) value of f on $[a, b]$. This situation occurs in Example 6 below.

EXAMPLE 6 For the camera dealer of Example 4, find the optimal lot size using the method suggested by the theorem of this section.

Solution From Equation (10), we want to find the lot size x that minimizes the total annual inventory cost

$$K(x) = 4x + \frac{160,000}{x}, \qquad (14)$$

assuming that $1 \leq x \leq 4000$.
We find the derivative,

$$K'(x) = 4 - \frac{160,000}{x^2}. \qquad (15)$$

To find the critical numbers of K we solve

$$0 = 4 - \frac{160,000}{x^2},$$

or as we have already seen,

$$x^2 = 40,000.$$

Thus, $x = 200$ is the only critical number of $K(x)$ in the interval $[1, 4000]$.
Next, we find the second derivative from (15):

$$K''(x) = \frac{d}{dx}(-160,000x^{-2})$$

$$= 320,000x^{-3}.$$

Since $K''(x) > 0$ when $x = 200$ we conclude by the second derivative test that K has a relative minimum at $x = 200$.

Therefore, by the theorem of this section, $K(x)$ has its absolute minimum for the interval $[1, 4000]$ at $x = 200$. ∎

4.7 EXERCISE SET ||||||||||||||||||||||||||||||||||||||||||||||||||||||||||||||||||||||||||||||||||||||||||

1. **(Advertising)** A record distributor has determined that the yearly profit $P(x)$ (in thousands of dollars) is related to the yearly advertising expenditures x (in thousands of dollars) by

$$P(x) = 50 + 60x - 0.5x^2.$$

If the distributor cannot spend more than $100,000 per year on advertising, how much should be spent on advertising to maximize the profit?

2. A manufacturer of bicycles finds that production of x bicycles per day entails a fixed cost of $1000, a labor cost of $10 per bicycle, and a cost of $25,000/x$ for advertising. How many bicycles should be made daily to minimize the total cost?

3. Suppose that 320 feet of fencing is available to enclose a rectangular field. Find the dimensions of the rectangle that will yield the largest possible area.

4. Suppose that in the fencing problem of Exercise 3, one side of the rectangular field must receive double fencing. Find the dimensions of the rectangle that will yield the largest possible area.

Figure 5

5. A farmer who wants to create a rectangular grazing field bordering on a straight river has 600 meters of fencing material (see Figure 5). If the side along the stream will not be fenced, what length and width will provide the maximum grazing area?

6. An experimental farm has 3600 feet of fencing to create a rectangular area that will be subdivided into four equal subregions as shown in Figure 6. What is the largest total area that can be enclosed?

Figure 6

7. The owner of a 100-room motel finds that when the rate is $40 per room, all rooms are occupied each night. For each $1 increase in the rate per room, 2 fewer rooms are rented. How much of an increase should be made to maximize daily revenue? How many rooms will then be occupied daily, and what will be the daily revenue?

8. A tour operator has chartered a 400-seat airplane for a trip to Paris. If 400 passengers agree to go, the price will be $500 per passenger. If the group is smaller than 400 passengers, the rate will rise $2 per passenger for each unoccupied seat. Naturally, it is to the passengers' advantage to travel with the plane completely full. But is that to the operator's advantage? How many passengers must go to give the

tour operator the greatest revenue? (*Hint:* Let x = number of unoccupied seats.)

9. An apple grower finds that the average yield is 60 bushels per tree when 80 or fewer trees are planted in an orchard. Each additional tree decreases the average yield per tree by 2 bushels. How many trees will maximize the total yield of the orchard? With this many trees, what will be the average yield per tree, and the total yield of the orchard?

10. (**Business**) A film processor has 10 pickup stations in Philadelphia and each of them produces an average weekly profit of $500. A marketing study has shown that opening additional pickup stations will reduce the average weekly profit for each location by $50 for each new station. Financial considerations allow the company to consider opening no more than 10 new stations. How many new stations should be opened to maximize total weekly profit?

11. (**Drug Concentration**) Suppose that t hours after a drug is administered to a patient, the concentration $k(t)$ of the drug in the body is given by

$$k(t) = \frac{2t}{t^2 + 4}.$$

How long after the drug has been administered is the drug concentration at a maximum?

12. (**Reaction to a drug**) Suppose it is known that when x milligrams of a certain drug is administered to a patient, the function

$$R(x) = \left(\frac{M}{2} - \frac{2x}{5}\right)x^2$$

is a reliable indicator of the patient's **reaction** to the dosage x. (M indicates the maximum allowable dosage, and is constant.) The rate of change in reaction relative to dosage, $\dfrac{dR}{dx}$, is called the patient's **sensitivity** to the drug at dosage x. Find the dos-

age at which the patient's sensitivity is greatest.

13. An airline finds that a cargo plane carrying a full load and flying x miles per hour consumes fuel at the rate of

$$F(x) = \frac{1}{500}\left(\frac{250,000}{x} + x\right)$$

gallons per mile. If the cost of jet fuel is $2 per gallon, what is the most economical speed for the plane on a 1000-mile trip? What is the minimum cost of fuel for the trip?

14. A truck traveling at the average speed of x miles per hour consumes fuel at the rate of

$$G(x) = \frac{1}{500}\left(\frac{1000}{x} + x\right)$$

gallons per mile. If fuel costs $1.50 per gallon, what speed will result in the minimum fuel cost for a 500-mile trip? What is the minimum cost of fuel for this trip?

15. (**Inventory Control**) A boat dealer in Florida expects to sell 1920 boats a year at a fairly uniform rate. The annual holding cost per boat is $200. Each time the dealer reorders, there is a basic $30 flat fee plus $10 per boat ordered. How many times per year and in what lot size should the dealer reorder to minimize the total annual inventory cost?

16. (**Inventory Control**) A moped dealer expects to sell 1200 mopeds during the year at a fairly constant rate. The annual holding cost is $10 per moped and the dealer can reorder any number of mopeds for a flat reorder fee of $60. How many times per year and in what lot size should the dealer reorder to minimize the total annual inventory cost?

17. A rectangular box with no top is to be made from a thin piece of cardboard that is 20 inches square by cutting a square from each corner. What size cutout will

give the box maximum volume? What is the maximum volume?

18. Starting with a rectangular piece of cardboard, 8 inches by 5 inches, a rectangular box with no top is to be made by cutting a square from each corner. What size cutout will give the box maximum volume? What is the maximum volume?

19. A piece of wire 40 centimeters long will be cut into two pieces. One piece will be bent into a circle and the other will be formed into a square. How should the wire be cut to minimize the sum of the areas?

20. The U.S. Postal Service will not accept rectangular fourth-class parcels if the perimeter of one end of the parcel plus the length is greater than 108 inches. What is the largest possible volume of an acceptable rectangular parcel whose ends are square? (See Figure 7.)

—Length—
Perimeter

Figure 7

21. A moving company needs to construct an open (topless) rectangular box with a square base. The box is to contain a volume of 96 cubic feet and the cost of the material for the bottom is $3 per square foot. The material for the sides costs $1 per square foot. Find the minimum cost of making the box.

22. An open rectangular box with a square base is to be made with 48 square feet of material. Find the dimensions of the box with the greatest possible volume.

23. A chemical manufacturer wants to produce a closed cylindrical can with a vol-

ume of 1024 cubic centimeters. What dimensions will minimize the surface area of the can? (*Hint:* Let r be the radius of the can and let h be its height. The volume V of the can is given by $V = \pi r^2 h$ and its surface area S by $S = 2\pi r^2 + 2\pi rh$.)

24. A producer of chemicals who makes a cylindrical steel can with a volume of 36 cubic inches, finds that it costs 6 cents per square inch for the top and bottom and 4 cents per square inch for the side. Find the minimal cost of producing a can. (*Hint:* See the *hint* in Exercise 23.)

25. A printer has to print a rectangular sheet containing 128 square inches of printed matter with a 2-inch margin at the top and bottom and a 1-inch margin along each side. Find the dimensions of the sheet that will minimize the total area of the sheet. (*Hint:* Refer to Figure 8.)

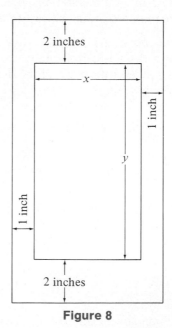

2 inches

—x—

1 inch

1 inch

y

2 inches

Figure 8

26. Suppose that it costs $6 + x/100$ dollars per mile for gasoline, maintenance, and depreciation to drive a moving van at the average rate of x miles per hour. If the driver's salary (including overhead and

fringe benefits) is $25 per hour, what driving speed will minimize the cost per mile?

27. A utility company is planning to run a power line from a generator located at point A (Figure 9) on a straight shoreline to an oil rig located at point B, which is 2 miles from the closest shore point C. The point C is 8 miles down the coast from A. The line is to run along the shore from A to a point D between A and C and from D directly to the oil rig at point B. If it costs $1000 per mile to run cable under water and $800 per mile under ground, where should the point D be located to minimize cost?

Figure 9

4.8 ‖‖‖‖‖ TANGENT LINE APPROXIMATIONS AND THE DIFFERENTIAL

In this section we shall see how to use the tangent line to the graph of f at x_0 to approximate the value of $f(x)$ when x is near x_0. We shall find this technique useful in estimating the change $\Delta f = f(x_0 + \Delta x) - f(x_0)$ in a function f, as x changes from x_0 to $x_0 + \Delta x$. We shall also define another useful concept, called *the differential*. Suppose that f is a function that is differentiable at $x = x_0$, so that

$$f'(x_0) = \lim_{h \to 0} \frac{f(x_0 + h) - f(x_0)}{h}.$$

It then follows that we can approximate $f'(x_0)$ by

$$f'(x_0) \approx \frac{f(x_0 + h) - f(x_0)}{h},$$

so

$$f'(x_0)h \approx f(x_0 + h) - f(x_0).$$

Hence, we can approximate $f(x_0 + h)$ as

$$f(x_0 + h) \approx f(x_0) + f'(x_0)h. \tag{1}$$

Recall from Section 3.1 that the equation of the tangent line to the graph of f at $(x_0, f(x_0))$ is

$$y = f(x_0) + f'(x_0)(x - x_0). \tag{2}$$

Thus, when $x = x_0 + h$, the corresponding value of y on the tangent line, obtained from (2), is

$$y = f(x_0) + f'(x_0)h. \tag{3}$$

Note that the right sides of Equations (1) and (3) are identical. This means that when $x = x_0 + h$, the approximate corresponding y-coordinate on the graph of f given by (1) is the y-coordinate at $x = x_0 + h$ on the line tangent to f at $(x_0, f(x_0))$. Thus, the approximation of $f(x_0 + h)$ given by (1) is called the **tangent line approximation.** Figure 1 is its graphic representation.

Figure 1

EXAMPLE 1   Approximate the value of $\sqrt{17}$.

Solution   Since we are approximating a square root, we let $f(x) = \sqrt{x}$. We want to find $f(17)$. Since

$$f(17) = f(16 + 1),$$

and we know $f(16) = \sqrt{16} = 4$, we take $x_0 = 16$ and $h = 1$, and apply the tangent line approximation (1). First,

$$f'(x) = \frac{1}{2\sqrt{x}},$$

so

$$f'(x_0) = f'(16) = \frac{1}{2\sqrt{16}} = \frac{1}{8}.$$

Substituting $x_0 = 16$ and $h = 1$ in Equation (1), we obtain

$$\sqrt{17} = f(16 + 1)$$
$$\approx f(16) + f'(16)(1)$$
$$= \sqrt{16} + \tfrac{1}{8}(1) = 4 + \tfrac{1}{8} = 4.125.$$

Hence, using the tangent line approximation, we estimate that

$$\sqrt{17} \approx 4.125.$$

The exact square root rounded to six decimal places is 4.123106. Thus, the error incurred by using the tangent line approximation is (to six decimal places)

$$4.125000 - 4.123106 = 0.001894.$$ ∎

Warning   When we use the tangent line approximation to estimate $f(x_0 + h)$, it is important that h be relatively small. As Figure 1 shows, we can expect the tangent line to closely approximate the curve only when $x_0 + h$ is close to x_0.
   Suppose that in Example 1 we had selected $f(x) = \sqrt{x}$, $x_0 = 1$, and $h = 16$. Then from (1) we would have

$$\sqrt{17} \approx f(1) + f'(1) \cdot 16 = \sqrt{1} + \frac{1}{2\sqrt{1}} \cdot 16,$$

or

$$\sqrt{17} \approx 9.$$

Why is this a poor approximation?

EXAMPLE 2   Approximate the value of $\sqrt[3]{26.5}$.

Solution   Observe that $\sqrt[3]{26.5} = \sqrt[3]{27 - 0.5}$ and that $\sqrt[3]{27} = 3$. Thus we take

$$f(x) = \sqrt[3]{x}, \qquad x_0 = 27, \quad \text{and} \quad h = -0.5.$$

Then

$$f'(x) = \frac{1}{3} x^{-2/3} = \frac{1}{3(\sqrt[3]{x})^2}$$

and

$$f'(27) = \frac{1}{3(\sqrt[3]{27})^2} = \frac{1}{3(3^2)} = \frac{1}{27}.$$

Substituting $x_0 = 27$ and $h = -0.5$ in Equation (1), we obtain

$$\begin{aligned}
\sqrt[3]{26.5} = f(27 - 0.5) &= f(x_0 + h) \\
&\approx f(x_0) + f'(x_0)h \\
&= f(27) + f'(27)(-0.5) \\
&= 3 + \frac{1}{27}(-0.5) = 2.981481.
\end{aligned}$$

Hence, using the tangent line approximation, we estimate that

$$\sqrt[3]{26.5} \approx 2.981481.$$

The exact cube root rounded to six decimal places is 2.981366. Thus, the error incurred by using the tangent line approximation is (to six decimal places)

$$2.981481 - 2.981366 = 0.000115. \qquad \blacksquare$$

Estimating Changes in a Function

When the input variable in a function $f(x)$ changes from $x = x_0$ to $x = x_0 + h$, the change in x is

$$\Delta x = (x_0 + h) - x_0 = h,$$

and the corresponding change in the function is

$$\Delta f = f(x_0 + h) - f(x_0).$$

Equation (1) tells us that for a differentiable function f, we can approximate the change in f by

$$\Delta f \approx f'(x_0)h. \qquad (4)$$

Differential Notation

The number $f'(x_0)h$ in Equation (4) is called the **differential of f at x_0** and is denoted by df:

$$df = f'(x_0)h. \qquad (5)$$

Observe that df is a function of both x_0 and the increment h.

When f is the function defined by $f(x) = x$, then df is denoted by dx. Since $f'(x_0) = 1$, we have

$$df = f'(x_0)h = 1h,$$

and thus

$$dx = h.$$

In summary, for a differentiable function $f(x)$, when x changes from x_0 to $x_0 + h$,

$$dx = \Delta x = h, \text{ the change in } x, \qquad (6)$$

and

$$df = f'(x_0)\, dx, \qquad (7)$$

which *approximately* equals the change in f,

$$\Delta f = f(x_0 + h) - f(x_0).$$

The differential of a function f is the derivative of f times the differential of its input variable.

In general, we write $y = f(x)$, and denote df by dy. Thus, we can write Equation (7) as

$$dy = f'(x_0)\,dx. \qquad (8)$$

This equation looks familiar to us because we already know that

$$\frac{dy}{dx} = f'(x_0).$$

Now, however, we can interpret the symbol dy/dx as a quotient of the differential dy by the differential dx. The geometric significance of Equation (8) is shown in Figure 2, where we see that $dx = PR$ and $dy = RS$. Hence the ratio $dy/dx = RS/PR$ is the slope $f'(x_0)$ of the tangent line to the graph of f at the point $P(x_0, f(x_0))$.

Figure 2

When computing the differential dy from the derivative dy/dx, we can conveniently rewrite Equation (8) by suppressing the subscript:

$$dy = f'(x)\,dx.$$

As an example, when $y = x^5$

$$\frac{dy}{dx} = 5x^4.$$

Multiplying both sides by dx, we obtain

$$dy = 5x^4\,dx.$$

EXAMPLE 3   If $y = \sqrt{x^2 + x}$, find dy.

Solution   We have

$$\frac{dy}{dx} = \frac{1}{2}(x^2 + x)^{-1/2}(2x + 1)$$

$$= \frac{2x + 1}{2\sqrt{x^2 + x}}.$$

Hence,

$$dy = \frac{2x + 1}{2\sqrt{x^2 + x}}\, dx.$$ ∎

|||||||||||| **Applications**

EXAMPLE 4
(Business)

A computer manufacturer has determined that when he spends x million dollars per year on research, his annual revenue is given (in millions of dollars) by

$$R(x) = 3 + 20x - x^2, \qquad 2 \le x \le 10.$$

Use differentials to approximate the increase in annual revenue when the annual research expenditure is increased from \$3 to \$3.1 million.

Solution

We have

$$\frac{dR}{dx} = 20 - 2x,$$

so

$$dR = (20 - 2x)\, dx.$$

We are given $x_0 = 3$ and $dx = 0.1$. Thus,

$$dR = (20 - (2 \cdot 3))(0.1)$$
$$= 1.4.$$

The increase in annual revenue is approximately \$1.4 million. ∎

EXAMPLE 5
(Sociology)

A sociologist has determined that when w million dollars are spent annually on work-incentive programs in a certain city, the annual amount spent on the prison system is given (in millions of dollars) by

$$P(w) = 16 - w^{2/3}, \qquad 6 \le w \le 27.$$

Use differentials to approximate the decrease in the amount spent annually on the city's prison system when the amount spent annually on the work-incentive program is increased from \$8 to \$8.5 million.

Solution

We have

$$\frac{dP}{dw} = -\frac{2}{3}w^{-1/3} = -\frac{2}{3w^{1/3}}.$$

So

$$dP = -\frac{2}{3w^{1/3}}\, dw.$$

We are given $w_0 = 8$ and $dw = 0.5$. Therefore

$$dP = -\frac{2}{3(8)^{1/3}}(0.5) = -\frac{1}{6}.$$

Thus, the amount spent annually on the city's prison system will decrease by 1/6 million dollars. ∎

|||||||||||| Error Estimation

Suppose that x_0 is a quantity that we want to measure and that $y = f(x)$ is a function that we will use to compute a corresponding value $f(x_0)$. Since measurements are seldom exact, suppose that when we attempt to measure x_0 we actually obtain $x_0 + dx$; that is, we make an error dx in measuring x_0. Then the computed quantity is $f(x_0 + dx)$ instead of $f(x_0)$. Thus, the error resulting in the computed quantity is $f(x_0 + dx) - f(x_0)$. We can approximate this error by using Equation (1):

$$f(x_0 + dx) - f(x_0) \approx f'(x_0)\,dx = dy.$$

(See also Figure 2.)

EXAMPLE 6 The side of a cube is measured as 20 centimeters with a possible error of 0.02 centimeters. What error might result when we compute the volume of the cube?

Solution Let $s =$ side length of the cube. Then the volume is given by

$$V(s) = s^3.$$

We have

$$\frac{dV}{ds} = V'(s) = 3s^2,$$

so

$$dV = 3s^2\,ds.$$

We are given $s_0 = 20$ and $ds = \pm 0.02$. Therefore,

$$dV = 3(20)^2(\pm 0.02)$$
$$= \pm 24.$$

Using the measured length $s_0 = 20$ centimeters, we find that

$$V = (20)^3 = 8000 \text{ cubic centimeters.}$$

The possible error in this computed volume is approximately 24 cubic centimeters. ∎

4.8 EXERCISE SET ||||||||||||||||||||||||||||||||||||||||||||||||||||||||||||||||||||||||||||||||

In Exercises 1–8, use the tangent-line approximation to approximate the given number.

1. $\sqrt{37}$ 2. $\sqrt{48.4}$ 3. $\sqrt[3]{29}$

4. $(63)^{2/3}$ 5. $\sqrt[5]{31.2}$ 6. $(2.02)^5$

7. $(15.5)^{-1/4}$ 8. $\dfrac{1}{1 + (1.023)^3}$

In Exercises 9–12, compute the differential df for the given values of x_0 and dx.

9. $f(x) = \sqrt{x}$, $x_0 = 9$, $dx = -0.3$

10. $f(x) = \sqrt[3]{x^2}$, $x_0 = 8$, $dx = 0.03$

11. $f(x) = \sqrt{1 + 2x^2}$, $x_0 = 2$, $dx = 0.2$

12. $f(x) = \dfrac{1}{1 + x^2}$, $x_0 = 3$, $dx = -0.02$

In Exercises 13–20, compute the differential dy.

13. $y = 2x + 3$

14. $y = 2x^2 - 3x$

15. $y = 2x^3 - 3x^2 + 5x + 2$

16. $y = \sqrt[3]{1 + x^2}$

17. $y = \dfrac{x - 1}{x + 1}$

18. $y = \dfrac{1}{(3x^2 + 2x - 1)^3}$

19. $y = (x^2 + 1)^5$

20. $y = (x^2 + 2)(x^2 - 3x)$

21. An ice cube measuring 3 centimeters on all sides melts down uniformly until each side measures 2.96 centimeters. Use differentials to estimate the decrease in volume.

22. **(Business)** A manufacturer of dolls finds that the total weekly cost of manufacturing x dolls (in dollars) is given by

$$C(x) = 0.01x^2 + 30x + 5000.$$

Use differentials to estimate the change in the total manufacturing cost when the weekly level of production changes from 200 to 210 dolls.

23. **(Medicine)** The volume of a spherical cancer tumor is given by

$$V = \tfrac{4}{3}\pi r^3,$$

where r is the radius. Radiation therapy reduces the radius from 2 centimeters to 1.8 centimeter. Use differentials to approximate the reduction in volume (use $\pi \approx 3.1416$).

24. **(Educational Psychology)** Suppose that the number $N(t)$ of words of Latin learned by the average student after t weeks of study is approximately given by

$$N(t) = 20t^{3/2}, \qquad 0 \le t \le 40.$$

Use differentials to approximate the increase in the number of words learned between weeks 16 and 17.

25. **(Business)** A manufacturer of refrigerators finds that the monthly cost and revenue functions (in dollars) are given by

$$C(x) = 40x + 8000$$
$$R(x) = 100x - \frac{x^2}{50} \qquad 0 \le x \le 5000,$$

where x is the number of refrigerators made per month. Find the approximate changes in revenue and profit if the monthly level of production changes from
(a) 2000 to 2010 refrigerators.
(b) 3000 to 2980 refrigerators.

26. **(Business)** A manufacturer of fertilizer finds that the annual demand in tons, when the price is x dollars per ton, is

$$D(x) = 2100 - \frac{x}{5} - \frac{x^2}{40}.$$

Use differentials to estimate the change in annual demand when the price changes

from
(a) $120 to $150 per ton.
(b) $140 to $130 per ton.

27. **(Medicine)** A physician measures the radius of a circular wound as 2 centimeters, with a possible error of 0.02 centimeters. Use differentials to estimate the possible error in the area of the wound. (*Hint:* The area A of a circle is $A = \pi r^2$; use $\pi \approx 3.1416$.)

28. A cube-shaped cargo container is to hold 1000 cubic feet. What is the possible error in the side of the cube if the possible error in volume cannot be more than 3 cubic feet.

29. **(Taxation)** A certain state's monthly income tax $T(x)$ (in dollars) is given by

$$T(x) = \sqrt{x},$$

where x is the monthly income in dollars. Use differentials to estimate the additional monthly tax due when a person's monthly salary increases from $1600 to $1650.

30. A hollow glass globe, made of quarter-inch-thick glass, is a sphere with an inner radius of 8 inches. Estimate the amount of glass, by volume, in the globe. (Recall that the volume of a sphere of radius r is $V = \frac{4}{3}\pi r^3$.)

4.9 |||||||| THE NEWTON-RAPHSON ALGORITHM

Applied mathematical problems often require that we solve an equation

$$f(x) = 0, \tag{1}$$

when f is some given function. If f is a first degree polynomial, solving for x is trivial; if f is a second degree polynomial we may use the quadratic formula. If f is a higher degree polynomial, a rational function, or a more complicated function, we may not have a way of solving Equation (1). The Newton-Raphson algorithm is a procedure that usually enables us to obtain an approximate solution, with excellent accuracy. The idea behind the algorithm is geometric, and makes use of tangent lines to the curve $y = f(x)$.

The problem of solving Equation (1) may be viewed geometrically as the problem of finding the x-intercepts of the curve $y = f(x)$ (See Figure 1).

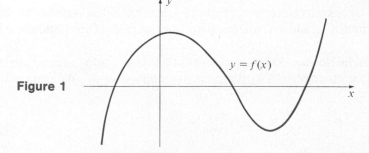

Figure 1

There may be several solutions to (1), but we shall focus our attention on finding *one* of them.

The first step in the Newton-Raphson algorithm is to make a "first guess" x_1. This is usually an integer that seems to be reasonably close to a solution. We then write the equation of the line tangent to the curve $y = f(x)$ at the point $(x_1, f(x_1))$:

$$y - f(x_1) = f'(x_1)(x - x_1). \tag{2}$$

We really want to know where the curve $y = f(x)$ crosses the x-axis, but since the curve and the tangent line are reasonably close for values of x close to x_1, we settle instead for finding where the *tangent line* intersects the x-axis (see Figure 2).

Figure 2

By Equation (2), that number x_2 must satisfy

$$0 - f(x_1) = f'(x_1)(x_2 - x_1)$$

or

$$\frac{-f(x_1)}{f'(x_1)} = x_2 - x_1,$$

so that

$$x_2 = x_1 - \frac{f(x_1)}{f'(x_1)}. \tag{3}$$

Equation (3) tells us how to calculate x_2. Figure 2 suggests that x_2 is closer to being a solution to (1) than x_1 is. Indeed, this happens in most cases. (See Example 4 for an exception.)

The next step in the Newton-Raphson algorithm is to use x_2 as a "second guess," and apply the same procedure to x_2 that was just applied to x_1. We thus obtain a "third guess" x_3:

$$x_3 = x_2 - \frac{f(x_2)}{f'(x_2)}. \tag{4}$$

(See Figure 3a.) We continue, finding a "fourth guess" x_4 (See Figure 3b), and so on until we are satisfied with the accuracy of the results.

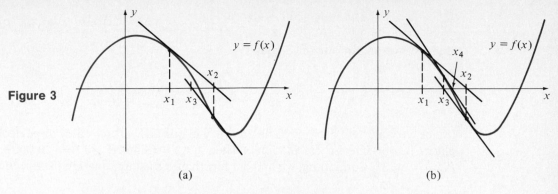

Figure 3

(a)   (b)

The general step in the Newton-Raphson algorithm is this:

> After the "nth guess" x_n has been determined, the $(n + 1)$th guess is
>
> $$x_{n+1} = x_n - \frac{f(x_n)}{f'(x_n)}. \tag{5}$$

We anticipate that the successive values of x_n will get closer and closer. When this happens, we terminate the procedure as soon as the value of x_{n+1} agrees with the value of x_n to the desired degree of accuracy. If this does not happen, either the first guess must be revised or the Newton-Raphson algorithm does not work with the function f.

In general, it will be necessary to use a calculator in working the examples and exercises in this section.

EXAMPLE 1   Find $\sqrt{12}$ to the nearest hundredth, using the Newton-Raphson algorithm.

Solution   We want to find $x = \sqrt{12}$. Thus, we want to solve $x^2 = 12$, or

$$x^2 - 12 = 0. \tag{6}$$

We let $f(x) = x^2 - 12$ and try to solve the equation $f(x) = 0$. Of course, $f'(x) = 2x$. A reasonable first guess for a solution to (6) is

$$x_1 = 3.$$

Then $f(x_1) = 9 - 12 = -3$, and $f'(x_1) = 2(3) = 6$. Applying Equation (3), we obtain our second guess,

$$x_2 = 3 - \frac{-3}{6} = \frac{7}{2} = 3.500.$$

Then $f(x_2) = \left(\dfrac{7}{2}\right)^2 - 12 = \dfrac{49 - 48}{4} = \dfrac{1}{4}$, and $f'(x_2) = 2\left(\dfrac{7}{2}\right) = 7$. Applying Equation (4), we obtain our third guess,

$$x_3 = \frac{7}{2} - \frac{\frac{1}{4}}{7} = \frac{7}{2} - \frac{1}{28} = 3.464.$$

(Since we want two decimal place accuracy in our answer, we carry three decimal places in our interim calculations.) Then $f(x_3) = (3.464)^2 - 12 = -0.0007$ and $f'(x_3) = 6.928$. Continuing, we find our fourth guess using Equation (5) with $n = 3$:

$$x_4 = 3.464 - \frac{-0.0007}{6.928} = 3.464.$$

Since x_3 and x_4 agree to the nearest thousandth, we are satisfied that we have found $\sqrt{12}$ to the desired accuracy. Thus,

$$\sqrt{12} \approx 3.46 \text{ (to the nearest hundredth).} \qquad \blacksquare$$

Locating Zeros of Continuous Functions

A value of x which makes $f(x)$ equal to 0 is called a *zero* of the function f. Thus, in attempting to solve the equation

$$f(x) = 0,$$

we are seeking all the zeros of f. The Newton-Raphson algorithm is a procedure for approximating the zeros of f to a specified degree of accuracy. One part of the algorithm that we have not discussed is the very first part: how to make a wise "first choice." We now turn our attention to that concern.

The following theorem describes an important property of continuous functions that is very helpful in making an educated first guess.

THEOREM

> **(The intermediate value property of continuous functions.)**
> Suppose f is continuous on an interval I containing two numbers, a and b. If $f(a)$ and $f(b)$ have opposite signs, then there is some number c between a and b such that
>
> $$f(c) = 0.$$

We shall not attempt to prove this theorem, but its plausibility is illustrated in Figure 4. The basic idea is that the graph of a continuous function cannot pass from a point on one side of the x-axis to a point on the other side without intersecting the x-axis at some intermediate point.

If we find two successive integers such that a continuous function f is positive at one integer and negative at the other, then, by the above theorem, we can be

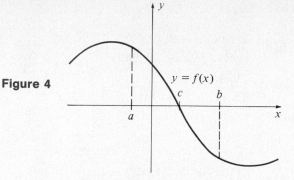

Figure 4

sure that there is a zero of f between these two integers. Either one of these integers is a wise first choice, x_1, for the Newton-Raphson algorithm.

EXAMPLE 2   Find a positive solution of the equation

$$2x^3 + 4x^2 = 7$$

to the nearest hundredth.

Solution   Let $f(x) = 2x^3 + 4x^2 - 7$; we seek a positive number that solves the equation $f(x) = 0$. We first calculate $f(0) = -7$, $f(1) = -1$, and $f(2) = 16 + 16 - 7 = 25$. Since $f(x)$ is negative at $x = 1$ and positive at $x = 2$, the intermediate value property assures us that there is a zero of f somewhere between $x = 1$ and $x = 2$. The above calculations suggest that the zero is closer to 1 than to 2. Thus, we let $x_1 = 1$ and apply the Newton-Raphson algorithm.

Now $f'(x) = 6x^2 + 8x$. Thus, Equation (5) becomes

$$x_{n+1} = x_n - \frac{2x_n^3 + 4x_n^2 - 7}{6x_n^2 + 8x_n}. \tag{7}$$

We first find x_2. Since $x_1 = 1$, we obtain

$$x_2 = 1 - \frac{2 + 4 - 7}{6 + 8} = 1 - \frac{-1}{14} = 1.071,$$

to the nearest thousandth. (In our calculations, we keep one more decimal place than specified in the problem.) We next use this value of x_2 in Equation (7) to find x_3:

$$x_3 = 1.071 - \frac{2(1.071)^3 + 4(1.071)^2 - 7}{6(1.071)^2 + 8(1.071)}$$

$$= 1.071 - \frac{.045}{15.45}$$

$$= 1.069.$$

We again use Equation (7) to find x_4:

$$x_4 = 1.069 - \frac{2(1.069)^2 + 4(1.069) - 7}{6(1.069)^2 + 8(1.069)}$$

$$= 1.069 - \frac{.014}{15.41}$$

$$= 1.068.$$

Computing x_5, we obtain

$$x_5 = 1.068 - \frac{2(1.068)^3 + 4(1.068) - 7}{6(1.068)^2 + 8(1.068)}$$

$$= 1.068 - \frac{-.001}{15.39}$$

$$= 1.068.$$

Since x_4 and x_5 agree to three decimal places, we conclude that our answer is reliable to at least the first two decimal places. Thus,

$$x = 1.07, \text{ to the nearest hundredth.}\qquad \blacksquare$$

EXAMPLE 3
(Educational
Psychology)

A certain standard test to measure students' performance in learning a foreign language indicates that after t weeks of study the average person's raw score $P(t)$ on the test is

$$P(t) = 12t^2 - t^3, \qquad 0 \le t \le 8.$$

(See Section 4.4, Exercise 31.) After how many days of study does the average person score 200 or higher?

Solution

We shall use the Newton-Raphson algorithm to solve the equation

$$12t^2 - t^3 = 200.$$

Since we want the answer to the nearest day, we shall be satisfied with an approximate solution to the nearest tenth of a week. Thus, we want to solve

$$f(t) = 0$$

to the nearest tenth, where $f(t) = 12t^2 - t^3 - 200$.
    Since $f'(t) = 24t - 3t^2$, Equation (5) becomes

$$t_{n+1} = t_n - \frac{12t_n^2 - t_n^3 - 200}{24t_n - 3t_n^2}. \tag{8}$$

Note that

$$f(5) = 12(25) - 125 - 200 = -25 < 0$$

and

$$f(6) = 12(36) - 216 - 200 = 16 > 0.$$

Thus, by the intermediate value property, f has a zero between $t = 5$ and $t = 6$. We take

$$t_1 = 5.$$

Using Equation (8) we obtain

$$t_2 = 5 - \frac{-25}{45} = 5.56.$$

Again, using Equation (8),

$$t_3 = 5.56 - \frac{-.92}{40.70} = 5.58.$$

Since t_2 and t_3 both round off the same, to the nearest tenth, the solution we are seeking is

$$t = 5.6 \text{ (weeks)}.$$

Converting $t = 5.6$ weeks to days, we obtain

$$t = 39.2 \text{ days}.$$

Thus, by the 40th day, the average person scores 200 or higher.   ■

EXAMPLE 4   Figure 5 shows the graph of a function f for which the Newton-Raphson algorithm fails.

Figure 5

In this case, we see that x_3 is *farther* from the zero of f than is x_1. When this happens we cannot use the Newton-Raphson algorithm. This situation can be detected by observing that the successive values of x_n $(n = 1, 2, 3, 4, \ldots)$ get farther apart rather than closer.

4.9 EXERCISE SET ||||||||||||||||||||||||||||||||||||||||||||||||||||||||||||||||||||||||||||||||||||

A calculator will be necessary for working these exercises.

In Exercises 1–6, use the Newton-Raphson algorithm to find the given root to the nearest hundredth.

1. $\sqrt{5}$

2. $\sqrt{30}$

3. $\sqrt[3]{20}$

4. $\sqrt[3]{100}$

5. $\sqrt[4]{100}$

6. $\sqrt[5]{500}$

In Exercises 7–14, use the Newton-Raphson algorithm to solve the given equation, under the given condition. First use the intermediate value property to locate the solution between successive integers. Estimate the solution to the nearest hundredth.

7. $x^3 - x - 5 = 0$, $x \geq 0$

8. $x^3 + 2x - 100 = 0$, $x \geq 0$

9. $x + 1 = \sqrt{x + 2}$, $x \geq 0$

10. $\sqrt{x + 3} = x^2$, $x \geq 0$

11. $3x^4 + 5x^3 = 200$, $x \geq 0$

12. $x^4 - 2x^3 + x^2 = 7$, $x \geq 2$

13. $\dfrac{1}{2x + 3} = x^2 + 1$, $x < 0$

14. $\dfrac{1}{\sqrt{1 - x}} = x^2$, $-1 < x < 1$

15. **(Motion)** An object is moving along a straight line in such a way that after t seconds ($0 \leq t \leq 10$) it has traveled

$$s = 10t^2 - t^3 \text{ feet.}$$

How long does it take to travel 80 feet (to the nearest tenth of a second)?

16. **(Business)** A chemical firm finds that the cost (in dollars) of producing x gallons of a special paint solvent weekly is

$$C(x) = 500 + 30x + 12\sqrt{x}.$$

How may gallons can be produced weekly, if the manufacturer must keep the weekly cost at $1800 or less (to the nearest hundredth of a gallon)?

17. **(Medicine)** One hour after x milligrams of a certain drug is administered, a patient's body temperature will change by

$$\Delta T = x^2\left(1 - \frac{x}{25}\right)$$

degrees Fahrenheit. How many milligrams must be administered to reduce the patient's temperature $2°F$ within one hour?

8. Consider the rectangular box described in Example 2 of Section 4.7. To the nearest tenth of an inch, how much must be cut out of each corner (that is, find x) so that the volume of the resulting box is exactly 80 cubic inches?

KEY IDEAS FOR REVIEW ||||||||||||||||||||||||||||||||||||||||||||||||||||||||||||||||||||||||||||

■ If $f'(x) > 0$ on (a, b), then f is increasing on (a, b); if $f'(x) < 0$, then f is decreasing on (a, b).

■ f has a relative maximum at $x = c$ if there is an open interval containing c such that for any x in this interval, $f(c) \geq f(x)$; f has a relative minimum at $x = c$ if there is an open interval containing c such that for any x in this interval, $f(c) \leq f(x)$. The value $f(c)$ is the relative extreme value.

■ If f is differentiable at c and f has a relative extremum at c, then $f'(c) = 0$.

■ If f is defined at c and either $f'(c) = 0$ or $f'(c)$ does not exist, then c is called a critical number of f.

■ The critical numbers of f give candidates for the values of x at which f may have relative extrema.

■ First derivative test: If the sign of f' changes from plus to minus at c, then f has a relative maximum at c; if the sign of f' changes from minus to plus at c, then f has a relative minimum at c.

- Extreme value theorem: If f is continuous on $[a, b]$, then f has an absolute maximum and an absolute minimum on $[a, b]$.
- To find the absolute extrema of a continuous function f over a closed interval, check the values of f at the critical numbers of f in the interval and at the endpoints of the interval.
- If $f''(x) > 0$ for all x in (a, b), then the curve $y = f(x)$ is concave upward over (a, b). If $f''(x) < 0$, then $y = f(x)$ is concave downward over (a, b).
- A point $(c, f(c))$ on a curve $y = f(x)$ where the curve changes from concave upward to concave downward or vice versa is called an inflection point of the curve.
- Second derivative test: Let c be a critical number of f such that $f'(c) = 0$. If $f''(c) > 0$, then f has a relative minimum at c. If $f''(c) < 0$, then f has a relative maximum at c.
- When the extreme value theorem is not applicable, the following theorem is useful in solving optimization problems in which the function has only one critical number in the interval. Theorem: Suppose f is continuous on the interval I, and I contains only one critical number c of f. If f has a relative maximum (relative minimum) at c, then $f(c)$, is the absolute maximum (absolute minimum) value of f on I.

- The basic procedures for sketching a continuous curve $y = f(x)$ are summarized in Table 1 of Section 4.5.
- If f is differentiable at $x = x_0$ and h is small, then

$$f(x_0 + h) \approx f(x_0) + f'(x_0)h.$$

- $df = f'(x_0)\, dx$, and if $y = f(x)$, $dy = f'(x_0)\, dx$.
- If $y = f(x)$, then

$$\Delta x = dx \qquad \text{and}$$
$$\Delta f = f(x_0 + dx) - f(x_0) \approx f'(x_0)\, dx = dy.$$

- If a function f is continuous on $[a, b]$ and has opposite signs at a and b, then it must have a zero somewhere in the interval (a, b).
- In the Newton-Raphson algorithm,

$$x_{n+1} = x_n - \frac{f(x_n)}{f'(x_n)}.$$

- The x-intercepts (where $y = 0$) and y-intercepts (where $x = 0$) are often important points of a curve to be sketched. Excluded regions and symmetry are also helpful.
- The points of discontinuity of an otherwise continuous function f separate the graph of f into separate "continuous" pieces.
- If $\lim\limits_{x \to \pm\infty} f(x)$ is a finite number, then the graph of f has a horizontal asymptote. If $\lim\limits_{x \to a+} f(x)$ or $\lim\limits_{x \to a-} f(x)$ is infinite, then the graph of f has a vertical asymptote at $x = a$.

REVIEW EXERCISES ||||||||||||||||||||||||||||||||||||||||||||||||||||||||||||||||||||||||||||||||||

Exercises 1–8 refer to the following figure.

1. Find all open intervals where f is increasing.

2. Find all open intervals where $f'(x) < 0$.

3. For what values of x does f have a relative maximum?

4. For what values of x does f have a relative minimum?

5. On what open intervals is the curve concave upward?

6. On what open intervals is $f''(x) < 0$?

7. Find all inflection points of f over (x_1, x_9).

8. For what values of x does the function have an absolute maximum and absolute minimum on the given interval?
   (a) $[x_1, x_5]$ (c) $[x_5, x_6]$
   (b) $[x_2, x_7]$ (d) $[x_4, x_9]$

In Exercises 9–12, determine where the given function is increasing and where it is decreasing. Also, for what values of x, if any, does the graph of f have a horizontal tangent?

9. $f(x) = 5 - 2x$

10. $f(x) = x - 2x^2$

11. $f(x) = 2x^3 - x^4$

12. $f(x) = \dfrac{x}{x - 2}$

In Exercises 13–16, locate the critical numbers of f and classify each as a relative maximum, relative minimum, or neither, based on the first derivative test.

13. $f(x) = 2x - x^2$

14. $f(x) = 9x - x^3$

15. $f(x) = (x - 2)^{1/3}$

16. $f(x) = \dfrac{x}{2x - 3}$

In Exercises 17–20, locate the absolute maximum and absolute minimum for each function on the given interval. If an absolute maximum or an absolute minimum does not exist, say so.

17. $f(x) = 2x - x^2$, $[-3, 2]$

18. $f(x) = x^3 - 12x + 8$, $[-4, 3]$

19. $f(x) = 3x^{1/3} - 4x$, $[0, 8]$

20. $f(x) = \dfrac{2}{x - 1}$, $[-2, 2]$

In Exercises 21–24, find the inflection points and determine where the given curve is concave upward and where it is concave downward.

21. $f(x) = x^2 + 5x - 4$

22. $f(x) = x^3 - x^2 - x + 3$

23. $f(x) = 2x^3 + 3x^2 - 12x + 3$

24. $f(x) = \dfrac{1}{x^2}$

In Exercises 25–28, find the relative extrema of f by using the second derivative test.

25. $f(x) = x^3 - 12x + 8$

26. $f(x) = \dfrac{1}{x^2 - 1}$

27. $f(x) = \dfrac{3}{2x + 1}$

28. $f(x) = \sqrt{x - 1}$

In Exercises 29–36, sketch the graph of each function.

29. $f(x) = (x + 1)^3$

30. $f(x) = \frac{1}{3}x^3 + \frac{1}{2}x^2 - 3$

31. $f(x) = \frac{1}{2}x^4 - 4x^2$

32. $f(x) = 2x^{3/2}$

33. $f(x) = \sqrt{x + 4}$ 34. $f(x) = \dfrac{1}{(x - 1)^2}$

35. $f(x) = \dfrac{1}{x^2 - 1}$ 36. $f(x) = \dfrac{x}{x^2 + 1}$

37. Use differentials to approximate
    (a) $\sqrt{48.3}$ (b) $(27.04)^{2/3}$

38. If $f(x) = 3\sqrt{x} - x$, $x_0 = 4$, and $dx = 0.2$, compute df.

In Exercises 39 and 40, compute dy.

39. $y = x^3 - x$ 40. $y = \sqrt{1 + x^3}$

41. A manufacturer of typewriters finds that the profit (in dollars) received from the

manufacture and sale of x typewriters per week is given by

$$P(x) = -\frac{x^2}{5} + 3000x - 180.$$

(a) For what levels of production is the profit increasing? For what level is it decreasing?

(b) For what levels of production is the profit neither increasing nor decreasing?

42. **(Medicine)** The concentration k of a drug in the bloodstream t hours after it has been administered is given by $k(t) = \dfrac{t}{t^2 + 3}$.

(a) When is the concentration of the drug increasing?

(b) When is the concentration of the drug decreasing?

43. **(Business)** A manufacturer of calculators finds that the profit (in millions of dollars) received from the manufacture and sale of x thousand calculators per week is given by

$$P(x) = \frac{x^4}{12} - \frac{7}{6}x^3 + 6x^2.$$

Find the level(s) of production for which the marginal profit is increasing.

44. A manufacturer of CB receivers finds that the weekly profit $P(x)$ (in dollars) is given by

$$P(x) = -120x^2 + 6000x - 2000$$

where x is the number of receivers made and sold each week. If production is limited to 100 receivers per week, how many receivers should be made each week to maximize the profit?

45. An appliance dealer expects to sell 2500 stoves during the year at a fairly constant rate. The annual holding cost is $10 per

stove and the dealer can reorder any number of stoves for a flat reorder fee of $20 per lot. How many times per year and in what lot size should the dealer reorder in order to minimize the total annual inventory cost?

46. A tool rental shop finds that it can rent 30 floor sanders per day at a daily rental of $12 per sander. For each additional $2 a day charged per sander, one less sander is rented. If the shop wants to maximize its daily revenue, what should it charge to rent a sander for one day?

47. An oil producer designs a closed barrel in the form of a right circular cylinder whose capacity is 32 cubic feet. Find the dimensions that will minimize the amount of material required to manufacture the barrel. (*Hint:* Refer to figure below. The area of the side is $2\pi rh$ and the total area of the top and bottom is $2\pi r^2$.)

48. A square measuring 5 centimeters on each side is enlarged uniformly so that each side measures 5.02 centimeters. Use differentials to estimate the increase in area.

49. A spherical satellite that has a radius of 10 feet is to be coated with a plastic shield that is 0.005 feet thick. Use differentials to estimate the volume of material required. (*Hint:* Estimate the increase in volume when the radius of the sphere changes from 10 feet to 10.005 feet, and

recall that the volume V of a sphere of radius r is given by $V = \frac{4}{3}\pi r^3$.)

50. Use the Newton-Raphson algorithm to find $\sqrt[3]{50}$ to the nearest hundredth.

51. Use the Newton-Raphson algorithm to find a positive solution to the equation

$$2x^3 + x = 40,$$

to the nearest hundredth.

CHAPTER TEST ||||||||||||||||||||||||||||||||||||||||||||||||||||||||||||||||||||||||||||||||||

1. Consider the function $f(x) = x^3 + 3x^2 - 9x - 7$.
   (a) Find the interval(s) over which f is increasing.
   (b) Over what interval(s) is f decreasing?
   (c) At what point(s) (x, y) does the graph of f have a horizontal tangent line?
   (d) Use the first derivative test to determine the relative maxima and the relative minima of f.
   (e) Find the interval(s) over which the graph of f is concave upward.
   (f) Does the graph of f have any inflection points? If so, where?

2. Sketch the graph of the function given in 1 above.

3. Consider the function $g(x) = x + \dfrac{1}{x}$. Locate all critical numbers of g, and use the *second* derivative test to determine whether g has a relative maximum, a relative minimum, or neither, at those numbers.

4. Find the absolute maximum and the absolute minimum values of the function $f(x) = \dfrac{x}{3} - \sqrt[3]{x}$ on the interval $[0, 8]$.

5. For the function $y = \sqrt{5x^2 + 6}$, find the differential dy.

6. A producer finds that when the selling price is x dollars per ton, the demand for their commodity is

$$D(x) = 4500 + 10x - \frac{x^2}{50} \text{ tons per year.}$$

Use differentials to estimate the change in annual demand when the price changes from \$400 per ton to \$450 per ton.

7. Suppose that t hours after a drug is administered to a patient the concentration $k(t)$ of the drug in the patient's body is given by

$$k(t) = \frac{0.1t}{4t^2 + 3t + 9}.$$

How long after the drug has been administered is the concentration at its maximum?

8. A rectangular box with no top is to be made from a square piece of cardboard, 10 inches by 10 inches, by cutting out and discarding a square

from each corner and then folding up the corners, as shown here. Find the maximum volume that can be obtained.

9. Consider the function $f(x) = x^3 - x - 20$.
   (a) Show that there must be a number c between 2 and 3 such that $f(c) = 0$.
   (b) Let $x_1 = 3$ be a first approximation for this number c. Use the Newton-Raphson algorithm to obtain a second approximation x_2, and then a third approximation x_3.

10. Obtain the graph of the function $f(x) = \dfrac{x}{x^2 - x - 2}$. First, however,
   (a) determine the x and y intercept(s).
   (b) tell how many continuous pieces the graph will have, and why.
   (c) find (and shade) excluded regions, using a sign analysis.
   (d) find all vertical and horizontal asymptotes.
   (e) determine where the function is increasing and where it is decreasing.
   Finally, sketch the graph.

THE EXPONENTIAL AND LOGARITHMIC FUNCTIONS

So far, our study of the calculus has focused primarily on functions that are polynomials or rational powers of polynomials, or sums, differences, products, and quotients of these two types of functions. In this chapter we shall introduce two new functions, exponential and logarithmic functions, both of which occur in a wide variety of applications. After defining these functions, discussing their graphs, studying their basic properties, and finding their derivatives, we shall turn to applications in which the exponential function arises, including growth and decay problems.

5.1 |||||||| THE EXPONENTIAL FUNCTION

In Section A.4 (Appendix A) we reviewed the definition of a^n for any real number a and positive integer n:

$$a^n = \underbrace{a \cdot a \cdot \cdots \cdot a.}_{n \text{ factors}}$$

Then we defined a^{-n}, where n is a positive integer, and finally $a^{m/n}$, where m and n are integers. Thus, using the material in Section A.4 (Appendix A), we can raise the base a to any rational exponent.

We now want to work with exponents that are irrational numbers. The first obstacle to overcome is the problem of *defining* a^x, where x is an irrational number. The approach we use here is based on the idea of continuity. We shall see that the requirement that the exponential function $f(x) = a^x$ be continuous requires that it satisfy a certain definition.

Every real number has a decimal representation. For an irrational number x, that decimal representation is infinitely long, and is nonrepeating. Thus,

$$x = N.d_1 d_2 d_3 \ldots d_n \ldots$$

where N is some integer and the $d_1, d_2, d_3, \ldots,$ are "digits," 0, 1, 2, . . . , 9.

Note that each of the finite decimals

$$x_1 = N.d_1$$
$$x_2 = N.d_1d_2$$
$$x_3 = N.d_1d_2d_3$$
$$\vdots$$
$$x_n = N.d_1d_2d_3 \ldots d_n$$
$$\vdots$$

is a rational number ($x_1 = N + \dfrac{d_1}{10}$, $x_2 = N + \dfrac{d_1}{10} + \dfrac{d_2}{100}$, and so on). Moreover, each of the finite decimals x_n (1, 2, 3, . . .) may be regarded as an **approximation** to x. These approximations get closer and closer to x as n gets larger and larger. Using limit notation we can say

$$x = \lim_{n \to \infty} x_n.$$

We want to define a^x in such a way that the function $f(x) = a^x$ is continuous at each x. Thus, we want to have

$$\lim_{n \to \infty} f(x_n) = f\left(\lim_{n \to \infty} x_n \right)$$

or

$$\lim_{n \to \infty} a^{x_n} = f(x).$$

Therefore, our definition is

$$a^x = \lim_{n \to \infty} a^{x_n}. \qquad (1)$$

In terms of concrete applications, we are saying that to calculate a^x for an irrational number x we may approximate x by a finite decimal x_n and then calculate a^{x_n}. If x_n is a good approximation to x, then a^{x_n} is a good approximation to a^x.

For example, to calculate $1024^{\sqrt{2}}$, approximate $\sqrt{2}$ by $1.4 = 14/10$. Thus,

$$1024^{\sqrt{2}} \approx 1024^{14/10} = (\sqrt[10]{1024})^{14} = 2^{14} = 16{,}384.$$

We should note that this is not a very good approximation; a calculator yields $1024^{\sqrt{2}} \approx 18{,}080.358$. By improving our approximation of x, we improve the accuracy of our approximation of a^x. We should also point out that the actual procedure used by a calculator or computer to calculate a^x is somewhat different, and based on more advanced techniques of calculus.

An **exponential function** f is defined by the equation

$$f(x) = a^x \qquad a > 0, a \neq 1, \tag{2}$$

where a is a constant real number called the **base.** We have chosen $a > 0$, $a \neq 1$, first because if $a = 1$, then $a^x = 1^x = 1$ for every x. Second, if $a = 0$, and $x = -\frac{1}{2}$, then

$$a^x = a^{-1/2} = \frac{1}{a^{1/2}} = \frac{1}{0^{1/2}},$$

which is not defined. Third, if $a < 0$ and $x = \frac{1}{2}$, then $a^x = a^{1/2}$ is not a real number.

The independent variable x, which is the exponent, can take on any rational or irrational value, so the domain of f is the set of all real numbers.

The most important algebraic properties of exponential functions are summarized in Section A.4 (Appendix A). You should review them at this time.

EXAMPLE 1   Sketch the graph of the exponential function

$$f(x) = 2^x.$$

Solution   We construct a table of values:

x	-4	-3	-2	-1	0	1	2	3	4
$y = f(x) = 2^x$	0.0625	0.125	0.25	0.5	1	2	4	8	16

Plotting the values, we obtain the points shown in Figure 1(a).

It can be shown that the exponential function (1) is continuous everywhere, so we may join the plotted points in Figure 1(a) to get the smooth curve shown in Figure 1(b).

Figure 1

(a)   (b)

To further illustrate the concept of exponential functions, we have sketched in Figure 2 the exponential functions $f(x) = 3^x$ and $f(x) = (\frac{3}{2})^x$ on the same coordinate axes.

Figure 2

Figures 1 and 2 illustrate the following general properties of the exponential functions $f(x) = a^x$ for $a > 1$:

1. The domain of $f(x)$ is the entire real line $(-\infty, +\infty)$.
2. $f(x) > 0$ for all x. In fact, the range of f is the interval $(0, +\infty)$.
3. f is an increasing function for all x (as we shall show in Section 5.3).
4. The graph of f passes through the point $(0, 1)$, regardless of the value of a (since $a^0 = 1$ for every $a > 1$).
5. $\lim\limits_{x \to +\infty} f(x) = +\infty$
6. The graph of f is concave upward for all x (as we shall show in Section 5.3).

EXAMPLE 2 Sketch the graph of the exponential function

$$f(x) = (\tfrac{1}{2})^x.$$

Solution We construct a table of values:

x	-4	-3	-2	-1	0	1	2	3	4
$y = f(x) = (\tfrac{1}{2})^x$	16	8	4	2	1	0.5	0.25	0.125	0.0625

Since the function is continuous, we may join the plotted points by a smooth curve shown in Figure 3.

Figure 3

Figure 4 shows the graphs of the exponential functions $f(x) = (\frac{1}{3})^x$ and $f(x) = (\frac{3}{4})^x$ on the same coordinate axes.

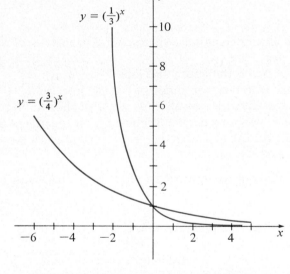

Figure 4

The graphs in Figures 3 and 4 illustrate the following general properties of the exponential functions $f(x) = a^x$ for $0 < a < 1$.

1. The domain of $f(x)$ is the entire real line $(-\infty, +\infty)$.
2. $f(x) > 0$ for all x. In fact, the range of f is the interval $(0, +\infty)$.
3. f is a decreasing function for all x (as we shall show in Section 5.3).
4. The graph of f passes through the point $(0, 1)$ regardless of the value of a (since $a^0 = 1$ for every a, $0 < a < 1$).
5. $\lim\limits_{x \to \infty} f(x) = 0$
6. The graph of f is concave upward for all x (as we shall shown in Section 5.3).

∎

||||||||||||| **The Number e**

In a surprisingly large number of applications, one particular value of a occurs often. This value of a is denoted by **e** and is an irrational number given by

$$e = \lim_{m \to \infty} \left(1 + \frac{1}{m}\right)^m. \tag{3}$$

Approximations of e have been carried to several thousand decimal places. Since e is an irrational number, its decimal representation is infinitely long and non-repeating. To twelve decimal places,

$$e = 2.718281828459\ldots$$

For our work, however, it will suffice to approximate e to three decimal places:

$$\boxed{e \approx 2.718.}$$

Table 1, constructed with a calculator, illustrates this result.

TABLE 1

m	1	5	10	20	100	1000	5000
$\left(1 + \dfrac{1}{m}\right)^m$	2	2.488	2.594	2.653	2.705	2.717	2.718

Many calculators can compute the values of e^x and e^{-x}, but before the availability of calculators, it was necessary to consult tables for these values. Appendix B, Table I, is an abbreviated form of such a table.

In the pages that follow we shall see that the number e is a natural base to use for an exponential function. Moreover, in applications involving exponential functions, the function e^x is used predominantly. Thus, the function e^x is usually called *the* exponential function. In Section 5.3 we shall see that the derivative of e^x is very simple to obtain.

EXAMPLE 3   Using Appendix B, Table I, or a calculator, sketch the graph of

$$y = 5e^{0.4x}, \qquad -4 \leq x \leq 4.$$

Solution   We construct a table of values and plot the corresponding points. The graph is shown in Figure 5.

x	-4	-3	-2	-1	0	1	2	3	4
$y = 5e^{0.4x}$	1.009	1.506	2.247	3.352	5	7.459	11.128	16.601	24.765

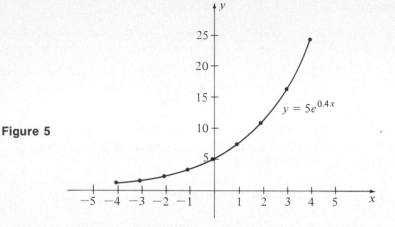

Figure 5

■

|||||||||||| **Compound Interest**

In Section 1.4 we studied simple interest as an application of linear functions. Recall that if we invest the principal P at a simple annual interest rate r for t years, then the **amount** or **sum** that we have on hand is given by

$$S = P + Prt$$

or

$$S = P(1 + rt). \tag{4}$$

In Equation (4) we refer to the principal P as the **present value** of S, and S is called the **future value** of P.

In many business transactions the interest that is added to the principal at regular time intervals earns interest itself. This process is called the **compounding** of interest. The time period between successive additions of interest is called the **conversion period.** Thus, if interest is compounded quarterly, the conversion period is three months; if interest is compounded semiannually, the conversion period is six months. In this book we shall adopt the common business practice of using an interest year that has 360 days.

Suppose now that we invest a principal P at an annual rate r, compounded k times a year. Thus, each conversion period lasts $1/k$ years, and the future value S_1 of P at the end of the first conversion period is

$$S_1 = P + P \cdot r \cdot \frac{1}{k} = P\left(1 + \frac{r}{k}\right). \tag{5}$$

The future value S_2 of P at the end of the second conversion period is

$$S_2 = S_1 + S_1 \cdot r \cdot \frac{1}{k}$$

$$= S_1\left(1 + \frac{r}{k}\right)$$

$$= P\left(1 + \frac{r}{k}\right)\left(1 + \frac{r}{k}\right). \qquad \text{[from (5)]}$$

Thus,

$$S_2 = P\left(1 + \frac{r}{k}\right)^2.$$

Continuing, we see that the future value S of P after n conversion periods is given by

$$S = P\left(1 + \frac{r}{k}\right)^n. \tag{6}$$

In summary,

If a principal P is invested at an annual rate r, compounded k times per year, then the future value S of P after n conversion periods is

$$S = P(1 + i)^n \tag{7}$$

where $i = r/k$.

Appendix B, Table III, gives $(1 + i)^n$ and $(1 + i)^{-n}$ for a number of values of i and n. Appendix B, Tables I and III, will be used for the computation of values of e^x, e^{-x}, $(1 + i)^n$, and $(1 + i)^{-n}$ that occur in the illustrative examples. The corresponding computations that occur in the exercises should be carried out with a calculator and the answers in the back are based on this assumption.

EXAMPLE 4 Suppose that we invest $6000 at an annual interest rate of 8%, compounded quarterly. What is the future value of the investment after 3 years?

Solution We are given $P = \$6000$, $r = 0.08$, $k = 4$, and $n = 12$ (4 conversion periods per year for 3 years). Then,

$$i = \frac{r}{k} = \frac{0.08}{4} = 0.02.$$

Using Equation (7), we obtain

$S = 6000(1 + 0.02)^{12}$

$\quad = 6000(1.26824179)$ (Appendix B, Table III, with $i = .02$ and $n = 12$)

$\quad = \$7609.45.$ ■

Often, instead of wanting to know the future value S of some principal P, we want to know the principal P that should be invested now to yield a certain amount S at some future time. The answer is obtained by solving Equation (7) for P:

$$P = S(1 + i)^{-n} \tag{8}$$

The amount P is called the **present value** of the (future) amount S.

EXAMPLE 5 How much money should a thirty-five-year-old person invest now at 12% per year, compounded semiannually, to have a lump sum of $40,000 when retiring at sixty years of age?

Solution We are given $S = \$40,000$, $r = 0.12$, $k = 2$, $n = 50$ (25 years, 2 semiannual conversion periods per year). Then,

$$i = \frac{0.12}{2} = 0.06.$$

Using Equation (8), we find

$$
\begin{aligned}
P &= 40,000(1 + 0.06)^{-50} \\
&= 40,000(0.05428836) \qquad \text{(Appendix B, Table III, with } i = .06, n = 50) \\
&= \$2171.53.
\end{aligned}
$$
∎

IIIIIIIIIIIII Continuous Compounding

When P, r, and t are fixed and the frequency of compounding interest is increased, the return on the investment increases. We want to determine the effect of increasing the number of conversions per year without bound; that is, we want to see what happens to S as $k \to +\infty$.

Suppose a principal P is invested at an annual interest rate r, compounded k times per year. After t years, the number of conversions is $n = tk$. The future value of the investment after t years is given by Equation (6), so

$$S = P\left(1 + \frac{r}{k}\right)^{tk}. \tag{9}$$

Letting $m = k/r$ $(k = mr)$, we can rewrite (9) as

$$S = P\left(1 + \frac{1}{m}\right)^{tmr},$$

or

$$S = P\left[\left(1 + \frac{1}{m}\right)^{m}\right]^{rt}.$$

If the number k of conversions per year increases without bound, then m also increases without bound (letting $k \to +\infty$ is equivalent to letting $m \to +\infty$). Consequently, as $k \to +\infty$,

$$
\begin{aligned}
S &= \lim_{m \to +\infty} P\left[\left(1 + \frac{1}{m}\right)^{m}\right]^{rt} \\
&= P\left[\lim_{m \to +\infty} \left(1 + \frac{1}{m}\right)^{m}\right]^{rt}. \tag{10}
\end{aligned}
$$

Here is where the importance of the number e becomes apparent. You will recall that by our definition (3), the limit in Equation (10) is e. Therefore, Equation (10) tells us that as $k \to +\infty$,

$$S = Pe^{rt}. \tag{11}$$

Equation (11) shows that increasing the frequency of compounding without bound does not cause the return on the principal to increase without bound. Because we obtained Equation (11) by letting the frequency of compounding increase without bound, we say that interest is **compounded continuously** by Equation (11).

In summary,

If a principal P is invested at an annual rate r, compounded continuously, then the future value S of P after t years is

$$S = Pe^{rt}. \tag{11}$$

EXAMPLE 6 What is the future value of $20,000 invested at 7% per year, compounded continuously, after 4 years?

Solution We have $P = \$20{,}000$, $r = 0.07$, and $t = 4$. From Equation (11),

$$S = 20{,}000e^{0.07(4)}$$
$$= 20{,}000e^{0.28}$$
$$= 20.000(1.3231) \qquad \text{(Appendix B, Table I)}$$
$$= \$26{,}462.00. \qquad \blacksquare$$

By solving Equation (11) for P, we can determine the principal P that must be invested at continuous compounding to have a certain amount S at some future time.

$$P = Se^{-rt}. \tag{12}$$

The amount P is called the **present value** of the (future) amount S.

EXAMPLE 7 How much money must be invested now at 9% continuous compound interest so that in 10 years the investment is worth $20,000?

Solution We are given $S = \$20{,}000$, $r = 0.09$ and $t = 10$. We use Equation (12):

$$P = Se^{-rt},$$

obtaining

$$P = 20{,}000e^{(-0.09)(10)}$$
$$= 20{,}000e^{-0.9}$$
$$= 20{,}000(.4066)$$
$$= \$8132.$$ ∎

|||||||||||||| ## Nominal Versus Effective Interest Rates

Interest rates are usually stated on a per-year basis. In compound interest, however, interest is computed periodically rather than at the end of the year only. The stated annual rate of interest, called the **nominal rate,** is not used in computing the interest at the end of each period. Instead, the nominal rate r is converted to the periodic rate $\dfrac{r}{k}$ [see Equation (7)]. For example, if money is invested at 8% per year, compounded quarterly, then the nominal rate is 0.08 and the period rate is $\dfrac{0.08}{4} = 0.02$.

If a principal P draws interest at the nominal rate r, compounded k times a year, then the **effective rate** of interest V is defined to be the rate that, if P were compounded *annually* at this rate, would yield the same interest. We shall derive a formula for V. By Equation (7), the compound amount at the nominal rate r, compounded k times a year, after n years is

$$A_1 = P\left(1 + \frac{r}{k}\right)^{nk}.$$

If we invest the same principal P for n years at the effective interest rate V (compounded annually), then after n years the compound amount is

$$A_2 = P(1 + V)^n.$$

By definition of effective interest rate V, A_1 and A_2 must be equal. Thus, we must have

$$P\left(1 + \frac{r}{k}\right)^{nk} = P(1 + V)^n. \tag{13}$$

Dividing both sides by P, we obtain

$$\left[\left(1 + \frac{r}{k}\right)^k\right]^n = (1 + V)^n.$$

Taking the nth root of both sides,

$$\left(1 + \frac{r}{k}\right)^k = 1 + V,$$

so

$$V = \left(1 + \frac{r}{k}\right)^k - 1.$$

Thus,

For the nominal interest rate r, compounded k times a year, the effective rate of interest is

$$V = \left(1 + \frac{r}{k}\right)^k - 1. \tag{14}$$

EXAMPLE 8 A bank is paying its vacation savings accounts 9% interest, compounded monthly. Find the effective rate of interest.

Solution We have here $r = 0.09$ and $k = 12$. Thus, by Equation (14),

$$V = \left(1 + \frac{0.09}{12}\right)^{12} - 1.$$

Using a calculator, we obtain

$$V = (1.0075)^{12} - 1$$
$$\approx 1.0938 - 1$$
$$= 0.0938.$$

Thus, the effective rate of interest is approximately 9.38%. ∎

When a principal P is invested at the nominal rate r, compounded continuously, the **effective rate** V is defined analogously, so that in place of (13) we have

$$Pe^{rn} = P(1 + V)^n$$

or

$$(e^r)^n = (1 + V)^n,$$

so

$$e^r = 1 + V$$

or

$$V = e^r - 1.$$

Therefore,

For the nominal interest rate r, compounded continuously, the effective rate of interest is

$$V = e^r - 1. \tag{15}$$

EXAMPLE 9 If a bank is paying interest at the nominal rate of 9%, compounded continuously, what is the effective rate of interest?

Solution   We have $r = 0.09$. Thus, by Equation (15), the effective rate is

$$V = e^{0.09} - 1.$$
$$\approx 1.0942 - 1 \qquad \text{(from Appendix B, Table 1)}$$
$$= .0942.$$

Therefore, the effective rate is approximately 9.42%.   ∎

Notice that in both formulas (14) and (15) the variables P and n do not appear. That is, the effective rate of interest does not depend on either the amount of money invested or the number of years the money is invested.

Section 5.4 is devoted entirely to further applications of the exponential functions.

5.1 EXERCISE SET ||||||||||||||||||||||||||||||||||||||||||||||||||||||||||||||||||||||||||||||||||||

1. Use the equation $P = 200e^{0.04t}$ to complete the following table.

t	1	10	20	120
P				

2. Use the equation $Q = 400e^{-0.02t}$ to complete the following table.

t	10	50	100	2000
Q				

In Exercises 3–16, sketch the graphs of the given functions. If the base is e, use a calculator.

3. $f(x) = 4^x$

4. $f(x) = (\frac{1}{4})^x$

5. $f(x) = (\frac{5}{2})^x$

6. $f(x) = (\frac{1}{10})^x$

7. $f(x) = 5^{-x} = (\frac{1}{5})^x$

8. $f(x) = 2^{-x^2}$

9. $f(x) = e^x$

10. $f(x) = e^{-x}$

11. $f(x) = 2^{x-1}$

12. $f(x) = 20e^{2x}$

13. $f(x) = 50e^{0.4x}$

14. $f(x) = 100e^{-0.5x}$

15. $f(x) = e^{-x^2}$

16. $f(x) = 50(1 - e^{-0.5x})$

17. Find the interest on $6000 invested for 4 years at 6% per year, compounded

   (a) annually.   (d) semiannually.
   (b) monthly.   (e) continuously.
   (c) quarterly.

18. Find the interest on $8000 invested for
   (a) 5 years at 8% per year, compounded semiannually.
   (b) 3 years at 6% per year, compounded monthly.
   (c) 15 years at 5% per year, compounded annually.
   (d) 9 months at 8% per year compounded continuously

19. Find the future value of a $12,000 savings certificate invested for 8 years at 9% per year, compounded
   (a) semiannually.   (b) continuously.

20. Find the future value of $10,000 invested for 12 years at 6% per year, compounded
   (a) quarterly.   (b) continuously.

21. An account that draws 10% interest, compounded quarterly, will be worth $12,000 in 6 years. Find the present value of this account.

22. If the account in Exercise 21 pays compound interest continuously, instead of quarterly, what is its present value?

23. In order to have the money to replace a machine 20 years from now, a manufacturer invests a lump sum of money at 8% per year, compounded semiannually. If it will take $20,000 to purchase the new equipment, how much should be invested now?

24. How much money should be invested now at 9% per year, compounded continuously, to obtain $10,000 in 6 years?

25. Find the effective rate of interest when the nominal rate is 8%, compounded
    (a) semiannually.   (c) monthly.
    (b) quarterly.   (d) continuously.

26. Find the effective rate of interest when the nominal rate is 6%, compounded
    (a) semiannually.   (c) weekly.
    (b) every 4 months.   (d) continuously.

27. Which pays more, 6% compounded continuously or $6\frac{1}{4}$% compounded semiannually?

28. Which pays more, 12% compounded continuously or $12\frac{1}{4}$% compounded semiannually?

29. **(Population Growth)** Suppose that the population P of a bacteria colony is given by $P(t) = 2000e^{0.04t}$, where t is the number of days that the colony has been growing. Find the size of the colony after
    (a) 10 days.   (b) 80 days.

30. **(Atmospheric Pressure)** The atmospheric pressure P in pounds per square inch is approximately given by the formula $P = 14.7e^{-0.21h}$, where h is the altitude (in miles) above sea level. Calculate the atmospheric pressure
    (a) in Denver, which has an altitude of 1 mile above sea level.
    (b) at the top of Mount Everest, the highest point on earth, where the altitude is 5.50 miles above sea level.
    (c) at the Dead Sea, the lowest point on earth, where the altitude is -0.247 miles (it is below sea level).

31. **(Light Absorption)** When a beam of light having an initial intensity of I_0 (in lumens) passes through a medium of thickness s (in centimeters), the resulting intensity I (in lumens) is given by the equation
    $$I = I_0 e^{-ks},$$
    where k is a constant depending upon the medium. Consider a medium for which $k = 3$, and find the resulting intensity of an 80-lumen beam passing through 4.2 centimeters of the medium.

32. **(Learning)** The number N of calculators tested daily by a worker on an assembly line after t hours of training is given by
    $$N = 60(1 - e^{-0.2t}).$$
    Find the number of calculators tested daily by a worker who has had 12 hours of training.

5.2 ||||||| THE LOGARITHMIC FUNCTION

John Napier* developed logarithms to simplify the tedious computational work encountered in the astronomical investigations of his times. The remarkable capabilities of modern, inexpensive calculators have diminished the use of logarithms in tedious hand computational work. Nevertheless, logarithmic functions

---

* John Napier (1550–1617) was born in Merchiston, near Edinburgh, Scotland. He graduated from St. Andrews University, studied in Paris, and traveled in Europe. Early in his career he became active in politics as an advocate of the Protestant cause. In 1593, his book *A Plaine Discovery of the Whole Revelation of St. John*, the first Scottish interpretation of scripture, was published. He invented a number of military devices in addition to devising the system of logarithms and computing logarithm tables. He also developed a system of arithmetic computation using counting rods, which were called "Napier's bones."

are useful in many science and engineering applications and study of them is important.

The notion of logarithm arises when we attempt to solve the equation

$$x = a^y \tag{1}$$

for y; that is, when we want to determine the exponent y to which we must raise the base a in order to yield x. The answer is formally called **the logarithm of x to the base a**, and is denoted

$$y = \log_a x. \tag{2}$$

The base a is required to be a positive number different from 1. As Equation (1) shows, y may be any real number, but x must be positive. Thus,

---

For a given base $a > 0$, $a \neq 1$,

$$y = \log_a x \qquad \text{if and only if } x = a^y. \tag{3}$$

EXAMPLE 1   (a) $\log_{10} 1000 = 3$, since $10^3 = 1000$;
(b) $\log_3 9 = 2$, since $3^2 = 9$;
(c) $\log_{25} 5 = \frac{1}{2}$, since $25^{1/2} = 5$;
(d) $\log_2 \frac{1}{8} = -3$, since $2^{-3} = \frac{1}{8}$.

Two bases are of special importance in applications: base 10. which yields **common logarithms,** and base e, which yields **natural logarithms.** It is customary to write

$$\log x \text{ rather than } \log_{10} x,$$

and

$$\ln x \text{ rather than } \log_e x. \qquad \blacksquare$$

EXAMPLE 2   (a) $\log 100 = 2$, since $10^2 = 100$;
(b) $\log (1/1000) = -3$, since $10^{-3} = 1/1000$;
(c) $\log 1 = 0$, since $10^0 = 1$;
(d) $\ln e = 1$, since $e^1 = e$;
(e) $\ln (1/e^4) = -4$, since $1/e^4 = e^{-4}$;
(f) $\ln 1 = 0$, since $e^0 = 1$. $\qquad \blacksquare$

EXAMPLE 3   For every $a > 0$, $a \neq 1$, we have

$$\log_a a = 1 \quad \text{and} \quad \log_a 1 = 0,$$

since $a^1 = a$ and $a^0 = 1$, respectively. $\qquad \blacksquare$

The following two identities are immediate and useful consequences of the definition of logarithm:

$$a^{(\log_a x)} = x \qquad \text{for } x > 0, \text{ and} \tag{4}$$

$$\log_a a^x = x \qquad \text{for any real number } x. \tag{5}$$

Both of these equations follow from the fact that $\log_a$ x *is the power to which we raise* a *to yield* x.

EXAMPLE 4   In each of the following, solve for x:
    (a) $\log_3 x = 4$ (c) $\log_2 x = -3$
    (b) $\log_x 100 = 2$ (d) $\log_x (1/9) = -\frac{1}{2}$

Solution   (a) Since $\log_3 x = 4$ is equivalent to $3^4 = x$, we see that $x = 81$.
    (b) Since $\log_x 100 = 2$ is equivalent to $x^2 = 100$, and $x > 0$, we see that $x = 10$.
    (c) Since $\log_2 x = -3$ is equivalent to $2^{-3} = x$, we have $x = 1/8$.
    (d) Since $\log_x (1/9) = -\frac{1}{2}$ is equivalent to $x^{-1/2} = 1/9$, or $1/\sqrt{x} = 1/9$, we have $\sqrt{x} = 9$, so that $x = 81$.   ■

||||||||||||| Logarithm Functions

If a is a positive number, $a \neq 1$, then the **logarithmic function to the base a** is defined by

$$f(x) = \log_a x. \tag{6}$$

The **graph** of the logarithm function (6) is the graph of the equation $y = \log_a x$, which by definition is equivalent to the graph of the equation $x = a^y$. This graph may be easily obtained from the already familiar graph of

$$y = a^x$$

by merely interchanging the roles of x and y. The results are shown in Figure 1.

Figure 1

We can see from Figure 1 that

   (a) the curve $y = \log_a x$ is the mirror image of the curve $y = a^x$ about the line $y = x$; thus, if we were to fold the paper along the dotted line in Figure 1, the two graphs would coincide exactly.

   (b) the domain of the logarithm function $f(x) = \log_a x$ is the interval $(0, +\infty)$, which is the range of the exponential function $g(x) = a^x$.

   (c) the logarithm function is continuous throughout its domain.

   (d) both of the functions $f(x) = \log_a x$ and $g(x) = a^x$ are increasing functions throughout their respective domains.

   (e) the range of the logarithm function $f(x) = \log_a x$ is the entire real line $(-\infty, +\infty)$, which is the domain of the exponential function $g(x) = a^x$.

If $f(x) = \log_a x$ and $g(x) = a^x$, then from Equation (5), we obtain

$$f(g(x)) = x \qquad \text{for any real number } x, \tag{7}$$

and from Equation (4), we obtain

$$g(f(x)) = x \qquad \text{for } x > 0. \tag{8}$$

Two functions that satisfy Equations (7) and (8) are said to be **inverses** of each other, so that $\log_a x$ and a^x are **inverses** of each other. Thus, each of these functions reverses or undoes the action of the other. That is, when $x = 8$ the function $\log_2 x$ gives the value 3. If we now apply the function a^x to the answer, we get $2^3 = 8$, our original value of x.

    Table 1 below reviews the basic properties required for the algebra of logarithms. Since logarithms are exponents, each of the properties listed is a fact

TABLE 1   **Properties of Logarithms**

$(a > 0, a \neq 1, x > 0, y > 0)$	
Property	Examples
$\log_a 1 = 0$	$\log 1 = 0;\ \ln 1 = 0$
$\log_a a^x = x$	$\log_3 81 = \log_3 3^4 = 4$
	$\log 1000 = \log 10^3 = 3$
$\log_a xy = \log_a x + \log_a y$	$\log 12 = \log(4 \cdot 3) = \log 4 + \log 3$
	$\ln 28 = \ln(4 \cdot 7) = \ln 4 + \ln 7$
$\log_a \dfrac{x}{y} = \log_a x - \log_a y$	$\log \frac{7}{5} = \log 7 - \log 5$
	$\ln \frac{2}{3} = \ln 2 - \ln 3$
	$\ln \frac{1}{7} = \ln 1 - \ln 7 = 0 - \ln 7 = -\ln 7$
$\log_a \dfrac{1}{x} = -\log_a x$	$\log_2 \frac{1}{16} = -\log_2 16 = -4$
$\log_a x^k = k \log_a x$	$\log 2^3 = 3 \log 2$
	$\log \sqrt{12} = \log 12^{1/2} = \frac{1}{2} \log 12$
	$\ln \sqrt[3]{(4.1)^2} = \ln(4.1)^{2/3} = \frac{2}{3} \ln 4.1$

about exponents, and can be proved from familiar laws of exponents. We shall not prove these properties here, but shall give examples showing how they are used.

EXAMPLE 5 Given that $\log_a u = b$, $\log_a v = c$ and $\log_a w = d$, express each of the following in terms of b, c, and d:

(a) $\log_a (uv^2w^3)$ (b) $\log_a \dfrac{uv}{w}$ (c) $\log_a \dfrac{u^4}{\sqrt[3]{vw^2}}$

Solution Using the properties given in Table 1,

(a) $\log_a (uv^2w^3) = \log_a u + \log_a v^2 + \log_a w^3$

$\qquad\qquad\qquad = \log_a u + 2 \log_a v + 3 \log_a w$

$\qquad\qquad\qquad = b + 2c + 3d.$

(b) $\log_a \left(\dfrac{uv}{w}\right) = \log_a (uv) - \log_a w$

$\qquad\qquad\qquad = \log_a u + \log_a v - \log_a w$

$\qquad\qquad\qquad = b + c - d.$

(c) $\log_a \dfrac{u^4}{\sqrt[3]{vw^2}} = \log_a u^4 - \log_a (vw^2)^{1/3}$

$\qquad\qquad\qquad = 4 \log_a u - \tfrac{1}{3} \log_a (vw^2)$

$\qquad\qquad\qquad = 4b - \tfrac{1}{3}[\log_a v + \log_a w^2]$

$\qquad\qquad\qquad = 4b - \tfrac{1}{3}[c + 2 \log_a w]$

$\qquad\qquad\qquad = 4b - \tfrac{1}{3}c - \tfrac{2}{3}d.$ ∎

In this book we shall be concerned mainly with natural logarithms. Appendix B, Table II, gives the natural logarithms for a number of values of x. We can use this table or a calculator to solve exponential equations, as the following examples illustrate.

EXAMPLE 6 If $300e^t = 600$, find t.

Solution $\qquad\qquad 300e^t = 600$

$\qquad\qquad\quad e^t = 2$ (divide both sides by 300)

$\qquad\qquad\quad t = \ln 2$ (remember that $y = e^x$ if and only if $x = \ln y$)

$\qquad\qquad\quad t = 0.6931.$ (Appendix B, Table II)   ∎

EXAMPLE 7 If \$2000 is invested at 8% per year continuous compound interest, how long will it take for the investment to grow to \$3000?

Solution We have seen in Section 5.1 that, after t years, the compound amount is

$$S = 2000e^{.08t}.$$

Thus, we seek the value of t for which

$$3000 = 2000e^{.08t}$$

or

$$\tfrac{3}{2} = e^{.08t}.$$

In logarithmic form, and using Appendix B, Table II,

$$\ln\left(\tfrac{3}{2}\right) = .08t$$
$$.4055 = .08t$$
$$t = 5.069 \text{ years}$$
$$\approx 5 \text{ years, } 25 \text{ days.}$$ ∎

EXAMPLE 8 For comparison with Example 7, determine how long it will take $2000 invested at 8% per year, compounded quarterly, to grow to $3000.

Solution From Section 5.1, after t years the compound amount is

$$S = 2000\left(1 + \frac{.08}{4}\right)^{4t}.$$

Thus, we seek the value of t for which

$$3000 = 2000(1.02)^{4t}$$
$$\tfrac{3}{2} = (1.02)^{4t}.$$

In logarithmic form,

$$\ln 1.5 = 4t \ln 1.02$$
$$\frac{\ln 1.5}{4 \ln 1.02} = t.$$

Using a calculator we find

$$t = \frac{.4055}{.0792} \approx 5.119 \text{ years}$$
$$\approx 5 \text{ years, } 43 \text{ days.}$$

Of course, if interest is paid only at the end of the quarter, then it will take 5 years, 90 days for the investment to reach (and exceed) $3000. Thus, at 8% quarterly compound interest, the $2000 investment will require 18 more days to reach $3000 than it would at 8% continuous compound interest. (If interest is paid only at the end of a quarter, then it takes 65 more days rather than 18 more days.)

∎

The following example utilizes a mathematical model known as a **logistic equation**. The general logistic equation, its origins and its applications will be discussed in Section 10.3.

EXAMPLE 9
(Medicine)

The spread of an epidemic of long-lasting flu is described by the equation

$$P(t) = \frac{32{,}000}{1 + 50e^{-0.1t}},$$

where $P(t)$ is the number of people infected t weeks after the outbreak of the disease. How long does it take for 2000 people to become infected?

Solution   We must solve for t in the equation

$$\frac{32{,}000}{1 + 50e^{-0.1t}} = 2000.$$

We have

$$2000(1 + 50e^{-0.1t}) = 32{,}000$$
$$2000 + 100{,}000e^{-0.1t} = 32{,}000$$
$$100{,}000e^{-0.1t} = 30{,}000$$
$$e^{-0.1t} = 0.30$$
$$-0.1t = \ln 0.30 \qquad \text{(since } y = e^x \text{ if and only if } x = \ln y)$$
$$-0.1t = -1.2040 \qquad \text{(Appendix B, Table II)}$$
$$t = \frac{1.2040}{0.1} = 12.04.$$

Thus, after 12 weeks approximately 2000 people will have been infected.   ∎

It is possible to change logarithms from one base to another. This is helpful in a number of circumstances. Suppose a and b are positive numbers different from 1, and that

$$y = \log_a x.$$

Then,

$$a^y = x.$$

Taking the logarithm of both sides, to the base b, we obtain

$$\log_b (a^y) = \log_b x$$
$$y \log_b a = \log_b x$$

and so,

$$y = \frac{\log_b x}{\log_b a}.$$

Summarizing,

If $a > 0$, $b > 0$, $a \neq 1$, $b \neq 1$, then for all $x > 0$,

$$\log_a x = \frac{\log_b x}{\log_b a}. \tag{9}$$

In fact, logarithms to any base a can be found by using only natural logarithms, or only common logarithms.

$$\log_a x = \frac{\ln x}{\ln a} = \frac{\log x}{\log a} \tag{10}$$

5.2 EXERCISE SET ||||||||||||||||||||||||||||||||||||||||||||||||||||||||||||||||||||||||||||

1. Write in exponential form:
   (a) $\log_3 9 = 2$ (c) $\log_2 (\frac{1}{4}) = -2$
   (b) $\log_4 64 = 3$

2. Write in logarithmic form:
   (a) $4^2 = 16$ (c) $7^{-2} = \frac{1}{49}$
   (b) $5^3 = 125$

3. Without tables or calculators, compute the following:
   (a) $\log_2 16$ (c) $\log_2 128$
   (b) $\log_2 (\frac{1}{2})$ (d) $\log_2 \sqrt[3]{32}$

4. Without tables or calculators, compute the following:
   (a) $\log_3 243$ (c) $\log_3 \sqrt{27}$
   (b) $\log_3 \frac{1}{9}$ (d) $\log_3 27^4$

5. Without tables or calculators, compute the following:
   (a) $\log 10,000$ (e) $10^{\log 10}$
   (b) $\log .01$ (f) $10^{\log 45}$
   (c) $\log .00001$ (g) $\log \sqrt[3]{100}$
   (d) $\log 1,000,000$ (h) $\log \dfrac{1}{\sqrt{10}}$

6. Without tables or calculators, compute the following:

7. (a) $\ln (e^2 \cdot e^5)$ (e) $e^{\ln 1.8}$
   (b) $\ln \left(\dfrac{e^8}{e^5}\right)$ (f) $e^{3 \ln 2}$
   (c) $\ln e^{3t}$ (g) $\ln \sqrt[4]{e}$
   (d) $\ln e^{-.01}$ (h) $\ln \dfrac{1}{\sqrt[3]{e}}$

7. Using only the values $\log 3 = 0.4771$ and $\log 5 = 0.6990$, compute
   (a) $\log 15$ (c) $\log 3000$ (e) $\log 75$
   (b) $\log \frac{5}{3}$ (d) $\log \frac{1}{500}$ (f) $\log \sqrt{45}$

8. Using only the values $\log 2 = .3010$ and $\log 3 = .4771$, compute
   (a) $\log 12$ (c) $\log \frac{8}{9}$ (e) $\log \sqrt{2}$
   (b) $\log \frac{1}{6}$ (d) $\log 0.5$ (f) $\log 6\frac{3}{4}$

9. Solve each of the following equations for x:
   (a) $\log_4 x = 3$ (c) $\log_x 9 = 2$
   (b) $\log_5 x = -1$ (d) $\log_x 4 = -2$

10. Solve each of the following equations for x:
    (a) $\log_3 x = -2$ (c) $\log_x 100 = -2$
    (b) $\log_x 9 = 2$ (d) $\log_x 4 = \frac{1}{3}$

In Exercises 11–18, sketch the graph of the given function. Use a calculator where necessary.

11. $f(x) = \log_2 x$ 12. $f(x) = \log_3 x$

13. $f(x) = \ln x$ 14. $f(x) = \ln\left(\dfrac{1}{x}\right)$

15. $f(x) = 1 + \ln x$ 16. $f(x) = \ln(x + 1)$

17. $f(x) = \ln\dfrac{x}{2}$ 18. $f(x) = 4 \ln 2x$

19. Solve for t. Use a calculator where necessary.
   (a) $e^t = 4$
   (b) $500e^{2t} = 3000$
   (c) $200e^{0.2t} = 100$
   (d) $50e^{-2t} = 60$
   (e) $100(1 - e^{-0.2t}) = 40$
   (f) $4 \ln x = 5$

20. Solve for t by using a calculator.
   (a) $e^{-t} = 8$
   (b) $3e^{2t} = 15$
   (c) $60e^{-0.3t} = 150$
   (d) $80 - 80e^{-0.4t} = 24$
   (e) $2 \ln t = 8$
   (f) $2 \ln(t + 1) = 8$

21. Given that $\log_a u = b$, $\log_a v = c$, and $\log_a w = d$, find
   (a) $\log_a\left(\dfrac{u^3}{v^2}\right)$
   (b) $\log_a\sqrt{\dfrac{u}{vw}}$
   (c) $\log_a u\sqrt[3]{v}$

22. Given that $\log_b x = r$, $\log_b y = s$, and $\log_b z = t$, find
   (a) $\log_b\left(\dfrac{x^3y^4}{z^2}\right)$ (c) $\log_b\left(\dfrac{1}{x^2yz^3}\right)$
   (b) $\log_b\left(\dfrac{y}{\sqrt{xz}}\right)$

23. Compute the following to four decimal place accuracy, using Equation (10):
   (a) $\log_3 70$ (b) $\log_2 10$ (c) $\log_{1/2} 20$

24. Compute the following to four decimal place accuracy, using Equation (10):
   (a) $\log_6 30$ (b) $\log_5\left(\frac{1}{4}\right)$ (c) $\log_{1/4}(50)$

25. How long does it take $8000 to grow to $20,000 if it is invested at 8% per year, compounded continuously?

26. How long will it take $5000 invested at 10% per year, compounded continuously, to grow to $12,000?

27. How long will it take $8000 to grow to $20,000 if it is invested at 8% per year, compounded quarterly? Compare with the answer to Exercise 25.

28. How long will it take $5000 invested at 10% per year, compounded semiannually, to grow to $12,000? Compare with the answer to Exercise 26.

29. (**Learning**) The number N of radios that an assembly-line worker can assemble daily after t days of training is given by

$$N(t) = 60 - 60e^{-0.04t}.$$

After how many days of training does the worker assemble 40 radios daily?

30. (**Air Pressure**) The atmospheric pressure P in pounds per square inch is approximately given by $P = 14.7e^{-0.21h}$, where h is the altitude (in miles) above sea level. At what altitude is the pressure 7.83 pounds per square inch?

31. (**Advertising**) Suppose that the number N of mopeds sold when x thousands of dollars are spent on advertising is given by

$$N = 4000 + 500 \ln(x + 2).$$

How much advertising money must be spent to sell 6000 mopeds?

32. (**Logistic Growth**) Consider a population whose growth is governed by a logistic equation, so that t days after some initial time the population is

$$P(t) = \dfrac{10,000}{5 + 20e^{-.1t}}.$$

(a) Find the initial population.

(b) How long does it take $P(t)$ to reach 500?

(c) How long does it take $P(t)$ to reach 800?

33. **(Earthquake Intensity)** The magnitude R on the Richter scale of an earthquake of intensity I is given by

$$R = \log \frac{I}{I_0},$$

where I_0 is a standard intensity used for comparison. Thus, a magnitude of $R = 4$ on the Richter scale means that

$$4 = \log \frac{I}{I_0}$$

or

$$\frac{I}{I_0} = 10^4$$

so

$$I = 10{,}000 I_0$$

which means that the earthquake being measured is 10,000 times more intense than the standard earthquake used for comparison.

(a) The San Francisco earthquake of 1906 had an intensity of 8.25 on the Richter scale. How much more intense was it than the standard earthquake?

(b) The devastating 1978 earthquake in Iran had an intensity of $10^{7.7} I_0$. What was its magnitude on the Richter scale?

5.3 |||||||| THE DERIVATIVES OF THE EXPONENTIAL AND LOGARITHMIC FUNCTIONS

In this section we shall obtain formulas for the derivative of the exponential function e^x and the logarithm function $\ln x$. In the next section we shall use these results in several diverse applications.

|||||||||||||| Differentiating Exponential Functions

Let f be the exponential function defined by

$$f(x) = e^x.$$

We compute $f'(x)$ by the definition of the derivative. Thus,

$$f'(x) = \lim_{h \to 0} \frac{f(x + h) - f(x)}{h}$$

$$= \lim_{h \to 0} \frac{e^{x+h} - e^x}{h}$$

$$= \lim_{h \to 0} \frac{e^x(e^h - 1)}{h} \qquad (e^{x+h} = e^x e^h \text{ and factoring})$$

$$= \lim_{h \to 0} e^x \lim_{h \to 0} \frac{e^h - 1}{h}$$

$$= e^x \lim_{h \to 0} \frac{e^h - 1}{h}. \qquad \left(\text{since } e^x \text{ does not involve } h,\ \lim_{h \to 0} e^x = e^x \right)$$

Hence,

$$f'(x) = e^x \lim_{h \to 0} \frac{e^h - 1}{h}. \tag{1}$$

To compute $\lim_{h \to 0} (e^h - 1)/h$, we construct Tables 1 and 2 with the aid of a calculator.

TABLE 1 $h \to 0$ from the right

h	1	0.5	0.25	0.1	0.01	0.001
$\dfrac{e^h - 1}{h}$	1.718	1.297	1.136	1.052	1.005	1.0005

TABLE 2 $h \to 0$ from the left

h	-1	-0.5	-0.25	-0.1	-0.01	-0.001
$\dfrac{e^h - 1}{h}$	0.632	0.787	0.885	0.952	0.995	0.9995

We conclude from the data in Tables 1 and 2 that

$$\lim_{h \to 0} \frac{e^h - 1}{h} = 1. \tag{2}$$

Hence, substituting the result of (2) into Equation (1), we have

$$f'(x) = e^x \cdot 1 = e^x,$$

or

$$\frac{d}{dx}[e^x] = e^x. \tag{3}$$

Thus, the derivative of the exponential function is the exponential function itself.

To differentiate more complicated exponential functions, such as

$$e^{5x} \qquad e^{x^2} \qquad e^{\sqrt{x^2 + 1}}$$

we must use the chain rule. To differentiate

$$y = e^{g(x)}$$

we first let $u = g(x)$. Then $y = e^u$, and by the chain rule,

$$\frac{dy}{dx} = \frac{dy}{du} \cdot \frac{du}{dx}.$$

But

$$\frac{dy}{du} = \frac{d}{du}[e^u] = e^u. \qquad \text{[by (3)]}$$

Thus,

$$\frac{dy}{dx} = e^u \frac{du}{dx}.$$

In summary,

If $u = g(x)$ is differentiable, then

$$\frac{d}{dx}[e^u] = e^u \frac{du}{dx}, \tag{4}$$

or

$$\frac{d}{dx}[e^{g(x)}] = e^{g(x)}g'(x). \tag{5}$$

EXAMPLE 1   Using Equation (5), we write

$$\frac{d}{dx}[e^{5x}] = e^{5x}\frac{d}{dx}[5x] = 5e^{5x},$$

$$\frac{d}{dx}[e^{x^2}] = e^{x^2}\frac{d}{dx}[x^2] = 2xe^{x^2}.$$

EXAMPLE 2   If $h(x) = e^{\sqrt{x^2+1}}$, find $h'(x)$.

Solution   Using (5), we write

$$h'(x) = \frac{d}{dx}[e^{\sqrt{x^2+1}}] = e^{\sqrt{x^2+1}}\frac{d}{dx}[\sqrt{x^2+1}].$$

We now compute $\dfrac{d}{dx}[\sqrt{x^2+1}]$ according to the chain rule, as follows:

$$\frac{d}{dx}[\sqrt{x^2+1}] = \frac{d}{dx}[x^2+1]^{1/2} = \frac{1}{2}(x^2+1)^{-1/2}\frac{d}{dx}[x^2+1]$$

$$= \frac{1}{2}(x^2+1)^{-1/2}(2x) = \frac{x}{\sqrt{x^2+1}}.$$

Hence,

$$h'(x) = \frac{x}{\sqrt{x^2+1}}e^{\sqrt{x^2+1}}.$$

■

EXAMPLE 3   Find the slope of the tangent line to the curve $y = xe^x$ at the point where $x = 0$.

Solution   The slope of this tangent line is dy/dx when $x = 0$. By the product rule,

$$\frac{dy}{dx} = \frac{d}{dx}[xe^x] = x\frac{d}{dx}[e^x] + e^x\frac{d}{dx}[x]$$

$$= xe^x + e^x \cdot 1 = xe^x + e^x.$$

When $x = 0$, we obtain

$$\frac{dy}{dx} = 0(1) + 1 = 1.$$ ∎

|||||||||||| **General Exponential Functions**

Suppose a is a positive real number, with $a \neq 1$. We shall show that the function

$$f(x) = a^x$$

is differentiable, and derive a formula for its derivative. Since $a = e^{\ln a}$, we have $a^x = (e^{\ln a})^x$. Thus, for all real numbers x,

$$a^x = e^{x \ln a}. \tag{6}$$

Substituting $u = x \ln a$ into Equation (4) we have

$$\frac{d}{dx}[a^x] = \frac{d}{dx}[e^{x \ln a}]$$

$$= e^{x \ln a}\frac{d}{dx}(x \ln a)$$

$$= e^{x \ln a} \ln a$$

$$= a^x \ln a.$$

Therefore, for all real numbers x,

$$\frac{d}{dx}[a^x] = a^x \ln a. \tag{7}$$

Moreover, if $u = g(x)$ is differentiable, the chain rule used in conjunction with (7) tells us that

$$\frac{d}{dx}[a^u] = a^u \ln a \frac{du}{dx} \tag{8}$$

$$\frac{d}{dx}[a^{g(x)}] = a^{g(x)} (\ln a)g'(x). \tag{9}$$

EXAMPLE 4 (a) $\dfrac{d}{dx}[5^x] = 5^x \ln 5$.

(b) $\dfrac{d}{dx}[3^{x^2}] = 3^{x^2} \ln 3 \cdot 2x$.

(c) $\dfrac{d}{dx}[12^{x^3-x}] = 12^{x^3-x} \ln 12 \cdot (3x^2 - 1)$.

||||||||||||| **Application to Graphs**

In Section 5.1 we displayed the graphs of $y = a^x$ and $y = a^{-x}$, but could not use our familiar calculus techniques because we did not yet know how to differentiate these functions. We are now able to supply the details.

Consider the function $f(x) = e^x$. Since its derivative is $f'(x) = e^x > 0$, the function is *increasing everywhere*. Since its second derivative is $f''(x) = e^x > 0$, the graph of f must be *concave upward everywhere*.

Similarly, the function $g(x) = e^{-x}$ has derivative $y'(x) = -e^{-x}$, which is negative everywhere, and second derivative $g''(x) = e^{-x}$, which is positive everywhere. Thus, the function g is decreasing everywhere and its curve is concave upward everywhere.

Figure 1 shows the graphs of $f(x) = e^x$ and $g(x) = e^{-x}$.

Figure 1

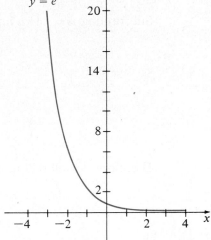

EXAMPLE 5 In the next section we shall consider many physical processes that are described by the function $Q(t) = Q_0 e^{kt}$ where Q_0 and k are constants and the variable t represents time. Find dQ/dt.

Solution

$$\frac{dQ}{dt} = \frac{d}{dt}[Q_0 e^{kt}]$$

$$= Q_0 \frac{d}{dt}[e^{kt}] \qquad (Q_0 \text{ is a constant})$$

$$= Q_0 k e^{kt} \qquad [\text{by Equation (4)}].$$

Now observe that

$$\frac{dQ}{dt} = k(Q_0 e^{kt}) = kQ(t).$$

Thus,

$$\frac{dQ}{dt} = kQ. \tag{10}$$

∎

Equation (10) says that the instantaneous rate of change of Q at any time t, is proportional to the value of Q itself at time t. Such a function $Q(t)$ is a very special function, indeed.

Differentiating the Natural Logarithm Function

We turn next to the derivative of the logarithm function

$$f(x) = \ln x, \qquad x > 0. \tag{11}$$

Since the graph of $\ln x$ is the mirror image of the graph of e^x across the line $y = x$ (see Section 5.2), the graph of $f(x) = \ln x$ is as shown in Figure 2.

Figure 2

We may thus *assume* that f is differentiable, and proceed to find its derivative.
Recall from Section 5.2 that Equation (11) can also be rewritten in equivalent exponential form as

$$x = e^{f(x)}. \tag{12}$$

We apply the method of implicit differentiation. Differentiating both sides of (12) and assuming that $f'(x)$ exists we have

$$\frac{d}{dx}[x] = \frac{d}{dx}[e^{f(x)}].$$

Using Equation (5) to simplify the right hand side, we have

$$1 = e^{f(x)}f'(x). \tag{13}$$

Substituting (12) into (13), we obtain

$$1 = xf'(x).$$

Therefore,

$$f'(x) = \frac{1}{x}.$$

We have obtained the derivative of $\ln x$:

$$\frac{d}{dx}[\ln x] = \frac{1}{x}, \qquad x > 0. \tag{14}$$

To differentiate more complicated expressions, such as $\ln (5x)$, $\ln (x^2)$, and $\ln \sqrt{x^2 + 1}$, requires use of the chain rule. To differentiate

$$y = \ln g(x),$$

where $g(x)$ is a *positive* differentiable function, we first let $u = g(x)$. Then $y = \ln u$, and by the chain rule,

$$\frac{dy}{dx} = \frac{dy}{du} \cdot \frac{du}{dx}.$$

But $\dfrac{dy}{du} = \dfrac{1}{u}$ by (14). Thus,

$$\frac{dy}{dx} = \frac{1}{u} \cdot \frac{du}{dx}.$$

In summary,

$$\frac{d}{dx}[\ln u] = \frac{1}{u}\frac{du}{dx}, \tag{15}$$

or

$$\frac{d}{dx}[\ln g(x)] = \frac{g'(x)}{g(x)}. \tag{16}$$

EXAMPLE 6 Using Equation (15), we write

(a) $\dfrac{d}{dx}[\ln (5x)] = \dfrac{1}{5x}\dfrac{d}{dx}[5x] = \dfrac{5}{5x} = \dfrac{1}{x}.$

(b) $\dfrac{d}{dx}[\ln (x^2)] = \dfrac{1}{x^2}\dfrac{d}{dx}[x^2] = \dfrac{2x}{x^2} = \dfrac{2}{x}.$

(c) $\dfrac{d}{dx} [\ln \sqrt{x^2 + 1}] = \dfrac{1}{\sqrt{x^2 + 1}} \dfrac{d}{dx} [\sqrt{x^2 + 1}]$

$\quad = \dfrac{1}{\sqrt{x^2 + 1}} \dfrac{1}{2} (x^2 + 1)^{-1/2} \dfrac{d}{dx} [x^2 + 1]$

$\quad = \dfrac{1}{\sqrt{x^2 + 1}} \cdot \dfrac{2x}{\sqrt{x^2 + 1}} \cdot \dfrac{2x}{2}$

$\quad = \dfrac{x}{\sqrt{x^2 + 1}\sqrt{x^2 + 1}} = \dfrac{x}{x^2 + 1}.$ ∎

Occasionally, we can simplify the computation of the derivative of $\ln g(x)$ by using the properties of logarithmic functions (Section 5.2) *before* differentiating. Let us use this approach to redo Example 6.

EXAMPLE 7   (a) $\dfrac{d}{dx} [\ln (5x)] = \dfrac{d}{dx} [\ln 5 + \ln x] = \dfrac{d}{dx} [\ln 5] + \dfrac{d}{dx} [\ln x] = 0 + \dfrac{1}{x} = \dfrac{1}{x}.$

(b) $\dfrac{d}{dx} [\ln (x^2)] = \dfrac{d}{dx} [2 \ln x] = 2 \cdot \dfrac{d}{dx} [\ln x] = 2 \cdot \dfrac{1}{x} = \dfrac{2}{x}.$

(c) $\dfrac{d}{dx} [\ln \sqrt{x^2 + 1}] = \dfrac{d}{dx} [\ln (x^2 + 1)^{1/2}] = \dfrac{d}{dx} \left[\dfrac{1}{2} \ln (x^2 + 1) \right]$

$\quad = \dfrac{1}{2} \cdot \dfrac{1}{x^2 + 1} \dfrac{d}{dx} [x^2 + 1]$

$\quad = \dfrac{1}{2} \cdot \dfrac{1}{x^2 + 1} \cdot 2x = \dfrac{x}{x^2 + 1}.$

EXAMPLE 8   Find the slope of the tangent line to the curve $y = x^2/\ln x$ at the point where $x = e$.

Solution   The slope of this tangent line is dy/dx when $x = e$. By the quotient rule,

$$\dfrac{d}{dx} \left[\dfrac{x^2}{\ln x} \right] = \dfrac{\ln x \cdot \dfrac{d}{dx} [x^2] - x^2 \cdot \dfrac{d}{dx} [\ln x]}{(\ln x)^2}$$

$$= \dfrac{2x \ln x - \dfrac{x^2}{x}}{(\ln x)^2}$$

$$= \dfrac{2x \ln x - x}{(\ln x)^2}.$$

When $x = e$, we obtain

$$\dfrac{dy}{dx} = \dfrac{2e \ln e - e}{(\ln e)^2} = e \qquad (\text{since } \ln e = 1).$$ ∎

||||||||||||| **Differentiating General Logarithm Functions**

Suppose b is any positive real number different from 1. The function

$$f(x) = \log_b x$$

is easily differentiated, as we shall now see. From Equation (10) of Section 5.2,

$$\log_b x = \frac{\ln x}{\ln b} = \frac{1}{\ln b} \cdot \ln x.$$

Therefore,

$$\frac{d}{dx} [\log_b x] = \frac{1}{\ln b} \cdot \frac{d}{dx} [\ln x] \qquad \left(\text{since } \frac{1}{\ln b} \text{ is a constant}\right)$$

$$= \frac{1}{\ln b} \cdot \frac{1}{x}. \qquad \text{[from Equation (14)]}$$

That is,

If b is any positive constant different from 1, then for all $x > 0$,

$$\frac{d}{dx} [\log_b x] = \frac{1}{x \ln b}. \tag{17}$$

Further, by applying the chain rule to (17),

If $g(x)$ is a positive differentiable function, then

$$\frac{d}{dx} [\log_b g(x)] = \frac{1}{g(x) \ln b} \cdot g'(x). \tag{18}$$

EXAMPLE 9   (a) $\dfrac{d}{dx} [\log_4 x] = \dfrac{1}{x \ln 4}.$ [by (17)]

(b) $\dfrac{d}{dx} [\log_3 x^2] = \dfrac{1}{x^2 \ln 3} \cdot 2x.$ [by (18)]

$$= \frac{2}{x \ln 3}$$

(c) $\dfrac{d}{dx} [\log_5 \sqrt{3x + 1}] = \dfrac{1}{\sqrt{3x + 1} \ln 5} \cdot \dfrac{1}{2} (3x + 1)^{-1/2}(3)$ [by (18)]

$$= \frac{3}{\sqrt{3x + 1} \ln 5 \cdot 2 \cdot \sqrt{3x + 1}}$$

$$= \frac{3}{2(3x + 1) \ln 5}.$$

5.3 EXERCISE SET ||||||||||||||||||||||||||||||||||||||||||||||||||||||||||||||||||||||||||||||

In Exercises 1–40, find dy/dx.

1. $y = e^{2x}$

2. $y = e^{-x}$

3. $y = e^{-1.08x}$

4. $y = 7e^{0.02x}$

5. $y = e^{x^3}$

6. $y = e^{2x^2 - 3x}$

7. $y = e^{\sqrt{3x+2}}$

8. $y = \sqrt{e^x + 3}$

9. $y = 3x^2 + e^{-x^3} + 4e^{x^2}$

10. $y = e^{(4-x)^5}$

11. $y = (e^x + e^{-x})^5$

12. $y = e^{x\sqrt{x+3}}$

13. $y = x^3 e^{4x}$

14. $y = x^2 e^{-x}$

15. $y = \dfrac{x}{e^x}$

16. $y = \dfrac{1}{1 + e^x}$

17. $y = \dfrac{e^x - e^{-x}}{e^x + e^{-x}}$

18. $y = \dfrac{1}{2x} e^{-3x^2}$

19. $y = \ln(x - 2)$

20. $y = \ln(3x - 5)$

21. $y = \ln(2x^2 + 3x - 1)$

22. $y = \ln(x^3 - 2x)$

23. $y = (\ln x)^5$

24. $y = \ln(x^5)$

25. $y = \ln \dfrac{1}{x}$

26. $y = \dfrac{1}{\ln x}$

27. $y = \ln \sqrt{x}$

28. $y = \ln \sqrt{9 - x^2}$

29. $y = \sqrt{\ln x}$

30. $y = \ln(x^2 + 9)^4$

31. $y = 3 \ln x^2 + 4x^2 - 5$

32. $y = \ln \left(\dfrac{x - 1}{x + 1} \right)$

33. $y = x^2 \ln x$

34. $y = \dfrac{\ln x}{x^2}$

35. $y = \dfrac{(\ln x)^2}{x^2}$

36. $y = \ln(\ln x)$

37. $y = \ln(e^x + x)$

38. $y = e^x \ln x$

39. $y = \dfrac{e^x}{\ln x}$

40. $y = e^{-x^2} \ln x$

Exercises 41–56 involve bases other than e. Find the derivative of the given function.

41. $y = 3^{2x}$

42. $y = x2^x$

43. $y = 4^{-3x}$

44. $y = 5^{3-4x}$

45. $y = x^2 2^x$

46. $y = 3^{x^2}$

47. $y = 500(1.02)^{x/3}$

48. $y = \dfrac{x}{3^x}$

49. $y = \log_3 x$

50. $y = 7 \log_4 x$

51. $y = \log_2(x^2 - 5x)$

52. $y = \log_3 \left(\dfrac{1}{x^2} \right)$

53. $y = \log \sqrt{x}$

54. $y = \log \left(\dfrac{x + 1}{x - 1} \right)$

55. $y = x^3 \log x$

56. $y = \dfrac{x^2}{\log x}$

57. Using information obtained from the first and second derivatives, show that $f(x) = \ln x$ is an increasing function for all $x > 0$, the graph of which is concave downward for all $x > 0$.

58. Repeat Exercise 57 for the common logarithm function $f(x) = \log x$.

59. Find the slope of the tangent line to the curve $y = x + e^x$ at the point $(0, 1)$.

60. Find the slope of the tangent line to the curve $y = \dfrac{e^x}{1 + \ln x}$ at the point where $x = 1$.

In Exercises 61–68, use information obtained from the first and second derivatives (Section 4.5) to sketch the graphs of the given functions. Find relative maxima and minima, and inflection points.

61. $y = e^{x/2}$

62. $y = \dfrac{e^x}{4}$

63. $y = e^x + e^{-x}$

64. $y = e^x - e^{-x}$

65. $y = xe^x$ 66. $y = e^x - x$

67. $y = x \ln x$ 68. $y = \ln x - x$

69. **(Business)** The cost (in hundreds of thousands of dollars) of manufacturing x diesel engines per week is given by

$$C(x) = \ln (x^2 + 5x) + 40,000.$$

Find an expression for the marginal cost.

70. **(Learning)** Suppose that the number of cards per hour N that an average person can keypunch after t hours of training is given by

$$N(t) = 80 - 60e^{-0.046t}.$$

Find the rate at which $N(t)$ is increasing with time

(a) after 5 hours of training.

(b) after 30 hours of training.

5.4 |||||||| APPLICATIONS OF THE EXPONENTIAL FUNCTION

There are many physical phenomena for which the rate of increase or decrease of a quantity is proportional to the amount of the quantity that is present. We discussed this concept in Example 5 of Section 5.3. Examples of such phenomena include growth of human population, growth of bacteria and other organisms, investment problems, radioactive decay, carbon dating, and concentration of a drug in the body. In this section we shall examine several of these phenomena. We shall discuss other important applications of the exponential function in Chapter 10.

|||||||||||||| Exponential Growth

A physical quantity is said to **grow exponentially,** or to show **exponential growth,** if at every instant, the rate of increase of the quantity is proportional to the amount of the quantity that is present at that instant.

Let $Q = Q(t)$ be the amount of a quantity that is present at time t, where t is the time that has elapsed from some initial observation. Thus, if $t = 2$ seconds, then 2 seconds have elapsed after the initial observation. The value of t at the initial observation is $t = 0$.

Suppose now that Q is growing exponentially. By definition, this means that the rate of increase at time t, dQ/dt, is proportional to the amount Q present at time t. Thus,

$$\frac{dQ}{dt} = kQ, \qquad k > 0, \tag{1}$$

where k is a constant of proportionality.

Equation (1) is an example of a **differential equation,** a topic which we shall study in more detail in Section 6.5. A **solution** to such a differential equation is a function $Q = Q(t)$ that satisfies the equation. As we have seen in Example 5 of Section 5.3, one solution to (1) is the function

$$Q = Q_0 e^{kt}, \tag{2}$$

where Q_0 is a constant. We shall also show in Section 10.2 that the *only* functions that satisfy the differential equation (1) must be of the form (2).

If $t = 0$ in (2), then

$$Q = Q_0 e^{k(0)} = Q_0 e^0 = Q_0 \cdot 1 = Q_0.$$

Thus, the initial quantity is Q_0. Figure 1 shows the graph of Equation (2) for $k > 0$.

Figure 1

The constant k in (1) and (2) is called the **growth constant,** because it determines the rate at which Q grows with time. Thus, the rate at which Q *grows* with time, at time t, is given by (1) as kQ.

EXAMPLE 1 Consider a quantity Q, the exponential growth of which is described by the equation

$$Q = 100 e^{0.04t},$$

where t is measured in days, and Q is measured in grams. Then the amount present initially is

$$Q_0 = 100 \text{ grams,}$$

and the growth constant is

$$k = 0.04.$$

The rate at which Q grows with time is given by

$$\frac{dQ}{dt} = kQ = (0.04)(100 e^{0.04t}).$$

Initially, $t = 0$, and $Q = 100$ grams. Thus, the rate at which Q is growing initially is

$$\frac{dQ}{dt} = (0.04)(100) = 4 \text{ grams/day.}$$

When $Q = 500$ grams the rate at which Q is growing is

$$\frac{dQ}{dt} = (0.04)(500) = 20 \text{ grams/day},$$

so the rate at which Q is growing does not remain constant. ∎

The constant k in (1) and (2) is often expressed as a percentage. For example, a growth constant of 5% means that $k = 0.05$ and a growth constant of 200% means that $k = 2$. A complication arises from the fact that k is also often called the **growth rate.** Observe that the "growth rate" k is *not* the rate at which Q is growing (which is kQ).

|||||||||||||| ## World Population Growth

We have all heard of the universal concern about the explosive growth of the world's population. This concern dates, principally, back to 1798, when the great British economist and social philosopher Thomas Robert Malthus* published *An Essay on the Principle of Population as It Affects Future Improvement of Society*. In this book, Malthus argued that human population grows exponentially, but the available food supply grows linearly, so that eventually the population must exceed the food supply. According to Malthus, the race between population and food supply was responsible for war, poverty, and disease. He opposed birth control measures, but he did favor "normal restraint" and the postponement of marriage as methods of limiting human reproduction. The Malthusian model of population growth is most appropriate for nonindustrial countries and for world population over long periods of time (30 to 40 years).

EXAMPLE 2 The world's population grows exponentially at the growth rate of 2% per year. If it was 4.8 billion at the beginning of 1986, estimate the world's population at the beginning of 2006.

Solution Let Q denote world population in billions. If the initial observation was made at the beginning of 1986, then

$$Q_0 = 4.8.$$

Since the growth rate is 2% per year, $k = 0.02$ and Equation (2), describing Q, becomes

$$Q = 4.8e^{0.02t},$$

---

* Thomas Robert Malthus (1766–1834), British economist and social philosopher, was born to a prosperous middle-class family and received his early schooling at home from his father and private tutors. He graduated from Cambridge University, where he won a number of prizes, with a degree in mathematics. For several years, he served as an ordained Anglican clergyman. The last twenty-nine years of his life were spent as a professor of history and political economy at the East India Company's Haileybury College.

where t is in years. At the beginning of 2006, the time t that will have elapsed since the initial observation is $t = 20$ years. Thus, the population in 2006 will be

$$Q = 4.8e^{0.02(20)} = 4.8e^{0.4}$$
$$= 4.8(1.4918) \qquad \text{(Appendix B, Table I)}$$
$$= 7.1606 \text{ billion.}$$

Hence, there will be approximately 7.2 billion people in the world in 2006. ■

EXAMPLE 3   Suppose Q is a quantity that is growing exponentially with growth rate k. How long will it take the initial quantity Q_0 to double?

Solution   If Q_0 doubles in time t, then at the end of that period, the quantity $2Q_0$ will be present. Substituting in Equation (2), we see that at time t,

$$2Q_0 = Q_0e^{kt}.$$

Dividing by Q_0, we have

$$2 = e^{kt}.$$

In exponential form,

$$\ln 2 = kt, \qquad \text{(since } y = e^x \text{ if and only if } \ln y = x)$$

so

$$t = \frac{\ln 2}{k}.$$

The value of t just obtained, that is,

$$T = \frac{\ln 2}{k} \qquad\qquad (3)$$

is called the **doubling time.** Observe that the doubling time does not depend on the initial quantity Q_0.

From Equation (3) we obtain

$$kT = \ln 2$$

or, in exponential form,

$$e^{kT} = 2. \qquad\qquad (4)$$

Thus, for all n, when we substitute $t = nT$ into Equation (2) we obtain $Q(nT) = Q_0e^{knT} = Q_0(e^{kT})^n = Q_02^n$.

That is,

For all n,

$$Q(nT) = 2^n Q_0. \tag{5}$$

Using Equation (5) we construct the following table of values for the function $Q = Q_0 e^{kt}$.

t	0	T	$2T$	$3T$	$4T$	$\cdots$	nT
Q	Q_0	$2Q_0$	$4Q_0$	$8Q_0$	$16Q_0$	$\cdots$	$2^n Q_0$

We can use this table in obtaining a graph of this function.

Figure 2

Figure 2 shows the graph of Equation (2), as well as doubling times, graphically. Notice that the quantity Q, which is growing exponentially, is growing at an extremely rapid rate. ∎

EXAMPLE 4   World population is growing exponentially at the growth rate of 2% per year. How long does it take the world's population to double?

Solution   Since $k = 0.02$, Equation (3) gives

$$T = \frac{\ln 2}{0.02} = \frac{0.6931}{0.02}$$
$$= 34.655 \text{ years.}$$

Hence, it takes approximately 35 years for the world's population to double. It is this phenomenal growth rate that alarms many demographers. ∎

In many applied exponential-growth models the constant k is unknown. We can determine it, however, if we have two values of Q at different times.

EXAMPLE 5
(Biology)

Consider a colony of *E. coli* bacteria that is growing exponentially. A microbiologist finds that, initially, 10,000 bacteria are present and 30 minutes later there are 18,000 bacteria. Find an expression for the number of bacteria $Q(t)$ after t minutes.

Solution

Given $Q_0 = 10,000$, all we need to find is the growth constant k. Since $Q(30) = 18,000$, by (2) we have

$$Q(30) = 18,000 = 10,000e^{k \cdot 30}.$$

Thus,

$$e^{k \cdot 30} = \frac{18,000}{10,000} = 1.8.$$

Taking natural logarithms of both sides, we obtain

$$30k = \ln 1.8 = 0.5878. \qquad \text{(Appendix B, Table II)}$$

Hence,

$$k = \frac{0.5878}{30} \approx 0.020.$$

Therefore,

$$Q(t) \approx 10,000e^{0.020t}. \qquad\qquad ∎$$

||||||||||||| **Exponential Decay**

A physical quantity is said to be **decaying exponentially,** or to **show exponential decay,** if at every instant, the rate of decrease of the quantity is proportional to the amount of the quantity that is present at that instant.

Let $Q = Q(t)$ be the amount of a quantity that is present at time t, where t is the time that has elapsed after some initial observation. Suppose that Q is decaying exponentially:

$$\frac{dQ}{dt} = -kQ, \qquad k > 0, \tag{6}$$

where the minus sign indicates that Q is decreasing. If a function Q satisfies the differential equation (6), then it can be proved that Q must be of the form

$$Q = Q_0 e^{-kt}, \tag{7}$$

where Q_0 is the amount present initially (see Section 10.2 for proof). In Equations (6) and (7) the contant k is called the **decay constant**. The rate at which Q decays with time at time t is given by kQ. Figure 3 shows the graph of Equation (7).

Figure 3

$$Q = Q_0 e^{-kt}$$

$$(k > 0)$$

EXAMPLE 6   Consider a quantity Q, the exponential decay of which is described by the equation

$$Q = 200e^{-0.03t},$$

where t is measured in hours and Q is measured in grams. The amount present initially is

$$Q_0 = 200 \text{ grams,}$$

and the decay constant is

$$k = 0.03.$$

The rate at which Q is decaying initially is

$$kQ = (0.03)(200) = 6 \text{ grams/hour.}$$

The rate at which Q is decaying when $Q = 150$ is

$$kQ = (0.03)(150) = 4.5 \text{ grams/hour.} \qquad \blacksquare$$

The constant k in Equations (6) and (7) is often expressed as percentage. For example, a decay constant of 5.2% means that $k = 0.052$ and a decay constant of 120% means that $k = 1.2$. As in the case of exponential growth, the constant k is also called the **decay rate.**

EXAMPLE 7   Consider a quantity Q that is decaying exponentially with time. How long does it take the initial quantity Q_0 to shrink to half its value?

Solution   Suppose that after a time period t, the quantity $\frac{1}{2}Q_0$ is present. Substituting in Equation (7), we write

$$\tfrac{1}{2}Q_0 = Q_0 e^{-kt}.$$

Dividing by Q_0 gives

$$\tfrac{1}{2} = e^{-kt},$$

or taking natural logarithms of both sides yields

$$\ln \tfrac{1}{2} = -kt$$
$$-\ln 2 = -kt$$
$$t = \frac{\ln 2}{k}.$$

∎

The value of t just obtained, that is,

$$T = \frac{\ln 2}{k}, \tag{8}$$

is called the **halving time,** or the **half-life.** It is exactly the same expression as the doubling time for a quantity that is growing exponentially [Equation (3)]. Figure 4 shows half-lives graphically.

Figure 4

EXAMPLE 8
(Radioactive Decay)

Flourine-17 has a decay rate of approximately 1.05% per second.
(a) If 200 grams of flourine-17 are present initially, how much will be left after 20 seconds?
(b) What is the half-life of flourine-17?

Solution

(a) We have $Q_0 = 200$ and $k = 0.0105$. According to Equation (7), the amount left after 20 seconds will be

$$Q = 200e^{-0.0105(20)}$$
$$= 200e^{-0.21}$$
$$= 200(0.8106) \qquad \text{(Appendix B, Table I)}$$
$$= 162.12 \text{ grams.}$$

(b) From Equation (8), the half-life is

$$T = \frac{\ln 2}{k} = \frac{0.6931}{0.0105} = 66 \text{ seconds.}$$ ∎

|||||||||||||| **Radiocarbon Dating**

When nitrogen in the earth's upper atmosphere is bombarded by cosmic rays, radioactive carbon-14 is produced. Because carbon-14 is radioactive, it decays with time and reforms nitrogen-14, from which it was made. During this process, the decaying carbon atoms emit energy in the form of radiation, which can be measured with a Geiger counter. Carbon-14 combines with oxygen to form carbon dioxide, which is consumed by plants, which, in turn, are eaten by animals. Thus, all organic forms of life contain radioactive carbon-14. In 1947, W. F. Libby* found that the percentage of carbon-14 in the atmosphere equals the percentage found in the living tissues of all organic forms of life. When an organism dies, it stops replacing carbon-14 in its living tissues. Yet the carbon-14 continues decaying. The half-life of carbon-14 is 5730 years and the decay rate is 0.012% per year. Thus, by measuring the amount of carbon-14 in the remains of an organism, it is possible to estimate fairly accurately when the organism died. This method of radiocarbon dating has been an invaluable tool in archaeology, geology, and anthropology. The following application illustrates the scientific usefulness of this technique.

EXAMPLE 9
(The Latest
Glacial Age)
In the late 1940s, radiocarbon dating was used to date the last ice sheet to cover the North American and European continents. Parts of trees in the Two Creeks Forest in northern Wisconsin were found and tested for their carbon-14 content. It was determined that they had retained 25.4% of the original quantity Q_0 of carbon-14 that was present when the descending ice sheet felled the trees. To find the age t of the wood, we substitute in Equation (7):

$$25.4\% \, Q_0 = Q_0 e^{-0.00012t} \qquad (k = 0.012\% = 0.00012)$$

or

$$0.254 Q_0 = Q_0 e^{-0.00012t}$$
$$0.254 = e^{-0.00012t}$$
$$\ln 0.254 = \ln e^{-0.00012t}$$
$$-1.3704 = -0.00012t.$$

---

* Willard Frank Libby (1908–80) was born in Colorado and received his Ph. D. at the University of California, where he taught prior to World War II. During the war he worked on the Manhattan Project, which developed the atomic bomb, and after the war he was a member of the U.S. Atomic Energy Commission. In 1959 he returned to teaching and research at the University of California, Los Angeles. In 1960 he was awarded the Nobel Prize in chemistry for his work on radiocarbon dating.

(ln 0.254 was obtained by a calculator or by a more extensive table than Appendix B, Table II.) Therefore,

$$t = -\frac{1.3704}{0.00012} = 11,420.$$

Thus, the ice age occurred approximately 11,420 years before the measurements were taken.

5.4 EXERCISE SET ||||||||||||||||||||||||||||||||||||||||||||||||||||||||||||||||||||||||||||||||||||||

1. Consider a quantity Q the exponential growth of which is given by the equation

$$Q = 5000e^{0.4t} \qquad (9)$$

where t is measured in days.
   (a) Find the amount present initially.
   (b) What is the growth constant?
   (c) Write the differential equation of which (9) is a solution.
   (d) Find the rate at which Q is growing when $Q = 15,000$.
   (e) How much is there after 5 days of growth?

2. Consider a quantity Q the exponential decay of which is given by the equation

$$Q = 400e^{-0.5t} \qquad (10)$$

where t is measured in weeks.
   (a) Find the amount present initially.
   (b) What is the decay constant?
   (c) Write the differential equation of which (10) is a solution.
   (d) Find the rate at which Q is decaying when $Q = 1200$.
   (e) How much is left after 6 weeks?

3. A quantity begins at 8000 lb and decays exponentially at the decay rate of 5% per year.
   (a) Write the differential equation that must be satisfied by Q.
   (b) How much of the quantity remains after t years?
   (c) How much of the quantity decays during the first 30 years?

4. A quantity Q, which is initially 600, is growing exponentially at the growth rate of 20% per hour.
   (a) Write the differential equation that must be satisfied by Q.
   (b) How much is there after t hours of growth?
   (c) How much is there after 20 hours of growth?

5. The amount of a quantity Q that is growing exponentially at the growth rate of 10% per hour is 500 after 10 hours of growth. Find the amount present initially.

6. **(Bacteria Growth)** To measure a bacterial colony that is growing exponentially, a scientist takes two readings. The first reading shows that 8000 bacteria are present. The second reading, taken 40 minutes later, shows that there are 12,000 bacteria. Find an expression for the number of bacteria present t minutes after the first reading.

7. **(Demography)** In 1983 Nigeria had a population of approximately 85 million, and was growing exponentially at the growth rate of 3.4% per year.
   (a) Write the differential equation that was satisfied by the population of Nigeria in 1983.
   (b) If this growth rate continues, what will be the population of Nigeria in the year 2013? (Use a calculator.)
   (c) At this growth rate, in what year will the population of Nigeria reach 300 million?

8. **(Demography)** The growth rate of the population of a certain country was 2% per year from 1920 to 1980.
   (a) If the population was 2.1 million in 1950, what was it in 1980?
   (b) Using the results of (a), estimate the population in 1990.

9. Find the doubling time for the population of Nigeria, assuming the growth rate given in Exercise 7.

10. **(Population Growth)** In 1975, the growth rate of Mexico's population was 3.3% per year. How long will it take the country's population to double?

11. A colony of fruit flies that is growing exponentially doubles in eight hours. What is the growth rate?

12. **(Radioactive Decay)** Carbon-14 has a half-life of 5730 years. Find the decay constant.

13. **(Medicine)** When insulin is injected in a patient's bloodstream, its concentration in the bloodstream decreases at a rate proportional to the existing concentration.
   (a) If the constant of proportionality is 25% per hour, how long does it take the concentration to decrease to $\frac{1}{5}$ of its original concentration.
   (b) If the initial concentration is 0.08 milligrams/milliliter of blood, what is the concentration after 4 hours?

14. **(Radioactive Decay)** The decay rate of krypton-85 is 6.3% per year.
   (a) What is the half-life?
   (b) How long does it take to decay to $\frac{1}{4}$ of the amount present initially?

15. **(Radioactive Decay)** Tritium has a half-life of 12.3 years. How much time is required for 60% of the tritium to decay?

16. **(Carbon Dating)** Grain discovered in an Egyptian pyramid was found to contain 60% of its carbon-14. How old was the grain?

17. **(Carbon Dating)** A fossilized skeleton was found to have lost $\frac{1}{5}$ of its original carbon-14 content. How old was the fossil?

18. **(Nuclear Energy Supply)** The power supply of a lunar quake detector is a radio-isotope (that is, a radioactive substance) the power output P (in watts) of which decreases with time according to the equation

$$P = 80e^{-0.005t}$$

   where t is in days.
   (a) What is the power output after 250 days?
   (b) What is the half-life of the power supply?
   (c) If the detector requires 20 watts of power for its operation, how long will it stay in operation?

19. **(Oil Consumption)** During the years 1950–77, the yearly consumption of oil in the United States was growing exponentially. In 1950, the demand was 1 billion barrels, but in 1977, it was 2.6 billion barrels.
   (a) Find the growth constant.
   (b) Estimate oil consumption for the year 2000, at this growth rate.

20. **(Radioactive Decay)** After 18 years, 30% of a radioactive substance decays.
   (a) Find its (annual) decay constant.
   (b) What is its half-life?
   (c) How long does it take for 90% of the substance to decay?

21. **(Nuclear Power Accident)** As a result of an accident, a nuclear power plant has contaminated the environment with strontium-90 and nearby residents have been evacuated. The half-life of strontium-90 is 28 years. Public officials will not allow the residents to return to their homes until 90% of the radioactivity has disappeared. Assuming that no cleanup effort is made, how long will it be before the area is once again fit for human habitation?

22. **(Nuclear Power Accident)** After a nuclear reactor accident, radioactive iodine-131 was detected in the pasture of a nearby dairy farm, at a level 20 times higher than the maximum level deemed "safe" for cows that produce milk for human consumption. Iodine-131 has a half-life of 8 days. How many days should the farmer keep his cows from grazing on this pasture, assuming there is no additional source of contamination?

KEY IDEAS FOR REVIEW ||||||||||||||||||||||||||||||||||||||||||||||||||||||||||||||||||||||||||

■ An exponential function is one defined by
$$f(x) = a^x \qquad a > 0, \ a \neq 1.$$

■ $e = \lim\limits_{n \to \infty} \left(1 + \dfrac{1}{n}\right)^n \approx 2.718.$

■ **The exponential function** is $f(x) = e^x$.

■ If $f(x) = a^x$, with $a > 1$, then
  . the domain of f is $(-\infty, +\infty)$.
  . $f(x) > 0$ for all x.
  . f is an increasing function for all x.
  . The graph of f passes through $(0, 1)$.
  . $\lim\limits_{x \to \infty} f(x) = \infty$.
  . The graph of f is concave upward for all x.

■ If $f(x) = a^{-x}$, with $a > 1$, then
  . the domain of f is $(-\infty, +\infty)$.
  . $f(x) > 0$ for all x.
  . f is a decreasing function for all x.
  . The graph of f passes through $(0, 1)$.
  . $\lim\limits_{x \to \infty} f(x) = 0$.
  . The graph of f is concave upward for all x.

■ If a principal P is invested at an annual rate r, compounded k times a year, and S is the value of the investment after n conversion periods, then
$$S = P\left(1 + \frac{r}{k}\right)^n \quad \text{and} \quad P = S\left(1 + \frac{r}{k}\right)^{-n}.$$

■ If a principal P is invested at an annual rate r, compounded continuously, and S is the value of the investment after t years, then
$$S = Pe^{rt} \quad \text{and} \quad P = Se^{-rt}.$$

■ $y = \log_a x$ if and only if $x = a^y$. The base a is always a positive number different from 1.

■ The **natural logarithm** of x is $\ln x = \log_e x$.

The **common logarithm** of x is $\log x = \log_{10} x$.

■ $a^{\log_a x} = x$; $\log_a (a^x) = x$. $e^{\ln x} = x$; $\ln e^x = x$.

■ $\log_a a = 1$; $\log_a 1 = 0$. $\ln e = 1$; $\ln 1 = 0$.

■ If $f(x) = \log_a x$, with $a > 1$, then
  . the domain of f is $(0, +\infty)$ and the range of f is $(-\infty, +\infty)$.
  . $f(x) > 0$ if $x > 1$; $f(x) < 0$ if $x < 1$.
  . f is an increasing function for all x.
  . the graph of f passes through the point $(1, 0)$.
  . $\lim\limits_{x \to +\infty} f(x) = +\infty$; $\lim\limits_{x \to 0^+} f(x) = -\infty$.
  . the graph of f is the reflection of the graph of $y = a^x$ across the line $y = x$.
  . the graph of f is concave downward.

■ $\log_a xy = \log_a x + \log_a y$.

■ $\log_a \dfrac{x}{y} = \log_a x - \log_a y$; $\log_a \dfrac{1}{x} = -\log_a x$.

■ $\log_a x^k = k \log_a x$.

■ $\log_a x = \dfrac{\ln x}{\ln a} = \dfrac{\log_b x}{\log_b a}$.

■ $\dfrac{d}{dx}[e^x] = e^x$.

■ $\dfrac{d}{dx}[e^{g(x)}] = e^{g(x)} g'(x)$.

■ $\dfrac{d}{dx}[e^u] = e^u \dfrac{du}{dx}$.

■ $\dfrac{d}{dx}[a^x] = a^x \ln a$; $\dfrac{d}{dx}[a^{g(x)}] = a^{g(x)} g'(x) \ln a$.

■ $\dfrac{d}{dx}[\ln x] = \dfrac{1}{x}$ (assuming $x > 0$).

■ $\dfrac{d}{dx}[\ln g(x)] = \dfrac{1}{g(x)} g'(x)$ (assuming $g(x) > 0$).

■ $\dfrac{d}{dx}[\log_b x] = \dfrac{1}{x \ln b}$ (assuming $x > 0$).

■ $\dfrac{d}{dx}[\log_b g(x)] = \dfrac{g'(x)}{g(x) \ln b}$
(assuming $g(x) > 0$).

■ If $Q = Q_0 e^{kt}$, then $dQ/dt = kQ$.

■ The solution to the differential equation $dQ/dt = kQ$, where k is a constant, is $Q = Q_0 e^{kt}$, where Q_0 is the constant $Q(0)$.

■ Q is growing exponentially if $Q = Q_0 e^{kt}$,

$k > 0$. The amount initially present is Q_0, and the constant k is called the growth constant (or growth rate).

■ Q is decaying exponentially if $Q = Q_0 e^{-kt}$, $k > 0$. The amount initially present is Q_0, and the constant k is called the decay constant (or rate of decay).

■ If $Q = Q_0 e^{kt}$, then the number $T = \dfrac{\ln 2}{|k|}$
is called the doubling time if $k > 0$, or the half-life if $k < 0$.

REVIEW EXERCISES ||||||||||||||||||||||||||||||||||||||||||||||||||||||||||||||||||||||||||||||||||

1. If $P(t) = 400e^{-0.04t}$, compute $P(10)$.

2. Sketch the graph of $f(x) = 2e^{4x}$.

3. Sketch the graph of $f(x) = 20e^{-2x}$.

4. Find the interest on $5000 invested at 7% per year, compounded semiannually, after 4 years.

5. Find the future value of $4000 invested at 6% per year, compounded quarterly, after 6 years.

6. How much money invested now at 10% per year, compounded continuously, will yield $12,000 in eight years?

7. What principal, if invested now at 8% per year, compounded semiannually, would grow to $5000 in 4 years?

8. Find the effective rate of interest, when the nominal rate is 6%, compounded monthly.

9. Find the effective rate of interest, when the nominal rate is 9%, compounded continuously.

10. **(Atmospheric Pressure)** The atmospheric pressure P (in pounds per square inch) is approximately given by $P = 14.7e^{-0.21h}$, where h is the altitude (in miles) above sea level.
    (a) What is the pressure at sea level?
    (b) At what altitude is the pressure twice what it is at sea level?

11. Compute
    (a) $\log 0.0001$ (g) $\ln e^{-2/3}$
    (b) $\log_6 36$ (h) $\ln e^{0.02t}$
    (c) $\log_9 \sqrt[4]{81}$ (i) $\ln \dfrac{1}{e^5}$
    (d) $\log_2 \left(\dfrac{1}{8}\right)$ (j) $2^{\log_2 7}$
    (e) $\log_{1/3}(9)$ (k) $e^{\ln 1.02}$
    (f) $\log 10,000$ (l) $5^{-\log_5 2}$

12. Using only the values $u = \log 2$, $v = \log 3$, and $w = \log 5$, find
    (a) $\log 24$.   (b) $\log \sqrt{15}$.   (c) $\log \frac{81}{5}$.

13. Using the values of u, v and w in Exercise 12, find
    (a) $\log_2 3$.   (b) $\log_5 6$.

14. Solve for t using a calculator:
    (a) $40e^{-4t} = 20$.
    (b) $50(1 - e^{-0.02t}) = 20$.

15. **(Business)** A computer manufacturer finds that when x millions of dollars are spent on research, the profit P (in millions of dollars) is given by

    $$P(x) = 20 + 5 \ln(x + 3).$$

    How much should be spent on research to make a profit of 40 million dollars?

In Exercises 16–32, find dy/dx.

16. $y = e^{-x^2}$ 17. $y = e^{1.03x}$

18. $y = e^{\sqrt{x}}$ 19. $y = e^{x\sqrt{2x+1}}$

20. $y = x^3 e^{-2x}$

21. $y = \dfrac{e^x}{x^2}$

22. $y = \dfrac{1}{1 - e^x}$

23. $y = \ln (x + 5)$

24. $y = \ln x^6$

25. $y = \ln \sqrt[3]{x^2}$

26. $y = \ln (x^2 + 2)^3$

27. $y = \ln (x^3 + 5x + 3)$

28. $y = \dfrac{\ln x^2}{x^2}$

29. $y = 3^{2x}$

30. $y = 4^{-5x/2}$

31. $y = \log_2 5x$

32. $y = \log_3 (x^2 + 4)$

33. Find the slope of the tangent line to the curve $y = x^2 + 5e^x$ at the point $(0, 5)$.

34. Use the first and second derivatives to analyze the graph of the function $f(x) = xe^{-x}$. Locate all relative maxima and minima and inflection points.

35. (**Business**) The revenue received from manufacturing and selling x stereos is given by

$$R(x) = x \ln (x + 2).$$

Find an expression for the marginal revenue.

36. How long does it take for a population to triple that is growing exponentially with growth rate k?

37. (**Nuclear Energy Accident**) A damaged nuclear reactor has discharged cesium-137, whose decay rate is 2.3% per year. Write the differential equation that is satisfied, and find its solution. How long will it take half of the material originally discharged to decay (assuming no cleanup operation)?

38. A bacterial colony is growing at a rate proportional to the number of bacteria present. If the initial colony consisted of 2000 bacteria and the constant of proportionality is 40% per hour, find an expression for the number of bacteria after t hours of growth.

39. Cobalt-60 has a decay rate of 13.3% per year.
    (a) Write the appropriate differential equation, and its solution.
    (b) How long does it take half the amount present initially to decay?
    (c) How long does it take $\frac{3}{4}$ the amount present initially to decay?

40. A certain country's population is growing exponentially at a growth rate of 2.5% per year. How many years does it take for the population to double?

41. Xenon-133 has a half-life of 5.27 days. Find its decay rate (per day).

42. A dart tip was found to contain 70% of its original carbon-14. How old was the dart tip?

CHAPTER TEST ||||||||||||||||||||||||||||||||||||||||||||||||||||||||||||||||||||||||||||

(A calculator with e^x and $\ln x$ keys will be needed.)

1. Sketch the graphs of $y = 3^x$ and $y = \log_3 x$ in the same coordinate system.

2. A principal of $5000 is invested at 8% per year compound interest. Find the compound value of the investment after 3 years if
   (a) the interest is compounded quarterly.
   (b) the interest is compounded continuously.

3. If $4000 is invested at 9% per year interest compounded continuously, how long will it take for the investment to grow to $7000?

4. Find
   (a) $\log_2 16$.   (c) $e^{2 \ln 7}$.   (e) $\log 100{,}000$.
   (b) $\log_3 \left(\dfrac{1}{9}\right)$.   (d) $\ln \left(\dfrac{1}{e^2}\right)$.   (f) $2^{-\log_2 5}$.

5. Given $r = \log_b x$, $s = \log_b y$, and $t = \log_b z$, express

$$\log_b \sqrt{\frac{xy^3}{z^4}}$$

   in terms of r, s, and t.

6. Find the slope of the tangent line to the curve $y = x^2 e^x$ at the point where $x = 1$.

7. For each of the following functions $f(x)$, find $f'(x)$.

   (a) $f(x) = 50e^{0.1x}$ (c) $f(x) = \dfrac{x^4}{e^{3x}}$

   (b) $f(x) = \ln (4x^3 + x)$ (d) $f(x) = e^{\sqrt{x}}$

8. (a) $\dfrac{d}{dx} [4^{x^2}] =$

   (b) $\dfrac{d}{dx} [x^2 \log_5 x] =$

   Alternate Question 8, for those classes that did not study bases other than e and 10:

   (a) $\dfrac{d}{dx} [\ln (x^2 + 1)]^5 =$

   (b) $\dfrac{d}{dx} [x^3 \ln 2x] =$

9. Starting with an initial population of 4000, a bacterial colony grows hourly at a rate that is always 3% of the number of bacteria present in the colony.
   (a) Write the differential equation that must be satisfied by the population, and write its solution.
   (b) How many bacteria are present 10 hours after the initial time?
   (c) How long does it take for the population to double?

10. After 17 years, 40% of a radioactive substance decays.
    (a) Find the decay rate.
    (b) What is its half-life?
    (c) How long does it take for 90% of the substance to decay?

ANTIDIFFERENTIATION

Calculus is divided into two general categories: **differential calculus,** which we studied in Chapters 3 and 4, and **integral calculus,** which we shall study in this chapter and in Chapter 7.

To most readers, the word integration means "putting together." In Chapter 7, we shall study integration from that point of view and we shall consider a wide variety of applications. In this chapter, we shall view integration as the reverse process of differentiation. Moreover, we shall develop some elementary methods of integration and discuss the use of integral tables.

6.1 ||||||| ANTIDERIVATIVES

In Chapters 3 and 4 we obtained properties of a given function by studying its derivative. In this section we shall tackle the reverse problem: given the derivative function at each value of x, we must find the original function at each value of x. This type of problem arises in many practical situations. For example, if we know the instantaneous velocity of an object at each instant, can we find the position of the object at each instant? Or, if we know a manufacturer's marginal revenue function, can we find the revenue function itself?

Given a function $f(x)$, we are interested in finding a function $F(x)$ whose derivative is $f(x)$; that is, a function $F(x)$ for which

$$F'(x) = f(x). \tag{1}$$

Such a function $F(x)$ is called an **antiderivative** of $f(x)$.

Caution Many students will allow sloppy notation, or inattention to proper notation, to get in their way at this point. It will be very important, in this chapter, to pay strict attention to correct symbolism. Begin by observing that both a lowercase f and a capital F have been used above in Equation (1). We are using capital F to denote an antiderivative of lower case f: $F'(x) = f(x)$. This notation is customary, and should be followed carefully.

As an example, we shall suppose that $f(x) = 2x$ and try to find an antiderivative of $f(x)$; that is, a function $F(x)$ whose derivative $F'(x)$ is $f(x) = 2x$. From our previous work with derivatives, we know that the function

$$F(x) = x^2$$

has the desired derivative. But other functions also have the same derivative. For example, we could choose

$$F(x) = x^2 + 8, \qquad F(x) = x^2 - 3, \quad \text{or} \quad F(x) = x^2 + C,$$

where C is any constant, because, for all the above choices, $F'(x) = 2x$.

We have shown that any function of the form $F(x) = x^2 + C$, where C is an arbitrary constant, is an antiderivative of the function $f(x) = x^2$. By an "arbitrary constant C" we mean that we can assign to C any real number. In general, if $F(x)$ is any antiderivative of $f(x)$ [so that $F'(x) = f(x)$] then $F(x) + C$ is also an anti-derivative of $f(x)$, where C is an arbitrary constant. This is because

$$\frac{d}{dx}[F(x) + C] = \frac{d}{dx}[F(x)] + \frac{d}{dx}[C] = F'(x) + 0$$

$$= F'(x)$$

$$= f(x).$$

It can be shown that *every* antiderivative of $f(x)$ can be written in the form $F(x) + C$, where $F(x)$ is a particular antiderivative of $f(x)$ and C is an arbitrary constant. This fact is not obvious, but is a consequence of the following argument, concluding with the corollary to Theorem 2.

THEOREM 1

(Zero Derivative Principle)
If f is differentiable on an interval I, and if $f'(x) = 0$ for all x in I, then f is constant throughout the interval I.

The proof of this theorem is beyond the scope of this book. It is the *converse* of a known fact, the "constant rule," which says that the derivative of a constant is 0, and which was proved in Section 3.2. From Theorem 1, the following is an immediate consequence.

THEOREM 2

(Equal Derivatives Principle)
If f and g are differentiable and have equal derivatives on an interval I, then they differ by a constant amount throughout I.

That is, if $f'(x) = g'(x)$ for all x in I, then there is a constant C such that

$$f(x) - g(x) = C,$$

or

$$f(x) = g(x) + C,$$

for all x in I.

Proof Suppose $f'(x) = g'(x)$ for all x in I. Let $h(x) = f(x) - g(x)$. Then, for all x in I,

$$h'(x) = \frac{d}{dx}[f(x) - g(x)]$$

$$= f'(x) - g'(x) \qquad \text{(the "difference rule", Section 3.2)}$$

$$= 0. \qquad [\text{since } f'(x) = g'(x)]$$

Therefore, by Theorem 1, $h(x)$ is constant on I. That is, there is some constant C such that for all x in I,

$$h(x) = C.$$

Thus,

$$f(x) - g(x) = C, \quad \text{or} \quad f(x) = g(x) + C,$$

for all x in I.

COROLLARY

(Antiderivatives)
Any two antiderivatives of the same function differ by a constant. More precisely, if $F(x)$ is one antiderivative of $f(x)$ over an interval I and $G(x)$ is any other, then there is a constant C such that for all x in I,

$$F(x) = G(x) + C.$$

Proof Suppose $F(x)$ and $G(x)$ are antiderivatives of the same function $f(x)$ on an interval I. Then $F'(x) = f(x)$ and $G'(x) = f(x)$ for all x in I. Hence, by Theorem 2, there is some constant C such that

$$F(x) = G(x) + C$$

for all x in I.

We have thus determined that the *only* antiderivatives of a function are of the form $F(x) + C$, where $F(x)$ is any specific antiderivative of $f(x)$.

A function $F(x)$ is called an **antiderivative** of $f(x)$ if $F'(x) = f(x)$ for all x in some interval. The process of finding $F(x)$ once we know $f(x)$ is called **antidifferentiation** or **integration**. The **most general antiderivative of $f(x)$** is denoted

$$\int f(x)\,dx = F(x) + C, \qquad (2)$$

where $F(x)$ is some particular antiderivative of $f(x)$.

EXAMPLE 1   (a) $\int 2x\,dx = x^2 + C$, because $\dfrac{d}{dx}[x^2 + C] = 2x$.

(b) $\int 3x^2\,dx = x^3 + C$, because $\dfrac{d}{dx}[x^3 + C] = 3x^2$.

(c) $\int (3x^2 + 2x)\,dx = x^3 + x^2 + C$, because $\dfrac{d}{dx}[x^3 + x^2 + C] = 3x^2 + 2x$.

There is a considerable amount of terminology associated with Equation (2). The symbol $\int$ is called the **integral sign,** x is called the **variable of integration,** the arbitrary constant C is called the **constant of integration,** and the function f is called the **integrand.** Moreover, the differential dx helps to keep track of the variable of the function being integrated. Finally, the symbol $\int f(x)\,dx$ is called the **indefinite integral of $f(x)$.**

Since integration or antidifferentiation is the reverse of differentiation, we can reverse the differentiation formulas given in Sections 3.2 and 5.3 to obtain integration formulas. Thus, we have the following integration formulas.

|||||||||||||| **Integrating Powers of x**

If k is a constant, then

$$\int k\,dx = kx + C, \tag{3}$$

because

$$\frac{d}{dx}[kx + C] = \frac{d}{dx}[kx] + \frac{d}{dx}[C]$$

$$= k + 0 = k.$$

Note that (3) yields

$$\int 0\,dx = 0 \cdot x + C = C.$$

The antiderivative of x^r is given by

$$\int x^r\,dx = \frac{x^{r+1}}{r+1} + C, \qquad \text{if } r \neq -1, \tag{4}$$

because

$$\frac{d}{dx}\left[\frac{x^{r+1}}{r+1} + C\right] = \frac{1}{r+1}\frac{d}{dx}[x^{r+1}] + \frac{d}{dx}[C]$$

$$= \frac{1}{r+1}(r+1)x^r + 0 = x^r.$$

Equation (4) says that

> to integrate x to a power (different from -1), raise x to the power increased by one and divide by the increased power.

The integral $\int 1 \, dx$ can be written simply as $\int dx$. By (4), we have

$$\int dx = \int 1 \, dx = \int x^0 \, dx = \frac{x^{0+1}}{0+1} + C,$$

or, directly from (3),

$$\int dx = x + C.$$

Next we want to integrate $1/x$; that is, we want to find the antiderivatives of $1/x$. This information will plug the gap in Equation (4), finding the antiderivative of x^r when $r = -1$. Recall from Section 5.3 that

$$\frac{d}{dx} [\ln x] = \frac{1}{x} \qquad \text{for } x > 0.$$

We are then led to conclude that

$$\int \frac{1}{x} \, dx = \ln x + C, \qquad \text{if } x > 0. \tag{5}$$

This formula holds only when x is positive, because $\ln x$ is not defined for $x < 0$. However, Equation (5) holds for all $x \neq 0$ if $\ln x$ on the right side is replaced by $\ln |x|$. To confirm this result, we proceed as follows. If $x < 0$, $|x| = -x$. Thus, when $x < 0$,

$$\frac{d}{dx} [\ln |x|] = \frac{d}{dx} [\ln (-x)]$$

$$= \frac{1}{-x} \frac{d}{dx} [-x] \qquad \text{[by the chain rule or by Equation (16) of Section 5.3]}$$

$$= -\frac{1}{x} (-1) = \frac{1}{x}.$$

We have just shown that

$$\int \frac{1}{x} \, dx = \ln |x| + C, \qquad \text{if } x < 0.$$

Moreover, if $x > 0$, then $|x| = x$ and

$$\frac{d}{dx}[\ln |x|] = \frac{d}{dx}[\ln x] = \frac{1}{x}.$$

Hence, for all values of x, except zero,

$$\frac{d}{dx}[\ln |x|] = \frac{1}{x}.$$

We can therefore write the following integration formula:

$$\int \frac{1}{x}\, dx = \ln |x| + C, \qquad x \neq 0. \tag{6}$$

Summarizing,

$$\int x^r\, dx = \begin{cases} \dfrac{x^{r+1}}{r+1} + C, & \text{if } r \neq -1; \\[2mm] \ln |x| + C, & \text{if } r = -1 \text{ and } x \neq 0. \end{cases}$$

EXAMPLE 2   (a) $\int 5\, dx = 5x + C.$ [Equation (3)]

(b) $\int x^3\, dx = \dfrac{x^4}{4} + C.$ [Equation (4)]

(c) $\int x^{-1}\, dx = \ln |x| + C,$ if $x \neq 0.$ [Equation (6)]

EXAMPLE 3   (a) $\int \sqrt{x}\, dx = \int x^{1/2}\, dx = \dfrac{x^{3/2}}{3/2} + C = \dfrac{2}{3} x^{3/2} + C.$

(b) $\int x^{-6}\, dx = \dfrac{x^{-5}}{-5} + C = -\dfrac{x^{-5}}{5} + C = -\dfrac{1}{5x^5} + C.$

|||||||||||||| **Integrating Exponential Functions**

Since $\dfrac{d}{dx}[e^x] = e^x$, we have

$$\int e^x\, dx = e^x + C. \tag{7}$$

Also, for any positive base different from 1, $\dfrac{d}{dx}[a^x] = a^x \ln a$, so

$$\frac{d}{dx}\left[\frac{1}{\ln a}\,a^x\right] = \frac{1}{\ln a}\,\frac{d}{dx}[a^x]$$

$$= \frac{1}{\ln a}\,a^x \ln a$$

$$= a^x.$$

Therefore,

$$\int a^x\,dx = \frac{a^x}{\ln a} + C, \tag{8}$$

for any positive base $a \neq 1$.

EXAMPLE 4
$$\int 10^x\,dx = \frac{10^x}{\ln 10} + C.$$

|||||||||||| **Integrating Algebraic Combinations of Functions**

By reversing the corresponding differentiation rules, we obtain the following integration rules:

The Constant Times a Function Rule
If k is a constant, then

$$\int kf(x)\,dx = k \int f(x)\,dx. \tag{9}$$

That is, the integral of a constant times a function is the constant times the integral of the function.

EXAMPLE 5 (a) $\displaystyle\int 5x^6\,dx = 5\int x^6\,dx = 5\left(\frac{x^7}{7} + C_1\right) = \frac{5}{7}x^7 + 5C_1$

$$= \frac{5}{7}x^7 + C \qquad \text{(where } C = 5C_1\text{)}.$$

(b) $\displaystyle\int \frac{3}{x}\,dx = 3\int \frac{1}{x}\,dx = 3(\ln|x| + C_1)$

$$= 3\ln|x| + C \qquad \text{(where } C = 3C_1\text{)}.$$

(c) $\displaystyle\int 10e^x\,dx = 10\int e^x\,dx = 10[e^x + C_1] = 10e^x + 10C_1$

$$= 10e^x + C \qquad \text{(where } C = 10C_1\text{)}.$$

Technically, each answer in Example 5 contained a constant C_1, which was then multiplied by a constant to produce still another constant C. This is superfluous. For all practical purposes, use of C_1 is unnecessary; we may skip that step. The next example shows how to do it.

EXAMPLE 6 (a) $\int 5\sqrt[3]{x}\,dx = 5\int x^{1/3}\,dx = 5\dfrac{x^{4/3}}{4/3} + C = \dfrac{15}{4}x^{1/3} + C.$

(b) $\int \dfrac{-2}{x}\,dx = -2\int \dfrac{1}{x}\,dx = -2\ln|x| + C.$

(c) $\int 8e^x\,dx = 8\int e^x\,dx = 8e^x + C.$

The Sum and Difference Rules

$$\int [f(x) + g(x)]\,dx = \int f(x)\,dx + \int g(x)\,dx \tag{10}$$

$$\int [f(x) - g(x)]\,dx = \int f(x)\,dx - \int g(x)\,dx \tag{11}$$

That is, the integral of a sum (difference) is the sum (difference) of the integrals. We shall demonstrate the use of these rules in the next three examples.

EXAMPLE 7

$$\int (x^2 + x^3)\,dx = \int x^2\,dx + \int x^3\,dx$$

$$= \frac{x^3}{3} + C_1 + \frac{x^4}{4} + C_2$$

$$= \frac{x^3}{3} + \frac{x^4}{4} + (C_1 + C_2),$$

where C_1 and C_2 are arbitrary constants. Since $C_1 + C_2 = C$ is another arbitrary constant, we may write our answer simply as

$$\frac{x^3}{3} + \frac{x^4}{4} + C. \qquad\blacksquare$$

Starting with the next example, we shall add an arbitrary constant C after we have completed *all* the antidifferentiating, rather than after each antidifferentiation.

EXAMPLE 8

$$\int \left(x^{2/3} + 2e^x - \frac{1}{x}\right)\,dx = \int x^{2/3}\,dx + \int 2e^x\,dx - \int \frac{1}{x}\,dx$$

$$= \int x^{2/3}\,dx + 2\int e^x\,dx - \int \frac{1}{x}\,dx$$

$$= \frac{x^{5/3}}{5/3} + 2e^x - \ln|x| + C$$

$$= \frac{3}{5}x^{5/3} + 2e^x - \ln|x| + C.$$

EXAMPLE 9

$$\int \left(\frac{3}{\sqrt{x}} + 5x^4 \right) dx = \int \frac{3}{\sqrt{x}} dx + \int 5x^4 dx$$

$$= 3 \int x^{-1/2} dx + 5 \int x^4 dx$$

$$= 3 \frac{x^{1/2}}{1/2} + 5 \frac{x^5}{5} + C$$

$$= 6x^{1/2} + x^5 + C.$$

EXAMPLE 10 When the variable of integration is denoted by a letter other than x, the integration formulas are modified accordingly. Thus,

$$\int s^4 \, ds = \frac{s^5}{5} + C$$

$$\int \frac{1}{t} \, dt = \ln |t| + C, \qquad t \neq 0$$

$$\int 5B^3 \, dB = 5 \int B^3 \, dB = 5 \frac{B^4}{4} + C = \frac{5}{4} B^4 + C$$

$$\int e^u \, du = e^u + C.$$

|||||||||||||| **Checking Answers**

When we think we have found the indefinite integral of a given function, we can always check the correctness of our work by differentiating the answer to obtain the original integrand. Thus, in Example 8,

$$\frac{d}{dx} \left[\frac{3}{5} x^{5/3} + 2e^x - \ln |x| + C \right] = \frac{3}{5} \frac{d}{dx} [x^{5/3}] + 2 \frac{d}{dx} [e^x] - \frac{d}{dx} [\ln |x|] + \frac{d}{dx} [C]$$

$$= \frac{3}{5} \cdot \frac{5}{3} x^{2/3} + 2e^x - \frac{1}{x} + 0$$

$$= x^{2/3} + 2e^x - \frac{1}{x}.$$

||||||||||||| **Integrating Derivatives**

Observe that an antiderivative of $f'(x)$ is $f(x)$, because $\frac{d}{dx} [f(x)] = f'(x)$. Hence,

$$\int f'(x) \, dx = f(x) + C. \tag{12}$$

The next two examples make use of Equation (12), as does Section 7.3. In many applied problems we know enough information to be able to compute a specific value for the arbitrary constant of integration. In the next two examples we shall do this.

EXAMPLE 11
(Marginal Cost)

A manufacturer of toys knows that the marginal cost MC of producing x toys (in dollars per toy) is given by

$$MC = x^2 + 5x.$$

Find the cost function if the fixed cost (the cost of producing no toys) is $100. (This condition on fixed cost is called an **initial condition** and it will come up again in Section 6.5.)

Solution

Since

$$MC = C'(x) = x^2 + 5x,$$

we can find $C(x)$ by computing the indefinite integral of MC [Equation (12)]:

$$\begin{aligned} C(x) &= \int MC\, dx = \int C'(x)\, dx \\ &= \int (x^2 + 5x)\, dx \\ &= \frac{x^3}{3} + \frac{5}{2}x^2 + K. \end{aligned} \tag{13}$$

(K is used as the constant of integration here to avoid confusion.)

The fixed cost is the cost when $x = 0$, that is, when no toys are made. So

$$C(0) = 100, \quad \text{or} \quad \tfrac{1}{3}(0)^3 + \tfrac{5}{2}(0)^2 + K = 100.$$

Thus,

$$K = 100.$$

We then substitute this value of K into (13). Hence, the cost function, when the marginal cost is $x^2 + 5x$ and the fixed cost is $100, is given by

$$C(x) = \frac{x^3}{3} + \frac{5}{2}x^2 + 100. \qquad \blacksquare$$

EXAMPLE 12
(Learning)

Let $N(t)$ denote the number of radios assembled daily by an average worker on an assembly line after having completed t hours of training ($0 \le t \le 10$). Suppose that the rate of change in production (in radios per day) at time t is given by $N'(t) = 20 - 3t + \tfrac{1}{2}t^2$, and that without any training an average worker cannot assemble any radios at all.
   (a) Find the number of radios $N(t)$ assembled daily by an average worker after t hours of training.
   (b) Find the number of radios assembled daily by an average worker after 6 hours of training.

Solution

(a) We have

$$\begin{aligned} N(t) &= \int N'(t)\, dt \\ &= \int (20 - 3t + \tfrac{1}{2}t^2)\, dt \\ &= 20t - \tfrac{3}{2}t^2 + \tfrac{1}{6}t^3 + C. \end{aligned}$$

Since $N(0) = 0$, we obtain

$$20(0) - \tfrac{3}{2}(0)^2 + \tfrac{1}{6}(0)^3 + C = 0.$$

So $C = 0$. Hence,

$$N(t) = 20t - \tfrac{3}{2}t^2 + \tfrac{1}{6}t^3.$$

(b) When $t = 6$,

$$N(6) = 20(6) - \tfrac{3}{2} \cdot 6^2 + \tfrac{1}{6} \cdot 6^3$$
$$= 120 - 54 + 36 = 102.$$

Thus, an average worker assembles 102 radios daily after 6 hours of training.

∎

6.1 EXERCISE SET ||||||||||||||||||||||||||||||||||||||||||||||||||||||||||||||||||||||||||||||||

In Exercises 1–26, find the indefinite integral.

1. $\displaystyle\int 3\,dx$

2. $\displaystyle\int 7\,dx$

3. $\displaystyle\int s^6\,ds$

4. $\displaystyle\int x^9\,dx$

5. $\displaystyle\int x^{1/5}\,dx$

6. $\displaystyle\int t^{-2/3}\,dt$

7. $\displaystyle\int 3u^7\,du$

8. $\displaystyle\int -5x^3\,dx$

9. $\displaystyle\int \frac{-6}{x^2}\,dx$

10. $\displaystyle\int \frac{3}{t^5}\,dt$

11. $\displaystyle\int 3e^u\,du$

12. $\displaystyle\int 8e^x\,dx$

13. $\displaystyle\int \frac{10}{x}\,dx$

14. $\displaystyle\int \frac{2}{3s}\,ds$

15. $\displaystyle\int \sqrt[3]{x}\,dx$

16. $\displaystyle\int \frac{1}{\sqrt[3]{x}}\,dx$

17. $\displaystyle\int 3^t\,dt$

18. $\displaystyle\int 2 \cdot 7^u\,du$

19. $\displaystyle\int \left(x + \frac{1}{x}\right)dx$

20. $\displaystyle\int \left(e^x - \frac{3}{x}\right)dx$

21. $\displaystyle\int (4t^3 - 3t^{1/4} + 3)\,dt$

22. $\displaystyle\int \left(x^3 + e^x - \frac{1}{x}\right)dx$

23. $\displaystyle\int \left(2t^{2/3} - 3e^t + \frac{2}{t}\right)dt$

24. $\displaystyle\int (2A^2 - 3A + 2)\,dA$

25. $\displaystyle\int \left(5x^4 - 3x^{5/2} + \frac{2}{5x^4}\right)dx$

26. $\displaystyle\int (\sqrt[3]{x^4} - \sqrt[4]{x^3})\,dx$

In Exercises 27–32, state whether F is an antiderivative of f.

27. $f(x) = x^4$, $F(x) = x^3$

28. $f(u) = u^{4/3}$, $F(u) = \tfrac{3}{7}u^{7/3}$

29. $f(x) = e^{3x}$, $F(x) = e^{3x}$

30. $f(t) = \dfrac{2}{t^3}$, $F(t) = \dfrac{1}{t^2}$

31. $f(x) = \ln x$, $F(x) = x \ln x - x$, for $x > 0$

32. $f(s) = \dfrac{3}{s^{2/5}}$, $F(s) = \dfrac{5}{s^{3/5}}$

In Exercises 33–36, state whether the integration is correct or incorrect.

33. $\displaystyle\int (x^3 + 2x - 1)\,dx = \frac{x^4}{4} + x^2 + C$

34. $\displaystyle\int (x^3 + 5)^4(3x^2)\,dx = \frac{(x^3 + 5)^5}{5} + C$

35. $\displaystyle\int e^{5t}\,dt = e^{5t} + C$

36. $\int \left(\dfrac{1}{x} - x^{3/2} + \dfrac{2}{x^2} \right) dx =$

$$\ln |x| - \dfrac{2}{5} x^{5/2} - \dfrac{2}{x} + C, \ x \neq 0$$

In Exercises 37–42, some algebraic simplification is required before the integration is begun.

37. $\int \dfrac{x^3 + 2x^2}{3x^5} \, dx$

38. $\int \dfrac{3 + 5x^2}{x} \, dx$

39. $\int (2x - 7)^2 \, dx$

40. $\int (x - 2)(x + 1)^2 \, dx$

41. $\int \sqrt{x}(x + 3) \, dx$

42. $\int \dfrac{3x - 1}{\sqrt{x}} \, dx$

43. **(Marginal Cost)** A manufacturer of mopeds knows that the marginal cost of producing x mopeds (in dollars per moped) is given by

$$C'(c) = 3\sqrt{x} + 2x + 5.$$

Find the cost function $C(x)$ if the fixed cost is $2000.

44. **(Marginal Profit)** A manufacturer of skis finds that producing x pairs of skis yields a marginal profit (in dollars per pair of skis) given by

$$P'(x) = 100 - 0.04x.$$

Assuming that every pair of skis that is manufactured is sold, find the profit function $P(x)$ if the profit on 100 pairs is $9600.

45. The slope of the tangent line to a curve at the point (x, y) on the curve is $2x^2 - 3x + 1$. If the point $(-2, 1)$ lies on the curve, find an equation for the curve.

46. The slope of the tangent line to a curve at the point (x, y) on the curve is $2/x^2 - 3x + 2$. If the point $(2, 3)$ lies on the curve, find an equation for the curve.

47. **(Population Growth)** Suppose that the rate of growth (in thousands per year) of the population of a newly incorporated city is given by

$$P'(t) = 10 + 6t^{1/5},$$

where t is the number of years since the city was incorporated. If the population were 20,000 when the city became incorporated, what would the population be 32 years after incorporation?

48. **(Motion)** A car, starting from rest, moves along a straight line so that its instantaneous velocity (in feet per second) at time t is given by

$$v(t) = t^2 - 3t + 5.$$

Find the total distance traveled (in feet) by the car in the first 6 seconds. [*Hint:* Let $s(t)$ denote the distance traveled in the first t seconds. Then $ds/dt = v(t)$.]

49. **(Natural Gas Development)** A newly developed natural-gas well is producing gas at the instantaneous rate given by

$$R'(t) = 20 + \tfrac{5}{2}t - \tfrac{1}{4}t^2,$$

where $R(t)$ is the number of millions of cubic feet of gas produced after t years of operation. How much gas will have been produced in the first 8 years of operation?

6.2 ||||||||| INTEGRATION BY SUBSTITUTION

In this section we shall discuss the method of integration by substitution, which will enable us to extend the basic integration formulas covered in Section 6.1 to many more integration problems than those formulas seem, at first, to cover.

In Section 3.5 we used the chain rule to develop the differentiation formula [Equation (6) in that section]:

$$\frac{d}{dx}[f(x)]^{r+1} = (r+1)[f(x)]^r f'(x). \tag{1}$$

Since every differentiation formula yields a corresponding integration formula, Equation (1) gives us

$$\int (r+1)[f(x)]^r f'(x)\,dx = [f(x)]^{r+1} + C, \tag{2}$$

or

$$\int [f(x)]^r f'(x)\,dx = \frac{1}{r+1}[f(x)]^{r+1} + C, \qquad r \neq -1. \tag{3}$$

Of course, the arbitrary constant C in Equation (3) is not the same as the constant in Equation (2), but we use the same letter for convenience. Equation (3) can be made to look simpler by making a **substitution.**
We let

$$u = f(x).$$

Then we compute the differential du [Section 4.8, Equation (8)] as

$$du = f'(x)\,dx.$$

Thus, Equation (3) becomes

$$\int u^r\,du = \frac{1}{r+1}u^{r+1} + C, \qquad r \neq -1, \tag{4}$$

where u is any differentiable function.

We shall illustrate the method of substitution with the following examples.

EXAMPLE 1   Find $\int \sqrt{\tfrac{1}{2}x^2 + 9}\, x\,dx$.

Solution   Suppose we try the substitution

$$u = f(x) = \tfrac{1}{2}x^2 + 9.$$

Then

$$\begin{aligned} du &= f'(x)\,dx \\ &= \tfrac{1}{2}(2x)\,dx = x\,dx. \end{aligned}$$

Substituting u for $\frac{1}{2}x^2 + 9$ and du for $x\,dx$, we have

$$\int \sqrt{\tfrac{1}{2}x^2 + 9}\, x\,dx = \int \sqrt{u}\, du$$
$$= \int u^{1/2}\, du$$
$$= \frac{u^{3/2}}{3/2} + C = \frac{2}{3}\, u^{3/2} + C.$$

Replacing u with $\frac{1}{2}x^2 + 9$, we obtain

$$\int \sqrt{\tfrac{1}{2}x^2 + 9}\, x\,dx = \tfrac{2}{3}(\tfrac{1}{2}x^2 + 9)^{3/2} + C. \qquad \blacksquare$$

EXAMPLE 2   Find $\int \sqrt{x^2 + 4}\, x\,dx$.

Solution   Suppose we try the substitution

$$u = f(x) = x^2 + 4.$$

Then

$$du = f'(x)\, dx = 2x\,dx.$$

In the given integral, we have $x\,dx$, but $du = 2x\,dx$. Multiplying the integrand by $\frac{1}{2} \cdot 2 = 1$, we rewrite the given integral as follows:

$$\int \sqrt{x^2 + 4}\, x\,dx = \int \sqrt{x^2 + 4} \cdot \tfrac{1}{2} \cdot 2x\,dx$$
$$= \tfrac{1}{2} \int \sqrt{x^2 + 4}\, 2x\,dx. \qquad \text{[Section 6.1, Equation (9)]}$$

Substituting for u and du, we now have

$$\int \sqrt{x^2 + 4}\, x\,dx = \frac{1}{2} \int \sqrt{u}\, du$$
$$= \frac{1}{2} \frac{u^{3/2}}{3/2} + C \qquad \text{[Equation (4)]}$$
$$= \frac{1}{3}\, u^{3/2} + C$$
$$= \frac{1}{3}\, (x^2 + 4)^{3/2} + C. \qquad \text{[replacing u by $x^2 + 4$]}$$

An alternate way of approaching this problem is as follows. Beginning with the same u and du as before, we solve the equation $du = 2x\,dx$ for $x\,dx$ to obtain $x\,dx = \frac{1}{2}du$. Then

$$\int \sqrt{x^2 + 4}\, x\,dx = \int \sqrt{u}\, \frac{1}{2}\, du = \frac{1}{2} \int u^{1/2}\, du.$$

[We have "factored out" the $\frac{1}{2}$ from the integral, as permitted by Equation (9) of Section 6.1.] Thus,

$$\int \sqrt{x^2 + 4}\, x\, dx = \frac{1}{2} \frac{u^{3/2}}{3/2} + C = \frac{1}{3} u^{3/2} + C$$

$$= \frac{1}{3} (x^2 + 4)^{3/2} + C. \qquad \blacksquare$$

The concrete examples we have seen so far are special cases of the general method of "integration by substitution," which is based on the following theorem.

THEOREM 1

(Integration by Substitution)
If $u = f(x)$ is a differentiable function of x, then

$$\int g(f(x))f'(x)\, dx = \int g(u)\, du. \qquad (5)$$

This theorem assures us that once we have substituted $u = f(x)$ and $du = f'(x)\, dx$, we can treat the integral as an antiderivative with respect to u. We can disregard x until after we have antidifferentiated, but then we replace u by $f(x)$ to give the final answer.

With the aid of the above theorem, we see that the three fundamental formulas of Section 6.1 are valid when the independent variable is replaced by a function. More precisely, we have the following theorem.

THEOREM 2

(Integration Formulas)
If $u = f(x)$ is a differentiable function, then

$$\int u^r \, du = \frac{u^{r+1}}{r+1} + C \qquad \text{if } r \neq -1 \qquad (6)$$

$$\int \frac{1}{u} \, du = \ln |u| + C \qquad \text{if } u \neq 0 \qquad (7)$$

$$\int e^u \, du = e^u + C. \qquad (8)$$

EXAMPLE 3 Find $\int e^{-5x}\, dx$.

Solution Let

$$u = -5x$$

so that

$$du = -5\, dx.$$

Then

$$dx = -\tfrac{1}{5}\,du,$$

and

$$\int e^{-5x}\,dx = \int e^u(-\tfrac{1}{5}\,du)$$

$$= -\tfrac{1}{5}\int e^u\,du = -\tfrac{1}{5}e^u + C \qquad \text{[by (8)]}$$

$$= -\tfrac{1}{5}e^{-5x} + C. \qquad \text{(replacing } u \text{ by } -5x)$$ ■

EXAMPLE 4 Find $\int e^{x^3}x^2\,dx$.

Solution Let

$$u = x^3$$

so that

$$du = 3x^2\,dx$$

and

$$x^2\,dx = \tfrac{1}{3}\,du.$$

Then

$$\int e^{x^3}x^2\,dx = \int e^u(\tfrac{1}{3})\,du$$

$$= \tfrac{1}{3}\int e^u\,du = \tfrac{1}{3}e^u + C \qquad \text{[by (8)]}$$

$$= \tfrac{1}{3}e^{x^3} + C.$$ ■

EXAMPLE 5 Find

$$\int \frac{4x^3}{1 + x^4}\,dx.$$

Solution Let

$$u = 1 + x^4$$

so that

$$du = 4x^3\,dx.$$

Then

$$\int \frac{4x^3}{1 + x^4}\,dx = \int \frac{1}{1 + x^4}\cdot 4x^3\,dx$$

$$= \int \frac{1}{u}\,du = \ln|u| + C \qquad \text{[by (7)]}$$

$$= \ln|1 + x^4| + C$$

$$= \ln(1 + x^4) + C. \qquad \text{(we may remove the absolute value signs because } 1 + x^4 > 0 \text{ for all } x)$$ ■

EXAMPLE 6   Find

$$\int \frac{e^x}{1 + e^x}\, dx.$$

Solution   Let

$$u = 1 + e^x$$

so that

$$du = e^x\, dx.$$

Then

$$\int \frac{e^x}{1 + e^x}\, dx = \int \frac{1}{1 + e^x} \cdot e^x\, dx$$

$$= \int \frac{1}{u}\, du = \ln |u| + C \qquad [\text{by (7)}]$$

$$= \ln |1 + e^x| + C$$

$$= \ln (1 + e^x) + C. \qquad (\text{since } 1 + e^x > 0 \text{ for all } x) \qquad \blacksquare$$

In the preceding examples we used the phrase "suppose we try the substitution," or "let $u = \ldots$." We must point out that the particular choice of u in making a substitution represents an intelligent guess. The method of substitution is essentially a trial-and-error technique. It might be necessary to try several substitutions before finding one that works or concluding that no appropriate substitution exists. Doing the exercises at the end of this section will help you develop a feel for choosing the right u.

EXAMPLE 7   Find

$$\int \frac{2x}{(x^2 + 9)^2}\, dx.$$

Solution   There are many conceivable substitutions: $u = x^2$, $u = 2x$, $u = x^2 + 9$, and so on. Suppose we try

$$u = x^2$$

so that

$$du = 2x\, dx.$$

Then

$$\int \frac{2x}{(x^2 + 9)^2}\, dx = \int \frac{1}{(u + 9)^2}\, du,$$

an integral that we still cannot find readily, since it does not fit one of the forms (6)–(8). Now try the substitution

$$u = x^2 + 9,$$

so that

$$du = 2x\,dx.$$

Then

$$\int \frac{2x}{(x^2 + 9)^2}\,dx = \int \frac{1}{u^2}\,du = \int u^{-2}\,du$$

$$= -\frac{1}{u} + C \qquad \text{[by (6)]}$$

$$= -\frac{1}{x^2 + 9} + C.$$ ■

Warning 1   *Calculating* du *correctly is extremely important.* This step must not be over-looked or taken lightly. For example,

$$\int \frac{1}{f(x)}\,dx \ne \ln |f(x)| + C$$

because the differential df is not present; dx and df are not the same! Formula (7) does not apply here at all. To give a specific example,

$$\int \frac{1}{x^2}\,dx \ne \ln |x^2| + C$$

because dx is not the differential of $\dfrac{1}{x^2}$. In fact, the correct answer is

$$\int \frac{1}{x^2}\,dx = \int x^{-2}\,dx$$

$$= \frac{x^{-1}}{-1} + C \qquad \text{[by (6)]}$$

$$= -\frac{1}{x} + C.$$

Warning 2   In calculating du we often get an unwanted constant. We compensate by factoring out the constant from the integral, according to Equation (9) of Section 6.1. For example,

$$\int (3x + 4)^4\,dx = \int u^4 \cdot \tfrac{1}{3}\,du = \tfrac{1}{3} \int u^4\,du.$$
$$(u = 3x + 4,\ du = 3\,dx,\ \tfrac{1}{3}\,du = dx.)$$

We have "factored out" the unwanted constant, $\tfrac{1}{3}$, from the integral. We can then finish the problem easily,

$$\int (3x + 4)^4\,dx = \frac{1}{3} \cdot \frac{u^5}{5} + C = \frac{1}{15}(3x + 4)^5 + C.$$

Beware   Never factor out variables from an integral; only constants may be factored out.

EXAMPLE 8   Suppose we wish to find $\int \sqrt{x^2 + 9}\, dx$. If we try the substitution

$$u = x^2 + 9$$

then

$$du = 2x\, dx$$

so that

$$\frac{1}{2x}\, du = dx.$$

It is *incorrect* to write

$$\int \sqrt{x^2 + 9}\, dx = \int \sqrt{u} \cdot \frac{1}{2x}\, du$$

$$= \frac{1}{2x} \int u^{1/2}\, du = \frac{1}{2x} \frac{u^{3/2}}{3/2} + C$$

$$= \frac{1}{3x} (x^2 + 9)^{3/2} + C.$$

This sequence of steps represents a common *mistake* and makes no sense at all. The error (and it is a very bad one) occurred in the step printed in color. It is *not permissible* to factor our $\dfrac{1}{2x}$, because $\dfrac{1}{2x}$ involves the variable, x.

Furthermore, we can check the answer ourselves, to see that it is wrong:

$$\frac{d}{dx} \left[\frac{1}{3x} (x^2 + 9)^{3/2} + C \right] \neq \sqrt{x^2 + 9}.$$

EXAMPLE 9
(Pollution)   A nuclear reactor dumps radioactive waste into a nearby stream at a rate (in tons per year) given by

$$A'(t) = \frac{\ln (t + 1)}{t + 1}, \qquad 0 \leq t \leq 20$$

where t is the number of years that the plant has been in operation. Find the amount $A(t)$ of radioactive waste that the plant has dumped into the stream during the first t years of operation.

Solution   We have

$$A(t) = \int A'(t)\, dt = \int \frac{\ln (t + 1)}{t + 1}\, dt.$$

Let

$$u = \ln (t + 1)$$

so that

$$du = \frac{1}{t + 1}\, dt.$$

Then

$$A(t) = \int \frac{\ln (t + 1)}{t + 1}\, dt$$

$$= \int u\, du$$

$$= \frac{u^2}{2} + C$$

$$= \tfrac{1}{2}[\ln (t + 1)]^2 + C.$$

When $t = 0$, the amount of radioactive waste that has gone into the stream is zero. Hence,

$$A(0) = 0$$
$$\tfrac{1}{2}[\ln (0 + 1)]^2 + C = 0$$
$$\tfrac{1}{2}[\ln 1]^2 + C = 0$$
$$0 + C = 0$$

so $C = 0$. Thus, the amount of radioactive waste that the plant has dumped into the stream during the first t years of operation is given by

$$A(t) = \tfrac{1}{2}[\ln (t + 1)]^2. \qquad \blacksquare$$

As we have already remarked, the method of integration by substitution involves some guessing and trial and error. As the preceding examples indicate, however, our objective is to change the integral into one of the forms

$$k \int u^r\, du, \qquad k \int e^u\, du, \quad \text{or} \quad k \int \frac{1}{u}\, du, \tag{9}$$

where $u = f(x)$ and $du = f'(x)\, dx$. The constant k arises when we differentiate u to get du.

||||||||||| Linear Substitution

The simplest substitution occurs when u is a first degree function of x. In fact, this substitution is so routine that one theorem supplies the answers for all cases that lead to one of the three basic forms (9). We state the theorem and then show how to apply it.

THEOREM 3

(Linear Substitution)

If

$$\int f(x)\, dx = F(x) + C$$

then, for all constants a and b, with $a \neq 0$,

$$\int f(ax + b)\, dx = \frac{1}{a} F(ax + b) + C. \tag{10}$$

Proof   Let $u = ax + b$, $a \neq 0$. Then $du = a\, dx$, or

$$\frac{1}{a}\, du = dx.$$

Thus,

$$\int f(ax + b)\, dx = \int f(u) \cdot \frac{1}{a}\, du$$

$$= \frac{1}{a} \int f(u)\, du \qquad (a \text{ is constant})$$

$$= \frac{1}{a} F(u) + C \qquad (\text{Why?})$$

$$= \frac{1}{a} F(ax + b) + C. \qquad \blacksquare$$

Theorem 3 can be applied to each of the standard forms (9). From equations (6)–(8) we obtain (assuming $a \neq 0$)

$$\int (ax + b)^n\, dx = \frac{1}{a} \frac{(ax + b)^{n+1}}{n + 1} + C, \qquad \text{if } n \neq -1 \tag{11}$$

$$\int \frac{1}{ax + b}\, dx = \frac{1}{a} \ln |ax + b| + C \tag{12}$$

$$\int e^{ax+b}\, dx = \frac{1}{a} e^{ax+b} + C. \tag{13}$$

EXAMPLE 10   (a) $\displaystyle \int (5x - 7)^3\, dx = \frac{1}{5} \cdot \frac{(5x - 7)^4}{4} + C$

$$= \frac{1}{20} (5x - 7)^4 + C.$$

(b) $\int \dfrac{1}{11x + 2}\, dx = \dfrac{1}{11} \ln |11x + 2| + C.$

(c) $\int e^{2 - 3x}\, dx = -\tfrac{1}{3}e^{2 - 3x} + C.$

6.2 EXERCISE SET ||||||||||||||||||||||||||||||||||||||||||||||||||||||||||||||||||||||||||||||

In Exercises 1–40, use the method of substitution to find the integrals.

1. $\int (5x - 3)^3\, dx$

2. $\int (4 - 2x)^5\, dx$

3. $\int s(s^2 + 1)^4\, dx$

4. $\int t(3t^2 - 5)^3\, dt$

5. $\int w^2 (1 - w^3)^7\, dw$

6. $\int (x^3 + 5)x^2\, dx$

7. $\int \dfrac{x}{(x^2 + 9)^3}\, dx$

8. $\int \dfrac{2x}{(x^2 - 1)^5}\, dx$

9. $\int \dfrac{1}{x + 5}\, dx$

10. $\int \dfrac{1}{12s - 5}\, ds$

11. $\int \dfrac{3x^2}{(x^3 + 2)^{10}}\, dx$

12. $\int \dfrac{w^3\, dw}{(w^4 - 3)^5}$

13. $\int (x^3 + 2)^{1/2}\, 3x^2\, dx$

14. $\int \sqrt{t^3 + 4}\ 3t^2\, dt$

15. $\int \sqrt{t^4 + 5}\ t^3\, dt$

16. $\int x(x^2 - 4)^{12}\, dx$

17. $\int e^{2x}\, dx$

18. $\int e^{t/5}\, dt$

19. $\int 2e^{4s - 1}\, ds$

20. $\int 3e^{2 - 8y}\, dy$

21. $\int 2xe^{x^2}\, dx$

22. $\int x^2 e^{x^3}\, dx$

23. $\int (3x^2 + 2x)e^{x^3 + x^2}\, dx$

24. $\int (x^3 + x^2 - 2x)^{15}(3x^2 + 2x - 2)\, dx$

25. $\int \dfrac{e^{\sqrt{t}}}{\sqrt{t}}\, dt$

26. $\int Q_0 e^{-kt}\, dt$

27. $\int e^y(1 + e^y)\, dy$

28. $\int \dfrac{3e^x}{\sqrt{1 - e^x}}\, dx$

29. $\int \dfrac{x + 1}{x^2 + 2x}\, dx$

30. $\int \dfrac{4x^3 + 3}{x^4 + 3x}\, dx$

31. $\int \dfrac{1}{2 - x}\, dx$

32. $\int \dfrac{1}{(t - 2)^3}\, dt$

33. $\int \dfrac{s}{3s^2 + 2}\, ds$

34. $\int \dfrac{w^2}{2w^3 - 1}\, dw$

35. $\int \dfrac{t^2}{\sqrt{t^3 + 4}}\, dt$

36. $\int \dfrac{5x^4 + 3x^2 - 2}{\sqrt{x^5 + x^3 - 2x}}\, dx$

37. $\int \dfrac{(\ln x)^2}{x}\, dx$

38. $\int \dfrac{\ln 2t}{t}\, dt$

39. $\int \dfrac{1}{x \ln x}\, dx$

40. $\int \dfrac{1}{x(\ln x)^2}\, dx$

41. **(Marginal Revenue)** A manufacturer's marginal revenue (in millions of dollars per thousand units) is given by

$$R'(x) = \dfrac{x}{\sqrt{x^2 + 16}}$$

where x is the number of units (in thousands) produced and sold. If the revenue is zero when the level of production is zero, find the revenue when the level of production is 3000 units, assuming that every unit that is manufactured is sold.

42. **(Public Health)** An epidemic of Russian flu has hit a certain city. The rate of growth of the disease (in new cases per day) is given by

$$N'(t) = 3t^2 \sqrt{t^3 + 9}$$

where t is the number of days after the start of the epidemic. On the day that the epidemic was discovered ($t = 0$), 30 people had the disease. How many people will have been infected during the first 3 days after the discovery of the epidemic?

43. **(Radioactive Decay)** A radioactive substance's rate of decay is given by

$$Q'(t) = -100e^{-0.2t} \text{ grams per week}$$

where t is in weeks. If there were 500 grams of the material when the decay process started, how much material will be present after 20 weeks?

6.3 |||||||| INTEGRATION BY PARTS

We now approach the problem of integrating a *product* of two functions,

$$\int f(x)g(x)\, dx.$$

In general, this is a quite difficult task. If we were trying to *differentiate* the product $f(x)\, g(x)$, our task would be much easier: we would merely apply the product rule of differentiation. Unfortunately, *there is no product rule for anti-differentiation*, a rule that gives us the antiderivative of $f(x)\, g(x)$ in terms of $f(x)$ and $g(x)$ and their antiderivatives. However, in case one of the functions is a derivative, our task may yield to a method known as "integration by parts."

Suppose we wish to obtain the integral

$$\int u(x)v'(x)\, dx.$$

In Section 3.3 we studied the product rule:

$$\frac{d}{dx}[u(x)v(x)] = u(x)v'(x) + v(x)u'(x). \tag{1}$$

Taking the indefinite integral of both sides of (1), we have

$$\int \frac{d}{dx}[u(x)v(x)]\, dx = \int [u(x)v'(x) + v(x)u'(x)]\, dx$$
$$= \int u(x)v'(x)\, dx + \int v(x)u'(x)\, dx.$$

That is,

$$u(x)v(x) = \int u(x)v'(x)\, dx + \int v(x)u'(x)\, dx.$$

Solving for $\int u(x)v'(x)\, dx$, we obtain the so-called **integration by parts** formula:

$$\int u(x)v'(x)\, dx = u(x)v(x) - \int v(x)u'(x)\, dx. \tag{2}$$

The arbitrary constant C does not appear in Equation (2) because we have agreed to insert C only after we have completed all integration.

Formula (2) may seen somewhat unusual. It does not provide the complete answer to the antiderivative of $u(x)v'(x)$, because it leaves another antiderivative to evaluate. The idea here is to finish with an integral on the right side that we can readily compute. If we cannot handle that integral, then the method may be of no use to us in the problem.

The following example illustrates the application of Equation (2).

EXAMPLE 1   Find $\int xe^x \, dx$.

Solution   We try to choose $u(x)$ and $v'(x)$ so that we can obtain $v(x)$ easily. Let

$$u(x) = x \quad \text{and} \quad v'(x) = e^x.$$

Then

$$u'(x) = 1$$

and

$$v(x) = \int v'(x) \, dx = \int e^x \, dx = e^x.$$

We have omitted the arbitrary constant that enters into the integration of $v'(x)$ because we want to combine all constants into one arbitrary constant at the end. So, from (2),

$$\int xe^x \, dx = \int u(x)v'(x) \, dx$$
$$= u(x)v(x) - \int v(x)u'(x) \, dx$$
$$= xe^x - \int e^x \cdot 1 \, dx$$
$$= xe^x - \int e^x \, dx$$
$$= xe^x - e^x + C. \qquad \blacksquare$$

The basic idea in using Equation (2) to find $\int f(x) \, dx$ is to factor the function f as the product

$$f(x) = u(x)v'(x)$$

so that the integral $\int v(x)u'(x) \, dx$ will be easier to find than $\int f(x) \, dx$. Note that in Example 1 it is much easier to evaluate

$$\int v(x)u'(x) \, dx = \int e^x \, dx$$

than to evaluate.

$$\int f(x) \, dx = \int u(x)v'(x) \, dx = \int xe^x \, dx.$$

The method of integration by parts is a trial-and-error procedure. There are often several candidates for $u(x)$ and $v'(x)$, and if a particular choice does not work, it is necessary to try another.

In Example 1, we might also have chosen $u(x)$ and $v'(x)$ as follows:

$$u(x) = e^x \quad \text{and} \quad v'(x) = x,$$

so that

$$u'(x) = e^x$$

and

$$v(x) = \int v'(x)\,dx = \int x\,dx = \frac{x^2}{2}.$$

Then, from (2),

$$\int xe^x\,dx = \int u(x)v'(x)\,dx$$

$$= u(x)v(x) - \int v(x)u'(x)\,dx$$

$$= \frac{x^2}{2}\,e^x - \int \frac{x^2}{2}\,e^x\,dx.$$

The integral $\int (x^2/2)e^x\,dx$ is more difficult to obtain than the original integral. This particular choice of $u(x)$ and $v'(x)$ was a rather poor one. Thus, the method of integration by parts must be used with care.

||||||||||||| **Differential Notation**

Formula (2) can be made to look easier and less confusing if we use the notation of differentials. Recall from Section 4.8 that if $u = u(x)$ and $v = v(x)$ are functions of x, then the *differentials* of u and v are defined as

$$du = u'(x)\,dx \quad \text{and} \quad dv = v'(x)\,dx.$$

Using these differentials, we can rewrite the integration by the parts formula as

$$\int u\,dv = uv - \int v\,du. \tag{3}$$

This formula is actually equivalent to (2), but its simpler form makes it much easier to remember. It is in the form (3) that we recommend the reader commit the integration by parts formula to memory.

EXAMPLE 2   Find $\int xe^x\,dx$, using Formula (3).

Solution   We are redoing Example 1, using (3) instead of (2). Let

$$u = x \quad \text{and} \quad dv = e^x\,dx.$$

Then

$$du = dx \quad \text{and} \quad v = e^x.$$

Thus, by (3),

$$\int xe^x \, dx = \int u \, dv$$

$$= uv - \int v \, du$$

$$= xe^x - \int e^x \, dx$$

$$= xe^x - e^x + C. \qquad \blacksquare$$

EXAMPLE 3   Find $\int x \ln x \, dx$, $x > 0$.

Solution   Let

$$u = \ln x \quad \text{and} \quad dv = x \, dx,$$

so that

$$du = \frac{1}{x} \, dx \quad \text{and} \quad v = \int dv = \int x \, dx = \frac{x^2}{2}.$$

Then, from (3),

$$\int x \ln x \, dx = \int u \, dv$$

$$= uv - \int v \, du$$

$$= \ln x \cdot \frac{x^2}{2} - \int \left(\frac{x^2}{2}\right)\left(\frac{1}{x}\right) dx$$

$$= \frac{x^2}{2} \ln x - \frac{1}{2} \int x \, dx$$

$$= \frac{x^2}{2} \ln x - \frac{1}{2}\left(\frac{x^2}{2}\right) + C$$

$$= \frac{x^2}{2} \ln x - \frac{x^2}{4} + C. \qquad \blacksquare$$

Sometimes, as in the next example, integration by parts must be done repeatedly in the same problem. When that happens, you are advised that a little extra patience may be necessary.

EXAMPLE 4   Find $\int x^2 e^{4x} \, dx$.

Solution   We let

$$u = x^2 \quad \text{and} \quad dv = e^{4x} \, dx.$$

Then

$$du = 2x \, dx \quad \text{and} \quad v = \int e^{4x} \, dx = \tfrac{1}{4}e^{4x}.$$

Then, by (3),

$$\int x^2 e^{4x}\,dx = \int u\,dv$$
$$= uv - \int v\,du$$
$$= x^2(\tfrac{1}{4}e^{4x}) - \int \tfrac{1}{4}e^{4x}e^x\,dx$$
$$= \tfrac{1}{4}x^2 e^{4x} - \tfrac{1}{2}\int xe^{4x}\,dx. \tag{4}$$

We first note that the integral on the right is "easier" than the original integral; the power on x has been reduced from 2 to 1. Thus, we are on the right track. In fact, the integral on the right reminds us of Examples 1 and 2. We shall have to use integration by parts one more time to find

$$\int xe^{4x}\,dx.$$

We let

$$u = x \quad \text{and} \quad dv = e^{4x}\,dx.$$

Then

$$du = dx \quad \text{and} \quad v = \int e^{4x}\,dx = \tfrac{1}{4}e^{4x}.$$

Thus,

$$\int xe^{4x}\,dx = \int u\,dv$$
$$= uv - \int v\,du$$
$$= x(\tfrac{1}{4}e^{4x}) - \int \tfrac{1}{4}e^{4x}\,dx$$
$$= \tfrac{1}{4}xe^{4x} - \tfrac{1}{4}\cdot\tfrac{1}{4}e^{4x} + C$$
$$= \tfrac{1}{4}xe^{4x} - \tfrac{1}{16}e^{4x} + C. \tag{5}$$

Substituting (5) into (4), we obtain

$$\int x^2 e^{4x}\,dx = \tfrac{1}{4}x^2 e^{4x} - \tfrac{1}{2}[\tfrac{1}{4}xe^{4x} - \tfrac{1}{16}e^{4x} + C]$$
$$= \tfrac{1}{4}x^2 e^{4x} - \tfrac{1}{8}xe^{4x} + \tfrac{1}{32}e^{4x} + C.$$

(Continuing our practice of using C as our standard constant of integration, we write C instead of $-\tfrac{1}{2}C$.) ∎

EXAMPLE 5
(Population Growth) The population of a newly incorporated city is growing at the rate (in people per year) given by

$$P'(t) = te^{t/10}, \qquad 0 \le t \le 20$$

where t is the number of years after incorporation. Find the population $P(t)$ of the city t years after incorporation, assuming that the population was 10,000 people at the time of incorporation.

Solution   We have

$$P(t) = \int P'(t) = \int te^{t/10} \, dt.$$

Let

$$u = t \quad \text{and} \quad dv = e^{t/10} \, dt.$$

Then

$$du = dt \quad \text{and} \quad v = \int e^{t/10} \, dt = 10e^{t/10}.$$

Then, from (3),

$$
\begin{aligned}
\int te^{t/10} \, dt &= \int u \, dv \\
&= uv - \int v \, du \\
&= t(10e^{t/10}) - \int 10e^{t/10} \, dt \\
&= 10te^{t/10} - 10 \int e^{t/10} \, dt \\
&= 10te^{t/10} - 100e^{t/10} + C.
\end{aligned}
$$

Thus,

$$P(t) = 10te^{t/10} - 100e^{t/10} + C.$$

When $t = 0$, the population is 10,000. Therefore,

$$
\begin{aligned}
P(0) &= 10,000 \\
(10)(0)e^{0/10} - 100e^{0/10} + C &= 10,000 \\
-100 + C &= 10,000.
\end{aligned}
$$

Thus, $C = 10,100$. Hence, the population of the city t years after incorporation is given by

$$P(t) = 10te^{t/10} - 100e^{t/10} + 10,100. \qquad \blacksquare$$

There are a few tricks that are often used in connection with integration by parts. One of them is illustrated in the following example.

EXAMPLE 6   Find $\int \ln x \, dx$, $x > 0$.

Solution   At first glance the method of integration by parts does not appear to apply in this example, since no product is indicated. But, if we let

$$u = \ln x \quad \text{and} \quad dv = dx,$$

then

$$du = \frac{1}{x} \, dx \quad \text{and} \quad v = x.$$

We can then apply (3) to obtain

$$\int \ln x \, dx = \int u \, dv$$

$$= uv - \int v \, du$$

$$= (\ln x) - \int x \left(\frac{1}{x} \, dx \right)$$

$$= x \ln x - \int dx$$

$$= x \ln x - x + C. \qquad \blacksquare$$

We have thus added one more formula to our list of integrals

If u is any positive differentiable function, then

$$\int \ln u \, du = u \ln u - u + C. \qquad (6)$$

6.3 EXERCISE SET ||||||||||||||||||||||||||||||||||||||||||||||||||||||||||||||||||||||||||

In Exercises 1–26, use the method of integration by parts to find the integrals. When $\ln x$ appears, assume that x is positive.

1. $\int 3xe^{3x} \, dx$

2. $\int xe^{-x} \, dx$

3. $\int t^2 e^t \, dt$

4. $\int x^2 \ln x \, dx$

5. $\int \sqrt{t} \ln t \, dt$

6. $\int 5xe^{4x} \, dx$

7. $\int x(\ln x)^2 \, dx$

8. $\int (\ln x)^2 \, dx$

9. $\int xe^{-5x} \, dx$

10. $\int (x - 2)e^{3x} \, dx$

11. $\int x\sqrt{x + 3} \, dx$

12. $\int t\sqrt[3]{3t + 1} \, dt$

13. $\int \frac{2x}{\sqrt{3x + 1}} \, dx$

14. $\int 7x \sqrt{3x - 2} \, dx$

15. $\int \ln (x + 1) \, dx$

16. $\int x \ln 2x \, dx$

17. $\int (x + 2) \ln x \, dx$

18. $\int \frac{x}{e^x} \, dx$

19. $\int x^3 \ln x \, dx$

20. $\int \frac{\ln x}{x^2} \, dx$

21. $\int x^3 e^x \, dx$

22. $\int x^3 e^{x^2} \, dx$

23. $\int \log x \, dx$

24. $\int x \log x \, dx$

25. $\int x3^x \, dx$

26. $\int x^2 3^x \, dx$

27. **(Marginal Revenue)** A manufacturer's marginal revenue (in millions of dollars per thousand units) is given by

$$R'(x) = 4x - xe^{-0.2x},$$

where x is the number of units (in thousands) manufactured and sold. If the revenue is zero when the level of production is zero, find the revenue when the level of production is 5000 units. Assume that every unit that is manufactured is sold.

6.4 |||||||| TABLES OF INTEGRALS

Thus far, we have studied only a few basic techniques of integration. There are many integrals that require techniques that we have not considered here, and that are beyond the scope of this book. In contrast to the problem of differentiation, in which a handful of formulas enables us to differentiate practically all functions we shall ever meet, the problem of integration requires a multitude of formulas. A function that appears quite simple may require a complicated integral formula; its antiderivative may involve functions that we have not even begun to study, such as inverse trigonometric functions.

There are many tables of integral formulas available to help solve more complicated integration problems. Some of these tables contain hundreds of integral formulas. Even a "short" table of integrals, such as one contained in a standard calculus textbook, may contain over a hundred formulas.

Our table of integrals, which can be found in Appendix B, includes sixty-one formulas. Forty-five of these are given at the end of this section; the remainder will be useful only after Chapter 9. To make the table easier to use, it is divided into categories, according to the *form* of the integrand. They are:

I   Basic forms (you should already have memorized these).

II   Integrals involving $au + b$.

III Integrals involving $\sqrt{au + b}$.

IV   Integrals involving $u^2 \pm a^2$, $a > 0$.

V   Integrals involving exponential and logarithmic functions.

VI   Integrals involving trigonometric functions.

It is to be understood that in all of the formulas, u may be a function of x. The other letters are constants. To use the table, one must manipulate the integrand algebraically, usually by making a substitution, until the integral matches one of the forms found in the table. The examples that follow will demonstrate the procedure.

EXAMPLE 1   Use the table of integrals to find

$$\int \frac{1}{x(5 - 2x)} \, dx.$$

Solution   We use Formula 14 in our table of integrals, with $u = x$:

$$\int \frac{1}{x(ax + b)} \, dx = \frac{1}{b} \ln \left| \frac{x}{ax + b} \right| + C,$$

where $a = -2$ and $b = 5$, and obtain

$$\int \frac{dx}{x(5 - 2x)} = \frac{1}{5} \ln \left| \frac{x}{5 - 2x} \right| + C.$$ ∎

EXAMPLE 2 Use the table of integrals to obtain

(a) $\int \dfrac{3x}{4x - 7} \, dx$ (b) $\int \dfrac{x^2}{(1 - 5x)^2} \, dx$ (c) $\int \dfrac{1}{6x^2 + 13x - 5} \, dx.$

Solution (a) $\int \dfrac{3x}{4x - 7} \, dx = 3 \int \dfrac{x}{4x - 7} \, dx.$ The latter integral matches Formula 10 when $a = 4$, $b = -7$, and $u = x$. Thus,

$$\int \frac{3x}{4x - 7} \, dx = 3 \left[\frac{x}{4} - \frac{-7}{16} \ln |4x - 7| \right] + C$$

$$= \frac{3x}{4} + \frac{21}{16} \ln |4x - 7| + C.$$

(b) $\int \dfrac{x^2}{(1 - 5x)^2} \, dx$ matches Formula 13, with $a = -5$, $b = 1$ and $u = x$. Thus,

$$\int \frac{x^2}{(1 - 5x)^2} \, dx = \frac{1}{-125} \left[(1 - 5x) - \frac{1}{1 - 5x} - 2 \ln |1 - 5x| \right] + C.$$

(c) $\int \dfrac{1}{6x^2 + 13x - 5} \, dx = \int \dfrac{1}{(3x - 1)(2x + 5)} \, dx$ matches Formula 17, with $a = 3$, $b = -1$, $c = 2$, $d = 5$, and $u = x$. Thus,

$$\int \frac{1}{6x^2 + 13x - 5} \, dx = \frac{1}{(-1)(2) - (3)(5)} \ln \left| \frac{2x + 5}{3x - 1} \right| + C$$

$$= -\frac{1}{17} \ln \left| \frac{2x + 5}{3x - 1} \right| + C.$$ ∎

EXAMPLE 3 Use the table of integrals to find

$$\int \frac{1}{\sqrt{4x^2 + 16}} \, dx.$$

Solution The integrand is almost of the form of Formula 31, with $u = x$:

$$\int \frac{1}{\sqrt{x^2 + a^2}} \, dx = \ln \left| x + \sqrt{x^2 + a^2} \right| + C. \tag{1}$$

To apply this formula, we must make the coefficient of x^2 equal to 1. Thus, we write the given integral as

$$\int \frac{1}{\sqrt{4x^2 + 16}} \, dx = \int \frac{1}{\sqrt{4(x^2 + 4)}} \, dx = \int \frac{1}{2\sqrt{x^2 + 4}} \, dx = \frac{1}{2} \int \frac{1}{\sqrt{x^2 + 4}} \, dx.$$

We can now apply Equation (1) with $a = 2$ to obtain

$$\int \frac{1}{\sqrt{4x^2 + 16}} \, dx = \frac{1}{2} \int \frac{1}{\sqrt{x^2 + 4}} \, dx = \frac{1}{2} \ln \left| x + \sqrt{x^2 + 4} \right| + C. \qquad \blacksquare$$

EXAMPLE 4   Use the table of integrals to find

$$\int \frac{1}{1 - 4t^2} \, dt.$$

Solution   The form of the integrand is almost the same as that of Formula 29, with u replaced by t:

$$\int \frac{1}{a^2 - t^2} \, dt = \frac{1}{2a} \ln \left| \frac{t + a}{t - a} \right| + C. \qquad (2)$$

To apply this formula, we must make the coefficient of t^2 equal to 1. Thus, we write the given integral as

$$\int \frac{1}{1 - 4t^2} \, dt = \int \frac{1}{4} \left(\frac{1}{(1/4) - t^2} \right) dt = \frac{1}{4} \int \frac{1}{(1/4) - t^2} \, dt.$$

We can now apply Equation (2) with $a = 1/2$ and $x = t$ to obtain

$$\int \frac{1}{1 - 4t^2} \, dt = \frac{1}{4} \int \frac{1}{(1/4) - t^2} \, dt = \frac{1}{4} \frac{1}{2(1/2)} \ln \left| \frac{t + \frac{1}{2}}{t - \frac{1}{2}} \right| + C$$

$$= \frac{1}{4} \ln \left| \frac{2t + 1}{2t - 1} \right| + C. \qquad \blacksquare$$

Alternate Solution   To find $\int \frac{1}{1 - 4t^2} \, dt$ we can begin by making the substitution $u = 2t$. Then $du = 2 \, dt$ or $\frac{1}{2} \, du = dt$, and

$$\int \frac{1}{1 - 4t^2} \, dt = \int \frac{1}{1 - (2t)^2} \, dt = \int \frac{1}{1 - u^2} \cdot \frac{1}{2} \, du$$

$$= \frac{1}{2} \int \frac{1}{1 - u^2} \, du.$$

We then apply Formula 29 with $a = 1$, so that

$$\int \frac{1}{1 - 4t^2} \, dt = \frac{1}{2} \cdot \frac{1}{2} \ln \left| \frac{u + 1}{u - 1} \right| + C$$

$$= \frac{1}{4} \ln \left| \frac{2t + 1}{2t - 1} \right| + C, \text{ since } u = 2t. \qquad \blacksquare$$

||||||||||||| **Reduction Formulas**

Some of the formulas in the table of integrals give only partial answers, leaving another integral to be evaluated. Formulas 42 and 43 are typical examples. It is anticipated that proper use of one of these formulas will leave an integral that is easier to obtain than the original. In this way, we use the formulas to "reduce" the problem to a simpler one.

EXAMPLE 5   Evaluate each of the following integrals:

(a) $\int 5x^2 e^{4x}\, dx$ (b) $\int 6x^3 e^x\, dx$

Solution   (a) We begin by using Formula 42, with $n = 2$, $k = 4$, and $u = x$:

$$\int 5x^2 e^{4x}\, dx = 5 \int x^2 e^{4x}\, dx$$

$$= 5 \left[\frac{x^2 e^{4x}}{4} - \frac{2}{4} \int xe^{4x}\, dx \right]$$

$$= \frac{5}{4} x^2 e^{4x} - \frac{5}{2} \int xe^{4x}\, dx.$$

The integral on the right may be computed using Formula 40, with $a = 4$ and $u = x$. Thus,

$$\int 5x^2 e^{4x}\, dx = \frac{5}{4} x^2 e^{4x} - \frac{5}{2} \left[\frac{e^{4x}}{16} (4x - 1) \right] + C$$

$$= \frac{5}{4} x^2 e^{4x} - \frac{5}{32} e^{4x}(4x - 1) + C.$$

(b) In this example, we shall have to use Formula 42 twice in succession, and then Formula 40.

$$\int 6x^3 e^x\, dx$$

$$= 6 \int x^3 e^x\, dx$$

$$= 6 \left[x^3 e^x - 3 \int x^2 e^x\, dx \right] \qquad \text{(Formula 42, with } n = 3, k = 1, u = x)$$

$$= 6x^3 e^x - 18 \int x^2 e^x\, dx$$

$$= 6x^3 e^x - 18 \left[x^2 e^x - 2 \int xe^x\, dx \right] \text{(Formula 42, with } n = 2, k = 1, u = x)$$

$$= 6x^3 e^x - 18x^2 e^x + 36 \int xe^x\, dx$$

$$= 6x^3 e^x - 18x^2 e^x + 36 \left[e^x(x - 1) \right] + C \qquad \text{(Formula 40, with } a = 1)$$

$$= 6x^3 e^x - 18x^2 e^x + 36xe^x - 36e^x + C$$

$$= 6e^x(x^3 - 3x^2 + 6x - 6) + C.$$

■

EXAMPLE 6 Compute $\int \dfrac{\sqrt{4x^2 - 4x + 10}}{2x - 1}\, dx.$

Solution First, we observe that

$$\sqrt{4x^2 - 4x + 10} = \sqrt{(4x^2 - 4x + 1) + 9}$$
$$= \sqrt{(2x - 1)^2 + 3^2}.$$

We are thus led to see that Formula 34 is the pertinent one. We let $u = 2x - 1$. Then $du = 2\, dx$, or $\frac{1}{2}\, du = dx$, and

$$\int \frac{\sqrt{4x^2 - 4x + 10}}{2x - 1}\, dx$$

$$= \int \frac{\sqrt{u^2 + 3^2}}{u} \cdot \frac{1}{2}\, du$$

$$= \frac{1}{2} \int \frac{\sqrt{u^2 + 3^2}}{u}\, du.$$

$$= \frac{1}{2} \left[\sqrt{u^2 + 9} - 3 \ln \left| \frac{3 + \sqrt{u^2 + 9}}{u} \right| \right] + C \qquad \text{(by Formula 34)}$$

$$= \frac{1}{2} \sqrt{(2x - 1)^2 + 9} - \frac{3}{2} \ln \left| \frac{3 + \sqrt{(2x - 1)^2 + 9}}{2x - 1} \right| + C \quad \text{(replace } u \text{ by } 2x - 1\text{)}$$

$$= \frac{1}{2} \sqrt{4x^2 - 4x + 10} - \frac{3}{2} \ln \frac{3 + \sqrt{4x^2 - 4x + 10}}{|2x - 1|} + C. \qquad \blacksquare$$

The example above shows, once again, the importance of substitution as a tool of antidifferentiation.

Unfortunately, our table of integrals is far from complete. There are many integrals left untouched by our table. In fact, *no* table of integrals can be exhaustive. A fairly comprehensive one can be found in the popular *CRC Standard Mathematics Tables**, which lists 596 formulas in some 50 pages. Indeed, there are entire *books* devoted to integral tables.

|||||||||||| **A BRIEF TABLE OF INTEGRALS**

Basic Forms

1. $\int du = u + C$

2. $\int af(u)\, du = a \int f(u)\, du$

3. $\int [f(u) \pm g(u)]\, du = \int f(u)\, du \pm \int g(u)\, du$

---

* *CRC Standard Mathematical Tables*, 27th Edition. Boca Raton: CRC Press, Inc., 1984.

4. $\int u^n \, du = \dfrac{u^{n+1}}{n+1} + C$ (if $n \neq -1$)

5. $\int \dfrac{1}{u} \, du = \ln |u| + C$

6. $\int e^u \, du = e^u + C$

7. $\int f(x)g'(x) \, dx = f(x)g(x) - \int g(x)f'(x) \, dx$ $\left(\int u \, dv = uv - \int v \, du \right)$

|||||||||||| **Integrals Involving $au + b$**

8. $\int (au + b)^n \, du = \dfrac{1}{a} \dfrac{(au+b)^{n+1}}{n+1} + C$ (if $n \neq -1$)

9. $\int \dfrac{1}{au+b} \, du = \dfrac{1}{a} \ln |au+b| + C$

10. $\int \dfrac{u}{au+b} \, du = \dfrac{u}{a} - \dfrac{b}{a^2} \ln |au+b| + C$

11. $\int \dfrac{u^2}{au+b} \, du = \dfrac{1}{a^3} \left[\dfrac{1}{2} (au+b)^2 - 2b(au+b) + b^2 \ln |au+b| \right] + C$

12. $\int \dfrac{u \, du}{(au+b)^2} = \dfrac{1}{a^2} \left[\dfrac{b}{au+b} + \ln |au+b| \right] + C$

13. $\int \dfrac{u^2 \, du}{(au+b)^2} = \dfrac{1}{a^3} \left[(au+b) - \dfrac{b^2}{au+b} - 2b \ln |au+b| \right] + C$

14. $\int \dfrac{du}{u(au+b)} = \dfrac{1}{b} \ln \left| \dfrac{u}{au+b} \right| + C$

15. $\int \dfrac{du}{u^2(au+b)} = -\dfrac{1}{bu} + \dfrac{a}{b^2} \ln \left| \dfrac{au+b}{u} \right| + C$

16. $\int \dfrac{du}{u(au+b)^2} = \dfrac{1}{b(au+b)} + \dfrac{1}{b^2} \ln \left| \dfrac{u}{au+b} \right| + C$

17. $\int \dfrac{du}{(au+b)(cu+d)} = \dfrac{1}{bc-ad} \ln \left| \dfrac{cu+d}{au+b} \right| + C$ (if $bc - ad \neq 0$)

18. $\int \dfrac{u \, du}{(au+b)(cu+d)} = \dfrac{1}{bc-ad} \left\{ \dfrac{b}{a} \ln |au+b| - \dfrac{d}{c} \ln |cu+d| \right\} + C$

   (if $bc - ad \neq 0$)

|||||||||||| **Integrals Involving $\sqrt{au + b}$**

19. $\int \sqrt{au+b} \, du = \dfrac{2}{3a} (au+b)^{3/2} + C$

20. $\int u\sqrt{au+b}\,du = \dfrac{2(3au-2b)(au+b)^{3/2}}{15a^2} + C$

21. $\int u^2\sqrt{au+b}\,du = \dfrac{2}{105a^3}(15a^2u^2 - 12abu + 8b^2)(au+b)^{3/2} + C$

22. $\int u^n\sqrt{au+b}\,du = \dfrac{2u^n(au+b)^{3/2}}{a(2n+3)} - \dfrac{2bn}{a(2n+3)}\int u^{n-1}\sqrt{au+b}\,du + C$

23. $\int \dfrac{u\,du}{\sqrt{au+b}} = \dfrac{2}{3a^2}(au-2b)\sqrt{au+b} + C$

24. $\int \dfrac{u^2\,du}{\sqrt{au+b}} = \dfrac{2}{15a^3}(3a^2u^2 - 4abu + 8b^2)\sqrt{au+b} + C$

25. $\int \dfrac{u^n\,du}{\sqrt{au+b}} = \dfrac{2u^n\sqrt{au+b}}{a(2n+1)} - \dfrac{2bn}{a(2n+1)}\int \dfrac{u^{n-1}\,du}{\sqrt{au+b}}$

26. $\int \dfrac{du}{u\sqrt{au+b}} = \dfrac{1}{\sqrt{b}}\ln\left|\dfrac{\sqrt{au+b}-\sqrt{b}}{\sqrt{au+b}+\sqrt{b}}\right| + C$ (if $b > 0$)

27. $\int \dfrac{\sqrt{au+b}\,du}{u} = 2\sqrt{au+b} + b\int \dfrac{du}{u\sqrt{au+b}}$

Integrals Involving $u^2 \pm a^2$, $a > 0$

Note When the symbol "$\pm$" is used several times in the same formula, it is understood to be "$+$" *throughout* the formula or "$-$" throughout the formula.

28. $\int \dfrac{du}{u^2-a^2} = \dfrac{1}{2a}\ln\left|\dfrac{u-a}{u+a}\right| + C$

29. $\int \dfrac{du}{a^2-u^2} = \dfrac{1}{2a}\ln\left|\dfrac{u+a}{u-a}\right| + C$

30. $\int \dfrac{u^2\,du}{\sqrt{u^2\pm a^2}} = \dfrac{u}{2}\sqrt{u^2\pm a^2} - \dfrac{\pm a^2}{2}\ln\left|u+\sqrt{u^2\pm a^2}\right| + C$

31. $\int \dfrac{du}{\sqrt{u^2\pm a^2}} = \ln\left|u+\sqrt{u^2\pm a^2}\right| + C$

32. $\int \sqrt{u^2\pm a^2}\,du = \dfrac{u}{2}\sqrt{u^2\pm a^2} \pm \dfrac{a^2}{2}\ln\left|u+\sqrt{u^2\pm a^2}\right| + C$

33. $\int u^2\sqrt{u^2\pm a^2}\,du = \dfrac{u}{8}(2u^2\pm a^2)\sqrt{u^2\pm a^2} - \dfrac{a^4}{8}\ln\left|u+\sqrt{u^2\pm a^2}\right| + C$

34. $\int \dfrac{\sqrt{a^2\pm u^2}\,du}{u} = \sqrt{a^2\pm u^2} - a\ln\left|\dfrac{a+\sqrt{a^2\pm u^2}}{u}\right| + C$

35. $\int \dfrac{\sqrt{u^2\pm a^2}\,du}{u^2} = -\dfrac{\sqrt{u^2\pm a^2}}{u} + \ln\left|u+\sqrt{u^2\pm a^2}\right| + C$

36. $\int \dfrac{du}{u\sqrt{a^2 \pm u^2}} = \dfrac{1}{a} \ln \left| \dfrac{u}{a + \sqrt{a^2 \pm u^2}} \right| + C$

37. $\int \dfrac{du}{u^2 \sqrt{u^2 \pm a^2}} = -\dfrac{\sqrt{u^2 \pm a^2}}{\pm a^2 u} + C$

38. $\int \dfrac{du}{u^2 \sqrt{a^2 - u^2}} = -\dfrac{\sqrt{a^2 - u^2}}{a^2 u} + C$

|||||||||||| **Integrals Involving Exponential and Logarithmic Functions**

39. $\int e^{au}\, du = \dfrac{1}{a} e^{au} + C$

40. $\int u e^{au}\, du = \dfrac{e^{au}}{a^2} (au - 1) + C$

41. $\int a^u\, du = \dfrac{a^u}{\ln a} + C, \qquad \text{for } a > 0,\, a \neq 1$

42. $\int u^n e^{ku}\, du = \dfrac{u^n e^{ku}}{k} - \dfrac{n}{k} \int u^{n-1} e^{ku}\, du \qquad (k \neq 0)$

43. $\int \dfrac{e^{ku}\, du}{u^n} = -\dfrac{e^{ku}}{(n-1)u^{n-1}} + \dfrac{k}{n-1} \int \dfrac{e^{ku}\, du}{u^{n-1}} \qquad (n \neq 1)$

44. $\int \ln u\, du = u \ln u - u + C$

45. $\int u^n \ln u\, du = \dfrac{u^{n+1}}{(n+1)^2} [(n+1) \ln u - 1] + C$

6.4 EXERCISE SET ||||||||||||||||||||||||||||||||||||||||||||||||||||||||||||||||||||||||||||||||

Compute each of the following integrals, using the table found before the exercise set. In each case, tell which formula you are using and what substitution(s) you are making.

1. $\int \dfrac{x}{5x + 10}\, dx$

2. $\int \dfrac{1}{\sqrt{x^2 + 25}}\, dx$

3. $\int \dfrac{1}{\sqrt{9x^2 - 25}}\, dx$

4. $\int \dfrac{t}{(5t - 3)^2}\, dt$

5. $\int \dfrac{1}{49x^2 - 4}\, dx$

6. $\int \dfrac{1}{x(3x - 5)^2}\, dx$

7. $\int \dfrac{1}{t\sqrt{9 + 16t^2}}\, dt$

8. $\int (-3x + 7)^5\, dx$

9. $\int \dfrac{1}{1 - 9x^2}\, dx$

10. $\int \dfrac{x}{(2x - 5)^2}\, dx$

11. $\int e^{5x/2}\, dx$

12. $\int \ln \left(\tfrac{2}{3} x \right) dx$

13. $\int \dfrac{1}{(4t + 3)^3}\, dt$

14. $\int \dfrac{1}{4t + 3}\, dt$

15. $\int \dfrac{1}{x\sqrt{9 - 4x^2}}\, dx$

16. $\int t e^{-t/3}\, dt$

17. $\int \dfrac{2x}{3x(x - 4)}\, dx$

18. $\int \dfrac{1}{5x\sqrt{4 + 9x^2}}\, dx$

19. $\int \dfrac{x^2}{3x - 5} \, dx$

20. $\int 3s\sqrt{2s + 1} \, ds$

21. $\int \dfrac{1}{(3t - 2)(t + 4)} \, dt$

22. $\int \dfrac{1}{2x^2 - x^3} \, dx$

23. $\int x^2 \sqrt{5x + 8} \, dx$

24. $\int \dfrac{t^2}{(3t - 2)^2} \, dt$

25. $\int \sqrt{4x^2 - 9} \, dx$

26. $\int \dfrac{1}{2y^2 - 5y - 3} \, dy$

27. $\int \dfrac{s}{2s^2 + s - 3} \, ds$

28. $\int \dfrac{3x}{\sqrt{2x - 5}} \, dx$

29. $\int \dfrac{1}{t\sqrt{t + 4}} \, dt$

30. $\int \dfrac{1}{s(3s + 4)} \, ds$

31. $\int \dfrac{\sqrt{2 - 9x^2}}{x} \, dx$

32. $\int \dfrac{x^2}{\sqrt{2 - 7x}} \, dx$

33. $\int t^2 e^{5t} \, dt$

34. $\int s^3 \ln s \, ds$

35. $\int \dfrac{1}{\sqrt{x^2 + 2x}} \, dx$

[Hint: $x^2 + 2x = (x^2 + 2x + 1) - 1$]

36. $\int \sqrt{9x^2 + 6x + 5} \, dx$

KEY IDEAS FOR REVIEW ||||||||||||||||||||||||||||||||||||||||||||||||||||||||||||||||||||||||

- $F(x)$ is an antiderivative of $f(x)$ if $F'(x) = f(x)$ for all x in the domain of f.
- Any two antiderivatives of the same function differ by a constant amount.
- If $F(x)$ is any antiderivative of $f(x)$, then every antiderivative of $f(x)$ is given by

$$\int f(x) \, dx = F(x) + C,$$

where C is an arbitrary constant.

- $\int k \, dx = kx + C$

- $\int x^r \, dx = \dfrac{x^{r+1}}{r + 1} + C, \qquad r \ne -1$

- $\int \dfrac{1}{x} \, dx = \ln |x| + C, \qquad x \ne 0$

- $\int e^x \, dx = e^x + C$

- $\int a^x \, dx = \dfrac{a^x}{\ln a} + C$

- $\int kf(x) \, dx = k \int f(x) \, dx$ (Beware: only *constants* "factor out" of an integral.)

- $\int [f(x) + g(x)] \, dx = \int f(x) \, dx + \int g(x) \, dx$

- $\int [f(x) - g(x)] \, dx = \int f(x) \, dx - \int g(x) \, dx$

- $\int [f(x)]^k f'(x) \, dx =$

$$\dfrac{1}{k + 1} [f(x)]^{k+1} + C, \qquad k \ne -1$$

or

$$\int u^k \, du = \dfrac{1}{k + 1} u^{k+1} + C,$$

where $u = f(x)$, $k \ne -1$

- $\int e^{f(x)} f'(x) \, dx = e^{f(x)} + C$

or

$\int e^u \, du = e^u + C$, where $u = f(x)$

- $\int \dfrac{f'(x)}{f(x)} \, dx = \ln |f(x)| + C, \qquad f(x) \ne 0$

or

$\int \dfrac{1}{u} \, du = \ln |u| + C$, where $u = f(x) \ne 0$

- Integration by substitution, in general:

$$\int g(f(x)) f'(x) \, dx = \int g(u) \, du$$

■ Linear substitution:

If $\qquad \int f(x)\,dx = F(x) + C,$

then $\quad \int f(ax + b)\,dx = \dfrac{1}{a}\,F(ax + b) + C$

■ Integration by parts:

$$\int u(x)v'(x)\,dx = u(x)v(x) - \int v(x)u'(x)\,dx$$

or

$$\int u\,dv = uv - \int v\,du$$

■ Before integration tables can be used, it may be necessary to algebraically manipulate the integrand.

REVIEW EXERCISES ||||||||||||||||||||||||||||||||||||||||||||||||||||||||||||||||||||||

In Exercises 1–26, find the indefinite integral without using the Table of Integrals. When $\ln x$ appears, assume that x is positive.

1. $\displaystyle\int x^5\,dx$

2. $\displaystyle\int x^{-1/4}\,dx$

3. $\displaystyle\int 3e^t\,dt$

4. $\displaystyle\int \frac{5}{x}\,dx$

5. $\displaystyle\int (3x + 4)^5\,dx$

6. $\displaystyle\int e^{2-5x}\,dx$

7. $\displaystyle\int \frac{1}{\sqrt[3]{x^2}}\,dx$

8. $\displaystyle\int (2x^2 + \sqrt[3]{x^4})\,dx$

9. $\displaystyle\int \left(\sqrt[3]{x^4} - \frac{4}{x} + e^x \right)dx$

10. $\displaystyle\int e^{3x/2}\,dx$

11. $\displaystyle\int \frac{1}{t + 3}\,dt$

12. $\displaystyle\int te^{t^2 + 1}\,dt$

13. $\displaystyle\int x^2(x^3 + 2)^8\,dx$

14. $\displaystyle\int \frac{4t^3 + 3}{\sqrt[3]{t^4 + 3t}}\,dt$

15. $\displaystyle\int \frac{3x^2 + 1}{x^3 + x}\,dx$

16. $\displaystyle\int \frac{x}{3x^2 - 7}\,dx$

17. $\displaystyle\int \frac{5e^{2x}}{e^{2x} + 1}\,dx$

18. $\displaystyle\int xe^{2x}\,dx$

19. $\displaystyle\int \frac{(\ln x)^2}{x}\,dx$

20. $\displaystyle\int \frac{1}{x(\ln x)^2}\,dx$

21. $\displaystyle\int t\sqrt{t - 3}\,dt$

22. $\displaystyle\int x^2 \ln 3x\,dx$

23. $\displaystyle\int x^3 e^{2x}\,dx$

24. $\displaystyle\int xe^{-x/2}\,dx$

25. $\displaystyle\int \ln (x + 5)\,dx$

26. $\displaystyle\int x(\ln x)^3\,dx$

In Exercises 27–36, use the Table of Integrals.

27. $\displaystyle\int \frac{1}{\sqrt{16 + 25x^2}}\,dx$

28. $\displaystyle\int \frac{1}{64 - u^2}\,du$

29. $\displaystyle\int \frac{x^2}{\sqrt{1 + 9x^2}}\,dx$

30. $\displaystyle\int \frac{5}{x(x - 5)}\,dx$

31. $\displaystyle\int \frac{t^2}{5 - 4t}\,dt$

32. $\displaystyle\int \frac{1}{t^2 - 2t - 8}\,dt$

33. $\displaystyle\int \frac{5x}{2x^2 + 5x - 3}\,dx$

34. $\displaystyle\int 7x^2\sqrt{2x + 3}\,dx$

35. $\displaystyle\int \frac{\sqrt{t^2 + 4}}{t}\,dt$

36. $\displaystyle\int \frac{8}{t\sqrt{t + 9}}\,dt$

37. The slope of the tangent line to a curve at the point (x, y) on the curve is $6x^3\sqrt{x^4 + 9}$. If the point $(2, 250)$ lies on the curve, find an equation for the curve.

38. **(Motion)** A car, starting from rest, moves along a straight line so that its acceleration after t seconds (in feet per second per second) is given by

$$a(t) = t^2 + 2t + 1.$$

Find the distance traveled by the car during the first 4 seconds. (*Hint:* Let $s(t)$ denote the distance traveled during the first t seconds. Then $v(t) = $ velocity $= ds/dt$, $a(t) = dv/dt = d^2s/dt^2$.)

39. **(Pollution)** A new plant discharges particulate matter into the atmosphere at a rate given by $Q'(t) = 10te^{-0.5t}$, where $Q(t)$ is the number of tons of particulate matter discharged during the first t years of operation (thus, $Q(0) = 0$). How much particulate matter will be discharged into the atmosphere during the first 8 years of operation?

CHAPTER TEST ||||||||||||||||||||||||||||||||||||||||||||||||||||||||||||||||||||||||||||||||||

In problems 1–8. find the indefinite integral without using the Table of Integrals. When ln x appears, assume that $x > 0$.

1. $\int (7x + 2)^5 \, dx$

2. $\int \left(4x^{1/3} - \dfrac{5}{\sqrt{x}} \right) dx$

3. $\int \dfrac{x}{(x^2 + 5)} \, dx$

4. $\int \dfrac{x^2}{(x^3 - 4)^3} \, dx$

5. $\int t\sqrt{t + 2} \, dt$

6. $\int te^{3t} \, dt$

7. $\int 3xe^{x^2} \, dx$

8. $\int \dfrac{\ln x}{x} \, dx$

In Problems 9–11, use the Table of Integrals.

9. $\int \dfrac{3x}{(5x - 4)^2} \, dx$

10. $\int \sqrt{x^2 + 4} \, dx$

11. $\int \dfrac{1}{x^2 + 2x - 3} \, dx$

12. For a manufacturer of surfboards the marginal profit in selling x surfboards is

$$P'(x) = 80 - 0.2x \text{ (dollars).}$$

Find the profit function $P(x)$, if the profit on 50 surfboards is known to be $1000.

THE DEFINITE INTEGRAL

An important type of calculus problem is finding the area of the region under the graph of a nonnegative function over an interval. The solution to this problem gives rise to another type of integral, called the definite integral. This integral is related to the indefinite integral defined in Chapter 6 by a central theorem of calculus. In this chapter we shall develop the definite integral, discuss its properties, and present a number of significant applications.

7.1 ||||||| THE AREA UNDER A CURVE

Consider the problem of finding the area of the region below the graph of a nonnegative function $y = f(x)$, and above the x-axis, between the vertical lines $x = a$ and $x = b$ (see Figure 1). We call this **the area under the graph of f over the interval $[a, b]$.** We assume, of course, that $a < b$ and that f is defined everywhere on $[a, b]$.

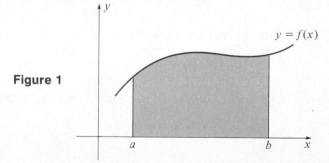

Figure 1

To set the stage for a powerful technique for finding the desired area, shaded in Figure 1, we change the problem somewhat. We shall find it advantageous to allow the right-hand endpoint of the interval to be a variable. Thus, we consider the apparently more difficult problem of finding the area under the graph of f over the interval $[a, x]$, where the right-hand endpoint is the variable x, $a \le x \le b$. The corresponding area is thus a function of x, which we denote by $A(x)$ (see Figure 2).

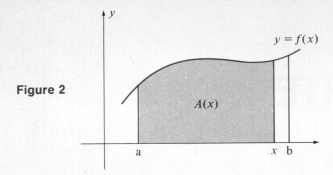

Figure 2

EXAMPLE 1   Let f be the constant function defined by $f(x) = 3$ on the interval $[0, 10]$. We want to find the area of the region under the curve $y = f(x) = 3$ over the interval

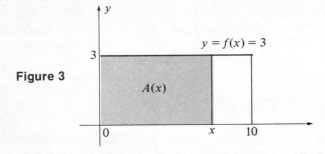

Figure 3

$[0, x]$ (Figure 3). The region whose area we want to compute is a rectangle with base width x and height 3. Hence, the area is

$$A(x) = 3x.$$

EXAMPLE 2   Consider the function f defined on the interval $[2, 8]$ by

$$f(x) = x - 2.$$

The graph of this function is the straight line shown in Figure 4. We want to find

Figure 4

the area of the region under the curve $y = f(x) = x - 2$ over the interval $[2, x]$. The region of interest is a triangle with base $x - 2$ and altitude $x - 2$. Hence, the area is given by

$$A(x) = \tfrac{1}{2} \cdot \text{altitude} \cdot \text{base} = \tfrac{1}{2}(x - 2)(x - 2)$$
$$= \tfrac{1}{2}(x^2 - 4x + 4) = \tfrac{1}{2}x^2 - 2x + 2.$$ ∎

Determining the function $A(x)$ in Examples 1 and 2 was easy because the regions involved were standard geometric regions for which area formulas were readily available. For most regions, however, no basic area formulas exist. It is the purpose of this chapter to develop techniques for obtaining these areas in many cases that are not covered by the area formulas of elementary geometry.

The first and most important step is to develop a connection between the problem of finding the area function $A(x)$ and the problem of antidifferentiation. This relationship is one of the cornerstones of calculus.

The First Fundamental Theorem of Calculus

Suppose that f is a function that is continuous and nonnegative on the interval $[a, b]$. Let $A(x)$ denote the area of the region under the curve $y = f(x)$ over the interval $[a, x]$, $a \le x \le b$. Then $A(x)$ is differentiable on $[a, b]$, and

$$A'(x) = f(x) \tag{1}$$

for each x in $[a, b]$. That is, $A(x)$ is an antiderivative of $f(x)$ on $[a, b]$.

Proof (Intuitive)   From the definition of the derivative,

$$A'(x) = \lim_{h \to 0} \frac{A(x + h) - A(x)}{h}. \tag{2}$$

Observe in Figure 5 that $A(x)$ is the area of the gray region and $A(x + h)$ is the sum of the area in gray and the area in color. Hence, $A(x + h) - A(x)$ is merely the area of the region shown in color. Notice that when h is small, this region is

Figure 5

"almost" a rectangle with base width h and height $f(x)$. Thus, the area in color is approximately $f(x) \cdot h$. That is,

$$A(x + h) - A(x) \approx h \cdot f(x),\tag{3}$$

or, equivalently,

$$\frac{A(x + h) - A(x)}{h} \approx f(x).\tag{4}$$

As h gets smaller and smaller, the difference between the colored region and the approximating rectangle shrinks. Consequently, the left side of (3) is a better approximation of the right side as h gets smaller and smaller. In other words, it appears (intuitively) that

$$\lim_{h \to 0} \frac{A(x + h) - A(x)}{h} = f(x).\tag{5}$$

Thus, the limit in (2) does exist, and this fact implies that A is differentiable and $A'(x) = f(x)$.   ∎

The First Fundamental Theorem of Calculus shows that $A'(x) = f(x)$, so $A(x)$ is one of the antiderivatives of $f(x)$. Hence, we can determine $A(x)$ from $\int f(x)\,dx$ once we have an appropriate numerical value for the integration constant. The following example uses this approach to determine $A(x)$.

EXAMPLE 3   Let $f(x) = x^2$, $1 \le x \le 4$.
  (a) Find the area $A(x)$ of the region under the curve $y = f(x)$ over the interval $[1, x]$ [Figure 6(a)].
  (b) Find the area of the region under the curve $y = f(x)$ over the interval $[1, 3]$ [Figure 6(b)].

Figure 6

(a)   (b)

Solution   (a) Since

$$\int x^2 \, dx = \frac{x^3}{3} + C,$$

we know, by the First Fundamental Theorem of Calculus, that

$$A(x) = \frac{x^3}{3} + C, \tag{6}$$

where C is an appropriately chosen constant. We can determine C by observing that if $x = 1$, the interval $[1, x]$ is simply a point and, in this case, the area under the curve over the interval is 0; that is, $A(1) = 0$. Then, from (6),

$$0 = A(1) = \frac{1^3}{3} + C.$$

Hence, $C = -\frac{1}{3}$ and

$$A(x) = \frac{x^3}{3} - \frac{1}{3}.$$

(b) Since

$$A(x) = \frac{x^3 - 1}{3},$$

we determine $A(3)$ by substituting $x = 3$ in the formula obtaining

$$A(3) = \frac{3^3 - 1}{3} = \frac{27 - 1}{3} = \frac{26}{3}. \qquad \blacksquare$$

Remembering Section 6.5, we see that the First Fundamental Theorem of Calculus shows that the problem of finding $A(x)$ for a given function f is equivalent to solving the initial-value problem: find $A(x)$ if

$A'(x) = f(x)$, and
$A(x) = 0$ when $x = a$.

The Second Fundamental Theorem of Calculus
If f is a function that is continuous and nonnegative on the interval $[a, b]$, then f has an antiderivative on $[a, b]$. Moreover, if F is *any* antiderivative of f on $[a, b]$, then the area under the curve $y = f(x)$ over $[a, b]$ is

$$\text{area} = F(b) - F(a). \tag{7}$$

Proof Suppose that f is a function that is continuous and nonnegative on the interval $[a, b]$. For each number x in the interval $[a, b]$, let $A(x)$ denote the area of the region under the curve $y = f(x)$ over the interval $[a, x]$ (see Figure 2).

Observe that if $x = a$, then the interval $[a, x]$ shrinks to a single point and the area under the curve $y = f(x)$ over this interval is zero; that is,

$$A(a) = 0.$$

Let $F(x)$ be any antiderivative of $f(x)$. By the First Fundamental Theorem of Calculus, $A(x)$ also is an antiderivative of $f(x)$. Because any two antiderivatives of the same function must differ by a constant C, we can write

$$F(x) = A(x) + C.$$

Hence,

$$
\begin{aligned}
F(b) - F(a) &= [A(b) + C] - [A(a) + C] \\
&= A(b) - A(a) \\
&= A(b) - 0 \\
&= A(b) \\
&= \text{area of the region under } y = f(x) \text{ over } [a, b].
\end{aligned}
$$

The Definite Integral of a Continuous Function over a Closed Interval

Suppose $f(x)$ is any function that is continuous over a closed interval $[a, b]$, not necessarily nonnegative there, and that $F(x)$ is any antiderivative of $f(x)$. The difference $F(b) - F(a)$ in Equation (7) leads to one of the fundamental notions of calculus.

If $f(x)$ is any continuous function on the interval $[a, b]$, then the **definite integral** (or just the **integral**) **of** $f(x)$ **from** a **to** b, denoted by

$$\int_a^b f(x)\,dx,$$

is defined by

$$\int_a^b f(x)\,dx = F(b) - F(a) \tag{8}$$

where $F(x)$ is any antiderivative of $f(x)$.

The numbers a and b in (8) are called the **limits of integration** and the expression on the right side of (8) is often written as

$$F(x)\Big]_a^b.$$

EXAMPLE 4   Evaluate the definite integral $\int_{-2}^{4} x^3 \, dx$.

Solution   Since $f(x) = x^3$ is continuous on $[-2, 4]$, we can evaluate the definite integral $\int_{-2}^{4} x^3 \, dx$ as $F(4) - F(-2)$, where $F(x)$ is any antiderivative of $f(x) = x^3$. One antiderivative of $f(x) = x^3$ is $F(x) = x^4/4$. Thus,

$$\int_{-2}^{4} x^3 \, dx = \frac{x^4}{4} \Bigg]_{-2}^{4} = \frac{4^4}{4} - \frac{(-2)^4}{4} = 64 - 4 = 60.$$

Note   Another antiderivative of $f(x)$ is

$$G(x) = \frac{x^4}{4} + 5.$$

Using this antiderivative, we obtain

$$\int_{-2}^{4} x^3 \, dx = \left(\frac{x^4}{4} + 5 \right) \Bigg]_{-2}^{4} = \left(\frac{4^4}{4} + 5 \right) - \left(\frac{(-2)^4}{4} + 5 \right)$$

$$= 64 + 5 - 4 - 5 = 60,$$

which agrees with the previously obtained value and illustrates the fact that the definite integral is independent of the particular antiderivative used in its evaluation.   ∎

Reminder

> If $f(x)$ is continuous and *nonnegative* over the closed interval $[a, b]$, then the area of the region under the curve $y = f(x)$ over the interval $[a, b]$ is the definite integral
>
> $$\int_{a}^{b} f(x) \, dx.$$

EXAMPLE 5   Use the definite integral to find the area of the region under the curve $y = x^2$ over the interval $[1, 3]$.

Solution   Since $f(x) = x^2 \geq 0$ on $[1, 3]$, the area of interest is $\int_{1}^{3} x^2 \, dx$. We can evaluate this definite integral as $F(3) - F(1)$, where $F(x)$ is any antiderivative of $f(x) = x^2$. Thus, the desired area is

$$\int_{1}^{3} x^2 \, dx = \frac{x^3}{3} \Bigg]_{1}^{3}$$

$$= \frac{3^3}{3} - \frac{1}{3} = \frac{26}{3},$$

which agrees with the answer obtained in Example 3(b).   ∎

EXAMPLE 6   Find the area of the region under the curve $y = 1/x$ over the interval $[1, 4]$ (Figure 7).

Figure 7

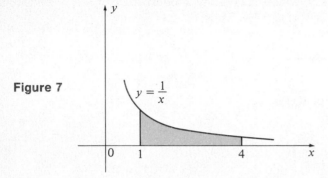

Solution   The desired area is

$$A(4) = \int_1^4 \frac{1}{x}\, dx = \ln |x| \Big]_1^4 = \ln 4 - \ln 1$$

$$= \ln 4 - 0$$

$$= \ln 4 \approx 1.3863. \qquad \text{(Appendix B, Table II)} \qquad \blacksquare$$

7.1 EXERCISE SET ||||||||||||||||||||||||||||||||||||||||||||||||||||||||||||||||||||||||||||||

In Exercises 1–8,

(a) sketch the region under the graph of the given function f over the given interval $[a, b]$;

(b) using only the graph and your knowledge of geometry (that is, without using the Second Fundamental Theorem of Calculus) find the area of this region;

(c) then use the Second Fundamental Theorem of Calculus to calculate this area.

1. $y = 4$, $[-1, 2]$

2. $y = 3$, $[2, 5]$

3. $y = 2x$, $[0, 3]$

4. $y = 1 + x$, $[-1, 5]$

5. $y = 2x$, $[2, 5]$

6. $y = -3x$, $[-2, -1]$

7. $y = x + 1$, $[2, 5]$

8. $y = 2x - 1$, $[1, 3]$

In Exercises 9–18, use a definite integral to find the area of the region under the curve over the designated interval. Note that each curve is a portion of the graph of a function that is nonnegative over the given interval. In each exercise, sketch the curve roughly and shade the appropriate region.

9. $y = x^2$, $[-1, 3]$

10. $y = x^2 - 2$, $[2, 4]$

11. $y = 3 - x^2$, $[-1, 1]$

12. $y = 4 - x^2$, $[0, 2]$

13. $y = 3x - x^2$, $[0, 3]$

14. $y = \sqrt{x}$, $[4, 9]$

15. $y = e^x$, $[0, 1]$

16. $y = e^{2x}$, $[0, 5]$

17. $y = \dfrac{3}{x}$, $[2, 5]$

18. $y = \dfrac{2}{x}$, $[1, 2]$

In Exercises 19–28, evaluate the definite integrals. Note that each integrand is continuous over the given interval.

19. $\displaystyle\int_{-1}^{2} dx$

20. $\displaystyle\int_{2}^{5} - dx$

21. $\displaystyle\int_{1}^{2} (x^4 - 10)\, dx$

22. $\displaystyle\int_{-1}^{2} 4 - x^5\, dx$

23. $\displaystyle\int_{0}^{2} e^t\, dt$

24. $\displaystyle\int_{-1}^{0} e^{2s}\, ds$

25. $\displaystyle\int_{1}^{3} \dfrac{1}{x^2}\, dx$

26. $\displaystyle\int_{2}^{5} \dfrac{1}{t}\, dt$

27. $\displaystyle\int_{1}^{4} \sqrt{x}\, dx$

28. $\displaystyle\int_{4}^{9} \dfrac{1}{\sqrt{t}}\, dt$

In Exercises 29–32, use definite integrals to find the area of the shaded regions.

29.

30.

31.

32.

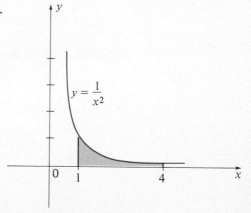

7.2 ‖‖‖‖‖‖‖ PROPERTIES OF THE DEFINITE INTEGRAL

In Section 7.1 we defined the definite integral $\int_a^b f(x)\, dx$ of a function that is continuous over the interval $[a, b]$. We have seen that if, in addition, $f(x)$ is nonnegative for all x in $[a, b]$, then $\int_a^b f(x)\, dx$ gives the area of the region under the curve $y = f(x)$ over the interval $[a, b]$.

In the definitions of 7.1 we have implicitly assumed that $a < b$. However, we can define $\int_a^b f(x)\,dx$ even when $a \geq b$. The general definition is as follows:

Let a and b be real numbers and suppose that f is continuous on an interval containing a and b. We define the definite integral $\int_a^b (x)\,dx$ by

$$\int_a^b f(x)\,dx = F(b) - F(a), \tag{1}$$

where $F(x)$ is any antiderivative of $f(x)$.

Many properties of the definite integral follow from this definition. We shall summarize some basic properties for a function f that is continuous on $[a, b]$.

Algebraic Properties of the Definite Integral

1. $\displaystyle\int_a^a f(x)\,dx = 0$

2. $\displaystyle\int_a^b kf(x)\,dx = k\int_a^b f(x)\,dx$ (k is a constant)

3. $\displaystyle\int_a^b f(x)\,dx = -\int_b^a f(x)\,dx$

4. $\displaystyle\int_a^b [f(x) \pm g(x)]\,dx = \int_a^b f(x)\,dx \pm \int_a^b g(x)\,dx$

5. $\displaystyle\int_a^b f(x)\,dx = \int_a^c f(x)\,dx + \int_c^b f(x)\,dx;\ a \leq c \leq b$

We can establish these properties as follows:

1. $\displaystyle\int_a^a f(x)\,dx = F(a) - F(a) = 0.$

2. If $F(x)$ is an antiderivative of $f(x)$, then $kF(x)$ is an antiderivative of $kf(x)$, so

$$\int_a^b kf(x)\,dx = kF(b) - kF(a) = k[F(b) - F(a)]$$
$$= k\int_a^b f(x)\,dx.$$

3. $\displaystyle\int_a^b f(x)\,dx = F(b) - F(a) = -[F(a) - F(b)]$
$$= -\int_b^a f(x)\,dx.$$

4. $\displaystyle\int_a^b [f(x) \pm g(x)]\,dx = [F(b) \pm G(b)] - [F(a) \pm G(a)]$
$$= [F(b) - F(a)] \pm [G(b) - G(a)]$$
$$= \int_a^b f(x)\,dx \pm \int_a^b g(x)\,dx$$

where $G(x)$ is an antiderivative of $g(x)$.

5. $\int_a^c f(x)\,dx + \int_c^b f(x)\,dx = [F(c) - F(a)] + [F(b) - F(c)]$

$$= F(b) - F(a)$$

$$= \int_a^b f(x)\,dx.$$

If f is a nonnegative function on $[a, b]$ and $a < c < b$, then property 5 states (Figure 1) that the area of the region under $y = f(x)$ over $[a, b]$ is the area over $[a, c]$ plus the area over $[c, b]$, as is intuitively clear.

Figure 1

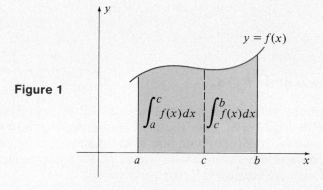

EXAMPLE 1   Use property 4 to evaluate

$$\int_2^5 \left(x^2 - e^x + \frac{1}{x} \right) dx.$$

Solution

$$\int_2^5 \left(x^2 - e^x + \frac{1}{x} \right) dx = \int_2^5 x^2\,dx - \int_2^5 e^x\,dx + \int_2^5 \frac{1}{x}\,dx$$

$$= \frac{x^3}{3} \Big]_2^5 - e^x \Big]_2^5 + \ln x \Big]_2^5$$

$$= \left(\frac{5^3}{3} - \frac{2^3}{3} \right) - (e^5 - e^2) + (\ln 5 - \ln 2)$$

$$= 39 - e^5 + e^2 + \ln \left(\frac{5}{2} \right).$$

EXAMPLE 2   Use property 5 to evaluate $\int_{-1}^3 |x|\,dx$.

Solution   Recall that

$$|x| = \begin{cases} x, & \text{if } x \geq 0 \\ -x, & \text{if } x \leq 0. \end{cases}$$

Thus, on the interval $[-1, 0]$, $|x| = -x$, while on the interval $[0, 3]$, $|x| = x$.

By property 5,

$$\int_{-1}^{3} |x|\, dx = \int_{-1}^{0} |x|\, dx + \int_{0}^{3} |x|\, dx$$

$$= \int_{-1}^{0} -x\, dx + \int_{0}^{3} x\, dx$$

$$= \frac{-x^2}{2}\Big]_{-1}^{0} + \frac{x^2}{2}\Big]_{0}^{3}$$

$$= \left(0 - \left(-\frac{1}{2}\right)\right) + \left(\frac{9}{2} - 0\right)$$

$$= \frac{10}{2}$$

$$= 5.$$

$\blacksquare$

||||||||||||| **Geometric Significance of $\int_{a}^{b} f(x)\, dx$**

We have seen that the definite integral $\int_{a}^{b} f(x)\, dx$ of a function f that is continuous and nonnegative on the interval $[a, b]$ gives the area of the region under the curve $y = f(x)$ over the interval $[a, b]$. Suppose now that f is merely continuous on $[a, b]$, so that f may have positive *and* negative values on $[a, b]$, as in Figure 2. We shall interpret the definite integral $\int_{a}^{b} f(x)\, dx$ geometrically.

Figure 2

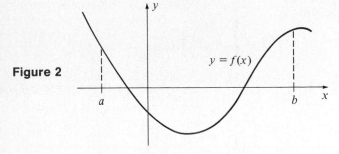

Case 1. $f(x) \leq 0$ for all x in $[a, b]$.

Consider the curve $y = f(x)$, where $f(x) \leq 0$ for all x in $[a, b]$, as in Figure 3.

Figure 3

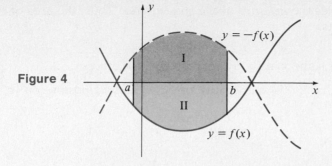

Figure 4

In Figure 4 the graphs of both $f(x)$ and $-f(x)$ are shown. Notice that $-f(x) \geq 0$ for all x in $[a, b]$. The region that lies above the interval $[a, b]$ and below the curve $y = -f(x)$ is labeled region I, while the region that lies below the interval $[a, b]$ and above the curve $y = f(x)$ is labeled II. Observe that regions I and II are mirror images of each other across the x-axis, and consequently they have the same area, A.

Since $-f(x) \geq 0$, the area A of region I is given by the definite integral

$$A = \int_a^b -f(x)\, dx = -\int_a^b f(x)\, dx.$$

This means that

$$\int_a^b f(x)\, dx = -A. \tag{2}$$

In summary, *if* f(x) $\leq$ 0 *for all* x *in* [a, b], *then* $\int_a^b$ f(x) dx *represents the negative of the area of the region that lies below the interval* [a, b] *and above the curve* y = f(x).

EXAMPLE 3   Evaluate and interpret $\int_0^2 (x^2 - 4)\, dx$.

Solution
$$\int_0^2 (x^2 - 4)\, dx = \left[\frac{x^3}{3} - 4x\right]_0^2 = \frac{8}{3} - 8 = -\frac{16}{3}.$$

The graph of $y = x^2 - 4$ is shown in Figure 5. Since $x^2 - 4 \leq 0$ for all x in $[0, 2]$,

Figure 5

the definite integral yields the *negative* of the shaded area. Thus, the area of the shaded region is $\frac{16}{3}$ units. ∎

Case 2. $f(x) \geq 0$ for some x in $[a, b]$, and $f(x) \leq 0$ for others.

For example, suppose that $f(x)$ is nonnegative on $[a, c]$ and nonpositive on $[c, b]$ (Figure 6).

Figure 6

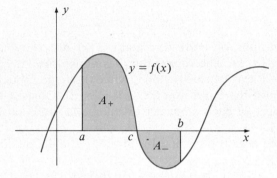

By Property 5,

$$\int_a^b f(x)\,dx = \int_a^c f(x)\,dx + \int_c^b f(x)\,dx. \tag{3}$$

Since $f(x)$ is nonnegative on $[a, c]$, $\int_a^c f(x)\,dx$ is the area A_+ of the upper region in Figure 6. Since $f(x)$ is nonpositive on $[c, b]$, $\int_c^b f(x)\,dx$ is the negative of the area A_- of the lower region in Figure 6. Hence, we can rewrite (3) as

$$\int_a^b f(x)\,dx = A_+ - A_-. \tag{4}$$

We can summarize our remarks as follows.

Geometric Interpretation of the Definite Integral
If f is a continuous function on the interval $[a, b]$, then the definite integral $\int_a^b f(x)\,dx$ represents the area of the region lying above $[a, b]$ and below $y = f(x)$ *minus* the area of the region lying below the interval $[a, b]$ and above $y = f(x)$.

Remark   The integral gives us a *net* area, which is not the same as the total area. The *total area* between the curve $y = f(x)$ and the x-axis, from $x = a$ to $x = b$, as in the combined shaded areas of Figure 6, is

$$A = A_+ + A_-. \tag{5}$$

EXAMPLE 4   Consider the function f defined by

$$f(x) = x^2 - 4$$

on the interval $[-2, 3]$ (Figure 7).

Figure 7

$$y = f(x) = x^2 - 4$$

We have

$$\int_{-2}^{3} (x^2 - 4)\, dx = \left(\frac{x^3}{3} - 4x\right)\Bigg]_{-2}^{3} = \left(\frac{27}{3} - 12\right) - \left(-\frac{8}{3} + 8\right)$$

$$= -\frac{25}{3}.$$

The negative value obtained for the integral means that the area of the shaded region below the x-axis is larger than the area of the shaded region above the x-axis.

To reconcile this conclusion with Equation (4), we calculate the area of each of the regions A_+ and A_-. The area of the shaded region above the x-axis is

$$A_+ = \int_{2}^{3} (x^2 - 4)\, dx = \left(\frac{x^3}{3} - 4x\right)\Bigg]_{2}^{3}$$

$$= \left(\frac{27}{3} - 12\right) - \left(\frac{8}{3} - 8\right) = \frac{7}{3}.$$

The area of the shaded region below the x-axis is

$$A_- = -\int_{-2}^{2} (x^2 - 4)\, dx = -\left(\frac{x^3}{3} - 4x\right)\Bigg]_{-2}^{2}$$

$$= -\left[\left(\frac{8}{3} - 8\right) - \left(-\frac{8}{3} + 8\right)\right] = -\left(-\frac{32}{2}\right) = \frac{32}{3}.$$

Thus, the area A_- of the shaded region below the x-axis is 32/3. The area of the region above the x-axis minus the area of the region below is

$$A_+ - A_- = \tfrac{7}{3} - \tfrac{32}{3} = -\tfrac{25}{3},$$

which agrees with the value obtained for the definite integral $\int_{-2}^{3} (x^2 - 4)\, dx$. From (5), the *total area* of the shaded region is

$$A = A_+ + A_- = \tfrac{7}{3} + \tfrac{32}{3} = \tfrac{39}{3} = 13.$$

EXAMPLE 5   Consider the function $f(x) = x$ over the interval $[-1, 1]$.

$$\int_{-1}^{1} f(x)\, dx = \int_{-1}^{1} x\, dx = \frac{x^2}{2} \Bigg]_{-1}^{1} = \frac{1}{2} - \frac{1}{2} = 0. \tag{6}$$

The graph of the function $f(x) = x$ is shown in Figure 8. In this graph we see that the definite integral $\int_{-1}^{1} f(x)\, dx$, evaluated in Equation (6) as zero, does not represent the total area of the region bounded by the curve $y = f(x)$, the x-axis, and the lines $x = 1$ and $x = -1$ (the region in color) because this area is clearly not zero.

Figure 8

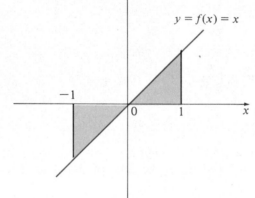

This definite integral is zero because the area of the region lying above the x-axis and below the curve is *equal to* the area of the region lying below the x-axis and above the curve.

The actual *total area* of the region shaded in Figure 8 is

$$A = -\int_{-1}^{0} x\, dx + \int_{0}^{1} x\, dx$$

$$= -\left[\frac{x^2}{2} \right]_{-1}^{0} + \left[\frac{x^2}{2} \right]_{0}^{1}$$

$$= -\left(0 - \frac{1}{2} \right) + \left(\frac{1}{2} - 0 \right) = \frac{1}{2} + \frac{1}{2}$$

$$= 1.$$

||||||||||||| **Area of a Region Bounded by Two Curves**

Suppose that f and g are continuous functions with $f(x) \geq g(x) \geq 0$ on $[a, b]$ and that we want to find the area of the region between the curves $y = f(x)$ and $y = g(x)$ over the interval $[a, b]$ [Figure 9(a)]. As we can see in Figures 9(b) and (c), the area A of the desired region can be written as

$$A = A_f - A_g \qquad (7)$$

Figure 9   (a)   (b)

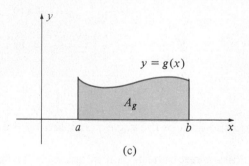

(c)

where A_f is the area of the region under $y = f(x)$ over $[a, b]$ and A_g is the area of the region under $y = g(x)$ over $[a, b]$. Using definite integrals, we can write (7) as

$$A = \int_a^b f(x)\,dx - \int_a^b g(x)\,dx$$

or

$$A = \int_a^b [f(x) - g(x)]\,dx. \qquad (8)$$

EXAMPLE 6   Find the area of the region between the curves $y = x + 3$ and $y = x^2 + 1$ over the interval $[0, 1]$ (the shaded region in Figure 10).

Figure 10

Solution   The desired area is obtained from (8) with

$$a = 0, \quad b = 1, \quad f(x) = x + 3, \quad \text{and} \quad g(x) = x^2 + 1.$$

We have

$$A = \int_0^1 [(x + 3) - (x^2 + 1)]\, dx$$

$$= \int_0^1 (x - x^2 + 2)\, dx$$

$$= \left(\frac{x^2}{2} - \frac{x^3}{3} + 2x \right)\Big]_0^1 = \frac{1}{2} - \frac{1}{3} + 2$$

$$= \frac{13}{6}. \qquad \blacksquare$$

EXAMPLE 7   Find the area of the region enclosed by the curves $y = x^2 + 1$ and $y = x + 3$.

Solution   We first sketch the region in question, as shown in Figure 11.

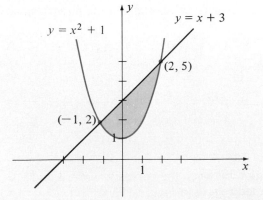

Figure 11

Next, we find the points where the two curves intersect by observing that the co-ordinates of any such point (x, y) must satisfy the equations of *both* curves. That is, (x, y) must satisfy both

$$y = x^2 + 1 \quad \text{and} \quad y = x + 3,$$

so that

$$x^2 + 1 = x + 3$$
$$x^2 - x - 2 = 0$$
$$(x - 2)(x + 1) = 0.$$

Hence,

$$x = 2 \quad \text{or} \quad x = -1.$$

When $x = 2$, $y = 2 + 3 = 5$, and when $x = -1$, $y = -1 + 3 = 2$.

   Thus, the points of intersection are $(2, 5)$ and $(-1, 2)$. Since the enclosed area contains only x-coordinates satisfying $-1 \le x \le 2$, the interval in question is $[-1, 2]$. The desired area is obtained from (8) with

$$a = -1, \quad b = 2, \quad f(x) = x + 3, \quad g(x) = x^2 + 1.$$

We have

$$A = \int_{-1}^{2} [(x + 3) - (x^2 + 1)] \, dx = \int_{-1}^{2} (x - x^2 + 2) \, dx$$

$$= \left(\frac{x^2}{2} - \frac{x^3}{3} + 2x \right) \Bigg]_{-1}^{2} = \left(\frac{4}{2} - \frac{8}{3} + 4 \right) - \left(\frac{1}{2} + \frac{1}{3} - 2 \right)$$

$$= \frac{9}{2}. \tag{9}$$

∎

   In developing Equation (8) for the area between two curves, we assumed that both functions lie above the x-axis over that interval. It turns out that this assumption is not necessary. Equation (8) remains true even when one or both of the curves falls below the x-axis somewhere (or everywhere) in the interval $[a, b]$.

EXAMPLE 8   Find the area enclosed between the curves $y = x^2 - 2$ and $y = x$.

Solution   The region in question is shown in Figure 12.

Figure 12

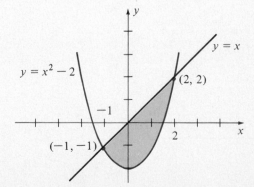

The x-coordinates of the points of intersection of the two curves are found by solving

$$x^2 - 2 = x$$
$$x^2 - x - 2 = 0$$
$$(x - 2)(x + 1) = 0,$$

so that $x = -1$ or $x = 2$. Thus, we shall integrate from -1 to 2. Using **Equation** (8), the area is

$$A = \int_{-1}^{2} [x - (x^2 - 2)] \, dx$$
$$= \int_{-1}^{2} (x - x^2 + 2) \, dx.$$

Observe that this integral was seen before, in (9), and was calculated to be $\frac{9}{2}$. Therefore,

$$A = \tfrac{9}{2}.$$

It is not a coincidence that this answer is identical to that of **Example 7**. The graphs of $y = x^2 - 2$ and $y = x$ are identical in shape, respectively, with the graphs of $y = x^2 + 1$ and $y = x + 3$. The former are merely 3 units lower than the latter. Since raising or lowering two curves by the same amount will not change the area they enclose, it is clear why Examples 7 and 8 have the same solution. ∎

Warning   When a region is bounded by curves that intersect somewhere *inside* the interval of integration (see Figure 13 below), then the integration must be broken up into a sum of separate integrals.

EXAMPLE 9   Find the area between the curves $y = x^2$ and $y = \sqrt{x}$ over the interval $[0, 2]$.

Solution   We first sketch the curves and shade the appropriate region in Figure 13.

Figure 13

We observe that these curves intersect when $x = 1$, which is inside the interval $[0, 2]$. Because of this, the upper and lower boundaries of the region must be given piecewise;

$$\text{upper boundary curve:} \quad y = \begin{cases} \sqrt{x} & \text{if} \quad 0 \le x \le 1 \\ x^2 & \text{if} \quad 1 \le x \le 2 \end{cases}$$

$$\text{lower boundary curve:} \quad y = \begin{cases} x^2 & \text{if} \quad 0 \le x \le 1 \\ \sqrt{x} & \text{if} \quad 1 \le x \le 2. \end{cases}$$

Thus, the integration will be performed over the intervals $[0, 1]$ and $[1, 2]$ separately. The desired area is

$$A = \int_0^2 (\text{upper boundary} - \text{lower boundary})\, dx$$

$$= \int_0^1 (\sqrt{x} - x^2)\, dx + \int_1^2 (x^2 - \sqrt{x})\, dx$$

$$= \left[\frac{2}{3} x^{3/2} - \frac{1}{3} x^3 \right]_0^1 + \left[\frac{1}{3} x^3 - \frac{2}{3} x^{3/2} \right]_1^2$$

$$= \left(\frac{2}{3} - \frac{1}{3} \right) + \left[\left(\frac{8}{3} - \frac{2}{3} (2)^{3/2} \right) - \left(\frac{1}{3} - \frac{2}{3} \right) \right]$$

$$= \frac{1}{3} + \frac{8}{3} - \frac{2}{3} \cdot 2\sqrt{2} - \left(-\frac{1}{3} \right)$$

$$= \frac{10}{3} - \frac{4\sqrt{2}}{3}. \qquad \blacksquare$$

7.2 EXERCISE SET ||||||||||||||||||||||||||||||||||||||||||||||||||||||||||||||||||||||||||||||

1. Use the Second Fundamental Theorem of calculus to evaluate $\int_2^2 (3x^2 + x)\, dx$, and verify property 1 for this particular case.

2. Calculate $\int_1^4 (x - 2)\, dx$ and $\int_1^4 (3x - 6)\, dx$ without using property 2. By comparing your results, verify property 2 in this case.

3. Use Equation (1) to evaluate directly

$$\int_2^3 (2x + x^2)\, dx \quad \text{and} \quad \int_3^2 (2x + x^2)\, dx.$$

By comparing their values, verify property 3 for this particular case.

4. (a) Evaluate $\int_0^3 x^2\, dx$ and $\int_0^3 6x\, dx$ separately.
   (b) Use one of the properties of definite integrals to write the sum $\int_0^3 x^2\, dx + \int_0^3 6x\, dx$ as a single definite integral, and evaluate this integral. Compare your answer with that found in (a).

5. (a) Evaluate

$$\int_0^2 x^3\, dx + \int_2^3 x^3\, dx$$

   by integrating each term.
   (b) Using one of the properties of definite integrals, write the sum in (a) as a single definite integral and evaluate this integral.

6. (a) Evaluate

$$\int_1^4 \frac{1}{\sqrt{x}}\, dx + \int_4^9 \frac{1}{\sqrt{x}}\, dx$$

   by integrating each term.
   (b) Using one of the properties of definite integrals, write the sum in (a) as a single definite integral and evaluate this integral.

In Exercises 7–24, evaluate the definite integrals using properties 1–5.

7. $\int_1^3 (1 + x^2)\, dx$

8. $\int_0^2 (2x - x^2)\, dx$

9. $\int_1^4 (\sqrt{x} + x^3)\, dx$

10. $\int_1^2 (3x^2 - 4x^3)\, dx$

11. $\int_{-1}^2 (3x^3 - 2x^2 + x)\, dx$

12. $\int_4^9 (3\sqrt{s} + s)\, ds$

13. $\int_3^3 \sqrt{1 + x^2}\, dx$

14. $\int_3^0 \frac{1}{x + 1}\, dx$

15. $\int_4^2 3x\, dx$

16. $\int_{-2}^{-2} \frac{x}{x^2 + 5}\, dx$

17. $\int_0^2 (e^{-2t} - 3t^2)\, dt$

18. $\int_1^2 \left(\frac{1}{x} + e^{2x} - \frac{2}{x^2}\right) dx$

19. $\int_1^8 (u^{-1/3} + 2u^3)\, du$

20. $\int_2^1 \left(2x^3 - \frac{3}{x^3} + 4\right) dx$

21. $\int_4^1 (2x^{-1/2} + 3x^{1/3})\, dx$

22. $\int_1^4 (3\sqrt{s} + s^2 - 3)\, ds$

23. $\int_1^4 \left(\frac{1}{2\sqrt{x}} + 5\sqrt{x}\right) dx$

24. $\int_1^3 \left(e^{-t/3} + \frac{4}{t^3} - 2t\right) dt$

In Exercises 25–32, find the areas of the shaded regions.

25.

$y = 4 - x^2$

26.

$y = x - 3$

27.

$y = \frac{1}{x}$ $y = x$

28.

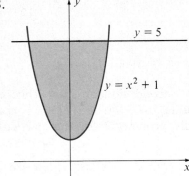

$y = 5$

$y = x^2 + 1$

29.

$y = 4 - x^2$

30.

31.

32.

In Exercises 33–40, find the area of the region enclosed by the given curves. First, sketch the

curves to determine which function is greater than the other.

33. $y = x^2$ and $y = x$

34. $y = 2x^2$ and $y = 2x$

35. $y = x^2$ and $y = 2x + 3$

36. $y = x^2$ and $y = x + 2$

37. $y = x^2 - 4$ and $y = 4 - x^2$

38. $y = x^2$ and $y = 18 - x^2$

39. $y = 9 - x^2$ and $y = 5$

40. $y = x^2 + 2$ and $y = 2x + 2$

41. Find the area of the region bounded by the curve $y = x^2 - 2x + 1$, the line $x = -2$, the x-axis, and the y-axis.

42. Find the area of the region bounded by the curve $y = 4 - x^2$, the line $x = 1$, the x-axis, and the y-axis.

In Exercises 43–46, find the area between the two given curves over the given interval. *Warning:* the curves intersect somewhere inside the interval.

43. $y = 2 - x^2$ and $y = x$ over the interval $[0, 2]$.

44. $y = \sqrt{x}$ and $y = \frac{1}{2}x$ over the interval $[0, 9]$.

45. $y = x^3$ and $y = x$ over the interval $[-1, 1]$.

46. $y = (x + 1)^2$ and $y = x + 3$ over the interval $[0, 2]$.

7.3 |||||||| SOME APPLICATIONS OF INTEGRATION

In this section we shall discuss applications of the definite integral to problems in medicine, economics, consumption of resources, and motion in physics. These applications are only a small sample of numerous important applications of the definite integral. Each of the applications that we shall examine is based on the change in the value of a function on an interval.

||||||||||||| **Change in the Value of a Function**

Suppose $f(x)$ has a continuous derivative $f'(x)$ over an interval $[a, b]$. Then, since $f(x)$ is an antiderivative of $f'(x)$, we have

$$\int_a^b f'(x)\,dx = f(b) - f(a). \tag{1}$$

The right side of Equation (1) is the change in the value of $f(x)$ as x varies from a to b (Figure 1).

Figure 1

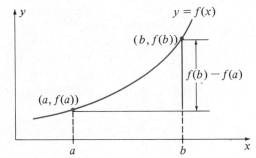

The left side is the integral from a to b of the rate of change of f with respect to x. Thus,

> If the rate of change $f'(x)$ of $f(x)$ with respect to x is continuous on $[a, b]$, then the definite integral $\int_b^a f'(x)\,dx$ gives the change in the value of $f(x)$ as x varies from a to b.

EXAMPLE 1   If $f(x) = x^2$, find $\int_1^2 f'(x)\,dx$.

Solution

$$\int_1^2 f'(x)\,dx = f(2) - f(1)$$
$$= 2^2 - 1^2 = 3. \qquad \blacksquare$$

In applied problems, we often know the rate of change of $f(x)$ and want to compute the change in the value of $f(x)$ as x varies from a to b.

EXAMPLE 2
(Medicine)

A certain drug causes the temperature of the human body to change at the rate given (in degrees Celsius per hour) by

$$r(t) = -0.06t^2 + 0.02t,$$

where t is the number of hours that have elapsed after the drug has been administered. Find the change in body temperature between the beginning of the third hour ($t = 2$) and the beginning of the sixth hour ($t = 5$).

Solution

We need to compute the change in body temperature from $t = 2$ to $t = 5$. This change is given by

$$\int_2^5 (-0.06t^2 + 0.02t)\, dt = (-0.02t^3 + 0.01t^2)\Big]_2^5$$
$$= (-2.5 + 0.25) - (-0.16 + 0.04)$$
$$= -2.13.$$

Thus, body temperature *decreases* by 2.13°C during the designated time period.

■

|||||||||||| **Marginal Analysis**

Recall from Section 3.4 that the marginal cost *MC*, marginal revenue *MR*, and marginal profit *MP* are the derivatives of the cost, revenue, and profit functions, respectively. Since

$$MC = C'(x), \quad MR = R'(x), \quad MP = P'(x),$$

integrating these functions over the interval $[a, b]$, we obtain from (1)

$$\int_a^b C'(x)\, dx = C(b) - C(a)$$
$$= \text{change in cost as } x \text{ varies from } a \text{ to } b,$$

$$\int_a^b R'(x)\, dx = R(b) - R(a)$$
$$= \text{change in revenue as } x \text{ varies from } a \text{ to } b, \quad \text{and}$$

$$\int_a^b P'(x)\, dx = P(b) - P(a)$$
$$= \text{change in profit as } x \text{ varies from } a \text{ to } b.$$

EXAMPLE 3

A manufacturer of computers finds that the marginal profit (in dollars) is given by

$$MP = 3x^2 - 2x$$

where x is the number of computers manufactured weekly. Assume that every computer that is manufactured is sold and that the profit is zero when no computers are sold.

(a) Find the profit when the level of production is 20 computers per week.
(b) Find the change in profit if the level of production changes from 30 to 60 computers per week.

Solution (a) As the level of production changes from $x = 0$ to $x = 20$, the change in weekly profit is

$$\int_0^{20} (3x^2 - 2x)\, dx = (x^3 - x^2)\Big]_0^{20} = (20^3 - 20^2) - (0 - 0)$$
$$= 8000 - 400 = \$7600.$$

(b) As the level of production changes from $x = 30$ to $x = 60$, the change in weekly profit is

$$\int_{30}^{60} (3x^2 - 2x)\, dx = (x^3 - x^2)\Big]_{30}^{60} = (60^3 - 60^2) - (30^3 - 30^2)$$
$$= 212{,}400 - 26{,}100 = \$186{,}300. \qquad \blacksquare$$

|||||||||||| **Depletion of Resources**

Suppose that $R(t)$ is a known rate at which a natural resource (such as coal, copper, zinc, gas, and oil) is being consumed. Select a reference point $t = 0$ in time and let $Q(t)$ denote the amount of the resource that was consumed during the time interval $[0, t]$; that is, $R(t) = Q'(t)$. If $R(t)$ is continuous for $0 \le t \le T$, then the amount $Q(T)$ of the resource that was consumed during the time interval $[0, T]$ can be determined from Equation (1) as

$$\int_0^T R(t)\, dt = \int_0^T Q'(t)\, dt = Q(T) - Q(0) = Q(T), \qquad (2)$$

since $Q(0) = 0$.

EXAMPLE 4 At the beginning of 1985 ($t = 0$), copper was being consumed in the United States at the rate of 1.203 million tons per year* and consumption was increasing 5% per year. Assume an exponential growth model for the rate of consumption so that

$$R(t) = 1.203e^{0.05t}.$$

Estimate the total amount of copper that will be used between 1985 and 2000.

Solution Consumption between $t = 0$ (1985) and $t = 15$ (2000) is given by (2) as

$$Q(15) = \int_0^{15} 1.203e^{0.05t}\, dt$$
$$= \frac{1.203}{0.05} e^{0.05t}\Big]_0^{15}$$
$$= 24.06(e^{0.75} - e^0)$$
$$= 24.06(2.117 - 1)$$
$$\approx 26.88 \text{ million tons.} \qquad \blacksquare$$

---

* *The World Almanac* (New York: Newspaper Enterprise Association, 1985).

||||||||||||| **Applications to Motion**

Recall that both velocity and acceleration are rates of change. If an object is moving along an axis with position coordinate x at time t, then its velocity at time t is

$$v = \frac{dx}{dt}$$

and its acceleration at time t is

$$a = \frac{dv}{dt}\left(= \frac{d^2x}{dt^2}\right)$$

Thus, Equation (1) tells us that over the time interval from $t = t_1$ to $t = t_2$ the **net change in velocity** is

$$v(t_2) - v(t_1) = \int_{t_1}^{t_2} v'(t)\, dt, \quad \text{or}$$

$$\Delta v = \int_{t_1}^{t_2} a(t)\, dt, \tag{3}$$

and that the net distance traveled (net change in position) is

$$x(t_2) - x(t_1) = \int_{t_1}^{t_2} x'(t)\, dt, \quad \text{or}$$

$$\Delta x = \int_{t_1}^{t_2} v(t)\, dt. \tag{4}$$

EXAMPLE 5   An object, starting at the point $x = 7$ with an initial velocity $v(0) = 5$ feet per second, is moving along the x-axis with acceleration $a = 2t$ feet per second per second. How far does it travel during the time interval from $t = 1$ to $t = 3$?

Solution   From Equation (3), the change in velocity over the first t seconds of motion is

$$v(t) - v(0) = \int_0^t a(t)\, dt$$
$$= \int_0^t 2t\, dt = t^2 \Big]_0^t$$
$$= t^2.$$

Thus,

$$v(t) = t^2 + v(0).$$

But we are given that $v(0) = 5$. Therefore, the velocity at time t is

$$v(t) = t^2 + 5 \text{ ft/sec}. \tag{5}$$

Now, using Equation (4), the distance traveled in the time interval from $t = 1$ to $t = 3$ is

$$x(3) - x(1) = \int_1^3 v(t)\,dt$$

$$= \int_1^3 (t^2 + 5)\,dt \qquad \text{[from (5)]}$$

$$= \left(\frac{t^3}{3} + 5t\right)\Bigg]_1^3$$

$$= \left(\frac{27}{3} + 15\right) - \left(\frac{1}{3} + 5\right)$$

$$= \frac{26}{3} + 10 = \frac{56}{3} \text{ feet.} \qquad \blacksquare$$

||||||||||||| Continuous Streams of Income

Imagine money flowing into a fund continuously, in a steady stream, although not necessarily at a constant rate. There are many examples, such as money flowing into the New York City subway system, income generated by a state lottery system, or income produced by a factory assembly line.

Choose a unit of time (hour, day, year, and so on), choose an initial time as $t = 0$, and suppose that at later time t, money is flowing into the fund with

$$\text{rate of flow} = r(t) \tag{6}$$

dollars per unit time. Let $A(t)$ denote the total amount of money in the fund at time t. Then (6) means that

$$\frac{dA}{dt} = r(t). \tag{7}$$

In a specific time interval, say from $t = t_1$ to $t = t_2$, the net amount of income that flows into the fund is the difference between $A(t_2)$ and $A(t_1)$. Using Equation (1),

$$A(t_2) - A(t_1) = \int_{t_1}^{t_2} A'(t)\,dt = \int_{t_1}^{t_2} r(t)\,dt.$$

In summary,

> If the rate of flow at time t is $r(t)$, then the total amount of flow from time t_1 to time t_2 is
>
> $$\text{total flow} = \int_{t_1}^{t_2} r(t)\,dt. \tag{8}$$

EXAMPLE 6   A company purchases a new machine for its assembly line. It is estimated that t years after it is purchased, the machine will be earning money at the rate of $150 - \frac{1}{3}t^2$ thousands of dollars per year. Find the total earnings of the machine during the first 3 years and during the next 3 years.

Solution   Let $R(t)$ denote the total revenue (in thousands of dollars) earned by the machine in the first t years. We are given the rate

$$R'(t) = 150 - \tfrac{1}{3}t^2, \tag{9}$$

and from this we want to obtain total revenues during the specified time intervals.

(a) During the first 3 years the earnings will be $R(3) - R(0)$. Of course, $R(0) = 0$, since at time $t = 0$ the machine will not yet have produced any revenue. Thus, we have

$$\begin{aligned} R(3) &= R(3) - R(0) \\ &= \int_0^3 R'(t)\, dt \\ &= \int_0^3 (150 - \tfrac{1}{3}t^2)\, dt \qquad \text{[from (9)]} \\ &= (150t - \tfrac{1}{9}t^3)\Big]_0^3 = 450 - 3 \\ &= 447 \text{ (thousand dollars)} \\ &= \$447{,}000. \end{aligned}$$

(b) During the next 3 years the earnings will be

$$\begin{aligned} R(6) - R(3) &= \int_3^6 R'(t)\, dt \\ &= \int_3^6 (150 - \tfrac{1}{3}t^2)\, dt = (150t - \tfrac{1}{9}t^3)\Big]_3^6 \\ &= (900 - \tfrac{216}{9}) - 447 = 876 - 447 \\ &= 429 \text{ (thousand dollars)} \\ &= \$429{,}000. \end{aligned}$$

∎

EXAMPLE 7   The company of Example 6 purchased the machine for \$350,000, and predicts that t years after it is purchased, the machine will have maintenance and other operating costs at the rate of $5 + \frac{14}{3}t$ thousand dollars per year. Predict the net profit earned by the machine during its first 3 years and during the next 3 years.

Solution   Let $C(x)$ denote the total cost of operating the machine during the first t years. We are given

$$C'(t) = 5 + \tfrac{14}{3}t. \tag{10}$$

The net profit produced by the machine during the first t years is

$$P(t) = R(t) - C(t).$$

Thus,

$$P'(t) = R'(t) - C'(t)$$
$$= (150 - \tfrac{1}{3}t^2) - (5 + \tfrac{14}{3}t) \qquad \text{[from (9) and (10)]}$$
$$= 145 - \tfrac{1}{3}t^2 - \tfrac{14}{3}t. \tag{11}$$

(a) During the first 3 years of operation the net profit from the machine will be $P(3) - P(0)$. Using Equation (1),

$$P(3) - P(0) = \int_0^3 P'(t)\, dt \tag{12}$$
$$= \int_0^3 (145 - \tfrac{1}{3}t^2 - \tfrac{14}{3}t)\, dt \qquad \text{[from (11)]}$$
$$= (145t - \tfrac{1}{9}t^3 - \tfrac{7}{3}t^2)]_0^3 = 435 - 3 - 21$$
$$= 411 \text{ (thousand dollars)}$$
$$= \$411{,}000. \tag{13}$$

This figure does not take into account the purchase price of the machine. This is because we have subtracted out $P(0)$. If we take $P(0) = -350$, then the purchase price of \$350,000 is taken as negative profit from the very beginning. Then, by the end of the third year, the net profit earned by the machine will be

$$P(3) = P(0) + \int_0^3 P'(t)\, dt \qquad \text{[from (12)]}$$
$$= -350 + 411 \qquad \text{[from (13)]}$$
$$= 61 \text{ (thousand dollars)}.$$

This figure tells us that by the end of the third year the machine will have "paid for itself" and earned an additional \$61,000 profit as well.

(b) During the next 3 years the machine will generate a profit of

$$P(6) - P(3) = \int_3^6 P'(t)\, dt$$
$$= \left(145t - \frac{t^3}{9} - \frac{7}{3}t^2\right)\Big]_3^6$$
$$= 762 - 411 = 351 \text{ (thousand dollars)}$$
$$= \$351{,}000. \tag{14}$$

The company will be concerned, of course, that the figure in (14) is lower than the figure in (13). As the machine ages its productivity diminishes while its costs of operation and maintenance rise. ∎

Total Realizable Profit

Suppose a company acquires a machine (or other income-producing resource) that produces revenue at the rate $R'(t)$ dollars per unit time, and incurs costs at the rate $C'(t)$ dollars per unit time, at time t. In order for the machine to operate profitably, it must produce revenue at a rate greater than the rate at which it incurs costs; that is,

$$R'(x) > C'(x). \tag{15}$$

Ordinarily, a machine operates most efficiently when it is new. As the machine ages its productivity diminishes and it becomes more and more costly to operate and maintain. Figure 2 shows graphs of typical revenue and cost rate functions.

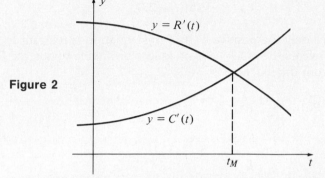

Figure 2

Eventually, at some time t_M shown in Figure 2, the cost $C'(t)$ of operating and maintaining the machine per unit time reaches and then exceeds $R'(t)$, the revenues produced by the machine per unit time. If the machine is used beyond that time it generates *loss* for the company, rather than profit. Hence, the *total realizable profit from the machine, ignoring* the purchase price, is

$$P(t_M) - P(0) = \int_0^{t_M} P'(t)\,dt$$
$$= \int_0^{t_M} (R'(t) - C'(t))\,dt$$
$$= \text{the area shaded in Figure 3.}$$

Figure 3

EXAMPLE 8   Find the total realizable profit for the machine described in Example 7.

Solution   We first find the time t_M by solving

$$R'(t) = C'(t),$$

$$150 - \frac{1}{3}t^2 = 5 + \frac{14t}{3}$$

$$450 - t^2 = 15 + 14t$$

$$0 = t^2 + 14t - 435$$

$$0 = (t - 15)(t + 29).$$

Therefore, $t_M = 15$ since only the positive root of this equation is relevant to our problem. Thus, after 15 years the machine is no longer profitable to operate. The total realizable profit from the machine is

$$P(15) - P(0) = \int_0^{15} P'(t)\,dt$$

$$= \int_0^{15} \left(145 - \frac{t^3}{3} - \frac{14}{3}t\right) dt$$

$$= \left(145t - \frac{t^3}{3} - \frac{7t}{3}\right)\Bigg]_0^{15}$$

$$= 2175 - 375 - 35 \text{ (thousand dollars)}$$

$$= \$1,765,000.$$

This figure ignores the $350,000 purchase cost of the machine. To take this cost into account, we set

$$P(15) = P(0) + \int_0^{15} P'(t)\,dt$$

$$= -350 + 1765 \text{ (thousand dollars)}$$

$$= \$1,415,000. \qquad \blacksquare$$

7.3 EXERCISE SET ||||||||||||||||||||||||||||||||||||||||||||||||||||||||||||||||||||||||||||||||||||||||

1. In each part, use Equation (1) to evaluate $\int_0^2 f'(x)\,dx$.
   (a) $f(x) = x^3$ (b) $f(x) = e^{-x^2/2}$

2. Use Figure 4 to evaluate $\int_3^7 f'(x)\,dx$.

3. **(Biology)** In an experiment to determine the rate of breeding of fruit flies, it was found that t days after the experiment started, the rate of change of the population was given by

$$P'(t) = 60e^{0.02t}.$$

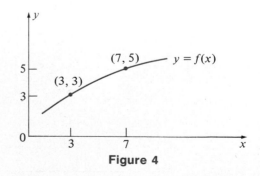

Figure 4

What was the change in the population from $t = 20$ to $t = 40$?

4. On a cold winter morning, Joe arrives at work and turns up the thermostat in the garage where he works. After the heating unit has been operating t minutes, the temperature in the garage is rising at the rate of

$$T'(t) = \frac{3}{2} + \frac{2}{(t+1)^2} \text{ degrees F per minute.}$$

How many degrees Fahrenheit does the temperature change during (a) the first 5 minutes? (b) the next 5 minutes?

5. **(Oil exploration)** A newly developed oil well is producing oil at the instantaneous rate given by

$$R'(t) = 40 + \tfrac{3}{2}t - \tfrac{1}{2}t^3,$$

where $R(t)$ is the number of millions of barrels produced during the first t years of operation. Determine the amount of oil produced
(a) in the first 4 years of operation.
(b) in the combined third and fourth years of operation.

6. **(Educational Psychology)** A psychologist has determined that the rate at which a subject learns a list of N meaningless words after t minutes of study is given by

$$N'(t) = 2\sqrt{t} \text{ words per minute,}$$
$$0 \le t \le 10.$$

Find the number of words learned in the time interval from $t = 4$ to $t = 9$ minutes of study.

7. **(Ecology)** As a result of a leak in the cooling system of a nuclear power plant, nuclear waste flows into a stream at a daily rate given by

$$A'(t) = -2t^3 + 30t^2,$$

where $A(t)$ is the number of gallons that flowed into the stream during the first t days after the leak occurred. Find the number of gallons of waste that flowed

into the stream during the fourth through sixth days after the leak occurred.

8. **(Business)** As a result of an advertising campaign, a firm's sales are growing at a rate given by

$$S'(t) = 100e^{-t},$$

where $S(t)$ is in thousands of dollars and t is the number of days after the campaign ended. Find the amount of sales made during the sixth through the tenth days after the campaign ended.

9. **(Marginal Analysis)** A ski manufacturer's marginal cost and marginal revenue functions (in dollars per pair) are given by

$$C'(x) = 400 - 0.2x$$
$$R'(x) = 400 + 0.1x,$$

where x is the level of production. Suppose that every pair of skis that is manufactured is sold.
(a) Find the change in revenue received when the level of production increases from 20 to 40 pairs of skis.
(b) Find the revenue received from the manufacture of 50 pairs of skis.
(c) If the fixed cost is $200, find the cost of producing 50 pairs of skis.
(d) Find the profit received from the manufacture of 50 pairs of skis. Assume that the fixed cost is $200.

10. **(Marginal Analysis)** A stereo manufacturer's marginal cost and marginal revenue functions (in dollars) are given by

$$C'(x) = 600 - 0.4x$$
$$R'(x) = 200 + 0.2x,$$

where x is the number of stereos manufactured. Assume that every stereo that is manufactured is sold.
(a) Find the change in cost when the level of production increases from 50 to 80 stereos.
(b) Find the change in revenue received when the level of production increases from 50 to 80 stereos.

(c) If the fixed cost is $2000, find the cost of producing 40 stereos.

(d) Find the change in profit when the level of production increases from 50 to 80 stereos.

11. **(Oil Consumption)** In a particular American city, the consumption of oil is growing exponentially at the rate of 4% per year. At the present time, the city is using oil at the rate of 2 million barrels per year. How much oil will the city require during the next 5 years? (*Hint:* You may want to refer to Section 5.4.)

12. **(Depletion of Coal)** At the beginning of 1985, coal production in the United States totaled 890 million short tons per year* and was increasing at the rate of 14% per year. Assuming an exponential growth model for the rate of coal production, estimate the total amount of coal produced in the decade from 1985 to 1995.

13. **(Production of Gold)** At the beginning of 1985, gold production in the United States was 2.06 million troy ounces per year, and increasing at the rate of about 5% per year. Assuming an exponential growth model for the rate of gold production, predict the number of troy ounces of gold produced in the United States in the years 1986 to 1990, inclusive.

14. **(Motion)** As a car moves along a straight track, its velocity (in feet per second) after t seconds is given by

$$f(t) = 4t^{1/2}, \qquad 0 \le t \le 100.$$

(a) How far did the car travel during the first 64 seconds?

(b) How far did the car travel between $t = 4$ and $t = 9$ seconds?

15. **(Motion)** When a stone is dropped from a tall building, t seconds later gravity im-

parts to it an acceleration of

$$a(t) = 32 \text{ ft. per sec. per sec.}$$

(a) Find its change in velocity during the first 3 seconds. The next 3 seconds.

(b) How far does it drop during the first 3 seconds. The next 3 seconds?

16. **(Motion)** An object starting with initial velocity $v(0) = 6$ ft./sec. from the initial position $x(0) = \frac{1}{3}$ ft. on the x-axis, moves along the x-axis in such a way that t seconds later its acceleration is

$$a(t) = 12t^2 + 2t \text{ ft. per sec. per sec.}$$

Find its velocity and position after 2 seconds.

17. **(Business)** When a small insurance broker retires, a large agency purchases his business and predicts that t years later revenues from the purchased business will be generated at the rate of $95 - \frac{1}{4}t - \frac{3}{2}t^2$ thousands of dollars per year. Predict the total earnings from the purchased business (a) over the first four years, (b) the next four years.

18. **(Business)** A photographic processing company installs some new equipment that is predicted to produce revenue at the rate of $33 - \dfrac{3t}{2} - \dfrac{t^2}{2}$ thousand dollars per year, and which will cost $3 + 2t$ thousand dollars per year to operate and maintain, t years after it is installed.

(a) Predict the total revenue and net profit earned by this equipment in the first 2 years.

(b) For how many years can the equipment be operated profitably?

(c) Find the total realizable profit that can be generated by this equipment.

(In Exercise 18 we are neglecting the purchase cost.)

19. **(Business)** A new machine is purchased by a company at a cost of $70,000. After it has been operating t years, it is expected

* The World Almanac (New York: Newspaper Enterprise Association, 1985).

to produce revenue at the rate of $180 - 2t - t^2$ thousand dollars per year, and to cost $50 + t$ thousand dollars per year to operate and maintain. Taking into account the purchase cost of the machine, answer the following.

(a) Predict the total revenue that the machine will produce over the first 6 years; over the next 6 years.

(b) Predict the total net profit the machine will produce over the first 6 years; over the next 6 years.

(c) After how many years will the machine no longer be profitable to operate?

(d) Find the total realizable net profit from the machine.

20. **(Business)** A large corporation is considering the purchase of a small manufacturer, whose rate of revenue is expected to be $16 - 2\sqrt{t}$ million dollars per year, at a cost of $1 + 3\sqrt{t}$ million dollars per year, t years later. How long can they expect to operate the small company profitably? Find the total net profit they can expect to earn from the small company over that length of time.

7.4 |||||||| THE RIEMANN INTEGRAL

In this section we shall develop an important alternate definition of the definite integral devised by Bernhard Riemann.* Suppose that f is a continuous nonnegative function on the interval $[a, b]$. Then the definite integral

$$\int_a^b f(x)\, dx$$

represents the area of the region under the curve $y = f(x)$ over the interval $[a, b]$.

Riemann devised an approach to the calculation of this area that is entirely different from the one that we have been using so far. His approach attacks the problem directly, obtaining the area as a limit without appealing to the notion of antiderivative at all. This approach has the advantage of showing that the area problem is completely distinct from the tangent line problem. An even greater advantage is that many applications of the integral derive their motivation from Riemann's approach. We shall see some of these in the next section.

We begin by examining the graph of a function $y = f(x)$, which is continuous and nonnegative over the interval $[a, b]$. Figure 1 shows the interval $[a, b]$ subdivided into n subintervals of equal width,

$$\Delta x = \frac{b - a}{n}. \tag{1}$$

---

* Georg Friedrich Bernhard Riemann (1826–66) was a German mathematician. The son of a Protestant minister, he obtained his elementary education from his father and showed brilliance in mathematics at an early age. At Göttingen University he studied theology and philology, but eventually he transferred to mathematics and studied under Gauss. In 1862 Riemann contracted pleuritis and was beset by seriously ill health for the rest of his life. He died in 1866 at the age of thirty-nine. Riemann's early death was unfortunate, for his mathematical work was exceptional and fundamentally important. His work in geometry helped Albert Einstein some fifty years later to formulate the theory of relativity.

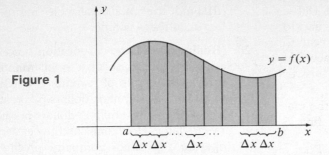

Figure 1

The area of the region under $y = f(x)$ over $[a, b]$, thus, is the sum of the areas of the vertical strips determined by the subintervals. Since the upper boundary of each strip may be curved, we cannot give a general formula for the area of such a strip. We can, however, approximate each strip by a rectangle whose base has width Δx and whose height is the value of the function at some point in the subinterval determining the strip [Figure 2(a)]. These points are denoted by $x_1, x_2, \ldots, x_n$. The subscript indicates the subinterval in which the point lies. Figure 2(b) shows the kth vertical strip, which is approximated by a rectangle with base width Δx and height $f(x_k)$, where x_k is the selected point in the kth subinterval.

Figure 2

(a)

(b)

Approximating the area of the region under $y = f(x)$ over $[a, b]$ by the sum of the rectangular areas, we have

$$R_n = f(x_1)\,\Delta x + f(x_2)\,\Delta x + f(x_3)\,\Delta x + \cdots + f(x_n)\,\Delta x. \qquad (2)$$

This expression is called a **Riemann sum.** As the number of subintervals becomes larger and larger, the width of each of the approximating rectangles becomes smaller and smaller, and the approximation given by (2) approaches the exact area $\int_a^b f(x)\,dx$. Moreover, this result is true no matter how the points $x_1, x_2 \ldots, x_n$ are chosen. Thus, the definite integral $\int_a^b f(x)\,dx$ is the limit of the Riemann sums as the number of subintervals becomes larger and larger.

Keeping technical details to a minimum, we can say that Riemann defined the integral of f over $[a, b]$ to be the *limit of the Riemann sums (2) as $n \to +\infty$.* That is,

$$\int_a^b f(x)\, dx = \lim_{n \to +\infty} [f(x_1) + f(x_2) + \cdots + f(x_n)]\, \Delta x, \qquad (3)$$

where $\Delta x = \dfrac{b-a}{n}$, provided that this limit exists and is independent of the precise choices of $x_1, x_2, \ldots, x_n$ in the n subintervals created in forming R_n.

A function f for which the limit (3) exists in the sense just described is said to be **Riemann integrable** over $[a, b]$. Such a function need not be continuous on $[a, b]$ or nonnegative there. A thorough study of Riemann's approach will reveal that for continuous functions, however, the value of Riemann's integral is identical to the value of the definite integral we have defined in Section 7.1.

EXAMPLE 1 Use a limit of Riemann sums to compute the area of the region under the curve $y = x$ over the interval $[0, 2]$.

Solution We divide the interval $[0, 2]$ into n subintervals, each of width

$$\Delta x = \frac{2-0}{n} = \frac{2}{n}$$

(See Figure 3).

Figure 3

In the kth subinterval, $k = 1, 2, \ldots, n$, we choose the right-hand endpoint as the point x_k. Thus,

$$x_1 = \frac{2}{n}, \quad x_2 = 2 \cdot \frac{2}{n} = \frac{4}{n}, \quad x_3 = 3 \cdot \frac{2}{n} = \frac{6}{n}, \ldots, x_n = n \cdot \frac{2}{n} = \frac{2n}{n}.$$

Then,

The height of the first rectangle is $f(x_1) = f(2/n) = 2/n$.

The height of the second rectangle is $f(x_2) = f(4/n) = 4/n$.

The height of the third rectangle is $f(x_3) = f(6/n) = 6/n$.

$$\vdots \qquad\qquad \vdots \qquad \vdots$$

The height of the nth rectangle is $f(x_n) = f(2n/n) = 2n/n$.

Therefore, the Riemann sum (2) consisting of the sum of the areas of the approximating rectangles is

$$R_n = f\left(\frac{2}{n}\right)\Delta x + f\left(\frac{4}{n}\right)\Delta x + f\left(\frac{6}{n}\right)\Delta x + \cdots + f\left(\frac{2n}{n}\right)\Delta x$$

$$= \frac{2}{n} \cdot \frac{2}{n} + \frac{4}{n} \cdot \frac{2}{n} + \frac{6}{n} \cdot \frac{2}{n} + \cdots + \frac{2n}{n} \cdot \frac{2}{n}, \quad \text{since } f(x) = x \text{ and } \Delta x = \frac{2}{n}$$

$$= \left[\frac{2}{n} + \frac{4}{n} + \frac{6}{n} + \cdots + \frac{2n}{n}\right]\left(\frac{2}{n}\right)$$

$$= \frac{2}{n}[1 + 2 + 3 + \cdots + n]\left(\frac{2}{n}\right)$$

$$= \frac{4}{n^2}[1 + 2 + 3 + \cdots + n].$$

Thus, for this problem, the Riemann sum (2) reduces to

$$R_n = \frac{4}{n^2}[1 + 2 + 3 + \cdots + n]. \tag{4}$$

We can show that the sum of the first n positive integers, the expression in the square brackets in (4), has the value

$$1 + 2 + 3 + \cdots + n = \frac{n(n + 1)}{2}. \tag{5}$$

Substituting (5) into (4), we obtain

$$R_n = \frac{4}{n^2}\frac{n(n + 1)}{2} = \frac{4n^2 + n}{2n^2}$$

or

$$R_n = 2 + \frac{2}{n}.$$

As the number of subintervals increases, $n \to +\infty$, we obtain

$$\lim_{n \to +\infty} \left(2 + \frac{2}{n} \right) = 2 + 0 = 2.$$

Hence, the limit of the Riemann sums is 2. We conclude that the area of the region under the curve $y = x$ over the interval $[0, 2]$ is 2.

As a check we compare with our previous method:

$$\int_0^2 x \, dx = \frac{x^2}{2} \bigg]_0^2 = \frac{2^2}{2} - \frac{0^2}{2} = 2 - 0 = 2,$$

which agrees with the result of the Riemann sum approach.

EXAMPLE 2   Use a limit of Riemann sums to compute the area of the region under the curve $y = x^2$ over the interval $[0, 3]$.

Solution   We divide the interval $[0, 3]$ into n subintervals of equal width, each of width

$$\Delta x = \frac{3 - 0}{n} = \frac{3}{n}.$$

(See Figure 4).

Figure 4

y

9

$y = x^2$

$x_3 = \frac{9}{n}$ $x_n = \frac{3n}{n} = 3$

x

$x_1 = \frac{3}{n}$ $x_2 = \frac{6}{n}$

In the kth subinterval, $k = 1, 2, \ldots, n$, we choose the right-hand endpoint as the point x_k. Thus,

$$x_1 = \frac{3}{n}, \quad x_2 = 2 \cdot \frac{3}{n}, \quad x_3 = 3 \cdot \frac{3}{n}, \ldots, x_n = n \cdot \frac{3}{n}.$$

Then,

The height of the first rectangle is $f\left(\dfrac{3}{n}\right) = \left(\dfrac{3}{n}\right)^2$.

The height of the second rectangle is $f\left(2 \cdot \dfrac{3}{n}\right) = 2^2\left(\dfrac{3}{n}\right)^2$.

The height of the third rectangle is $f\left(3 \cdot \dfrac{3}{n}\right) = 3^2\left(\dfrac{3}{n}\right)^2$.

$$\vdots \qquad\qquad \vdots \quad\; \vdots$$

The height of the nth rectangle is $f\left(n \cdot \dfrac{3}{n}\right) = n^2\left(\dfrac{3}{n}\right)^2$.

Therefore, the Riemann sum (2) consisting of the sum of the areas of the approximating rectangles is

$$R_n = f\left(\frac{3}{n}\right)\Delta x + f\left(2 \cdot \frac{3}{n}\right)\Delta x + f\left(3 \cdot \frac{3}{n}\right)\Delta x + \cdots + f\left(n \cdot \frac{3}{n}\right)\Delta x$$

$$= \left(\frac{3}{n}\right)^2\left(\frac{3}{n}\right) + 2^2\left(\frac{3}{n}\right)^2\left(\frac{3}{n}\right) + 3^2\left(\frac{3}{n}\right)^2\left(\frac{3}{n}\right) + \cdots + n^2\left(\frac{3}{n}\right)^2\left(\frac{3}{n}\right)$$

since $f(x) = x^2$ and $\Delta x = \dfrac{3}{n}$

$$= \left(\frac{3}{n}\right)^3[1 + 2^2 + 3^2 + \cdots + n^2].$$

Thus, in this problem, the Riemann sum (2) reduces to

$$R_n = \left(\frac{3}{n}\right)^3[1^2 + 2^2 + \cdots + n^2]. \tag{6}$$

We can show that the sum of the squares of the first n positive integers, the expression in the square brackets in (6), has the value

$$1^2 + 2^2 + 3^2 + \cdots + n^2 = \frac{n(n + 1)(2n + 1)}{6}. \tag{7}$$

Substituting (7) into (6), we obtain

$$R_n = \left(\frac{3}{n}\right)^3\frac{n(n + 1)(2n + 1)}{6} = \frac{27n}{6n} \cdot \frac{n + 1}{n} \cdot \frac{2n + 1}{n}$$

or (verify)

$$R_n = \frac{9}{2}\left(1 + \frac{1}{n}\right)\left(2 + \frac{1}{n}\right).$$

As the number of subintervals increases, $n \to +\infty$. We find

$$\lim_{n \to \infty}\left[\frac{9}{2}\left(1 + \frac{1}{n}\right)\left(2 + \frac{1}{n}\right)\right] = \frac{9}{2}(1)(2) = 9.$$

Hence, the limit of the Riemann sums is 9. We conclude that the area of the region under the curve $y = x^2$ over the interval $[0, 3]$ is 9.

As a check we compare with our previous method:

$$\int_0^3 x^2\,dx = \left.\frac{x^3}{3}\right]_0^3 = \frac{3^3}{3} - \frac{0^3}{3} = 9,$$

which agrees with the result of the Riemann sum approach. ■

EXAMPLE 3   Evaluate $\int_1^5 (12 - 2x)\,dx$ as a limit of Riemann sums. Choose each x_k to be the right-hand endpoint of the kth subinterval.

Solution   We subdivide $[1, 5]$ into n subintervals, each of length

$$\Delta x = \frac{5 - 1}{n} = \frac{4}{n}$$

(see Figure 5).

Figure 5

$y = 12 - 2x$

$x_1 = 1 + \frac{4}{n}$ $x_2 = 1 + \frac{8}{n}$

Since we are choosing x_k to be the right endpoint of the kth subinterval, we have

$$x_1 = 1 + \frac{4}{n}$$

$$x_2 = 1 + 2\left(\frac{4}{n}\right)$$

$$x_3 = 1 + 3\left(\frac{4}{n}\right)$$

$$\vdots$$

$$x_n = 1 + n\left(\frac{4}{n}\right).$$

Thus,

The height of the first rectangle is $f\left(1 + \frac{4}{n}\right) = 12 - 2\left(1 + \frac{4}{n}\right) = 10 - \frac{8}{n}$,

The height of the second rectangle is $f\left(1 + 2 \cdot \frac{4}{n}\right) = 12 - 2\left(1 + 2 \cdot \frac{4}{n}\right)$

$$= 10 - 2 \cdot \frac{8}{n},$$

The height of the third rectangle is $f\left(1 + 3 \cdot \frac{4}{n}\right) = 12 - 2\left(1 + 3 \cdot \frac{4}{n}\right)$

$$= 10 - 3 \cdot \frac{8}{n},$$

The height of the nth rectangle is $f\left(1 + n \cdot \frac{4}{n}\right) = 12 - 2\left(1 + n \cdot \frac{4}{n}\right)$

$$= 10 - n \cdot \frac{8}{n}.$$

Therefore, the Riemann sum (2) is

$$R_n = f\left(1 + \frac{4}{n}\right)\Delta x + f\left(1 + 2 \cdot \frac{4}{n}\right)\Delta x + f\left(1 + 3 \cdot \frac{4}{n}\right)\Delta x + \cdots$$

$$+ f\left(1 + n \cdot \frac{4}{n}\right)\Delta x$$

$$= \left(10 - \frac{8}{n}\right)\left(\frac{4}{n}\right) + \left(10 - 2 \cdot \frac{8}{n}\right)\left(\frac{4}{n}\right) + \left(10 - 3 \cdot \frac{8}{n}\right)\left(\frac{4}{n}\right) + \cdots$$

$$+ \left(10 - n \cdot \frac{8}{n}\right)\left(\frac{4}{n}\right)$$

$$= \frac{4}{n}\left[\left(10 - 1 \cdot \frac{8}{n}\right) + \left(10 - 2 \cdot \frac{8}{n}\right) + \left(10 - 3 \cdot \frac{8}{n}\right) + \cdots\right.$$

$$\left. + \left(10 - n \cdot \frac{8}{n}\right)\right]$$

$$= \frac{4}{n}\left[\underbrace{(10 + 10 + 10 + \cdots + 10)}_{n \ 10\text{'s}} - \frac{8}{n}(1 + 2 + 3 + \cdots + n)\right]$$

$$= \frac{4}{n}\left[10n - \frac{8}{n} \cdot \frac{n(n + 1)}{2}\right] \quad \text{[using Equation (5)]}$$

$$= \frac{40n}{n} - \frac{32(n^2 + n)}{2n^2}$$

$$= 40 - 16\left(1 + \frac{1}{n}\right).$$

Letting $n \to +\infty$, we have

$$\int_1^5 (12 - 2x)\,dx = \lim_{n \to +\infty} R_n$$

$$= \lim_{n \to +\infty} \left[40 - 16\left(1 + \frac{1}{n}\right)\right]$$

$$= 40 - 16(1 + 0)$$

$$= 24.$$

As a check, we use our previous method:

$$\int_1^5 (12 - 2x)\,dx = [12x - x^2]_1^5$$

$$= (60 - 25) - (12 - 1) = 35 - 11$$

$$= 24,$$

which agrees with the result of the Riemann sum approach. ∎

We conclude this section with an application showing the usefulness of the Riemann sum approach to integration in defining and calculating the average value of a function.

||||||||||||| **Average Value of a Function over an Interval**

The average value of the set of numbers $t_1, t_2, \ldots, t_n$ is

$$\frac{t_1 + t_2 + \cdots + t_n}{n}.$$

Suppose that f is a function that is continuous on the interval $[a, b]$. To obtain the average value of $f(x)$ over $[a, b]$, we proceed as follows. Divide the interval $[a, b]$ into n subintervals, each of width $\Delta x = (b - a)/n$, and choose an arbitrary point x_k in the kth subinterval $(k = 1, 2, \ldots, n)$. The average value of the corresponding function values $f(x_1), f(x_2), \ldots, f(x_n)$ is

$$\frac{f(x_1) + f(x_2) + \cdots + f(x_n)}{n}. \tag{8}$$

Since $\Delta x = (b - a)/n$, we may write (8) as

$$\frac{1}{b-a} [f(x_1) \Delta x + f(x_2) \Delta x + \cdots + f(x_n) \Delta x]. \tag{9}$$

Expression (9) represents an *approximation* to the average value of f over $[a, b]$. To obtain the exact average value we must take the limit as $n \to +\infty$.

As n becomes larger and larger, the Riemann sum $f(x_1) \Delta x + f(x_2) \Delta x + \cdots + f(x_n) \Delta x$ approaches the definite integral $\int_a^b f(x) \, dx$. Hence, the **average value of $f(x)$ over $[a, b]$,** denoted by $y_{\text{ave.}}$, is given by

$$y_{\text{ave.}} = \frac{1}{b-a} \int_a^b f(x) \, dx. \tag{10}$$

EXAMPLE 4   Find the average value of $f(x) = x^2$ over the interval $[0, 4]$.

Solution   Using Equation (10), we obtain this average value as

$$\frac{1}{4-0} \int_0^4 x^2 \, dx = \frac{1}{4} \frac{x^3}{3} \Big]_0^4 = \frac{1}{12} [4^3 - 0] = \frac{64}{12} = \frac{16}{3}. \qquad \blacksquare$$

We can interpret Equation (10), the average value of a function $f(x)$ over the interval $[a, b]$, geometrically as follows. We write Equation (10) as

$$\int_a^b f(x) \, dx = (b - a) y_{\text{ave.}}. \tag{11}$$

Suppose now that $f(x)$ is continuous and nonnegative on $[a, b]$. Then the left side of (11) is the area under the curve $y = f(x)$ over the interval $[a, b]$. The right side is the area of a rectangle whose base has width $b - a$ and whose height is $y_{\text{ave.}}$ (Figure 6).

Figure 6

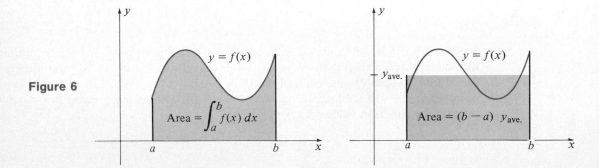

Thus, the average value of a nonnegative function f over $[a, b]$ is the height of the rectangle of base $[a, b]$ whose area is identical with the area under the curve $y = f(x)$ over the interval $[a, b]$.

EXAMPLE 5
(Demography)

As we observed in Chapter 5 (Example 2 of Section 5.4), the world population at the beginning of 1986 was 4.8 billion and it was growing exponentially at the rate of about 2% per year. Thus, t years after 1986, the population $Q(t)$ (in billions) is given by

$$Q(t) = 4.8e^{0.02t}.$$

Find the average population between the beginning of 1986 and the beginning of 2011.

Solution

If we represent 1986 by $a = 0$, then since $2011 - 1986 = 25$, $b = 25$. The average population is given by Equation (10), with $a = 0$ and $b = 25$. Thus, we have

$$Q_{ave.} = \frac{1}{25 - 0} \int_0^{25} 4.8e^{0.02t}\, dt = \frac{4.8}{25} \int_0^{25} e^{0.02t}\, dt$$

$$= \frac{4.8}{(25)(0.02)} e^{0.02t} \Big]_0^{25}$$

$$= \frac{4.8}{0.5}(e^{0.5} - 1) = \frac{4.8}{0.5}(1.6487 - 1)$$

$$\text{(Appendix B, Table I)}$$

$$\approx 6.2.$$

Hence, the average world population between 1986 and 2011 will be approximately 6.2 billion people. ■

EXAMPLE 6
(Economics)

To stabilize the dollar, the United States begins to sell gold. The price of gold (in dollars per ounce) t days after the policy has been implemented is given by

$$f(t) = 400 + 50t - 3t^2, \qquad 0 \le t \le 20.$$

Find the average price per ounce of gold between the beginning of the sixth day ($t = 5$) and the beginning of the eleventh day ($t = 10$).

Solution

The average price is given by Equation (10), with $a = 5$ and $b = 10$. Thus, we have

$$\frac{1}{10 - 5} \int_5^{10} (400 + 50t - 3t^2)\, dt = \frac{1}{5}(400t + 25t^2 - t^3) \Big]_5^{10}$$

$$= \frac{1}{5}[5500 - 2500] = \frac{3000}{5} = 600.$$

Hence, the average price of gold between the sixth and eleventh days is $600 per ounce. ■

7.4 EXERCISE SET ||||||||||||||||||||||||||||||||||||||||||||||||||||||||||||||||||||||||||||||||||||

In Exercises 1–8, find the area of the region under the curve $y = f(x)$ for the function f over the given intervals by using a limit of Riemann sums. In each exercise, choose the point x_k as the right-hand endpoint of the kth subinterval. Use formulas (5) and (7), when appropriate.

1. $f(x) = 2x$, $[0, 3]$

2. $f(x) = 3x$, $[0, 4]$

3. $f(x) = x + 2$, $[0, 3]$

4. $f(x) = 4x$, $[1, 4]$

5. $f(x) = 2x^2$, $[0, 2]$

6. $f(x) = \frac{1}{2}x^2$, $[0, 4]$

7. $f(x) = x^2 + 1$, $[0, 2]$

8. $f(x) = 4 - x^2$, $[0, 2]$

In Exercises 9–16, evaluate the definite integrals as limits of Riemann sums. In each exercise, choose the point x_k as the right-hand endpoint of the kth subinterval.

9. $\int_0^4 \frac{1}{2} x \, dx$

10. $\int_0^2 (-x + 2) \, dx$

11. $\int_0^1 5x \, dx$

12. $\int_1^3 (x + 1) \, dx$

13. $\int_0^1 3x^2 \, dx$

14. $\int_0^1 (x^2 + 2) \, dx$

15. $\int_0^2 (2x^2 + 1) \, dx$

16. $\int_0^1 (1 - x^2) \, dx$

In Exercises 17–20, find the average value of $f(x)$ over the designated interval.

17. $f(x) = \sqrt{x}$, $[0, 3]$

18. $f(x) = x^3$, $[2, 5]$

19. $f(x) = \frac{1}{x}$, $[1, 3]$

20. $f(x) = 3x - 2$, $[2, 6]$

21. **(Temperature)** At a certain ski resort, the temperature (in degrees Fahrenheit) is given by

$$h(t) = \frac{5}{36} t^2 - \frac{10}{3} t + 20, \qquad 0 \le t \le 10$$

where t is the number of hours after noon. What is the average temperature between 3 PM and 8 PM?

22. **(Pollution)** The amount of particulate matter discharged into the atmosphere by a steel plant is given by

$$A(t) = 5t e^{-0.2t}$$

where $A(t)$ is in tons and t is the number of years that the plant has been in operation. Find the average amount of particulate matter discharged into the atmosphere during the first 20 years of the plant's operation.

23. **(Oil Prices)** A new oil-price-monitoring agency finds that t weeks after its inception, the price of oil (in dollars per barrel) is given by

$$f(t) = 33 + t - \frac{1}{20} t^2, \qquad 0 \le t \le 10.$$

Find the average price per barrel of oil between the beginning of the third week ($t = 2$) and the beginning of the ninth week ($t = 8$).

24. **(Ecology)** As a result of an accident, the amount of krypton gas that a nuclear power plant has been discharging into the atmosphere (in thousands of cubic feet) is given by

$$A(t) = 3e^{-0.02t}, \qquad 0 \le t \le 30$$

where t is the number of weeks since the accident. Find the average amount of krypton gas that the plant has discharged into the atmosphere between the beginning of the sixth week ($t = 5$) and the beginning of the sixteenth week ($t = 15$).

7.5 |||||||| APPLICATIONS OF THE RIEMANN INTEGRAL TO BUSINESS

In this section we shall see the importance of the Riemann sum approach to integration. The applications that we shall discuss will all be explained from this approach.

||||||||||||| Present and Future Values of a Continuous Stream of Income

In Section 7.3 we used a definite integral to calculate the total earnings of an income stream over an interval of time. In practice a company will reinvest these earnings in order to earn even more revenue. When this happens, the total value of an income stream is greater than the direct income it produces. Let us suppose that the company is able to reinvest funds at an annual continuous compound interest rate r, which we shall call the company's **prevailing rate of return.** Using Reimann's approach we shall see how definite integrals can be used to calculate the accumulated (future) value of the income stream, as well as its present value.

Suppose a sequence of payments $P_1, P_2, \ldots, P_n$ (dollars) is received at times $t_1, t_2, \ldots, t_n$ (years) measured from some initial time $t = 0$, which we shall call "the present" or "now." Each payment is immediately invested at the prevailing rate of return, r. We shall find a formula for the accumulated (future) value of this sequence of payments T years from the present.

From Section 5.1, Equation (11), the future value of an investment P at continuous compound interest rate r per year after t years is

$$S = Pe^{rt}. \tag{1}$$

The first payment P_1, made t_1 years from now, after T years from now will have been invested $T - t_1$ years. Thus, its future value T years from now is

$$S_1 = P_1 e^{r(T - t_1)}.$$

Similarly, the future values of the other payments T years from now are $S_2 = P_2 e^{r(T - t_2)}$, $S_3 = P_3 e^{r(T - t_2)}, \ldots, S_n = P_n e^{r(T - t_n)}$. Therefore, the total accumulated (future) value of this sequence of payments, T years from now, is

$$S = P_1 e^{r(T - t_1)} + P_2 e^{r(T - t_2)} + \cdots + P_n e^{r(T - t_n)}. \tag{2}$$

Now suppose that we have a *continuous stream* of income over a time interval $[0, T]$, rather than a sequence of discrete payments. Suppose that at time t income is being received at the rate of $f(t)$ dollars per year, and that all income is invested immediately at the prevailing rate of return, r. We want to develop a formula analogous to (2) that will tell us the total accumulated (future) value of the stream, T years from now.

We proceed to set up a definite integral using Riemann sums, as in Section 7.4. First, we subdivide the interval $[0, T]$ into n subintervals, each of length

$\Delta t = \dfrac{T}{n}$ (years). In each of these n subintervals we select any point; the resulting points are denoted $t_1, t_2, \ldots, t_n$ (see Figure 1).

Figure 1

In general, the income received in the kth subinterval is approximately

$$f(t_k)\,\Delta t,$$

which is the rate of income at time t_k times the length of the time interval (dollars per year · number of years). The future value of this amount, T years from now, calculated as if earned at time t_k, is found using (1):

$$f(t_k)\,\Delta t e^{r(T - t_k)}.$$

The total future value of the n approximate amounts of income $f(t_1)\,\Delta t$, $f(t_2)\,\Delta t, \ldots, f(t_n)\,\Delta t$ made during their respective time intervals is approximately

$$f(t_1)\,\Delta t e^{r(T - t_1)} + f(t_2)\,\Delta t e^{r(T - t_2)} + \cdots + f(t_n)\,\Delta t e^{r(T - t_n)}. \tag{3}$$

The sum (3) is an approximation to the formula we are seeking. As the final step we take the limit as $n \to +\infty$. Then, instead of using future values of approximate incomes produced over discrete intervals of time, we will be using future values of exact incomes produced instantaneously. We therefore obtain

$$\text{total future value} = \lim_{n \to +\infty} [f(t_1)e^{r(T - t_1)} + f(t_2)e^{r(T - t_2)} + \cdots + f(t_n)e^{r(T - t_n)}]\,\Delta t.$$

By the Riemann sum approach to integrals, this limit is merely $\int_0^T f(t)e^{r(T - t)}\,dt$. Therefore, we make the following definition:

The total **future value** of an income stream producing income t years from time $t = 0$ at the rate $f(t)$ dollars per year, invested at the prevailing rate of return r, T years from time $t = 0$ is

$$S(T) = \int_0^T f(t)e^{r(T - t)}\,dt. \tag{4}$$

In the above integral, r and T are constant relative to the integration. The reader should be able to show that e^{rT} can be factored out of the integral, and that we can thus rewrite Equation (4) as

$$S(T) = e^{rT} \int_0^T f(t)e^{-rt} \, dt. \tag{5}$$

EXAMPLE 1 An industrial machine is expected to earn $20,000 per year in revenues for a company, which has an 8% prevailing rate of return. Find the total accumulated value of these earnings by the end of the fifth year.

Solution We are given that income is being produced at the constant rate $f(t) = 20,000$, and that the prevailing rate of return is $r = 0.08$. We want to find the future value $S(5)$. Using Equation (5),

$$S(5) = e^{0.08(5)} \int_0^5 20,000 e^{-0.08t} \, dt.$$

Recall that

$$\int e^{at} \, dt = \frac{e^{at}}{a} + C. \tag{6}$$

Thus,

$$S(5) = e^{0.40} \left[20,000 \frac{e^{-0.08t}}{-0.08} \right]_0^5$$

$$= \frac{20,000 e^{0.40}}{-0.08} \left[e^{-0.40} - e^0 \right]$$

$$= -250,000 e^{0.40} \left[e^{-0.40} - 1 \right]$$

$$\approx 123,000.$$

Therefore, the value of the machine's earnings to the company by the end of the fifth year is approximately $123,000, which is $23,000 more than the earnings directly produced by the machine itself during that time. ∎

EXAMPLE 2 A large corporation whose prevailing rate of return is 12% per year purchases a small company whose earnings t years from now are projected to be $600 - t$ thousand dollars per year. Ten years from now, what will be the total accumulated value of this purchase?

Solution The small company produces a stream of income at the rate $f(t) = 600 - t$ (thousands), and the large corporation has a prevailing rate of return $r = 0.12$. We want to find the future value $S(10)$. Using (5) we have

$$S(10) = e^{0.12(10)} \int_0^{10} (600 - t)e^{-0.12t} \, dt$$

$$= e^{1.2} \left[600 \int_0^{10} e^{-0.12t} \, dt - \int_0^{10} te^{-0.12t} \, dt \right]. \tag{7}$$

The first integral is evaluated using (6). The second integral is found by using the method of integration by parts or by Formula 40 in the Table of Integrals:

$$\int te^{-0.12t}\, dt = \frac{e^{-0.12t}}{(-0.12)^2}(-0.12t - 1)$$

$$= -\frac{e^{-0.12t}}{(0.12)^2}(0.12t + 1).$$

Thus, from (7),

$$S(10) = e^{1.2}\left[600\,\frac{e^{-0.12t}}{-0.12} + \frac{e^{-0.12t}}{(0.12)^2}(0.12t + 1)\right]_0^{10}$$

$$= e^{1.2}\left[e^{-0.12t}\left(-5000 + \frac{t}{0.12} + \frac{1}{(0.12)^2}\right)\right]_0^{10}$$

$$= e^{1.2}\left[e^{-1.2}(-5000 + 83.33) + 69.44 - e^0(-5000 + 69.44)\right]$$

$$= e^0(-4847.23) - e^{1.2}(-4930.56)$$

$$\approx -4847 + 16{,}370$$

$$\approx 11{,}523 \text{ (thousands)}.$$

Therefore, 10 years from now the total accumulated value of the purchase will be approximately \$11,523,000. ∎

We turn now to the notion of present value. As an alternative to a sequence of future payments or a stream of income, we may consider a single cash payment made now. For example, someone may offer to purchase the income stream from us. What single payment should satisfy us, considering that we can invest that payment at our prevailing rate of return? By the **present value** of an income stream $f(t)$ over T years we mean the principal $P(T)$ which, if invested *now* at the prevailing rate of return r, would result in the same total accumulated (future) value after T years as the income stream itself. That is, putting together Equations (1) and (5), the present value $P(T)$ must satisfy

$$P(T)e^{rT} = e^{rT}\int_0^T f(t)e^{-rt}\, dt.$$

Dividing both sides of this equation by e^{rT}, we obtain

$$P(T) = \int_0^T f(t)e^{-rt}\, dt.$$

Therefore, we have the following result:

The **present value of an income stream** producing income t years from now at the rate $f(t)$ dollars per year, assuming the prevailing rate of return r, over the first T years is

$$P(T) = \int_0^T f(t)e^{-rt}\, dt. \tag{8}$$

EXAMPLE 3 A certain industrial machine with a life expectancy of 8 years is expected to produce revenue at the rate $200 - 16t$ thousands of dollars per year, t years after installation. Find the present value of the revenue generated by the machine over its lifetime, given that the prevailing rate of return is 10%.

Solution Income is produced at the rate $f(t) = 200 - 16t$, and the prevailing rate of return is $r = 0.10$. We want to find the present value $P(8)$ of 8 years of revenue. We use Equation (8):

$$P(8) = \int_0^8 (200 - 16t)\, e^{-0.10t}\, dt$$
$$= \left[200 \int e^{-0.10t}\, dt - 16 \int t e^{-0.10t}\, dt \right]_0^8. \tag{9}$$

By Equation (6),

$$\int e^{-0.10t}\, dt = \frac{e^{-0.10t}}{-0.10},$$

and by Formula 40 of the Table of Integrals,

$$\int t e^{-0.10t}\, dt = \frac{e^{-0.10t}}{(-0.10)^2}(-0.10t - 1)$$
$$= 100 e^{-0.10t}(-0.10t - 1)$$
$$= -e^{-0.10t}(10t + 100).$$

Substituting these results into (9) we have

$$P(8) = \left[200 \frac{e^{-0.10t}}{-0.10} - 16(-e^{-0.10t})(10t + 100) \right]_0^8$$
$$= [-2000 e^{-0.10t} + 16 e^{-0.10t}(10t + 100)]_0^8$$
$$= [e^{-0.10t}(-2000 + 160t + 1600)]_0^8$$
$$= [e^{-0.10t}(-400 + 160t)]_0^8$$
$$= e^{-0.80}(-400 + 1280) - e^0(-400)$$
$$= \frac{880}{e^{0.80}} + 400$$
$$\approx 795 \text{ (thousand dollars)}.$$

Therefore, the present value of all revenues generated by the machine over its expected lifetime is $795,000. ∎

|||||||||||| Consumers' and Producers' Surplus

In Section 1.6 we discussed supply and demand curves. Recall that the p-coordinate p_E of the point of intersection of the supply and demand curves (Figure 2) is called the equilibrium price.

Figure 2

The demand function $p = D(x)$ for a product tells us that in order to sell x units of the product, the producer must set the price at $D(x)$ dollars per unit. Suppose that the producer is currently selling $\tilde{x}$ units of the product at $\tilde{p}$ dollars per unit, not necessarily the equilibrium price. Some consumers are willing to pay a price higher than $\tilde{p}$; those consumers save by being able to buy the product at price $\tilde{p}$. Using the Riemann sum approach, we shall show that a definite integral can be used to calculate a measure of the savings realized by these consumers.

Divide the quantity interval $[0, \tilde{x}]$ into n subintervals of equal length $\Delta x = \dfrac{\tilde{x}}{n}$. In each subinterval we choose the right endpoint, and label the resulting points $x_1, x_2, \ldots, x_n$ (Figure 3). Imagine that the $\tilde{x}$ units were sold in n small lots of Δx units each; first $\Delta x = x_1$ units, then $\Delta x = x_2 - x_1$ units, then $\Delta x = x_3 - x_2$ units, and so on to the last lot $\Delta x = x_n - x_{n-1}$ units. To sell the first lot of $\Delta x = x_1$ units the demand function tells us that the price can be set as high as $D(x_1)$ dollars per unit, thereby costing consumers

$$D(x_1)\,\Delta x \text{ dollars.}$$

Figure 3

To sell the next lot of $\Delta x = x_2 - x_1$ units, the demand function tells us that the price could be set as high as $D(x_2)$ dollars per unit, thereby costing consumers

$$D(x_2)\,\Delta x \text{ dollars.}$$

Continuing, we see that if the product were sold in this way the total cost to consumers would be

$$D(x_1)\Delta x + D(x_2)\Delta x + \cdots + D(x_n)\Delta x.$$

By taking smaller and smaller lot sizes (that is, $\Delta x \to 0$ and $n \to +\infty$) we arrive at the integral

$$\int_0^{\tilde{x}} D(x)\,dx,$$

which represents the total cost to the consumers of the first $\tilde{x}$ units produced, if every unit is sold to a consumer at the highest price he or she is willing to pay.

But at the uniform selling price $\tilde{p}$, the total cost to the consumers is $\tilde{p}\tilde{x}$. The difference,

$$\int_0^{\tilde{x}} D(x)\,dx - \tilde{p}\tilde{x} = \int_0^{\tilde{x}} [D(x) - \tilde{p}]\,dx,$$

represents the total savings to consumers when the product is sold at price $\tilde{p}$. Thus, we arrive at the following definition:

If $p = D(x)$ is the demand function for a certain product, then when $\tilde{x}$ units are sold at the (uniform) market price $\tilde{p}$, the integral

$$CS = \int_0^{\tilde{x}} [D(x) - \tilde{p}]\,dx \qquad (10)$$

is called the **consumers' surplus at market price** $\tilde{p}$. It is the area between the demand curve and the horizontal line $p = \tilde{p}$, over the interval $[0, \tilde{x}]$. (See Figure 4.)

Figure 4

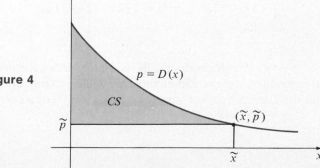

EXAMPLE 4   Suppose that the demand function for a certain commodity is $p = 200 - 4x - 3x^2$. Find the consumers' surplus

(a) when 5 units of the commodity are sold.

(b) when the market price is $68 per unit.

Solution   Observe that Equation (10) requires that we know both $\tilde{x}$ and $\tilde{p}$; when we are given one of them, we must use the demand equation to find the other.

(a) When 5 units of the commodity are sold, $\tilde{x} = 5$. From the demand equation, the corresponding market price is

$$\tilde{p} = 200 - 4(5) - 3(5)^2 = 200 - 20 - 75$$
$$= \$105.$$

Then by Equation (10) the consumers' surplus is

$$CS = \int_0^5 [(200 - 4x - 3x^2) - 105] \, dx$$
$$= \int_0^5 (95 - 4x - 3x^2) \, dx$$
$$= [95x - 2x^2 - x^3]_0^5 = 475 - 50 - 125$$
$$= \$300.$$

(b) When the market price is $68 per unit, $\tilde{p} = \$68$. From the demand equation, the number $\tilde{x}$ must satisfy

$$200 - 4x - 3x^2 = 68$$
$$3x^2 + 4x - 132 = 0$$
$$(3x + 22)(x - 6) = 0.$$

Thus, $\tilde{x} = 6$ units. (The "solution" $x = -\dfrac{22}{3}$ is rejected. Why?) Then, from Equation (10), the consumers' surplus is

$$CS = \int_0^6 [(200 - 4x - 3x^2) - 68] \, dx$$
$$= \int_0^6 (132 - 4x - 3x^2) \, dx$$
$$= [132x - 2x^2 - x^3]_0^6 = 792 - 72 - 216$$
$$= \$504. \qquad \blacksquare$$

The supply function $p = S(x)$ for a product tells us that in order for producers to supply x units of the product, the consumer must be willing to pay $S(x)$ dollars per unit. Suppose that the current market supply ($=$ sales level*) of a commodity

---

* We are assuming that all units produced are sold.

is $\tilde{x}$ units, and the selling price is $\tilde{p}$ dollars per unit. Some producers would be willing to supply the commodity at a price lower than $\tilde{p}$; these producers gain by being able to sell at price $\tilde{p}$. We shall use a Riemann integral to calculate a measure of this gain by these producers.

Divide $[0, \tilde{x}]$ into n subintervals of equal length $\Delta x = \dfrac{\tilde{x}}{n}$. Label the right endpoints of these subintervals $x_1, x_2, \ldots, x_n$. As before, we imagine selling the

Figure 5

$\tilde{x}$ units in lots of Δx units each. To assure that producers will supply the first $\Delta x = x_1$ units, the supply function tells us that the price could be set as low as $S(x_1)$ dollars per unit, bringing the producers

$$S(x_1)\,\Delta x \text{ dollars}$$

in revenue. To assure supply of the next lot of $\Delta x = x_2 - x_1$ units, the supply function tells us that the price could be as low as $S(x_2)$ dollars per unit, resulting in $S(x_2)\,\Delta x$ dollars in revenue for the producers. Continuing as before, we see that if the commodity were sold in this way, the total producers' revenues would be

$$S(x_1)\,\Delta x + S(x_2)\,\Delta x + \cdots + S(x_n)\,\Delta x.$$

Taking the limit of this Riemann sum, we obtain the integral

$$\int_0^{\tilde{x}} S(x)\,dx,$$

which represents the producers' total revenues for the first $\tilde{x}$ units, if each unit is supplied by a producer at the lowest price that that producer is willing to accept.

But at the uniform selling price $\tilde{p}$, the total revenues to producers is $\tilde{p}\tilde{x}$. The difference,

$$\tilde{p}\tilde{x} - \int_0^{\tilde{x}} S(x)\, dx = \int_0^{\tilde{x}} [\tilde{p} - S(x)]\, dx,$$

represents the total gain by the producers when the product is sold at price $\tilde{p}$. Therefore, we have the following definition:

If $p = S(x)$ is the supply function for a certain product, then when the market supply is $\tilde{x}$, selling at market price $\tilde{p}$, the integral

$$PS = \int_0^{\tilde{x}} [\tilde{p} - S(x)]\, dx \qquad (11)$$

is called the **producers' surplus at market price $\tilde{p}$**. It is the area between the horizontal line $p = \tilde{p}$ and the supply curve, over the interval $[0, \tilde{x}]$. (See Figure 6.)

Figure 6

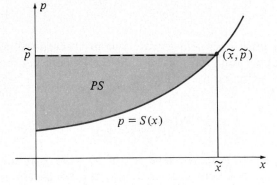

EXAMPLE 5   Suppose that the supply function for a commodity is $p = \dfrac{x^2}{4} + 50$. Find the producers' surplus when the market price is $75 per unit.

Solution   When $\tilde{p} = 75$, to find the corresponding sales level $\tilde{x}$ we solve

$$75 = \frac{x^2}{4} + 50$$
$$300 = x^2 + 200$$
$$100 = x^2.$$

Therefore,

$$\tilde{x} = 10.$$

By Equation (11) the producers' surplus is

$$PS = \int_0^{10} \left[75 - \left(\frac{x^2}{4} + 50 \right) \right] dx$$

$$= \int_0^{10} \left(25 - \frac{x^2}{4} \right) dx$$

$$= \left[25x - \frac{x^3}{12} \right]_0^{10} = 250 - \frac{1000}{12}$$

$$= \$166.67. \qquad\qquad\blacksquare$$

When the supply and demand functions are both known, and the market is in equilibrium, the consumers' and producers' surpluses may be depicted graphically, as shown in Figure 7.

Figure 7

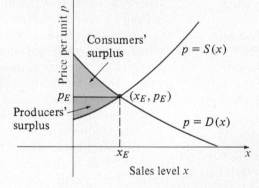

EXAMPLE 6   Consider the following supply and demand functions:

$$p = S(x) = x^2 + 2$$
$$p = D(x) = 8 - x.$$

Find the consumers' surplus and the producers' surplus when the market is in equilibrium.

Solution   We first find the equilibrium price by equating $S(x)$ and $D(x)$:

$$x^2 + 2 = 8 - x$$
$$x^2 + x - 6 = 0$$
$$(x + 3)(x - 2) = 0$$
$$x = -3 \quad \text{or} \quad x = 2.$$

Prices cannot be negative, so $x_E = 2$. Now, using either the supply or demand function, we obtain $p_E = 6$. Using Equations (10) and (11), we find

$$CS = \int_0^2 [(8 - x) - 6]\, dx = \int_0^2 (2 - x)\, dx$$

$$= \left[2x - \frac{x^2}{2} \right]_0^2 = 4 - \frac{4}{2} = 2,$$

and

$$PS = \int_0^2 [6 - (x^2 + 2)]\, dx = \int_0^2 (4 - x^2)\, dx$$

$$= \left[4x - \frac{x^3}{3} \right]_0^2 = \left(8 - \frac{8}{3} \right) = \frac{16}{3}.$$

Therefore, when the market is in equilibrium the consumers' surplus is 2 and the producers' surplus is $5\frac{1}{3}$. ∎

7.5 EXERCISE SET ||||||||||||||||||||||||||||||||||||||||||||||||||||||||||||||||||||||||||||||||||||||

1. **(Income Stream)** A continuous stream of income is being produced at the constant rate of $12,000 per year. For a company whose prevailing rate of return is 9%,
   (a) find the total accumulated (future) value of these earnings after 10 years.
   (b) find the present value of these earnings over the first 10 years.

2. **(Income Stream)** Find the present and future values of an income stream generated over the next 4 years at the rate of $60 + 5t$ thousand dollars per year at time t, assuming that the prevailing rate of return is $r = 10\%$.

3. **(Income Stream)** Find the total accumulated value of a continuous stream of revenue after 9 years, if revenues are received after t years at the rate of $1000 - 50t$ dollars per year and reinvested at the prevailing annual rate of 10% continuous compound interest.

4. **(Income Stream)** Engineers for a company predict that a newly acquired machine will produce income after t years at the rate of $50e^{-.4t}$ thousand dollars per year. Find the future value of these revenues after 5 years, if the company invests all revenues at the prevailing continuous compound interest rate of 6%.

5. **(Income Stream)** A manufacturer whose prevailing rate of return is 12% has the option to purchase a new assembly machine that is expected to produce a net profit of $15 - e^{-.3t}$ thousand dollars per year t years from now. The machine is expected to last 10 years, after which it will be scrapped. What is the largest price the manufacturer should be willing to pay for the machine in order to break even? (*Hint:* What is the present value of the total net profit expected from the machine?)

6. **(Income Stream)** A large corporation whose prevailing rate of return is 8% purchases a small company whose profits are $250,000 a year and growing exponentially at the rate of 5% per year.
   (a) Six years from now, what will the total accumulated value of this purchase be?
   (b) Find the present value of those 6 years' earnings.

7. **(Consumers' Surplus)** Suppose that the demand function for a certain commodity is $D(x) = 30 - \dfrac{x^2}{2}$. Find the consumers' surplus when the sales level is $\tilde{x} = 6$ units.

8. **(Consumers' Surplus)** The demand function for a certain product is $P = \sqrt{25 - 0.3x}$. Find the consumers' surplus when
   (a) the sales level is 70 units.
   (b) the market price is $\tilde{p} = 4$.

9. **(Producers' Surplus)** Suppose the supply function for a certain commodity is $S(x) = 3 + \dfrac{x^2}{8}$. Find the producers' surplus when the market price is $\tilde{p} = 21$.

10. **(Producers' Surplus)** The supply function for a certain product is $p = 4e^{0.4x}$.

Find the producers' surplus when
   (a) the market price is $\tilde{p} = 4e^4$.
   (b) the sales level is 5.

In Exercises 11–14, find the consumers' surplus and the producers' surplus for the given supply and demand functions, assuming that the market is in equilibrium.

11. $S(x) = x + 3$, $D(x) = 10 - x$

12. $S(x) = x^2 + x$, $D(x) = 15 - x$

13. $S(x) = x^2 + 1$, $D(x) = 9 - 2x$

14. $S(x) = 1 + 2x$, $D(x) = 16 - x^2$

7.6 |||||||| IMPROPER INTEGRALS

In Sections 7.1 and 7.4 we defined the definite integral $\int_a^b f(x)\,dx$ for a function f that is continuous on a closed interval $[a, b]$. In this section we shall extend the definition of a definite integral to infinite intervals of the form $[a, +\infty)$, $(-\infty, b]$, or $(-\infty, +\infty)$. Integrals of this type arise in a number of important applications. We begin with an example.

EXAMPLE 1   Consider the problem of computing the area of the region bounded by the curve $y = e^{-x}$, the y-axis, the x-axis, and the line $x = b$, $b > 0$ (Figure 1). Using the

Figure 1

definite integral, we can write the area of the region as

$$\int_0^b e^{-x}\,dx = -e^{-x}\Big]_0^b = -[e^{-b} - e^0]$$

$$= -e^{-b} + 1$$

$$= 1 - \frac{1}{e^b}.$$

As b increases without bound, we find that

$$\lim_{b \to +\infty} \int_0^b e^{-x}\, dx = \lim_{b \to \infty}\left[1 - \frac{1}{e^b}\right]$$

$$= 1 - \lim_{b \to +\infty}\frac{1}{e^b} = 1 - 0$$

$$= 1. \tag{1}$$

Equation (1) implies that as b becomes larger and larger, the area of the associated region moves closer and closer to 1. We rewrite Equation (1) as

$$\int_0^{+\infty} e^{-x}\, dx = 1.$$

This means that the total area under the curve $y = e^{-x}$ over the infinite interval $[0, +\infty)$ is 1 (see Figure 2). This is an example of a region that is infinitely long and yet has a finite area!

Figure 2

Total area = 1

Integrals of the form $\int_a^{+\infty} f(x)\, dx$ are called **improper integrals;** strictly speaking, we have not yet defined such an integral. We do so now.

Definition 1

If $f(x)$ is defined over $[a, +\infty)$, then

$$\int_a^{+\infty} f(x)\, dx = \lim_{b \to +\infty}\int_a^b f(x)\, dx, \tag{2}$$

provided this limit exists.

If the limit in (2) exists as a real number, we say that the improper integral $\int_a^{+\infty} f(x)\, dx$ **converges;** otherwise, we say that it **diverges.**

EXAMPLE 2   Evaluate $\int_1^{+\infty}\left(\frac{1}{x^3}\right)\, dx$ if it converges.

Solution   By definition, (2),

$$\int_1^{+\infty} \frac{1}{x^3}\, dx = \lim_{b \to +\infty} \int_1^b \frac{1}{x^3}\, dx.$$

Now,

$$\int_1^b \frac{1}{x^3}\, dx = -\frac{1}{2x^2}\Big]_1^b = -\frac{1}{2}\left[\frac{1}{b^2} - \frac{1}{1^2}\right] = \frac{1}{2} - \frac{1}{2b^2}.$$

Thus,

$$\int_1^{+\infty} \frac{1}{x^3}\, dx = \lim_{b \to +\infty}\left[\frac{1}{2} - \frac{1}{2b^2}\right] = \frac{1}{2}. \qquad\blacksquare$$

EXAMPLE 3   Evaluate $\int_1^{+\infty}\left(\dfrac{1}{\sqrt{x}}\right) dx$ if it converges.

Solution   By definition,

$$\int_1^{+\infty} \frac{1}{\sqrt{x}}\, dx = \lim_{b \to +\infty} \int_1^b \frac{1}{\sqrt{x}}\, dx.$$

Thus,

$$\int_1^b \frac{1}{\sqrt{x}}\, dx = 2\sqrt{x}\Big]_1^b$$

$$= 2\sqrt{b} - 2$$

and

$$\int_1^{\infty} \frac{1}{\sqrt{x}}\, dx = \lim_{b \to +\infty} [2\sqrt{b} - 2] = +\infty.$$

Because the limit on the right does not exist (is not a real number), we conclude that the given improper integral diverges. To be more precise, we may say that $\int_1^{+\infty}\left(\dfrac{1}{\sqrt{x}}\right) dx$ **diverges to $+\infty$.** $\blacksquare$

EXAMPLE 4
(Nuclear
Accident)   Suppose that, as a result of an accident, krypton gas is being discharged into the atmosphere by a nuclear power plant at a rate (in thousands of cubic feet per year) given by

$$A'(t) = 2e^{-0.03t},$$

where t is the number of years since the accident took place. Assume that the damaged reactor continues to discharge krypton gas indefinitely (forever) and that no cleanup effort is made. How much krypton gas will be discharged into the atmosphere during the indefinite lifetime of the reactor?

Solution The total amount of krypton gas that will be discharged into the atmosphere is given by

$$\int_0^\infty A'(t)\, dt = \int_0^\infty 2e^{-0.03t}\, dt$$

$$= \lim_{b\to\infty} 2 \int_0^b e^{-0.03t}\, dt$$

$$= -\frac{2}{0.03} \lim_{b\to\infty} e^{-0.03t} \Big]_0^b$$

$$= -\frac{2}{0.03} \lim_{b\to\infty} \left[\frac{1}{e^{0.03b}} - 1 \right]$$

$$= -\frac{2}{0.03} [0 - 1]$$

$$= \frac{2}{0.03} = 66.667.$$

Thus, the damaged reactor will release 66,667 cubic feet of krypton gas during its indefinite lifetime. ∎

||||||||||||| **Endless Streams of Income; Capitalization**

In certain applications an income stream may be expected to continue indefinitely, without end, as in earnings from an indestructible asset such as land, or earnings from stock in a stable corporation. In such cases it is practical to assume that the income stream will continue *forever*. This assumption actually facilitates calculations, as we shall see.

Suppose that t years from now an endless stream generates income at the rate of $f(t)$ dollars per year, and that the prevailing rate of continuous compound interest is r. From Equation (8) of Section 7.5, the present value of the first T years of income from this continuous stream is

$$P(T) = \int_0^T f(t)e^{-rt}\, dt.$$

Since the income stream is assumed to continue forever, the **present value of the endless stream of income** is defined to be the improper integral

$$P(+\infty) = \int_0^{+\infty} f(t)e^{-rt}\, dt. \tag{3}$$

The number (3) is sometimes called the **capitalized value** of the income stream.

It can be shown that the capitalized value of an income stream is the smallest amount of money that, if invested now at the continuous compound interest rate r, will guarantee that at any time at least as much money is generated by the investment as by the income stream. In short, it is the fair price for the income stream *today*.

EXAMPLE 5
(Perpetual Annuities)

A **perpetuity** is a sequence of equal payments paid at regular time intervals (that is, an annuity), continuing without end.

Find the capitalized value of a perpetuity paying $1000 per year, assuming that the available rate of continuous compound interest is fixed at 8%.

Solution

We treat the perpetuity as a continuous stream of income, producing income at the rate of $1000 per year, forever. Using Equation (3), the capitalized value of this income stream is

$$P(+\infty) = \int_0^{+\infty} 1000e^{-0.08t}\, dt$$

$$= \lim_{b \to +\infty} 1000 \int_0^b e^{-0.08t}\, dt$$

$$= 1000 \lim_{b \to +\infty} \left[\frac{e^{-0.08t}}{-0.08} \right]_0^b$$

$$= 1000 \lim_{b \to +\infty} \left[\frac{e^{-0.08b}}{-0.08} - \left(\frac{1}{-0.08} \right) \right]$$

$$= \frac{1000}{0.08} \lim_{b \to \infty} \left[1 - \frac{1}{e^{0.08b}} \right] = 12{,}500[1 - 0]$$

$$= \$12{,}500.$$

EXAMPLE 6
(College Endowments)

A loyal alumnus wishes to establish a fund that will contribute $1000 per year to his alma mater forever by investing a lump sum of money into an account that will pay continuous compound interest at the fixed rate of 8% per year. How much must he invest into the account?

Solution

Restating the question, what is the present value of a perpetuity paying $1000 per year if the prevailing rate of continuous compound interest is fixed at 8%? Of course, this is just the answer to Example 5. The loyal alumnus must invest $12,500 in one lump sum.

In the financial world, business assets are bought, sold, and evaluated. One of the methods used to appraise a business enterprise is to compute the present value of all its projected future profits, considered as an endless stream of income. In this method, called **capitalization,** the appraised value of the enterprise is the capitalized value of this endless stream.

EXAMPLE 7
(Oil Well Appraisal)

An independently owned oil well is expected to produce a net annual profit of $300{,}000e^{-0.2t}$ dollars, t years from now, indefinitely. Use the method of capitalization to appraise the oil well, given that the available rate of interest is 6% compounded continuously.

Solution   In the notation of this section, we are given $f(t) = 300{,}000e^{-0.2t}$ and $r = 0.06$. Using Equation (3), the capitalized value of the oil well is

$$P(+\infty) = \int_0^{+\infty} 300{,}000e^{-0.2t}e^{-0.06t}\,dt$$

$$= \lim_{b \to +\infty} 300{,}000 \int_0^b e^{-0.26t}\,dt$$

$$= 300{,}000 \lim_{b \to +\infty} \left[\frac{e^{-0.26t}}{-0.26} \right]_0^b$$

$$= \frac{300{,}000}{0.26} \lim_{b \to +\infty} [-e^{-0.26b} - (-1)]$$

$$\approx 1{,}154{,}000 \lim_{b \to \infty} \left[1 - \frac{1}{e^{0.26b}} \right]$$

$$= 1{,}154{,}000.$$

A reasonable appraisal would be $1,154,000.   ∎

It is sometimes necessary to integrate a function over an infinite interval of the form $(-\infty, b]$. This leads us to a second type of improper integral.

||||||||||||| **Other Improper Integrals**

Definition 2

> If $f(x)$ is defined over $(-\infty, b]$, then
>
> $$\int_{-\infty}^b f(x)\,dx = \lim_{a \to -\infty} \int_a^b f(x)\,dx, \qquad (4)$$
>
> provided this limit exists.

If the limit in (4) exists as a real number, we say that the improper integral $\int_{-\infty}^b f(x)\,dx$ **converges;** otherwise, we say that it **diverges.**

EXAMPLE 8   Evaluate $\int_{-\infty}^{-1} \dfrac{1}{x^4}\,dx$, if it converges.

Solution   By Definition 2,

$$\int_{-\infty}^{-1} \frac{1}{x^4}\,dx = \lim_{a \to -\infty} \int_a^{-1} x^{-4}\,dx$$

$$= \lim_{a \to -\infty} \left[\frac{x^{-3}}{-3} \right]_a^{-1}$$

$$= \lim_{a \to -\infty} \frac{1}{3} \left[-\frac{1}{x^3} \right]_a^{-1} = \frac{1}{3} \lim_{a \to -\infty} \left[\frac{-1}{(-1)^3} - \frac{-1}{a^3} \right]$$

$$= \frac{1}{3} \lim_{a \to -\infty} \left[1 + \frac{1}{a^3} \right] = \frac{1}{3} [1 + 0]$$

$$= \frac{1}{3}.$$

Therefore, $\int_{-\infty}^{-1} \frac{1}{x^4} \, dx$ converges to $\frac{1}{3}$.   ∎

The final type of improper integral that we shall consider is of the form $\int_{-\infty}^{+\infty} f(x) \, dx$.

Definition 3

If $f(x)$ is defined over the entire real number line $(-\infty, +\infty)$, then

$$\int_{-\infty}^{+\infty} f(x) \, dx = \int_{-\infty}^{0} f(x) \, dx + \int_{0}^{+\infty} f(x) \, dx, \qquad (5)$$

provided that each of the two improper integrals on the right side of (5) converges in the sense of definitions 1 and 2.

It is important to remember that an improper integral of the form $\int_{-\infty}^{+\infty} f(x) \, dx$ *must* be treated by breaking it down into two separate improper integrals of the previous types. If both integrals on the right side of Equation (5) converge, we say that $\int_{-\infty}^{+\infty} f(x) \, dx$ **converges;** otherwise, we say that it **diverges.**

EXAMPLE 9   Evaluate $\int_{-\infty}^{+\infty} e^x \, dx$, if it converges.

Solution   By Definition 3,

$$\int_{-\infty}^{+\infty} e^x \, dx = \int_{-\infty}^{0} e^x \, dx + \int_{0}^{+\infty} e^x \, dx,$$

provided that both improper integrals on the right converge. Now,

$$\int_{-\infty}^{0} e^x \, dx = \lim_{a \to -\infty} \int_{a}^{0} e^x \, dx$$

$$= \lim_{a \to -\infty} [e^x]_{a}^{0} = \lim_{a \to -\infty} [e^0 - e^a]$$

$$= 1 - \lim_{a \to -\infty} e^a = 1 - 0$$

$$= 1.$$

Thus, the first improper integral on the right converges to 1. However, the second integral diverges, since

$$\int_0^{+\infty} e^x \, dx = \lim_{b \to +\infty} \int_0^b e^x \, dx$$

$$= \lim_{b \to +\infty} [e^x]_0^b = \lim_{b \to +\infty} [e^b - e^0]$$

$$= +\infty.$$

Therefore, the improper integral $\int_{-\infty}^{+\infty} e^x \, dx$ diverges.   ∎

7.6 EXERCISE SET ||||||||||||||||||||||||||||||||||||||||||||||||||||||||||||||||||||||||||||||||||||||

In Exercises 1–18, evaluate each improper integral if it converges.

1. $\int_2^{+\infty} \frac{1}{x^2} \, dx$

2. $\int_2^{+\infty} \frac{1}{x} \, dx$

3. $\int_0^{+\infty} e^{-2x} \, dx$

4. $\int_0^{+\infty} e^x \, dx$

5. $\int_1^{+\infty} x^2 \, dx$

6. $\int_3^{+\infty} \frac{1}{x^3} \, dx$

7. $\int_1^{+\infty} \frac{1}{\sqrt[3]{x}} \, dx$

8. $\int_1^{+\infty} \frac{1}{x^{4/3}} \, dx$

9. $\int_1^{+\infty} \frac{1}{x^4} \, dx$

10. $\int_0^{+\infty} e^{-x/2} \, dx$

11. $\int_0^{+\infty} \frac{1}{x+1} \, dx$

12. $\int_1^{+\infty} \frac{1}{2x+3} \, dx$

13. $\int_0^{+\infty} x^3 \, dx$

14. $\int_1^{+\infty} 3x^{-2} \, dx$

15. $\int_2^{+\infty} \frac{1}{\sqrt{x-1}} \, dx$

16. $\int_0^{+\infty} \frac{1}{(x+1)^2} \, dx$

17. $\int_0^{+\infty} \frac{x}{(x^2+1)^{1/2}} \, dx$

18. $\int_2^{+\infty} \frac{x}{(x^2-1)^{5/2}} \, dx$

In Exercises 19 and 20, compute the area of the region bounded by the given curve and the x-axis, over the interval $[1, +\infty)$.

19. $y = \frac{1}{x^2}$

20. $y = \frac{1}{x^{3/2}}$

21. **(Harvesting)** In one forest the rate at which trees spring to life (in millions of trees per year) is given by

$$N'(t) = \frac{2t}{(t^2+4)^2}$$

where t is the number of years the forest has existed. Assuming that the forest will exist forever, determine how many trees it will produce during its lifetime.

22. **(Oil exploration)** A newly drilled oil well is producing oil at a rate (in billions of tons per year) given by

$$A'(t) = \frac{1}{(t+1)^3}$$

where t is the number of years the oil well has been producing oil. Assuming that the oil well will last forever, determine how much oil it will produce during its lifetime.

23. **(Perpetuity)** Find the capitalized value of a perpetuity paying $1800 per year, if the available rate of continuous compound interest is fixed at 6%.

24. **(Endowment to Charity)** How much money must be invested now into an account that will pay continuous compound interest at the fixed rate of 10%, so that from this account a contribution of $500 per year can be paid to a charity?

25. **(Appraisal of a Mine)** A small mine is expected to produce a profit of $700,000

per year, indefinitely. Assuming that the annual continuous compound interest rate is fixed at 9%, use the method of capitalization to appraise the value of this mine.

26. **(Perpetuities)** Show that the capitalized value of a perpetuity paying D dollars per year is D/r, if r is the available fixed rate of continuous compound interest.

27. **(Income Stream)** Find the present value of an endless stream of income that, after t years, produces income at the rate of $1200e^{-0.5t}$ dollars per year, assuming that the available rate of continuous compound interest is fixed at 8%.

28. **(Appraisal of a Company)** Use the method of capitalization to appraise the value of a company that, t years from now, is projected to produce a profit at the rate of $550,000e^{-0.1t}$ dollars per year,

given that the available rate of interest is 6% compounded continuously.

In Exercises 29–37, evaluate each improper integral, if it converges.

29. $\int_{-\infty}^{-2} \dfrac{1}{(1-x)^2}\, dx$ 30. $\int_{-\infty}^{-1} \dfrac{1}{1-x}\, dx$

31. $\int_{-\infty}^{0} \dfrac{1}{\sqrt{4-3x}}\, dx$ 32. $\int_{-\infty}^{-1} \dfrac{1}{x^{4/3}}\, dx$

33. $\int_{-\infty}^{+\infty} e^{x/5}\, dx$ 34. $\int_{-\infty}^{+\infty} e^{-0.1x}\, dx$

35. $\int_{-\infty}^{+\infty} \dfrac{x}{(x^2+1)^2}\, dx$

36. $\int_{-\infty}^{+\infty} \dfrac{2x}{\sqrt{x^2+1}}\, dx$

37. $\int_{-\infty}^{+\infty} xe^{-x^2}\, dx$

7.7 |||||||| NUMERICAL INTEGRATION TECHNIQUES

Riemann's approach suggests that a definite integral $\int_a^b f(x)\, dx$ may exist even when an antiderivative $\int f(x)\, dx$ cannot be found. Indeed, in Riemann's approach we calculate the definite integral as a limit of Riemann sums, without any reference to antiderivatives (see Examples 1–3 of Section 7.4). The key point to remember is that, unlike an indefinite integral, a definite integral is a *number*. When we evaluate a definite integral we seek not a function, but a numerical value.

EXAMPLE 1   Evaluate $\int_{-2}^{2} \sqrt{4-x^2}\, dx$.

Solution   This is an example of a situation in which it is impractical to seek an antiderivative. None of the methods of antidifferentiation that we have studied will work in this case. Even the formula given in tables of integrals is no help; it involves a function that we have not studied. Nevertheless, this integral is very easy to evaluate if we adopt the right perspective.

Recall that $\int_{-2}^{2} \sqrt{4-x^2}\, dx$ represents the area under the curve $y = \sqrt{4-x^2}$ from $x = -2$ to $x = 2$. Figure 1 shows this region. The trick is to recognize that the equation $y = \sqrt{4-x^2}$ is equivalent to

$$x^2 + y^2 = 4, \qquad y \geq 0, \tag{1}$$

and that the graph of (1) is the semicircle shown in Figure 1.

Figure 1

The area of the complete circle is 4π (πr^2, with $r = 2$). Thus, the area under the semicircle is $\frac{1}{2}(4\pi) = 2\pi$. Therefore,

$$\int_{-2}^{2} \sqrt{4 - x^2}\, dx = 2\pi. \qquad\blacksquare$$

Approximate Integration

We are interested in methods of evaluating

$$\int_{a}^{b} f(x)\, dx \tag{2}$$

when we cannot find an antiderivative of $f(x)$, or when it is inconvenient to attempt to find one. Example 1 illustrates an unusual case—one in which the exact value of (2) can be found using an elementary geometric formula. We shall describe methods that generally do not give us the exact value of (2), but that give us excellent, reliable approximations. For this reason we call these methods **approximate integration** methods. When used with the aid of a computer, these approximations can be made extremely accurate.

Rectangular Approximations

Consider an arbitrary function f that is continuous over the closed interval $[a, b]$. To obtain a numerical approximation of $\int_{a}^{b} f(x)\, dx$ we proceed as follows:
1. Choose a positive integer n.
2. Subdivide the interval $[a, b]$ into n subintervals of equal length. Each subinterval will have length

$$\Delta x = (b - a)/n. \tag{3}$$

Label the endpoints of these intervals

$$x_0,\ x_1,\ x_2,\ \ldots,\ x_n,$$

where $x_0 = a$ and $x_n = b$ (see Figure 2).

Figure 2

$$\begin{array}{ccccccccc} & a & \Delta x & \Delta x & \Delta x & & \Delta x & \Delta x & b \\ \hline & x_0 & x_1 & x_2 & x_3 & \cdots & x_{n-2} & x_{n-1} & x_n \end{array}$$

3. The n subintervals created in step 2 are $[x_0, x_1]$, $[x_1, x_2]$, $[x_2, x_3]$, . . . , $[x_{n-1}, x_n]$. In each of these intervals we choose a number, $\tilde{x}_i$. That is, for each i from 1 to n, $\tilde{x}_i$ belongs to the ith interval:

$$x_{i-1} \le \tilde{x}_i \le x_i. \tag{4}$$

4. The sum

$$S_n = \Delta x[f(\tilde{x}_1) + f(\tilde{x}_2) + f(\tilde{x}_3) + \cdots + f(\tilde{x}_n)] \tag{5}$$

is called a **rectangular approximation** of $\int_a^b f(x)\,dx$.

Sometimes Equation (5) is written in a slightly different form,

$$S_n = f(\tilde{x}_1)\,\Delta x + f(\tilde{x}_2)\,\Delta x + f(\tilde{x}_3)\,\Delta x + \cdots + f(\tilde{x}_n)\Delta_n\,\Delta x. \tag{6}$$

In Section 7.4 a sum such as (6) was called a Riemann sum. The notation we use here, however, is somewhat more elaborate: the x_i are the *endpoints* of the subintervals, and the $\tilde{x}_i$ are the points selected from *within* the respective subintervals.

Figure 3 shows the geometric interpretation of Equation (5). It also shows clearly that the sum S_n is an approximation to $\int_a^b f(x)\,dx$. We have drawn this figure so that $f(x) \ge 0$ throughout $[a, b]$, although our approximations remain valid even when that is not the case.

Figure 3

The first term $f(\tilde{x}_1)\,\Delta x$ in the sum (6) is equal to the area of the first (leftmost) shaded rectangle in Figure 3. That is because the height of the rectangle is $f(\tilde{x}_1)$ and the width is Δx.

Similarly, the second term $f(\tilde{x}_2)\,\Delta x$ in (6) equals the area of the second rectangle, which has height $f(\tilde{x}_2)$ and width Δx. Each term in the sum (6) represents the area of one of the rectangles in Figure 3, and S_n represents *the total area within all the rectangles.*

Therefore,

$$S_n \approx \int_a^b f(x)\,dx. \qquad (7)$$

Two factors influence the degree of accuracy of approximation (7).

(1) The size of n. For small values of n we should not expect the rectangular areas to be very close to the actual area. To improve the accuracy of approximation we must make n sufficiently large.

(2) The choice of the $\tilde{x}_i$'s. Remember that the rectangular approximation procedure allows us to choose them arbitrarily, as long as $x_{i-1} \le \tilde{x}_i \le x_i$ for each i. Some choices may be better than others, although that will depend somewhat on the function f.

We shall now show three common procedures for the implementation of rectangular approximation. Suppose $f(x)$ is continuous on $[a, b]$. Choose a positive integer n. Let $\Delta x = (b - a)/n$ and

$$
\begin{aligned}
x_0 &= a & &\text{and} & y_0 &= f(x_0) \\
x_1 &= a + \Delta x & &\text{and} & y_1 &= f(x_1) \\
x_2 &= a + 2\,\Delta x & &\text{and} & y_2 &= f(x_2) \\
&\;\;\vdots & & & &\;\;\vdots \\
x_n &= a + n\,\Delta x = b & &\text{and} & y_n &= f(x_n).
\end{aligned}
\qquad (8)
$$

The numbers $x_0, x_1, x_2, \ldots, x_n$ are the endpoints of the subintervals in Figure 3, and the numbers $y_0, y_1, y_2, \ldots, y_n$ are the heights of the curve above these endpoints. Then:

The **left endpoint approximation** of $\int_a^b f(x)\,dx$ is

$$L_n = \Delta x[y_0 + y_1 + y_2 + \cdots + y_{n-1}] \qquad (9)$$

(see Figure 4a), and the **right endpoint approximation** of $\int_a^b f(x)\,dx$ is

$$R_n = \Delta x[y_1 + y_2 + y_3 + \cdots + y_n] \qquad (10)$$

(see Figure 4b).

(a)   (b)

Figure 4

The midpoints of the subintervals in Figure 3 are

$$\tilde{x}_1 = \frac{x_0 + x_1}{2}, \quad \tilde{x}_2 = \frac{x_1 + x_2}{2}, \ldots, \tilde{x}_n = \frac{x_{n-1} + x_n}{2}.$$

We also let $\tilde{y}_1 = f(\tilde{x}_1), \tilde{y}_2 = f(\tilde{x}_2), \ldots, \tilde{y}_n = f(\tilde{x}_n)$. Then:

The **midpoint approximation** of $\int_a^b f(x)\,dx$ is

$$M_n = \Delta x[\tilde{y}_1 + \tilde{y}_2 + \cdots + \tilde{y}_n] \tag{11}$$

(see Figure 5).

Figure 5

EXAMPLE 2   Approximate $\int_0^2 x^2 \, dx$, using all three of the above methods, with $n = 4$.

Solution   First we let $\Delta x = (2 - 0)/4 = 1/2$; then

$$
\begin{aligned}
x_0 &= 0 & y_0 &= 0^2 = 0 \\
x_1 &= \tfrac{1}{2} & y_1 &= (\tfrac{1}{2})^2 = \tfrac{1}{4} \\
x_2 &= 1 & y_2 &= (1)^2 = 1 \\
x_3 &= \tfrac{3}{2} & y_3 &= (\tfrac{3}{2})^2 = \tfrac{9}{4} \\
x_4 &= 2 & y_4 &= (2)^2 = 4
\end{aligned}
$$

Thus,

$$
\begin{aligned}
L_4 &= \tfrac{1}{2}[0 + \tfrac{1}{4} + 1 + \tfrac{9}{4}] = (\tfrac{1}{2})(\tfrac{7}{2}) = \tfrac{7}{4} \\
&= 1.75
\end{aligned}
$$

and

$$
\begin{aligned}
R_4 &= \tfrac{1}{2}[\tfrac{1}{4} + 1 + \tfrac{9}{4} + 4] = (\tfrac{1}{2})(\tfrac{15}{2}) = \tfrac{15}{4} \\
&= 3.75.
\end{aligned}
$$

For the midpoint approximation,

$$
\tilde{x}_1 = \frac{0 + 1/2}{2} = \frac{1}{4}, \qquad \tilde{y}_1 = \left(\frac{1}{4}\right)^2 = \frac{1}{16}
$$

$$
\tilde{x}_2 = \frac{1/2 + 1}{2} = \frac{3}{4}, \qquad \tilde{y}_2 = \left(\frac{3}{4}\right)^2 = \frac{9}{16}
$$

$$
\tilde{x}_3 = \frac{1 + 3/2}{2} = \frac{5}{4}, \qquad \tilde{y}_1 = \left(\frac{5}{4}\right)^2 = \frac{25}{16}
$$

$$
\tilde{x}_4 = \frac{3/2 + 2}{2} = \frac{7}{4}, \qquad \tilde{y}_4 = \left(\frac{7}{4}\right)^2 = \frac{49}{16}
$$

Thus,

$$
\begin{aligned}
M_4 &= \tfrac{1}{2}[\tfrac{1}{16} + \tfrac{9}{16} + \tfrac{25}{16} + \tfrac{49}{16}] = (\tfrac{1}{2})(\tfrac{84}{16}) = \tfrac{21}{8} \\
&= 2.625.
\end{aligned}
$$

We may compare these approximations with the actual values of the integral:

$$
\int_0^2 x^2 \, dx = \tfrac{1}{3}x^3 \Big]_0^2 = \tfrac{8}{3} = 2.666 \ldots
$$

In this example it appears that the midpoint approximation is more reliable than either of the two endpoint approximations. ∎

EXAMPLE 3   Approximate $\int_0^2 x^2 \, dx$ using all three of the above methods, with $n = 8$.

Solution   First we let

$$
\Delta x = \frac{(2 - 0)}{8} = \frac{1}{4},
$$

and then we tabulate the following calculations.

$$x_0 = 0 \qquad y_0 = 0$$
$$x_1 = \tfrac{1}{4} \qquad y_1 = \tfrac{1}{16}$$
$$x_2 = \tfrac{1}{2} \qquad y_2 = \tfrac{1}{4}$$
$$x_3 = \tfrac{3}{4} \qquad y_3 = \tfrac{9}{16}$$
$$x_4 = 1 \qquad y_4 = 1$$
$$x_5 = \tfrac{5}{4} \qquad y_5 = \tfrac{25}{16}$$
$$x_6 = \tfrac{3}{2} \qquad y_6 = \tfrac{9}{4}$$
$$x_7 = \tfrac{7}{4} \qquad y_7 = \tfrac{49}{16}$$
$$x_8 = 2 \qquad y_8 = 4$$

$$\tilde{x}_1 = \tfrac{1}{8} \qquad \tilde{y}_1 = \tfrac{1}{64}$$
$$\tilde{x}_2 = \tfrac{3}{8} \qquad \tilde{y}_2 = \tfrac{9}{64}$$
$$\tilde{x}_3 = \tfrac{5}{8} \qquad \tilde{y}_3 = \tfrac{25}{64}$$
$$\tilde{x}_4 = \tfrac{7}{8} \qquad \tilde{y}_4 = \tfrac{49}{64}$$
$$\tilde{x}_5 = \tfrac{9}{8} \qquad \tilde{y}_5 = \tfrac{81}{64}$$
$$\tilde{x}_6 = \tfrac{11}{8} \qquad \tilde{y}_6 = \tfrac{121}{64}$$
$$\tilde{x}_7 = \tfrac{13}{8} \qquad \tilde{y}_7 = \tfrac{169}{64}$$
$$\tilde{x}_8 = \tfrac{15}{8} \qquad \tilde{y}_8 = \tfrac{225}{64}$$

Thus,

$$L_8 = \frac{1}{4} \cdot \frac{0 + 1 + 4 + 9 + 16 + 25 + 36 + 49}{16} = \frac{35}{16}$$

$$= 2.1875,$$

$$R_8 = \frac{1}{4} \cdot \frac{1 + 4 + 9 + 16 + 25 + 36 + 49 + 64}{16} = \frac{51}{16}$$

$$= 3.1875,$$

and

$$M_8 = \frac{1}{4} \cdot \frac{1 + 9 + 25 + 49 + 81 + 121 + 169 + 225}{64} = \frac{85}{32}$$

$$= 2.65625. \qquad \blacksquare$$

Observe that when $n = 8$ all three approximations are closer to the actual value $\int_0^2 x^2 \, dx = 2.6666 \ldots$ than when $n = 4$.

Trapezoidal Approximations

An alternative, and often better, approximation of $\int_a^b f(x) \, dx$ is achieved using trapezoids rather than rectangles. Consider a function $f(x)$, which is continuous on $[a, b]$. Subdivide $[a, b]$ into n subintervals of equal length $\Delta x = (b - a)/n$. Define $x_0, x_1, \ldots, x_n$, and $y_0, y_1, \ldots, y_n$ as in (8). Figure 6(a) shows the resulting trapezoidal approximation to the area under the curve $y = f(x)$ over $[a, b]$ when $n = 8$. A typical trapezoid is shown in Figure 6(b). Recall that the area of a trapezoid is one-half the sum of its "bases" times the distance between the bases. Thus the area of the ith trapezoid shown in Figure 6(b) is $\frac{1}{2}(y_{i-1} + y_i) \Delta x$. The sum of the areas of all the trapezoids is thus

$$T_n = \frac{1}{2}(y_0 + y_1)\Delta x + \frac{1}{2}(y_1 + y_2)\Delta x + \frac{1}{2}(y_2 + y_3)\Delta x + \cdots + \frac{1}{2}(y_{n-1} + y_n)\Delta x.$$

$$= \frac{\Delta x}{2}[(y_0 + y_1) + (y_1 + y_2) + (y_2 + y_3) + \cdots + (y_{n-1} + y_n)].$$

Figure 6

The trapezoidal approximation

$$T_n = \frac{\Delta x}{2}\,[y_0 + 2y_1 + 2y_2 + \cdots + 2y_{n-1} + y_n] \qquad (12)$$

is therefore a reasonable approximation of $\int_a^b f(x)\,dx$. It is often better than rectangular approximations for the same value of n.

EXAMPLE 4 Approximate $\int_0^2 x^2\,dx$ using the trapezoidal approximation with $n = 4$; then with $n = 8$.

Solution Using the values of Δx, x_i, and y_i found in Example 2,
($n = 4$)

$$T_4 = \frac{\Delta x}{2}\,[y_0 + 2y_1 + 2y_2 + 2y_3 + y_4]$$

$$= \frac{1/2}{2}\left[0 + 2\left(\frac{1}{4}\right) + 2(1) + 2\left(\frac{9}{4}\right) + 4\right]$$

$$= \frac{1}{4}\left[\frac{1}{2} + 2 + \frac{9}{2} + 4\right]$$

$$= \left(\frac{1}{4}\right)(11) = 2.75. \qquad \blacksquare$$

Solution Using the values of Δx, x_i, and y_i found in Example 3,
($n = 8$)

$$T_8 = \frac{\Delta x}{2} [y_0 + 2y_1 + 2y_2 + 2y_3 + 2y_4 + 2y_5 + 2y_6 + 2y_7 + y_8]$$

$$= \frac{1/4}{2} \left[0 + 2\left(\frac{1}{16}\right) + 2\left(\frac{1}{4}\right) + 2\left(\frac{9}{16}\right) + 2(1) + 2\left(\frac{25}{16}\right) + 2\left(\frac{9}{4}\right) + 2\left(\frac{49}{16}\right) + 4 \right]$$

$$= \frac{1}{8} \left[\frac{1 + 9 + 25 + 49}{8} + \frac{1 + 9}{2} + (0 + 2 + 4) \right]$$

$$= \frac{1}{8} \left[\frac{84}{8} + 5 + 6 \right] = \frac{1}{8} \left[\frac{21}{2} + 11 \right] = \frac{1}{8} \left(\frac{43}{2} \right) = \frac{43}{16}$$

$$= 2.6875.$$

Observe that T_8 is closer to the integral (2.666 . . .) than is T_4 ($= 2.75$). It is also better than the rectangular approximations we obtained with $n = 4$ found in Example 2. It is better than the rectangular approximations L_8 ($= 2.1875$) and R_8 ($= 3.18750$) found in Example 3, but not quite as good as M_8 ($= 2.65625$).

One final word of advice: considering both accuracy and the amount of effort required, the trapezoidal approximation is the recommended method. It is both effective and efficient.

7.7 EXERCISE SET ||||||||||||||||||||||||||||||||||||||||||||||||||||||||||||||||||||||||||||||||||||

In each of the following exercises, use the given value of n and approximate the given integral using all four methods: L_n, R_n, M_n, and T_n. A calculator will be helpful.

1. $\int_0^2 (2x + 1)\, dx$, $n = 4$

2. $\int_{-1}^5 (x + 3)\, dx$, $n = 6$

3. $\int_{-1}^1 (1 - x^2)\, dx$, $n = 4$

4. $\int_{-2}^2 \sqrt{4 - x^2}\, dx$, $n = 4$

5. $\int_1^7 x^2\, dx$, $n = 6$

6. $\int_2^6 (x^2 + x)\, dx$, $n = 4$

7. $\int_{-3}^3 x^3\, dx$, $n = 6$

8. $\int_0^3 \frac{1}{x + 2}\, dx$, $n = 6$

9. $\int_1^5 \frac{1}{x}\, dx$, $n = 8$

10. $\int_0^4 \frac{1}{x^2 + 1}\, dx$, $n = 8$

11. $\int_0^4 e^x\, dx$, $n = 8$

12. $\int_1^3 2^x\, dx$, $n = 4$

13. $\int_0^2 \sqrt{x^3 + 1}\, dx$, $n = 4$

14. $\int_0^5 \sqrt{x^2 + 2}\, dx$, $n = 5$

15. $\int_2^5 \ln x\, dx$, $n = 6$

16. $\int_0^3 \frac{1}{(1 + x^2)^2}\, dx$, $n = 6$

KEY IDEAS FOR REVIEW ||||||||||||||||||||||||||||||||||||||||||||||||||||||||||||||||||||||||||||

- The First Fundamental Theorem of Calculus: Suppose that f is a function that is continuous and nonnegative on the interval $[a, b]$. Let $A(x)$ denote the area of the region under the curve $y = f(x)$ over the interval $[a, x]$. Then $A'(x) = f(x)$ for each x in $[a, b]$.

- If $f(x)$ is a continuous function on the interval $[a, b]$, we define the definite integral of f from a to b by

$$\int_a^b f(x)\,dx = F(b) - F(a)$$

where $F(x)$ is any antiderivative of $f(x)$.

- If $f(x)$ is continuous and nonnegative on $[a, b]$, then $\int_a^b f(x)\,dx$ represents the area of the region under the curve $y = f(x)$ over the interval $[a, b]$.

- Properties of the definite integral:

$$\int_a^a f(x)\,dx = 0$$

$$\int_a^b kf(x)\,dx = k \int_a^b f(x)\,dx$$

$$\int_a^b f(x)\,dx = -\int_b^a f(x)\,dx$$

$$\int_a^b [f(x) \pm g(x)]\,dx = \int_a^b f(x)\,dx \pm \int_a^b g(x)\,dx$$

$$\int_a^b f(x)\,dx = \int_a^c f(x)\,dx + \int_c^b f(x)\,dx$$

- If f is a continuous function on $[a, b]$, then the definite integral $\int_a^b f(x)\,dx$ represents the area of the region lying above $[a, b]$ and below $y = f(x)$ minus the area of the region lying below $[a, b]$ and above $y = f(x)$.

- If f and g are continuous functions with $f(x) \geq g(x)$ on $[a, b]$, then the area of the region between the curves $y = f(x)$ and $y = g(x)$ over the interval $[a, b]$ is $\int_a^b [f(x) - g(x)]\,dx$.

- If the rate of change $f'(x)$ of $f(x)$ with respect to x is continuous on $[a, b]$, then the definite integral $\int_a^b f'(x)\,dx$ gives the change in the value of $f(x)$ as x varies from a to b.

- Divide the interval $[a, b]$ into n subintervals of equal width $\Delta x = (b - a)/n$. Let x_k be an arbitrary point in the kth subinterval. The expression

$$f(x_1)\,\Delta x + f(x_2)\,\Delta x + \cdots + f(x_n)\,\Delta x$$

is called a **Riemann sum.**

- If $f(x)$ is continuous and nonnegative on $[a, b]$, then the limit of the Riemann sums as $n \to +\infty$ is $\int_a^b f(x)\,dx$.

- The average value of $f(x)$ over $[a, b]$ is given by

$$y_{\text{ave.}} = \frac{1}{b - a} \int_a^b f(x)\,dx.$$

- Let f be a continuous function for $x \geq a$. The improper integral of f from a to $+\infty$, denoted by $\int_a^{+\infty} f(x)\,dx$, is defined by

$$\int_a^{+\infty} f(x)\,dx = \lim_{b \to +\infty} \int_a^b f(x)\,dx$$

if this limit exists. An improper integral is said to converge or diverge.

- Similarly, we define the improper integrals

$$\int_{-\infty}^b f(x)\,dx = \lim_{a \to -\infty} \int_a^b f(x)\,dx,$$

and

$$\int_{-\infty}^{+\infty} f(x)\,dx = \int_{-\infty}^0 f(x)\,dx + \int_0^{+\infty} f(x)\,dx.$$

- If a continuous stream of income generates income at the rate of $f(t)$ dollars per year t years from now, then
  - the total amount of the flow from time t_1 to time t_2 is

$$\text{total flow} = \int_{t_1}^{t_2} f(t)\,dt.$$

  - the total future value of the income stream after T years, invested at the continuous compound interest rate r, is

$$S(T) = \int_0^T f(t)e^{r(T-t)}\, dt$$

$$= e^{rT} \int_0^T f(t)e^{-rt}\, dt.$$

• the present value of the first T years of income is

$$P(T) = \int_0^T f(t)e^{-rt}\, dt,$$

assuming that the prevailing continuous compound interest rate is r.

■ Consumers' and Producers' Surplus: If $p = D(x)$ is the demand function and $p = S(x)$ is the supply function for a certain commodity, then when $\tilde{x}$ units are supplied (and sold) on the market and when the selling price is $\tilde{p}$,

$$CS = \int_0^{\tilde{x}} [D(x) - \tilde{p}]\, dx,$$

and

$$PS = \int_0^{\tilde{x}} [\tilde{p} - S(x)]\, dx.$$

■ The present value of an endless stream of income:

$$P(+\infty) = \int_0^{+\infty} f(t)e^{-rt}\, dt.$$

■ Rectangular approximations of $\int_a^b f(x)\, dx$:

$$L_n = \Delta x[y_0 + y_1 + y_2 + \cdots + y_{n-1}]$$
$$R_n = \Delta x[y_1 + y_2 + y_3 + \cdots + y_n]$$
$$M_n = \Delta x[\tilde{y}_1 + \tilde{y}_2 + \tilde{y}_3 + \cdots + \tilde{y}_n]$$

where $\Delta x = \dfrac{b-a}{n}$, $y_i = f(x_i)$ and

$$\tilde{y}_i = f\left(\frac{x_{i-1} + x_i}{2}\right)$$

■ Trapezoidal approximation of $\int_a^b f(x)\, dx$:

$$T_n = \frac{\Delta x}{2}[y_0 + 2y_1 + 2y_2 + \cdots + 2y_{n-1} + y_n]$$

where $\Delta x = \dfrac{b-a}{n}$ and $y_i = f(x_i)$.

REVIEW EXERCISES ||||||||||||||||||||||||||||||||||||||||||||||||||||||||||||||||||||||||||||||||||||||

In Exercises 1–6, use a definite integral to find the area under the curve over the given interval. Sketch the curve roughly and shade the appropriate region.

1. $y = \frac{1}{2}x$, $[1, 4]$ 2. $y = 2 - x$, $[0, 1]$

3. $y = \frac{1}{2}x^2$, $[1, 4]$ 4. $y = x^2 + 1$, $[1, 4]$

5. $y = e^{3x}$, $[0, 4]$ 6. $y = \sqrt{3x}$, $[0, 3]$

In Exercises 7–12, evaluate the definite integrals.

7. $\int_1^4 dx$ 8. $\int_2^3 (3 - x^2)\, dx$

9. $\int_1^{27} (x^{1/3} - 5)\, dx$ 10. $\int_8^{27} x^{-2/3}\, dx$

11. $\int_0^1 e^{t/2}\, dt$ 12. $\int_1^5 \frac{1}{x^3}\, dx$

In Exercises 13 and 14, find the area of each region by using definite integrals.

13.

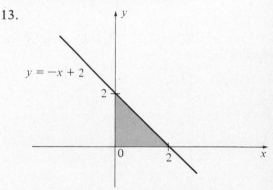

$y = -x + 2$

14.

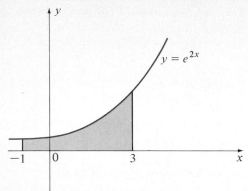

In Exercises 15–20, evaluate the definite integrals.

15. $\int_2^4 \left(3x + x^2 - \frac{1}{x} \right) dx$

16. $\int_0^1 (e^{3x} - 3x^2) \, dx$

17. $\int_1^4 (3\sqrt{x} + 2x^2 - 1) \, dx$

18. $\int_1^1 \left(2 - \frac{1}{x} + 3x^3 \right) dx$

19. $\int_4^9 \left(2e^{-x/2} + \frac{1}{\sqrt{x}} - 3x^2 \right) dx$

20. $\int_8^{27} (2x^{-1/3} + x^{1/3}) \, dx$

In Exercises 21 and 22, find the area of the shaded region.

21.

22.

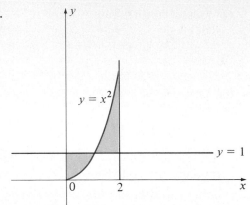

In Exercises 23 and 24, find the area of the region enclosed by the given curves. Graph the curves first.

23. $y = x^2$ and $y = \sqrt{x}$

24. $y = 4x - x^2$ and $y = x$

25. Find the total area of the region bounded by the curve $y = 9 - x^2$, the y-axis, the line $x = 1$, and the line $y = 1$.

In Exercises 26–29, find the area under the curve $y = f(x)$ over the given intervals by using limits of Riemann sums. Choose x_k as the right-hand endpoint of the kth subinterval. Use formulas (5) and (7) in Section 7.4 when appropriate.

26. $f(x) = \frac{1}{3}x$, $[0, 2]$

27. $f(x) = -x + 1$, $[0, 1]$

28. $f(x) = 4x^2$, $[0, 1]$

29. $f(x) = x^2 + 4$, $[0, 2]$

30. **(Marginal Analysis)** A moped manufacturer's marginal cost and marginal revenue functions (in hundreds of dollars per moped) are given by

$$C'(x) = 250 - 0.5x$$

$$R'(x) = 100 + 0.2x$$

where x is the number of mopeds manufactured. Assume that every moped that is manufactured is sold.

(a) Find the change in revenue received when the level of production increases from 40 to 100 mopeds.

(b) Find the revenue received from the manufacture and sale of 60 mopeds.

(c) Find the change in profit when the level of production increases from 40 to 100 mopeds.

31. **(Psychology)** An astronaut's reaction time to a flashing light changes at a rate given by

$$r'(t) = 0.003t^2 - 0.4t$$

where $r(t)$ is in milliseconds and t is the number of hours that the subject has been weightless. Find the decrease in reaction time between the fourth and eighth hours of weightlessness.

32. **(Pollution)** A steel plant discharges particulate matter into the atmosphere over a nearby town. The rate of discharge (in tons per year) is given by

$$A'(t) = te^{-t/10}, \qquad 0 \le t \le 10,$$

where t is the number of years that the plant has been in operation. Find the amount of particulate matter discharged into the atmosphere during the first 5 years of the plant's operation.

33. **(Motion)** An object starts from the initial position $x(0) = 0$ ft with initial velocity $v(0) = 5$ ft/sec, and moves along the x-axis with acceleration

$$a(t) = \frac{1}{\sqrt{t+1}} \text{ ft/sec}^2.$$

Find its velocity and position after t seconds.

34. **(Income Stream)** A small business account generates profit at the rate of $4500(1 + t)$ dollars per year after t years. Find the total profit earned over the first 5 years.

35. **(Total Realizable Profit)** A new assembly line is expected to produce revenue at the rate of $510 - t^2$ thousand dollars per year at a cost of $30 + 28t$ thousand dollars per year to operate and maintain, t years after it is installed.

(a) For how many years can the assembly line be operated profitably?

(b) Find the total realizable profit that can be generated by this equipment (ignoring the initial cost).

In Exercises 36 and 37, find the average value of $f(x)$ over the given interval.

36. $f(x) = \dfrac{1}{x^2}$, $[1, 3]$

37. $f(x) = e^{0.2x}$, $[5, 10]$

38. **(Population Growth)** Consider a recently incorporated city where the population (in tens of thousands) t months after incorporation is given by

$$Q(t) = 4e^{0.005t}.$$

Find the average population during the first 40 years after incorporation.

In Exercises 39–44, evaluate the improper integral if it converges.

39. $\displaystyle\int_1^{+\infty} \frac{3}{x^3}\, dx$

40. $\displaystyle\int_1^{+\infty} \frac{1}{\sqrt{2x+1}}\, dx$

41. $\displaystyle\int_0^{+\infty} 2xe^{-x^2}\, dx$

42. $\displaystyle\int_{-\infty}^0 e^{2x}\, dx$

43. $\displaystyle\int_3^{+\infty} \frac{1}{(x-2)^{1/3}}\, dx$

44. $\displaystyle\int_{-\infty}^{+\infty} \frac{x}{(3x^2+2)^2}\, dx$

In Exercises 45–47, compute the area bounded by the given curve and the x-axis to the right of the line $x = 1$.

45. $y = \dfrac{2x}{(1 + x^2)^2}$ 46. $y = e^{-3x}$

47. $y = \dfrac{1}{(1 + x)^2}$

48. **(Pollution)** An aluminum plant is dumping waste into a nearby stream at a rate given by

$$A'(t) = \frac{1}{(t + 4)^2}$$

where $A'(t)$ is in tons per year and t is the number of years that the plant has been in existence. Assuming that the plant will never close, determine how much waste will be dumped into the stream during the lifetime of the plant.

49. For the income stream of Exercise 34, find the total accumulated value of the account at the end of five years, if the profits have all been invested at the prevailing rate of return of 8%.

50. **(Income Stream)** A company whose prevailing rate of return is 5% per year pur-chases a machine that is expected to produce revenue at the rate of $80e^{-0.25t}$ thousand dollars per year after t years. Find the present value to the company of the first 10 years of revenue from this machine.

In Exercises 51 and 52, find the consumers' surplus and the producers' surplus for the given supply and demand functions, assuming that the market is in equilibrium.

51. $S(x) = x + 4$, $D(x) = 12 - x$

52. $D(x) = (x - 3)^2$, $0 \le x \le 3$, $S(x) = 4x^2$

53. **(Endless Stream of Income)** Find the present value of an endless stream of income that, after t years, produces income at the rate of $800e^{-0.2t}$ dollars per year, assuming that the available rate of continuous compound interest is fixed at 7%.

In Exercises 54 and 55, use the given value of n and approximate the given integral using all four methods: L_n, R_n, M_n, and T_n.

54. $\displaystyle\int_0^2 \frac{1}{x + 1}\, dx$, $n = 4$

55. $\displaystyle\int_1^5 \sqrt{x + 3}\, dx$, $n = 4$.

CHAPTER TEST ||||||||||||||||||||||||||||||||||||||||||||||||||||||||||||||||||||||||||

1. Use a definite integral to find the area under the curve $y = \sqrt{4x - 3}$ over the interval $[1, 3]$.

2. Find the area enclosed between the curve $y = 3 - x^2$ and the line $y = x + 1$.

3. Suppose that f is a continuous function on the interval $[0, 5]$, and

$$\int_0^5 f(x)\, dx = 10 \qquad \text{while} \qquad \int_2^5 f(x)\, dx = 4.$$

Find

(a) $\displaystyle\int_5^2 f(x)\, dx$ (b) $\displaystyle\int_0^2 f(x)\, dx$ (c) $\displaystyle\int_2^2 f(x)\, dx$.

4. Find $\int_0^2 x^2\, dx$ as a limit of Riemann sums. Choose each x_k as the right-hand endpoint of the kth subinterval. Show all the details.

5. To stabilize the dollar, the United States begins to sell gold. The rate at which gold is sold is given by

$$A'(t) = 50e^{-3t}$$

where $A(t)$ is in millions of ounces and t is the number of weeks that the policy of selling gold has been in effect. Find the change in the amount of gold sold between $t = 0$ and $t = 4$ weeks.

6. Find the average value of the function $f(x) = \dfrac{3}{(x + 2)^2}$ over the interval $[1, 4]$.

7. Evaluate each of the following improper integrals, or show that it does not converge.

   (a) $\displaystyle\int_1^{+\infty} \frac{1}{x^5}\, dx$ (b) $\displaystyle\int_2^{\infty} \frac{x}{x^2 - 1}\, dx$

8. For a company whose prevailing rate of continuous compound interest is 10% per year, find the present value of an income stream that will produce income at the rate of \$15,000 per year over the next 8 years. Also, find the future value of the income stream to the company at the end of the 8 years.

9. Find the consumers' surplus and the producers' surplus for a market that has supply function $S(x) = 4 + \dfrac{x^2}{3}$ and demand function $D(x) = 22 - \dfrac{x^2}{6}$, assuming that the market is in equilibrium.

10. Using $n = 4$, approximate the integral $\int_0^4 (x^2 + 3)\, dx$ using all four methods, L_n, R_n, M_n, and T_n.

FUNCTIONS OF SEVERAL VARIABLES

Up to this point, we have considered functions of only one variable. Yet many functions in applied problems depend on more than one variable. For example, the cost of manufacturing and marketing a unit of a certain product depends on the costs of material, labor, transportation, insurance, advertising, and research, and other such factors. In this chapter we shall introduce functions of several variables.

8.1 ‖‖‖‖‖ FUNCTIONS OF SEVERAL VARIABLES

A manufacturer of skis makes two models: A and B. Suppose that the manufacture and sale of x units of model A and y units of model B produce a total revenue R, given by

$$R = 100x + 60y. \qquad (1)$$

Equation (1) describes "R as a function of x and y." The quantities x and y are called **independent variables;** the quantity R is the **dependent variable** because its value depends on the selected values of x and y.

In general, a **function of two independent variables** x and y is a rule, or formula, f, which determines exactly one value of the dependent variable z for each ordered pair of values (x, y) for which the rule is defined. We denote this value of z by $f(x, y)$ and write

$$z = f(x, y).$$

In a similar manner, we can define a function g of the three independent variables x, y, and z. In this case, the rule g determines exactly one value of the dependent variable w for each ordered triple of values (x, y, z) for which the rule is defined. We denote this value of w by $g(x, y, z)$ and write

$$w = g(x, y, z).$$

EXAMPLE 1   The following are familiar functions of several variables.

$A = f(l, w) = lw$ A, the area of a rectangle, is a function of its length l and width w.

$V = f(l, w, h) = lwh$ V, the volume of a rectangular box, is a function of its length l, width w, and height h.

$S = f(P, r, t) = P(1 + rt)$ S, the amount repaid after t years when a principal P is borrowed at the annual simple interest rate r, is a function of the three variables P, r, and t.

EXAMPLE 2   If $f(x, y) = 2x^2 - 3xy - y^2$, find
(a) $f(2, -3)$ (b) $f(-3, 4)$.

Solution   (a) $f(2, -3) = 2(2)^2 - 3(2)(-3) - (-3)^2$
$$= 8 + 18 - 9$$
$$= 17$$

(b) $f(-3, 4) = 2(-3)^2 - 3(-3)(4) - (4)^2$
$$= 18 + 36 - 16$$
$$= 38. \qquad \blacksquare$$

EXAMPLE 3   If $f(x, y, z) = 3x^2 + 2yz - 5z^2$, find
(a) $f(4, 2, 3)$ (b) $f(2, 0, 5)$.

Solution   (a) $f(4, 2, 3) = 3(4)^2 + 2(2)(3) - 5(3)^2$
$$= 48 + 12 - 45$$
$$= 15$$

(b) $f(2, 0, 5) = 3(2)^2 + 2(0)(5) - 5(5)^2$
$$= 12 + 0 - 125$$
$$= -113. \qquad \blacksquare$$

As in Section 1.1, we may view a function as an input/output machine. A function of two variables

$$z = f(x, y) \qquad (2)$$

assigns to each input (x, y) a unique output $f(x, y)$. Notice that each input for (2) is an ordered pair of numbers. We define the **domain** of a function to be the set of all inputs for which the function is defined. For the function (2) the domain is the set of all ordered pairs of numbers (x, y) for which $f(x, y)$ exists.

EXAMPLE 4   For each of the following functions, find the domain.

(a) $f(x, y) = \dfrac{3x}{x - y}$ (b) $g(x, y) = \sqrt{2x + y}$ (c) $h(x, y) = \dfrac{x^2 + y^2}{xy}$

Solution   (a) The function $f(x, y)$ is defined for all (x, y) except when $x = y$, since we cannot divide by 0. Thus, the domain consists of all (x, y) for which $x \neq y$. As a set of points in the xy plane, the domain consists of all points in the coordinate system of Figure 1(a) except those points on the dashed line.

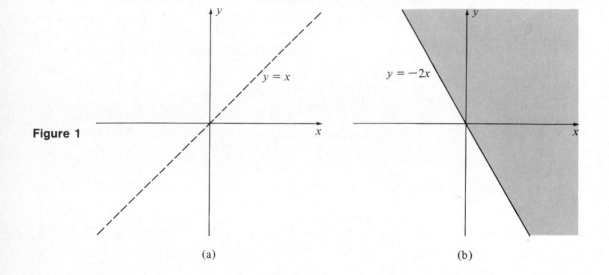

Figure 1

(a)   (b)

(b) The function $g(x, y)$ is defined for all (x, y) for which the number under the radical is nonnegative; that is,

$$2x + y \geq 0, \quad \text{or} \quad y \geq -2x.$$

Thus, the domain consists of all points in the coordinate system of Figure 1(b) that lie in the shaded area above or on the line $y = -2x$.

(c) The function $h(x, y)$ is defined for all (x, y) for which $xy \neq 0$. This occurs only when both $x \neq 0$ and $y \neq 0$. Therefore, the domain of h consists of all points in the xy-coordinate system except points that lie on either the x-axis or the y-axis. ∎

Three-Dimensional Rectangular Coordinate System

Before discussing the graphs of functions of several variables, we must extend the rectangular coordinate system of the plane to three-dimensional space.

We fix a three-dimensional coordinate system by choosing a point, called the **origin,** and three lines called the **coordinate axes,** which intersect at the origin in such a way that each line is perpendicular to the other two. These lines are called the **x-, y-, and z-axes.** On each of these axes we choose a point fixing the units of length and positive directions. Frequently, but not always, we use the same unit of length for all the coordinate axes. The x-, y-, and z-axes along with their units of length and positive directions constitute a **rectangular coordinate system,** or a **Cartesian coordinate system.** Figure 2 shows the rectangular coordinate system that we shall use in this book.

Figure 2

Each pair of coordinate axes determines a plane. The plane determined by the x- and y-axes is called the **xy-plane.** Similarly, the x- and z-axes determine the xz-plane, and the y- and z-axes determine the yz-plane. These three planes are called the **coordinate planes.** It takes some imagination to "see" textbook figures of three-dimensional coordinate systems because the page on which they appear is two-dimensional. In order to achieve a convincing three-dimensional effect on a two-dimensional page, we draw distances in the x-direction shorter than distances in the y- and z-directions. Also, distances that extend in the negative x-, y-, or z-directions are often shown as dashed lines (see Figure 5).

If P is any point in three-dimensional space, then the three planes through P that are parallel to the yz-, and xz-, and xy-planes intersect the x-, y-, and z-axes at points a, b, and c, respectively (Figure 3). Therefore, we can associate the ordered triple (a, b, c) with the point P, and we denote the point P by (a, b, c) or

Figure 3

by $P(a, b, c)$. The numbers a, b, and c are called the **x-, y-, and z-coordinates**, respectively, of P. Observe also that $|a|$ is the distance from P to the yz-plane, $|b|$ is the distance from P to the xz-plane, and $|c|$ is the distance from P to the xy-plane. Conversely, for each ordered triple of real numbers (a, b, c) there is a unique point P in three-dimensional space, the coordinates of which are (a, b, c) (Figure 4). Thus, we have established a one-to-one correspondence between points in three-dimensional space and ordered triples of real numbers.

Figure 4

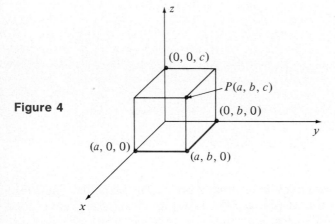

EXAMPLE 5 Figure 5 shows a number of points and their coordinates.

Figure 5

||||||||||||| **Graphs of Equations and Functions**

The **graph of an equation** in three-dimensional space is the set of all points (x, y, z) whose coordinates satisfy the given equation. The graph of an equation in three-dimensional space is also called a **surface.**

EXAMPLE 6   Figures 6(a)–(c) show the graphs of the equations $x = 3$, $y = 5$, and $z = 4$. Observe that the surfaces $x = 3$, $y = 5$, and $z = 4$ are planes that are perpendicular to the corresponding coordinate axes. Moreover, the graph of the equation $z = 0$ is the xy-plane. Similarly, the graphs of the equations $x = 0$ and $y = 0$ are the yz- and xz-planes, respectively.

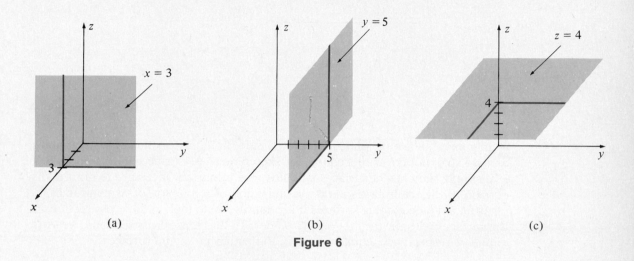

(a)   (b)   (c)

Figure 6

The graph of a **first-degree equation** in x, y, and z,

$$Ax + By + Cz + D = 0 \tag{3}$$

is always a **plane**. Conversely, every plane in the x-, y-, z-coordinate system has a first-degree equation.

The **graph of a function** f of the variables x and y is the graph of the equation $z = f(x, y)$. As noted above, the graph of a **first-degree function**

$$f(x, y) = ax + by + c$$

is always a plane.

EXAMPLE 7   Figure 7 shows a portion of the graph of the function

$$z = f(x, y) = 3 - x - y.$$

The surface is a plane intersecting the x-, y-, and z-axes at the points $(3, 0, 0)$, $(0, 3, 0)$, and $(0, 0, 3)$, respectively.

Figure 7

In general, it is quite difficult to sketch the graph of a function of two variables, and we shall not cover procedures for sketching these surfaces in this book. Several simple sketches of surfaces will illustrate important ideas needed in the discussion of the minimization and maximization of a function of two variables.

As an aid to sketching surfaces it is often useful to look at the cross sections of the surfaces with the xy-, xz-, and yz-planes. To find these cross sections, we set z, y, and x, respectively, equal to zero in the equation for the surface.

EXAMPLE 8 Figure 8 shows a portion of the graph of the function

$$z = f(x, y) = x^2 + y^2.$$

Figure 8

The cross section in the xy-plane, the intersection of the plane $z = 0$ with the surface $z = x^2 + y^2$, is merely the origin since $x^2 + y^2 = 0$ means that $x = 0$ and

$y = 0$. The cross section in the xz-plane, the intersection of the plane $y = 0$ with the surface $z = x^2 + y^2$, is the parabola $z = x^2$. The cross section in the yz-plane, the intersection of the plane $x = 0$ with the surface $z = x^2 + y^2$, is the parabola $z = y^2$. Note that the cross section in the plane $z = 4$, the intersection of the plane $z = 4$ with the surface $z = x^2 + y^2$, is the circle $x^2 + y^2 = 4$. It can be shown that the cross section in any horizontal plane is always a circle. It is clear that the smallest value of z on the surface is $z = 0$. The surface has a low point at $(0, 0, 0)$.

EXAMPLE 9   Figure 9 shows the graph of the function

$$z = f(x, y) = \sqrt{9 - x^2 - y^2}.$$

The domain of this function is the set of all (x, y) for which the expression inside the radical is nonnegative; that is,

$$9 - x^2 - y^2 \geq 0, \quad \text{or} \quad x^2 + y^2 \leq 9.$$

This is the set of all points in the xy-plane that lie inside or on the circle whose center is $(0, 0)$ and radius is 3, as shown in Figure 9.

Figure 9

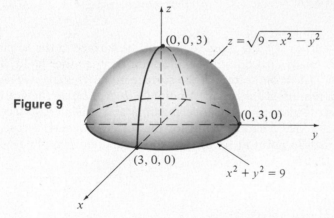

The cross section of the surface in the xy-plane is the curve obtained by setting $z = 0$:

$$\sqrt{9 - x^2 - y^2} = 0, \quad \text{or}$$
$$9 - x^2 - y^2 = 0, \quad \text{or}$$
$$x^2 + y^2 = 9,$$

which is a circle of radius 3, centered at the origin. The cross section of the surface in the xz-plane is the curve $z = \sqrt{9 - x^2}$, which is the upper half of a circle of radius 3, centered at the origin. The cross section of the surface in the yz-plane is the curve $z = \sqrt{9 - y^2}$, which is the upper half of a circle of radius 3, centered at the origin. The surface is the upper half of a sphere of radius 3 centered at the origin. It is clear that the largest value of z on the surface is $z = 3$, so the surface has a high point at $(0, 0, 3)$.

EXAMPLE 10   Figure 10 shows a portion of the graph of the function

$$z = f(x, y) = y^2 - x^2.$$

Figure 10

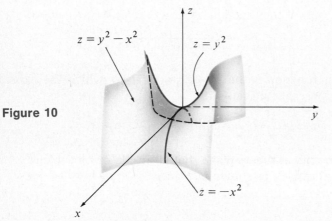

If we look at the curve $z = -x^2$, the cross section of the surface in the xz-plane, we notice that it has a relative maximum at the origin. On the other hand, if we look at the curve $z = y^2$, the cross section of the surface in the yz-plane, we notice that it has a relative minimum at the origin. A point $(a, b, f(a, b))$ on the graph of f at which some cross sections have relative maxima and other cross sections have relative minima is called a **saddle point** for the surface. In such a case, we also say that f has a **saddle point at (a, b)**. Thus, the function $f(x, y) = y^2 - x^2$ has a saddle point at $(0, 0)$.

8.1 EXERCISE SET ||||||||||||||||||||||||||||||||||||||||||||||||||||||||||||||||||||||||||||||

1. If $f(x, y) = 5 + 3x - 2y$, find
   (a) $f(1, 2)$ (d) $f(-1, -3)$
   (b) $f(-2, 3)$ (e) the domain of f
   (c) $f(0, 4)$

2. If $f(x, y) = x^2 + 3xy$, find
   (a) $f(2, 1)$ (d) $f(-1, -2)$
   (b) $f(1, -3)$ (e) the domain of f
   (c) $f(3, 0)$

3. If $g(x, y) = ye^x + xy + \ln y$, find
   (a) $g(0, 1)$ (c) the domain of g
   (b) $g(2, 2)$

4. If $h(x, y) = \dfrac{x + y}{x - y}$, find
   (a) $h(4, -1)$ (d) $h(-2, -3)$
   (b) $h(4, -4)$ (e) the domain of h
   (c) $h(0, 2)$

5. If $f(x, y, z) = x^2 + y^3 - xz + z^2$, find
   (a) $f(1, -1, 2)$ (d) $f(2, 1, -2)$
   (b) $f(0, 0, 4)$ (e) the domain of f
   (c) $f(-2, 3, 1)$

6. If $g(x, y, z) = (x^2 + y - z^2)/(x + z)$, find
   (a) $g(1, 2, 3)$ (b) $g(-4, -2, 3)$

(c) $g(0, 0, 3)$ (e) the domain of g
(d) $g(1, 1, 1)$

7. If $V(x, y, z) = xyz$, find
  (a) $V(1, 2, 3)$ (b) $V(2, 5, 7)$

8. If $T(A, B) = \sqrt{A} + \sqrt{B}$, find
  (a) $T(4, 9)$ (b) $T(25, 36)$

9. If $K(a, b, c) = ce^{a+b^2}$, find
  (a) $K(2, 3, 1)$ (b) $K(1, 0, 3)$

10. If $M(r, s) = \ln(r^2 + s^2)$, find
  (a) $M(1, 5)$ (b) $M(2, 3)$

11. For the function $f(x, y) = 2x + 3y$, find
  (a) $f(2 + h, 3)$
  (b) $f(x + h, y)$
  (c) $f(2 + h, 3) - f(2, 3)$
  (d) $f(x + h, y) - f(x, y)$
  (e) $\dfrac{f(x + h, y) - f(x, y)}{h}$

12. Follow the directions in Exercise 11 for the function
$$f(x, y) = 3x + y^2.$$

13. Plot the following points in a rectangular coordinate system:
  (a) $(1, 2, 3)$ (c) $(0, 4, 3)$
  (b) $(2, -2, 4)$ (d) $(3, 5, -1)$

14. Plot the following points in a rectangular coordinate system:
  (a) $(4, -1, 2)$ (c) $(-2, 3, -4)$
  (b) $(0, 0, -4)$ (d) $(1, 2, 0)$

15. **(Ecology)** The pollution index I of the air over a certain town is defined to be
$$I(x, y) = \sqrt{x} + \sqrt{y} + 2xy,$$
where x is the number of milligrams of nitrous oxide and y is the number of milligrams of particulate matter in a standard sample. Find
  (a) $I(4, 25)$ (b) $I(9, 4)$

16. **(Advertising)** An automobile manufacturer's profit P (in millions of dollars) is given by
$$P(x, y) = 2x^2 + 3y^2 - 2x + 2y - 6,$$
where x and y are the amounts spent (in millions of dollars) on TV and magazine advertising, respectively. Find
  (a) $P(3, 2)$ (b) $P(1, 1)$

17. **(Manufacturing Cost)** A bicycle manufacturer's production cost C (in thousands of dollars) is given by
$$C(x, y) = 10 + 2x^2 + 3y,$$
where x is the number of employees and y is the amount (in thousands of dollars) spent on materials. Find
  (a) $C(3, 12)$ (b) $C(15, 8)$

18. **(Supply and Demand)** A gasoline dealer discovers that the daily demand for regular gasoline (in dollars) is given by
$$D(x, y) = 5000 - 25x^2 + 50y$$
where x and y are the prices (in dollars per gallon) of regular and high-test gasoline, respectively. Find
  (a) $D(0.80, 0.90)$ (b) $D(1.0, 1.2)$

In Exercises 19–24, sketch the graph of the plane whose equation is given.

19. $3x + 2y + 4z = 12$

20. $2x - y + 3z = 6$

21. $x - y - 2z = 8$

22. $5x + y - 2z = 10$

23. $z = x - y + 2$

24. $z = y - 3x + 9$

8.2 |||||||| PARTIAL DERIVATIVES

In Section 3.1 we saw that if $y = f(x)$ is a function of one variable, then the derivative dy/dx represents the instantaneous rate of change of y with respect to x. Suppose now that $z = f(x, y)$ is a function of the variables x and y. In many applications, we are interested in determining the instantaneous rate of change of z with respect to x when the variable y is held constant and the instantaneous rate of change of z with respect to y when the variable x is held constant. Such rates of change are called **partial derivatives** and the process of obtaining them is called **partial differentiation.** The formal definitions of the partial derivatives are as follows.

If $z = f(x, y)$ is a function of the variables x and y, then the limit

$$\lim_{h \to 0} \frac{f(x + h, y) - f(x, y)}{h}$$

(if it exists) is called the **partial derivative of f with respect to x.** It is denoted by

$$f_x \quad \text{or} \quad \frac{\partial f}{\partial x} \quad \text{or} \quad \frac{\partial z}{\partial x}.$$

If $z = f(x, y)$ is a function of the variables x and y, then the limit

$$\lim_{k \to 0} \frac{f(x, y + k) - f(x, y)}{k}$$

(if it exists) is called the **partial derivative of f with respect to y.** It is denoted by

$$f_y \quad \text{or} \quad \frac{\partial f}{\partial y} \quad \text{or} \quad \frac{\partial z}{\partial y}.$$

As with functions of a single variable, we seldom use these definitions directly in computing partial derivatives. Rather, we use rules that are derived from these definitions. We observe that $\partial f / \partial x$ is the ordinary derivative of f with respect to x, if we consider y as a constant; that is,

$$\frac{\partial f}{\partial x} = \left(\frac{df}{dx} \right)_{y = \text{constant}}.$$

Similarly, $\partial f / \partial y$ is the ordinary derivative of f with respect to y, if we consider x as a constant.

EXAMPLE 1   For the function $f(x, y) = \dfrac{4\sqrt{x}}{y^3}$, find (a) $\dfrac{\partial f}{\partial x}$ (b) $\dfrac{\partial f}{\partial y}$.

Solution (a) $\dfrac{\partial f}{\partial x} = \dfrac{\partial}{\partial x}[4x^{1/2}y^{-3}]$

$$= 4y^{-3}\dfrac{\partial}{\partial x}[x^{1/2}] \qquad \text{(regarding } y \text{ as constant)}$$

$$= \dfrac{4}{y^3} \cdot \dfrac{1}{2}x^{-1/2}$$

$$= \dfrac{2}{y^3\sqrt{x}}.$$

(b) $\dfrac{\partial f}{\partial y} = \dfrac{\partial}{\partial y}[4x^{1/2}y^{-3}]$

$$= 4x^{1/2}\dfrac{\partial}{\partial y}[y^{-3}] \qquad \text{(regarding } x \text{ as constant)}$$

$$= 4x^{1/2} \cdot (-3y^{-4})$$

$$= \dfrac{-12\sqrt{x}}{y^4}. \qquad\qquad\qquad\qquad\qquad \blacksquare$$

If $z = f(x, y)$, then the values of the partial derivatives when $x = x_0$ and $y = y_0$ are denoted by

$$\left.\dfrac{\partial z}{\partial x}\right|_{(x_0, y_0)} \quad \text{and} \quad \left.\dfrac{\partial z}{\partial y}\right|_{(x_0, y_0)}$$

or

$$f_x(x_0, y_0) \quad \text{and} \quad f_y(x_0, y_0).$$

EXAMPLE 2 Let

$$z = f(x, y) = x^2 + xy^2 + y^3.$$

Find

(a) $\dfrac{\partial z}{\partial x}$ (b) $\dfrac{\partial z}{\partial y}$ (c) $f_x(1, -3)$ (d) $f_y(2, 1)$.

Solution (a) To find $\partial z/\partial x$ we treat y as a constant and differentiate z with respect to x, so that

$$\dfrac{\partial z}{\partial x} = \dfrac{\partial}{\partial x}[x^2 + xy^2 + y^3]$$

$$= \dfrac{\partial}{\partial x}[x^2] + y^2\dfrac{\partial}{\partial x}[x] + \dfrac{\partial}{\partial x}[y^3]$$

$$= 2x + y^2 \cdot 1 + 0$$

$$= 2x + y^2.$$

(b) To find $\partial z/\partial y$ we treat x as a constant and differentiate z with respect to y, so that

$$\frac{\partial z}{\partial y} = \frac{\partial}{\partial y} [x^2 + xy^2 + y^3]$$

$$= \frac{\partial}{\partial y} [x^2] + x \frac{\partial}{\partial y} [y^2] + \frac{\partial}{\partial y} [y^3]$$

$$= 0 + x \cdot 2y + 3y^2$$

$$= 2xy + 3y^2.$$

(c) From (a),

$$f_x(x, y) = \frac{\partial z}{\partial x} = 2x + y^2,$$

so

$$f_x(1, -3) = 2(1) + (-3)^2 = 11.$$

(d) From (b),

$$f_y(x, y) = \frac{\partial z}{\partial y} = 2xy + 3y^2,$$

so

$$f_y(2, 1) = 2 \cdot 2 \cdot 1 + 3(1)^2 = 7. \qquad \blacksquare$$

EXAMPLE 3 Let

$$z = xe^{2x^2 + y^2}.$$

Find

(a) $\dfrac{\partial z}{\partial x}$ (b) $\dfrac{\partial z}{\partial y}$ (c) $f_x(1, 0)$ (d) $f_y(1, -1)$.

Solution (a) To find $\partial z/\partial x$ we treat y as a constant and differentiate z with respect to x, so that

$$\frac{\partial z}{\partial x} = \frac{\partial}{\partial x} [xe^{2x^2 + y^2}]$$

$$= x \frac{\partial}{\partial x} [e^{2x^2 + y^2}] + e^{2x^2 + y^2} \frac{\partial}{\partial x} [x] \qquad \text{(by the product rule)}$$

$$= xe^{2x^2 + y^2} \cdot \frac{\partial}{\partial x} [2x^2 + y^2] + e^{2x^2 + y^2} \cdot 1 \qquad \text{(by the chain rule on } e^{2x^2 + y^2})$$

$$= xe^{2x^2 + y^2} \cdot 4x + e^{2x^2 + y^2}$$

$$= 4x^2 e^{2x^2 + y^2} + e^{2x^2 + y^2}$$

$$= (4x^2 + 1)e^{2x^2 + y^2}.$$

(b) To find $\partial z/\partial y$ we treat x as a constant and differentiate z with respect to y, so that

$$\frac{\partial z}{\partial y} = \frac{\partial}{\partial y}\,[xe^{2x^2+y^2}]$$

$$= x\frac{\partial}{\partial y}\,[e^{2x^2+y^2}] + e^{2x^2+y^2}\frac{\partial}{\partial y}\,[x] \quad \text{(by the product rule)}$$

$$= xe^{2x^2+y^2}\cdot\frac{\partial}{\partial y}\,[2x^2+y^2] + e^{2x^2+y^2}\cdot 0 \quad \text{(by the chain rule on } e^{2x^2+y^2}\text{)}$$

$$= xe^{2x^2+y^2}\cdot 2y + 0$$

$$= 2xy\,e^{2x^2+y^2}.$$

(c) From (a),

$$f_x(x,\,y) = \frac{\partial z}{\partial x} = (4x^2+1)e^{2x^2+y^2}$$

so

$$f_x(1,\,0) = (4\cdot 1^2 + 1)e^{2(1)^2+0^2} = 5e^2.$$

(d) From (b),

$$f_y(x,\,y) = \frac{\partial z}{\partial y} = 2xye^{2x^2+y^2}$$

so

$$f_y(1,\,-1) = 2(1)(-1)e^{2(1)^2+(-1)^2} = -2e^3. \qquad \blacksquare$$

EXAMPLE 4
(Business) A steel manufacturer finds that the total production cost C (in millions of dollars) depends on the amount x (in millions of units) of labor used and upon the amount y (in millions of tons) of raw materials used. Suppose that

$$C = 20 + 5x + 2y.$$

Thus,

$$\frac{\partial C}{\partial x} = 5 \quad \text{and} \quad \frac{\partial C}{\partial y} = 2.$$

The first result, $\partial C/\partial x = \$5$ million per 1 million units of labor, means that if the amount of raw materials used remains fixed, an increase of 1 million units in the amount of labor used will result in an increase of approximately $5 million in the total production cost. The second result, $\partial C/\partial y = \$2$ million per 1 million tons of raw materials, indicates that if the amount of labor used remains fixed, an additional 1 million tons of raw materials will yield an increase of approximately $2 million in the total production cost. The partial derivative $\partial C/\partial x$ is called the **marginal cost of labor** and $\partial C/\partial y$ is called the **marginal cost of raw materials**.

||||||||||||| **Geometric Interpretation**

If f is a function of a single variable x, then the derivative $f'(a)$ is the slope of the tangent line to the graph of f at the point $(a, f(a))$, as shown in Figure 1.

Figure 1

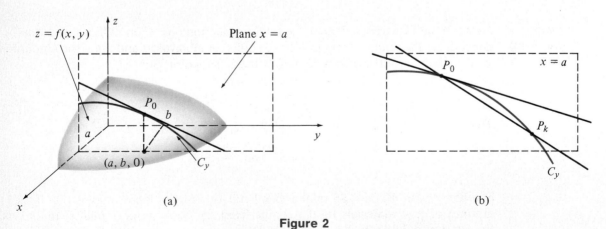

We can interpret partial derivatives geometrically as follows. Suppose $z = f(x, y)$ is a function of two variables and that $f_x(a, b)$ and $f_y(a, b)$ exist. From the formal definition of partial derivative we know that

$$f_y(a, b) = \lim_{k \to 0} \frac{f(a, b + k) - f(a, b)}{k}.$$

Consider what happens when we move from the point (a, b) to the point $(a, b + k)$ in the domain of f, holding x fixed, but allowing y to change from b to $b + k$. The points $P_0(a, b, f(a, b))$ and $P_k(a, b + k, f(a, b + k))$ both lie in the plane $x = a$. In fact, both points lie on the curve of intersection C_y of the plane $x = a$ with the surface $z = f(x, y)$. (See Figure 2.)

(a) (b)

Figure 2

The quantity

$$\frac{f(a, b + k) - f(a, b)}{k}$$

is the slope of the secant line (in the plane $x = a$) joining the points P_0 and P_k [Figure 2(b)]. As k approaches zero, this slope approaches the slope of the line (in the plane $x = a$) that is tangent to the curve C_y at the point P_0. Thus, $f_y(a, b)$ is the slope of the tangent line to the curve C_y at the point $P_0(a, b, f(a, b))$. Figure 2 depicts the curve and its tangent line.

Similarly, we let C_x denote the curve of intersection of the plane $y = b$ with the surface $z = f(x, y)$. The formal definition of $f_x(a, b)$ confirms that $f_x(a, b)$ is the slope of the tangent line to the curve C_x at the point $P_0(a, b, f(a, b))$ as Figure 3 shows.

Figure 3

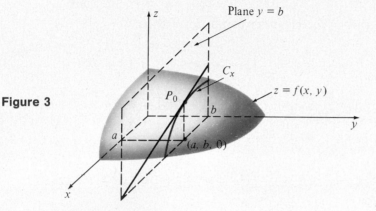

Second-Order Partial Derivatives

If $z = f(x, y)$, then

$$f_x = \frac{\partial z}{\partial x} \quad \text{and} \quad f_y = \frac{\partial z}{\partial y}$$

are both functions of x and y. Each of them may also have partial derivatives. In this case, we obtain the **second-order partial derivatives**

$$\frac{\partial}{\partial x}\left[\frac{\partial z}{\partial x}\right] = \frac{\partial^2 z}{\partial x^2} \quad \text{or} \quad (f_x)_x = f_{xx},$$

$$\frac{\partial}{\partial y}\left[\frac{\partial z}{\partial x}\right] = \frac{\partial^2 z}{\partial y\,\partial x} \quad \text{or} \quad (f_x)_y = f_{xy},$$

$$\frac{\partial}{\partial x}\left[\frac{\partial z}{\partial y}\right] = \frac{\partial^2 z}{\partial x\,\partial y} \quad \text{or} \quad (f_y)_x = f_{yx},$$

$$\frac{\partial}{\partial y}\left[\frac{\partial z}{\partial y}\right] = \frac{\partial^2 z}{\partial y^2} \quad \text{or} \quad (f_y)_y = f_{yy}.$$

The subscripts x and y in the notations f_{xy} and $\partial^2 z/\partial x\,\partial y$ indicate the order of differentiation. In both cases, x and y appear in the same order from left to right. However, f_{xy} means that we differentiate first with respect to x, then with respect to y, whereas $\partial^2 z/\partial x\,\partial y$ means that we differentiate first with respect to y, then with respect to x.

EXAMPLE 5   If

$$f(x, y) = x^2 e^y + ye^x + xy^2$$

find f_{xx}, f_{xy}, f_{yx}, and f_{yy}.

Solution   First, we form the partial derivatives:

$$f_x = 2xe^y + ye^x + y^2$$
$$f_y = x^2 e^y + e^x + 2xy.$$

Then we obtain the second-order partial derivatives:

$$f_{xx} = \frac{\partial}{\partial x}[f_x] = \frac{\partial}{\partial x}[2xe^y + ye^x + y^2] = 2e^y + ye^x$$

$$f_{yy} = \frac{\partial}{\partial y}[f_y] = \frac{\partial}{\partial y}[x^2 e^y + e^x + 2xy] = x^2 e^y + 2x$$

$$f_{xy} = \frac{\partial}{\partial y}[f_x] = \frac{\partial}{\partial y}[2xe^y + ye^x + y^2] = 2xe^y + e^x + 2y$$

$$f_{yx} = \frac{\partial}{\partial x}[f_y] = \frac{\partial}{\partial x}[x^2 e^y + e^x + 2xy] = 2xe^y + e^x + 2y. \qquad \blacksquare$$

In Example 5, it happened that

$$f_{xy} = f_{yx}.$$

Although this equation does not always hold, it is true for **almost all common** functions, as the following theorem implies.

THEOREM (Equality of Mixed Second-Order Partials)

If a function $f(x, y)$ and its first and second order partial derivatives f_x, f_y, f_{xy}, and f_{yx} are all continuous in a region around (x_0, y_0), then

$$f_{xy}(x_0, y_0) = f_{yx}(x_0, y_0).$$

It is beyond the scope of this book to give a rigorous proof of this theorem, or even a definition of "continuity" in this context. Virtually all functions the student encounters will have the requisite continuity.

EXAMPLE 6 (Agriculture)   In an agricultural experiment, tomato plants are being grown in a small fixed plot of land that is being fertilized with fertilizers **A** and **B**. The weight W of the tomatoes grown (in pounds) is given by

$$W(x, y) = 180 + x^2 - 3xy + y, \qquad 0 \le x \le 10, 0 \le y \le 8,$$

where x and y are the amounts (in gallons) of fertilizers A and B, respectively, that are being used.

(a) Find the instantaneous rate of change of W with respect to x as x varies and y is held constant.

(b) Find

$$\frac{\partial W}{\partial x}\bigg|_{(5,2)}$$

and interpret the result.

Solution   (a) $\dfrac{\partial W}{\partial x} = \dfrac{\partial}{\partial x} [180 + x^2 - 3xy + y]$

$$= 2x - 3y.$$

Thus, if y is held constant, the instantaneous rate of change of W with respect to x is $2x - 3y$.

(b) Here $y = 2$ and we want to find the instantaneous rate of change of W with respect to x when $x = 5$. From (a), we have

$$\frac{\partial W}{\partial x}\bigg|_{(5,2)} = 2 \cdot 5 - 3 \cdot 2 = 4.$$

Thus, after 5 gallons of fertilizer A and 2 gallons of fertilizer B have been used to treat the soil, each extra gallon of fertilizer A will increase the tomato yield by approximately 4 pounds. ∎

Functions of More Than Two Variables

A function may have more than two independent variables, as we saw in Section 8.1. For example, we may have

$$w = f(x, y, z).$$

In this case we have three first-order partial derivatives:

$$\frac{\partial w}{\partial x} = f_x(x, y, z),$$

$$\frac{\partial w}{\partial y} = f_y(x, y, z),$$

and

$$\frac{\partial w}{\partial z} = f_z(x, y, z).$$

We obtain these partial derivatives by holding all variables constant, except the one indicated in the notation.

EXAMPLE 7   For the function $f(x, y, z) = 5x^3y - \dfrac{x}{z} + (yz)^4$, we have the following first-order partial derivatives:

(a) Holding all variables constant except x, we obtain

$$f_x(x, y, z) = 5y\frac{\partial}{\partial x}[x^3] + \left(-\frac{1}{z}\right)\frac{\partial}{\partial x}[x] + 0$$

$$= 15yx^2 - \frac{1}{z}.$$

(b) Holding all variables constant except y, we obtain

$$f_y(x, y, z) = 5x^3\frac{\partial}{\partial y}[y] - 0 + 4(yz)^3\frac{\partial}{\partial y}[yz]$$

$$= 5x^3 + 4y^3z^4.$$

(c) Holding all variables constant except z, we obtain

$$f_z(x, y, z) = 0 - x\frac{\partial}{\partial z}[z^{-1}] + 4(yz)^3\frac{\partial}{\partial z}[yz]$$

$$= \frac{x}{z^2} + 4y^4z^3.$$

EXAMPLE 8   For the function $w = 3ux^3z^5 - 4v^2y^4z^2$, we have the following five first-order partial derivatives:

$$\frac{\partial w}{\partial x} = 3uz^5\frac{\partial}{\partial x}[x^3] - 0 = 9uz^5x^2;$$

$$\frac{\partial w}{\partial y} = 0 - 4v^2z^2\frac{\partial}{\partial y}[y^4] = -16v^2z^2y^3;$$

$$\frac{\partial w}{\partial z} = 3ux^3\frac{\partial}{\partial z}[z^5] - 4v^2y^4\frac{\partial}{\partial z}[z^2] = 15ux^3z^4 - 8v^2y^4z;$$

$$\frac{\partial w}{\partial u} = 3x^3z^5\frac{\partial}{\partial u}[u] - 0 = 3x^3z^5;$$

$$\frac{\partial w}{\partial v} = 0 - 4y^4z^2\frac{\partial}{\partial v}[v^2] = -8y^4z^2v.$$

8.2 EXERCISE SET ||||||||||||||||||||||||||||||||||||||||||||||||||||||||||||||||||||||||||||||||||||||||

In Exercises 1–12, find f_x, f_y, and f_{xy}.

1. $f(x, y) = x^3 + xy + 3y^2$

2. $f(x, y) = x^2y - y^2x + y^3$

3. $f(x, y) = (x^2 + 5y)^4$

4. $f(x, y) = \sqrt{xy}$

5. $f(x, y) = \dfrac{x^3}{y^2}$

6. $f(x, y) = \dfrac{x + y}{x - y}$

7. $f(x, y) = e^{xy^2}$

8. $f(x, y) = \dfrac{y}{x} - \dfrac{x}{y}$

9. $f(x, y) = xe^y + y^2 e^x - y^4$

10. $f(x, y) = xe^{2x + 3y}$

11. $f(x, y) = \ln(xy)$

12. $f(x, y) = (x + y) \ln(xy)$

In Exercises 13–16, calculate $f_x(x_0, y_0)$, $f_y(x_0, y_0)$, and $f_{yx}(x_0, y_0)$ for the specified functions at the given points.

13. $f(x, y) = xy^2 + 2xy - y^3$, $(x_0, y_0) = (-1, 2)$

14. $f(x, y) = x^3 + xy^2 - x^2 y + y^4$, $(x_0, y_0) = (2, 1)$

15. $f(x, y) = xe^y + y^2 e^x$, $(x_0, y_0) = (0, 1)$

16. $f(x, y) = x \ln(xy)$, $(x_0, y_0) = (1, 1)$

In Exercises 17–22, calculate f_{xx}, f_{yy}, f_{xy}, and f_{yx}.

17. $f(x, y) = 2x^2 + xy - y^3$

18. $f(x, y) = 3x^3 - 4xy + y^4 x^2$

19. $f(x, y) = 3x - 5y$

20. $f(x, y) = e^{x^2 y}$

21. $f(x, y) = xe^{x + y^2}$

22. $f(x, y) = y \ln x$

23. Let $f(x, y) = \ln(x^2 + y^2)$. Show that $f_{xx} + f_{yy} = 0$.

In Exercises 24–27, evaluate f_{xx}, f_{yy}, f_{xy}, and f_{yx} at the point (x_0, y_0).

24. $f(x, y) = 3xy^2 + xy - y^3$, $(x_0, y_0) = (1, -1)$

25. $f(x, y) = 5x^2 y - y^3$, $(x_0, y_0) = (-3, 2)$

26. $f(x, y) = e^{3x + 2y}$, $(x_0, y_0) = (0, 1)$

27. $f(x, y) = x \ln y$, $(x_0, y_0) = (3, 5)$

In Exercises 28–33, find all first-order partial derivatives at the indicated point.

28. $w = f(x, y, z) = x^3 y^2 - \dfrac{y}{z^2}$, at $(2, 3, -1)$.

29. $w = f(x, y, z) = xy - yz + xz$, at $(3, -4, 2)$.

30. $w = f(u, v, x, y) = (3u - v)(x^2 - y^3)$, at $(-4, 3, 2, 1)$.

31. $w = f(u, v, x, y) = \dfrac{u^2 - 4v}{x + y^2}$, at $(1, 2, 2, -1)$.

32. $w = f(u, v, x, y, z) = u^2 v(x^2 - 3y + 5z)$, at $(2, 3, -2, -1, -3)$.

33. $w = f(u, v, x, y, z) = \dfrac{3u}{v} - \dfrac{x^2 y}{z}$ at $(1, -1, -2, 2, -1)$.

34. **(Business)** A car manufacturer makes gasoline engines and diesel engines. The cost C (in millions of dollars) is given by

$$C = 20 + 5x^2 + 2x + 2y^2 + xy,$$

where x is the number (in thousands) of gasoline engines produced and y is the number (in thousands) of diesel engines produced. Suppose that the current level of production is 8000 gasoline engines and 5000 diesel engines.
(a) Find $\partial C / \partial x$ and $\partial C / \partial y$.
(b) Find the marginal cost (in millions of dollars per thousand engines) of gasoline engines at the current level of production.
(c) Find the marginal cost (in millions of dollars per thousand engines) of diesel engines at the current level of production.
(d) Interpret your answers to parts (b) and (c).

35. **(Production)** The production function $Q = 60x^{2/3} y^{1/3}$ gives the number of cameras produced by a manufacturer with x units of labor and y units of capital.
(a) Find $\partial Q / \partial x$ and $\partial Q / \partial y$.

(b) Evaluate

$$\left.\frac{\partial Q}{\partial x}\right|_{(8,27)} \quad \text{and} \quad \left.\frac{\partial Q}{\partial y}\right|_{(8,27)}$$

(c) Interpret the results of part (b).

36. **(Pollution)** The total annual amount A of particulate matter present in the atmosphere in a certain town is traceable to two nearby steel factories. It has been found that

$$A = 2x^3 + 3xy + y^2,$$

where A is in thousands of pounds and x and y are the annual production levels (in millions of tons) at the respective factories.
(a) Find $\partial A/\partial x$ and $\partial A/\partial y$.
(b) If the annual production levels at the two factories are 5 and 8 million tons, compute

$$\left.\frac{\partial A}{\partial x}\right|_{(5,8)} \quad \text{and} \quad \left.\frac{\partial A}{\partial y}\right|_{(5,8)}$$

(c) Interpret the results of part (b).

37. **(Business)** A computer manufacturer finds that the annual revenue R (in millions of dollars) is given by

$$R(x, y) = 50 + 2x^2 + 3xy,$$

where x is the annual amount (in millions of dollars) spent on research and y is the annual amount (in millions of dollars) spent on advertising.
(a) Find the instantaneous rate of change of R with respect to x.
(b) Find the instantaneous rate of change of R with respect to y.

38. **(Medicine)** In many medical applications it is necessary to know how the surface area of the average person's body varies with height and weight. The following empirical formula is commonly used to relate body surface area A (in square meters) to weight W (in kilograms) and height H (in meters):

$$A = 2.024 W^{0.425} H^{0.725}.$$

(a) Find $\partial A/\partial W$ and $\partial A/\partial H$ when $W = 90$ and $H = 2$. Use these approximations:

$$90^{0.425} \approx 6.769 \qquad 90^{-0.575} \approx 0.0752$$
$$2^{0.725} \approx 1.653 \qquad 2^{-0.275} \approx 0.826.$$

(b) Give a physical interpretation of the results of part (a).

39. **(Business)** A roofing contractor has a work force of 20 roofers and 12 apprentices. The number of roofs N completed annually by x roofers and y apprentices is given by

$$N = 20x + 3xy + y^2.$$

(a) Use partial derivatives to obtain an approximation to the number of additional roofs that can be completed per year if the contractor hires one more roofer.
(b) Use partial derivatives to obtain an approximation to the number of additional roofs that can be completed per year if the contractor hires one more apprentice.

8.3 |||||||| EXTREME VALUES

In Sections 4.2 and 4.3 we studied the problem of maximizing or minimizing a function of one variable. Functions of several variables provide an analogous problem. In this section we shall limit our attention to the problem of maximizing or minimizing a function of two variables.

The function f is said to have a **relative maximum at (a, b)** if there is a circular disk in the xy-plane centered at (a, b), throughout which f is defined, such that

for all points (x, y) in this disk

$$f(x, y) \le f(a, b).$$

[See Figure 1(a).]

Figure 1

$(a, b, f(a, b))$

$z = f(x, y)$

$(a, b, 0)$

(a)

f has a relative
maximum at (a, b)

$z = f(x, y)$

$(a, b, f(a, b))$

$(a, b, 0)$

(b)

f has a relative
minimum at (a, b)

The function *f* is said to have a **relative minimum at (a, b)** if there is a circular disk in the *xy*-plane centered at (a, b), throughout which *f* is defined, such that for all points (x, y) in this disk

$$f(a, b) \le f(x, y).$$

[See Figure 1(b).]

We say that *f* has a **relative extremum at (a, b)** if *f* has a relative maximum or a relative minimum at (a, b). The relative maxima and relative minima of *f* correspond, respectively, to the high points and low points on the surface $z = f(x, y)$. For example, the graph of the function

$$z = f(x, y) = x^2 + y^2$$

is the surface in Figure 8 of Section 8.1. This surface has a low point at $(0, 0, 0)$ since the function $f(x, y) = x^2 + y^2$ has a relative minimum at $(0, 0)$. Similarly, the graph of the function

$$z = f(x, y) = \sqrt{9 - x^2 - y^2}$$

is the surface in Figure 9 of Section 8.1. This surface has a high point at $(0, 0, 3)$ since the function $f(x, y) = \sqrt{9 - x^2 - y^2}$ has a relative maximum at $(0, 0)$. The graph of the function

$$z = f(x, y) = y^2 - x^2$$

is the saddle-shaped surface in Figure 10 of Section 8.1. This surface has neither a high point nor a low point at the saddle point $(0, 0, 0)$ since $f(x, y) = y^2 - x^2$ has no relative extremum at $(0, 0)$. It becomes clear that no relative extremum of f occurs at $(0, 0)$ if we note that in every circular disk in the xy-plane centered at $(0, 0)$ there are points (x, y) at which

$$f(x, y) \leq 0 = f(0, 0),$$

and other points (x, y) at which

$$f(x, y) \geq 0 = f(0, 0).$$

Thus, neither inequality holds for *all* points within the disk.

From our experience with functions of a single variable, we would expect to be able to say something about the partial derivatives $f_x(a, b)$ and $f_y(a, b)$ if f has a relative extremum at (a, b). To determine whether we can, let us suppose that $f(x, y)$ has a relative maximum at (a, b) and that both partials $f_x(a, b)$ and $f_y(a, b)$ exist. Figure 2 shows the curve C_y obtained by intersecting the surface $z = f(x, y)$ with the plane $x = a$, and it also shows the curve C_x obtained by intersecting the surface $z = f(x, y)$ with the plane $y = b$. The curve C_y has a relative maximum at $y = b$, so $f_y(a, b) = 0$. Similarly, the curve C_x has a relative maximum at $x = a$, so $f_x(a, b) = 0$.

Figure 2

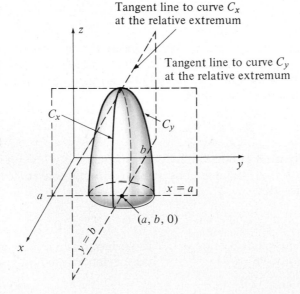

Tangent line to curve C_x
at the relative extremum

Tangent line to curve C_y
at the relative extremum

We can generalize this discussion as follows.

THEOREM 1

> If f has a relative extremum at (a, b), and $f_x(a, b)$ and $f_y(a, b)$ both exist, then
>
> $$f_x(a, b) = 0 \quad \text{and} \quad f_y(a, b) = 0. \qquad (1)$$

The converse of this theorem is false. That is, if $f_x(a, b) = 0$ and $f_y(a, b) = 0$, it need not follow that f must have a relative extremum at (a, b). For example, consider the saddle surface of Example 10 and Figure 10 in Section 8.1:

$$z = f(x, y) = y^2 - x^2.$$

For that function,

$$f_x = -2x \quad \text{and} \quad f_y = 2y.$$

Thus, $f_x(0, 0) = 0$ and $f_y(0, 0) = 0$, but f has neither a relative maximum nor a relative minimum at $(0, 0)$.

Critical Points

We say that a point (a, b) is a **critical point** of f if (a, b) is the center of a circular disk throughout which f is defined and if either

$$f_x(a, b) = 0 \quad \text{and} \quad f_y(a, b) = 0,$$

or one of the partials $f_x(a, b)$, $f_y(a, b)$ fails to exist.

Thus, by Theorem 1, if f has a relative extremum at (a, b), then (a, b) is a critical point of f. Once again the converse of this statement is not true. The critical points of f are only *candidates* for the points at which f has relative extrema. Of course, this is the same situation for functions of one variable, as we saw in Section 4.2. In summary, if f has a relative extremum at (a, b), then (a, b) is a critical point of f; if (a, b) is a critical point of f, then f may or may not have a relative extremum at (a, b).

EXAMPLE 1 Find the critical points of

$$f(x, y) = 2x^2 + xy + y^2 + 2x - 3y.$$

Solution To find the critical points, we determine f_x and f_y and set them equal to zero. In this case,

$$f_x(x, y) = 4x + y + 2$$
$$f_y(x, y) = x + 2y - 3.$$

Consequently, to find the critical points, we must solve the system of equations

$$
\begin{aligned}
4x + \ y + 2 &= 0 \\
x + 2y - 3 &= 0
\end{aligned}
\quad \text{or} \quad
\begin{aligned}
4x + \ y &= -2 \\
x + 2y &= 3.
\end{aligned}
$$

Solving the first equation for y, we have

$$y = -2 - 4x.$$

Substituting this value of y into the second equation and simplifying, we obtain (verify)

$$-7x = 7$$
$$x = -1.$$

Then

$$y = -2 - 4x$$
$$= -2 - 4(-1) = 2.$$

Thus, $(-1, 2)$ is the only critical point of f. ■

EXAMPLE 2   Find the critical points of

$$f(x, y) = 2xy + x^2 - y^3.$$

Solution   We first determine f_x and f_y and then set them equal to zero. In this case,

$$f_x(x, y) = 2x + 2y$$
$$f_y(x, y) = 2x - 3y^2.$$

Consequently, to find the critical points, we must solve the system of equations

$$2x + 2y = 0$$
$$2x - 3y^2 = 0.$$

We solve the first equation for y in terms of x, obtaining

$$y = -x \tag{2}$$

and substitute this value of y in the second equation,

$$2x - 3(-x)^2 = 2x - 3x^2 = 0$$

or

$$x(2 - 3x) = 0.$$

Thus,

$$x = 0 \quad \text{or} \quad x = \tfrac{2}{3}.$$

From (2), we obtain the corresponding values of y,

$$y = 0, \quad y = -\tfrac{2}{3}.$$

Hence, the critical points of f are

$$(0, 0) \quad \text{and} \quad (\tfrac{2}{3}, -\tfrac{2}{3}).$$ ∎

||||||||||||| **Testing Critical Points**

For functions of a single variable, we used the second derivative test to determine the critical numbers that correspond to relative extrema. The following more complicated test, which we state without proof, is available for functions of two variables.

THEOREM 2 (Second Partials Test for Relative Extrema)

Let (a, b) be a critical point of f where both partial derivatives of f are 0, and let

$$M = f_{xx}(a, b) \cdot f_{yy}(a, b) - [f_{xy}(a, b)]^2.$$

(a) If $M > 0$ and $f_{xx}(a, b) < 0$, then f has a relative maximum at (a, b).
(b) If $M > 0$ and $f_{xx}(a, b) > 0$, then f has a relative minimum at (a, b).
(c) If $M < 0$, then f has no relative extremum at (a, b).
(d) If $M = 0$, then f may or may not have a relative extremum at (a, b). (When $M = 0$ this test is useless, and some other criterion must be used.)

We thus have the following method for finding the relative extrema of a function of two variables.

Finding the Relative Extrema of $f(x, y)$
Step 1.   Compute f_x, f_y, f_{xx}, f_{yy}, and f_{xy}.
Step 2.   Find the critical points of $f(x, y)$, where both partial derivatives are 0.
Step 3.   Apply the second partials test to each critical point found in step 2.
*Step 4.** Critical points where one or both partial derivatives fail to exist must be considered separately.

EXAMPLE 3   In Example 1 we found that $(-1, 2)$ is the only critical point of

$$f(x, y) = 2x^2 + xy + y^2 + 2x - 3y.$$

To apply the second partials test, we must calculate

$$f_{xx}(-1, 2), \quad f_{yy}(-1, 2), \quad \text{and} \quad f_{xy}(-1, 2).$$

---

* In this textbook we do not encounter critical points of this type. Step 4 will not be used.

We saw in Example 1 that

$$f_x(x, y) = 4x + y + 2$$
$$f_y(x, y) = x + 2y - 3.$$

Thus,

$$f_{xx}(x, y) = 4 \quad \text{so} \quad f_{xx}(-1, 2) = 4$$
$$f_{xy}(x, y) = 1 \quad \text{so} \quad f_{xy}(-1, 2) = 1$$
$$f_{yy}(x, y) = 2 \quad \text{so} \quad f_{yy}(-1, 2) = 2.$$

Then,

$$M = f_{xx}(-1, 2)f_{yy}(-1, 2) - [f_{xy}(-1, 2)]^2$$
$$= 4 \cdot 2 - 1^2 = 7.$$

Since $M > 0$ and $f_{xx}(-1, 2) = 4 > 0$, it follows that $f(x, y)$ has a relative minimum at $(-1, 2)$.

EXAMPLE 4   In Example 2 we found that $(0, 0)$ and $(\frac{2}{3}, -\frac{2}{3})$ are the only critical points of

$$f(x, y) = 2xy + x^2 - y^3.$$

We find (verify) that

$$f_{xx}(x, y) = 2, \quad f_{yy}(x, y) = -6y, \quad \text{and} \quad f_{xy}(x, y) = 2.$$

Thus, at the critical point $(0, 0)$,

$$M = f_{xx}(0, 0)f_{yy}(0, 0) - [f_{xy}(0, 0)]^2$$
$$= 2 \cdot 0 - 2^2$$
$$= -4.$$

Since $M < 0$, there is no relative extremum at $(0, 0)$. At the critical point $(\frac{2}{3}, -\frac{2}{3})$,

$$M = f_{xx}(\tfrac{2}{3}, -\tfrac{2}{3})f_{yy}(\tfrac{2}{3}, -\tfrac{2}{3}) - [f_{xy}(\tfrac{2}{3}, -\tfrac{2}{3})]^2$$
$$= 2 \cdot 4 - 2^2 = 4 > 0.$$

Since $M > 0$ and $f_{xx}(\frac{2}{3}, -\frac{2}{3}) > 0$, we conclude that $f(x, y)$ has a relative minimum at $(\frac{2}{3}, -\frac{2}{3})$.

Absolute Extrema

In applications we often want to find the absolute extrema of a function over a closed region R, rather than its relative extrema.

By a **region** R we generally mean the interior of some common two-dimensional geometric figure, such as a circle or rectangle. Sometimes the region R may be more complicated, but a rigorous definition of a region is beyond the scope of this book. By a **closed region** we mean a region, together with all of its boundary.

The function $f(x, y)$ is said to have an **absolute maximum** at (a, b) for a region R if (a, b) is in R, and

$$f(x, y) \leq f(a, b) \text{ for all } (x, y) \text{ in } R.$$

The function $f(x, y)$ is said to have an **absolute minimum** at (a, b) for a region R if (a, b) is in R, and

$$f(a, b) \leq f(x, y) \text{ for all } (x, y) \text{ in } R.$$

Before finding an absolute extremum of a given function, we must determine whether or not the function has an absolute extremum. In general, this can be a very difficult question to answer. There are two-dimensional analogues of the Extreme Value Theorem for functions of a single variable (Section 4.3) that ensure the existence of an absolute extremum. The following theorem justifies the procedure we shall be following.

THEOREM 3 (Extreme Value Theorem)

If $f(x, y)$ is continuous on a closed, bounded region R, then $f(x, y)$ has an absolute maximum and an absolute minimum on R.

In applied problems, the existence or nonexistence of an absolute extremum is often clear from the nature of the problem. Of course, this absolute extremum may occur on the boundary of the domain or in its interior. In this simplified presentation we shall assume that the absolute extrema of f always exist and that the absolute extrema do not occur at points on the boundary if the domain of f has a boundary. Thus, an absolute extremum will be found among the critical points.

EXAMPLE 5 (Economics)

A foreign manufacturer that has a monopoly on a particular transistor radio exports the radio to the United States and sells it at home. Let x be the number of radios sold each week in the United States and let y be the number sold each week at home. Market research has led the manufacturer to set a price (in dollars) of $100 - x/2$ for each radio sold in the United States and a price (in dollars) of $120 - y/4$ for each radio sold at home. The weekly cost (in dollars) of producing these radios is $2000 + 20x + 20y$. How many radios should be sold in each country to maximize the weekly profit?

Solution

The revenue (in dollars) is given by

$$R(x, y) = \left(100 - \frac{x}{2}\right)x + \left(120 - \frac{y}{4}\right)y,$$

so the profit is

$$P(x, y) = \left(100 - \frac{x}{2}\right)x + \left(120 - \frac{y}{4}\right)y - [2000 + 20x + 20y]$$

$$= 80x - \frac{x^2}{2} + 100y - \frac{y^2}{4} - 2000.$$

Since x and y denote the number of radios sold, it must be that $x \geq 0$ and $y \geq 0$. Also, since prices cannot be negative, it must be that

$$100 - \frac{x}{2} \geq 0 \quad \text{and} \quad 120 - \frac{y}{4} \geq 0.$$

Hence, $x \leq 200$ and $y \leq 480$. We therefore want to maximize the function $P(x, y)$ over the closed rectangle shown in Figure 3.

Figure 3

To find the critical points of P we set the first partial derivatives equal to zero. That is,

$$P_x(x, y) = 80 - x = 0$$

$$P_y(x, y) = 100 - \frac{y}{2} = 0,$$

so

$$x = 80, \qquad y = 200.$$

To test the critical point $(80, 200)$, we use the second partials test for relative extrema. We have

$$P_{xx}(80, 200) = -1$$
$$P_{yy}(80, 200) = -\tfrac{1}{2}$$
$$P_{xy}(80, 200) = 0,$$

so

$$M = P_{xx}(80, 200)P_{yy}(80, 200) - [P_{xy}(80, 200)]^2$$
$$= (-1)(-\tfrac{1}{2}) - 0^2 = \tfrac{1}{2} > 0.$$

Since $M > 0$ and $P_{xx}(80, 200) < 0$, we conclude that $P(x, y)$ has a relative maximum at $(80, 200)$. Thus, to maximize the weekly profit, the manufacturer should sell 80 radios in the United States and 200 radios at home. The maximum profit made each week from the sale of these quantities would be

$$P(80, 200) = 80 \cdot 80 - \frac{80^2}{2} + 100 \cdot 200 - \frac{200^2}{2} - 2000$$

$$= \$1200. \qquad \blacksquare$$

**EXAMPLE 6
(Advertising)**

The annual profit of an automobile manufacturer (in millions of dollars) is given by

$$P(x, y) = -3x^2 - 2y^2 + 4x + 2y - 2xy + 20$$

where x and y are the amounts spent annually (in hundreds of thousands of dollars) on TV and magazine advertising, respectively. Find the values of x and y that maximize the profit and determine the maximum profit.

Solution

To find the critical points of P, we set the first partial derivatives equal to zero; that is,

$$P_x(x, y) = -6x + 4 - 2y = 0$$
$$P_y(x, y) = -4y + 2 - 2x = 0.$$

Solving this linear system, we obtain (verify)

$$x = \tfrac{3}{5}, \qquad y = \tfrac{1}{5}.$$

Applying the second partials test for relative extrema, we find (verify)

$$M = P_{xx}(\tfrac{3}{5}, \tfrac{1}{5})P_{yy}(\tfrac{3}{5}, \tfrac{1}{5}) - [P_{xy}(\tfrac{3}{5}, \tfrac{1}{5})]^2$$
$$= -6(-4) - (-2)^2 = 20 > 0$$

and

$$P_{xx}(\tfrac{3}{5}, \tfrac{1}{5}) = -6 < 0,$$

so $P(x, y)$ has a relative maximum and hence an absolute maximum at $x = \tfrac{3}{5}$, $y = \tfrac{1}{5}$. Thus, \$60,000 should be spent annually on TV advertising and \$20,000 on magazine advertising to maximize annual profit. The resulting maximum profit would be

$$P(\tfrac{3}{5}, \tfrac{1}{5}) = -3(\tfrac{3}{5})^2 - 2(\tfrac{1}{5})^2 + 4(\tfrac{3}{5}) + 2(\tfrac{1}{5}) - 2(\tfrac{3}{5})(\tfrac{1}{5}) + 20$$
$$= \$21.4 \text{ million.} \qquad \blacksquare$$

Least Squares

In Section 1.4 we discussed the method of least squares for finding the line of best fit for the data points (x_1, y_1), (x_2, y_2), ..., (x_r, y_r). Denoting the line of best fit by

$$y = mx + b,$$

the method of least squares determines m and b so that the sum of the squares of the vertical distances from the data points to the line is as small as possible. The problem is to minimize the function

$$E(m, b) = (y_1 - mx_1 - b)^2 + (y_2 - mx_2 - b)^2 + \cdots + (y_r - mx_r - b)^2.$$

To find the minimum of E, we set $\partial E/\partial b$ and $\partial E/\partial m$ equal to zero:

$$\frac{\partial E}{\partial b} = 2(y_1 - mx_1 - b)(-1) + 2(y_2 - mx_2 - b)(-1) + \cdots$$

$$+ 2(y_r - mx_r - b)(-1) = 0 \tag{3}$$

$$\frac{\partial E}{\partial m} = 2(y_1 - mx_1 - b)(-x_1) + 2(y_2 - mx_2 - b)(-x_2) + \cdots$$

$$+ 2(y_r - mx_r - b)(-x_r) = 0. \tag{4}$$

Dividing Equations (3) and (4) by 2 and rearranging the terms, we have

$$rb + (x_1 + x_2 + \cdots + x_r)m = y_1 + y_2 + \cdots + y_r$$
$$(x_1 + x_2 + \cdots + x_r)b + (x_1^2 + x_2^2 + \cdots + x_r^2)m = x_1 y_1 + x_2 y_2 + \cdots + x_r y_r, \tag{5}$$

which are precisely the same as Equations (16) in Section 1.4. The solution to the system of equations in (5) gives the desired values for m and b.

8.3 EXERCISE SET ||||||||||||||||||||||||||||||||||||||||||||||||||||||||||||||||||||||||||||||||||

In Exercises 1–8, find the critical points of f.

1. $f(x, y) = x^2 + y^2 - 6x + 3y + 5$

2. $f(x, y) = -2x^2 - 5y^2 + 5x - 2y + 8$

3. $f(x, y) = 3x^2 + 2y^2 + 2xy - 4x + y - 1$

4. $f(x, y) = 5x^2 + 3y^2 - 4xy + 2x - 3y + 12$

5. $f(x, y) = -4x^2 + 3y^2 - 3xy + 2x - 5y + 3$

6. $f(x, y) = x^3 + y^3 + 3xy$

7. $f(x, y) = 2x^3 - 3x^2 - y^2$

8. $f(x, y) = x^3 + y^3 - 3x - 3y + 2$

In Exercises 9–17, find the relative extrema, if any exist.

9. $f(x, y) = x^2 + xy + y^2 - 4x - 5y$

10. $f(x, y) = x^2 + y^2$

11. $f(x, y) = -x^2 - y^2 + 6x + 2y + 12$

12. $f(x, y) = x^2 - y^2$

13. $f(x, y) = x^3 + xy - y^3$

14. $f(x, y) = x^2 + y^2 + 2x + 6y + 8$

15. $f(x, y) = 2x^3 - 6xy + 3y^2 + 6x - 18y$

16. $f(x, y) = -2x^2 - 2xy - y^2 + 4x + 2y$

17. $f(x, y) = x^2 - 2xy + y^2$

18. Find the dimensions of the closed rectangular box of least surface area with a volume of 64 cubic centimeters. (*Hint:* Let x, y, and z denote the dimensions of the box; express the surface area in terms of x and y alone.)

19. **(Business)** A computer manufacturer finds that outlays of x million dollars each year for research and y million dollars for development are related to the annual profit $P(x, y)$ (in millions of dollars) by

$$P(x, y) = 3000 - \tfrac{1}{2}x^2 + 50x - \tfrac{1}{3}y^3 + 16y.$$

How much money should be spent each year for research and how much for development to maximize the annual profit? What is the maximum annual profit?

20. Find three positive numbers whose sum is 30 and whose product is a maximum.

21. **(Business)** A shopping center's daily profit P (in dollars) is given by

$$P(x, y) = 120,000 - (60 - x)^2 - (10 - y)^2,$$

where x is the number of sales people and y is the number of stores. Find the values of x and y that will maximize the daily profit. What is the maximum daily profit?

22. A closed rectangular box having a volume of 1400 cubic feet is to be made from three different materials. The material for the top costs $8 per square foot, the material for the sides costs $5 per square foot, and the material for the bottom costs $6 per square foot. Find the dimensions of the box that will minimize the total cost of the materials.

23. In bidding on a housing development, a firm estimates that the total plumbing cost (including materials) is given by

$$C(x, y) \cdot= 12{,}000 + 5x^2 - 6xy \\ - 18x - 14y + 8y^2,$$

where x and y are the number of master plumbers and the number of apprentice plumbers, respectively, and the cost is in dollars. How many plumbers of each type should the firm use to minimize the total plumbing cost? What is the minimum cost?

24. A manufacturer makes calculators in models A and B, which sell for $60 and $80 each, respectively. The cost (in dollars) of making x units of model A and y units of model B is given by

$$C(x, y) = x^2 - 80x + y^2 - 20y + 5500.$$

Assuming that every unit that is made can be sold, find the values of x and y that will maximize the profit. What is the maximum profit?

25. A zoo plans to construct an uncovered rectangular aquarium that will have a volume of 32 cubic meters. Find the dimensions that will result in the smallest surface area. What is the smallest surface area?

26. **(Medicine)** A patient being treated for a certain disease is receiving x grams of Curine I and y grams of Curine II daily. The amount (in milligrams) of Trophane, a beneficial enzyme, produced by the medications being administered is given by

$$A(x, y) = -x^2 + 2xy - 2y^2 + 8x + 4y.$$

How much Curine I and how much Curine II should be administered to the patient to maximize the amount of Trophane produced? What is the maximum amount of Trophane produced?

8.4 |||||||| LAGRANGE MULTIPLIERS

Many applications involve **constrained optimization problems.** In these problems we must maximize or minimize a function whose variables satisfy an additional equation called a **constraint.** A simple example of such a problem is a computer manufacturer whose profit is a function of the amount of money spent for research and the amount spent for development. The manufacturer's problem is to maximize the profit subject to the constraint that the total amount of money available for research and development is fixed. We solved problems of this type earlier. In Example 5 of Section 4.7, we were designing a closed rectangular cargo box having a square base and a volume of 1024 cubic meters. The material for the top and bottom cost $4 per square meter while the material for the sides cost $2 per square meter. We had to choose the dimensions of the box to minimize the total cost. We let x denote the length (in meters) of a side of the base and y the height (in meters) of the box. Our task was to minimize the function

$$C(x, y) = 8x^2 + 8xy$$

where the variables x and y satisfy the constraint

$$x^2y = 1024 \quad \text{or} \quad x^2y - 1024 = 0.$$

We can state the general constrained optimization problem as follows: We want to maximize or minimize a function $z = f(x, y)$ whose variables satisfy the constraint $g(x, y) = 0$. If we can solve the equation $g(x, y) = 0$ for one variable in terms of the other and substitute the resulting expression in $f(x, y)$, then we have obtained z as a function of one variable. We can maximize or minimize z by the methods developed in Chapter 4. Indeed, this was the procedure we used to solve Example 5 in Section 4.7 and other problems of this type.

The main drawback to this method of solving a constrained optimization problem is that it may be difficult, or even impossible, to solve the constraint $g(x, y) = 0$ for one variable in terms of the other. We shall now describe a very powerful technique for solving a constrained optimization problem. Developed by the French mathematician Joseph L. Lagrange,* the technique is called the **method of Lagrange multipliers.** This method eliminates the necessity of solving $g(x, y) = 0$ for one variable in terms of the other. Moreover, it is easy to generalize this method to handle functions of three or more variables. We shall briefly discuss this generalization to the three-variable case, after we have seen several examples of the two-variable case.

The following theorem establishes the foundation for Lagrange's method. Although we are careful to state the theorem correctly, its proof is beyond the scope of this book.

THEOREM

Suppose $f(x, y)$ and $g(x, y)$ have continuous partial derivatives. If, among all points (x, y) satisfying the constraint equation

$$g(x, y) = 0, \tag{1}$$

the function $f(x, y)$ has a relative extremum at a point (a, b) where the partial derivatives $g_x(a, b)$ and $g_y(a, b)$ are not both zero, then there is a real number k such that

$$f_x(a, b) = kg_x(a, b), \quad \text{and} \tag{2}$$
$$f_y(a, b) = kg_y(a, b). \tag{3}$$

---

* Joseph L. Lagrange (1736–1813) was born in Turin, Italy, of French and Italian descent. His talents in mathematics were recognized early, and before he was nineteen he had been appointed professor at the Turin Artillery School. In 1766 he took over Euler's position as director of the scientific section of the Royal Academy in Berlin. In 1786 Lagrange accepted a position at the distinguished Paris Academy of Sciences. He became a French citizen and spent the rest of his life in France where he taught at several major institutions. During the French Revolution, he helped to develop new weights and measures as well as a calendar. He was greatly admired by Napoleon and was highly honored during his lifetime. In mathematics, Lagrange laid the foundations for the calculus of variations and contributed to many fields of pure mathematics, including probability theory and the theory of equations. In physics, he worked on wave theory, sound propagation, and hydrodynamics. He also made important contributions to astronomy.

In order to see how this theorem enables us to locate the relative maxima and minima of $f(x, y)$ we introduce a new variable λ (*lambda*) and form a new function:

$$F(x, y, \lambda) = f(x, y) + \lambda g(x, y).$$

Then

$$F_x(x, y, \lambda) = f_x(x, y) + \lambda g_x(x, y),$$
$$F_y(x, y, \lambda) = f_y(x, y) + \lambda g_y(x, y), \quad \text{and}$$
$$F_\lambda(x, y, \lambda) = g(x, y).$$

If we let $\lambda = -k$, then Equation (2) becomes

$$F_x(a, b, \lambda) = 0,$$

while Equation (3) becomes

$$F_y(a, b, \lambda) = 0$$

and Equation (1) becomes

$$F_\lambda(a, b, \lambda) = 0.$$

We are now ready to summarize the method of Lagrange multipliers.

The Method of Lagrange Multipliers
To locate the (relative) maxima and minima of the function $f(x, y)$ subject to the constraint $g(x, y) = 0$, proceed as follows:

Step 1. Introduce a new variable λ (lambda), called a **Lagrange multiplier,** and form the new function

$$F(x, y, \lambda) = f(x, y) + \lambda g(x, y).$$

Step 2. Solve the system

$$F_x(x, y, \lambda) = 0$$
$$F_y(x, y, \lambda) = 0$$
$$F_\lambda(x, y, \lambda) = 0.$$

(We assume that these partial derivatives exist.)

Step 3. For each solution (a, b, λ_0) obtained in step 2, evaluate $f(a, b)$. The constrained relative maxima or minima of f, if they exist, will be among these values.

In this method, note that λ always multiplies the constraint function $g(x, y)$. Also, the constraint must be written as $g(x, y) = 0$.

The method of Lagrange multipliers has its limitations. It gives only *candidate* points (a, b) at which f may or may not have a constrained maximum or minimum. At a particular point (a, b) obtained in step 2, f may have no constrained extremum. For simplicity, all Lagrange multiplier problems considered in this book have a constrained extremum. A second limitation of the method is that it detects only *relative* extreme values. An absolute extreme value of a function over some subset of this domain will not be detected by the method of Lagrange multipliers unless it is also a relative extreme value. Difficulties of this type are avoided in this book.

EXAMPLE 1   Find the maximum and minimum values of

$$f(x, y) = x^2 + y$$

subject to the constraint

$$x^2 + y^2 = 4.$$

Solution   We first rewrite the constraint as

$$g(x, y) = x^2 + y^2 - 4 = 0.$$

Step 1. Form the new function

$$F(x, y, \lambda) = x^2 + y + \lambda(x^2 + y^2 - 4).$$

Step 2. Solve the system

$$F_x(x, y, \lambda) = 2x + 2\lambda x = 0 \tag{4}$$
$$F_y(x, y, \lambda) = 1 + 2\lambda y = 0 \tag{5}$$
$$F_\lambda(x, y, \lambda) = x^2 + y^2 - 4 = 0. \tag{6}$$

Equation (4) gives us

$$2x(1 + \lambda) = 0$$

so

$$x = 0 \quad \text{or} \quad \lambda = -1. \tag{7}$$

We first consider $x = 0$. Then Equation (6) yields

$$y = \pm 2.$$

Equation (5) can be rewritten as

$$\lambda = -1/(2y). \tag{8}$$

Thus, when $y = 2$, $\lambda = -\frac{1}{4}$, and when $y = -2$, $\lambda = \frac{1}{4}$. Therefore, $(0, 2, -\frac{1}{4})$ and $(0, -2, \frac{1}{4})$ are solutions to the system obtained in step 2. Hence, $(0, 2)$ and $(0, -2)$ are candidates for points at which f has a constrained maximum or minimum value.

We next consider $\lambda = -1$, from (7). Then from (8),

$$y = -\frac{1}{2\lambda} = \frac{1}{2},$$

and from (6),

$$x^2 + \frac{1}{4} - 4 = 0$$

$$x = \pm\frac{\sqrt{15}}{2}.$$

Thus, $(\sqrt{15}/2, 1/2, -1)$ and $(-\sqrt{15}/2, 1/2, -1)$ also are solutions to the system obtained in step 2. Hence, $(\sqrt{15}/2, 1/2)$ and $(-\sqrt{15}/2, 1/2)$ also are candidates for points at which f has a constrained maximum or minimum value.

Step 3. We now tabulate the values of f at these four candidate points.

(x, y)	$(0, 2)$	$(0, -2)$	$\left(\dfrac{\sqrt{15}}{2}, \dfrac{1}{2}\right)$	$\left(-\dfrac{\sqrt{15}}{2}, \dfrac{1}{2}\right)$
$f(x, y) = x^2 + y$	2	-2	$\dfrac{17}{4}$	$\dfrac{17}{4}$

The largest value tabulated for $f(x, y)$ is $\frac{17}{4}$, and the smallest is -2. Hence, subject to the constraint $x^2 + y^2 = 4$, $f(x, y)$ has a maximum value of $17/4$ at the points $(\sqrt{15}/2, 1/2)$ and $(-\sqrt{15}/2, 1/2)$ and a minimum value of -2 at the point $(0, -2)$. The function $f(x, y)$ has neither a constrained maximum nor a constrained minimum value at $(0, 2)$. (Why?) ∎

EXAMPLE 2   Solve the problem posed at the beginning of this section: Minimize

$$C(x, y) = 8x^2 + 8xy$$

subject to the constraint

$$x^2 y = 1024.$$

Solution   We first rewrite the given constraint as

$$g(x, y) = x^2 y - 1024 = 0.$$

Step 1. Form the new function

$$F(x, y, \lambda) = 8x^2 + 8xy + \lambda(x^2 y - 1024).$$

Step 2. Solve the system

$$F_x(x, y, \lambda) = 16x + 8y + 2\lambda xy = 0 \qquad (9)$$
$$F_y(x, y, \lambda) = 8x \qquad\quad + \lambda x^2 \;\; = 0 \qquad (10)$$
$$F_\lambda(x, y, \lambda) = x^2 y - 1024 \qquad = 0. \qquad (11)$$

Equation (10) gives us $x(8 + \lambda x) = 0$ so that

$$x = 0 \quad \text{or} \quad \lambda x = -8.$$

Since x denotes the length of a side of the base, we reject the solution $x = 0$, for it makes no sense physically. We thus take

$$\lambda x = -8. \qquad (12)$$

Substituting -8 for λx in (9) yields

$$16x + 8y - 16y = 0$$
$$16x - 8y = 0$$
$$y = 2x. \qquad (13)$$

Substituting (13) into (11), we have

$$x^2(2x) = 1024,$$
$$2x^3 = 1024,$$

or

$$x = 8.$$

[and (12) gives us $\lambda = -\dfrac{8}{x} = -\dfrac{8}{8} = -1$, but we do not need the value of λ in this problem.] Since $y = 2x$, we obtain the pair

$$x = 8, \; y = 16$$

for the constrained minimum. Thus, the dimensions of the box having minimum cost subject to the volume constraint are 8 meters by 8 meters by 16 meters. The minimum cost is $C(8, 16) = \$1536$. These answers agree with those obtained in Section 4.7, Example 5. ∎

||||||||||||| **Three-Variable Problems**

We can easily extend the method of Lagrange multipliers to maximize or minimize a function $f(x, y, z)$ whose variables satisfy the constraint $g(x, y, z) = 0$. In this case, we form the new function

$$F(x, y, z, \lambda) = f(x, y, z) + \lambda g(x, y, z)$$

where λ is a Lagrange multiplier. We next solve the system

$$F_x(x, y, z, \lambda) = 0$$
$$F_y(x, y, z, \lambda) = 0$$
$$F_z(x, y, z, \lambda) = 0$$
$$F_\lambda(x, y, z, \lambda) = 0.$$

We assume that these partial derivatives exist. Finally, for each solution (a, b, c, λ_0) to this system we evaluate $f(a, b, c)$. The constrained maximum or minimum value of f will be among these values.

EXAMPLE 3   An uncovered rectangular box has a volume of 32 cubic meters. Find the dimensions that will minimize the total surface area.

Figure 1

Solution   We denote the lengths (in meters) of the sides of the box by x, y, and z as shown in Figure 1. Thus, the surface area is given by

$$S(x, y, z) = xy + 2xz + 2yz.$$

Our problem is to minimize S subject to the constraint

$$xyz = 32.$$

We first write the constraint as

$$g(x, y, z) = xyz - 32 = 0.$$

Form the new function

$$F(x, y, z, \lambda) = xy + 2xz + 2yz + \lambda(xyz - 32).$$

Solve the system

$$y + 2z + \lambda yz = 0 \tag{14}$$
$$x + 2z + \lambda xz = 0 \tag{15}$$
$$2x + 2y + \lambda xy = 0 \tag{16}$$
$$xyz - 32 = 0. \tag{17}$$

Equation (14) gives us

$$z\lambda = -\frac{y + 2z}{y}$$

and Equation (15) gives us

$$z\lambda = -\frac{x + 2z}{x}.$$

Equating these two expressions for $z\lambda$, we have

$$\frac{y + 2z}{y} = \frac{x + 2z}{x}$$
$$xy + 2xz = xy + 2yz$$
$$2xz = 2yz. \tag{18}$$

Since z denotes the length of a side, we know that $z \neq 0$. Thus, we may divide both sides of (18) by z. We have

$$x = y. \tag{19}$$

Substituting (19) into (16) gives us

$$2x + 2x + \lambda x^2 = 0$$
$$4x + \lambda x^2 = 0$$
$$x(4 + \lambda x) = 0$$
$$x = 0 \quad \text{or} \quad \lambda x = -4.$$

Since x denotes a length, we reject the solution $x = 0$. We must thus take

$$\lambda x = -4. \tag{20}$$

Substituting (20) into (15), we have

$$x + 2z - 4z = 0$$
$$x - 2z = 0$$
$$z = \frac{x}{2}. \tag{21}$$

Substituting (19) and (21) into (17) gives us

$$\frac{x^3}{2} = 32$$

$$x^3 = 64$$

$$x = 4.$$

Finally, when $x = 4$, Equations (19), (21), and (20) yield

$$y = 4, \quad z = 2, \quad \text{and} \quad \lambda = -1.$$

Hence, the dimensions of the box that will minimize the total surface area subject to the volume constraint are 4 meters by 4 meters by 2 meters. Note that the base is a square and the minimum surface area is 48 cubic meters. ∎

8.4 EXERCISE SET ||||||||||||||||||||||||||||||||||||||||||||||||||||||||||||||||||||||||||||||

In Exercises 1–8, use the method of Lagrange multipliers to find the extreme values.

1. Minimize $f(x, y) = x^2 + y^2$ subject to $x + 2y = 3$.

2. Find the extreme values of $f(x, y) = xy$ subject to $x^2 + y^2 = 32$.

3. Find the extreme values of $f(x, y) = x + y$ subject to $x^2 + y^2 = 72$.

4. Maximize $f(x, y) = xy^2$ subject to $x^2 + y^2 = 1$.

5. Maximize $f(x, y) = x^2 - y^2$ subject to $2x + y - 5 = 0$.

6. Minimize $f(x, y) = x^2 + y^2$ subject to $xy = 1$.

7. Minimize $x^2 + y^2 + z^2$ subject to $x + y + 2z = 1$.

8. Find the extreme values of $x + y + z$ subject to $x^2 + y^2 + z^2 = 1$.

In Exercises 9–16, use the method of Lagrange multipliers.

9. Find three positive numbers whose sum is 24 and whose product is as large as possible.

10. Solve Exercise 5 in Section 4.7.

11. Solve Exercise 6 in Section 4.7.

12. Solve Exercise 21 in Section 4.7.

13. Solve Exercise 20 in Section 4.7.

14. Solve Exercise 18 in Section 8.3.

15. Solve Exercise 20 in Section 8.3.

16. Solve Exercise 22 in Section 8.3.

17. **(Business)** A manufacturer of air conditioners makes x units of model A and y units of model B each week. The total weekly cost of making these units (in dollars) is given by

$$C(x, y) = 50 + 3x^2 + 4y^2.$$

Union contracts and storage considerations require that the total number of air conditioners made each week be exactly 70. How many units of each type should be made to minimize the total weekly manufacturing cost? What is the minimum weekly cost subject to the manufacturing constraint?

18. **(Business)** A solar-heater manufacturer has a plant in San Antonio and a plant in Denver. The firm is committed to the manufacture of 90 heaters per month. If x and y are the monthly levels of

production in San Antonio and Denver, respectively, then the total monthly manufacturing cost (in dollars) is given by

$$C(x, y) = 30x^2 + 20xy + 50y^2.$$

How many heaters should be made at each plant to minimize the total monthly manufacturing cost? What is the minimum constrained monthly manufacturing cost?

19. **(Business)** In order to assemble x compact cars and y standard cars, an automobile assembly plant needs the number of units of labor given by

$$N(x, y) = 3x^2 - 2xy - 10y + 1,000,000.$$

If the level of production is a total of 2000 cars, how many cars of each type should be made to minimize the number of units of labor required?

20. **(Economics)** The production function $Q = 20x^{1/2}y$ gives the number of radios produced by a manufacturer per month with x units of labor and y units of raw materials. A unit of labor costs $12 and a unit of raw materials costs $50. If $12,000 is available monthly for labor and raw materials, how much money should be allocated each month for labor and for raw materials to maximize monthly production?

21. A closed rectangular cargo container is to be built at a cost of $4800 from three different materials. The material for the top costs $9 per square meter, the material for the bottom costs $7 per square meter and the material for the sides costs $4 per square meter. Find the dimensions that will maximize the volume. What is the maximum volume?

22. Find three numbers x, y, and z, whose sum is 24, for which the product xy^2z^3 is maximum.

23. A company purchases x, y, and z units of commodities X, Y, and Z weekly, at unit costs of $3, $5, and $2, respectively. Its satisfaction is measured by the "utility index"

$$u(x, y, z) = \sqrt{xyz}.$$

When spending a total of $1800 weekly on these commodities, how many units of each commodity should the company purchase weekly, in order to maximize its utility index?

24. A factory produces $100z$ units of a commodity daily, which it sells for $24 per unit. To produce $100z$ units of the commodity requires $100x$ units of raw material A and $100y$ units of raw material B, which cost $12 per unit and $11 per unit, respectively. The variables x, y, and z are related by the equation

$$12z = 3x + 4y - \frac{12}{x} - \frac{24}{y} + 86.$$

The factory has additional daily production costs of $3500. Show that the daily profit function for this commodity is

$$P(x, y, z) = 2400z - 1200x - 1100y - 3500$$

and determine the values of x, y, and z which maximize this profit.

8.5 |||||||| TOTAL DIFFERENTIALS AND APPROXIMATIONS

In Section 4.8 we defined the differential of a function $y = f(x)$ by the equation

$$dy = f'(x)\, dx. \tag{1}$$

We found that the differential dy can be used as an approximation to the change in y when x changes by an amount $dx = \Delta x$. More precisely, if x changes from

x_0 to $x_0 + \Delta x$, the quantity

$$\Delta y = f(x_0 + \Delta x) - f(x_0) \tag{2}$$

represents the exact change in y, while the differential

$$dy = f'(x)\, dx \qquad (dx = \Delta x)$$

represents the approximate change in y. Thus,

$$dy \approx \Delta y, \tag{3}$$

and the approximation approaches complete accuracy as $\Delta x \to 0$.

Differentials of Functions of Two Variables

Suppose $z = f(x, y)$ is a function of x and y. We define the **total differential** of f at (x_0, y_0) to be

$$dz = f_x(x_0, y_0)\, dx + f_y(x_0, y_0)\, dy. \tag{4}$$

Observe that dz is a function of four independent variables: x_0, y_0, dx, and dy. We often write (4) in one of the alternate forms

$$dz = \frac{\partial z}{\partial x}\, dx + \frac{\partial z}{\partial y}\, dy, \quad \text{or} \tag{5}$$

$$df = \frac{\partial f}{\partial x}\, dx + \frac{\partial f}{\partial y}\, dy. \tag{6}$$

EXAMPLE 1 For each of the following functions, find the total differential dz.
(a) $z = 3x^2 - \sqrt{y}$
(b) $z = x^3 e^{2y}$.

Solution (a) For $z = 3x^2 - \sqrt{y}$ we have $\dfrac{\partial z}{\partial x} = 6x$ and $\dfrac{\partial z}{\partial y} = -\dfrac{1}{2\sqrt{y}}$. Thus, from Equation (5) we have

$$dz = (6x)\, dx + \left(-\frac{1}{2\sqrt{y}}\right) dy$$

$$= 6x\, dx - \frac{1}{2\sqrt{y}}\, dy.$$

(b) For $z = x^3 e^{2y}$ we have $\dfrac{\partial z}{\partial x} = 3x^2 e^{2y}$ and $\dfrac{\partial z}{\partial y} = 2x^3 e^{2y}$. Thus, from Equation (5),

$$dz = 3x^2 e^{2y}\, dx + 2x^3 e^{2y}\, dy. \qquad \blacksquare$$

EXAMPLE 2 For each of the following functions, find dz at the specified point (x_0, y_0), given that $dx = 0.01$ and $dy = -0.02$.

(a) $z = 3x^2 - \sqrt{y}$, at the point $(x_0, y_0) = (1, 4)$.

(b) $z = x^3 e^{2y}$, at the point $(x_0, y_0) = (2, 0)$.

Solution (a) From Example 1(a), $dz = 6x\,dx - \dfrac{1}{2\sqrt{y}}\,dy$. At the point $(1, 4)$ we have

$$dz = 6 \cdot 1\,dx - \frac{1}{2\sqrt{4}}\,dy$$

$$= 6\,dx - \frac{1}{4}\,dy.$$

When $dx = 0.01$ and $dy = -0.02$ we have

$$dz = 6(0.01) - \tfrac{1}{4}(-0.02) = 0.06 + 0.005$$

$$= 0.065.$$

(b) From Example 1(b), $dz = 3x^2 e^{2y}\,dx + 2x^3 e^{2y}\,dy$. At the point $(2, 0)$ we have

$$dz = 3(4)e^0\,dx + 2(8)e^0\,dy$$

$$= 12\,dx + 16\,dy.$$

When $dx = 0.01$ and $dy = -0.02$ we have

$$dz = 12(0.01) + 16(-0.02) = 0.12 - 0.32$$

$$= -0.20. \qquad\blacksquare$$

|||||||||||| Differentials as Approximations

For functions of two variables, differentials provide an approximation analogous to (3). This is expressed in the following theorem.

THEOREM

Suppose $z = f(x, y)$. If the partial derivatives $\dfrac{\partial z}{\partial x}$ and $\dfrac{\partial z}{\partial y}$ exist and are continuous* in some region surrounding (x_0, y_0), then within this region,

$$dz \approx \Delta z. \tag{7}$$

That is, for all $(x_0 + \Delta x, y_0 + \Delta y)$ within this region,

$$f(x_0 + \Delta x, y_0 + \Delta y) - f(x_0, y_0) \approx f_x(x_0, y_0)\,\Delta x + f_y(x_0, y_0)\,\Delta y. \tag{8}$$

Moreover, this approximation approaches complete accuracy as both $\Delta x \to 0$ and $\Delta y \to 0$.

---

* We do not define continuity in this context.

EXAMPLE 3   Consider the function $z = f(x, y) = 5x^2 + 3xy + y^3$. Estimate the change in z as x changes from 2 to 1.98 and y changes from 1 to 1.03.

Solution   The actual change in z is

$$\Delta z = f(1.98, 1.03) - f(2, 1). \tag{9}$$

To estimate this change we shall use dz. By Equation (4), we have

$$dz = f_x(x_0, y_0)\, dx + f_y(x_0, y_0)\, dy$$
$$= (10x_0 + 3y_0)\, dx + (3x_0 + 3y_0^2)\, dy.$$

In our case we have $x_0 = 2$, $y_0 = 1$, $\Delta x = -0.02$, and $\Delta y = 0.03$. Thus,

$$dz = [10(2) + 3(1)](-0.02) + [3(2) + 3(1^2)](0.03)$$
$$= 23(-0.02) + 9(0.03)$$
$$= -0.19.$$

Therefore, z changes by an amount approximately equal to -0.19.

NOTE   From Equation (9), the actual change in z is

$$\Delta z = 5(1.98)^2 + 3(1.98)(1.03) + (1.03)^3 - 5(2)^2 - 3(2) - 1^3$$
$$= -0.187.$$

When rounded off to two decimal places, dz and Δz agree.   ■

EXAMPLE 4
(Agriculture)   In the agricultural experiment of Example 6, Section 8.2, a field of tomato plants, when fed x gallons of fertilizer A and y gallons of fertilizer B, produced a weight yield of

$$w(x, y) = 180 + x^2 - 3xy + y \text{ pounds.}$$

Find the approximate change in weight yield if the amount of fertilizer A is changed from 6 to 7 gallons and the amount of fertilizer B used is changed from 1 gallon to $\frac{3}{2}$ gallons.

Solution   The total differential of w is

$$dw = (2x - 3y)\, dx + (-3x + 1)\, dy.$$

We want to calculate dw when $x = 6$, $\Delta x = 1$, $y = 1$ and $\Delta y = 0.5$. Then

$$dw = [2(6) - 3(1)](1) + [-3(6) + 1](0.5)$$
$$= 9 + (-17)(0.5)$$
$$= 0.5.$$

Therefore, the weight yield will increase by about one-half pound.   ■

When we apply the theorem of this section to obtain approximations, we often use Equation (8) in the following alternate form:

$$f(x_0 + \Delta x, y_0 + \Delta y) \approx f(x_0, y_0) + f_x(x_0, y_0)\, \Delta x + f_y(x_0, y_0)\, \Delta y. \qquad (10)$$

EXAMPLE 5 Use differentials to approximate $\sqrt{(3.01)^2 + (3.98)^2}$.

Solution We shall use the function $f(x, y) = \sqrt{x^2 + y^2}$. If we let $x_0 = 3$, $\Delta x = 0.01$, $y_0 = 4$, and $\Delta y = -0.02$, then our problem is to find

$$f(3.01, 3.98) = f(x_0 + \Delta x, y_0 + \Delta y). \qquad (11)$$

We shall use Equation (10), but first we must calculate the partial derivatives of f:

$$\frac{\partial f}{\partial x} = \frac{1}{2}(x^2 + y^2)^{-1/2}(2x) = \frac{x}{\sqrt{x^2 + y^2}}, \quad \text{and}$$

$$\frac{\partial f}{\partial y} = \frac{1}{2}(x^2 + y^2)^{-1/2}(2y) = \frac{y}{\sqrt{x^2 + y^2}}.$$

Thus, from (10) we have

$$f(3.01, 3.98) \approx f(3, 4) + f_x(3, 4)(0.01) + f_y(3, 4)(-0.02)$$

$$= \sqrt{3^2 + 4^2} + \frac{3}{\sqrt{3^2 + 4^2}}(0.01) + \frac{4}{\sqrt{3^2 + 4^2}}(-0.02)$$

$$= 5 + \frac{3}{5}(0.01) + \frac{4}{5}(-0.02)$$

$$= 4.99.$$

Therefore, $\sqrt{(3.01)^2 + (3.98)^2} \approx 4.99$.

For comparison, the correct value of $\sqrt{(3.01)^2 + (3.98)^2}$ to five decimal places is 4.99004. ∎

Functions of Three or More Variables

We can define differentials of functions of three or more variables by generalizing Equations (4) − (6). Thus, if $w = f(x, y, z)$, then

$$dw = f_x(x, y, z)\, dx + f_y(x, y, z)\, dy + f_z(x, y, z)\, dz,$$

or

$$dw = \frac{\partial w}{\partial x}\, dx + \frac{\partial w}{\partial y}\, dy + \frac{\partial w}{\partial z}\, dz.$$

EXAMPLE 6   For each of the following functions calculate dw.
(a) $w = x^3y - x^2z^2 + y/z$,
(b) $w = 3uv^2\sqrt{x + y^2}$.

Solution   (a) We first calculate the partial derivatives,

$$\frac{\partial w}{\partial x} = 3x^2y - 2xz^2, \qquad \frac{\partial w}{\partial y} = x^3 + \frac{1}{z}, \quad \text{and} \quad \frac{\partial w}{\partial z} = -2x^2z - \frac{y}{z^2}.$$

Then,

$$dw = \frac{\partial w}{\partial x}\,dx + \frac{\partial w}{\partial y}\,dy + \frac{\partial w}{\partial z}\,dz$$

$$= (3x^2y - 2xz^2)\,dx + \left(x^3 + \frac{1}{2} \right)dy + \left(-2x^2z - \frac{y}{z^2} \right)dz.$$

(b) Calculating the partial derivatives, we have

$$\frac{\partial w}{\partial u} = 3v^2\sqrt{x + y^2}, \qquad \frac{\partial w}{\partial v} = 6uv\sqrt{x + y^2},$$

$$\frac{\partial w}{\partial x} = 3uv^2 \cdot \frac{1}{2}(x + y^2)^{-1/2} = \frac{3uv^2}{2\sqrt{x + y^2}}, \quad \text{and}$$

$$\frac{\partial w}{\partial y} = 3uv^2 \cdot \frac{1}{2}(x + y^2)^{-1/2}(2y) = \frac{3uv^2y}{\sqrt{x + y^2}}.$$

Then,

$$dw = 3v^2\sqrt{x + y^2}\,du + 6uv\sqrt{x + y^2}\,dv + \frac{3uv^2}{2\sqrt{x + y^2}}\,dx + \frac{3uv^2y}{\sqrt{x + y^2}}\,dy. \quad \blacksquare$$

EXAMPLE 7   A rectangular box that has a height of 10 cm, a width of 6 cm, and a height of 4 cm is made of material that is 0.05 cm thick. Estimate the amount of material (by volume) that is present in the box.

Solution   The volume of a rectangular box of length l, width w, and height h is

$$V = lwh.$$

Taking $l = 10$, $w = 6$, and $h = 4$, we want to find Δv when $\Delta l = \Delta w = \Delta h = 0.05$. Using dV to approximate ΔV we have

$$dV = wh\,dl + lh\,dw + lw\,dh$$
$$= 6 \cdot 4(0.05) + 10 \cdot 4(0.05) + 10 \cdot 6(0.05)$$
$$= 6.2.$$

Thus, the box contains approximately 6.2 cm^3 of material. $\blacksquare$

EXAMPLE 8
(Error
Estimation)

A rectangular tropical fish aquarium is advertised as having length 45 cm, width 20 cm, and height 30 cm, with a volume of 27,000 cm³. If the length and width measurements each have a maximum error of 0.25 cm, and the height measurement has a maximum error of 0.50 cm, estimate the maximum possible error in the advertised volume of the tank.

Solution

As in Example 7, we have

$$V = lwh \tag{12}$$

and

$$dV = wh\,dl + lh\,dw + lw\,dh.$$

The given error estimates tell us that $|dl| \le 0.25$, $|dw| \le 0.25$, and $|dh| \le 0.50$. Thus, when we take $l = 45$, $w = 20$, and $h = 30$, the maximum error in the value of V computed using (12) is

$$|dV| \le (20)(30)(0.25) + (45)(30)(0.25) + (45)(20)(0.50)$$
$$= 150 + 337.5 + 450$$
$$= 937.5 \text{ cm}^3.$$

Thus, the maximum error in the advertised volume of 27,000 cm³ is approximately 937.5 cm³. ∎

8.5 EXERCISE SET ||||||||||||||||||||||||||||||||||||||||||||||||||||||||||||||||||||||||||||||

In Exercises 1–10 find the total differential df.

1. $f(x, y) = 5x^3y + 7x - 8y^2$

2. $f(x, y) = 4xy^5 - 5x^3 + \sqrt{y}$

3. $f(x, y) = \sqrt{x^2 + y^3}$

4. $f(x, y) = 4x\sqrt{y}$

5. $f(x, y) = x^2e^{3y}$

6. $f(x, y) = e^{3x-4y}$

7. $f(x, y) = \dfrac{x + y}{x - y}$

8. $f(x, y) = \dfrac{3x + 5y}{x^2 - y^2}$

9. $f(x, y) = \ln(4x^2y^5)$

10. $f(x, y) = x^6 \ln y$

In Exercises 11–14, find the total differential of the given function at the given point, with the given values of dx and dy.

11. The function f of Exercise 1, at $(-1, 2)$, with $dx = 0.02$ and $dy = 0.01$.

12. The function f of Exercise 2, at $(1, 1)$, with $dx = 0.03$ and $dy = 0.02$.

13. The function f of Exercise 3, at $(-1, 2)$, with $dx = 0.06$ and $dy = -0.04$.

14. The function f of Exercise 4, at $(-4, 9)$, with $dx = 0.01$ and $dy = 0.12$.

15. For the function $f(x, y) = x^3 - x^2y^2 + 11$, estimate the change in $f(x, y)$ as x changes from 1 to 1.02 and y changes from 3 to 2.97.

16. For the function $f(x, y) = \dfrac{x^2 + 7}{y - 3}$, estimate the change in $f(x, y)$ as x changes from 5 to 4.9 and y changes from 7 to 7.15.

17. For the function $f(x, y)$ of Exercise 15, estimate $f(1.02, 2.97)$.

18. For the function $f(x, y)$ of Exercise 16, estimate $f(4.90, 7.15)$.

19. For the function $f(x, y) = \sqrt[3]{xy^4}$, estimate $f(7.940, 1.030)$.

20. For the function $f(x, y) = 8xe^y$, estimate $f(2.95, 0.02)$.

21. **(Production)** A manufacturer of cameras produces
$$Q(x, y) = 60x^{2/3}y^{1/3}$$
units of cameras when using x units of labor and y units of capital (see Exercise 35 of Section 8.2). Suppose the manufacturer increases its utilization of labor from 8 to $\frac{35}{4}$ units, and increases its utilization of capital from 27 to 30 units. Estimate the resulting change in the production function $Q(x, y)$.

22. **(Pollution)** In Exercise 36 of Section 8.2, when the annual production levels (in millions of tons) at two steel factories are x and y, respectively, the total amount of particulate matter in the atmosphere over a nearby town is
$$A(x, y) = 2x^3 + 3xy + y^2$$
(in thousands of pounds). In that exercise the two factories had annual production levels of 5 and 8 million tons, respectively. Estimate the change in $A(x, y)$ if the first factory increases its annual production level to 5.4 million tons, while the second factory reduces its production level to 7.2 million tons.

23. **(Medicine)** The formula
$$A = 2.024 \, W^{0.425} H^{0.725},$$
expressing the human body surface area (in square meters) as a function of its weight W (in kilograms) and height H (in meters), was introduced in Exercise 38 of Section 8.2. Estimate the change in body surface area of a youth who gains 2 kg in weight and 3 cm in height, from an initial weight of 60 kg and height of 1.75 m. Use the decimal approximations
$$60^{0.425} \approx 5.698 \qquad 60^{-0.575} \approx 0.095$$
$$1.75^{0.725} \approx 1.500 \qquad 1.75^{-0.275} \approx 0.857.$$

24. The volume of a cylindrical can is given by the formula
$$v = \pi r^2 h.$$
Estimate the amount of material (by volume) in a sheet metal can that has a radius of 2 inches, a height of 6 inches, and a uniform thickness of 0.02 inches.

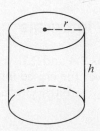

In Exercises 25–34, find the differential df.

25. $f(x, y, z) = 4xy^3 - 10yz + 8x^2z^3$.

26. $f(x, y, z) = \dfrac{5x}{y} + \dfrac{4y}{z^2}$.

27. $f(x, y, z) = \dfrac{x^2z^4}{1 + y^3}$.

28. $f(x, y, z) = x^3\sqrt{y^2 - 3z}$.

29. $f(x, y, z) = \dfrac{5x^2 - 3y}{\sqrt{z}}$.

30. $f(x, y, z) = xy^2e^{yz}$.

31. $f(u, v, x, y) = (x^2 + y^3)e^{u^2v}$.

32. $f(u, v, x, y) = (5x^2u)^8 - 7yv^3$.

33. $f(u, v, w, x, y) = \dfrac{2u + v - 3w}{x - y}$.

34. $f(u, v, w, x, y) = (u + v^2)e^{wx} \ln y$.

35. Estimate the change in the function

$$f(x, y, z) = \frac{x\sqrt{y}}{z}$$

as x changes from 3 to 3.01, y changes from 4 to 3.97, and z changes from 5 to 5.02.

36. For the function $f(x, y, z) = x^2 e^y z$, estimate the change in f as (x, y, z) changes from $(3, 0, 2)$ to $(2.97, 0.04, 1.98)$.

37. Use differentials to approximate $(1.03)^{10}\sqrt{(3.05)^2 + (3.96)^2}$.

38. Use differentials to approximate $\dfrac{(2.02)^3(2.97)^2}{\sqrt{8.94}}$.

39. **(Error Estimation)** The dimensions of a closed rectangular box are measured to be 2 feet by 3 feet by 1.5 feet, with a maximum error of $\frac{1}{4}$ inch. Use differentials to estimate the maximum error in using these measurements to calculate
    (a) the surface area of the box;
    (b) the volume of the box.

40. **(Error Estimation)** Approximate the amount of error that could occur in calculating $f(x, y, z) = x^2 y^3 z^4$ if the input variable x is measured to be 2.50 with an error no greater than 0.03, y is measured to be 3.00 with an error no greater than 0.15, and z is measured to be 2.00 with an error no greater than 0.10.

8.6 ‖‖‖‖ Double Integrals

In Sections 7.4 and 7.7 we defined the definite integral of a function f that is continuous over the closed interval $[a, b]$ as follows. Choose a positive integer n and select $n + 1$ arbitrary points

$$x_0, x_1, \ldots, x_{n-1}, x_n$$

in $[a, b]$ with $x_0 = a$ and $x_n = b$, so that $[a, b]$ is divided into n subintervals of equal length. Thus, the length of the ith subinterval $[x_{i-1}, x_i]$ is $\Delta x = (b - a)/n$. (See Figure 2 in Section 7.7.) In each of these subintervals we now select an arbitrary point $\tilde{x}_i$ and form the Riemann sum

$$S_n = f(\tilde{x}_1)\Delta x + f(\tilde{x}_2)\Delta x + \cdots + f(\tilde{x}_n)\Delta x. \tag{1}$$

If $f(x) \geq 0$ over $[a, b]$, then as shown in Figure 3 of Section 7.7, S_n is the sum of areas of rectangles approximating the area under the curve $y = f(x)$ over the interval $[a, b]$. If the Riemann sums in (1) approach a limiting value as $n \to +\infty$ and $\Delta x \to 0$, then this value is defined as $\int_a^b f(x)\,dx$, the definite integral of f over $[a, b]$. Recall that when $f(x)$ is *nonnegative* on $[a, b]$, then the definite integral $\int_a^b f(x)\,dx$ is the area under the curve $y = f(x)$ over the interval $[a, b]$.

‖‖‖‖‖‖‖‖ Double Integrals over Rectangles

Suppose now that f is a function of the variables x and y which is defined over the rectangle R consisting of all points (x, y) such that

$$a \leq x \leq b \qquad \text{and} \qquad c \leq y \leq d.$$

See Figure 1a.

Figure 1

(a)

(b)

We first draw lines parallel to the x- and y-axes, thereby partitioning the region R into n subrectangles $R_1, R_2, \ldots, R_n$. See Figure 1b. Now choose an arbitrary point $(\tilde{x}_i, \tilde{y}_i)$ in each subrectangle R_i as shown in Figure 1b. The sum

$$S_n = f(\tilde{x}_1, \tilde{y}_1)\,\Delta x \Delta y + f(\tilde{x}_2, \tilde{y}_2)\,\Delta x \Delta y + \cdots + f(\tilde{x}_n, \tilde{y}_n)\,\Delta x \Delta y \tag{2}$$

is called a **Riemann sum for f over the rectangle R.** If the Riemann sums in (2) approach a limit I as $n \to +\infty$ and as Δx and Δy both approach zero, then I is defined as the **double integral of f over the rectangle R** and we write

$$I = \iint\limits_R f(x, y)\,dx\,dy. \tag{3}$$

If $f(x, y)$ is *nonnegative* over the region R, then the double integral term gives the volume of the solid below the surface $z = f(x, y)$ over the region R in the xy-plane. This can be seen by observing, as in Figure 2, that each $f(\tilde{x}_i, \tilde{y}_i)\,\Delta x \Delta y$ in Equation 2 is the volume of a rectangular parallelepiped whose base R_i has area $\Delta x \Delta y$ and whose height is $f(\tilde{x}_i, \tilde{y}_i)$, so that the Reimann sum is (2) gives an

Figure 2

approximation to the volume of the solid below the surface $z = f(x, y)$ and over the region R in the xy-plane.

If $f(x, y)$ assumes both positive and negative values over the region R, then the double integral $\iint_R f(x, y)\, dx\, dy$ is interpreted as the *net volume* between the surface $z = f(x, y)$ and the xy-plane, counting that portion of the solid which lies above the xy-plane as having positive volume, and that portion which lies below the xy-plane as having negative volume.

In more advanced courses, it is shown that if the function f is *continuous* over the rectangle R, then the Riemann sums in (2) do approach a limit as $n \to +\infty$ and Δx and Δy both approach zero. That is, the double integral of a function $z = f(x, y)$ exists on any rectangle over which f is continuous.

‖‖‖‖‖‖‖‖‖‖‖‖ Double Integrals over More General Regions

The assumption that the region R is a rectangle is unnecessarily restrictive. Suppose that g_1 and g_2 are functions that are continuous and that $g_1(x) \le g_2(x)$ for all x in $[a, b]$. Then we may consider the region R, shown in Figure 3a, as consisting of all points (x, y) such that

$$a \le x \le b \qquad \text{and} \qquad g_1(x) \le y \le g_2(x).$$

Figure 3

Type I region

(a)

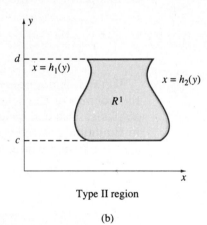

Type II region

(b)

Such a region is called a **Type I region.** Similarly, let h_1 and h_2 be functions that are continuous and $h_1(y) \le h_2(y)$ for all y in $[c, d]$. We can then define the region R, shown in Figure 3b, consisting of all points (x, y) such that

$$c \le y \le d \qquad \text{and} \qquad h_1(y) \le h_2(y).$$

Such a region is called a **Type II region.**

In the case of a general region R of either Type I or Type II, we partition R by means of lines parallel to the coordinate axes into rectangles, each having area $\Delta x \Delta y$. Some of these rectangles are entirely contained in R, while others are only partly contained in R. See Figure 4. In each rectangle R_i that is entirely

Figure 4

contained in R we choose an arbitrary point $(\tilde{x}_i, \tilde{y}_i)$ and form the Riemann sum

$$f(\tilde{x}_1, \tilde{y}_1)\,\Delta x\,\Delta y + f(\tilde{x}_2, \tilde{y}_2)\,\Delta x\,\Delta y + \cdots + f(\tilde{x}_n, \tilde{y}_n)\,\Delta x\,\Delta y. \tag{4}$$

An important result in calculus, proved in more advanced courses, is that if f is continuous over a region R of Type I or Type II, then the limit of the Riemann sums (4) as $n \to +\infty$ and Δx and Δy approach zero, exists and is independent of the choice of points $(\tilde{x}_i, \tilde{y}_i)$. This limit is called the **double integral of f over the region R,** and is denoted by

$$\iint\limits_{R} f(x, y)\,dx\,dy. \tag{5}$$

The double integral (5) may be interpreted geometrically as the net volume between the surface $z = f(x, y)$ and the xy-plane, counting the portion of the solid lying above the xy-plane as having positive volume, and the portion lying below the xy-plane as having negative volume.

Iterated Integrals

The evaluation of a double integral as a limit of Riemann sums is too burdensome a task, except for very simple functions. Thus, we now turn to more practical ways of evaluating double integrals.

Let f be a continuous function of the variables x and y, and let g_1 and g_2 be functions of x which are continuous over the closed interval $[a, b]$. Then an expression of the form

$$\int_a^b \left(\int_{g_1(x)}^{g_2(x)} f(x, y)\,dy \right) dx \tag{6}$$

is called an **iterated integral.**

To evaluate the expression in (6) we first evaluate the inner integral $\int_{g_1(x)}^{g_2(x)} f(x, y)\,dy$ by considering x as a constant, so that $f(x, y)$ is a function only of y. Thus, we need to find an antiderivative of f, which we evaluate between the limits $y = g_1(x)$ and $y = g_2(x)$. The result is a function of x only, which we now integrate between the limits $x = a$ and $x = b$.

Example 1   Evaluate the iterated integral

$$\int_2^4 \left(\int_1^6 (2x + y)\, dy \right) dx.$$

Solution   In this case, $g_1(x) = 1$, $g_2(x) = 6$, $a = 2$, and $b = 4$. We first evaluate the inner integral

$$\int_1^6 (2x + y)\, dy$$

by regarding x as a constant. Thus,

$$\int_1^6 (2x + y)\, dy = \left[2xy + \frac{y^2}{2} \right]_{y=1}^{y=6}$$

$$= \left(2x(6) + \frac{6^2}{2} \right) - \left(2x(1) + \frac{1^2}{2} \right)$$

$$= 10x + \frac{35}{2}.$$

We now evaluate the outer integral

$$\int_2^4 \left(10x + \frac{35}{2} \right) dx = \left[5x^2 + \frac{35}{2} x \right]_{x=2}^{x=4}$$

$$= 5(4^2) + \frac{35}{2}(4) - \left[5(2^2) + \frac{35}{2}(2) \right]$$

$$= 95.$$

Thus

$$\int_2^4 \left(\int_1^6 (2x + y)\, dy \right) dx = 95. \qquad \blacksquare$$

Example 2   Evaluate the iterated integral

$$\int_0^2 \left(\int_{x^2}^{2x} (x^2 + 2y)\, dy \right) dx.$$

Solution   In this case, $g_1(x) = x^2$, $g_2(x) = 2x$, $a = 0$, and $b = 2$. We first evaluate the inner integral

$$\int_{x^2}^{2x} (x^2 + 2y)\, dy$$

by regarding x as a constant. Thus,

$$\int_{x^2}^{2x} (x^2 + 2y)\, dy = \left[x^2 y + y^2 \right]_{y=x^2}^{y=2x}$$

$$= [x^2(2x) + (2x)^2] - [x^2(x^2) + (x^2)^2]$$

$$= 2x^3 + 4x^2 - 2x^4.$$

Now we evaluate the outer integral

$$\int_0^2 (2x^3 + 4x^2 - 2x^4)\, dx = \left[\frac{2}{4} x^4 + \frac{4}{3} x^3 - \frac{2}{5} x^5 \right]_{x=0}^{x=2}$$

$$= \frac{1}{2} 2^4 + \frac{4}{3} 2^3 - \frac{2}{5} 2^5 - 0 = \frac{88}{15}.$$

Thus,

$$\int_0^2 \left(\int_{x^2}^{2x} (x^2 + 2y)\, dy \right) dx = \frac{88}{15}. \qquad \blacksquare$$

So far we have considered only iterated integrals in which the inner integration was performed with respect to y, and the outer integration with respect to x. If we "switch the order of integration," another type of iterated integral is obtained. Suppose that f is a continuous function of the variables x and y, and h_1 and h_2 are functions of y which are continuous over the closed interval $[c, d]$. Then an expression of the form

$$\int_c^d \left(\int_{h_1(y)}^{h_2(y)} f(x, y)\, dx \right) dy \qquad (7)$$

is also called an **iterated integral**.

To evaluate the expression in (7) we first evaluate the inner integral $\int_{h_1(y)}^{h_2(y)} f(x, y)\, dx$ by considering y as a constant. The resulting function of y only is then integrated between $y = c$ and $y = d$.

Example 3 Evaluate the iterated integral

$$\int_1^6 \left(\int_2^4 (2x + y)\, dx \right) dy.$$

Solution This iterated integral is of the form given in (7) with $h_1(y) = 2$, $h_2(y) = 4$, $c = 1$, and $d = 6$. Evaluating the inner integral first by treating y as a constant yields

$$\int_2^4 (2x + y)\, dx = [x^2 + xy]_{x=2}^{x=4}$$

$$= (4^2 + 4y) - (2^2 + 2y)$$

$$= 12 + 2y.$$

Now we evaluate the outer integral,

$$\int_1^6 (12 + 2y)\, dy = [12y + y^2]_{y=1}^{y=6}$$
$$= [12(6) + 6^2] - (12(1) + 1^2)$$
$$= 95.$$

Examples 1 and 3 illustrate an important principle. In both examples, the integration may be described as taking place over a rectangle given by the inequalities

$$2 \le x \le 4 \qquad \text{and} \qquad 1 \le y \le 6. \qquad \text{(See Figure 5.)}$$

Figure 5

In Example 1, the inequality $2 \le x \le 4$ governs the outer integration while the inequality $1 \le y \le 6$ governs the inner integration. In Example 3, the roles played by the inequalities are interchanged. We say that we have "interchanged the order of integration." The answer is the same in both examples.

Examples 1 and 3 illustrate how iterated integrals may be used to evaluate a double integral. In Example 1 we viewed the rectangle R as a Type I region, whereas in Example 3 we viewed it as a Type II region.

The following theorem states the underlying principle.

Theorem

(a) If f is a function of x and y which is continuous over a region R of Type I, then

$$\iint_R f(x, y)\, dy\, dx = \int_a^b \left(\int_{g_1(x)}^{g_2(x)} dy \right) dx.$$

(b) If f is a function of x and y which is continuous over a region R of Type II, then

$$\iint_R f(x, y)\, dy\, dx = \int_c^d \left(\int_{h_1(y)}^{h_2(y)} dx \right) dy.$$

Note: If R is a rectangular region in the xy-plane, then R is both of Type I and Type II so the double integral $\iint\limits_R f(x, y) \, dx \, dy$ can be evaluated as in either part (a) or (b).

Example 4   Determine the volume of the solid below the surface $z = 2x + y$ and over the rectangular region R consisting of all points (x, y) in the xy-plane satisfying

$$2 \le x \le 4 \quad \text{and} \quad 1 \le y \le 6. \quad \text{(See Figure 6.)}$$

Figure 6

Solution   Since $f(x, y) = 2x + y \ge 0$ over R, the volume of the solid is given by

$$\iint\limits_R (2x + y) \, dx \, dy.$$

The answer is either of the iterated integrals evaluated in Examples 1 and 3. Thus, the volume of the solid is 95.   ■

Example 5   Compute the volume of the solid lying under the surface $z = 4 - x - y$ and over the region R in the first qudrant bounded by the lines $y = 4 - x$, $y = 0$, and $x = 0$. This is a tetrahedron.

Solution   We first sketch the region R in Figure 7(a) and the solid tetrahedron in Figure 7(b). The volume of the solid is $\iint\limits_R (4 - x - y) \, dx \, dy$. This double integral is now evaluated as the interated integral

$$\int_0^4 \left(\int_0^{4-x} (4 - x - y) \, dy \right) dx.$$

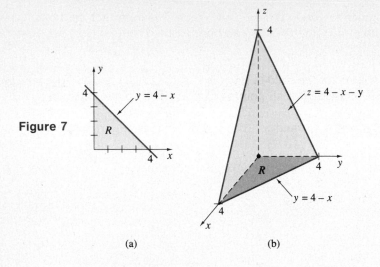

Figure 7

First, the inner integral is

$$\int_0^{4-x} (4 - x - y)\, dy = \left[4y - xy - \frac{y^2}{2} \right]_0^{4-x}$$

$$= 4(4 - x) - x(4 - x) - \frac{1}{2}(4 - x)^2$$

$$= 16 - 4x - 4x + x^2 - 8 + 4x - \frac{1}{2}x^2$$

$$= 8 - 4x + \frac{x^2}{2}.$$

Then the outer integral is

$$\int_0^4 \left(8 - 4x + \frac{x^2}{2} \right) dx = \left[8x - 2x^2 + \frac{x^3}{6} \right]_0^4$$

$$= 32 - 32 + \frac{4^3}{6} = \frac{64}{6}.$$

Hence, the volume of the tetrahedron is $\frac{64}{6}$. ∎

|||||||||||| **Average Value of a Function of Two Variables**

In Section 7.4 we defined the average value of a function f that is continuous over the interval $[a, b]$ as

$$y_{\text{ave.}} = \frac{1}{b - a} \int_a^b f(x)\, dx.$$

We now wish to extend this idea to a function f of two variables x and y that is continuous over a region R in the xy-plane. We define the **average value of f over R** by

$$y_{\text{ave.}} = \frac{1}{\text{area of } R} \iint\limits_{R} f(x, y)\, dx\, dy. \tag{8}$$

Example 6   Compute the average value of

$$f(x) = xe^y$$

over the region R bounded by $y = x^2$ and $y = x$.

Solution   We first sketch the region R in Figure 8. We compute the area of R by proceeding as in Section 7.2. We find the x-coordinates of the points of intersection of the boundary curves by equating their y-coordinates:

$$x^2 = x$$
$$x(x - 1) = 0$$
$$x = 0 \quad \text{or} \quad x = 1.$$

Figure 8

Using Equation (8) of Section 7.2 with

$$a = 0, \ b = 1, \ f(x) = x, \qquad \text{and} \qquad g(x) = x^2$$

we have

$$\text{Area of region } R = \int_0^1 (x - x^2)\, dx$$

$$= \left[\frac{x^2}{2} - \frac{x^3}{3} \right]_0^1$$

$$= \left(\frac{1}{2} - \frac{1}{3} \right) - 0 = \frac{1}{6}.$$

Next, we compute the double integral in Equation (8) as the iterated integral

$$\int_0^1 \left(\int_{x^2}^x xe^y \, dy \right) dx.$$

First,

$$\int_{x^2}^x xe^y \, dy = [xe^y]_{y=x^2}^x \qquad \text{(Remember that } x \text{ is constant.)}$$
$$= xe^x - xe^{x^2}.$$

Now

$$\int_0^1 (xe^x - xe^{x^2}) \, dx = \int_0^1 xe^x \, dx - \int_0^1 xe^{x^2} \, dx. \tag{9}$$

The first integral on the right side of Equation (9) is evaluated by the method of integration by parts (Section 6.3). In fact, the corresponding indefinite integral was evaluated in Example 2 of Section 6.3. We have

$$\int_0^1 xe^x \, dx = [xe^x - e^x]_0^1$$
$$= (e^1 - e^1) - (0e^0 - e^0)$$
$$= 1. \tag{10}$$

The second integral in the right side of Equation (9) is evaluated by the method of substitution (Section 6.2); Let $u = x^2$, so $du = 2x \, dx$ or $x \, dx = \frac{1}{2} \, du$. Then

$$\int xe^{x^2} \, dx = \frac{1}{2} \int e^u \, du = \frac{1}{2} e^u = \frac{1}{2} e^{x^2}$$

so that

$$\int_0^1 xe^{x^2} = \frac{1}{2} [e^{x^2}]_0^1$$
$$= \frac{1}{2} [e^1 - e^0] = \frac{1}{2} e - \frac{1}{2}. \tag{11}$$

Substituting the results of (10) and (11) into Equation (9), we find that the double integral over R is

$$\iint_R xe^y \, dx \, dy = 1 - \left(\frac{1}{2} e - \frac{1}{2} \right)$$
$$= \frac{3}{2} - \frac{1}{2} e.$$

Hence, the average value of f over R is

$$y_{\text{ave.}} = \frac{1}{\frac{1}{6}} \left(\frac{3}{2} - \frac{1}{2} e \right) = 6 \left(\frac{3}{2} - \frac{1}{2} e \right) = 9 - 3e$$
$$\approx 0.845. \qquad \blacksquare$$

8.6 EXERCISE SET ||||||||||||||||||||||||||||||||||||||||||||||||||||||||||||||||||||||||||||

In Exercises 1–12, evaluate the given iterated integral.

1. $\int_2^3 \left(\int_{-1}^2 (2xy + xy^2) \, dy \right) dx$

2. $\int_{-2}^5 \left(\int_2^3 (3x^2 + y^4) \, dy \right) dx$

3. $\int_0^1 \left(\int_0^1 e^{x-y} \, dx \right) dy$

4. $\int_0^1 \left(\int_1^2 ye^x \, dx \right) dy$

5. $\int_1^e \left(\int_1^2 \frac{x}{y} \, dy \right) dx$

6. $\int_1^2 \left(\int_1^{e^2} \frac{x}{y} \, dy \right) dx$

7. $\int_1^2 \left(\int_{x+1}^{-x+3} (x - y) \, dy \right) dx$

8. $\int_1^2 \left(\int_{y-2}^{y+2} xy \, dx \right) dy$

9. $\int_1^3 \left(\int_0^{\ln y} 2y \, dx \right) dy$

10. $\int_0^1 \left(\int_{x^2}^2 xe^y \, dy \right) dx$

11. $\int_{-1}^1 \left(\int_0^{\sqrt{1-y^2}} x \, dx \right) dy$

12. $\int_1^2 \left(\int_1^x \frac{y}{x} \, dy \right) dx$

In Exercises 13–20, evaluate the double integral of f over the region R.

13. $f(x, y) = x + 2y$ R is the rectangle defined by $0 \le x \le 2$ and $1 \le y \le 3$.

14. $f(x, y) = 2xy$ R is the rectangle bounded by the lines $x = 1$, $x = 5$, $y = 2$, and $y = 4$.

15. $f(x, y) = x$ R is the triangle with vertices $(1, 2)$, $(4, -1)$, and $(4, 5)$.

16. $f(x, y) = 2x + y$ R is the triangle with vertices, $(0, 1)$, $(0, 3)$, and $(1, 1)$.

17. $f(x, y) = x$ R is the triangle bounded by $y = 2x$, $y = -x$, and $x = 1$.

18. $f(x, y) = x^2$ R is the region bounded by $y = x^2$ and $y = \sqrt{x}$.

19. $f(x, y) = x^2 + y$ R is the region in the first quadrant bounded by $y = x^3$ and $y = x$.

20. $f(x, y) = e^{x-y}$ R is the triangle bounded by the lines $x = 0$, $y = 0$ and $y = 1 - x$.

In Exercises 21–26, compute the volume of the solid lying under the surface $z = f(x, y)$ and above the region R.

21. $z = 2x + 5y$ R is the rectangle defined by $2 \le x \le 4$, $1 \le y \le 3$.

22. $z = 2x + 3y$ R is the rectangle whose vertices are $(1, 2)$, $(3, 2)$, $(1, 5)$, and $(3, 5)$.

23. $z = 10 - x - y$ R is the rectangle bounded by the lines $x = 1$, $x = 4$, $y = 2$, and $y = 6$.

24. $z = 4 - x^2 - y^2$ R is the triangle bounded by the lines $y = x$, $y = 2 - x$, and the y-axis.

25. $z = xy$ *R* is the region bounded by the line $y = 3x$, the curve $y = 4 - x^2$, and the *y*-axis.

26. $z = x^2 + 2y$ *R* is the region bounded by the curves $y = 4 - x^2$ and $y = 2$.

27. Use a double integral to compute the volume of the rectangular parallelepiped whose height is 4 and whose base is the rectangle bounded by $x = 1$, $x = 5$, $y = 2$, and $y = 4$.

28. The double integral $\iint\limits_{R} dy\, dx$ gives the area of the region *R* in the *xy*-plane. (Why?) Compute the area of the region *R* bounded by the curves $y = x^2 + 1$ and $y = x + 3$.

In Exercises 29–34, compute the average value of *f* over the region *R*.

29. $f(x, y) = x^2 y$ *R* is the rectangle with vertices $(0, 0)$, $(4, 0)$, $(0, 2)$ and $(4, 2)$

30. $f(x, y) = x^2 + y$ *R* is the region in the first quadrant bounded by $y = 4 - x^2$, $y = 3$, and the *y*-axis.

31. $f(x, y) = e^x$ *R* is the triangle in the first quadrant bounded by the line $y = 1 - x$, the *x*-axis, and the *y*-axis.

32. $f(x, y) = e^{x-y}$ *R* is the triangle with vertices $(0, 0)$, $(4, 4)$ and $(8, 0)$.

33. $f(x, y) = x^2 - 2y$ *R* is the region bounded by the curves $y = 4 - x^2$ and $y = 3x^2$.

34. $f(x, y) = x^3 + y^2$ *R* is the rectangle bounded by $x = 1$, $x = 3$, $y = 2$, and $y = 5$.

35. (Production). Suppose that a manufacturer of a certain chemical product finds out that the number of gallons (in thousands) produced weekly when *x* units of labor and *y* units of capital are used is given by

$$f(x, y) = x^{2/3} \, y^{1/3}.$$

Determine the average number of units produced if $20 \le x \le 50$ and $30 \le y \le 80$.

36. (Pollution). Suppose that the pollution (in grams of mercury per liter) in a rectangular lake bounded by the lines $x = 0$, $x = 50$, $y = 0$, and $y = 100$ is given by $z = 2x^2 + y$. Determine the average pollution in the lake.

37. (Average Height). Suppose that the floor of a tent being set up for an antique show is a rectangle measuring 100 feet by 50 feet. If the floor of the tent is in the first quadrant of the *xy*-plane with one corner at the origin, and the longer side along the *y*-axis, then the roof of the tent is a plane whose equation is $z = 20 - \dfrac{x}{10} - \dfrac{y}{25}$. Determine the average height of the tent.

KEY IDEAS FOR REVIEW ||||||||||||||||||||||||||||||||||||||||||||||||||||||||||||||||||||||||||||||||||

■ A function of the independent variables *x* and *y* is a rule, or formula, that determines exactly one value of the dependent variable *z* for each ordered pair of values (x, y) for which the rule is defined. We denote this value of *z* by $f(x, y)$ and write

$$z = f(x, y).$$

- A three-dimensional Cartesian coordinate system sets up a one-to-one correspondence between points in three-dimensional space and ordered triples of real numbers.

- The graph of a function f of x and y, called a surface, is the set of all points (x, y, z) in three-dimensional space for which $z = f(x, y)$.

- The graph of a first degree equation in x, y, and z is a plane.

- If $z = f(x, y)$, then the partial derivative of f with respect to x—denoted by f_x, $\partial f/\partial x$, or $\partial z/\partial x$—is obtained by taking the ordinary derivative of f with respect to x, while the variable y remains constant. Similarly, the partial derivative of f with respect to y—denoted by f_y, $\partial f/\partial y$, or $\partial z/\partial y$—is obtained by taking the ordinary derivative of f with respect to y, while the variable x remains constant.

- If $z = f(x, y)$, then the second-order partial derivatives are defined by

$$\frac{\partial}{\partial x}\left(\frac{\partial z}{\partial x}\right) = \frac{\partial^2 z}{\partial x^2} \quad \text{or} \quad f_{xx}$$

$$\frac{\partial}{\partial y}\left(\frac{\partial z}{\partial y}\right) = \frac{\partial^2 z}{\partial y^2} \quad \text{or} \quad f_{yy}$$

$$\frac{\partial}{\partial x}\left(\frac{\partial z}{\partial y}\right) = \frac{\partial^2 z}{\partial x\,\partial y} \quad \text{or} \quad f_{yx}$$

$$\frac{\partial}{\partial y}\left(\frac{\partial z}{\partial x}\right) = \frac{\partial^2 z}{\partial y\,\partial x} \quad \text{or} \quad f_{xy}$$

(the mixed partials f_{xy} and f_{yx} are equal whenever f_x, f_y, f_{xy}, and f_{yx} are continuous.)

- The notion of partial derivative extends to functions of three or more variables.

- $f(x, y)$ has a relative maximum (or relative minimum) at (a, b) if there is a circular disk in the xy-plane centered at (a, b), throughout which f is defined, such that for all points (x, y) in this disk $f(x, y) \le f(a, b)$ (or $f(a, b) \le f(x, y)$).

- (a, b) is a critical point of f if (a, b) is the center of a circular disk throughout which f is defined and either $f_x(a, b) = 0$ and $f_y(a, b) = 0$, or $f_x(a, b)$ and $f_y(a, b)$ do not both exist.

- If f has a relative extremum at (a, b), then (a, b) is a critical point of f. Critical points are *candidates* for points where f has relative extrema.

- If (a, b) is a critical point of f, then f may have a relative maximum, a relative minimum, or no relative extremum at (a, b). A critical point of f at which f does not have a relative extremum is called a saddle point.

- Second partials test for relative extrema: Let (a, b) be a critical point of f and let

$$M = f_{xx}(a, b) \cdot f_{yy}(a, b) - [f_{xy}(a, b)]^2.$$

  (a) If $M > 0$ and $f_{xx}(a, b) < 0$, then f has a relative maximum at (a, b).
  (b) If $M > 0$ and $f_{xx}(a, b) > 0$, then f has a relative minimum at (a, b).
  (c) If $M < 0$, then f has no relative extremum at (a, b).
  (d) If $M = 0$, then f may or may not have a relative extremum at (a, b).

- The method of Lagrange multipliers: To maximize or minimize $f(x, y)$ subject to the constraint $g(x, y) = 0$, proceed as follows:

  Step 1. Form $F(x, y, \lambda) = f(x, y) + \lambda g(x, y)$.
  Step 2. Solve the system

$$F_x(x, y, \lambda) = 0$$
$$F_y(x, y, \lambda) = 0$$
$$F_\lambda(x, y, \lambda) = 0.$$

  Step 3. For each solution (a, b, λ_0) to the system obtained in step 2, evaluate $f(a, b)$. The constrained maximum or minimum value of f, if it exists, will be among these values.

- The method of Lagrange multipliers extends to the problem of maximizing or minimizing a function $f(x, y, z)$ subject to the constraint $g(x, y, z) = 0$.

- The total differential of a function $z = f(x, y)$ is defined by

$$dz = \frac{\partial z}{\partial x}\,dx + \frac{\partial z}{\partial y}\,dy.$$

This definition extends to functions of three or more variables.

- $df \approx \Delta f$.

- A region R of Type I in the xy-plane is the set of all points (x, y) such that $a \leq x \leq b$, and $g_1(x) \leq y \leq g_2(x)$, where g_1 and g_2 are continuous functions over the interval $[a, b]$.

- A region R of Type II in the xy-plane is the set of all points (x, y) such that $c \leq y \leq d$ and $h_1(y) \leq x \leq h_2(y)$, where h_1 and h_2 are continuous functions over the interval $[c, d]$.

- Suppose f is a function of x and y that is continuous over a region R of Type I or Type II. Partition R into n rectangles whose sides are parallel to the coordinate axes, and each having area $\Delta x \Delta y$. In each rectangle R_i that is entirely contained in R choose any arbitrary point $(\tilde{x}_i, \tilde{y}_i)$ and form the **Riemann sum**

$$f(\tilde{x}_1, \tilde{y}_1)\,\Delta x\Delta y + f(\tilde{x}_1, \tilde{y}_2)\,\Delta x\Delta y + \cdots + f(\tilde{x}_n, \tilde{y}_n)\,\Delta x\Delta y.$$

Then, the limit of the Riemann sums, as $n \to +\infty$ and Δx and Δy approach zero,

exists and is independent of the choice of points $(\tilde{x}_i, \tilde{y}_i)$. This limit is called the double integral of f over R and is denoted by

$$\iint_R f(x, y)\,dy\,dx.$$

- An iterated integral is an expression of the form

$$\int_a^b \left(\int_{g_1(x)}^{g_2(x)} f(x, y)\,dy \right) dx$$

or

$$\int_c^d \left(\int_{h_1(y)}^{h_2(y)} f(x, y)\,dx \right) dy,$$

where g_1 and g_2 are continuous over $[a, b]$ and h_1 and h_2 are continuous over $(c, d]$.

- If f is a function of x and y which is continuous over a region R of Type I, then

$$\iint_R f(x, y)\,dy\,dx = \int_a^b \left(\int_{g_1(x)}^{g_2(x)} f(x, y)\,dy \right) dx.$$

- If f is a function of x and y which is continuous over a region R of Type II, then

$$\iint_R f(x, y)\,dy\,dx = \int_c^d \left(\int_{h_1(y)}^{h_2(y)} f(x, y)\,dx \right) dy.$$

REVIEW EXERCISES ||||||||||||||||||||||||||||||||||||||||||||||||||||||||||||||||||||||||||||

1. If $f(x, y) = 3x + x^2y - 2y^2$, find
   (a) $f(4, 2)$ (c) $f(0, 3)$
   (b) $f(1, -1)$ (d) $f(-2, -3)$

2. If $f(x, y) = \dfrac{e^{\sqrt{x^2+y^2}}}{3y}$ find
   (a) $f(0, 1)$ (d) $f(1, -2)$
   (b) $f(4, 3)$ (e) the domain of f
   (c) $f(-2, -3)$

3. If $S(P, r, t) = P(1 + rt)$, find
   (a) $S(4000, 0.08, 3)$
   (b) $S(5000, 0.10, 5)$

4. Plot the following points in a rectangular coordinate system:
   (a) $(2, 3, 1)$ (c) $(1, -2, 3)$
   (b) $(2, 2, -1)$ (d) $(0, 4, 0)$

5. Sketch the graph of the plane $x + 3y - 2z = 6$.

In Exercises 6 and 7, find f_x, f_y, and f_{xy}.

6. $f(x, y) = x^2y + xy^3 - y^2$

7. $f(x, y) = x^2 e^{x-y^2}$

8. If $f(x, y) = x^{3/2} + xy^2 - x^2y + y^3$, calculate $f_x(4, 1)$, $f_y(4, -1)$, and $f_{yx}(4, 2)$.

9. If $f(x, y) = 4x^3 - xy^2 + xe^{x^2-y^2}$, calculate f_{xx}, f_{yy}, f_{xy}, and f_{yx}.

10. If $f(x, y) = 2x^3y + 3xy^2 - ye^{xy}$, evaluate $f_{xx}(1, 1)$, $f_{yy}(0, 1)$, and $f_{xy}(-1, -1)$.

11. Calculate all the first-order partial derivatives of the function $f(x, y, z) = \dfrac{3x^5 - xy^4}{z^3}$.

12. Calculate all the first-order partial derivatives of the function $f(w, x, y, z) = w^3 e^{x^2 y} \ln z$.

In Exercises 13 and 14, find the critical points of f.

13. $f(x, y) = x^2 + 3xy + 6y^2 + x - 3y + 2$

14. $f(x, y) = x^3 + y^3 - 3xy$

In Exercises 15–18, find the relative extrema of f, if any exist.

15. $f(x, y) = 3x^2 + 2xy - y^2$

16. $f(x, y) = x^3 + y^3 - 3xy$

17. $f(x, y) = -x^3 - y^3 + 6xy$

18. $f(x, y) = x^3 + y^3 + 18xy$

In Exercises 19–21, use the method of Lagrange multipliers to find the extreme values.

19. Minimize $f(x, y) = x^2 + y^2$ subject to $x + y = 20$.

20. Maximize $f(x, y) = xy + 4$ subject to $x^2 + y^2 = 2$.

21. Minimize $f(x, y, z) = x^2 + y^2 + z^2$ subject to $x + y + z = 8$.

In Exercises 22 and 23, find the total differential of the given function.

22. $f(x, y) = \dfrac{5x^3}{\sqrt{y}}$

23. $f(x, y, z) = x^3 e^{y^2 z}$

24. Find the total differential of the function $f(x, y) = 2x^4 y + x\sqrt{y}$ at the point $(-2, 1)$, when $dx = -0.01$ and $dy = 0.02$.

25. Find the total differential of the function

$$f(x, y, z) = x^2 y^5 - 4yz^3$$

at the point $(3, -1, 2)$, when $dx = 0.03$, $dy = 0.02$, and $dz = -0.10$.

26. **(Profit)** A stereo manufacturer's profit P (in dollars) is given by

$$P(x, y, z) = 100x + 600y + 500z,$$

where x is the number of cassette recorders sold, y is the number of receivers sold, and z is the number of speakers sold. Find
(a) $P(20, 30, 50)$ (b) $P(80, 20, 100)$

27. **(Advertising)** A car-rental firm's weekly revenue R (in millions of dollars) is given by

$$R(x, y) = 2 + 3x^2 + 2xy$$

where x is the number of times that its commercial is shown on television and y is the number of people (in millions) who see the commercial.
(a) Find the instantaneous rate of change of R with respect to x.
(b) Find the instantaneous rate of change of R with respect to y.

28. **(Production)** The production function $Q = 50x^{3/4} y^{1/4}$ gives the number of typewriters produced by a manufacturer with x units of labor and y units of capital.
(a) Find $\partial Q / \partial x$ and $\partial Q / \partial y$. These partial derivatives are called the **marginal productivity of labor** and the **marginal productivity of capital,** respectively.
(b) Evaluate

$$\left. \frac{\partial Q}{\partial x} \right|_{(16, 256)} \quad \text{and} \quad \left. \frac{\partial Q}{\partial y} \right|_{(16, 256)}.$$

(c) Interpret the results of part (b).

29. A closed rectangular box having a surface area of 24 square meters is to be constructed. Find the dimensions that will yield a box of maximum volume.

30. When a corn farmer spends x hundred dollars for labor and y hundred dollars for fertilizers, the farm's profit (in thousands of dollars per acre) is given by

$$P(x, y) = 6x - x^2 - y^2 + 4y - 4.$$

How much money should the farmer

spend for labor and fertilizers to maximize the profit? What is the maximum profit?

31. **(Economics)** A manufacturer has a monopoly on two types of solar engines: A and B. Market research indicates that if x engines of model A are made, they can all be sold at $150 - 3x$ dollars each and, if y engines of type B are made, they can all be sold at $160 - 4y$ dollars each. The total cost (in dollars) of manufacturing x engines of type A and y engines of type B is $46x + 30y + 10xy$. How many engines of each type should be made to maximize the profit? What is the maximum profit?

In Exercises 32 and 33, use the method of Lagrange multipliers.

32. Maximize the function $P(x, y) = 800x + 500y + 10xy - 5320$ subject to the constraint $100x + 50y = 2200$.

33. **(Business)** A manufacturer has $60,000 to spend monthly for labor, interest, and raw materials. If x, y, and z are the amounts (in dollars) spent monthly on these resources, respectively, then the monthly profit (in thousands of dollars) is given by

$$P(x, y, z) = xz + yz + xy.$$

How much should be allocated monthly for each resource to maximize the profit?

34. For the function $f(x, y) = 9x^{1/3}y^{4/3}$, estimate $f(27.5, -8.2)$ using differentials.

35. Use differentials to approximate $5.96\sqrt{5.02 - (.97)^2}$.

36. **(Sociology)** A large city has determined that the number N of public assistance clients per year (in thousands) is given by

$$N = 4 + 2x^2 + 3y^3 - 4xy,$$

where x is the amount (in millions of dollars) spent per year on work-incentive programs and y is the amount (in millions of dollars) spent per year on remedial educa-

tion. At the present time, the city is spending $3 million annually on work-incentive programs and $2 million annually on remedial education. Use a differential to estimate the increase or decrease in the number of public assistance clients per year if the city were to spend an additional 1 million dollars per year on work-incentive programs, and an additional half million dollars per year on remedial education.

37. **(Error Estimation)** The volume of a cylinder is given by the formula $v = \pi r^2 h$, where r is the radius of the base and h is the height of the cylinder. The radius is measured to be 3 inches, with a maximum error of $\frac{1}{8}$ inch, and the height is measured to be 1 foot, with a maximum error of $\frac{1}{4}$ inch. Estimate the maximum error in calculating the volume of the cylinder from these measurements.

In Exercises 38–41, evaluate the given iterated integral.

38. $\int_1^4 \left(\int_0^2 x^2 y \, dx \right) dy$

39. $\int_{-2}^1 \left(\int_x^{2x} (x - 2y) \, dy \right) dx$

40. $\int_{-2}^2 \left(\int_{-\sqrt{4-x^2}}^{\sqrt{4-x^2}} y \, dx \right) dy$

41. $\int_1^3 \int_0^x e^{y/x} \, dy \, dx$

In Exercises 42–44, evaluate the double integral of f over R.

42. $f(x, y) = xe^{x-y}$, R is the region bounded by the lines $y = x$, $y = 1$, and the y-axis.

43. $f(x, y) = e^{2x+y}$, R is the triangle bounded by the lines $y = 2x$, $y = 1 - x$, and the y-axis.

44. $f(x, y) = 2y$, R is the region bounded by $y = 4/x$, $x = 1$, and $y = x$.

In Exercises 45 and 46, compute the volume of the solid below the surface $z = f(x, y)$ and above the region R in the xy-plane.

45. $z = xe^y$, R is the region bounded by the lines $y = x$, $y = 1$, $y = 3$ and the y-axis.

46. $z = x + y^2$, R is the triangle with vertices $(2, 1)$, $(5, 1)$, and $(4, 2)$.

In Exercises 47–49, compute the average value of f over the region R.

47. $f(x, y) = x^2y$, R is the region in the first quadrant bounded by $y = 4 - x^2$, the x-axis, and the y-axis.

48. $f(x, y) = e^y$, R is the triangle bounded by the lines $y = 1 - x$, $y = 1 + x$, and $x = 2$.

49. $f(x, y) = 3x - y$, R is the region bounded by $y = \sqrt{x}$ and $y = x$.

50. A manufacturer of two types of film developer, Sharp and Ultra Sharp, finds that the weekly cost to make x gallons of Sharp and y gallons of Ultra Sharp (both in thousands) is given (in dollars) by

$$C(x, y) = 200 + 2x^2 - 0.05x + 5y^2 - 0.02y.$$

If x varies between 10 and 30 gallons and y varies between 20 and 50 gallons, find the average weekly cost.

51. Suppose that the region bounded by the lines $y = x - 2$, $y = x + 2$, $x = 0$, and $x = 1$ describes a lake. If the salt content at any point (x, y) is given (in milligrams per liter) by $f(x, y) = 2x + y^2 + y$, find the average salt content in the lake.

CHAPTER TEST ||||||||||||||||||||||||||||||||||||||||||||||||||||||||||||||||||||||||||||||||||||||||

1. For the function $f(x, y) = \dfrac{x + 2y}{x - y}$, find

    (a) $f(-1, 2)$ (c) $f(3, -5)$
    (b) $f(0, -1)$ (d) the domain of f

2. **(Ecology)** The number of bald eagles on a preserve is given by

$$N(x, y) = 32x + 10y + 20$$

    where x is the number of squirrels and y is the number of rats on the preserve. Rats and squirrels constitute the eagles' basic food supply. Find
    (a) $N(50, 30)$ (b) $N(40, 60)$

3. For the function $f(x, y) = ye^{-2x} + 4x^2$, find f_x, f_y, f_{xx}, f_{xy}, f_{yx} and f_{yy}.

4. For the function $f(x, y, z) = 4xz - e^{y^2}z$, find all the first-order partial derivatives.

5. Find the total differential of the function $f(x, y) = x^3(3y^2 - x)$ at the point $(1, -2)$ when $dx = -0.01$ and $dy = 0.02$.

6. Find all relative extrema of the function $f(x, y) = x^3 - 3x + y^2$.

7. Use the method of Lagrange multipliers to find the extreme values of $x - y$ subject to $x^2 + 3y^2 = 3$.

8. **(Psychology)** In an experiment designed to study the effects of weightlessness and exercise on puzzle-solving ability, a returning astronaut is required to solve a certain puzzle. The time T (in minutes) it takes to solve the puzzle is given by

$$T = 2 + 0.1x + 0.2xy^2$$

where x is the number of days that the astronaut was weightless and y is the number of hours that the astronaut exercised. Find
   (a) the instantaneous rate of change of T with respect to x, when $x = 4$ and $y = 2$.
   (b) the instantaneous rate of change of T with respect to y, when $x = 4$ and $y = 2$.

9. **(Production)** A garment manufacturer has two plants producing the same coat. When plant A is producing x hundred coats weekly, and plant B is producing y hundred coats weekly, the company realizes a weekly profit of

$$P(x, y) = 4xy + 20x + 16y - 3x^2 - 2y^2$$

(in thousands of dollars) from these coats. Determine the weekly production levels at the two plants that will maximize the company's weekly profit from these coats.

10. **(Business)** A manufacturer makes three products—A, B, and C—on which the profit per unit is $6, $8, and $10, respectively. The daily levels of production are x, y, and z, respectively. The research department has decided that x, y, and z must be subject to the constraint

$$x^2 + y^2 + z^2 = 800.$$

Find the levels of production of each product that will maximize daily profit. (Use the method of Lagrange multipliers.)

11. A rectangular concrete vault has inside dimensions 12 feet by 8 feet by 4 feet, and walls of uniform thickness 6 inches. Estimate the amount of concrete used in making the vault, using differentials.

12. Evaluate $\int_0^1 \int_x^{\sqrt{x}} 2y \, dy \, dx$.

13. Compute the volume of the solid below the surface $z = 3x + y$ and over the rectangular region R with vertices $(0, 0)$, $(3, 0)$, $(0, 4)$, and $(3, 4)$.

THE TRIGONOMETRIC FUNCTIONS

It should come as no surprise to read that many real phenomena show a cyclical or repetitive mode of behavior. For example, business cycles, sunspots, earthquakes, the motion of the planets, heartbeats, respiration, and sound waves all occur in cyclical fashion. In this chapter we shall study an important class of functions that are useful in describing cyclical phenomena. They are called the **trigonometric functions**, and their graphs exhibit cyclical behavior.

9.1 IIIIIII THE TRIGONOMETRIC FUNCTIONS

Angles and Degree Measure

We shall view an angle as the result of rotating a ray (or half-line) about its endpoint. In Figure 1(a), the ray initially on the x-axis (called the **initial side**) is rotated in a counterclockwise direction until it coincides with its final position (called the **terminal side**) forming the angle θ.

Figure 1

(a) Counterclockwise rotation
for a positive angle

(b) Clockwise rotation
for a negative angle

When the rotation is counterclockwise we say that θ is a **positive angle.** When a ray is rotated in a clockwise direction to yield the angle ϕ (as in Figure 1(b)) we say that ϕ is a **negative angle.** The end point of the two rays is called the **vertex** of the angle. An angle is said to be in **standard position** when it is placed so that its vertex is at the origin and its initial side coincides with the positive x-axis.

Note that an angle has an initial side, a terminal side, a vertex, and a direction. The direction of the angle is positive when the angle is obtained by a counterclockwise rotation. It is negative when the angle is the result of a clockwise rotation.

Degrees and radians are the two units commonly used to measure an angle. An angle is said to have a measure of **one degree** (written $1°$) if it is the result of rotating the initial side $1/360$ of a complete revolution in the counterclockwise direction. Thus, an angle obtained by a complete counterclockwise revolution of the initial ray has a measure of $360°$ and an angle obtained by $1/4$ of a complete counterclockwise rotation has a measure of $(1/4)360° = 90°$ (see Figure 2).

$\theta = 360°$
(1 complete revolution)

(a)

$\theta = 180°$
($\frac{1}{2}$ revolution)

(b)

$\theta = 90°$
($\frac{1}{4}$ revolution)

(c)

Figure 2

An angle measuring $90°$ is called a **right angle;** an angle measuring $180°$ is called a **straight angle.** An angle between $0°$ and $90°$ is called an **acute angle,** while an angle between $90°$ and $180°$ is called an **obtuse angle.** When placed in standard position, an acute angle has its terminal side in the first quadrant, while an obtuse angle has its terminal side in the second quadrant.

Some elementary facts about angles in triangles should be reviewed at this time. First, the sum of the three angles of a triangle is always $180°$ (see Figure 3).

Figure 3

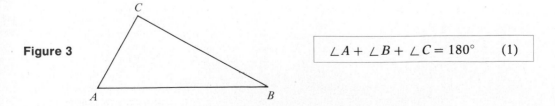

$$\angle A + \angle B + \angle C = 180° \qquad (1)$$

An **equilateral triangle** is a triangle in which all three sides are the same; equivalently, all three angles are the same. Since the angles must add up to $180°$ by (1), they must each be $60°$ (see Figure 4).

Figure 4   (an equilateral triangle)

An **isosceles triangle** is a triangle in which (at least) two sides are the same, or equivalently, (at least) two of the angles are the same (see Figure 5).

Figure 5

(an isosceles triangle)
$$a = b$$
$$\angle A = \angle B$$

A **right triangle** is a triangle in which there is one right angle. The *Pythagorean theorem* tells us that in any right triangle, the sum of the squares of the two shorter sides equals the square of the largest side (the **hypotenuse**), as shown in Figure 6.

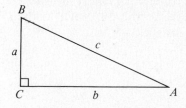

Figure 6

(a right triangle)

Pythagorean Theorem:
$$a^2 + b^2 = c^2 \qquad (2)$$

In an isosceles right triangle the two equal angles must add up to 90° (why?); hence, they must each be 45°. If the hypotenuse is taken to be 1 unit long, then the Pythagorean theorem tells us that the two equal sides must satisfy

$$1^2 = a^2 + a^2$$
$$1 = 2a^2$$
$$a^2 = \frac{1}{2}$$
$$a = \pm\frac{\sqrt{2}}{2}$$

(see Figure 7).

Figure 7   (an isosceles right triangle)

Another important right triangle is exactly half of an equilateral triangle, shaded in Figure 8.

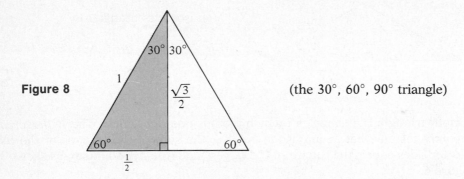

Figure 8   (the 30°, 60°, 90° triangle)

Suppose the side of the equilateral triangle in Figure 8 were chosen as 1 unit. In the shaded triangle, the side opposite the 30° angle is then $\frac{1}{2}$, and by the Pythagorean theorem the third side satisfies

$$1^2 = \left(\frac{1}{2}\right)^2 + y^2$$

$$1 = \frac{1}{4} + y^2$$

$$y^2 = \frac{3}{4}$$

$$y = \frac{\sqrt{3}}{2}. \qquad \text{(see Figure 8)}$$

‖‖‖‖‖‖‖‖‖‖‖ Radian Measure of Angles

If we consider a circle of radius 1 centered at the origin, the positive angle in standard position that cuts off an arc of length 1 is said to measure **one radian** (Figure 9). Geometry tells us that a circle of radius 1 has an arc length (or circumference) of $2\pi r = 2\pi \cdot 1 = 2\pi$. Hence, the radian measure of the angle obtained by making one complete counterclockwise revolution of the initial ray is 2π radians.

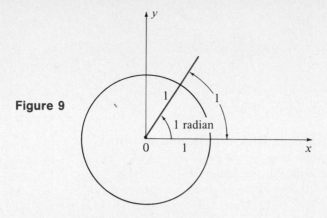

Figure 9

Thus, we conclude that

$$2\pi \text{ radians} = 360°. \tag{3}$$

It then follows that

$$\pi \text{ radians} = 180°, \tag{4}$$

and since $\pi \approx 3.1416$,

$$1 \text{ radian} \approx \frac{180}{3.1416} \approx 57°18'.$$

The following proportion, obtained from (4), gives a simple way of converting an angular measure from radians to degrees and vice versa.

$$\frac{\text{radian measure of angle } \theta}{\pi \text{ radians}} = \frac{\text{degree measure of angle } \theta}{180°}. \tag{5}$$

We shall find that in the study of calculus, radian measure is more convenient than degree measure. Formulas that would be awkward in degree measure turn out to have elegant simplicity in radian measure. For this reason, when no unit of measure is mentioned, we shall presume that radian measure is being used. We shall write $\theta = t$ to mean that the measure of angle θ is t radians, and $\theta = t°$ to mean that the measure of angle θ is t degrees.

EXAMPLE 1   An angle θ measures 150°. What is its radian measure?

Solution   Using the conversion formula (5), we find

$$\frac{\theta}{\pi} = \frac{150°}{180°},$$

or

$$\theta = \frac{150°}{180°}\pi = \frac{5}{6}\pi.$$

Thus, $\theta = \frac{5}{6}\pi$ radians, or simply $\theta = \frac{5\pi}{6}$. ∎

EXAMPLE 2   An angle θ measures $\frac{2}{3}\pi$ radians. What is its degree measure?

Solution   Using the conversion formula (5), we find

$$\frac{(2/3)\pi}{\pi} = \frac{\theta}{180°},$$

or

$$\theta = \frac{(2/3)\pi}{\pi}(180°) = 120°.$$

Thus, $\theta = 120°$. ∎

Table 1 gives the radian and degree measures of some important angles:

TABLE 1

Radian	0	$\dfrac{\pi}{6}$	$\dfrac{\pi}{4}$	$\dfrac{\pi}{3}$	$\dfrac{\pi}{2}$	$\dfrac{2}{3}\pi$	$\dfrac{5}{6}\pi$	π	$\dfrac{3}{2}\pi$	2π
Degree	0°	30°	45°	60°	90°	120°	150°	180°	270°	360°

Figure 10 shows that two angles in standard position may share the same terminal side. In fact, because additional complete revolutions do not affect the position of the terminal side, each position of the terminal side determines infinitely many angles.

Figure 10

Figure 11 shows other angles that have the same terminal side as the angles in Figure 10. These angles are determined by additional complete revolutions. Thus, given any real number θ, we can always construct an angle with radian measure θ, although the process may require many revolutions.

Figure 11

(a)   (b)

############# **The Trigonometric Functions**

Corresponding to every angle there are defined six special functions called **trigonometric functions.**

Let θ be a fixed angle. We begin by placing θ in standard position. Construct a unit circle (radius = 1) centered at the origin (Figure 12). Let $P(x, y)$ denote the point at which the terminal side of θ intersects the unit circle (see Figure 12). Table 2 defines the six trigonometric functions of the angle θ.

Figure 12

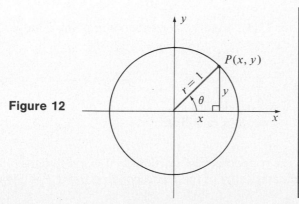

TABLE 2
Definition of the Trigonometric Functions

$$\text{sine } \theta = \sin \theta = y$$
$$\text{cosine } \theta = \cos \theta = x$$
$$\text{tangent } \theta = \tan \theta = \frac{y}{x} \qquad \text{if } x \neq 0$$
$$\text{cosecant } \theta = \csc \theta = \frac{1}{\sin \theta} = \frac{1}{y} \qquad \text{if } y \neq 0$$
$$\text{secant } \theta = \sec \theta = \frac{1}{\cos \theta} = \frac{1}{x} \qquad \text{if } x \neq 0$$
$$\text{cotangent } \theta = \cot \theta = \frac{1}{\tan \theta} = \frac{x}{y} \qquad \text{if } y \neq 0$$

Observe that $\tan \theta = \sin \theta/\cos \theta$. Because the cosecant, secant, and cotangent are the reciprocals of the sine, cosine, and tangent, respectively, we shall limit our attention to the latter three functions. Also, because the sine is a function of the angle, we should denote the value of the sine of the angle θ by sine (θ). But we shall simply write it as sine θ or merely $\sin \theta$.

Using the definitions in Table 2, we can obtain values of the trigonometric functions for any angle.

Let $\theta = 0$ (radians). Then, as Figure 13(a) shows, the corresponding point P has coordinates $(1, 0)$. Hence

$$\sin 0 = y = 0,$$
$$\cos 0 = x = 1, \quad \text{and}$$
$$\tan 0 = \frac{y}{x} = \frac{0}{1} = 0.$$

Figure 13

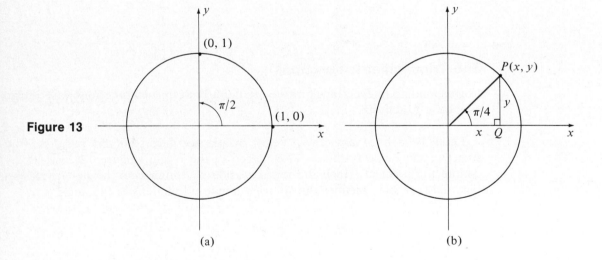

(a)　　　　　　　　　　　　(b)

If $\theta = \pi/2$ (radians), as in Figure 13(a), then the corresponding point P has coordinates $(0, 1)$. Hence,

$$\sin \pi/2 = y = 1,$$
$$\cos \pi/2 = x = 0, \quad \text{and}$$
$$\tan \pi/2 = y/x = 1/0 \text{ is undefined.}$$

To determine the values of the trigonometric functions at $\theta = \pi/4$ radians $(45°)$, we must determine the coordinates (x, y) of the corresponding point P in Figure

13(b). The triangle OQP is a right triangle and one angle is $\pi/4$, so OQP is an isosceles right triangle and $x = y$. From Figure 7 we can see that

$$x = \frac{1}{\sqrt{2}} = \frac{\sqrt{2}}{2} \quad \text{and} \quad y = \frac{\sqrt{2}}{2}.$$

Consequently,

$$\sin\frac{\pi}{4} = y = \frac{\sqrt{2}}{2},$$

$$\cos\frac{\pi}{4} = x = \frac{\sqrt{2}}{2}, \quad \text{and}$$

$$\tan\frac{\pi}{4} = \frac{y}{x} = 1.$$

We now determine the trigonometric functions of $\pi/6$ radians (30°) and $\pi/3$ radians (60°). Figure 14(a) shows an angle of $\pi/6$ in standard position. From Figure 8 we can see that $x = \sqrt{3}/2$ and $y = \frac{1}{2}$, so that

$$\sin\frac{\pi}{6} = y = \frac{1}{2},$$

$$\cos\frac{\pi}{6} = x = \frac{\sqrt{3}}{2}, \quad \text{and}$$

$$\tan\frac{\pi}{6} = \frac{y}{x} = \frac{1}{2} \div \frac{\sqrt{3}}{2} = \frac{1}{\sqrt{3}} = \frac{\sqrt{3}}{3}.$$

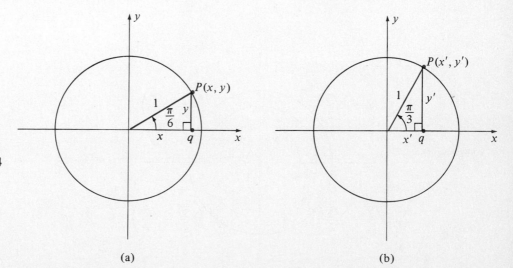

Figure 14

(a)   (b)

Figure 14(b) shows an angle of $\pi/3$ in standard position. Again, Figure 8 shows that $x' = \frac{1}{2}$ and $y' = \sqrt{3}/2$, so that

$$\sin \frac{\pi}{3} = y' = \frac{\sqrt{3}}{2},$$

$$\cos \frac{\pi}{3} = x' = \frac{1}{2}, \quad \text{and}$$

$$\tan \frac{\pi}{3} = \frac{y'}{x'} = \frac{\sqrt{3}}{2} \div \frac{1}{3} = \sqrt{3}.$$

The most efficient way to find the sine, cosine, and tangent of arbitrary angles is to use a calculator. Inexpensive scientific calculators with built-in trigonometric function keys are widely available. Most of them will accept both degrees and radians. Before the availability of calculators, numerical tables and slide rules were the common tools used.

Once we know the trigonometric functions of angles whose terminal sides are in the first quadrant (acute angles), we can use them to find the trigonometric functions of angles whose terminal sides fall in other quadrants.

EXAMPLE 3   As Figure 15 shows, it follows from the symmetries of the circle that

$$\sin \frac{3}{4}\pi = \frac{\sqrt{2}}{2}, \qquad \cos \frac{3}{4}\pi = -\frac{\sqrt{2}}{2}$$

$$\sin \frac{5}{4}\pi = -\frac{\sqrt{2}}{2}, \qquad \cos \frac{5}{4}\pi = -\frac{\sqrt{2}}{2}$$

$$\sin \frac{7}{4}\pi = -\frac{\sqrt{2}}{2}, \qquad \cos \frac{7}{4}\pi = \frac{\sqrt{2}}{2}.$$

Figure 15

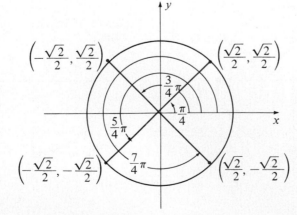

Using the approach we used in Example 3, we can determine the values of the trigonometric functions at $\theta = 5\pi/6$, $7\pi/6$, and $11\pi/6$, which are all related by circular symmetries to $\pi/6$, and at $2\pi/3$, $4\pi/3$, and $5\pi/3$, which are all related by circular symmetries to $\pi/3$. Table 3 lists these values together with the trigonometric functions of a number of other common angles.

TABLE 3   Trigonometric Functions of Common Angles

θ				
Radians	Degrees	$\sin\theta$	$\cos\theta$	$\tan\theta$
0	0	0	1	0
$\dfrac{\pi}{6}$	30	$\dfrac{1}{2}$	$\dfrac{\sqrt{3}}{2}$	$\dfrac{1}{\sqrt{3}}$
$\dfrac{\pi}{4}$	45	$\dfrac{\sqrt{2}}{2}$	$\dfrac{\sqrt{2}}{2}$	1
$\dfrac{\pi}{3}$	60	$\dfrac{\sqrt{3}}{2}$	$\dfrac{1}{2}$	$\sqrt{3}$
$\dfrac{\pi}{2}$	90	1	0	—
$\dfrac{2}{3}\pi$	120	$\dfrac{\sqrt{3}}{2}$	$-\dfrac{1}{2}$	$-\sqrt{3}$
$\dfrac{3}{4}\pi$	135	$\dfrac{\sqrt{2}}{2}$	$-\dfrac{\sqrt{2}}{2}$	-1
$\dfrac{5}{6}\pi$	150	$\dfrac{1}{2}$	$-\dfrac{\sqrt{3}}{2}$	$-\dfrac{1}{\sqrt{3}}$
π	180	0	-1	0
$\dfrac{7}{6}\pi$	210	$-\dfrac{1}{2}$	$-\dfrac{\sqrt{3}}{2}$	$\dfrac{1}{\sqrt{3}}$
$\dfrac{5}{4}\pi$	225	$-\dfrac{\sqrt{2}}{2}$	$-\dfrac{\sqrt{2}}{2}$	1
$\dfrac{4}{3}\pi$	240	$-\dfrac{\sqrt{3}}{2}$	$-\dfrac{1}{2}$	$\sqrt{3}$
$\dfrac{3}{2}\pi$	270	-1	0	—
$\dfrac{5}{3}\pi$	300	$-\dfrac{\sqrt{3}}{2}$	$\dfrac{1}{2}$	$-\sqrt{3}$
$\dfrac{7}{4}\pi$	315	$-\dfrac{\sqrt{2}}{2}$	$\dfrac{\sqrt{2}}{2}$	-1
$\dfrac{11}{6}\pi$	330	$-\dfrac{1}{2}$	$\dfrac{\sqrt{3}}{2}$	$-\dfrac{1}{\sqrt{3}}$
2π	360	0	1	0

As we noted earlier, if θ is an angle in standard position, then $\theta + 2\pi$ designates the same point on the unit circle because it is the result of adding a complete revolution to the terminal side of θ. It then follows that

$$\sin(\theta + 2\pi) = \sin \theta \qquad \cos(\theta + 2\pi) = \cos \theta. \tag{6}$$

For this reason we say that the sine and cosine functions are periodic (or cyclical), with period 2π or $360°$. The values of the sine and cosine functions repeat every 2π radians or every $360°$.

From Figure 16 we obtain the following identities, which enable us to compute the trigonometric functions of negative angles.

$$\sin(-\theta) = -\sin \theta \qquad \cos(-\theta) = \cos \theta \tag{7}$$

Figure 16

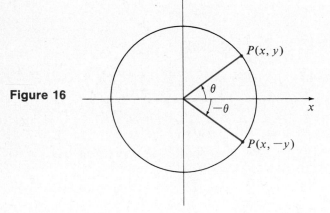

EXAMPLE 4   Evaluate

(a) $\cos \dfrac{5}{2}\pi$ (b) $\sin\left(-\dfrac{\pi}{3}\right)$

Solution   (a) From (6), we have

$$\cos \frac{5}{2}\pi = \cos\left(\frac{\pi}{2} + 2\pi\right) = \cos \frac{\pi}{2} = 0.$$

(b) From (7), we have

$$\sin\left(-\frac{\pi}{3}\right) = -\sin \frac{\pi}{3} = -\frac{\sqrt{3}}{2}. \qquad \blacksquare$$

|||||||||||| **Trigonometry in Right Triangles**

It is not difficult to show (Exercise 49) that if θ is an acute angle, then each trigonometric function can be expressed as a ratio of the sides of a right triangle (Figure 17). The ratios that represent the sine, cosine, and tangent functions are shown in Table 4.

TABLE 4

$$\sin \theta = \frac{\text{length of side opposite } \theta}{\text{length of hypotenuse}}$$

$$\cos \theta = \frac{\text{length of side adjacent to } \theta}{\text{length of hypotenuse}}$$

$$\tan \theta = \frac{\text{length of side opposite } \theta}{\text{length of side adjacent to } \theta}$$

Figure 17

EXAMPLE 5   Find the values of the sine, cosine, and tangent functions of the angle θ in Figure 18.

Figure 18

Solution   We have

$$\sin \theta = \frac{4}{5}, \qquad \cos \theta = \frac{3}{5}, \qquad \tan \theta = \frac{4}{3}. \qquad \blacksquare$$

EXAMPLE 6   A ladder leaning against a building makes an angle of 60° with the ground. If the base of the ladder is 6 meters from the building, how long is the ladder? To what height does it rise along the building?

Solution   We seek the length d of the ladder and the height h along the building shown in Figure 19.
We have

$$\cos 60° = \frac{6}{d}$$

$$d = \frac{6}{\cos 60°} = \frac{6}{0.500} = 12 \text{ meters.}$$

Figure 19

Thus, the ladder is 12 meters long. Also,

$$\tan 60^\circ = \frac{h}{6}$$

$$h = 6 \tan 60^\circ$$

$$\approx 6(1.732) \approx 10.4 \text{ meters.}$$

Thus, the ladder rises approximately 10.4 meters along the building.   ■

|||||||||||| **Basic Trigonometric Identities**

There are equations involving trigonometric functions that are true *for all values of* θ for which both sides of the equation are defined. Equations (6) and (7) are examples. Such equations are called **identities.** We collect a few basic examples here.

From their definitions in Table 2, we see that

$$\tan \theta = \frac{\sin \theta}{\cos \theta}, \qquad \cot \theta = \frac{\cos \theta}{\sin \theta} \tag{8}$$

$$\sec \theta = \frac{1}{\cos \theta}, \qquad \csc \theta = \frac{1}{\sin \theta} \tag{9}$$

$$\cot \theta = \frac{1}{\tan \theta}. \tag{10}$$

Referring to Figure 12, observe that

$$x^2 + y^2 = 1 = (\cos \theta)^2 + (\sin \theta)^2.$$

It is more convenient to write $(\cos \theta)^2$ as $\cos^2 \theta$ and $(\sin \theta)^2$ as $\sin^2 \theta$. Thus, we have the identity

$$\cos^2 \theta + \sin^2 \theta = 1. \tag{11}$$

It is also easy to establish (Exercise 50) the identities

$$\tan^2 \theta + 1 = \sec^2 \theta \quad \text{and} \tag{12}$$

$$\cot^2 \theta + 1 = \csc^2 \theta. \tag{13}$$

||||||||||||| ## Graphs of Sine, Cosine, and Tangent Functions

Using a calculator, we can sketch the graphs of $y = \sin x$, $y = \cos x$, and $y = \tan x$, as shown in Figures 20, 21, and 22, respectively. Observe that the graph of the sine function repeats every 2π radians. Similarly, the graph of the cosine function repeats every 2π radians and that of the tangent function repeats every π radians.

Figure 20

Figure 21

Figure 22

EXAMPLE 7
(Motion)

Consider an object moving along a straight line. The distance (in feet) it travels during the first t seconds is given by

$$s(t) = 2 \sin t, \qquad 0 \le t \le \frac{\pi}{2},$$

where t is given in seconds. Find the distance traveled in the first $\pi/4$ seconds.

Solution

We have

$$s\left(\frac{\pi}{4}\right) = 2 \sin\left(\frac{\pi}{4}\right) = 2\frac{\sqrt{2}}{2} = \sqrt{2} \text{ feet.} \qquad \blacksquare$$

EXAMPLE 8
(Predator-Prey Interaction)

Two plant or animal species interact in their environment in such a manner that one species (the prey) serves as the primary food supply for the second species (the predator). Examples of such interaction are the relationships between trees (prey) and insects (predators) and between rabbits (prey) and lynxes (predators). As the population of the prey increases, the additional food supply causes an increase in the population of the predators. More predators consume more food, so the population of the prey will decrease, which, in turn, will lead to a decrease in the population of the predators. This reduction in the predator population will now cause an increase in the number of prey and the cycle with which we started our discussion will begin again. Figure 23, adapted from *Mathematics: Ideas and Applications*, by Daniel D. Benice, copyright © 1978 by Harcourt Brace Jovanovich, Inc. (reprinted by permission of the publisher), shows the interaction between lynxes and rabbits. Both curves can be described by trigonometric functions. To construct a mathematical model of this interaction, a system of differential equations is used. For further details, see Exercise 13 of Section 10.6.

Figure 23

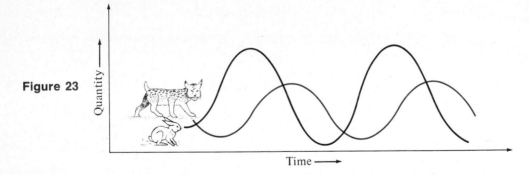

9.1 EXERCISE SET ||||||||||||||||||||||||||||||||||||||||||||||||||||||||||||||||||||||||||||||||||||||

In Exercises 1–6, assume that the angle θ is in standard position and determine the quadrant in which the terminal side of the angle lies.

In Exercises 7–12, convert from degree measure to radian measure.

1. $\theta = 315°$

2. $\theta = 192°$

3. $\theta = 12°$

4. $\theta = -340°$

5. $\theta = 140°$

6. $\theta = -167°$

7. $80°$

8. $200°$

9. $-150°$

10. $-330°$

11. $75°$

12. $570°$

In Exercises 13–18, convert from radian measure to degree measure.

13. $\dfrac{4}{3}\pi$ 14. $-\dfrac{\pi}{2}$ 15. $-\dfrac{7\pi}{12}$

16. $\dfrac{2}{3}\pi$ 17. $\dfrac{5}{2}\pi$ 18. $-\dfrac{5}{3}\pi$

In Exercises 19 and 20, give the radian measure of the angles shown.

19.

20.

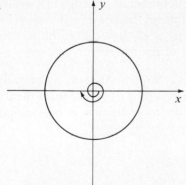

In Exercises 21–26, use Equations (6) and (7) along with Table 3 to evaluate the trigonometric functions.

21. $\sin(-\pi)$ 22. $\cos \frac{7}{2}\pi$

23. $\tan\left(-\dfrac{\pi}{4}\right)$ 24. $\tan(10\pi)$

25. $\cos\left(\dfrac{8\pi}{3}\right)$ 26. $\sin\left(-\dfrac{11\pi}{6}\right)$

In Exercises 27–30, use Tables 3 and 4 to obtain the desired quantities.

27. Find the lengths of the sides of a 30°, 60°, 90° triangle whose hypotenuse has length 10.

28. Find the length of the hypotenuse of an isosceles right triangle, if one of the shorter sides has length 6.

29. Find the length of the sides of an isosceles right triangle, if the hypotenuse has length 8.

30. Find the lengths of the sides of a 30°, 60°, 90° triangle, if the shortest side has length 2.

In Exercises 31–34, find the values of $\sin\theta$, $\cos\theta$, and $\tan\theta$ for θ in each right triangle.

31.

32.

33.

34.

In Exercises 35–40, find the values of sin θ, cos θ, and tan θ if the point P lies on the terminal side of θ and θ is in standard position.

35. $P(-5, 12)$ **36.** $P(-3, 4)$

37. $P(12, -5)$ **38.** $P(-3, -2)$

39. $P(-2, -4)$ **40.** $P(2, -1)$

In Exercises 41–44, express the length h in each right triangle in terms of a trigonometric function of the angle θ.

41.

42.

43.

44.

45. An airborne kite is staked to the ground by its string. The string makes an angle of 35° with the ground. If 150 meters of string has been let out, how high is the kite?

46. A monument is 500 feet high. What is the length of the shadow cast by the monument when the sun is 66° above the horizontal?

47. A forest ranger is in a tower 80 feet above the ground. If the ranger spots a fire at an angle of depression (the angle between the horizontal and the line of sight when looking down) of 8°, how far is the fire from the base of the tower?

48. (**Business**) Suppose that the average daily revenue received by a ski shop during a fixed month (in hundreds of dollars) is given by

$$R(t) = 18 + 12 \sin \frac{\pi}{6} t,$$

where t is the number of months since July 1. Find the average daily revenue received the following January.

49. In this exercise we establish the relationships expressed in Table 4. Figure 24(a) shows an arbitrary acute angle θ in a right triangle, with hypotenuse r, adjacent side x, and opposite side y. In Figure 24(b) the same triangle is placed in a coordinate system so that θ is in standard position. Then triangle OMN is similar to triangle OQP.

(a)

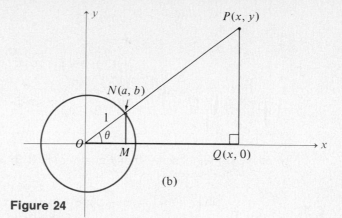

(b)

Figure 24

Use these similar triangles to prove that

$$\sin \theta = \frac{y}{r}, \qquad \cos \theta = \frac{x}{r}, \qquad \tan \theta = \frac{y}{x}.$$

50. Establish the identities

$$\tan^2 \theta + 1 = \sec^2 \theta \quad \text{and}$$
$$1 + \cot^2 \theta = \csc^2 \theta.$$

(*Hint:* Begin with Equation (11), which we know to be an identity, and divide both sides by the same quantity.)

9.2 |||||||| DERIVATIVES OF TRIGONOMETRIC FUNCTIONS

In this section we shall present formulas for the derivatives of the sine, cosine, and tangent functions.

To obtain formulas for the derivatives of the sine and cosine functions, we need several trigonometric identities and special limit results. The fundamental trigonometric identity we shall need is

$$\sin (x + y) = \sin x \cos y + \cos x \sin y. \tag{1}$$

This remarkable identity is proved in every standard course in trigonometry. To prove it here would take us too far afield, so we simply state it as an accepted fact. Observe that (1) tells us that $\sin (x + y) \neq \sin x + \sin y$. It is likewise true that $\cos (x + y) \neq \cos x + \cos y$, and $\tan (x + y) \neq \tan x + \tan y$; these common mistakes must be avoided.

From Equation (1) we derive two other important identities.

For any θ,

$$\sin\left(\frac{\pi}{2} - \theta\right) = \cos\theta, \quad \text{and} \tag{2}$$

$$\cos\left(\frac{\pi}{2} - \theta\right) = \sin\theta. \tag{3}$$

To prove (2) we observe that Equation (1) with $x = \pi/2$ and $y = -\theta$ will yield

$$\sin\left(\frac{\pi}{2} - \theta\right) = \sin\frac{\pi}{2}\cos\theta + \cos\frac{\pi}{2}\sin(-\theta)$$

$$= 1 \cdot \cos\theta + 0 \cdot \sin(-\theta) \quad \text{(from Table 3)}$$

$$= \cos\theta.$$

From Equation (2) it follows that

$$\sin\left[\frac{\pi}{2} - \left(\frac{\pi}{2} - \theta\right)\right] = \cos\left(\frac{\pi}{2} - \theta\right), \quad \text{or}$$

$$\sin\theta = \cos\left(\frac{\pi}{2} - \theta\right),$$

which proves (3).

In the language of traditional trigonometry, Equations (2) and (3) say that if two angles are **complementary** (that is, their sum is $\pi/2$ or $90°$) then the sine of either angle is the cosine of the other.

We shall need two special limit results:

$$\lim_{\theta \to 0}\frac{\sin\theta}{\theta} = 1, \quad \text{and} \tag{4}$$

$$\lim_{\theta \to 0}\frac{\cos\theta - 1}{\theta} = 0, \tag{5}$$

where θ is in radians (*not* degrees).

To verify the first limit we construct Table 1 with the aid of a calculator.

TABLE 1 $\theta \to 0$ (*h* in radians)

θ	± 1	± 0.5	± 0.25	± 0.1	± 0.01	± 0.001
$\dfrac{\sin\theta}{\theta}$	.8414710	.9588511	.9896158	.9983342	.9999833	.9999998

The second limit can be derived from the first, as follows:

$$\lim_{\theta \to 0} \frac{\cos \theta - 1}{\theta} = \lim_{\theta \to 0} \frac{\cos \theta - 1}{\theta} \cdot \frac{\cos \theta + 1}{\cos \theta + 1}$$

$$= \lim_{\theta \to 0} \frac{\cos^2 \theta - 1}{\theta(\cos \theta + 1)}$$

$$= \lim_{\theta \to 0} \frac{-\sin^2 \theta}{\theta(\cos \theta + 1)} \qquad (\text{since } \sin^2 \theta + \cos^2 \theta = 1)$$

$$= -\lim_{\theta \to 0} \left(\frac{\sin \theta}{\theta} \cdot \frac{\sin \theta}{\cos \theta + 1} \right)$$

$$= -\left(\lim_{\theta \to 0} \frac{\sin \theta}{\theta} \right)\left(\lim_{\theta \to 0} \frac{\sin \theta}{\cos \theta + 1} \right).$$

From the graphs of the sine and cosine functions [Figure (20) and (21) of Section 9.1] we know that $\lim_{\theta \to 0} \sin \theta = 0$ and $\lim_{\theta \to 0} \cos \theta = 1$. Therefore, using (4),

$$\lim_{\theta \to 0} \frac{\cos \theta - 1}{\theta} = -(1) \cdot \frac{0}{1 + 1} = 0,$$

and we have proved Equation (5).

We are now ready to state and prove the main theorems on the derivatives of trigonometric functions.

THEOREM 1

$$\frac{d}{dx} [\sin x] = \cos x \qquad (6)$$

(for x in radians).

Proof   Using the definition of the derivative (Equation (1) of Section 3.1) we have

$$\frac{d}{dx} [\sin x] = \lim_{h \to 0} \frac{\sin (x + h) - \sin x}{h}$$

$$= \lim_{h \to 0} \frac{1}{h} [\sin x \cos h + \cos x \sin h - \sin x] \qquad [\text{by (1)}]$$

$$= \lim_{h \to 0} \frac{1}{h} [\sin x (\cos h - 1) + \cos x \sin h]$$

$$= \lim_{h \to 0} \frac{\sin x (\cos h - 1)}{h} + \lim_{h \to 0} \frac{\cos x \sin h}{h}$$

$$= \sin x \left(\lim_{h \to 0} \frac{\cos h - 1}{h} \right) + \cos x \left(\lim_{h \to 0} \frac{\sin h}{h} \right)$$

$$= (\sin x) \cdot 0 + (\cos x) \cdot 1 \qquad [\text{by (4) and (5)}]$$

$$= \cos x.$$

Corollary 1

If u is a differentiable function of x, then

$$\frac{d}{dx}[\sin u] = \cos u \cdot \frac{du}{dx}. \qquad (7)$$

Proof   Apply the chain rule (Section 3.5) to Theorem 1.

EXAMPLE 1   Differentiate
(a) $\sin 5x$ (b) $\sin x^2$ (c) $\sin^2 x$

Solution   (a) $\dfrac{d}{dx}[\sin 5x] = \cos 5x \cdot \dfrac{d}{dx}[5x]$

$$= 5 \cos 5x$$

(b) $\dfrac{d}{dx}[\sin x^2] = \cos x^2 \dfrac{d}{dx}[x^2] = 2x \cos x^2$

(c) $\dfrac{d}{dx}[\sin^2 x] = \dfrac{d}{dx}[(\sin x)^2] = 2 \sin x \dfrac{d}{dx}[\sin x]$

$$= 2 \sin x \cos x. \qquad\blacksquare$$

THEOREM 2

$$\frac{d}{dx}[\cos x] = -\sin x \qquad (8)$$

(for x in radians)

Proof

$$\frac{d}{dx}[\cos x] = \frac{d}{dx}\left[\sin\left(\frac{\pi}{2} - x\right)\right] \qquad \text{[by Identity (2)]}$$

$$= \cos\left(\frac{\pi}{2} - x\right) \cdot (-1) \qquad \text{[by (7)]}$$

$$= \sin x \cdot (-1) \qquad \text{[by (3)]}$$

$$= -\sin x.$$

Corollary 2

If u is a differentiable function of x, then

$$\frac{d}{dx}[\cos u] = -\sin u \cdot \frac{du}{dx}. \qquad (9)$$

EXAMPLE 2   Differentiate

(a) $(x + \cos x)^2$ (b) $\sqrt{\cos x}$ (c) $\cos(e^x)$

Solution   (a) $\dfrac{d}{dx}[(x + \cos x)^2] = 2(x + \cos x)\dfrac{d}{dx}[x + \cos x]$

$$= 2(x + \cos x)(1 - \sin x)$$

(b) $\dfrac{d}{dx}[\sqrt{\cos x}] = \dfrac{d}{dx}[(\cos x)^{1/2}] = \dfrac{1}{2}(\cos x)^{-1/2}\dfrac{d}{dx}[\cos x]$

$$= \dfrac{1}{2}(\cos x)^{-1/2}(-\sin x)$$

$$= -\dfrac{\sin x}{2\sqrt{\cos x}}$$

(c) $\dfrac{d}{dx}[\cos(e^x)] = -\sin(e^x)\dfrac{d}{dx}[e^x]$

$$= -e^x \sin(e^x). \qquad \blacksquare$$

THEOREM 3

$$\frac{d}{dx}[\tan x] = \frac{1}{\cos^2 x} = \sec^2 x \tag{10}$$

(for x in radians).

Proof   Recall that $\tan x = \dfrac{\sin x}{\cos x}$. Thus, $\dfrac{d}{dx}[\tan x]$ may be found using Equations (6) and (8) along with the quotient rule.

$$\frac{d}{dx}[\tan x] = \frac{d}{dx}\left[\frac{\sin x}{\cos x}\right]$$

$$= \frac{\cos x \dfrac{d}{dx}[\sin x] - \sin x \dfrac{d}{dx}[\cos x]}{\cos^2 x}$$

$$= \frac{\cos x \cdot \cos x - \sin x \cdot (-\sin x)}{\cos^2 x}$$

$$= \frac{\cos^2 x + \sin^2 x}{\cos^2 x} = \frac{1}{\cos^2 x} = \left(\frac{1}{\cos x}\right)^2 = \sec^2 x.$$

Corollary 3   If u is a differentiable function of x, then

$$\frac{d}{dx}[\tan u] = \frac{du/dx}{\cos^2 u} = \sec^2 u \cdot \frac{du}{dx}. \tag{11}$$

EXAMPLE 3   Find the slope of the tangent line to the graph of $y = \tan 2x$ at the point where $x = \pi/8$.

Solution   We want to know dy/dx when $x = \pi/8$. We have

$$\frac{dy}{dx} = \frac{d}{dx} [\tan 2x] = \sec^2 2x \cdot \frac{d}{dx} [2x]$$

$$= 2 \sec^2 2x.$$

When $x = \pi/8$, we have

$$\left. \frac{dy}{dx} \right|_{x = \pi/8} = 2 \sec^2 \left[2 \left(\frac{\pi}{8} \right) \right] = 2 \sec^2 \left(\frac{\pi}{4} \right)$$

$$= 2 \cdot \left(\frac{2}{\sqrt{2}} \right)^2 = 4. \quad \left(\cos \frac{\pi}{4} = \frac{\sqrt{2}}{2}, \text{ so } \sec \frac{\pi}{4} = \frac{1}{\cos \dfrac{\pi}{4}} = \frac{2}{\sqrt{2}} \right) \quad \blacksquare$$

EXAMPLE 4   Recall the predator-prey interaction between lynxes and rabbits discussed in Sec-
(Predator   tion 9.1, and suppose that in a certain area the number of rabbits (in thousands)
-Prey) t months after 1 January 1987 is given by

$$N(t) = 10 + \sin \left(\frac{\pi}{24} t \right).$$

Find the instantaneous rate of change of the number of rabbits on 1 July 1988.

Solution   We have

$$N'(t) = \frac{d}{dt} \left[10 + \sin \left(\frac{\pi}{24} t \right) \right]$$

$$= \cos \left(\frac{\pi}{24} t \right) \frac{d}{dt} \left[\frac{\pi}{24} t \right]$$

$$= \frac{\pi}{24} \cos \left(\frac{\pi}{24} t \right).$$

On 1 July 1988, $t = 18$ months after 1 January 1987. Thus,

$$N'(18) = \frac{\pi}{24} \cos \left(\frac{\pi}{24} \cdot 18 \right) = \frac{\pi}{24} \cos \left(\frac{3}{4} \pi \right)$$

$$= \frac{\pi}{24} \left(-\frac{\sqrt{2}}{2} \right) = -\frac{\pi \sqrt{2}}{48} \approx -0.0926.$$

Hence, on 1 July 1988 the number of rabbits is decreasing at the rate of approximately 93 ($0.0926 \cdot 1000 = 92.6$) rabbits per month. $\blacksquare$

We can easily obtain the derivatives of the **remaining trigonometric functions** (Exercises 31–33) from the derivative formulas for the sine, cosine, and tangent functions.

9.2 EXERCISE SET ||||||||||||||||||||||||||||||||||||||||||||||||||||||||||||||||||||||||||||||||||||||||

In Exercises 1–30, find dy/dx.

1. $y = \cos 3x$

2. $y = \sin x^3$

3. $y = \sin (x^2 + 1)$

4. $y = x \cos x$

5. $y = (\sin x)^3$

6. $y = \sin x + \cos x$

7. $y = x^2 \cos x$

8. $y = \dfrac{\sin x}{x^2}$

9. $y = \sin x \cos x$

10. $y = \dfrac{\sin x}{\cos^2 x}$

11. $y = \cos x^2 + \cos x^3$

12. $y = \cos (x^2 + x^3)$

13. $y = \tan^2 x$

14. $y = \tan x^2$

15. $y = x^3 \tan x$

16. $y = \dfrac{\tan x}{x}$

17. $y = \sqrt{\sin x}$

18. $y = \sqrt{\tan x}$

19. $y = \cos (e^{2x})$

20. $y = \sin (\ln x)$

21. $y = e^x \sin x$

22. $y = e^{\cos x}$

23. $y = \sin \sqrt{x - 1}$

24. $y = \cos (1 + \sqrt{x})$

25. $y = \sin^2 x \cos x$

26. $y = \sin x \tan x$

27. $y = \ln (\sin x)$

28. $y = \ln (\tan x)$

29. $y = \dfrac{x}{\sin x}$

30. $y = \dfrac{x^2 + 1}{\cos x}$

31. Show that $\dfrac{d}{dx} [\csc x] = -\csc x \cot x$.

   (*Hint:* $\csc x = 1/\sin x$.)

32. Show that $\dfrac{d}{dx} [\sec x] = \sec x \tan x$. (*Hint:* $\sec x = 1/\cos x$.)

33. Show that $\dfrac{d}{dx} [\cot x] = -\csc^2 x$.

   (*Hint:* $\cot x = \cos x/\sin x$ and $\sin^2 x + \cos^2 x = 1$.)

In Exercises 34–39, find $\partial z/\partial x$ and $\partial z/\partial y$.

34. $z = \sin x + \cos y$

35. $z = \cos x \sin y$

36. $z = \dfrac{\cos (3x)}{\sin (2y)}$

37. $z = \sin \left(\dfrac{x}{y}\right)$

38. $z = \sin x \tan y$

39. $z = \tan (x^3 y^4)$

40. Find the slope of the tangent line to the graph of $f(x) = \sin x$ at the point where $x = \pi/3$.

41. Find the slope of the tangent line to the graph of $f(x) = \cos 2x$ at the point where $x = \pi/6$.

42. Find an equation of the tangent line to the graph of $f(x) = \cos 3x$ at the point where $x = \pi/2$.

43. Find an equation of the tangent line to the graph of $f(x) = \tan x$ at the point where $x = \pi$.

44. **(Marginal Cost)** A stove manufacturer finds that the cost (in hundreds of dollars) of making x stoves is given by

$$C(x) = x\left[200 + 20 \sin \left(\frac{\pi}{12} x\right)\right]$$

where x is in thousands. Find the marginal cost when the level of production is 18,000 stoves.

45. **(Pollution)** An aluminum-manufacturing plant that went into operation on 1 July 1987, is dumping waste into a nearby stream. The amount of waste dumped

every day (in thousands of gallons) is given by

$$A(t) = 18 + 12 \sin\left(\frac{\pi}{6} t\right)$$

where t is the number of months the plant has been in operation. Find the instantaneous rate at which the amount of waste dumped each day is changing on 1 January 1988 ($t = 6$).

46. **(Temperature)** During 1987, it was found that the temperature (in degrees Fahrenheit) at a certain Caribbean resort is given by

$$F(t) = 60 + \sin\left(\frac{\pi}{4} t\right)$$

where t is the number of months that have passed since 1 January 1987. Find the instantaneous rate at which the temperature is changing on 1 October 1987 ($t = 9$).

47. **(Length of Shadow)** The angle of elevation (the angle between the horizontal and the line of sight) of the sun is decreasing at the rate of 15° per hour. When the angle of elevation is 30°, find the rate at which the length of the shadow cast by a 20-foot tree is increasing. (*Hint:* Change all degrees to radians.)

48. A balloon is rising straight up into the air. Fifty feet from the point on the ground directly beneath the balloon, a TV camera is videotaping its ascent. How fast is the camera angle θ (see Figure 1) changing when the balloon is 100 feet off the ground and rising at the rate of 8 feet per second?

h

θ

50

Figure 1

9.3 ‖‖‖‖‖‖   INTEGRATION OF TRIGONOMETRIC FUNCTIONS

Because every differentiation formula yields a corresponding integration formula, we have the following integration formulas for the sine and cosine functions (see Theorems 1 and 2 of Section 9.2):

THEOREM 1

$$\int \sin u \, du = -\cos u + C \qquad (1)$$

$$\int \cos u \, du = \sin u + C \qquad (2)$$

EXAMPLE 1   Find the indefinite integral of $f(x) = \cos 3x$.

Solution   Using the method of substitution (Section 6.2), we let

$$u = 3x.$$

Then $du/dx = 3$ and so

$$du = 3 \, dx.$$

Substituting, we have

$$\int \cos 3x \, dx = \tfrac{1}{3} \int \cos 3x \, (3 \, dx)$$
$$= \tfrac{1}{3} \int \cos u \, du$$
$$= \tfrac{1}{3} \sin u + C \qquad \text{[from (2)]}$$
$$= \tfrac{1}{3} \sin 3x + C.$$

EXAMPLE 2   Evaluate

$$\int_{\pi/4}^{\pi/2} \sin x \, dx.$$

Solution   We first obtain the indefinite integral,

$$\int \sin x \, dx = -\cos x + C.$$

By definition of the definite integral (Section 7.1), we have

$$\int_{\pi/4}^{\pi/2} \sin x \, dx = -\cos x \Big]_{\pi/4}^{\pi/2} = -\cos \frac{\pi}{2} - \left(-\cos \frac{\pi}{4} \right)$$
$$= 0 - \left(-\frac{\sqrt{2}}{2} \right)$$
$$= \frac{\sqrt{2}}{2}.$$

EXAMPLE 3   Find $\int \sin^3 x \cos x \, dx$.

Solution   Again using the method of substitution, we let

$$u = \sin x.$$

Then $du/dx = \cos x$ and so

$$du = \cos x \, dx.$$

Hence,

$$\int \sin^3 x \cos x \, dx = \int u^3 \, du$$
$$= \frac{u^4}{4} + C$$
$$= \frac{\sin^4 x}{4} + C.$$

EXAMPLE 4   Evaluate $\int_0^{\pi/2} x \cos x \, dx$.

Solution   We use the method of integration by parts (Section 6.3), as follows. Let

$$u = x \quad \text{and} \quad dv = \cos x,$$

so that

$$du = dx \quad \text{and} \quad v = \int v'(x)\,dx = \int \cos x\,dx$$
$$= \sin x.$$

Then we obtain the indefinite integral.

$$\int x \cos x\,dx = \int u\,dv$$
$$= uv - \int v\,du$$
$$= x \sin x - \int \sin x\,dx$$
$$= x \sin x - (-\cos x) + C$$
$$= x \sin x + \cos x + C.$$

Finally,

$$\int_0^{\pi/2} x \cos x\,dx = (x \sin x + \cos x)\Big]_0^{\pi/2}$$
$$= \left(\frac{\pi}{2} \sin \frac{\pi}{2} + \cos \frac{\pi}{2}\right) - \left(0 \cdot \sin 0 + \cos 0\right)$$
$$= \left(\frac{\pi}{2} + 0\right) - \left(0 + 1\right) = \frac{\pi}{2} - 1. \qquad \blacksquare$$

EXAMPLE 5   Find the area of the region bounded by the curve $y = \sin x$ and the x-axis from $x = 0$ to $x = \pi$ (Figure 1).

Figure 1

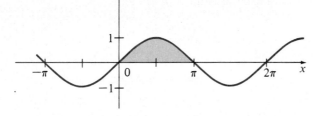

Solution   The area we want to find is the definite integral $\int_0^\pi \sin x\,dx$, which we evaluate as follows:

$$\int_0^\pi \sin x\,dx = -\cos x\Big]_0^\pi = -\cos \pi - (-\cos 0)$$
$$= -(-1) - (-1) = 2. \qquad \blacksquare$$

THEOREM 2
$$\int \tan x\,dx = -\ln |\cos x| + C \qquad (3)$$

Proof We represent $\tan x$ as $\sin x/\cos x$ and use the method of substitution as follows. Let

$$u = \cos x.$$

Then $du/dx = -\sin x$, so that

$$du = -\sin x \, dx.$$

Substituting, we have

$$\int \tan x \, dx = \int \frac{\sin x}{\cos x} \, dx$$

$$= -\int \frac{-\sin x}{\cos x} \, dx$$

$$= -\int \frac{1}{u} \, du$$

$$= -\ln |u| + C$$

$$= -\ln |\cos x| + C.$$

Corollary If u is a differentiable function then

$$\int \tan u \, du = -\ln |\cos u| + C. \qquad (4)$$

EXAMPLE 6 Find $\int x \tan x^2 \, dx$.

Solution Using the method of substitution, we let

$$u = x^2.$$

Then $du = 2x \, dx$, so that $\frac{1}{2} du = x \, dx$. Thus,

$$\int x \tan x^2 \, dx = \int \tan u \cdot \frac{1}{2} \, du$$

$$= \frac{1}{2} \int \tan u \, du$$

$$= -\frac{1}{2} \ln |\cos u| + C \qquad \text{[by (4)]}$$

$$= -\frac{1}{2} \ln |\cos x^2| + C. \qquad \blacksquare$$

EXAMPLE 7 (Marginal Revenue) Suppose that the marginal revenue for a manufacturer of sports cars (in millions of dollars per thousand cars) is given by

$$R'(x) = 5x - \sin \pi x,$$

where x is the number of cars (in thousands) manufactured and sold. Suppose that every car that is made is sold. If the revenue is zero when the level of production is zero, find the revenue when the level of production is 4000 cars.

Solution   We can obtain the revenue function $R(x)$ by integrating the marginal revenue $R'(x)$. That is,

$$
\begin{aligned}
R(x) &= \int R'(x)\, dx \\
&= \int [5x - \sin \pi x]\, dx \\
&= \frac{5}{2} x^2 - \frac{1}{\pi} \int \sin \pi x (\pi\, dx) \qquad (u = \pi x;\ du = \pi\, dx) \\
&= \frac{5}{2} x^2 - \frac{1}{\pi} \int \sin u\, du \\
&= \frac{5}{2} x^2 - \frac{1}{\pi} [-\cos u] + C \\
&= \frac{5}{2} x^2 + \frac{1}{\pi} \cos \pi x + C.
\end{aligned}
$$

Thus,

$$
R(x) = \frac{5}{2} x^2 + \frac{1}{\pi} \cos \pi x + C.
$$

We now determine C by using the initial condition $R(0) = 0$. We have

$$
0 = R(0) = \frac{5}{2} (0)^2 + \frac{1}{\pi} \cos 0 + C = \frac{1}{\pi} + C.
$$

Hence,

$$
C = -\frac{1}{\pi}
$$

and

$$
R(x) = \frac{5}{2} x^2 + \frac{1}{\pi} \cos \pi x - \frac{1}{\pi}.
$$

Then

$$
R(4) = \frac{5}{2} \cdot (4)^2 + \frac{1}{\pi} \cos 4\pi - \frac{1}{\pi}
$$

$$
= 40 + \frac{1}{\pi} - \frac{1}{\pi} = 40.
$$

Thus, the revenue is \$40 million when the level of production is 4000 cars.   ∎

EXAMPLE 8
(Predator-
Prey)

Consider a predator-prey interaction in a forest between a certain type of sapling and a predator insect. Suppose that the rate of change in the number of saplings (in thousands per month) is given by

$$N'(t) = -\frac{5\pi}{18} \sin\left(\frac{\pi}{18} t\right)$$

where t is the number of months after 1 May. If the forest contained 8000 saplings on 1 May, how many saplings were there 36 months later?

Solution

The number of saplings in the forest t months after 1 May is

$$N(t) = \int N'(t)\, dt$$

$$= \int -\frac{5\pi}{18} \sin\left(\frac{\pi}{18} t\right) dt$$

$$= -\frac{5\pi}{18} \int \sin\left(\frac{\pi}{18} t\right) dt$$

$$= 5 \cos\left(\frac{\pi}{18} t\right) + C.$$

Thus,

$$N(t) = 5 \cos\left(\frac{\pi}{18} t\right) + C.$$

We determine C by using the initial condition $N(0) = 8$.

$$8 = N(0) = 5 \cos\left(\frac{\pi}{18} \cdot 0\right) + C$$

$$= 5 + C$$

so

$$C = 3.$$

Hence,

$$N(t) = 5 \cos\left(\frac{\pi}{18} t\right) + 3.$$

After 36 months, the number of saplings in the forest is given by

$$N(36) = 5 \cos\left(\frac{\pi}{18} \cdot 36\right) + 3$$

$$= 5 \cos 2\pi + 3$$

$$= 8.$$

Hence, the forest again has 8000 saplings after 36 months. ∎

9.3 EXERCISE SET ||||||||||||||||||||||||||||||||||||||||||||||||||||||||||||||||||||||||||||||||||||

In Exercises 1–22, find the indefinite integral.

1. $\int \sin 5x \, dx$

2. $\int \cos \frac{1}{2} x \, dx$

3. $\int \cos (x + 4) \, dx$

4. $\int \sin (-3x) \, dx$

5. $\int (\sin x + \cos x) \, dx$

6. $\int (1 - \cos x) \, dx$

7. $\int \sin x \cos x \, dx$

8. $\int \sin^2 x \cos x \, dx$

9. $\int \sin x \cos^3 x \, dx$

10. $\int \sqrt{\cos x} \sin x \, dx$

11. $\int x^2 \sin (x^3) \, dx$

12. $\int x \cos (x^2) \, dx$

13. $\int \frac{\cos x}{\sin^2 x} \, dx$

14. $\int \frac{\sin x}{\cos^3 x} \, dx$

15. $\int \frac{\cos x}{\sqrt{\sin x}} \, dx$

16. $\int \sin \left(x + \frac{\pi}{4} \right) dx$

17. $\int x \sin x \, dx$

18. $\int e^{\sin x} \cos x \, dx$

19. $\int \tan 3x \, dx$

20. $\int x^2 \tan x^3 \, dx$

21. $\int \tan x \sec^2 x \, dx$

22. $\int \frac{\sec^2 x}{\tan x} \, dx$

In Exercises 23–32, evaluate the definite integral.

23. $\int_{\pi/6}^{\pi/3} \cos x \, dx$

24. $\int_0^\pi \sin \frac{1}{3} x \, dx$

25. $\int_{\pi/4}^{3\pi/4} \sin 2x \, dx$

26. $\int_0^{\pi/2} x \sin x \, dx$

27. $\int_0^{\pi/4} \sin \left(x + \frac{\pi}{4} \right) dx$

28. $\int_0^{\pi/2} \sin 2x \cos 2x \, dx$

29. $\int_0^\pi x^3 \sin (x^4) \, dx$

30. $\int_{\pi/4}^{\pi/2} \sin x \cos x \, dx$

31. $\int_0^{\pi/4} x \cos 2x \, dx$

32. $\int_0^\pi \cos x \sqrt{\sin x} \, dx$

33. Referring to Figure 2, find the area of the region below the curve $y = \sin (x/2)$ and above the x-axis from $x = 0$ to $x = \pi/2$.

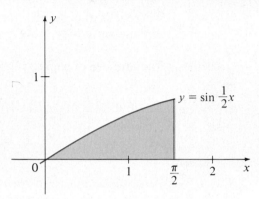

Figure 2

34. Referring to Figure 3, find the area of the region below the curve $y = x + \sin x$ and above the x-axis from $x = 0$ to $x = \pi/4$.

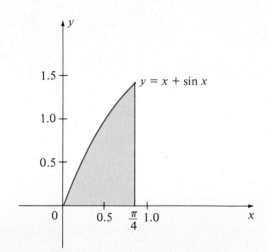

Figure 3

35. Referring to Figure 4, find the area of the region bounded by the y-axis and the curves $y = \sin x$ and $y = \cos x$ from $x = 0$ to $x = \pi/4$.

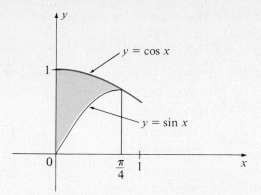

Figure 4

36. Referring to Figure 5, find the area of the region below the curve $y = x$ and above the curve $y = \sin x$ from $x = 0$ to $x = \pi$.

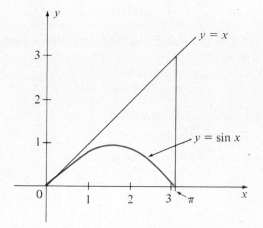

Figure 5

37. The slope of the tangent line to a curve at the point (x, y) on the curve is $1 - 2 \cos 2x$. If the point $(\pi/2, \pi/2)$ lies on the curve, find an equation for the curve.

38. **(Population Growth)** The rate of change in an animal population is given by

$$P'(t) = 40\pi \sin{(\pi t/2)}$$

where t is the number of days after the start of an experiment. What is the change in population from $t = 1$ to $t = 7$?

39. **(Motion)** Starting from rest, an object moves along a straight line so that its velocity after t seconds of motion is given by

$$v(t) = 1 - \cos{(\pi t/2)}$$

Find the distance traveled
(a) during the first 10 seconds
(b) between $t = 5$ and $t = 12$ seconds

40. **(Business)** Suppose that the rate of change in the profit of a toy manufacturer (in thousands of dollars per month) is given by

$$P'(t) = -\pi \sin{(\pi t/6)}$$

where t is the number of months since 1 December. If the profit on 1 March ($t = 3$) was $28,000, find the profit on the following 1 September ($t = 9$).

41. Show that $\int \sec^2 x\, dx = \tan x + C.$

42. Show that $\int \tan^2 x\, dx = \tan x - x + C.$

[*Hint:* use Equation (12) of Section 9.1]

KEY IDEAS FOR REVIEW ||||||||||||||||||||||||||||||||||||||||||||||||||||||||||||||||||||||||||||

- Angles are measured in degrees and in radians.

- 2π radians $= 360°$.

- $\dfrac{\text{radian measure of } \theta}{\pi \text{ radians}} = \dfrac{\text{degree measure of } \theta}{180°}.$

- The isosceles right triangle:

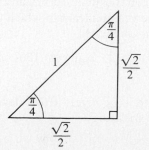

■ The 30°, 60°, 90° triangle:

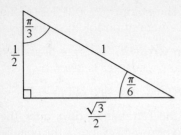

■ Let θ be a fixed angle that is in standard position. Let $P(x, y)$ be the point at which the terminal side of θ intersects the unit circle centered at the origin. Then

$$\sin \theta = y, \qquad \cos \theta = x,$$

$$\tan \theta = \frac{y}{x}, \qquad x \neq 0$$

$$\csc \theta = \frac{1}{y}, \qquad y \neq 0$$

$$\sec \theta = \frac{1}{x}, \qquad x \neq 0$$

$$\cot = \frac{x}{y} = \frac{1}{\tan \theta}, \qquad y \neq 0.$$

■ $\tan \theta = \dfrac{\sin \theta}{\cos \theta};$ $\sec \theta = \dfrac{1}{\cos \theta};$ $\csc \theta = \dfrac{1}{\sin \theta};$

$\cot \theta = \dfrac{1}{\tan \theta}$

■ $\sin (\theta + 2\pi) = \sin \theta; \cos (\theta + 2\pi) = \cos \theta;$
$\tan (\theta + \pi) = \tan \theta$

■ $\sin (-\theta) = -\sin \theta; \cos (-\theta) = \cos \theta;$
$\tan (-\theta) = -\tan \theta$

■ $\sin^2 \theta + \cos^2 \theta = 1$

■ If θ is an acute angle in a right triangle, then

$$\sin \theta = \frac{\text{opposite}}{\text{hypotenuse}}$$

$$\cos \theta = \frac{\text{adjacent}}{\text{hypotenuse}}$$

$$\tan \theta = \frac{\text{opposite}}{\text{adjacent}}.$$

■ If $y = \sin u$ and u is a differentiable function of x, then

$$\frac{dy}{dx} = \frac{d}{dx} [\sin u] = \cos u \cdot \frac{du}{dx}.$$

■ If $y = \cos u$ and u is a differentiable function of x, then

$$\frac{dy}{dx} = \frac{d}{dx} [\cos u] = -\sin u \cdot \frac{du}{dx}.$$

■ If $y = \tan u$ and u is a differentiable function of x, then

$$\frac{dy}{dx} = \frac{d}{dx} [\tan u] = \sec^2 u \cdot \frac{du}{dx}.$$

■ $\int \sin u \, du = -\cos u + C$

■ $\int \cos u \, du = \sin u + C$

■ $\int \tan u \, du = -\ln |\cos u| + C$

REVIEW EXERCISES ||||||||||||||||||||||||||||||||||||||||||||||||||||||||||||||||||||||||||||

In Exercises 1 and 2, if the angle θ is in standard position, determine the quadrant in which the terminal side of the angle lies.

1. $\theta = 112°$ 2. $\theta = -130°$

In Exercises 3 and 4, convert from degree measure to radian measure.

3. $315°$ 4. $-220°$

In Exercises 5 and 6, convert from radian measure to degree measure.

5. $\dfrac{11}{6} \pi$

6. $-\dfrac{9}{4} \pi$

7. Give the radian measure of the angle shown in Figure 6.

Figure 6

In Exercises 8 and 9, use (6) and (7) from Section 9.1 to evaluate the trigonometric functions.

8. $\sin (5\pi)$

9. $\tan \left(-\frac{2}{3}\pi\right)$

10. Use trigonometry to find the hypotenuse of an isosceles right triangle the legs of which are 6 units in length.

In Exercises 11 and 12, find the values of $\sin \theta$, $\cos \theta$, and $\tan \theta$ for the indicated angle θ in each right triangle.

11.

12.

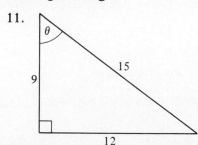

13. Find the values of $\sin \theta$, $\cos \theta$, and $\tan \theta$ if the point $P(-6, -8)$ lies on the terminal side of θ and θ is in standard position.

In Exercises 14–21, find dy/dx.

14. $y = \sin \left(\dfrac{\pi}{2} - 2x\right)$

15. $y = \sin x - \cos 2x$

16. $y = \cos (e^{x^2})$

17. $y = \cos^2 x \sin 2x$

18. $y = \sin^3 x + 2 \cos^2 x$

19. $y = \dfrac{3x}{\sin x}$

20. $y = x \tan x$

21. $y = \sqrt{\tan x}$

In Exercises 22 and 23, find $\partial z/\partial x$ and $\partial z/\partial y$.

22. $z = \sin 2x \cos \frac{1}{2}y$

23. $z = 3 \sin 2x - 2 \cos 3y$

24. Find the slope of the tangent line to the graph of $f(x) = \cos (x/2)$ at the point where $x = \pi/2$.

25. Find an equation of the tangent line to the graph of $f(x) = \sin x + \cos 2x$ at the point where $x = \pi/2$.

In Exercises 26–29, find the indefinite integrals.

26. $\displaystyle\int \frac{\sin x}{\sqrt{\cos x}}\, dx$

27. $\displaystyle\int e^{\cos x} \sin x\, dx$

28. $\displaystyle\int \frac{\sin x}{\cos^2 x}\, dx$

29. $\displaystyle\int x \sin \frac{1}{2} x\, dx$

In Exercises 30 and 31, evaluate the definite integrals.

30. $\displaystyle\int_{\pi/4}^{3\pi/4} \sin 2x\, dx$

31. $\displaystyle\int_{\pi/2}^{3\pi/2} x \sin 2x\, dx$

32. The slope of the tangent line to a curve at the point (x, y) on the curve is $\sin \frac{1}{2}x$. If

the point $(\pi, 1)$ lies on the curve, find an equation for the curve.

33. Find the area of the region bounded by the curve $y = \sin x$ and the x-axis from $x = -\pi$ to $x = \pi$.

34. Referring to Figure 7, find the area of the region below the curve $y = \sin x$ and above the curve $y = \cos x$ from $x = \pi/4$ to $x = \pi/2$.

Figure 7

35. A ladder that is 20 feet long touches the wall against which it leans at a point 16 feet above the ground. Find the angle that the ladder makes with the ground.

36. Find the approximate angle of elevation (the angle between the horizontal and the line of sight) of the sun when a tower 50 meters in height casts a shadow 25 meters in length.

37. **(Medicine)** In a town that has been struck by a certain disease, public health officials have determined that the number $N(t)$ of people who have the disease at time t (in days) is approximately

$$N(t) = 100 + 100 \sin\left(\frac{\pi t}{10}\right), \qquad 0 \le t \le 15.$$

At what rate is the number of people who have the disease changing when
(a) $t = \frac{5}{2}$ days   (b) $t = \frac{40}{3}$ days

38. **(Marginal Cost)** A manufacturer's marginal cost (in millions of dollars per one thousand units) is given by

$$C'(x) = 2 + 3x - \sin\left(\frac{\pi x}{20}\right),$$

where x is the number of thousands of units made. If the cost is \$4 million when the level of production is zero, find the cost when the level of production is 10,000 units.

CHAPTER TEST |||||||||||||||||||||||||||||||||||||||||||||||||||||||||||||||||||||||||||||||||||

1. (a) Convert $150°$ to radian measure.

   (b) Convert $\dfrac{7\pi}{3}$ radians to degree measure.

2. Find the sine, cosine, and tangent of the angle θ in standard position, if the point $(-3, 1)$ lies on its terminal side.

In Exercises 3–6, find $\dfrac{dy}{dx}$.

3. $y = \cos\left(3x + \dfrac{\pi}{2}\right)$.

4. $y = 2 \cos x + 3 \tan x$

5. $y = x^3 \sin(x^2)$

6. $y = (\tan x)^3$

In Exercises 7–9, find the integral.

7. $\int \sqrt{\sin x} \cos x \, dx$ 8. $\int x \cos (x^2) \, dx$ 9. $\int \tan (3x - 4) \, dx$

10. Due to the interaction between a species of plant and its environment, the rate of change in the number of plants per month (in millions per month) in a certain forest is

$$N'(t) = \frac{5\pi}{3} \cos \left(\frac{\pi}{6} t \right),$$

where t is the number of months after 1 April. If the forest contains 12 million plants of this species on 1 April, how many plants of this species are there t months later?

DIFFERENTIAL EQUATIONS

10.1 |||||| FUNDAMENTAL CONCEPTS

A **differential equation** is an equation that involves an unknown function and its derivatives. In Chapter 5 we saw an example of a simple differential equation: $dQ/dt = kQ$, where k is a constant. Other examples of differential equations are

$$\frac{dy}{dx} + 2y = 4e^x \tag{1}$$

$$\frac{d^2y}{dx^2} - 3\frac{dy}{dx} - 4y = 0 \tag{2}$$

$$xy^3 \frac{d^2y}{dx^2} + 3x^2y = xe^y. \tag{3}$$

Differential equations appear in mathematical models of many phenomena in the real world. Applications occur in engineering, the physical sciences, biology, the medical sciences, psychology, and business and economics.

To solve many of the exercises of this chapter, the student will need a good quality scientific calculator with the functions e^x, $\ln x$, $\sin x$ and $\cos x$ (in radians); parentheses and memory will also be helpful.

|||||||||||| **Higher Order Derivatives**

We are already familiar with the second derivative of a function $y = f(x)$:

$$y'' = \frac{d}{dx}[y'] = \frac{d^2}{dx^2}[y] = \frac{d^2}{dx^2}[f(x)] = \frac{d}{dx}\left[\frac{d}{dx}f(x)\right]. \tag{4}$$

We can continue, with the third derivative

$$y''' = \frac{d}{dx}[y''] = \frac{d^3}{dx^3}[y] = \frac{d^3}{dx^3}[f(x)] = \frac{d}{dx}\left[\frac{d^2}{dx^2}f(x)\right] \tag{5}$$

and the fourth derivative

$$y^{(4)} = y'''' = \frac{d}{dx}[y'''] = \frac{d^4}{dx^4}[y] = \frac{d^4}{dx^4}[f(x)] = \frac{d}{dx}\left[\frac{d^3}{dx^3}f(x)\right], \qquad (6)$$

and so on, as many times as the function is differentiable.

As in Equation (6), when we wish to indicate an nth derivative, we may use the notation $y^{(n)}$; the parentheses are used to distinguish the nth derivative from the nth power. This is particularly useful when $n \geq 4$, as the prime notation then becomes more awkward. In general, an nth derivative is called an **nth order derivative**.

EXAMPLE 1   For the function $f(x) = x^4 - 5x^3 + \dfrac{3}{x}$,

$$f'(x) = 4x^3 - 15x^2 - \frac{3}{x^2},$$

$$f''(x) = 12x^2 - 30x + \frac{6}{x^3},$$

$$f'''(x) = 24x - 30 - \frac{18}{x^4},$$

$$f^{(4)}(x) = 24 + \frac{72}{x^5},$$

$$f^{(5)}(x) = -\frac{360}{x^6},$$

and so on. ∎

Be careful to note the difference between $\left(\dfrac{d}{dx}f(x)\right)^n$ and $\dfrac{d^n}{dx^n}[f(x)]$; that is, between y'^n and $y^{(n)}$. For example,

$$\left(\frac{d}{dx}x^4\right)^3 = (4x^3)^3 = 64x^9, \qquad (7)$$

whereas

$$\frac{d^3}{dx^3}[x^4] = \frac{d^2}{dx^2}[4x^3] = \frac{d}{dx}[12x^2] = 24x. \qquad (8)$$

Thus, in Equation (8) we have a third derivative, but in Equation (7) we have the third power of a first derivative.

The **order** of a differential equation is the order of the highest derivative appearing in the equation. Thus, the differential equation $dQ/dt = kQ$ and Equation (1) are first-order equations. Equations (2) and (3) are second-order differential equations.

EXAMPLE 2   State the order of each of the following differential equations:

(a) $x^5 \dfrac{d^3y}{dx^3} + 5x \dfrac{d^2y}{dx^2} - 8y^4 = x^7$.

(b) $x^3 y^{(4)} - x^8 y''' + 25y'' = x^{15}$.

(c) $\sin x \left(\dfrac{dy}{dx}\right)^5 + (\cos y) \dfrac{d^2y}{dx^2} = x^3 y^6$.

Solution   (a) The order is 3, since the highest order derivative occurring in the equation is the third derivative.

(b) The order is 4.

(c) The order is 2, since the highest order derivative occurring in the equation is $\dfrac{d^2y}{dx^2}$. ∎

|||||||||||| **Solutions of a Differential Equation**

A function $y = f(x)$ is called a **solution** of a differential equation if it satisfies the given equation. Thus, a solution is a function rather than a number. Moreover, a differential equation may have *many* solutions, even infinitely many.

EXAMPLE 3   Verify that every function of the form $y = Ce^{-3x}$, where C is an arbitrary constant, is a solution of the differential equation

$$y' + 3y = 0. \qquad (9)$$

Solution   If $y = Ce^{-3x}$, then

$$y' = \frac{d}{dx}[Ce^{-3x}]$$
$$= -3Ce^{-3x}.$$

Substituting in the given equation, we obtain

$$y' + 3y = -3Ce^{-3x} + 3(Ce^{-3x}) = 0.$$

Therefore, $y = Ce^{-3x}$ is a solution of Equation (9). ∎

Note that differential equation (9) has infinitely many solutions. Each value of the arbitrary constant C determines another solution of the given differential equation. For example,

$$5e^{-3x}, \quad \tfrac{2}{3}e^{-3x}, \quad \text{and} \quad \sqrt{2}e^{-3x}$$

are all solutions of $y' + 3y = 0$.

On the other hand, $y = e^{4x}$ is not a solution of differential equation (9) because $y' = 4e^{4x}$ and

$$y' + 3y = 4e^{4x} + 3e^{4x} = 7e^{4x} \neq 0.$$

In Section 10.2 it will be shown that *every* solution of the given differential equation (9) can be obtained from the solution $y = Ce^{-3x}$ by assigning a value to the arbitrary constant C.

EXAMPLE 4   Verify that $y = x^2 + 2x + 1$ is a solution of the differential equation

$$(x^2 - 1)y'' + 2y' - 2y = 0. \tag{10}$$

Solution   First, we calculate y' and y'' for the given function $y = x^2 + 2x + 1$:

$$y' = 2x + 2 \qquad \text{and} \qquad y'' = 2.$$

Then

$$\begin{aligned}
(x^2 - 1)y'' + 2y' - 2y &= (x^2 - 1)(2) + 2(2x + 2) - 2(x^2 + 2x + 1) \\
&= 2x^2 - 2 + 4x + 4 - 2x^2 - 4x - 2 \\
&= 0.
\end{aligned}$$

Therefore, $y = x^2 + 2x + 1$ is a solution of Equation (10).   ∎

The simplest type of differential equation is one of the form

$$\frac{dy}{dx} = f(x). \tag{11}$$

Clearly, this equation is "solved" by the function

$$y = \int f(x)\, dx. \tag{12}$$

By definition of the indefinite integral, the function given by Equation (12) is the most general solution of Equation (11); there are no others.

EXAMPLE 5   Find the most general solution of

$$\frac{d^2 y}{dx^2} = x^3 + \cos x. \tag{13}$$

Solution   From Equation (13), one integration shows that

$$\begin{aligned}
\frac{dy}{dx} &= \int (x^3 + \cos x)\, dx \\
&= \tfrac{1}{4}x^4 + \sin x + C.
\end{aligned}$$

A second integration shows that

$$y = \int (\tfrac{1}{4}x^4 + \sin x + C)\, dx$$
$$y = \tfrac{1}{20} x^5 - \cos x + Cx + D. \tag{14}$$

■

The **general solution** of a differential equation is a formula from which all particular solutions can be obtained by substituting values for the unspecified constants contained in that formula. In Example 5, every solution of the differential equation (13) can be obtained from the general solution (14) by substituting numerical values for C and D.

||||||||||||| **Initial-Value Problems**

As we have seen, a differential equation may have infinitely many solutions. Very often in applications, only a certain specific one of these solutions is of interest to us. Data specified in the problem may require us to reject the others.

EXAMPLE 6   Find the solution of the differential equation

$$\frac{dy}{dx} = x + e^{-x} \tag{15}$$

for which $y(0) = 1$.

Solution   Integrating (15) we obtain

$$y = \tfrac{1}{2}x^2 - e^{-x} + C. \tag{16}$$

Equation (16) is to be regarded as an identity; that is, it is true for all values of x. The required condition $y(0) = 1$ means that $y = 1$ when $x = 0$, or

$$1 = \tfrac{1}{2}(0)^2 - e^{-0} + C$$
$$1 = -1 + C$$
$$2 = C.$$

Thus, from the general solution (16) we obtain the particular solution

$$y = \tfrac{1}{2}x^2 - e^{-x} + 2.$$

This is the *only* solution of (15) that satisfies the required condition $y(0) = 1$.   ■

In Example 6 the condition $y(0) = 1$ is called an **initial condition.** An **initial-value problem** is a differential equation with one or more initial conditions specified. To "solve" an initial-value problem means to find the specific solution(s) of the given differential equation that also satisfy the given initial condition(s). Very often, as in Example 6, there is only one such solution.

10.1 EXERCISE SET ||||||||||||||||||||||||||||||||||||||||||||||||||||||||||||||||||||||||||||||||||||

In Exercises 1–8, find the indicated derivative of the given function.

1. $y = 5x^3 + 21x - 7$; y'''

2. $y = 18x^2 + 4x^5 - 4x^3$; $y^{(4)}$

3. $f(x) = e^{-3x}$; $f^{(4)}(x)$

4. $f(x) = x^4 + \sin\left(\dfrac{x}{2}\right)$; $f^{(6)}(x)$

5. $\dfrac{d^4}{dx^4}\left(\dfrac{1+x}{1-x}\right)$

6. $\dfrac{d^3}{dx^3}(x \ln x)$

7. $y = \sqrt{x}$; $\dfrac{d^5 y}{dx^5}$

8. $y = e^{-x} - \dfrac{3}{x}$; $y^{(5)}$

In Exercises 9–14, state the order of the given differential equation.

9. $x^2 y^5 + x^3 y^{(3)} - \dfrac{x}{y} = 10$

10. $xy^{(4)} - x^3 y^{(2)} + y^5 = x^6$

11. $8x\left(\dfrac{dy}{dx}\right)^4 - \dfrac{y^5}{\sqrt{x}} - \dfrac{d^2 y}{dx^2} = 0$

12. $\left(\dfrac{dy}{dx}\right)^3 = x^3 + 3\dfrac{dy}{dx}$

13. $\dfrac{1}{x^2}\dfrac{dy}{dx} + \dfrac{1}{x}\dfrac{d^4 y}{dx^4} = x\left(\dfrac{d^2 y}{dx^2}\right)^3$

14. $\sqrt{x}\, y' + \dfrac{y''}{x} - y^4 = 0.$

In Exercises 15–22, verify that the given function $y = f(x)$ is a solution of the given differential equation.

15. $y = x^2 - 1$; $(y')^2 - 4y = 4$

16. $y = Ce^{x^2}$; $y' = 2xy$

17. $y = 3 \sin x - 2 \cos x$; $y'' + y = 0$

18. $y = xe^x$; $y'' - y' = \dfrac{y}{x}$

19. $y = e^{3x}$ and $y = e^{-x}$; $y'' - 2y' - 3y = 0$

20. $y = e^{2x}$ and $y = e^{-5x}$; $y'' + 3y' = 10y$

21. $y = \sqrt{x}$; $x^2 y''' - xy'' - y' = \dfrac{1}{8y}$

22. $y = x \ln x$; $x^2 y^{(4)} - 2y'' + xy' - y = x$

In Exercises 23–30, find the general solution of the given differential equation.

23. $y' = 3x - \dfrac{1}{x}$

24. $y' = x^3 + \dfrac{1}{x^3}$

25. $y'' = x + 4$

26. $y'' - 2x = 3$

27. $x^2 y' = 3x + 5$

28. $xy' + x^2 = 4$

29. $y'' = 6 \sin 2x$

30. $x^2(y'' + e^x) = 1$

In Exercises 31–36, solve the given initial-value problem.

31. $y' = e^{3x}$, $y(0) = 1$

32. $y' = \sin x + \cos x$, $y(0) = 2$

33. $y' = \cos 2x - \sin 3x$, $y(0) = 0$

34. $y' = x^3 - \dfrac{1}{x}$, $y(1) = 3$

35. $y'' = \sin x$, $y(0) = 4$, $y'(0) = 2$

36. $y'' = \sec^2 x + 2$, $y(0) = -1$, $y'(0) = 3$.

10.2 |||||| SEPARABLE, FIRST-ORDER DIFFERENTIAL EQUATIONS

We shall now look at first-order differential equations that can be written in the form

$$\frac{dy}{dx} = f(x)g(y), \tag{1}$$

where $f(x)$ is a function of x only and $g(y)$ is a function of y only. Such differential equations are called **separable**, because we can rewrite the equation as

$$\frac{1}{g(y)} \, dy = f(x) \, dx. \tag{2}$$

That is, we can rewrite the equation so that all terms involving y and dy appear on one side of the equation and all terms involving x and dx appear on the other side. Separable differential equations (1) are easily solved, provided we can integrate both sides of (2).

EXAMPLE 1   (a) The following differential equations are separable:

(1) $y' + x^2y = 0$ since we can rewrite this equation as

$$\frac{dy}{dx} = -x^2y, \text{ or } \frac{dy}{y} = -x^2 \, dx.$$

(2) $2x - 3yy' = 5$ since we can rewrite it as

$$\frac{dy}{dx} = (2x - 5)\left(\frac{1}{3y}\right), \text{ or } 3y \, dy = (2x - 5) \, dx.$$

(3) $y'e^{-y} = x$ since we can rewrite it as

$$\frac{dy}{dx} = xe^y, \text{ or } \frac{1}{e^y} \, dy = x \, dx.$$

(b) The following differential equations are not separable:

$$y' = 2(x + y) \quad \text{and} \quad y' = x^2y^2 + 2.$$

They are not in form (1) and cannot be rewritten in that form.   ■

We shall turn now to a method for finding the general solution of a separable first-order differential equation. The following example illustrates the technique.

EXAMPLE 2   Find all the solutions of the separable first-order differential equation

$$y' = 3x^2e^{-y}. \tag{3}$$

Solution   We write the given equation as

$$\frac{dy}{dx} = 3x^2 e^{-y}.$$

Next we separate the variables; that is, we write all terms involving x and dx on one side of the equals sign and all terms involving y and dy on the other side. Thus,

$$e^y\,dy = 3x^2\,dx.$$

Integrating both sides, we obtain

$$\int e^y\,dy = \int 3x^2\,dx,$$

or

$$e^y + C_1 = x^3 + C_2,$$

or

$$e^y = x^3 + C_2 - C_1,$$

where C_1 and C_2 are arbitrary constants. Since $C = C_2 - C_1$ is an arbitrary constant, we can write the last equation as

$$e^y = x^3 + C.$$

Whenever possible, we want to express y explicitly as a function of x. In this case, we can do it by taking natural logarithms of both sides of the last equation to obtain

$$y = \ln(x^3 + C). \qquad (4)$$

Equation (4) gives an expression for the general solution of Equation (3). ■

We can now summarize the procedure for finding the general solution of a first-order separable differential equation.

Solving a Separable First-Order Differential Equation
Step 1. Express y' as a quotient of two differentials: $y' = dy/dx$.
Step 2. Separate the variables. That is, move dx and all terms involving x to one side of the equal sign, and dy and all terms involving y to the other side.
Step 3. Integrate both sides of the equation obtained in step 2.
Step 4. Solve the resulting equation for y as an explicit function of x, if possible.

EXAMPLE 3 Find the general solution of the separable first-order differential equation

$$y' = \frac{x^3}{y^2}. \tag{5}$$

Solution We write the given equation as

Step 1. $\dfrac{dy}{dx} = \dfrac{x^3}{y^2}.$

Step 2. Separating the variables, we have $y^2 \, dy = x^3 \, dx.$

Step 3. Integrating both sides, we obtain

$$\int y^2 \, dy = \int x^3 \, dx,$$

or

$$\frac{y^3}{3} = \frac{x^4}{4} + C,$$

or

$$y^3 = 3\left(\frac{x^4}{4} + C\right)$$
$$= \tfrac{3}{4}x^4 + K,$$

where $K = 3C$ is an arbitrary constant.

Step 4. In this case, we are able to solve for y explicitly:

$$y = (\tfrac{3}{4}x^4 + K)^{1/3}, \tag{6}$$

which gives the general solution of Equation (5). ∎

EXAMPLE 4 Solve the initial-value problem

$$y' = \frac{x^3}{y^2}, \qquad y = 2 \text{ when } x = 0.$$

Solution The differential equation is the same as in Example 3 where we found that the general solution is given by Equation (6). Substituting $y = 2$ and $x = 0$ in Equation (6), we obtain

$$2 = [\tfrac{3}{4}(0)^4 + K]^{1/3}$$

or

$$2 = K^{1/3}$$
$$K = 8.$$

Thus, the (only) solution of the given initial-value problem is

$$y = (\tfrac{3}{4}x^4 + 8)^{1/3}.$$

∎

THEOREM 1

The only solutions of the differential equation

$$\frac{dy}{dx} = Ky \tag{7}$$

are the functions of the form

$$y = Ce^{Kx}, \tag{8}$$

for an arbitrary constant C.

Proof Suppose $y = f(x)$ is a solution of (7). Then

$$\frac{dy}{dx} = Ky.$$

Separating the variables, we have

$$\frac{dy}{y} = K\,dx.$$

Integrating both sides, we obtain

$$\ln|y| = Kx + C_1.$$

We can solve for $|y|$ explicitly:

$$|y| = e^{Kx + C_1} = e^{Kx} \cdot e^{C_1}.$$

That is,

$$|y| = C_2 e^{Kx},$$

where C_2 is an arbitrary positive constant ($C_2 = e^{C_1} > 0$). Therefore,

$$y = \pm C_2 e^{Kx},$$

or

$$y = Ce^{Kx},$$

where C is an arbitrary constant, which may be positive or negative.

Therefore, all solutions of Equation (7) must have the form of Equation (8).

∎

THEOREM 2

The only solution of the initial-value problem

$$\frac{dy}{dx} = Ky, \qquad y = y_0 \text{ when } x = 0, \tag{9}$$

is

$$y = y_0 e^{Kx}. \tag{10}$$

Proof To solve the initial-value problem (9) we must first solve the differential equation (7). By Theorem 1, every solution must be of the form

$$y = Ce^{Kx}. \tag{11}$$

We can find the constant C by letting $x = 0$:

$$y_0 = Ce^{K(0)},$$

or

$$y_0 = C.$$

Substituting this value of C into Equation (11), we obtain

$$y = y_0 e^{Kx}.$$

Therefore, the initial-value problem (9) has only one solution, and it is (10). ∎

Unrestricted Growth or Decay

In Section 5.4 we discussed the exponential or unrestricted growth model, in which a quantity Q is growing or decaying at a rate that is at each instant proportional to the amount of Q present at that instant. If the constant of proportionality is k, then the amount Q present at time t must satisfy the differential equation

$$\frac{dQ}{dt} = kQ. \tag{12}$$

For growth models, the constant k is positive; for decay models k is negative.
According to Theorem 1, the general solution of (12) is

$$Q(t) = Ce^{kt}, \tag{13}$$

where C is an arbitrary constant.
Given the initial condition that $Q = Q_0$ when $t = 0$, or

$$Q(0) = Q_0, \tag{14}$$

Theorem 2 tells us that the only solution of differential equation (12) that satisfies the initial condition (14) is

$$Q = Q_0 e^{kt}. \qquad (15)$$

In Section 5.4 we applied the unrestricted growth model to problems dealing with the growth of populations (human, bacteria, and so on), and to decay problems such as radioactive decay. Other applications occur in business and economics, such as continuous compound interest and depreciation.

In Section 1.4 we introduced the notion of depreciation and considered the method of straight-line (or linear) depreciation. When an asset depreciates more rapidly at some times than it does at others, the depreciation is nonlinear. For nonlinear depreciation, differential equations are the natural tool to use.

EXAMPLE 5
(Depreciation)

A certain television set depreciates at a rate that is always proportional to its value. One of these sets sells for $350 and is worth $250 after 3 years. Calculate its worth
(a) 1 year after it was bought.
(b) 2 years after it was bought.
(c) 5 years after it was bought.

Solution

We are given that the value V of the television set is changing at a rate proportional to V; that is,

$$\frac{dV}{dt} = kV.$$

We are also given that $V(0) = \$350$. By Theorem 2, therefore, the value of the television set t years after it was bought is

$$V = 350 e^{kt}.$$

We can calculate k by using the given information, $V(3) = \$250$:

$$250 = 350 e^{k(3)},$$

or

$$\tfrac{250}{350} = \tfrac{5}{7} = e^{3k}.$$

Then,

$$\ln \left(\tfrac{5}{7} \right) = 3k,$$

or

$$\tfrac{1}{3} \ln \left(\tfrac{5}{7} \right) = k.$$

Therefore, the value after t years is

$$V = 350 e^{\frac{1}{3} \ln (5/7) t}.$$

Finally, we calculate

(a) $V(1) = 350e^{\frac{1}{3}\ln(5/7)1} = \$312.87.$

(b) $V(2) = 350e^{\frac{1}{3}\ln(5/7)2} = \$279.67.$

(c) $V(5) = 350e^{\frac{1}{3}\ln(5/7)5} = \$199.77.$ ∎

|||||||||||| ## The Differential Equation $\dfrac{dy}{dt} = ay + b$

Given any constants a and b, the differential equation

$$\frac{dy}{dt} = ay + b \tag{16}$$

is separable. We can solve it, as follows:

$$dy = (ay + b)\, dt$$

$$\frac{dy}{ay + b} = dt$$

$$\int \frac{dy}{ay + b} = \int dt$$

$$\frac{1}{a}\ln|ay + b| = t + C_0$$

$$\ln|ay + b| = at + aC_0$$

$$|ay + b| = e^{at + aC_0}$$

$$ay + b = C_1 e^{at}, \qquad \text{where } C_1 = \pm e^{aC_0}.$$

By taking $C = \dfrac{C_1}{a}$, we obtain the general solution

$$y = Ce^{at} - \frac{b}{a}. \tag{17}$$

Moreover, if we are given $y(0) = y_0$ then we can evaluate the constant C in Equation (17):

$$y_0 = Ce^0 - \frac{b}{a},$$

$$C = y_0 + \frac{b}{a}. \tag{18}$$

We summarize these results in the following theorem.

THEOREM 3

The general solution of the differential equation

$$\frac{dy}{dt} = ay + b \tag{16}$$

is

$$y = Ce^{at} - \frac{b}{a}. \tag{17}$$

Moreover, if $y(0) = y_0$, then

$$C = y_0 + \frac{b}{a}. \tag{18}$$

**EXAMPLE 6
(Exponential
Growth with
Migration)**

Consider a population Q which has natural exponential growth (with annual growth rate k) but which is also affected by migration at a constant net gain of b individuals per year. In this case a differential equation that models the population at time t more accurately than Equation (12) is

$$\frac{dQ}{dt} = kQ + b. \tag{19}$$

This model is effective both when annual net immigration is positive ($b > 0$) and when it is negative ($b < 0$). It is likewise effective in decay problems ($k < 0$) as well as in growth problems ($k > 0$). ∎

EXAMPLE 7

An economically troubled city has a natural growth rate (birth rate minus death rate) of 2%, but 5000 more people per year are moving out of the city than are moving in. At these rates, if the population of the city is now 300,000, how many people will live in the city 10 years from now?

Solution

Let $P(t) =$ the population of the city (in thousands) after t years. Then $P(0) = 300$ and

$$\frac{dP}{dt} = 0.02P - 5.$$

By Theorem 3, the solution of this initial-value problem is

$$P = Ce^{0.02t} - \frac{(-5)}{0.02},$$

where $C = 300 + \dfrac{(-5)}{0.02} = 50$. Thus,

$$P = 50e^{0.02t} + 250.$$

Then

$$P(10) = 50e^{0.2} + 250$$
$$\approx 311.$$

The population of the city 10 years from now will thus be approximately 311,000.

∎

EXAMPLE 8
(Mixing Problems)

A 30-gallon tank is filled with a water/alcohol mixture that is 20% alcohol by volume. A water/alcohol mixture that is 60% alcohol is added at the rate of 2 gallons per minute, and the resulting mixture is drained out at the same rate.
(a) How much alcohol is in the tank t minutes later?
(b) After how many minutes will the solution in the tank be 50% alcohol?

Solution

(a) Let $y(t)$ denote the amount of alcohol in the tank after t minutes. The rate at which the amount of alcohol in the tank changes is

$$\frac{dy}{dt} = (\text{rate in}) - (\text{rate out}), \tag{20}$$

where "rate in" denotes the rate in gallons per minute at which alcohol enters the tank, and "rate out" denotes the rate at which it is drained out. From the given information,

$$\text{rate in} = (2 \text{ gal/min}) \times (60\%)$$
$$= 1.2 \text{ gal/min},$$

and

$$\text{rate out} = (2 \text{ gal/min}) \times \left(\frac{y \text{ gal alcohol}}{30 \text{ gal mix}}\right)$$
$$= \frac{y}{15} \text{ gal/min}.$$

Thus from (20),

$$\frac{dy}{dt} = 1.2 - \frac{y}{15}.$$

By Theorem 3, the solution of this differential equation is

$$y = Ce^{-t/15} - \frac{1.2}{-1/15} = Ce^{-t/15} + 18,$$

where $C = y_0 + \dfrac{1.2}{-1/15}$. Now $y_0 = 30(.20) = 6$ gal. Thus,

$$C = 6 - 18 = -12.$$

Therefore, the amount of alcohol in the tank after t minutes is

$$y(t) = 18 - 12e^{-t/15} \text{ gal.}$$

(b) To answer the second question, we want to solve the equation

$$y(t) = 0.5 \times 30 = 15,$$
$$18 - 12e^{-t/15} = 15$$
$$-12e^{-t/15} = -3$$
$$e^{-t/15} = \tfrac{1}{4}$$
$$-t/15 = -\ln 4$$
$$t = 15 \ln 4 \approx 20.8.$$

Thus, after about 21 minutes, the solution in the tank will be 50% alcohol.   ■

|||||||||||||| **Lotka–Volterra (Predator–Prey) Models (Optional)**

In Example 8 of Section 9.3 we introduced the concept of predator–prey inter-acting populations. In this interaction one species, called the **predator,** feeds on another, called the **prey.** The Lotka*–Volterra† differential equations provide a useful mathematical model of this interaction. Let $x(t)$ denote the population of the predator at time t, and $y(t)$ the population of the prey. Without the predator, the prey would increase with some natural growth rate $r > 0$; $dy/dt = ry$. On the other hand, with the predator and prey together in the same environment, it is reasonable to postulate that the rate of kill would be jointly proportioned to both x and y. We then have the

$$\textbf{prey equation:} \quad \frac{dy}{dt} = ry - sxy \quad (r > 0, s > 0). \tag{21}$$

Similarly, without the prey the predators tend to die out at some death rate $a > 0$, $dx/dt = -ax$; and in the presence of the prey it is reasonable to postulate that the predator population would rebound at a rate jointly proportional to both x and y. Thus we have the

$$\textbf{predator equation:} \quad \frac{dx}{dt} = -ax + bxy \quad (a > 0, b > 0). \tag{22}$$

---

* Alfred James Lotka (1880–1949) was born in Austria to American parents and educated in France, England, Germany, and the United States. His development of mathematical demography, achieved while he was working for many years in the statistical bureau of the Metropolitan Life Insurance Company of New York, earned him worldwide respect. He served as president of the American Statistical Association and of the Population Association of America. He also made significant con-tributions to mathematical developments in other areas of social science and in biology.

† Vito Volterra (1860–1940) was born in Ancona, Italy, earned his doctorate in mathematical physics at the University of Pisa, and established his research career at the University of Rome. After serving with the Allies in World War I, he began his long creative period of research in mathematical biology and ecology. For refusing allegiance to the Fascist government of Italy in 1931, he was stripped of all official positions. He continued active mathematical research, nonetheless, so that by the time of his death he had published nearly 300 papers.

Taken together, the predator and prey equations yield a system of two differential equations,

$$\begin{cases} \dfrac{dx}{dt} = -ax + bxy \\[2mm] \dfrac{dy}{dt} = ry - sxy \end{cases} \tag{23}$$

which is called the Lotka–Volterra Model of the predator–prey interaction.

Unfortunately, we are unable to solve the system (23) to obtain the functions $x = x(t)$ and $y = y(t)$ using techniques of this course. Instead, we shall show how to derive an equation *relating* the functions $x(t)$ and $y(t)$. In Section 10.6 we shall show how to solve completely a system similar to (23).

Suppose x and y are related by some function, $y = f(x)$. Then, by the chain rule,

$$\frac{dy}{dt} = \frac{dy}{dx} \frac{dx}{dt}.$$

Thus,

$$\frac{dy}{dx} = \frac{dy}{dt} \div \frac{dx}{dt} \tag{24}$$

whenever $dx/dt \neq 0$. For the Lotka–Volterra model, Equation (24) becomes

$$\frac{dy}{dx} = \frac{ry - sxy}{-ax + bxy}. \tag{25}$$

This is a separable differential equation, and we proceed to solve it.

$$\frac{dy}{dx} = \frac{y(r - sx)}{x(-a + by)}$$

$$\frac{-a + by}{y} dy = \frac{r - sx}{x} dx$$

$$\int \left(-\frac{a}{y} + b \right) dy = \int \left(\frac{r}{x} - s \right) dx$$

$$\boxed{-a \ln y + by = r \ln x - sx + C. \tag{26}}$$

In general, we cannot solve Equation (26) to obtain an explicit function $y = f(x)$ which satisfies the differential equation (25), much less obtain explicit functions $x = x(t)$, $y = y(t)$ which satisfy the system (23). We must be satisfied here with an equation relating x and y. We say that Equation (26) represents an **implicit solution** of (25).

EXAMPLE 9
Parasites and
Hosts
(optional)

Suppose that a population of x (in millions) parasites live on a host population of y (in hundreds) animals, and that the two populations interact according to the Lotka–Volterra model (23). The natural annual growth rate of the host, in the absence of the parasite, is $r = 300\%$, and when the parasite is present the kill rate is given by $s = 20\%$. Without the host, the parasites will die out at the annual rate of $a = 90\%$, but with the host present the parasites revive with $b = 50\%$. If there are 100 million parasites living on a population of 1000 hosts, how many parasites are present when there are 2000 hosts?

Solution

Using the given information, the Lotka–Volterra differential equations (23) which model the populations $x(t)$ and $y(t)$ are

$$\begin{cases} \dfrac{dx}{dt} = -0.9x + 0.5xy \\[2mm] \dfrac{dy}{dt} = 3y - 0.2xy \end{cases} \tag{27}$$

According to Equation (26), the general solution of (27) is given implicitly by

$$-0.9 \ln y + 0.5y = 3 \ln x - 0.2x + C$$

or

$$-9 \ln y + 5y = 30 \ln x - 2x + D. \tag{28}$$

We solve for D by substituting the given data $x = 100$, $y = 10$ into Equation (28):

$$-9 \ln 10 + 50 = 30 \ln 100 - 200 + D$$
$$250 - 9 \ln 10 - 60 \ln 10 = D \qquad (\ln 100 = 2 \ln 10)$$
$$250 - 69 \ln 10 = D$$
$$D \approx 91.$$

Substituting this value of D into (28), we see that at any given time t, the populations $x(t)$ and $y(t)$ are related by the equation

$$-9 \ln y + 5y = 30 \ln x - 2x + 91. \tag{29}$$

To finish the problem, we want to find x when $y = 20$. That is, we want to solve the equation

$$-9 \ln 20 + 5(20) = 30 \ln x - 2x + 91,$$

or

$$2x - 30 \ln x + 9 - 9 \ln 20 = 0.$$

Using a calculator we find $9 - 9 \ln 20 \approx -18$. Thus, we want to solve the equation

$$2x - 30 \ln x - 18 = 0. \tag{30}$$

We shall use the Newton–Raphson algorithm (Section 4.9) to find an approximate solution. We let

$$f(x) = 2x - 30 \ln x - 18.$$

We are seeking a value of x for which $f(x) = 0$. Since we are already given some information about $x = 100$, it is reasonable to take as our "first guess"

$$x_1 = 100.$$

Then $f(x_1) = 200 - 30 \ln 100 - 18 \approx 44$, and $f'(x_1) = 2 - \dfrac{30}{x_1} = 1.7$. The Newton–Raphson procedure yields the next approximation:

$$x_2 = x_1 - \frac{f(x_1)}{f'(x_1)}$$

$$= 100 - \frac{44}{1.7}$$

$$\approx 74.$$

An iteration of the Newton–Raphson algorithm will produce $x_3 \approx 73$, and another iteration will yield the same value. Therefore, there are about 73 million parasites living on 20 thousand hosts.

Notice that in this example the hosts double in number while there are *fewer* parasites present. This is a reasonable expectation in the parasite–host interaction. ∎

10.2 EXERCISE SET ||||||||||||||||||||||||||||||||||||||||||||||||||||||||||||||||||||||||||||

In Exercises 1–22, find all the solutions of each separable first-order differential equation.

1. $y' = 2y$

2. $y' = \dfrac{2}{y}$

3. $y' = xy$

4. $y' = \dfrac{x}{y}$

5. $y' = \dfrac{2y}{x}$

6. $2y^2y' = 3x$

7. $y' = 3x^2y$

8. $y' = xe^y$

9. $y' = 5 + y$

10. $y' = \dfrac{x^2}{y}$

11. $y' = x - 2xy$

12. $y' = 2x + xy$

13. $y' = \dfrac{x}{y^2}$

14. $y' = e^{x+y}$ (Hint: $e^{x+y} = e^xe^y$)

15. $2xy' = x^2 + 1$

16. $xy' = 3y$

17. $y' = y + 5xy$

18. $2xy' = 4x^2 + 3$

19. $(4 + x^2)y' = x$

20. $(1 + x^3)y' - 3x^2 = 0$

21. $dN/dt = 5(50 - N)$

22. $dQ/dt = (2t - 3)(Q + 4)$

In Exercises 23–30, solve the initial-value problems.

23. $y' = \dfrac{5}{y}$; $y = 3$ when $x = 0$.

24. $y' = -3y$; $y = 5$ when $x = 0$.

25. $y' = -2xy$; $y = 5$ when $x = 0$.

26. $y = xe^{-y}$; $y = 2$ when $x = 0$.

27. $y' = 3x/y$; $y = 6$ when $x = 0$.

28. $y' = xy^2$; $y = 1$ when $x = 0$.

29. $(x^2 + 2)y' = x$; $y = 3$ when $x = 2$.

30. $e^{x^2}y' = x$; $y = e$ when $x = 3$.

31. **(Business)** Suppose that $5000 is invested at 8% per year, compounded continuously. Then the rate of growth of the investment is proportional to the amount of money present at any time t and the proportionality constant is 0.08. Find an expression for the amount of the investment after (a) t years, (b) 10 years.

32. **(Depreciation)** A manufacturing company purchases an industrial machine for $85,000. It depreciates in such a way that t years after purchase its rate of depreciation is

    $$300t - 250 \text{ dollars per year.}$$

    Find the value of the machine after (a) t years, (b) 10 years, (c) 20 years.

33. **(Medicine)** After a certain drug is administered, the patient's body eliminates the drug from the blood stream at a rate proportional to the amount present in the blood stream.
    (a) One hour after an initial dose of 80 milligrams (mg) of this drug has been administered, 65 mg of the drug remains in the blood stream. How much remains after t hours?
    (b) A physician wishes to prescribe a "booster" dose to be administered every 6 hours, so that with each booster the original amount of 80 mg of the drug will be restored in the blood stream. Determine this booster dose, to the nearest milligram.

34. **(Medicine)** Suppose that t hours after a certain drug is administered, the patient's body is eliminating the drug from the blood stream at the rate of 10% of the amount present, per hour. Determine the

booster dose to be administered every T hours, where T is a constant, so that with each booster the original amount A_0 mg will be restored to the blood stream. (See Exercise 33.)

35. A cylindrical tank of height 25 feet and radius 12 feet is being emptied by gravity through a drain at the bottom. The rate at which the water level goes down is always proportional to the square root of the depth of the water remaining in the tank. Starting with a full tank, the water level goes down 9 feet in the first hour.
    (a) Find the depth of the water in the tank after t hours.
    (b) How long does it take to empty the tank?

$$V = \pi r^2 h$$

36. A cylindrical tank of height 9 feet and radius 4 feet is being emptied by gravity, as in Exercise 35. (The water level goes down at a rate proportional to the square root of the depth of the water remaining in the tank.)

$$V = \pi r^2 h$$

(a) Prove that the rate of flow (dV/dt) of water from the tank is always proportional to the square root of the volume (V) of the water remaining in the tank.

(b) If it takes 3 hours to empty a full tank, how much water is in the tank after t hours ($0 \leq t \leq 3$)?

Exercises 37–40 are concerned with the **stimulus–response relationship.** Psychologists and physiologists have studied the mathematical relationship between a physical stimulus of magnitude s and the intensity of the resulting sensation r perceived by an individual. Of particular concern is the ability of an individual to perceive a *change* in the stimulus; that is, the relationship between ds and dr. An adequate mathematical model must account for the observed fact that small changes in s are more easily perceived when s is small. For example, it is easier to hear a clock ticking when the background is relatively quiet than when it is quite noisy. Mathematically, this means dr decreases as s increases.

37. **(Stimulus–Response)** Early experimental evidence seemed to suggest that dr is directly proportional to ds and inversely proportional to s. That is,

$$dr = \frac{k\,ds}{s} \qquad (31)$$

for some constant k. This is known as the **Weber–Fechner Law,** and was proposed about 1850. Solve this differential equation to express r as a function of s. Assume that the response r is 0 when the stimulus s is $s_0 > 0$. (The number s_0 is called the *threshold of detection.*)

38. **(Stimulus–Response)** Further studies of the stimulus–response relationship suggested that a more accurate mathematical model is

$$dr = \frac{kr\,ds}{s}. \qquad (32)$$

Proposed in the 1950s, this is known as **Stevens' Power Law.** Solve this differential equation to express r as a function of s, and explain why this is called a "power" law.

39. **(Stimulus–Response)** Suppose that a person's perception of the illumination given off by a w-watt light bulb follows the Weber–Fechner law

$$dx = \frac{k\,dw}{w},$$

where x is a measure of the person's perception and k is a constant. Suppose that the threshold of detection is $w_0 = 1$ watt and, for convenience, that the perceived illumination from a 10-watt light bulb is 1 unit ($x = 1$ when $w = 10$). Calculate the perceived illumination from a light bulb rated at 40 watts; at 60 watts; at 75 watts; at 100 watts.

40. **(Stimulus–Response)** In a stimulus–response relationship that satisfies Stevens' Power Law, a stimulus of 3 units produces a response of 2 units, while a stimulus of 4 units produces a response of 6 units.
    (a) Find the solution of the differential equation (32) that satisfies the given data.
    (b) When the stimulus is 5 units, what is the response?

41. **(Income Distribution)** In a stable economy, let $y = y(x)$ denote the number of people who have an annual income of at least x dollars. Of course, y decreases as x increases. One mathematical model of this relationship, called **Pareto's Law,** is

$$\frac{dy}{dx} = -k\frac{y}{x}$$

where k is some positive constant. Suppose that in a given community, 200 people earn at least \$100,000 per year, while 1200 people earn at least \$50,000 per year. According to Pareto's Law, how many people in this community earn at least \$30,000 per year?

42. **(Human Growth)** The **allometric law** of proportional growth says that if $x(t)$ and $y(t)$ are measurements of the sizes of different organs (or other parts) of a growing

child at time t, then the relative growth rates of these parts are proportional:

$$\frac{y'}{y} = k\frac{x'}{x} \tag{33}$$

for some positive constant k. Use the chain rule as in Equation (24) to transform (33) into a differential equation involving only x, y, and dy/dx. Find its general solution.

43. **(Population Growth with Migration)** The population of a country has a natural growth rate of 2.5% per year and experiences an average annual net immigration of 50,000 people. Assuming that these trends continue and that the present population is 30 million, determine the population after 20 years.

44. **(Animal Population with Harvesting)** A breeder of rabbits maintains a natural growth rate of 40% per year in his rabbit stock and sells 250 rabbits per year. If his present stock is 900 rabbits, how many will he have after 5 years?

45. **(Mixing)** A 100-liter tank is filled with brine (a solution of salt in water) that contains 5 grams of salt per liter. A stronger brine that is 20 grams of salt per liter is added at the rate of 8 liters per minute, and the resulting well-mixed brine is drawn off at the same rate.
    (a) How many grams of salt are in the tank after 10 minutes?
    (b) After how many minutes will there be 1800 grams of salt in the tank?

46. **(Absorption of a Drug)** Suppose that a drug is administered intravenously to a patient at the constant rate of 0.5 gram per hour, and that the patient's system removes it from the blood stream at the rate of 80% per hour.
    (a) How much of the drug is in the patient's blood stream t hours after treatment is begun?
    (b) How much of the drug is in the bloodstream after 1 hour; 2 hours; 5 hours?

(c) Is there a limit to the amount of the drug in the patient's blood stream as $t \to +\infty$?

47. **(Air Conditioning)** Into a 3000 cubic foot room full of "stale" air containing 0.2% carbon dioxide, an air conditioner circulates a "fresh" air supply containing 0.03% carbon dioxide at the rate of 600 cubic feet per minute, and draws out the resulting mixture at the same rate. How many cubic feet of carbon dioxide will be in the room after 5 minutes? After 10 minutes? After 15 minutes? After 20 minutes? Compare these figures with the amount of carbon dioxide that would be in the room if it were filled with "fresh" air.

48. **(Trust Fund)** A trustee establishes a $90,000 savings account paying 8% continuous compound interest, for the benefit of a young child. He makes no additional deposits but makes frequent and regular withdrawals for the benefit of the child, averaging $5000 per year. Approximately how much money remains in the account after 10 years?

Exercises 49–52 pertain to optional material in this section.

49. **(Predator–Prey)** A population of x predators feeds on a population of y prey, according to the Lotka–Volterra model (23). The natural annual growth rate of the prey is $r = 100\%$, but in the presence of the predator the kill rate is given by $s = 30\%$. Without the prey, the predator will die at an annual rate $a = 50\%$, but with the prey present the predator rebounds with $b = 20\%$.
    (a) Write the Lotka–Volterra differential equations that model this situation.
    (b) Obtain an equation relating x and y, as in Example 9, given that there are 100 predators when the prey population is 1000.

50. **(Predator–Prey)** A predator population of x (in hundreds) feeds on a prey population of y (in thousands) according to the

Lotka–Volterra model. If the two populations were kept separated, the prey would have an annual growth rate of 80% while the predator would have an annual death rate of 60%. When the two populations are placed together, however, the kill rate is given by $s = 10\%$ while the predator rebound rate is given by $b = 20\%$.

(a) Write the Lotka–Volterra differential equations that model this situation.

(b) Obtain an equation relating x and y, as in Example 9, given that there are 1000 predators when the prey population is 20,000.

51. For the predator–prey model of Exercise 49, determine the prey population when the predator population is 200. Use the Newton–Raphson algorithm, as in Example 9, with $y_1 = 1000$.

52. For the predator–prey model of Exercise 50, how many predators will be present when the prey population is 10,000? Use the Newton–Raphson algorithm, as in Example 9, with $x_1 = 50$.

Exercises 53–59 are concerned with **bimolecular chemical reactions,** in which chemicals A and B are combined in the ratio of a units of A to b units of B to form a chemical "product" C. Let $x = x(t)$, $y = y(t)$, and $z = z(t)$ denote the numbers of units of A, B, and C present at time t, with $x_0 = x(0)$, $y_0 = y(0)$, and $z_0 = z(0)$. We take $z_0 = 0$. (Why?) In a bimolecular reaction,

$$x(t) = x_0 - \frac{a}{a + b} z(t), \quad y(t) = y_0 - \frac{b}{a + b} z(t),$$

$$\tag{34}$$

and the *rate of reaction*, dz/dt, is jointly proportional to the amounts of A and B present at time t:

$$\frac{dz}{dt} = kxy \tag{35}$$

for some positive constant k. We want to determine how many units of C have been produced through time t. Combining (34) and (35), we

have

$$\frac{dz}{dt} = k(x_0 - a_1 z)(y_0 - b_1 z) \tag{36}$$

where $a_1 = \dfrac{a}{a + b}$ and $b_1 = \dfrac{b}{a + b}$. For example, if a particular bimolecular reaction requires 2 units of chemical A to every 3 units of chemical B, then

$$\frac{dz}{dt} = k\left(x_0 - \frac{2}{5} z\right)\left(y_0 - \frac{3}{5} z\right).$$

53. Separate the variables in the differential equation (36) and, assuming that Formula 17 in our table of integrals (Appendix B, Table IV) can be used, show that the general solution $z = z(t)$ must satisfy the equation

$$\frac{b_1 z - y_0}{a_1 z - x_0} = M e^{k(a_1 y_0 - b_1 x_0)t}$$

for some constant M. Use the initial condition $z_0 = 0$ to show that $M = y_0/x_0$, and thus

$$\frac{b_1 z - y_0}{a_1 z - x_0} = \frac{y_0}{x_0} e^{k(a_1 y_0 - b_1 x_0)t}. \tag{37}$$

54. Suppose that the ratio of the initial amounts of A and B is exactly the same as the ratio in which A and B combine to produce C. That is,

$$\frac{x_0}{y_0} = \frac{a}{b}.$$

Show that in this case,

(a) $\dfrac{a_1}{b_1} = \dfrac{a}{b}$ and $\dfrac{a_1}{x_0} = \dfrac{b_1}{y_0}$.

(b) Integral Formula 17 (Appendix B, Table IV) cannot be used to solve Equation (36).

(c) Equation (36) takes the form

$$\frac{dz}{dt} = \frac{ky_0}{x_0} (x_0 - a_1 z)^2. \tag{38}$$

55. Separate the variables in (38) and show that the general solution $z = z(t)$ of this

differential equation must satisfy the equation

$$\frac{1}{a_1(x_0 - a_1 z)} = \frac{ky_0}{x_0} t + M \qquad (39)$$

for some constant M. Use the initial condition $z_0 = 0$ to show that $M = \dfrac{1}{a_1 x_0}$. Use this and (39) to show that the desired solution is

$$z = \frac{x_0 y_0 kt}{a_1 y_0 kt + 1}. \qquad (40)$$

56. Chemicals A and B combine in the ratio 2-to-3 to yield C. Starting with 50 grams each of A and B, 20 grams of C are produced in 3 minutes.
    (a) How many grams of C are produced in t minutes?
    (b) How many grams of C are formed in 6 minutes? In 15 minutes?
    [*Hint:* substitute the given data into Equation (37) and solve for k and z.]

57. Chemicals A and B combine in 1-to-1 ratio to yield C. Starting with 20 grams of A and 30 grams of B, 10 grams of C are produced in 5 minutes.
    (a) How many grams of C are produced in t minutes?
    (b) How many grams of C are formed in 3 minutes? In 10 minutes?
    [See hint in Exercise 56.]

58. Chemicals A and B combine in 1-to-1 ratio to yield C. Starting with 40 grams each of A and B, 8 grams of C are produced in 30 seconds. How many grams of C are produced in t seconds? In 1 minute?
    [*Hint:* substitute the given data into Equation (40) and solve for k and z.]

59. Two units of A and one unit of B combine to produce one unit of C. Starting with 60 grams of A and 30 grams of B, 12 grams of C are produced in 1 minute. How many grams of C are produced in t minutes? In 5 minutes?
    [See hint in Exercise 58.]

10.3 ⅠⅠⅠⅠⅠⅠ BOUNDED GROWTH MODELS

When a quantity undergoes unrestricted exponential growth, its size increases without any bound at an ever faster rate. Figure 1 shows the graph of a quantity Q whose growth satisfies the differential equation

$$\frac{dQ}{dt} = kQ. \qquad (1)$$

Figure 1

As suggested by the graph, this type of growth is unbounded. We considered examples of this type of growth in Sections 5.4 and 10.2. However, few quantities can continue to grow in this manner over an indefinite period of time because the necessary resources, such as space, raw materials, and so on, would eventually be depleted. In this section we introduce several mathematical models that are more appropriate for studying a quantity whose growth has some limiting value, or upper bound, as $t \to +\infty$.

Let $Q(t)$ be the amount of a quantity present at time t and let L be an upper bound for quantity. That is, $Q(t) \leq L$ for all t. We further assume that $\lim_{t \to +\infty} Q(t) = L$. We can construct several alternative mathematical models that describe the growth of a quantity Q satisfying this restricted case. To select the appropriate model it is necessary to know a little more about the behavior of the rate of growth, dQ/dt, as time t increases.

Simple Restricted Growth

First, observe that as $Q(t)$ approaches L, we expect the rate of growth dQ/dt to decrease. The simplest mathematical model for such restricted growth is that in which the rate of growth dQ/dt of Q is proportional to the difference $L - Q$. The differential equation for this mathematical model is thus

$$\frac{dQ}{dt} = k(L - Q), \qquad (2)$$

where k is a positive constant. For a quantity Q satisfying Equation (2), as Q approaches L, the growth rate dQ/dt approaches zero. Thus, when Q is near L, dQ/dt is small, and when Q is far from L, dQ/dt is large. A quantity Q satisfying the differential equation (2) is said to have **simple restricted growth.**

In mathematical terms, the essential difference between unrestricted growth and simple restricted growth is this: for unrestricted growth the rate of growth is always proportional to the quantity present, whereas for simple restricted growth the rate of growth is always proportional to the amount the quantity has left to grow. In unrestricted growth, the rate of growth grows without bound; in simple restricted growth, the rate of growth decreases to 0 as the amount grows to its limiting value L.

To find all the solutions of the separable differential equation (2), we proceed as follows. First, note that the constant function $Q(t) = L$ is a solution of Equation (2). Second, assuming that Q is not the constant function L, we can write (2) as

$$\frac{1}{L - Q} \, dQ = k \, dt,$$

and integrate both sides to obtain

$$\int \frac{1}{L - Q} \, dQ = \int k \, dt,$$

so that

$$-\ln |L - Q| = kt + C_1 \qquad \text{(Integral Table, Formula 9, with } a = -1, \, b = L)$$

or

$$\ln |L - Q| = -kt - C_1, \tag{3}$$

where C_1 is an arbitrary constant. In (3), we do not need to use absolute value because we are given that $Q \le L$, or $L - Q \ge 0$. Rewriting (3) in exponential form, we have

$$L - Q = e^{-C_1}e^{-kt}.$$

Letting $C = e^{-C_1}$, we have

$$Q(t) = L - Ce^{-kt}. \tag{4}$$

Since the constant function $Q(t) = L$ also has this form (pick $C = 0$), Equation (4) yields all the solutions of the restricted growth differential equation (2). Thus, (4) is the general solution.

If Q_0 is the amount of the quantity present when $t = 0$, then substituting $t = 0$ and $Q = Q_0$ into (4), we obtain

$$Q_0 = L - Ce^{-k0},$$

or

$$C = L - Q_0.$$

Substituting this value of C into (4) we obtain the solution to our initial-value problem:

$$Q(t) = L - (L - Q_0)e^{-kt}, \tag{5}$$

or

$$Q(t) = L(1 - e^{-kt}) + Q_0 e^{-kt}. \tag{6}$$

It is the only solution of the differential equation (2) that also satisfies the initial condition $Q = Q_0$ when $t = 0$.

We can summarize our discussion with the following formal definition and theorem.

Definition

A positive quantity Q is said to have **simple restricted growth** if there is an upper bound L on the size of Q, and the rate of growth of Q is always proportional to the difference between L and Q; that is,

$$\frac{dQ}{dt} = k(L - Q) \tag{2}$$

for some constant k.

THEOREM 1   For given positive constants k, L, and Q_0, the initial-value problem

$$\frac{dQ}{dt} = k(L - Q), \qquad (Q \leq L)$$

$$Q(0) = Q_0,$$

has the unique solution

$$Q(t) = L - (L - Q_0)e^{-kt}, \tag{5}$$

or

$$Q(t) = L(1 - e^{-kt}) + Q_0 e^{-kt}. \tag{6}$$

To sketch the graph of (5) we examine the first and second derivatives of Q. We have

$$\frac{dQ}{dt} = k(L - Q_0)e^{-kt},$$

which is positive for all values of t. Hence, $Q(t)$ is always increasing. We also have

$$\frac{d^2Q}{dt^2} = -k^2(L - Q_0)e^{-kt},$$

which is negative for all values of t. Hence, the graph of Q is concave downward for all values of t. Moreover,

$$\lim_{t \to +\infty} Q(t) = \lim_{t \to +\infty} [L - (L - Q_0)e^{-kt}]$$
$$= L - (L - Q_0) \lim_{t \to +\infty} e^{-kt}$$
$$= L,$$

since $\lim_{t \to +\infty} e^{-kt} = 0$. Figure 2 shows the graph of Q.

Figure 2

$$Q(t) = L - (L - Q_0)e^{-kt}$$

EXAMPLE 1
(Psychology)

Psychologists have discovered that there is a limit L to the number of meaningless words that a person can learn, no matter how much study time is available, and that the rate of learning is proportional to the number of words not yet learned. Suppose that $L = 100$ words and that at the start of an experiment the number of words known by a subject is zero. Let $N(t)$ denote the number of words mastered in the first t minutes of study. Assume that a subject learns 30 words in the first 5 minutes.

(a) Determine an expression for $N(t)$.

(b) How many words are learned in the first 3 minutes of study?

Solution

(a) The number of words not yet learned is $100 - N$, so the rate of learning dN/dt satisfies the differential equation

$$\frac{dN}{dt} = k(100 - N).$$

This is a restricted growth differential equation. By Theorem 1, Equation (6) gives all solutions and we obtain

$$N(t) = 100(1 - e^{-kt}) + N_0 e^{-kt}. \tag{7}$$

Since $N(0) = 0$, by substituting $t = 0$ into (7) we find that

$$N(t) = 100(1 - e^{-kt}).$$

We next find the constant k using data given in the problem. Because $N(5) = 30$, we substitute $t = 5$ and $N(5) = 30$ to obtain

$$30 = 100(1 - e^{-5k}),$$

or

$$0.30 = 1 - e^{-5k},$$

or

$$e^{-5k} = 1 - 0.30 = 0.70.$$

Taking the natural logarithm of both sides we have

$$-5k = \ln 0.70.$$

Then

$$k = -\frac{\ln 0.70}{5} = \frac{0.3567}{5},$$

or

$$k \approx 0.07134,$$

or

$$k \approx 0.07.$$

Hence, the number of words learned in the first t minutes is given approximately by

$$N(t) = 100(1 - e^{-0.07t}).$$

(b) In the first 3 minutes, the number of words learned is

$$\begin{aligned}
N(3) &= 100(1 - e^{-0.07(3)}) \\
&= 100(1 - e^{-0.21}) \\
&= 100(1 - 0.8106) = 100(0.1894) \\
&= 18.94.
\end{aligned}$$

Hence, after 3 minutes, the subject will know approximately 19 meaningless words.   ∎

**EXAMPLE 2
(Terminal
Velocity)**

In Section 7.3 we noted that the acceleration of an object moving along a straight line is the instantaneous rate of change of its velocity:

$$a = \frac{dv}{dt}. \tag{8}$$

When an object falls through the air to the earth, its acceleration is determined by two forces: a *downward* force due to gravity and an *upward* force due to air resistance. The downward component of acceleration is known to be constant (32 ft/sec²) and, in our mathematical model, the upward component is assumed to be directly proportional to the velocity and inversely proportional to the weight w of the object. Thus, the downward acceleration is

$$a = 32 - \frac{kv}{w} \text{ ft/sec}^2 \tag{9}$$

for some constant k.

We would like to know how fast the object is falling t seconds after some initial time. Using (8) we can rewrite (9) as a differential equation

$$\frac{dv}{dt} = 32 - \frac{kv}{w} = \frac{k}{w}\left(\frac{32w}{k} - v\right). \tag{10}$$

Comparing this equation with (2) we see that the problem of determining the velocity of an object falling through the air is a simple restricted growth problem. By Theorem 1, the solution of (10) is

$$v = \frac{32w}{k}(1 - e^{-kt/w}) + v_0 e^{-kt/w}. \tag{11}$$

If the object is simply dropped, with initial velocity $v_0 = 0$, then Equation (11) becomes

$$v = \frac{32w}{k}(1 - e^{-kt/w}). \qquad (12)$$

When an object falls according to this model, its velocity does *not* increase without bound. For example, the speed of a falling skydiver tends to level off rather than increase indefinitely. Indeed, Equation (12) tells us that the limiting value of v, as $t \to +\infty$, is

$$L = \frac{32w}{k}. \qquad (13)$$

As the object falls, its velocity approaches, and will not exceed, $32w/k$ ft/sec, which is called its **terminal velocity.**

The value of k is determined by the particular falling object. For example, k would be much larger for a feather than for a pin of the same weight, since the air resistance is greater for the feather than for the pin. ∎

EXAMPLE 3 (Skydiving) A skydiver whose weight, including parachute, is 200 pounds jumps from an airplane. His parachute opens at the instant when he is falling 96 ft/sec. With the chute open, the value of k in Equation (9) is 400.
(a) How fast is the skydiver falling t seconds later?
(b) How fast is he falling after 1 second? After 2 seconds? After 3 seconds?
(c) Find his terminal velocity.

Solution (a) We are given that $w = 200$ and $k = 400$. Thus, Equation (10) becomes

$$\frac{dv}{dt} = \frac{400}{200}\left[\frac{32(200)}{400} - v\right]$$
$$= 2(16 - v).$$

By Theorem 1, the solution (5) of this initial value problem with $v_0 = 96$ is

$$v = 16 - (16 - 96)e^{-2t}$$
$$= 16 + 80e^{-2t}.$$

This is the skydiver's speed t seconds after his parachute opens.
(b) $v(1) = 16 + 80e^{-2} \approx 26.8$ ft/sec.
$v(2) = 16 + 80e^{-4} \approx 17.5$ ft/sec.
$v(3) = 16 + 80e^{-6} \approx 16.2$ ft/sec.
(c) By Equation (13) the terminal velocity is $L = \dfrac{32(200)}{400} = 16$ ft/sec. For the curious, this is about 11 miles per hour. ∎

|||||||||||||| **Logistic Growth**

It is clear that a population cannot grow exponentially over an indefinite period of time, because the living space and all other resources would eventually be depleted. In a realistic population model, growth is initially nearly exponential, yet there is a limit L to how large the population can become. We shall develop a model (called the **logistic growth model**) that combines the best features of both the exponential and the simple restricted growth models.

Let $P(t)$ denote the size of the population at time t. Then $P(t) < L$ for all t. For small values of P, the population should undergo approximately exponential growth. That is, the rate of growth dP/dt of the population at time t should be approximately proportional to the size of the population $P(t)$ at time t. As the size of the population approaches L, the rate of growth dP/dt levels off and eventually approaches zero. We shall see that both of these properties are satisfied if P satisfies the separable differential equation

$$\frac{dP}{dt} = kP(L - P). \tag{14}$$

We proceed to solve (14). Separating variables,

$$\frac{1}{P(L - P)}\, dP = k\,dt. \tag{15}$$

We then integrate both sides,

$$\int \frac{1}{P(L - P)}\, dP = \int k\, dt.$$

When we apply Integral Formula 14, with $a = -1$, $b = L$, and $u = P$, Equation (15) becomes

$$\frac{1}{L} \ln \left| \frac{P}{L - P} \right| = kt + C_1.$$

Since $P > 0$ and $P < L$, we may drop the absolute value bars. Thus,

$$\frac{1}{L} \ln \frac{P}{L - P} = kt + C_1,$$

or

$$\ln \frac{P}{L - P} = Lkt + C_2, \tag{16}$$

where $C_2 = LC_1$.

In exponential form, Equation (16) is equivalent to

$$\frac{P}{L-P} = e^{Lkt+C_2}$$
$$= e^{C_2}e^{Lkt}.$$

But e^{C_2} is a constant, so we may let $C_3 = e^{C_2}$. Then

$$\frac{P}{L-P} = C_3e^{Lkt}. \tag{17}$$

Continuing, we solve (17) for P:

$$P = C_3e^{Lkt}(L-P)$$
$$= C_3Le^{Lkt} - C_3e^{Lkt}P$$
$$P + C_3e^{Lkt}P = C_3Le^{Lkt}$$
$$P(1 + C_3e^{Lkt}) = C_3Le^{Lkt}$$
$$P = \frac{C_3Le^{Lkt}}{1 + C_3e^{Lkt}}. \tag{18}$$

We can simplify (18) by multiplying the numerator and the denominator of the fraction by $(C_3e^{Lkt})^{-1}$ obtaining

$$P = \frac{L}{\dfrac{1}{C_3}e^{-Lkt} + 1}.$$

Finally, letting $C = 1/C_3$, we have

$$P = \frac{L}{1 + Ce^{-Lkt}}. \tag{19}$$

The constant solution $P(t) = L$ has the form given by (19) (pick $C = 0$). The constant solution $P(t) = 0$ does not have this form. Thus, all solutions of (14) are given by (19) and the zero solution.

We summarize our discussion with the following formal definition and theorem.

Definition

A positive quantity P is said to have **logistic growth** if there is an upper bound L on the size of P and the rate of growth of P is jointly proportional to both P and the difference $L - P$; that is,

$$\frac{dP}{dt} = kP(L-P), \tag{14}$$

for some constant k.

THEOREM 2

For positive constants k and L, the general nonzero solution of the differential equation (14) is

$$P = \frac{L}{1 + Ce^{-Lkt}}. \tag{19}$$

Equation (19) is called the **logistic equation.** Figure 3 shows a graph of the logistic equation for $C > 0$. It can be shown (See Exercise 35) that this curve has an inflection point at $(\ln C/Lk, L/2)$. Before P reaches the value $L/2$ its growth rate is increasing; after that it is decreasing.

Figure 3

P. F. Verhulst* introduced the logistic equation in 1838 to facilitate the study of human population growth. Applications modeled by the logistic equation include the growth of the population of fish in a lake, the spread of an epidemic throughout a population, the spread of a rumor throughout a population, the growth of bacteria in a culture, autocatalytic chemical reactions, and advertising problems.

EXAMPLE 4
(Ecology)

An ecological study shows that a newly developed lake can support a maximum population of 2400 fish. Suppose that the lake is initially stocked with 400 fish and that there are 1200 fish in the lake after 4 months. Assuming that the logistic equation applies, find an expression for the number $P(t)$ of fish that are present in the lake t months after it has been stocked.

Solution

The desired expression is given by the logistic equation [Equation (19)]. Since $L = 2400$, we obtain

$$P(t) = \frac{2400}{1 + Ce^{-2400kt}}. \tag{20}$$

---

* Pierre-Francois Verhulst (1804–1849) was born in Brussels, Belgium, and was educated at the University of Ghent. After some disappointing experiences in politics, in 1833 he became a professor at the University of Brussels. His contributions to mathematics include work in the calculus of variations, number theory, and applications of probability. In 1838 he introduced a model of population growth that he called "logistic growth."

We can use $P(0) = 400$ to find the constant C:

$$400 = \frac{2400}{1 + Ce^0} = \frac{2400}{1 + C}.$$

Thus,

$$1 + C = 6,$$

or

$$C = 5.$$

We can then use $P(4) = 1200$ to find the constant k:

$$1200 = \frac{2400}{1 + 5e^{-2400k(4)}} = \frac{2400}{1 + 5e^{-9600k}}.$$

Then

$$1 + 5e^{-9600k} = 2,$$

or

$$e^{-9600k} = \tfrac{1}{5} = 0.2.$$

Taking the natural logarithm of both sides, we find

$$-9600k = \ln{(0.2)} = -1.6094,$$

or

$$k = \frac{1.6094}{9600}.$$

We substitute the values of C and k that we have obtained into (20). Hence, t months after the lake has been stocked, the number of fish is given by

$$P(t) = \frac{2400}{1 + 5e^{-2400(1.6094/9600)t}}$$

$$= \frac{2400}{1 + 5e^{-(1.6094/4)t}}$$

$$= \frac{2400}{1 + 5e^{-0.402t}}. \tag{21}$$

∎

EXAMPLE 5 How long will it take for the fish population of the lake described in Example 4 to reach 80% of its limit?

Solution We must find the value of t when $P(t) = 0.8(2400) = 1920$. Using Equation (21), we want to solve for t:

$$1900 = \frac{2400}{1 + 5e^{-0.402t}},$$

or

$$1 + 5e^{-0.402t} = \frac{2400}{1920} = 1.25$$

$$5e^{-0.402t} = 0.25$$

$$e^{-0.402t} = 0.05.$$

In logarithmic form,

$$-0.402t = \ln 0.05$$

or

$$t = \frac{\ln 0.05}{-0.402}.$$

Therefore,

$$t \approx 7.45 \text{ months.} \qquad \blacksquare$$

||||||||||||| **Gompertz Growth (Optional)**

There is an alternative to the logistic growth model that has similar behavior but that is occasionally more suitable for applications in biology, ecology, or economics. It is a more sophisticated model which in certain circumstances may reflect tabulated data with greater accuracy than the logistic model.

Definition

> A positive quantity P is said to have **Gompertz* growth** if there is an upper bound L on the size of P, and if the rate of growth of P is jointly proportional to both P and the difference $\ln L - \ln P$; that is,
>
> $$\frac{dP}{dt} = kP(\ln L - \ln P), \qquad P < L \qquad (22)$$
>
> or
>
> $$\frac{dP}{dt} = kP \ln\left(\frac{L}{P}\right), \qquad P < L \qquad (23)$$
>
> for some positive constant k.

The distinction between logistic growth and Gompertz growth is that in the latter, $L - P$ is replaced by $\ln L - \ln P$.

We solve the differential equation (22) by separating the variables:

$$\frac{dP}{P(\ln L - \ln P)} = k\,dt$$

$$\int \frac{1}{\ln L - \ln P} \frac{dP}{P} = \int k\,dt. \qquad (24)$$

---

* Benjamin Gompertz (1779–1865) was an English mathematician and actuary.

If we make the substitution $u = \ln L - \ln P$, then $du = (-1/P)\, dP$ and so the left side of Equation (24) is merely $\int (-1/u)\, du$, or $-\ln u$. (Note that $u > 0$, since $L > P$). Thus, (24) becomes

$$-\ln (\ln L - \ln P) = kt + C_1$$
$$\ln (\ln L - \ln P) = -kt + C_2 \qquad (C_2 = -C_1)$$
$$\ln L - \ln P = e^{-kt + C_2}$$
$$\ln \left(\frac{L}{P}\right) = e^{C_2} e^{-kt}$$
$$= Ce^{-kt} \qquad \text{(Where } C = e^{C_2}.)$$

Then

$$\ln \left(\frac{P}{L}\right) = -\ln \left(\frac{L}{P}\right) = -Ce^{-kt}$$

so that

$$\frac{P}{L} = e^{-Ce^{-kt}}$$

and finally,

$$P = Le^{-Ce^{-kt}}. \tag{25}$$

Equation (25) gives us the general solution of (22). As we have come to expect, the general solution involves a constant of integration, C. We can evaluate this constant if we know the initial value $P_0 = P(0)$. Substituting $t = 0$ into (25) we obtain

$$P_0 = Le^{-Ce^0} = Le^{-C}, \qquad \text{or}$$
$$P_0 e^C = L$$
$$e^C = \frac{L}{P_0},$$

so that

$$C = \ln \left(\frac{L}{P_0}\right). \tag{26}$$

Finally, we substitute this value of C back into the general solution (25) and obtain

$$P = Le^{-\ln \left(\frac{L}{P_0}\right) e^{-kt}}$$

or

$$P = L\, [e^{\ln \left(\frac{P_0}{L}\right)}]^{e^{-kt}}. \tag{27}$$

Therefore,

$$P = L\left(\frac{P_0}{L}\right)^{e^{-kt}}. \tag{28}$$

Equation (28) represents the specific solution of the Gompertz differential equation (22) which satisfies the initial condition $P(0) = P_0$.

We summarize the above derivation in the following theorem.

THEOREM 3
(optional)

For given positive constants K, L, and P_0, the initial-value problem

$$\frac{dP}{dt} = kP(\ln L - \ln P), \qquad (P < L) \tag{22}$$

$$P(0) = P_0,$$

has the unique solution

$$P = L\left(\frac{P_0}{L}\right)^{e^{-kt}}. \tag{28}$$

Figure 4 shows the graph of a Gompertz growth curve. It can be shown that this curve has an inflection point when $t = \frac{1}{k}\ln\left(\ln\left(\frac{L}{P_0}\right)\right)$, and $P = \frac{L}{e}$. Before P reaches the value $\frac{L}{e}$, its growth rate is increasing; after that, its growth rate is decreasing.

By comparing Figures 3 and 4, one can easily see similarities between logistic growth and Gompertz growth.

Figure 4

EXAMPLE 6
(optional)

Consider the newly developed lake of Example 4, which can support a maximum of 2400 fish. Stocked initially with 400 fish, the lake contains 1200 fish after 4 months. Suppose that a Gompertz growth model is used, instead of the logistic growth model, to describe the fish population of the lake.

(a) Find an expression for the number $P(t)$ of fish in the lake after t months.

(b) How long will it take for the fish population to reach 80% of its limit?

(c) Compare the number of fish in the lake after 1 month as predicted by the Gompertz growth model, with the number as predicted by the logistic model. Compare the predicted populations after 8 months.

Solution (a) We substitute $P_0 = 400$ and $L = 2400$ into (28) to obtain

$$P = 2400(\tfrac{400}{2400})^{e^{-kt}}$$
$$= 2400(\tfrac{1}{6})^{e^{-kt}}.$$

Thus,

$$P = 2400(6)^{-e^{-kt}} \qquad (29)$$

To find the constant k, we can use the given data $P(4) = 1200$. Substituting these values into (29) we have

$$1200 = 2400(6)^{-e^{-k(4)}}$$
$$\tfrac{1}{2} = 6^{-e^{-4k}}.$$

Taking reciprocals,

$$2 = 6^{e^{-4k}}.$$

Taking the natural logarithm of both sides, we have

$$\ln 2 = e^{-4k}\ln 6$$
$$\frac{\ln 2}{\ln 6} = e^{-4k}.$$

Again, taking the natural logarithm of both sides,

$$\ln\left(\frac{\ln 2}{\ln 6}\right) = -4k,$$

so that

$$k = -\frac{1}{4}\ln\left(\frac{\ln 2}{\ln 6}\right)$$
$$k \approx 0.2374.$$

Therefore, t months after the lake is stocked, the population of fish in the lake will be

$$P = 2400(6)^{-e^{-0.2374t}}. \qquad (30)$$

(b) Since 80% of 2400 is 1920, we want to determine the value of t for which

$$1920 = 2400(6)^{-e^{-0.2374t}}$$

or

$$0.8 = (6)^{-e^{-0.2374t}}.$$

Taking the natural logarithm of both sides,

$$\ln 0.8 = -e^{-0.2374t} \ln 6$$

$$\frac{-\ln 0.8}{\ln 6} = e^{-0.2374t}.$$

Again, taking the natural logarithm,

$$\ln\left(\frac{-\ln 0.8}{\ln 6}\right) = -0.2374t.$$

Thus,

$$t = \frac{\ln\left(\dfrac{-\ln 0.8}{\ln 6}\right)}{-0.2374}$$

$$t \approx 8.77 \text{ months.}$$

Therefore, after about 8.77 months, the fish population of the lake will (according to the Gompertz model) reach 80% of its limit.

(c) Using Equation (30), the Gompertz model predicts that after 1 month the population of the fish in the lake will be

$$P_G(1) = 2400(6)^{-e^{-0.2374(1)}} \approx 584$$

and after 8 months the population will be

$$P_G(8) = 2400(6)^{-e^{-0.2374(8)}} \approx 1835.$$

On the other hand, using Equation (21), the logistic model predicts that after 1 month the fish population will be

$$P_l(1) = \frac{2400}{1 + 5e^{-0.402(1)}} \approx 552$$

and after 8 months it will be

$$P_l(8) = \frac{2400}{1 + 5e^{-0.402(8)}} \approx 1999.$$

For small values of t, Gompertz growth rate is larger than logistic growth rate. Thus, $P_G(1) > P_l(1)$.
Later on, however, the reverse is true: $P_G(8) < P_l(8)$. ∎

|||||||||||| **Critique of Mathematical Modeling**

The conflicting results of Examples 4 and 6 afford an excellent opportunity to reflect on the nature of the mathematical modeling process. This process was outlined in Section 1.2. After carefully formulating a real-life problem, we select "an appropriate mathematical model." In Examples 4 and 6, two different differential equations were used as alternative models of the same real life fish population. How is one to know which model is better? One cannot always guarantee in advance that a particular model adequately reflects all significant aspects of the problem. That is why a later step of the modeling process requires that the model be evaluated "to determine whether it is realistic and workable." In the case of the fish population of a lake, experiments would be made and data collected to help determine which of the alternative models is more reliable. Of course, experience with other lakes or other species of fish would also help us choose the best model.

Perhaps no available mathematical model will provide a perfect "fit" to a given real-life problem. In such a case we try to come as close as we can. The point is that we must continue to be critical of the mathematical models we use and strive to improve them.

10.3 EXERCISE SET |||||||||||||||||||||||||||||||||||||||||||||||||||||||||||||||||||||||||||||||||||

In Exercises 1–6, find the general solution to the given differential equation.

1. $\dfrac{dP}{dt} = 4(80 - P)$

2. $\dfrac{dP}{dt} = 75 - 25P$

3. $\dfrac{dP}{dt} = 10P(100 - P)$

4. $\dfrac{dP}{dt} = 48P - 6P^2$

5. $\dfrac{dP}{dt} = 5P(\ln 30 - \ln P)$

6. $\dfrac{dP}{dt} = 8P \ln\left(\dfrac{20}{P}\right)$

In Exercises 7–12, solve the given initial-value problem.

7. $\dfrac{dP}{dt} = 5(50 - P)$

   $P = 12$ when $t = 0$

8. $\dfrac{dP}{dt} = 1 - 0.01P$

   $P = 1.7$ when $t = 0$

9. $\dfrac{dP}{dt} = 3P(80 - P)$

   $P = 10$ when $t = 0$

10. $\dfrac{dP}{dt} = 100P(500 - P)$

    $P = 25$ when $t = 0$

11. $\dfrac{dP}{dt} = 3P(\ln 16 - \ln P)$

    $P = 2$ when $t = 0$

12. $\dfrac{dP}{dt} = 0.02 P \ln\left(\dfrac{5}{P}\right)$

    $P = \frac{1}{2}$ when $t = 0$

13. **(Learning)** Suppose that a factory apprentice produces 80 units of work daily before the first day of his training period and by the end of the fifth day produces 100 units of work daily. His productivity

continues to improve so that on the tth day the rate of change in his daily production quantity $Q(t)$ is

$$\frac{dQ}{dt} = k(200 - Q).$$

(a) Approximately how many units of work will he produce on the tth day? The 15th day? The 30th day?
(b) On what day will he produce 140 units of work?

14. **(Sociology)** Suppose that a rumor starts to spread through a city with a fixed population of 10,000. Sociologists have discovered that the rate of growth of the number of people $P(t)$ who have heard a rumor t hours after it has been planted is proportional to the number of people who have not yet heard it. That is, the spread of the rumor is governed by a restricted growth model. Suppose that, initially, 400 people know the information and that 6160 people have heard the rumor 4 hours later.

(a) Approximately how many people will have heard the rumor 10 hours after it has been planted?
(b) How long will it be before 8000 people have heard the rumor?

15. **(Marketing)** A computer game company has been studying its weekly sales as a function of the number of dollars it spends weekly on advertising. When it spends $\$x$ (thousand) weekly on advertising, it sells

$$S = f(x),$$

thousand units weekly. Their studies show that the market will support a maximum of 20,000 units sold weekly, and that the rate of change in their sales is directly proportional to $20 - S$, the number of potential sales (in thousands) *not* made.

(a) Write the differential equation that must be satisfied.

(b) Find the solution of this differential equation, assuming that 1000 units are sold weekly when no money at all is spent on advertising.
(c) Under all of the above assumptions, assume further that a weekly advertising budget of $\$2000$ results in a weekly sales level of 4000 units. Find the weekly sales level as a function of x.
(d) Estimate the weekly sales when the weekly advertising budget is $\$5000$.

16. **(Medicine) Fick's law of diffusion** says that the rate at which a solute diffuses through a thin membrane of fixed surface area is proportional to the difference between the concentrations of the solute on the two sides of the membrane.

A cell of fixed surface area is suspended in a solution in which the concentration of a certain drug is 0.030 milligrams per cubic centimeter. At time $t = 0$ the cell contains none of the drug, but 5 minutes later it contains 0.003 milligrams per cubic centimeter.

(a) Find the concentration of the drug in the cell at time t minutes.
(b) How long does it take for the concentration of the drug in the cell to reach 0.010 mg/cc?

Exercises 17 and 18 use **Newtons law of temperature change:** when an object is placed in a medium of constant temperature M, the object changes temperature at a rate proportional to the *difference* between the temperature T of the object and that of the surrounding medium. That is,

$$\frac{dT}{dt} = k(M - T).$$

17. **(Temperature Change)** A roast at a temperature of 80°F is placed in a 400°F oven. After 3 hours, the temperature of the roast is 240°F. What is the temperature of the roast after t hours? After 5 hours?

18. **(Temperature Change)** A pot of hot coffee 200°F is set out in a room that is 70°F. Ten minutes later its temperature is 180°F. Find its temperature after 10 more minutes.

19. **(Terminal Velocity)** A skydiver and her parachute together weigh 220 pounds. After jumping from the plane she falls freely for 10 seconds before opening her chute. While free-falling, the value of k in Equation (9) is approximately 33.
    (a) How fast is she falling t seconds after she jumps? ($t \leq 10$)
    (b) How fast is she falling 10 seconds after she jumps?
    (c) Find her terminal velocity while free-falling.

20. **(Terminal Velocity)** An object dropped from a great height reaches half of its terminal velocity in 8 seconds.
    (a) Calculate its terminal velocity. (*Hint:* use the given data along with Equation (12) to determine the value of k/w.)
    (b) How fast is the object falling after 10 seconds?
    (c) How long does it take the object to reach 80% of its terminal velocity?

21. Prove that if $P = P_0$ when $t = 0$, then the value of the constant C in the logistic equation (19) is

$$C = \frac{L}{P_0} - 1. \qquad (31)$$

This equation may be useful in the next few exercises.

22. Prove that if $P = P_0$ when $t = 0$, and $P = P_1$ when $t = t_1$, then the value of k in the logistic equation (19) is

$$k = \frac{1}{Lt_1} \ln \left[\frac{P_1(L - P_0)}{P_0(L - P_1)} \right]. \qquad (32)$$

23. **(Biology)** A bacteria culture is growing in a medium that can support a maximum population of 21,000 bacteria. Initially, there are 1000 bacteria present. The rate of growth of the culture at time t is proportional to the size of the population $P(t)$ and to $21,000 - P(t)$. The size of the population is 3000 bacteria after 4 hours of growth.
    (a) Using the logistic growth model, find an expression for the size of the population after t hours of growth. After 5 hours.
    (b) How long will it take the population to grow to 10,000 bacteria?

24. **(Medicine)** An epidemic of Victoria flu has spread throughout a city of 80,200 people at a rate that is proportional to both the number of people who have been infected and to the number of people who have not been infected. Suppose that the disease had infected 200 people by the time the epidemic was discovered and that 2005 had it 20 days later.
    (a) Using the logistic growth model, find an expression for the number of people infected t days after the discovery of the disease.
    (b) How long will it take for half the population of the city to become infected?

25. **(Advertising)** A market research firm studying the effectiveness of a radio advertising campaign for a new product in a community of 48,000 people finds that the rate at which people hear about the product is proportional to both the number of people who have heard about it and the number of people who have not yet heard about it. When the advertising campaign started, 3000 people had already heard about the product. Three weeks after the advertising campaign had started, 12,000 people knew about the product. Using the logistic equation, find an expression for the number of people who knew about the product t days after the advertising campaign started.

26. **(Sociology)** A fad spreads through a college of 8000 students at a rate proportional to both the number of student adherents of the fad and the number of susceptible students who are not yet adherents. The fad starts with 10 students, and after 1 week the fad has 100 adherents. If only half the student body is susceptible, after how long will $\frac{1}{4}$ of the student body be adherents of the fad?

27. **(Medicine)** After winter vacation, 5% of the students at an isolated college return with a contagious virus. One week later a total of 10% of the students have been infected by the virus. Assuming a logistic model, what percentage of the students will have been infected by the end of the second week? By the end of the third week?

28. **(Chemistry)** In an **autocatalytic reaction,** one substance changes into another at a rate proportional to both the amount of the new substance present and the amount of the original substance not yet changed.

    In a certain autocatalytic reaction the constant of proportionality is 0.2% per second. Starting with an initial amount of 100 units of the first substance and 20 units of the new one, how much of the new substance is present after t seconds? After 10 seconds?

29. **(Sociology)** In a 1948 paper, L. F. Richardson* applied the methods of calculus to the study of "War-moods" during periods of international tension. Let P denote the fraction of a society which favors war at time t; then $1 - P$ denotes the fraction of the society that does not favor war. In Richardson's model, this fraction P is assumed to change at a rate that is jointly proportional to P and $1 - P$.
    (a) Write the differential equation for Richardson's "War-moods" model.

(b) Suppose that at the beginning of a period of worsening relations with a foreign enemy, 5% of the members of a society favor going to war, and that 3 months later 12% favor going to war. After how many months will 50% of the members of that society favor going to war, according to Richardson's model?

30. **(Plant growth)** Suppose that the height of a growing plant follows a logistic growth model. One day a plant whose limiting height is 30 centimeters is observed to be 6 centimeters high. A day later it is 7 centimeters high. How high will it be after 3 more days?

Exercises 31–40 pertain to optional material in this section.

31. **(Gompertz Growth)** Suppose that the growth of the bacteria culture of Exercise 23 follows a Gompertz model rather than a logistic model.
    (a) Find the population after t hours of growth. After 5 hours of growth.
    (b) Show that at time $t = 4$ hours the Gompertz model and the logistic model predict virtually the same population.
    (c) How long will it take the population to reach 10,000 bacteria?

32. **(Gompertz Growth)** Suppose that the fad of Exercise 26 spreads according to a Gompertz model rather than a logistic model. Use the data given in Exercise 26 to determine
    (a) the number of student adherents of the fad by the end of the tth week.
    (b) how long it will take before one-fourth of the student body are adherents of the fad.

33. **(Gompertz Growth)** Suppose that in the advertising campaign of Exercise 25, the number of people $P(t)$ who know about

---

* See footnote, page 622.

the product after t days follows a Gompertz model rather than a logistic one. Use the data of Exercise 25 to obtain the function $P(t)$. How many people know about the product after 15 weeks?

34. **(Gompertz Growth)** Use the data given in Exercise 30 to answer the same question given there, assuming that the growth of the plant follows a Gompertz model rather than a logistic model.

35. Show that the graph of the logistic equation (19) has no relative extreme values, but has a point of inflection when

$$t = \frac{\ln C}{Lk}.$$

Find the value of P at this point of inflection.

36. Use the results of Exercise 35 to sketch the graph of

$$P = \frac{80}{1 + 10e^{-0.4t}},$$

showing the point of inflection.

37. (a) Differentiate both sides of Equation (22) to obtain an equation which must be satisfied by the second derivative of the Gompertz function

$$P = L\left(\frac{P_0}{L}\right)^{e^{-kt}}.$$

(b) Use this result to show that the Gompertz curve has only one inflection point, and find its coordinates.
[*Hint:* first, show that if (t, P) is a point of inflection, then $P = L/e$; then find t.]

38. Show that the general solution (25) of the Gompertz differential equation (22) is also a solution of the differential equation

$$\frac{dP}{dt} = CkPe^{-kt}. \qquad (33)$$

[Note that for any solution $P = P(t)$ of (33), $dP/dt = CkP_0$ when $t = t_0$. For this reason, the quantity Ck is called the *initial growth rate*.]

39. Solve the differential equation (33) by separating variables and using the methods of Section 10.2. Show that the general solution is

$$P = Le^{-Ce^{-kt}}$$

for some constant L. [For this reason, (22) and (33) may both be called "the Gompertz differential equation." Here is an example of two quite different equations which have the same general solution.]

40. **(Growth of a Tumor)** Medical researchers have found that the growth of a solid tumor follows closely a Gompertz model. While the tumor is small its growth rate is nearly exponential, and as it grows larger its growth rate slows down in a pattern that is more nearly Gompertzian than logistic.

Suppose that at time $t = 0$ a solid tumor consists of 2 million cells and has an initial growth rate of 12% per month. [See note in Exercise 38.] The value of the *retarding constant* k in Equation (33) is 4%.
(a) Use Equation (26) to find L, the limiting number of cells in the tumor.
(b) How many cells are contained in the tumor after t months? After 6 months?

10.4 ▍▍▍▍▍▍ LINEAR FIRST-ORDER DIFFERENTIAL EQUATIONS

A first-order differential equation that can be expressed in the form

$$\frac{dy}{dx} + u(x)y = v(x) \qquad (1)$$

where $u(x)$ and $v(x)$ are functions of x, is said to be **linear.** There is a standard procedure for solving equations of this type, which we now describe.

Given specific functions* $u(x)$ and $v(x)$, we seek all functions $y = y(x)$ which satisfy the differential equation (1). First, however, observe that if $a(x)$ is any differentiable function, then by the product rule,

$$\frac{d}{dx} [e^{a(x)}y] = e^{a(x)} \frac{dy}{dx} + ya'(x)e^{a(x)},$$

or

$$\frac{d}{dx} [e^{a(x)}y] = e^{a(x)} \left[\frac{dy}{dx} + a'(x)y \right]. \tag{2}$$

Suppose that we can find a function $a(x)$ such that

$$a'(x) = u(x) \tag{3}$$

$$\left(\text{that is, } a(x) = \int u(x)\, dx \right). \tag{4}$$

Multiplying both sides of (1) by $e^{a(x)}$, we transform Equation (1) into the form

$$e^{a(x)} \left[\frac{dy}{dx} + a'(x)y \right] = e^{a(x)}v(x) \tag{5}$$

which, in view of Equation (2), is equivalent to

$$\frac{d}{dx} [e^{a(x)}y] = e^{a(x)}v(x). \tag{6}$$

We have thus transformed the problem of solving (1) into the equivalent problem of solving (6).

By integrating both sides of Equation (6), we have

$$e^{a(x)}y = \int e^{a(x)}v(x)\, dx + C. \tag{7}$$

Finally, we multiply both sides of (7) by $e^{-a(x)}$ to obtain the general solution

$$y = e^{-a(x)} \int e^{a(x)}v(x)\, dx + Ce^{-a(x)}. \tag{8}$$

While Equation (8) does indeed provide, in theory, the general solution of the linear first-order differential equation (1), the reader is discouraged from memo-

---

* We shall assume, throughout this section and without further comment, that $u(x)$ and $v(x)$ are continuous functions.

rizing this solution, or even using it directly. Instead, you are encouraged to re-member the *procedure* used in obtaining it. This procedure is summarized below.

To solve a linear first-order differential equation

$$\frac{dy}{dx} + u(x)y = v(x), \tag{1}$$

Step 1. Find $a(x) = \int u(x)\,dx$. (Ignore the constant of integration.)

Step 2. Let $A(x) = e^{a(x)}$. (Note that $A(x) > 0$.)

Step 3. Multiple both sides of (1) by $A(x)$, obtaining (5), which is equivalent to

$$\frac{d}{dx}[A(x)y] = A(x)v(x). \tag{6}$$

Step 4. Integrate both sides of (6), obtaining

$$A(x)y = \int A(x)v(x)\,dx + C. \tag{7}$$

Step 5. Obtain the general solution of (1) by dividing both sides of (7) by $A(x)$.

The function $A(x) = e^{a(x)}$ described above is called an **integrating factor.** When we multiply both sides of the linear first-order differential equation by the appropriate integrating factor, we are left with a differential equation which can be solved, at least in theory, by "integrating both sides."

EXAMPLE 1   Solve the differential equation

$$\frac{dy}{dx} + 3y = 5x. \tag{9}$$

Solution   *Step 1.* $a(x) = \int 3\,dx = 3x$.

Step 2. $A(x) = e^{3x}$.

Step 3. Multiplying both sides of (9) by the integrating factor $A(x)$, we have

$$e^{3x}\frac{dy}{dx} + 3e^{3x}y = 5xe^{3x}$$

$$\frac{d}{dx}[e^{3x}y] = 5xe^{3x}.$$

Step 4. Integrating both sides, we have

$$e^{3x}y = 5\int xe^{3x}\,dx.$$

To obtain the integral on the right-hand side of this equation, we use Integral Formula 40 (Appendix B, Table IV). We then have

$$e^{3x}y = 5 \cdot \tfrac{1}{9}e^{3x}(3x - 1) + C.$$

Step 5. $y = e^{-3x}[\tfrac{5}{9}e^{3x}(3x - 1) + C],$

or

$$y = \tfrac{5}{9}(3x - 1) + Ce^{-3x}, \tag{10}$$

which is the general solution of the differential equation (9).   ∎

EXAMPLE 2   Solve the differential equation

$$x^2 \frac{dy}{dx} + xy - x^3 + 4 = 0. \qquad (x > 0) \tag{11}$$

Solution   We must first write the equation in the standard form (1) by writing

$$x^2 \frac{dy}{dx} + xy = x^3 - 4,$$

and then dividing both sides by x^2,

$$\frac{dy}{dx} + \frac{y}{x} = x - \frac{4}{x^2}. \tag{12}$$

Step 1. $a(x) = \displaystyle\int \frac{1}{x}\, dx = \ln |x| = \ln x,$ since $x > 0$.

Step 2. $A(x) = e^{a(x)} = e^{\ln x} = x.$

Step 3. Multiplying both sides of (12) by the integrating factor $A(x)$, we have

$$x\left(\frac{dy}{dx} + \frac{y}{x}\right) = x\left(x - \frac{4}{x^2}\right)$$

$$x\frac{dy}{dx} + y = x^2 - \frac{4}{x}$$

$$\frac{d}{dx}[xy] = x^2 - \frac{4}{x}.$$

Step 4. Integrating both sides, we have

$$xy = \tfrac{1}{3}x^3 - 4 \ln x + C.$$

Step 5. Dividing both sides by x, we obtain the general solution

$$y = \tfrac{1}{3}x^2 - \frac{4 \ln x}{x} + \frac{C}{x}. \tag{13}$$

∎

EXAMPLE 3
(Population
Growth with
Migration)

A city with natural growth rate (birth rate minus death rate) of 2% per year is experiencing economic trouble, so that in the tth year after 1985, $5000 + 100t$ more people move out of the city than move in. If the population of the city was 300,000 in 1985, what will it be in the year 2000?

Solution

Let $P(t)$ = the population of the city (in thousands) t years after 1985. From the given information,

$$\frac{dP}{dt} = 0.02P - (5 + 0.1t) \tag{14}$$

and $P(0) = 300$. Rewriting Equation (14) in the form

$$\frac{dP}{dt} - 0.02P = -5 - 0.1t, \tag{15}$$

we see that our problem is to solve a linear first-order differential equation.

Step 1. $a(t) = \int (-0.02)\, dt = -0.02t$.

Step 2. $A(t) = e^{-0.02t}$.

Step 3. Multiplying both sides of (15) by $A(t)$, we have

$$e^{-0.02t}\frac{dP}{dt} - 0.02Pe^{-0.02t} = (-5 - 0.1t)e^{-0.02t}$$

$$\frac{d}{dt}[e^{-0.02t}P] = -5e^{-0.02t} - 0.1te^{-0.02t}. \tag{16}$$

Step 4. Integrating both sides of (16), and using Integral Formulas 39 and 40,

$$e^{-0.02t}P = \frac{-5}{-0.02}e^{-0.02t} - 0.1\left[\frac{e^{-0.02t}}{(-0.02)^2}(-0.02t - 1)\right] + C$$

$$= 250e^{-0.02t} + 250[0.02te^{-0.02t} + e^{-0.02t}]$$

$$= 500e^{-0.02t} + 5te^{-0.02t}.$$

Step 5. Multiplying both sides by $e^{0.02t}$, we have

$$P = [500e^{-0.02t} + 5te^{-0.02t} + C]\, e^{0.02t}$$

$$P = 500 + 5t + Ce^{0.02t}. \tag{17}$$

Equation (17) is the general solution of our differential equation (14). To find the constant C we substitute the initial condition $P(0) = 300$ into Equation (17):

$$300 = 500 + 0 + Ce^0$$

$$-200 = C.$$

Therefore, the population after t years is

$$P = 500 + 5t - 200e^{0.02t}.$$

To find the population in the year 2000, we set $t = 15$. Then,

$$P = 500 + 5(15) - 200e^{(0.02)(15)}$$
$$= 575 - 200e^{0.3}$$
$$\approx 305.$$

Therefore, the city will have a population of about 305,000 in the year 2000.   ∎

The reader should compare the answer of Example 3 above with that of Example 7 in Section 10.2. What is different about these examples?

EXAMPLE 4
(Mixing)

A tank initially contains 30 gallons of a water/alcohol mixture that is 20% alcohol by volume. A water/alcohol mixture that is 60% alcohol is added at the rate of 3 gallons per minute, and the resulting mixture is pumped out at the rate of 2 gallons per minute. How much alcohol is in the tank after t minutes?

Solution

Let $y(t)$ denote the volume of the alcohol in the tank after t minutes. As in Example 8 of Section 10.2,

$$\frac{dy}{dt} = (\text{rate in}) - (\text{rate out}). \tag{18}$$

From the given information,

$$\text{rate in} = (3 \text{ gal/min}) \times 60\%$$
$$= 1.8 \text{ gal/min}, \tag{19}$$

and

$$\text{rate out} = (2 \text{ gal/min}) \times \left(\frac{y \text{ gal alcohol}}{V(t)} \right) \tag{20}$$

where $V(t) =$ the total volume of the mixture in the tank at time t. Now liquid is entering the tank at the rate of 3 gal/min and leaving at the rate of 2 gal/min. Thus,

$$V(t) = 30 + (3 - 2)t$$
$$= 30 + t \text{ (gal)}.$$

Substituting this result into (20), we have

$$\text{rate out} = (2 \text{ gal/min}) \times \left(\frac{y}{30 + t} \right)$$
$$= \frac{2y}{30 + t} \text{ gal/min}. \tag{21}$$

Combining Equations (18), (19) and (21) we have

$$\frac{dy}{dt} = 1.8 - \frac{2y}{30 + t}.$$

Rewriting this differential equation in the form

$$\frac{dy}{dt} + \left(\frac{2}{30 + t}\right)y = 1.8 \tag{22}$$

we see that our problem is to solve a linear first-order differential equation.

Step 1. $a(t) = \int \frac{2}{30 + t}\, dt = 2 \ln (30 + t).$

Step 2. $A(t) = e^{2 \ln (30+t)} = (30 + t)^2.$

Step 3. Multiplying both sides of (22) by $A(t)$, we have

$$(30 + t)^2 \frac{dy}{dt} + 2(30 + t)y = 1.8(30 + t)^2$$

$$\frac{d}{dt}[(30 + t)^2 y] = 1.8(30 + t)^2.$$

Step 4. Integrating both sides,

$$(30 + t)^2 y = 1.8 \int (30 + t)^2\, dt$$

$$= 1.8 \frac{(30 + t)^3}{3} + C$$

$$= 0.6(30 + t)^3 + C.$$

Step 5. $y = \dfrac{0.6(30 + t)^3 + C}{(30 + t)^2}$

$$= 0.6(30 + t) + \frac{C}{(30 + t)^2}. \tag{23}$$

Equation (23) is the general solution of our differential equation (22).

To find the constant C, we substitute the initial condition $y(0) = 30 \times 20\% = 6$ into Equation (23):

$$6 = 0.6(30 + 0) + \frac{C}{(30 + 0)^2}$$

$$6 = 18 + \frac{C}{900}$$

$$-10,800 = C.$$

Substituting this value of C into (23), we find that

$$y = 0.6(30 + t) - \frac{10,800}{(30 + t)^2} \text{ gal.} \qquad \blacksquare$$

The reader should compare Example 4 above with Example 8 of Section 10.2. How do these examples differ?

|||||||||||| Supply–Demand Equilibrium Price

In Sections 1.6 and 7.5 we have considered the supply and demand of a commodity as functions of the selling price,

$$S = S(x) \text{ and } D = D(x),$$

where x represents the selling price per unit of the commodity. In more sophisticated economics models, the price $x = x(t)$ is not assumed to be constant but is allowed to fluctuate with time t. Supply and demand are then considered to be functions of *both* the price and the rate of change of the price. In the notation of Chapter 8,

$$S = S(x, x') = S\left(x(t), \frac{dx}{dt}\right),$$

$$D = D(x, x') = D\left(x(t), \frac{dx}{dt}\right).$$

The market for the commodity is in equilibrium when $S = D$. That is, when

$$S(x, x') = D(x, x'). \qquad (24)$$

Equation (24) is actually a differential equation, since it involves both the function $x = x(t)$ and its derivative $x' = dx/dt$. When we solve the differential equation (24) we obtain a function $x = x(t)$ which represents the **equilibrium price** as a function of time.

If the limit

$$\bar{x} = \lim_{t \to +\infty} x(t)$$

exists, then $\bar{x}$ is called the **long-range equilibrium price.**

EXAMPLE 5   Suppose that the supply and demand functions for a given commodity are

$$S = -120 + 4x(t) + 7x'(t),$$
$$D = 90e^{-t} - 2x(t) + 4x'(t),$$

and the initial equilibrium price is $x(0) = \$35$ per unit.

(a) Find the equilibrium price as a function of time.

(b) Find the long-range equilibrium price, $\bar{x}$.

Solution   (a) For the equilibrium price $x(t)$ we must have

$$S = D,$$

or

$$-120 + 4x + 7x' = 90e^{-t} - 2x + 4x'$$

$$3x' + 6x = 90e^{-t} + 120$$

$$\frac{dx}{dt} + 2x = 30e^{-t} + 40. \tag{25}$$

We proceed to solve this linear first-order differential equation.

Step 1. $a(t) = \displaystyle\int 2\, dt = 2t.$

Step 2. $A(t) = e^{2t}.$

Step 3. Multiplying both sides of (25) by $A(t)$, we have

$$e^{2t}\frac{dx}{dt} + 2e^{2t} = (30e^{-t} + 40)e^{2t}$$

$$\frac{d}{dt}[e^{2t}x] = 30e^{t} + 40e^{2t}. \tag{26}$$

Step 4. Integrating both sides of (26), we have

$$e^{2t}x = 30e^{t} + \tfrac{40}{2}e^{2t} + C.$$

Step 5. Dividing both sides of this equation by e^{2t}, we have

$$x = (30e^{t} + 20e^{2t} + C)e^{-2t}$$

$$x = 30e^{-t} + 20 + Ce^{-2t}. \tag{27}$$

Equation (27) is the general solution of the differential equation (25). To find the constant C we substitute the initial condition $x(0) = 35$ into (27).

$$35 = 30e^{0} + 20 + Ce^{0}$$

$$35 = 30 + 20 + C$$

$$-15 = C.$$

Thus, the function

$$x(t) = 30e^{-t} + 20 - 15e^{-2t} \tag{28}$$

represents the equilibrium price at time t.

(b) The long-range equilibrium price of the commodity is found by taking the limit

$$\bar{x} = \lim_{t \to +\infty} [30e^{-t} + 20 - 15e^{-2t}]$$
$$= 0 + 20 - 0$$
$$= 20.$$

Thus, the long-range equilibrium price is $20 per unit.   ∎

Note: In Example 5, the long-range equilibrium price could have been calculated from Equation (27) just as well as from (28):

$$\bar{x} = \lim_{t \to +\infty} [30e^{-t} + 20 + Ce^{-2t}]$$
$$= 20.$$

The long-range equilibrium price $\bar{x}$, in this case, is totally independent of the initial price x_0.

10.4 EXERCISE SET ||||||||||||||||||||||||||||||||||||||||||||||||||||||||||||||||||||||||||||||||||||||||

In Exercises 1–10, find the general solution of the given differential equation by the methods of this section. In any exercise where ln x arises, assume $x > 0$.

1. $\dfrac{dy}{dx} + 4y = 7$

2. $\dfrac{dy}{dx} = 3 + 8y$

3. $\dfrac{dy}{dx} = 3y - 2x$

4. $\dfrac{dy}{dx} - 5y = x + 1$

5. $xy' + y = 2x$

6. $y' = xy$

7. $y' - 2y = e^{5x}$

8. $y' = 4x + 2xy$

9. $x^2y' + xy = 1 + x^2$

10. $xy' - y = x \ln x$

In Exercises 11–16, solve the given initial-value problem by the methods of this section. When ln x occurs, assume $x > 0$.

11. $\dfrac{dy}{dx} = y + e^x,\ y(0) = 1$

12. $\dfrac{dy}{dx} = 3y + 5,\ y(0) = 2$

13. $\dfrac{dy}{dx} + 2xy = 4x,\ y(0) = 10$

14. $y' + 3x^2y = 6x^2,\ y(0) = 3$

15. $x^2y' + xy = 1,\ y(1) = 2$

16. $xy' + 2y = e^x - 3x,\ y(1) = 1$

17. In Section 10.2 we solved the differential equation

$$\frac{dy}{dx} = ay + b,$$

where a and b are constants, by separation of variables. Solve this equation by the methods of the present section. [Your answer should agree with Theorem 3 of Section 10.2.]

18. In Section 10.3 we solved the differential equation

$$\frac{dQ}{dt} = k(L - Q),$$

where K and L are constants, by separation of variables. Solve this equation by the methods of the present section. [Your answer should agree with Equation (4) of Section 10.3. Explain the slight difference.]

19. **(Population Growth with Migration)** A suburban county on the east coast has a natural growth rate (birth rate minus death rate) of 1% per year. Its population is 400,000 in 1988, and its immigration is increasing steadily, being approximately $2000t$ per year t years after 1988. Predict its population in the year 2000.

20. **(Population Growth with Migration)** An emerging third-world nation has a national growth rate of 3% per year. If its present population is 12 million, and t years from now its annual net immigration rate is $0.7e^{-0.04t}$ million people, what will its population be after 25 years?

21. **(Trust Fund)** A trustee establishes a $100,000 savings account paying 8% annual continuous compound interest, for the benefit of a young child. She makes no additional deposits, but makes frequent and regular withdrawals for the benefit of the child. The expenses increase over the years, so that t years after the account is set up, these withdrawals are at the rate of $5000 + 500t$ dollars per year. How much money remains in the account after 10 years? After 15 years?

22. **(Animal Population with Harvesting)** A rabbit breeder presently has 900 rabbits. Suppose that by aggressive sales promotions he is able to sell $250 + 50t$ rabbits per year every year over the next 10 years. Assuming that his rabbit stock maintains an annual growth rate of 40% (birth rate minus death rate), how many rabbits will he have after 10 years?

23. **(Mixing)** A tank initially contains 100 liters of a salt/water brine that is 5 grams of salt per liter. A stronger brine that is 20 grams of salt per liter is added at the rate of 8 liters per minute, and the result-ing well-mixed brine is drawn off at the rate of 6 liters per minute. How many grams of salt are in the tank after t minutes? After 10 minutes?

24. **(Mixing)** A tank initially contains 5000 gallons of contaminated water which is $\frac{1}{8}$ ounce pollutant per gallon of water. A more highly contaminated water containing $\frac{3}{4}$ ounce pollutant per gallon of water is pumped into the tank at the rate of 100 gallons per hour. The resulting mixture is drained off at the rate of 50 gallons per hour. How many ounces of the pollutant are in the tank after t hours? After 10 hours?

25. **(Affirmative Action)** An industry currently has 20,000 employees, of whom 30% are women. Its management adopts a new hiring policy stipulating that 60% of all new employees will be women. If the industry loses 100 employees per month and replaces them according to this policy, how many women will be employed in this industry after t months? After 12 months? After 2 years? After 5 years? (*Hint:* treat this as a mixture problem.)

26. **(Affirmative Action)** Suppose that the industry of Exercise 25 decides to replace only half of the employees who leave, but stipulates that 80% (rather than 60%) of all new employees will be women. How many women will it employ after t months? After 12 months? After 2 years? After 5 years? (See *Hint* to Exercise 25.)

In Exercises 27–29, supply and demand functions $S(x)$ and $D(x)$ for a commodity are given in terms of the unit price $x = x(t)$ at time t. An initial market price $x(0)$ is also given. Determine the equilibrium price function $x(t)$, and find the long-range equilibrium price $\bar{x}$ if it exists.

27. **(Supply–Demand Equilibrium)**
$S(x) = 80 - 9x + x'$,
$D(x) = 60 - 7x + 3x'$, $x(0) = \$23$

28. **(Supply–Demand Equilibrium)**
$S(x) = 35 + 3x + 7x'$,
$D(x) = 65 - 2x + 5x'$, $x(0) = \$15$

29. **(Supply–Demand Equilibrium)**
$S(x) = -36 + 7x + 8x'$,
$D(x) = 42e^{-0.2t} - 2x + 5x'$, $x(0) = \$3$

30. **(Sales Growth)** Suppose that the annual sales $y = y(t)$ (in millions of dollars) of a

product satisfies the growth equation

$$\frac{dy}{dt} = 2.6e^{0.1t} + 0.03(100 - y),$$

where time t is measured in years. If the sales this year amount to 150 million dollars, what will they amount to 8 years from now?

10.5 |||||| LINEAR SECOND-ORDER DIFFERENTIAL EQUATIONS WITH CONSTANT COEFFICIENTS (OPTIONAL)

A second-order differential equation that can be expressed in the form

$$a(x)\frac{d^2y}{dx^2} + b(x)\frac{dy}{dx} + c(x)y = d(x) \tag{1}$$

is said to be **linear**. By dividing both sides by $a(x)$, we can always rewrite (1) so that the coefficient of d^2y/dx^2 is 1. Thus, the general linear second-order differential equation can always be written in the form

$$\frac{d^2y}{dx^2} + b(x)\frac{dy}{dx} + c(x)y = d(x). \tag{2}$$

|||||||||||| **The Homogeneous Case**

If the function $d(x)$ in Equation (2) is the constant 0, the resulting equation is said to be **homogeneous**. The problem of solving the linear second-order differential equation is greatly simplified by considering the homogeneous case first. We begin by defining a few terms.

Two functions $f_1(x)$ and $f_2(x)$ are said to be **independent** if neither of them is a constant multiple of the other; that is, there is no constant c such that for all x in the common domain of f_1 and f_2,

$$f_1(x) = cf_2(x),$$

or

$$f_2(x) = cf_1(x).$$

A function of $g(x)$ is said to be a **linear combination** of functions $f_1(x)$ and $f_2(x)$ if there exist constants c_1 and c_2 such that

$$g(x) = c_1f_1(x) + c_2f_2(x)$$

for all x in the common domain of f_1 and f_2. For example, the function $g(x) = 5x^3 - 8x + 8$ is a linear combination of the functions $f_1(x) = x^3$ and $f_2(x) = x - 1$, since

$$5x^3 - 8x + 8 = 5(x^3) - 8(x - 1),$$

or

$$g(x) = 5f_1(x) + (-8)f_2(x).$$

We are now in a position to state the first important theorem in the theory of linear second-order differential equations.

THEOREM 1

Consider the general homogeneous second-order linear differential equation

$$\frac{d^2y}{dx^2} + b(x)\frac{dy}{dx} + c(x)y = 0. \tag{3}$$

(a) If $y_1(x)$ and $y_2(x)$ are solutions of (3), then any linear combination of them

$$y(x) = c_1 y_1(x) + c_2 y_2(x)$$

is also a solution of (3).

(b) If $y_1(x)$ and $y_2(x)$ are *independent* solutions of (3), then *every* solution of (3) is a linear combination of them.

We can easily prove (a) of Theorem 1, but the proof of (b) is much more difficult and is beyond the level of this book.

Proof of (a)   Suppose that $y_1(x)$ and $y_2(x)$ are both solutions of the differential equation (3). Let c_1 and c_2 be any constant real numbers, and let $y(x) = c_1 y_1(x) + c_2 y_2(x)$. Then, applying the differentiation rules of Section 3.2,

$$\frac{d}{dx}[y(x)] = \frac{d}{dx}[c_1 y_1(x) + c_2 y_2(x)]$$

$$= \frac{d}{dx}[c_1 y_1(x)] + \frac{d}{dx}[c_2 y_2(x)]$$

$$= c_1 \frac{d}{dx}[y_1(x)] + c_2 \frac{d}{dx}[y_2(x)].$$

Similarly,

$$\frac{d^2}{dx^2}[y(x)] = \frac{d}{dx}\left[\frac{d}{dx}y(x)\right]$$

$$= \frac{d}{dx}\left\{c_1 \frac{d}{dx}[y_1(x)] + c_2 \frac{d}{dx}[y_2(x)]\right\}$$

$$= \frac{d}{dx}\left\{c_1 \frac{d}{dx}[y_1(x)]\right\} + \frac{d}{dx}\left\{c_2 \frac{d}{dx}[y_2(x)]\right\}$$

$$= c_1 \frac{d^2}{dx^2}[y_1(x)] + c_2 \frac{d}{dx}[y_2(x)].$$

Thus,

$$\frac{d^2}{dx^2}[y(x)] + b(x)\frac{d}{dx}[y(x)] + c(x)y(x)$$

$$= \left\{ c_1 \frac{d^2}{dx^2}[y_1(x)] + c_2 \frac{d^2}{dx^2}[y_2(x)] \right\} + b(x)\left\{ c_1 \frac{d}{dx}[y_1(x)] + c_2 \frac{d}{dx}[y_2(x)] \right\}$$

$$+ c(x)[c_1 y_1(x) + c_2 y_2(x)]$$

$$= \left\{ c\frac{d^2}{dx^2}[y_1(x)] + b(x)c_1\frac{d}{dx}[y_1(x)] + c(x)c_1 y_1(x) \right\}$$

$$+ \left\{ c_2 \frac{d^2}{dx^2}[y_2(x)] + b(x)c_2\frac{d}{dx}[y_2(x)] + c(x)c_2 y_2(x) \right\}$$

$$= c_1 \left\{ \frac{d^2}{dx^2}[y_1(x)] + b(x)\frac{d}{dx}[y_1(x)] + c(x)y_1(x) \right\}$$

$$+ c_2 \left\{ \frac{d^2}{dx^2}[y_2(x)] + b(x)\frac{d}{dx}[y_2(x)] + c(x)y_2(x) \right\}$$

$$= c_1 \cdot 0 + c_2 \cdot 0 \qquad \text{since } y_1(x) \text{ and } y_2(x) \text{ are solutions of (3)}$$

$$= 0.$$

This proves that every linear combination $y(x)$ of $y_1(x)$ and $y_2(x)$ is a solution of (3). ■

Any two independent functions which are solutions of (3) are said to be **basic solutions** of (3). Theorem 1 tells us that *to solve a homogeneous linear second-order differential equation we look for a pair of basic solutions*. The general solution is always expressible as a linear combination of a pair of basic solutions.

EXAMPLE 1 Show that the functions $y_1(x) = x^2$ and $y_2(x) = x^3$ are solutions of the differential equation

$$x^2 y'' - 4xy' + 6y = 0 \tag{4}$$

and obtain the general solution of (4).

Solution For $y_1(x) = x^2$ we have $y_1'(x) = 2x$ and $y_1''(x) = 2$. Thus,

$$x^2 y_1'' - 4xy_1' + 6y_1 = x^2(2) - 4x(2x) + 6(x^2)$$
$$= x^2(2 - 8 + 6)$$
$$= 0.$$

For $y_2(x) = x^3$ we have $y_2'(x) = 3x^2$ and $y_2''(x) = 6x$. Thus,

$$x^2 y_2'' - 4xy_2' + 6y_2 = x^2(6x) - 4x(3x^2) + 6x^3$$
$$= x^3(6 - 12 + 6)$$
$$= 0.$$

Therefore, $y_1(x) = x^2$ and $y_2(x) = x^3$ are solutions of (4). Now, x^2 and x^3 are independent functions, since neither is a constant multiple of the other. Therefore, by Theorem 1, the general solution of (4) is a linear combination of x^2 and x^3,

$$y(x) = c_1 x^2 + c_2 x^3.$$ ∎

Observe that the general solution of (4) involves *two* unknown constants, whereas the general solution of a first-order linear differential equation involves only one. It is plausible to expect that the general solution of a second-order differential equation will always involve two constants, since the presence of the second derivative suggests that integration will somehow be performed twice in the process of obtaining the solution. One constant enters into the solution each time integration is performed.

The problem of solving the general linear second-order differential equation (2) is a profound one, even in the homogeneous case, requiring advanced mathematical techniques that are well beyond the level of this book. We shall now consider a less general version of (2) which we can always solve.

||||||||||||| ## Equations with Constant Coefficients

Throughout the remainder of this section we confine our attention to linear second-order differential equations with constant coefficients:

$$a\frac{d^2 y}{dx^2} + b\frac{dy}{dx} + cy = d$$

or equivalently,

$$ay'' + by' + cy = d \tag{5}$$

where a, b, c, and d are constants.

We first consider only the homogeneous case,

$$ay'' + by' + cy = 0. \tag{6}$$

By Theorem 1, solving this differential equation requires that we find two "basic solutions." Since functions of the form e^{mx} have been so helpful in solving previous differential equations, we are tempted to take a wild guess for one basic solution:

$$y = e^{mx},$$

where m is a constant. For this function we have

$$y' = me^{mx} \quad \text{and} \quad y'' = m^2 e^{mx}.$$

Thus,

$$ay'' + by' + cy = am^2 e^{mx} + bme^{mx} + ce^{mx}$$
$$= (am^2 + bm + c)e^{mx}.$$

From this we see that $y = e^{mx}$ is a solution of (6) if and only if m is a solution of the quadratic equation

$$am^2 + bm + c = 0, \tag{7}$$

which is called the **auxiliary equation** of (6). Our wild guess turned out to be not so wild, after all. It leads us to the following theorem.

THEOREM 2

> A homogeneous linear second-order differential equation with constant coefficients
>
> $$ay'' + by' + cy = 0 \tag{6}$$
>
> has the function $y = e^{mx}$ as a solution if and only if m satisfies the auxiliary equation
>
> $$am^2 + bm + c = 0. \tag{7}$$

We now begin a sequence of three theorems that will tell us exactly how to solve the differential equation (6).

THEOREM 3

> If the auxiliary equation (7) has two (different) real number solutions, m_1 and m_2, then $e^{m_1 x}$ and $e^{m_2 x}$ are basic solutions, and the general solution of the differential equation (6) is
>
> $$y = c_1 e^{m_1 x} + c_2 e^{m_2 x} \tag{8}$$
>
> where c_1 and c_2 are unspecified constants.

Proof   Suppose the auxiliary equation (7) has two different real number roots, m_1 and m_2. Then, by Theorem 2, the functions $y_1 = e^{m_1 x}$ and $y_2 = e^{m_2 x}$ are both solutions of (6). We now show that $y_1(x)$ and $y_2(x)$ are independent functions. Suppose that $y_1(x)$ and $y_2(x)$ are *not* independent. Then there exists a constant c such that for all x,

$$e^{m_1 x} = c e^{m_2 x}$$

or

$$\frac{e^{m_1 x}}{e^{m_2 x}} = c.$$

But the left-hand side of this equation cannot be constant, since $m_1 \neq m_2$. Therefore, no such constant c exists. That is, y_1 and y_2 are independent functions.

By Theorem 1, therefore $e^{m_1 x}$ and $e^{m_2 x}$ are basic solutions and the general solution is

$$y = c_1 e^{m_1 x} + c_2 e^{m_2 x}. \qquad \blacksquare$$

EXAMPLE 2 Solve the differential equation $2y'' + 5y' - 3y = 0$.

Solution The given differential equation is homogeneous, linear, second-order, with constant coefficients. Its auxiliary equation is

$$2m^2 + 5m - 3 = 0$$
$$(2m - 1)(m + 3) = 0.$$

Since the auxiliary equation has $m = \frac{1}{2}$ and $m = -3$ as roots, Theorem 3 tells us $e^{\frac{2}{3}x}$ and $xe^{\frac{2}{3}x}$ are basic solutions, and the general solution is

$$y = c_1 e^{\frac{1}{2}x} + c_2 e^{-3x}$$

where c_1 and c_2 are unspecified constants. ∎

THEOREM 4

If the auxiliary equation (7) has only one solution, m, then e^{mx} and xe^{mx} are basic solutions and the general solution of the differential equation (6) is

$$y = c_1 e^{mx} + c_2 x e^{mx}. \tag{9}$$

Proof The quadratic formula (see Appendix A, Section A.6) tells us that the solutions of the auxiliary equation (7) are

$$m = \frac{-b \pm \sqrt{b^2 - 4ac}}{2a}.$$

Since (7) has only one solution we must have $b^2 - 4ac = 0$, and so

$$m = \frac{-b \pm 0}{2a},$$

or

$$2am + b = 0. \tag{10}$$

We already know, by Theorem 2, that $y_1(x) = e^{mx}$ is a solution of (6). We now consider the function

$$y_2(x) = xe^{mx}.$$

Taking the first and second derivatives of $y_2(x)$, we have

$$y_2'(x) = xme^{mx} + e^{mx},$$
$$y_2''(x) = xm^2 e^{mx} + 2me^{mx}.$$

Then

$$ay_2'' + by_2' + cy_2 = a(xm^2 e^{mx} + 2me^{mx}) + b(xme^{mx} + e^{mx}) + cxe^{mx}$$
$$= e^{mx}[axm^2 + 2am + bmx + b + cx]$$
$$= e^{mx}[x(am^2 + bm + c) + (2am + b)].$$

Now $am^2 + bm + c = 0$ (why?) and by (10), $2am + b = 0$. Therefore,

$$ay_2'' + by_2' + cy_2 = e^{mx}[x \cdot 0 + 0]$$
$$= 0.$$

This shows that $y_2(x) = xe^{mx}$ is also a solution of (6).

We thus have two independent functions, $y_1(x) = e^{mx}$ and $y_2(x) = xe^{mx}$, which are solutions of (6). By Theorem 1, they are basic solutions of (6) and the general solution is

$$y = c_1 e^{mx} + c_2 x e^{mx}. \qquad \blacksquare$$

EXAMPLE 3 Solve the differential equation $9y'' - 12y' + 4y = 0$.

Solution The given differential equation is homogeneous, linear, second-order, with constant coefficients. Its auxiliary equation is

$$9m^2 - 12m + 4 = 0$$
$$(3m - 2)^2 = 0,$$

which has only one root, the real number $m = \frac{2}{3}$. By Theorem 4, the functions $e^{\frac{2}{3}x}$ and $xe^{\frac{2}{3}x}$ are basic solutions, and the general solution is

$$y = c_1 e^{\frac{2}{3}x} + c_2 x e^{\frac{2}{3}x}. \qquad \blacksquare$$

Theorems 3 and 4 tell us how to find the general solution of a homogeneous linear second-order differential equation with constant coefficients, in case the auxiliary equation has 1 or 2 real number solutions. But what can we do if the auxiliary equation has no real number solution? By Theorem 2, we know that in this situation no function of the form e^{mx} will be a solution. We must look for another kind of basic solution.

The reader should be forwarned that the trigonometric functions sin x and cos x will appear throughout the remainder of this section and the next section.

EXAMPLE 4 Solve the differential equation $y'' + y = 0$. $\qquad\qquad$ (11)

Solution The quadratic formula applied to the auxiliary equation $m^2 + 1 = 0$ yields

$$m = \frac{-1 \pm \sqrt{0 - 4}}{2} = \frac{-1 \pm \sqrt{-4}}{2}.$$

The discriminant $D = b^2 - 4ac$ (see Appendix A, Section A.6) in this case is negative, $D = -4$. This tells us that the auxiliary equation has no real number roots. Thus, e^{mx} cannot be used as one of the basic solutions.

Our given differential equation can be rewritten in the form

$$y'' = -y. \tag{12}$$

We shall use our knowledge of trigonometry (Chapter 9) to find two functions which satisfy Equation (12). Let

$$y_1(x) = \sin x \qquad \text{and} \qquad y_2(x) = \cos x.$$

From Chapter 9 we know that

$$y_1'(x) = \cos x \text{ and } y_2'(x) = -\sin x.$$

Moreover,

$$y_1''(x) = -\sin x = -y_1(x)$$

and

$$y_2''(x) = -\cos x = -y_2(x).$$

Thus $\sin x$ and $\cos x$ are solutions of (12). Furthermore, $\sin x$ and $\cos x$ are independent functions. By Theorem 1, therefore, these functions are basic solutions of (11) and the general solution of (11) is

$$y = c_1 \sin x + c_2 \cos x. \qquad \blacksquare$$

The reader might find it surprising to suddenly encounter the trigonometric functions, $\sin x$ and $\cos x$, in the study of differential equations which do not seem to involve trigonometry at all. But Example 4 is no accident. We are going to find that whenever the auxiliary equation of a homogeneous linear second-order differential equation with constant coefficients has no real number roots, the general solution will *always* involve the sine or cosine functions.

Let us take a closer look at the general quadratic equation,

$$ax^2 + bx + c = 0. \tag{13}$$

As we know (see Appendix A, Section A.6) the discriminant

$$D = b^2 - 4ac \tag{14}$$

is negative precisely when (13) has no real number solutions. In this case we let

$$r = \frac{-b}{2a} \qquad \text{and} \quad s = \frac{\sqrt{4ac - b^2}}{2a}. \tag{15}$$

We shall need two identities in our work below. First, from (15),

$$2ar + b = 0. \tag{16}$$

Second,

$$r^2 - s^2 = \frac{b^2}{4a^2} - \frac{4ac - b^2}{4a^2} = \frac{2b^2 - 4ac}{4a^2}$$

$$= \frac{b^2}{2a^2} - \frac{c}{a}$$

$$= \frac{-b}{a}\left(\frac{-b}{2a}\right) - \frac{c}{a}$$

$$= \frac{-b}{a} r - \frac{c}{a}.$$

Thus,

$$a(r^2 - s^2) = -br - c,$$

or

$$a(r^2 - s^2) + br + c = 0. \tag{17}$$

We are now ready to state and prove the following important result.

THEOREM 5

If the auxiliary equation of the differential equation

$$ay'' + by' + cy = 0 \tag{18}$$

has no real number roots, and if we let

$$r = -\frac{b}{2a} \quad \text{and} \quad s = \frac{\sqrt{-D}}{2a}, \quad \text{where } D = b^2 - 4ac, \tag{19}$$

then the functions

$$y_1 = e^{rx} \cos sx \quad \text{and} \quad y_2 = e^{rx} \sin sx$$

are basic solutions of (18), and the general solution is

$$y = e^{rx}[c_1 \cos sx + c_2 \sin sx]. \tag{20}$$

where c_1 and c_2 are unspecified constants.

Proof   Suppose that the auxiliary equation of (18) has no real number roots. We define r and s as in (19), and we define the functions

$$y_1(x) = e^{rx} \cos sx \quad \text{and} \quad y_2(x) = e^{rx} \sin sx.$$

Then, using the product rule of differentiation,

$$
\begin{aligned}
y_1'(x) &= e^{rx}(-s \sin sx) + re^{rx} \cos sx \\
&= e^{rx}[r \cos sx - s \sin sx]
\end{aligned}
$$

and

$$
\begin{aligned}
y_1''(x) &= e^{rx}[-rs \sin sx - s^2 \cos sx] + re^{rx}[r \cos sx - s \sin sx] \\
&= e^{rx}[(r^2 - s^2) \cos sx - 2rs \sin sx].
\end{aligned}
$$

Then,

$$
\begin{aligned}
ay_1'' + by_1' + cy_1 &= ae^{rx}[(r^2 - s^2) \cos sx - 2rs \sin sx] \\
&\quad + be^{rx}[r \cos sx - s \sin sx] + ce^{rx} \cos sx \\
&= e^{rx}\{[a(r^2 - s^2) + br + c] \cos sx - (2ar + b)s \sin sx\}.
\end{aligned}
$$

Using the identities (16) and (17) we see that

$$
\begin{aligned}
ay_1'' + by_1' + cy_1 &= e^{rx}\{0 \cdot \cos sx - 0 \cdot s \cdot \sin sx.\} \\
&= 0.
\end{aligned}
$$

Therefore, $y_1(x)$ is a solution of the given differential equation (18).

We leave it to the reader to show (See Exercise 41) that $y_2(x)$ is also a solution of (18).

Since $y_1(x)$ and $y_2(x)$ are two independent solutions of (18), Theorem 1 assumes us that they are basic solutions, and that the general solution of (18) is

$$
\begin{aligned}
y &= c_1 y_1(x) + c_2 y_2(x) \\
&= c_1 e^{rx} \cos sx + c_2 e^{rx} \sin sx \\
&= e^{rx}[c_1 \cos sx + c_2 \sin sx]
\end{aligned}
$$

where c_1 and c_2 are unspecified constants.   ∎

EXAMPLE 5   Solve the differential equation $25y'' + 4y = 0$.

Solution   The auxiliary equation is $25m^2 + 4 = 0$. Since the discriminant is $D = 0^2 - 4(25)(4) = -400 < 0$, the auxiliary equation has no real number roots. Thus, Theorem 5 applies. We let

$$r = -\frac{b}{2a} = -\frac{0}{50} = 0,$$

and

$$s = \frac{\sqrt{400}}{50} = \frac{20}{50} = 0.4.$$

By Theorem 5, the functions

$$y_1(x) = e^0 \cos 0.4x = \cos 0.4x$$

and

$$y_2(x) = e^0 \sin 0.4x = \sin 0.4x$$

are basic solutions, and the general solution is

$$y = c_1 \cos 0.4x + c_2 \sin 0.4x.$$ ■

EXAMPLE 6 Solve the differential equation $2y'' + 3y' + 2y = 0$.

Solution The auxiliary equation is $2m^2 + 3m + 2 = 0$. Since the discriminant is $D = 3^2 - 4(2)(2) = -7 < 0$, the auxiliary equation has no real number roots. To apply Theorem 5 we let

$$r = -\frac{b}{2a} = -\frac{3}{4} \quad \text{and} \quad s = \frac{\sqrt{-D}}{2a} = \frac{\sqrt{7}}{4}.$$

Then the functions

$$y_1(x) = e^{-\frac{3}{4}x} \cos \frac{\sqrt{7}}{4} x$$

and

$$y_2(x) = e^{-\frac{3}{4}x} \sin \frac{\sqrt{7}}{4} x$$

are basic solutions, and the general solution is

$$y = e^{-\frac{3}{4}x} \left[c_1 \cos \frac{\sqrt{7}}{4} x + c_2 \sin \frac{\sqrt{7}}{4} x \right].$$ ■

We summarize our results in the following table.

Homogeneous Linear Second-Order Differential Equations with Constant Coefficients
$ay'' + by' + cy = 0$

Discriminant $D = b^2 - 4ac$	Real roots of auxiliary equation	Basic solutions	General solution
$D > 0$	r_1, r_2 $(r_1 \neq r_2)$	$y_1 = e^{r_1 x}$ $y_2 = e^{r_2 x}$	$y = c_1 e^{r_1 x} + c_2 e^{r_2 x}$
$D = 0$	r	$y_1 = e^{rx}$ $y_2 = xe^{rx}$	$y = c_1 e^{rx} + c_2 xe^{rx}$
$D < 0$ $\left(\text{Take } r = -\dfrac{b}{2a}, s = \dfrac{\sqrt{-D}}{2a} \right)$	none	$y_1 = e^{rx} \cos sx$ $y_2 = e^{rx} \sin sx$	$y = e^{rx} [c_1 \cos sx + c_2 \sin sx]$

|||||||||||| **Initial-Value Problems and Boundary-Value Problems**

Since the general solution of a second-order differential equation involves two constants, c_1 and c_2, two conditions must be given in order to completely specify a particular solution. The problem of finding the particular solution which satisfies two "initial conditions"

$$y(x_0) = y_0, \ y'(x_0) = y_0'$$

is called an **initial-value problem**, and the problem of finding the particular solution which satisfies two "boundary conditions"

$$y(x_1) = y_1, \ y(x_2) = y_2$$

is called a **boundary-value problem.**

EXAMPLE 7 Solve the initial-value problem

$$3y'' - 11y' - 4y = 0, \qquad y(0) = 8, \ y'(0) = 6.$$

Solution The auxiliary equation is

$$3m^2 - 11m - 4 = 0$$
$$(3m + 1)(m - 4) = 0,$$

and the general solution is

$$y = c_1 e^{-\frac{1}{3}x} + c_2 e^{4x}. \tag{21}$$

To find the particular solution which satisfies the given initial conditions, we first note that

$$y' = -\tfrac{1}{3}c_1 e^{-\frac{1}{3}x} + 4c_2 e^{4x}. \tag{22}$$

Substituting $y(0) = 5$ into (21) and $y'(0) = 2$ into (22) we have

$$\begin{cases} 8 = c_1 + c_2 \\ 6 = -\tfrac{1}{3}c_1 + 4c_2. \end{cases}$$

Solving these two equations simultaneously we find that

$$c_1 = 6, \qquad c_2 = 2.$$

Therefore, the particular solution we are seeking is

$$y = 6e^{-\frac{1}{3}x} + 2e^{4x}. \qquad \blacksquare$$

EXAMPLE 8   Solve the boundary-value problem $y'' + 9y = 0$, $y\left(\dfrac{\pi}{2}\right) = 4$, $y(\pi) = 8$.

Solution   The auxiliary equation $m^2 + 9 = 0$ has no real number roots, so we let

$$r = -\frac{b}{2a} = 0 \quad \text{and} \quad s = \frac{\sqrt{4ac - b^2}}{2a} = \frac{\sqrt{36 - 0}}{2} = 3.$$

The functions

$$\begin{cases} y_1 = e^0 \cos 3x = \cos 3x \\ y_2 = e^0 \sin 3x = \sin 3x \end{cases}$$

are basic solutions, and the general solution is

$$y = c_1 \cos 3x + c_2 \sin 3x. \tag{23}$$

Substituting the boundary conditions into (23), we have

$$\begin{cases} 4 = c_1 \cos \dfrac{3\pi}{2} + c_2 \sin \dfrac{3\pi}{2} = 0 - c_2 \\ 8 = c_1 \cos 3\pi + c_2 \sin 3\pi = -c_1 + 0. \end{cases}$$

Therefore, $c_1 = -8$, $c_2 = -4$, and the particular solution we are seeking is

$$y = -8 \cos 3x - 4 \sin 3x. \qquad\blacksquare$$

|||||||||||| The Nonhomogeneous Case

Now that we have learned how to solve homogeneous linear second-order differential equations with constant coefficients, it is a relatively easy task to extend our techniques to solve the **nonhomogeneous** equation

$$ay'' + by' + cy = d. \tag{24}$$

THEOREM 6

If $c \neq 0$, the nonhomogeneous equation

$$ay'' + by' + cy = d$$

can be transformed into a homogeneous equation

$$au'' + bu' + cu = 0$$

by making the substitution

$$u = y - \frac{d}{c}. \tag{25}$$

Rather than give a formal proof of this theorem, we shall simply give an example.

EXAMPLE 9   Solve the differential equation

$$y'' + 2y' + 3y = 5. \tag{26}$$

Solution   Since this is a nonhomogeneous equation, we follow the advice of Theorem 6 and make the substitution

$$u = y - \tfrac{5}{3}.$$

Then

$$y = u + \tfrac{5}{3} \tag{27}$$

and

$$y' = u', \qquad y'' = u''.$$

Substituting these expressions for y, y', and y'' into (26), our differential equation becomes

$$u'' + 2u' + 3(u + \tfrac{5}{3}) = 5$$

or

$$u'' + 2u' + 3u = 0. \tag{28}$$

The auxiliary equation $m^2 + 2m + 3 = 0$ has discriminant $D = 4 - 12 = -8 < 0$, so we let

$$r = -\frac{2}{2} = -1 \qquad \text{and} \qquad s = \frac{\sqrt{8}}{2} = \sqrt{2}.$$

By Theorem 5, the general solution of (28) is

$$u = e^{-x}[c_1 \cos \sqrt{2}\, x + c_2 \sin \sqrt{2}\, x].$$

Therefore, by (27), the general solution of our given differential equation (26) is

$$y = e^{-x}[c_1 \cos \sqrt{2}\, x + c_2 \sin \sqrt{2}\, x] + \tfrac{5}{3}. \qquad \blacksquare$$

If $c = 0$ in Equation (24), then Theorem 6 cannot be used. In that case, the following theorem applies.

THEOREM 7

The nonhomogeneous differential equation

$$ay'' + by' = d$$

can be transformed into a first-order linear differential equation

$$au' + bu = d$$

by making the substitution $u = y'$.

EXAMPLE 10   Solve the differential equation $3y'' - 4y' = 15$.

Solution   Following the advice of Theorem 7, we let $u = y'$ and solve the first-order differential equation

$$3u' - 4u = 15,$$

or

$$u' = \tfrac{4}{3}u + 5. \tag{29}$$

In Theorem 3 of Section 10.2 we found that the general solution of the differential equation $u' = au + b$ is $u = Ce^{ax} - \dfrac{b}{a}$, where C is a constant. Thus the general solution of (29) is

$$u = Ce^{\frac{4}{3}x} - \frac{5}{(4/3)}$$

$$= Ce^{\frac{4}{3}x} - \tfrac{15}{4}.$$

But $u = y'$, so we have

$$y' = Ce^{\frac{4}{3}x} - \tfrac{15}{4}.$$

To solve for y we integrate both sides of this equation, obtaining

$$y = \tfrac{3}{4}Ce^{\frac{4}{3}x} - \tfrac{15}{4}x + C_2.$$

Finally, letting $C_1 = \tfrac{3}{4}C$, we see that the general solution of our differential equation is

$$y = C_1 e^{\frac{4}{3}x} - \tfrac{15}{4}x + C_2. \qquad \blacksquare$$

Geometry of Solutions When $D < 0$.

When the auxiliary equation of a homogeneous linear second-order differential equation with constant coefficients

$$ay'' + by' + cy = 0 \tag{30}$$

has no real roots, the graph of a solution is one of two basic types. Each of these two types of curves has a significant shape, and will be given a corresponding name. The following theorem is technical, but crucial in our development. Its proof may be omitted.

THEOREM 8

If s, c_1, and c_2 are arbitrary real numbers, then there is a unique angle θ (in radians, $0 \leq \theta < 2\pi$) such that for all x,

$$c_1 \cos sx + c_2 \sin sx = \sqrt{c_1^2 + c_2^2} \, \sin(sx + \theta).$$

Moreover, the angle θ can be found from the equations

$$\sin \theta = \frac{c_1}{\sqrt{c_1^2 + c_2^2}}, \qquad \cos \theta = \frac{c_2}{\sqrt{c_1^2 + c_2^2}}.$$

Proof (Optional)

First, observe that

$$c_1 \cos sx + c_2 \sin sx = \sqrt{c_1^2 + c_2^2} \left[\frac{c_1}{\sqrt{c_1^2 + c_2^2}} \cos sx + \frac{c_2}{\sqrt{c_1^2 + c_2^2}} \sin sx \right]$$

$$= \sqrt{c_1^2 + c_2^2} \, [u \cos sx + v \sin sx]$$

where $u = \dfrac{c_1}{\sqrt{c_1^2 + c_2^2}}$ and $v = \dfrac{c_2}{\sqrt{c_1^2 + c_2^2}}$. Observe that $-1 \leq u \leq 1$, $-1 \leq v \leq 1$,

and $u^2 + v^2 = \dfrac{c_1^2 + c_2^2}{c_1^2 + c_2^2} = 1$. Hence, there is a unique angle θ (in radians, $0 \leq \theta < 2\pi$)

such that $\sin \theta = u$ and $\cos \theta = v$. Thus,

$$c_1 \cos sx + c_2 \sin sx = \sqrt{c_1^2 + c_2^2} \, [\sin \theta \cos sx + \cos \theta \sin sx].$$

By the remarkable identity found as Equation (1) in Section 9.2, the quantity inside the brackets equals $\sin(\theta + sx)$. Therefore, we have

$$c_1 \cos sx + c_2 \sin sx = \sqrt{c_1^2 + c_2^2} \, \sin(\theta + sx). \qquad \blacksquare$$

Theorem 8 shows that every function of the form $c_1 \cos sx + c_2 \sin sx$ can be written in the form

$$y = A \sin(sx + \theta) \qquad (A > 0, 0 \leq \theta < 2\pi). \qquad (31)$$

Functions of this type are called sinusoids; their graphs have the shape of a sine curve. Figure 1 shows the graph of a sinusoidal function (or **sinusoid**).

The number A is called the **amplitude** of the sinusoidal function (31). Recall that the function $y = \sin x$ is periodic (or cyclic) with period 2π. Thus, the sinusoidal function (31) is also **periodic** and has **period** $\dfrac{2\pi}{|s|}$. The number $\dfrac{\theta}{s}$ is sometimes called the (leftward) **phase shift** of the sinusoid (31).

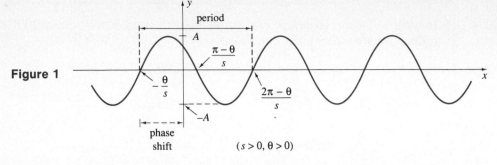

Figure 1

$(s > 0, \theta > 0)$

When the independent variable x represents time (t), the period is the length of time it takes the sinusoidal function to complete one cycle. The reciprocal of the period is then called the **frequency** of the sinusoid, and indicates how many cycles there are in one unit of time. Thus,

$$\text{frequency} = \text{cycles per unit time} = \frac{|s|}{2\pi}. \tag{32}$$

EXAMPLE 11   Find the numbers A and θ such that for all x,

$$-3 \cos 5x + 2 \sin 5x = A \sin (5x + \theta).$$

Determine the amplitude, period, frequency and phase shift of the resulting sinusoid.

Solution   From Theorem 8, $A = \sqrt{(-3)^2 + 2^2} = \sqrt{13}$, and $\sin \theta = \dfrac{-3}{\sqrt{13}}$, $\cos \theta = \dfrac{2}{\sqrt{13}}$. Since $\sin \theta < 0$ and $\cos \theta > 0$, the terminal side of the angle θ must be in the fourth quadrant. Using a calculator we find the reference angle $\theta \approx 0.2308$ radians, and so

$$\theta \approx 2\pi - 0.2308 \approx 6.0524.$$

Therefore,

$$-3 \cos 5x + 2 \sin 5x = \sqrt{13} \sin (5x + 6.0524).$$

The amplitude is $\sqrt{13}$, the period is $\dfrac{2\pi}{5}$, the frequency is $\dfrac{5}{2\pi} \approx 0.796$ cycles per unit time, and the (leftward) phase shift is $\dfrac{6.0524}{5} = 1.2105$.   ∎

A function of the form

$$y = e^{rx} A \sin (sx + \theta)$$

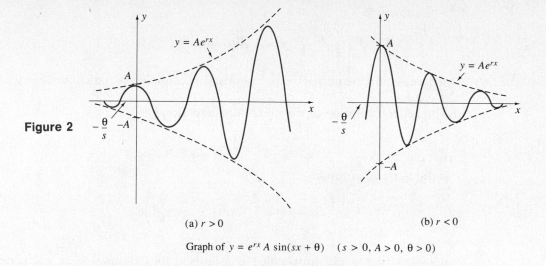

Figure 2

(a) $r > 0$ (b) $r < 0$

Graph of $y = e^{rx} A \sin(sx + \theta)$ $(s > 0, A > 0, \theta > 0)$

where r, s, A, and θ are constants, with $A > 0$ and $0 \le \theta < 2\pi$, is called a **damped sinusoid**. Figure 2 shows graphs of typical damped sinusoids. Notice that damped sinusoidal functions are not periodic, but are nevertheless oscillating.

We now have the necessary insight and vocabulary to describe the graphs of solution functions of the homogeneous linear second-order differential equation with constant coefficients, when $D < 0$. Our analysis shows that these solutions are always oscillating functions, and must be either sinusoidal (when $b = 0$) or damped sinusoidal (when $b \ne 0$). We summarize our results in the following theorem.

THEOREM 9

If the auxiliary equation of the differential equation

$$ay'' + by' + cy = 0 \qquad (30)$$

has no real number roots (i.e., $D < 0$) then the general solution can be written in the form

$$y = Ae^{rx} \sin (sx + \theta)$$

where A and θ are unspecified constants, $r = -\dfrac{b}{2a}$ and $s = \dfrac{\sqrt{-D}}{2a}$.

Thus, when $D < 0$ the solutions of (30) are
(1) sinusoidal (if $b = 0$), or
(2) damped sinusoidal (if $b \ne 0$).

EXAMPLE 12   (a) The general solution of the differential equation

$$25y'' + 4y = 0$$

is the sinusoid

$$y = A \sin(0.4x + \theta)$$

where A and θ are unspecified constants. (In Example 5 we calculated $r = 0$ and $s = 0.4$.)

(b) The general solution of the differential equation

$$2y'' + 3y' + 2y = 0$$

is the damped sinusoid

$$y = e^{-\frac{3}{4}x} A \sin\left(\frac{\sqrt{7}}{4} x + \theta\right)$$

where A and θ are unspecified constants. $\left(\text{In Example 6 we calculated}\right.$

$r = -\frac{3}{4}$ and $s = \dfrac{\sqrt{7}}{4}.\Big)$ ∎

10.5 EXERCISE SET ||||||||||||||||||||||||||||||||||||||||||||||||||||||||||||||||||||||||

In Exercises 1–4, show that the functions $y_1(x)$, $y_2(x)$ are solutions of the given differential equation, and write the general solution of the differential equation.

1. $x^2 y'' + xy' - y = 0$;

   $y_1(x) = x,\ y_2(x) = \dfrac{1}{x}$ $(x \neq 0)$

2. $x^2 y'' + xy' - 9y = 0$;

   $y_1(x) = x^3,\ y_2(x) = \dfrac{1}{x^3}$ $(x \neq 0)$

3. $2x^2 y'' - 3xy' - 3y = 0$;

   $y_1(x) = \dfrac{1}{\sqrt{x}},\ y_2(x) = x^3$ $(x > 0)$

4. $x^2 y'' - xy' + y = 0$;

   $y_1(x) = x,\ y_2(x) = x \ln x$ $(x > 0)$

In Exercises 5–22, find the general solution of the given differential equation.

5. $y'' - 3y' - 10y = 0$

6. $3y'' + 11y' - 4y = 0$

7. $y'' - 4y = 0$

8. $y'' - 6y + 9 = 0$

9. $y'' + 25y = 0$

10. $12y'' - y' - 6y = 0$

11. $4y'' + 4y' + y = 0$

12. $9y'' + y = 0$

13. $y'' + y' + y = 0$

14. $y'' + 4y' + 13y = 0$

15. $3y'' - y' = 0$

16. $2y'' + 4y' = 7$

17. $10y'' + 11y' - 6y = 30$

18. $16y'' - 24y' + 9y = 45$

19. $y'' - 0.8y' + 0.16y = 1$

20. $y'' - 2y = 3$

21. $4y'' + 4y' + 17y = 5$

22. $y'' + 3y' + 5y = -8$

In Exercises 23–30, find the solution of the given initial-value or boundary-value problem.

23. $3y'' + 8y' - 3y = 0$, $y(0) = 8$, $y'(0) = -4$

24. $y'' + 2y' - 8y = 0$, $y(0) = 4$, $y'(0) = -1$

25. $y'' + 2y' + y = 0$, $y(0) = 2$, $y(1) = \dfrac{5}{e}$

26. $y'' + 6y' + 9y = 5$, $y(0) = 1$, $y'(0) = \frac{4}{3}$

27. $y'' + 2y' + 3y = 12$, $y(0) = 1$, $y'(0) = 1$

28. $y'' + 4y = 8$, $y(0) = 5$, $y\left(\dfrac{\pi}{4}\right) = -5$

29. $5y'' - 2y' = 4$, $y(0) = 10$, $y\left(\dfrac{5}{2}\right) = 5e$

30. $4y'' + y' = 1$, $y(0) = 5$, $y'(0) = 1$

In Exercises 31–34, find constants A, B, and θ such that the given expression can be written as $A \sin(Bx + \theta)$. Also determine the amplitude, period, frequency and phase shift of the resulting sinusoid.

31. $3 \cos 4x + \sin 4x$

32. $-\cos 3x + 4 \sin 3x$

33. $-2 \cos \dfrac{x}{3} + 5 \sin \dfrac{x}{3}$

34. $3 \cos 0.2x - 5 \sin 0.2x$

In Exercises 35–40, take the solution of the differential equation found in the given exercise and express it in the form $y = Ae^{rx} \sin(sx + \theta)$.

35. Exercise 9   36. Exercise 12

37. Exercise 13   38. Exercise 14

39. Exercise 21   40. Exercise 22

41. Prove that the function $y_2(x)$ given in Theorem 5 is a solution of (18).

10.6 |||||| APPLICATIONS OF LINEAR SECOND-ORDER DIFFERENTIAL EQUATIONS (OPTIONAL)

In this section we consider a variety of applications of linear second-order differential equations with constant coefficients. After investigating several applications in economics, we show how our techniques for solving linear second-order differential equations can be applied fruitfully in solving some systems of first-order differential equations which model significant ecological and sociological phenomena.

|||||||||||| **Applications to Supply–Demand**

In Section 10.4 we noted that the supply and demand of a commodity are functions, not only of the selling price of the commodity, but also of the rate of change of this selling price. Buyers and sellers of the commodity base their decisions on the *price trends* as well as on the actual price itself. That is where differential equations come into play. As in Section 10.4, we let $x(t)$ denote the selling price at time t.

The second derivative d^2x/dt^2 of the price yields significant information about price trends which affect buyer and seller behavior. In addition to being concerned about whether the price is rising or falling, buyers and sellers might well be concerned about whether it is doing so at an increasing or decreasing rate. Figures 1 and 2 show various possible price trends in an interval of time around time t_0.

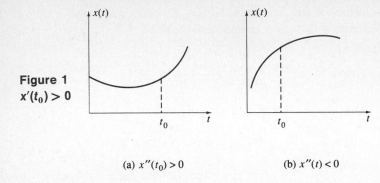

Figure 1
$x'(t_0) > 0$

(a) $x''(t_0) > 0$ (b) $x''(t) < 0$

Figure 1 shows two price trend functions that are increasing in a time interval around t_0. In Figure 1(a) the price is increasing with increasing derivative; buyers will want to buy now before the price becomes prohibitively high, and producers will hasten to supply more of the commodity. In Figure 1(b) the price is also rising, but with decreasing derivative; buyers will still want to buy now, but with less urgency, and producers will still be motivated to produce, but may show some caution.

Figure 2
$x'(t_0) < 0$

(a) $x''(t_0) < 0$ (b) $x''(t_0) > 0$

Figure 2 shows two price trend functions that are decreasing in a time interval around t_0. In Figure 2(a) the price is decreasing, with decreasing derivative. Producers will have no incentive to produce, because the trend suggests that "the bottom is dropping out of the market"; buyers will prefer to postpone buying until the price comes down dramatically as expected. In Figure 2(b), producers see an upturn in the future, so they will be less conservative in production, while buyers will anticipate a future increase in price and be less inclined to postpone buying.

**EXAMPLE 1
(Supply–
Demand
Equilibrium
Price)**

Suppose that when the second derivative of the selling price function $x(t)$ is taken into account, the supply and demand functions for a given commodity (in thousands of units) are

$$S = 100 + 7x + 6x' + 8x''$$
$$D = 100 + 9x + 5x' - 2x''.$$

At time $t = 0$ the equilibrium price is \$87 per unit and decreasing at the rate of \$3 per year. Find the equilibrium price t years later. What is the equilibrium price after 1 year?

Solution For the equilibrium price $x(t)$ we must have $S = D$, or

$$100 + 7x + 6x' + 8x'' = 100 + 9x + 5x' - 2x''$$
$$10x'' + x' - 2x = 0.$$

The auxiliary equation of this linear second-order differential equation is

$$10m^2 + m - 2 = 0$$
$$(2m + 1)(5m - 2) = 0,$$

which has roots $m = -\frac{1}{2}$ and $\frac{2}{5}$. By Theorem 5 of Section 10.5 the general solution is thus

$$x = c_1 e^{-0.5t} + c_2 e^{0.4t}. \tag{1}$$

Then

$$x' = -0.5c_1 e^{-0.5t} + 0.4c_2 e^{0.4t}. \tag{2}$$

Substituting the initial conditions $x(0) = 87$ and $x'(0) = -3$ into Equations (1) and (2), we have

$$\begin{cases} 87 = c_1 + c_2 \\ -3 = -0.5c_1 + 0.4c_2. \end{cases}$$

When we solve this pair of equations simultaneously, we obtain $c_1 = 42$ and $c_2 = 45$. We then substitute these constants into Equation (1). Therefore, after t years the equilibrium price is

$$x = 42e^{-0.5t} + 45e^{0.4t}.$$

Substituting $t = 1$ into this equation, we find that the equilibrium price after 1 year is $x(1) \approx \$93$ per unit.

Observe that, although the equilibrium price at time $t = 0$ is decreasing at the rate of \$3 per year, the equilibrium price actually *increases* by \$6 over the first year. ∎

The Role of the Middleman: A Mathematical Model

Up to this point, our mathematical models of supply and demand have ignored the role of the merchant, the "middle-man" who buys the commodity from the producer, stockpiles it, and sets the price $x(t)$ which the consumer must pay for the commodity. We now set up a simplified mathematical model which takes his role into account. We speak of "the merchant" in the singular, because it simplifies our analysis to regard all merchants collectively as a single unit.

Let $M(t)$ denote the merchant's inventory, the amount of the commodity held by the merchant at time t. This inventory fluctuates with time, increasing when there is excess supply and decreasing when there is excess demand. One mathematical model of this fluctuation is

$$\frac{dM}{dt} = k_1(S - D) \tag{3}$$

where k_1 is a positive constant. Assume that there is a certain "normal" inventory M_0, and that the merchant raises (or lowers) his selling price $x(t)$ at a rate that is proportional to the amount by which the inventory falls below (or rises above) M_0:

$$\frac{dx}{dt} = k_2(M_0 - M), \tag{4}$$

where k_2 is a positive constant. Differentiating both sides of Equation (4) yields

$$\frac{d^2x}{dt^2} = -k_2 \frac{dM}{dt}$$
$$= -k_2 k_1(S - D), \qquad \text{by (3)}$$

or merely,

$$\frac{d^2x}{dt^2} = -k(S - D) \tag{5}$$

where k is a positive constant. This second-order differential equation is a model of the effect of the supply and demand functions on the selling price set by the merchant.

**EXAMPLE 2
(Supply,
Demand, and
Merchant's
Selling Price)**

Suppose that the supply and demand functions for a certain commodity are

$$S = 2200 + 15x + 7x'$$
$$D = 3000 + 5x + 3x'$$

where $x = x(t)$ is the selling price set by the merchant at time t (in years after some initial time $t = 0$), and assume that $x(t)$ satisfies the mathematical model (5) with $k = 0.5$. At time $t = 0$ the selling price is $70 per unit and falling at the rate of $6 per year. Find the selling price after t years. In particular, find the selling price after 1 year, 2 years, and 3 years.

Solution   We first find

$$S - D = (2200 + 15x + 7x') - (3000 + 5x + 3x')$$
$$= -800 + 10x + 4x'.$$

Then Equation (5) becomes

$$\frac{d^2x}{dt^2} = -0.5(-800 + 10x + 4x')$$

$$= 400 - 5x - 2x',$$

or

$$x'' + 2x' + 5x = 400. \tag{6}$$

To solve this nonhomogeneous differential equation we apply Theorem 6 of Section 10.5. We let $u = x - \dfrac{400}{5} = x - 80$; then $u' = x'$ and $u'' = x''$. Equation (6) becomes

$$u'' + 2u' + 5(u + 80) = 400$$
$$u'' + 2u' + 5u = 0.$$

The auxiliary equation is $m^2 + 2m + 5 = 0$. Since $D = 4 - 20 = -16 < 0$, we let

$$r = \frac{-2}{2} = -1 \quad \text{and} \quad s = \frac{\sqrt{16}}{2} = 2.$$

By Theorem 5 of Section 10.5, the general solution is

$$u = e^{-t}[c_1 \cos 2t + c_2 \sin 2t].$$

Since $x = u + 80$, we have

$$x = e^{-t}[c_1 \cos 2t + c_2 \sin 2t] + 80. \tag{7}$$

To finish solving our initial-value problem we need to find the derivative,

$$x' = e^{-t}[-2c_1 \sin 2t + 2c_2 \cos 2t] - e^{-t}[c_1 \cos 2t + c_2 \sin 2t]$$
$$= e^{-t}[(2c_2 - c_1) \cos 2t - (2c_1 + c_2) \sin 2t]. \tag{8}$$

Substituting the initial conditions $x(0) = 70$, $x'(0) = -6$ into Equations (7) and (8), we have

$$\begin{cases} 70 = c_1 + 80 \\ -6 = 2c_2 - c_1 \end{cases}$$

from which we obtain $c_1 = -10$ and $c_2 = -8$. We then substitute these constants into Equation (7). Therefore, after t years the merchant's selling price is

$$x = -e^{-t}[10 \cos 2t + 8 \sin 2t] + 80.$$

Using a calculator we find the selling prices:

$$x(1) = \$78.85$$
$$x(2) = \$81.70$$
$$x(3) = \$79.63.$$

These selling prices fluctuate above and below \$80, since the solution function $x = x(t)$ is a damped sinuosid.   ∎

|||||||||||| Applications to Systems of First-Order Differential Equations

Many mathematical models involve several differential equations in several variables. We consider here mathematical models in which the rates of change of two variables are each given as a first degree function of those variables:

$$\begin{cases} \dfrac{dx}{dt} = ax + by + c \\[2mm] \dfrac{dy}{dt} = dx + ey + f. \end{cases} \tag{9}$$

To "solve" such a system means to find two functions, $x = x(t)$ and $y = y(t)$ which satisfy both differential equations in (9) simultaneously. We shall show how to solve (9) by transforming the system into a single second-order linear differential equation.

EXAMPLE 3   Solve the system

$$\begin{cases} \dfrac{dx}{dt} = 4x - y & (10) \\[2mm] \dfrac{dy}{dt} = 6x - 3y & (11) \end{cases}$$

for functions $x = x(t)$, $y = y(t)$ satisfying the initial conditions $x(0) = 7$, $y(0) = -8$.

Solution   Differentiating both sides of (10), we have

$$x'' = 4x' - y'. \tag{12}$$

Then, substituting (11) into (12),

$$x'' = 4x' - (6x - 3y)$$

or

$$x'' = 4x' - 6x + 3y. \tag{13}$$

We shall obtain a differential equation that contains only one function and its derivatives, by eliminating y from Equation (13). We solve Equation (10) for y,

$$y = 4x - \frac{dx}{dt}, \tag{14}$$

and substitute the result into (13):

$$x'' = 4x' - 6x + 3(4x - x')$$
$$x'' = x' + 6x$$
$$x'' - x' - 6x = 0. \tag{15}$$

The auxiliary equation of this second-order linear differential equation is $m^2 - m - 6 = 0$, which has roots $m = -2, 3$. Thus, the general solution of (15) is

$$x = c_1 e^{-2t} + c_2 e^{3t}. \tag{16}$$

To find the general solution for the function $y = y(t)$ we use Equation (14). Differentiating (16) we have

$$x' = -2c_1 e^{-2t} + 3c_2 e^{3t}. \tag{17}$$

Substituting (16) and (17) into (14) we have

$$y = 4(c_1 e^{-2t} + c_2 e^{3t}) - (-2c_1 e^{-2t} + 3c_2 e^{3t})$$
$$y = 6c_1 e^{-2t} + c_2 e^{3t}. \tag{18}$$

Putting (16) and (18) together, we see that the general solution of our system (10) and (11) is

$$\begin{cases} x = c_1 e^{-2t} + c_2 e^{3t} \\ y = 6c_1 e^{-2t} + c_2 e^{3t}. \end{cases}$$

Using the given initial conditions $x(0) = 7$, $y(0) = -8$, we must have

$$\begin{cases} 7 = c_1 + c_2 \\ -8 = 6c_1 + c_2. \end{cases}$$

Solving these equations simultaneously we find that $c_1 = -3$ and $c_2 = 10$. Therefore, the solution of our given initial-value problem is

$$\begin{cases} x = -3e^{-2t} + 10e^{3t} \\ y = -18e^{-2t} + 10e^{3t}. \end{cases}$$

∎

EXAMPLE 4
(Richardson's
Arms Race
Model)

In the 1930s, L. F. Richardson* developed a mathematical model of an arms race between two nations (or blocs of nations). In his model, x and y represent the respective annual expenditures on armaments of countries X and Y in some standard monetary unit, say dollars. These numbers are nonnegative, and are functions of time t, measured in years from some initial time $t = 0$. Richardson's research suggested the model

$$\frac{dx}{dt} = ay - bx + e \tag{19}$$

$$\frac{dy}{dt} = cx - dy + f \tag{20}$$

where a, b, c, and d are positive constants, while e and f are any constants. Richardson called the constants a and c **defense coefficients;** they reflect the tendency for each side to escalate its military arming in proportion to the armaments possessed by the other side. The constants b and d are called **restraint coefficients;** they reflect the tendency of high armament expenditures to have a braking or fatiguing effect on the buildup of arms within the same country. Finally, the constants e and f may be called **attitude constants;** they reflect underlying feelings in each country of grievance (when positive) or good will (when negative) toward the other country.

Briefly, Richardson's model asserts that the rate of increase in arms expenditures by one country depends positively on the arms expenditure level of the other, negatively on its own arms expenditures, and positively or negatively on its internal attitudes. The model is quite flexible, since the constants may be chosen in a way that best reflects the nations under consideration. As we have seen in Example 3, Richardson's model can be solved by transforming the system into a single second-order differential equation. The solution will tell us the annual arms expenditures of both countries after t years.

An arms race is said to have reached an **equilibrium** at time $t = t_0$ if neither country is escalating its armaments at time $t = t_0$; that is,

$$x'(t_0) = y'(t_0) = 0.$$

In view of Equations (19) and (20), such an equilibrium point (x_0, y_0) must satisfy the following pair of equations simultaneously:

$$\begin{cases} ay - bx + e = 0 & \text{(21)} \\ cx - dy + f = 0 & \text{(22)} \end{cases}$$

When we solve these two equations algebraically we find that

$$x_0 = \frac{af + de}{ac - bd}, \qquad y_0 = \frac{bf + ce}{ac - bd} \tag{23}$$

---

* Lewis Fry Richardson (1881–1953) was born at Newcastle-upon-Tyne, England, and educated at Cambridge University. As a physicist he made outstanding contributions to the mathematical development of meteorology. A Quaker with strong convictions against war, he served in ambulance duty in World War I, and after the war he pursued extensive research into the causes of war.

provided that $ac - bd \neq 0$. To make sense, we must have $x_0 \geq 0$, $y_0 \geq 0$. Thus, if $ac - bd = 0$ or $x_0 < 0$ or $y_0 < 0$, there will be no equilibrium point in the arms race.

An equilibrium point (x_0, y_0) will be called **stable** if, once it is achieved, it is maintained. An unstable equilibrium, which is achieved only momentarily and then abandoned, is not especially satisfactory. To see whether a stable equilibrium is attainable in a given arms race satisfying Richardson's model we take a closer look at Equations (21) and (22).

We rewrite Equation (21) in the form

$$y = \frac{b}{a} x - \frac{e}{a}. \tag{24}$$

When we graph this equation in the xy plane we obtain the straight line L_x shown in Figure 3. Notice that its slope is positive, since both a and b are positive.

Figure 3

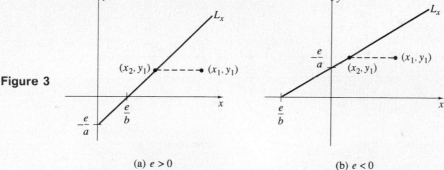

(a) $e > 0$ (b) $e < 0$

Consider any point (x_1, y_1) not on line L_x, and the corresponding point (x_2, y_1) on L_x, as shown in Figure 3. From (19) the derivative dx/dt at (x_1, y_1) is

$$x'(x_1, y_1) = ay_1 - bx_1 + e. \tag{25}$$

Since (x_2, y_1) lies on L_x, Equation (21) tells us that

$$e = bx_2 - ay_1. \tag{26}$$

Substituting (26) into (25) we have

$$x'(x_1, y_1) = ay_1 - bx_1 + (bx_2 - ay_1)$$
$$x'(x_1, y_1) = b(x_2 - x_1). \tag{27}$$

Equation (27) tells us that if (x_1, y_1) is to the right of L_x (i.e., $x_1 > x_2$) then $dx/dt = x'(x_1, y_1) < 0$; that is, x is decreasing and the point (x, y) representing the arms expenditures of countries X and Y is moving to the left from its present position (x_1, y_1) toward the line L_x. Similarly, if (x_1, y_1) is to the left of L_x (i.e., $x_1 < x_2$) then by (27) $dx/dt = x'(x_1, y_1) > 0$; that is, x is increasing and the point (x, y) moves to the right from (x_1, y_1) toward the line L_x.

In summary, Richardson's model predicts that country X responds to *any* behavior of country Y in a way that moves the point (x, y) horizontally toward the line L_x. For this reason we call the line L_x the **optimal line** for country X. A similar analysis shows that country Y responds to any behavior of X in a way that moves the point (x, y) vertically toward the line L_y, which we call the optimal line for Y.

From the above analysis we see that an arms race which follows Richardson's model will have a stable equilibrium (x_0, y_0) if and only if the two lines L_x and L_y intersect at (x_0, y_0), in the first quadrant. (See Figure 4.) Moreover, since two lines cannot intersect at more than one point, an arms race cannot have more than one stable equilibrium point.

Figure 4

(a) Stable equilibrium at (x_0, y_0) (b) No equilibrium

Two other possibilities are worth mentioning here. If both $\lim\limits_{t \to +\infty} x(t) = +\infty$ and $\lim\limits_{t \to +\infty} y(t) = +\infty$, then the arms race is said to be a **runaway arms race.** If there exist finite numbers $\bar{x}$ and $\bar{y}$ such that $\lim\limits_{t \to +\infty} x(t) = \bar{x}$ and $\lim\limits_{t \to +\infty} y(t) = \bar{y}$, then the arms race is said to be a **stabilizing arms race.** If these limits are both ≤ 0, then we say that **total disarmament** is possible between these two nations.

A more detailed and extensive analysis of Richardson's model is beyond the scope of this book.* In the decades since it was first proposed it has been refined, extended and debated. It remains today a subject of some discussion and controversy. ∎

10.6 EXERCISE SET ||||||||||||||||||||||||||||||||||||||||||||||||||||||||||||||||||||||||||||||||||||||||||||

(Supply–Demand Equilibrium Price) In Exercises 1–3, supply and demand functions for a commodity (in thousands of units) are given

as functions of the unit selling price $x = x(t)$. The initial selling price $x(0)$ and its initial rate of change $x'(0)$ are also given. Find the equili-

---

* See *An Introduction to Mathematical Models in the Social and Life Sciences*, by Michael Olinick, © 1978 by Addison-Wesley Publishing Company, Inc.

brium price after t years and after 5 years. Find the long-range equilibrium price if it exists.

1. $S = 35 + 2x + 9x' + 5x''$
$D = 45 - 3x + x' + 2x''$
$x(0) = 3, \ x'(0) = 1$

2. $S = 80 + 9x + 11x' + 8x''$
$D = 120 + 2x - 5x' + 4x''$
$x(0) = 7, \ x'(0) = 3$

3. $S = 80 + 5x + 6x' + 15x''$
$D = 72 + 6x - 2x' - 5x''$
$x(0) = 3, \ x'(0) = 1$

(Supply, Demand, and Merchant's Selling Price) In Exercises 4–6, supply and demand functions for a commodity (in thousands of units) are given as functions of the unit selling price $x = x(t)$. Suppose that the selling price is set by the merchant, according to the model described in this section. Use the value of k and the initial values given in each exercise to determine the selling price after t years and after 1 year.

4. $S = 3200 + 80x + 45x'$
$D = 4700 + 30x + 25x'$
$k = 0.4, \ x(0) = 40, \ x'(0) = -4$

5. $S = 4850 + 70x + 14x'$
$D = 6100 + 20x + 4x'$
$k = 0.2, \ x(0) = 20, \ x'(0) = -1$

6. $S = 5500 + 17x + 20x'$
$D = 6400 + \frac{1}{3}x + 10x'$
$k = 0.6, \ x(0) = 63, \ x'(0) = 1$

In Exercises 7–12, find the general solution of the given system.

7. $\dfrac{dx}{dt} = 4x + 2y$

$\dfrac{dy}{dt} = -3x + 5y$

8. $\dfrac{dx}{dt} = x + y$

$\dfrac{dy}{dt} = -x + 3y$

9. $\dfrac{dx}{dt} = 2x - y$

$\dfrac{dy}{dt} = 8x - 2y$

10. $\dfrac{dx}{dt} = x - y$

$\dfrac{dy}{dt} = 2x + 3y$

11. $\dfrac{dx}{dt} = 4x - y + 7$

$\dfrac{dy}{dt} = 6x - 3y - 1$

12. $\dfrac{dx}{dt} = 3x - y - 2$

$\dfrac{dy}{dt} = 7x - 5y + 4$

13. **(Predator–Prey)** In a simple model sometimes used in studying certain predator–prey relations, the population $x = x(t)$ of the predator species is assumed to grow at a rate proportional to the population of the prey species, while the population $y = y(t)$ of the prey is assumed to decline at a rate proportional to the population of the predator. Then,

$$\begin{cases} \dfrac{dx}{dt} = ay \\ \dfrac{dy}{dt} = -bx \end{cases}$$

for positive constants a and b. Suppose that in a given predator–prey relation t is measured in weeks and $a = 0.2$ while $b = 0.8$. Assuming that $x(0) = 500$ and $y(0) = 800$, find the populations $x(t)$ and $y(t)$ after t weeks. Show that the population of each species is sinusoidal. (See Figure 23, page 520.)

14. **(Competing Species)** Consider two species with populations $x(t)$ and $y(t)$, which are competing for the resources of a common environment. An increase in the population of either species causes a decline in the growth rate of the other. A simple model of this competition is

$$\begin{cases} \dfrac{dx}{dt} = ax - by \\ \dfrac{dy}{dt} = cy - dx \end{cases}$$

where a, b, c, and d are positive constants. The coefficients a and c are the natural growth rates of the two species, while the coefficients b and d reflect the tendency of the presence of either species to retard the growth rate of the other. Given that two species satisfy the system

$$\begin{cases} \dfrac{dx}{dt} = 2x - y \\[2mm] \dfrac{dy}{dt} = -2x + 3y \end{cases}$$

where t is measured in years, $x(0) = 5600$ and $y(0) = 3200$, find the population of each species after t years. Show that the first species will never be eliminated. What about the second species?

(Richardson's Arms Race Models) In Exercises 15–18, a system of differential equations modeling an arms race between two nations is given. In each exercise, (a) solve the system to determine the level of expenditures on arms by each nation at time t, (b) graph the optimal lines for the two nations, (c) determine whether the arms race has a stable equilibrium, and (d) check to see if the arms race is runaway or stabilizing as $t \to +\infty$, and if total disarmament is possible.

15. $\begin{cases} \dfrac{dx}{dt} = 4y - 3x + 1 \\[2mm] \dfrac{dy}{dt} = 2x - 5y + 4 \\[2mm] x(0) = 18, \ y(0) = 14 \end{cases}$

16. $\begin{cases} \dfrac{dx}{dt} = 3y - 2x + 8 \\[2mm] \dfrac{dy}{dt} = 2x - y + 2 \\[2mm] x(0) = 15, \ y(0) = 11 \end{cases}$

17. $\begin{cases} \dfrac{dx}{dt} = 2y - 3x - 1 \\[2mm] \dfrac{dy}{dt} = x - 2y - 2 \\[2mm] x(0) = 4, \ y(0) = 6 \end{cases}$

18. $\begin{cases} \dfrac{dx}{dt} = 3y - 2x - 4 \\[2mm] \dfrac{dy}{dt} = 2x - y - 1 \\[2mm] x(0) = 2, \ y(0) = 2 \end{cases}$

19. Show that in Exercise 15, changing the initial values $x(0)$ and $y(0)$ will not affect the conclusion about whether the arms race is stabilizing or is a runaway.

20. Show that in the arms race of Exercise 17, total disarmament is possible regardless of the initial values $x(0)$ and $y(0)$.

21. Show that the arms race of Exercise 18 can be changed into a runaway arms race by suitably changing the initial values $x(0)$ and $y(0)$. [Try $x(0) = 10$, $y(0) = 7$, and explain how such a change affects the outcome.]

22. **(Romeo and Juliet*)** In this non-Shakesperian version of the story, Romeo is in love with Juliet, but she is fickle. The more he loves her, the more she dislikes him; but as he begins to dislike her, she warms up and begins to love him. Romeo's love for Juliet, on the other hand, increases in proportion to her love for him, and similarly decreases in proportion to her dislike of him. If we let $R = R(t)$ denote Romeo's love for Juliet and $J = J(t)$ denote Juliet's love for Romeo, at time t, then a simple model of their relationship is

$$\begin{cases} \dfrac{dJ}{dt} = -aR \\[2mm] \dfrac{dR}{dt} = bJ \end{cases}$$

for positive constants a and b.

Using the methods of this section and Theorem 9 of Section 10.5, find the gen-

---

* Exercises 22–24 are suggested by the article, *Love Affairs and Differential Equations*, by Steven H. Strogatz, in *Mathematics Magazine*, Vol. 61 No. 1, February, 1988.

eral solutions $J = J(t)$ and $R = R(t)$. Give a mathematical argument to support your hunch that the "love" Romeo and Juliet have for each other will forever vacillate between two extremes.

23. **(Romeo and Juliet)** Suppose that Juliet has "mixed feelings" about Romeo. Part of her behaves as described in Exercise 22, but another part of her has a growing admiration for Romeo, moving her toward love for him. If Romeo's behavior remains as described in Exercise 22, then a reasonable model of their relationship now becomes

$$\begin{cases} \dfrac{dJ}{dt} = -aR + c \\ \dfrac{dR}{dt} = bJ \end{cases}$$

for positive constants a, b, and c. Show mathematically that, despite the part of Juliet which is moving toward love for Romeo, their feelings for each other vacillate between two extremes with the same frequency as before.

24. **(Romeo and Juliet)** Let us suppose that Juliet, in addition to behaving as described in Exercise 23, finds herself spurred on by her own romantic feelings (positive *or* negative) but that Romeo's response to her remains as described. Then a better model of their relationship is

$$\begin{cases} \dfrac{dJ}{dt} = eJ - aR + c \\ \dfrac{dR}{dt} = bJ \end{cases}$$

for positive constants a, b, c, and e. Show mathematically that their feelings toward each other still vacillate, but now between ever-widening extremes!

KEY IDEAS FOR REVIEW ||||||||||||||||||||||||||||||||||||||||||||||||||||||||||||||||||||||||

- A separable differential equation is one that can be written in the form $dy/dx = f(x)g(y)$. Such an equation is solved by separating the variables and integrating both sides.

- The general solution of the differential equation $dy/dx = ky$ is $y = Ce^{kx}$. Further, $C = y(0)$.

- The general solution of $dy/dx = ay + b$, where a and b are constants, is $y = Ce^{at} - b/a$. Further, $C = y(0) + b/a$.

- The differential equation $dQ/dt = k(L - Q)$ describes simple restricted growth. The expression $Q(t) = L - Ce^{-kt}$ yields all the solutions. Moreover, $C = L - Q_0$, where $Q_0 = Q(0)$.

- The differential equation $dP/dt = kP(L - P)$ describes logistic growth. It behaves like unrestricted growth for small values of P and like simple restricted growth as P approaches L. The expression $P(t) = L/(1 + Ce^{-Lkt})$, called the logistic equa-

tion, together with the constant function $P(t) = 0$, yields all the solutions.

- **(Optional)** The differential equation $dP/dt = kP(\ln L - \ln P)$, where k and L are constants and $P < L$; describes Gompertz growth. The unique solution such that $P(0) = P_0$ is

$$P = L\left(\frac{P_0}{L}\right)^{e^{-kt}}.$$

Its graph is similar to the graph of the logistic equation.

- To solve the linear first-order differential equation

$$\frac{dy}{dx} + u(x)y = v(x),$$

first multiply both sides by the integrating factor $A(x) = e^{a(x)}$, where $a(x) = \int u(x)\, dx$. Then integrate both sides, and finish by dividing both sides by $A(x)$.

- In mixing problems, let $y(t)$ denote the amount of one ingredient present at time t, and then use the principle

$$\frac{dy}{dt} = \text{rate in} - \text{rate out.}$$

- When the selling price $x(t)$ of a commodity varies with time, then to find the equilibrium price at time t we must solve the differential equation $S = D$.

The remaining items pertain to optional material in this chapter.

- To solve a homogeneous linear second-order differential equation we look for a pair of basic solutions. The general solution is a linear combination of them.

- To solve the homogeneous differential equation $ay'' + by' + cy = 0$ we solve the auxiliary equation $am^2 + bm + c = 0$ and apply Theorem 3, Theorem 4, or Theorem 5 of Section 10.5. The table on page 606 summarizes the results.

- The nonhomogeneous linear second-order differential equation $ay'' + by' + cy = d$ can be transformed into a homogeneous equation by substituting $u = y - d/c$ (if $c \neq 0$)

or into a first-order linear differential equation by substituting $u = y'$ (if $c = 0$).

- The expression $c_1 \cos sx + c_2 \sin sx$ can always be expressed as a sinusoid

$$A \sin(sx + \theta).$$

- If the auxiliary equation of $ay'' + by' + cy = 0$ has no real number roots then the general solution is either a sinusoid (if $b = 0$) or a damped sinusoid (if $b \neq 0$).

- In our model, if $x = x(t)$ is the price set by the merchant of a commodity with supply S and demand D, then

$$\frac{d^2x}{dt^2} = -k(S - D).$$

- To solve a system of differential equations,

$$\begin{cases} \dfrac{dx}{dt} = ax + by + c \\[2mm] \dfrac{dy}{dt} = dx + ey + f, \end{cases}$$

differentiate both sides of one of the equations and use substitution to replace the system by a single linear second-order differential equation with constant coefficients.

REVIEW EXERCISES ||||||||||||||||||||||||||||||||||||||||||||||||||||||||||||||||||||||

The exercises marked with an asterisk (*) pertain to the optional material in this chapter.

1. For $y = x^6 - 4x^3 + \sqrt{x}$, find $\dfrac{d^4y}{dx^4}$.

2. State the order of each of the following differential equations:

    (a) $x^5y^2 + x\left(\dfrac{dy}{dx}\right)^4 + 4x^2\dfrac{d^3y}{dx^3} = 10.$

    (b) $\dfrac{y'}{x^2} + x^3y'' = x^4y^4.$

3. Verify that $y = 1 - \dfrac{1}{x}$ is a solution of the differential equation $x\dfrac{dy}{dx} + (x + 1)y = x.$

4. Verify that for every constant C, $y = \dfrac{1}{Cx - x^2}$ is a solution of the differential equation $\dfrac{dy}{dx} + \dfrac{y}{x} = xy^2.$

In Exercises 5–19, solve the given differential equation.

5. $y' = 4xy^2$

6. $\dfrac{dy}{dx} = 3y - 7$

7. $y' + 2y = e^{-3x}$

8. $y' = \dfrac{2x + 3}{y}$

*9. $3y'' + 20y' - 7y = 0$

*10. $7y'' + 3y' = 0$

11. $\frac{1}{2}y' - y = 3xy$, $y > 0$

*12. $y'' - 4y + 6 = 0$

13. $y' - 5y = 2$

*14. $25y'' + 10y' + y = 0$

15. $(x^2 + 1)y' + 2xy = x^3 + x$

*16. $y'' - 16y = 5$

17. $\dfrac{dy}{dx} + y^2 e^x = y^2$

*18. $4y'' + 9y = 0$

*19. $2y'' - y' + y = 3$

In Exercises 20–28, solve the given initial-value or boundary-value problem.

20. $\dfrac{dP}{dt} = 3(80 - P)$

$P(0) = 50$

21. $\dfrac{dP}{dt} = 5P(30 - P)$

$P(0) = 20$

*22. $\dfrac{dP}{dt} = 0.5P \ln\left(\dfrac{50}{P}\right)$

$P(0) = 10$

23. $\dfrac{dy}{dx} = (y + 1)(x - 3)$,

$y > -1$, $y(0) = 1$

*24. $4y'' + y = 3$

$y(0) = 1$, $y(\pi) = 8$

*25. $5y'' + 3y' - 2y = 0$

$y(0) = 8$, $y'(0) = 6$

26. $\dfrac{dy}{dx} = 8y + 9$

$y(0) = \frac{3}{8}$

27. $y' + \dfrac{2}{x} y = x + 1$,

$x > 0$, $y(1) = \frac{1}{12}$

*28. $2y'' - 6y' + 5y = 0$

$y(0) = -2$, $y'(0) = 1$

*29. Find numbers A, B, and θ such that

$$5 \cos 3x - 2 \sin 3x = A \sin(Bx + \theta)$$

where $A > 0$ and $0 \le \theta < 2\pi$.

*30. Express the general solution of Exercise 28 in the form $Ae^{rx} \sin(sx + \theta)$.

31. **(Demand vs. Selling Price)** The Weber–Fechner Law, $dD = k\,dx/x$, is sometimes used to model the relationship between the unit price x of a product and the demand $D = D(x)$. Suppose that the average consumer is willing to rent 2 videotapes per month when the cost is $5 each, and 12 videotapes per month when the cost is only $1. Use the Weber–Fechner Law to predict how many videos per month the average consumer is willing to rent when the cost is $2 each; when the cost is $3 each. [To an individual consumer a fraction of a video is nonsense, but to a supplier it is significant.]

32. **(Continuous Annuity)** A deposit of $50,000 is made into a savings account earning 8% annual continuous compound interest. From this account a beneficiary is paid continuously at the rate of $3000 per year. Write the differential equation that is satisfied by A, the amount of money in the savings account after t years. Solve this differential equation to find A as a function of t. How much money is left in the account after 5 years?

33. **(Mixture)** For a prank, a college student puts 10 pounds of detergent into a rival college's fountain, which contains 500 gallons of water. If fresh water runs into the fountain at the rate of 5 gallons per minute, and the water–detergent mixture runs out at the same rate, how many pounds of detergent are in the fountain after t minutes? After 1 hour? How long does it take before there is only $\frac{1}{2}$ pound of detergent left in the fountain?

34. **(Population with Migration)** A rural county has a natural growth rate (birth minus death) of 3% per year, but 1800 more people per year move out of the county than move in. Write the differential equation that the population must satisfy at time t. If the population is now 100,000 what will it be 12 years from now?

*35. **(Symbiotic Relations)** The Lotka–Volterra model is useful in modeling the growth rates of two species which *depend* on each other for coexistence. Suppose that $x = x(t)$ and $y = y(t)$ represent the populations of the two species at time t, and that their growth rates satisfy

$$\begin{cases} \dfrac{dx}{dt} = -2x + 5xy \\ \dfrac{dy}{dt} = -3y + 4xy. \end{cases}$$

(The positive coefficients of xy indicate that this relation is symbiotic, and the negative coefficients indicate that each species would die out in the absence of the other.) Derive an equation relating the populations x and y.

36. **(Bimolecular Chemical Reaction)** Chemicals A and B combine in the ratio 3 to 4 to yield chemical C. Starting with 100 grams each of A and B, 30 grams of C are produced in 2 minutes.
   (a) Letting z represent the number of grams of C present after t minutes, write the differential equation that z must satisfy.
   (b) Solve this differential equation for z and evaluate the constant.
   (c) How many grams of C are formed in the first 10 minutes?
   [See Exercise 53, Section 10.2.]

37. **(Bimolecular Chemical Reaction)** Chemicals A and B combine in the ratio 3 to 4 to yield chemical C. Starting with 42 grams of A and 56 grams of B, 14 grams of C are produced in 2 minutes. Answer

(a), (b), and (c) of Exercise 36. (See Exercises 54 and 55, Section 10.2.)

38. **(Advertising)** An advertising campaign for a new soap product is directed at a community of 8100 people. When the campaign starts, 100 people already know about the product. Two weeks after the start of the campaign, 1700 people have heard about the product. Suppose that the rate at which people hear about the product is proportional to the number of people who have not heard about the product. Using a simple restricted growth model, find an expression for the number of people who know about the product t weeks after the beginning of the campaign.

39. **(Terminal Velocity)** From high above the earth a 10-pound object is thrown downward with an initial velocity of 9 ft/sec. Due to air resistance, the value of k in Equation (9) of Section 10.3 is 20.
   (a) Write the differential equation that must be satisfied by the velocity at time t.
   (b) Solve this differential equation.
   (c) How fast is the object falling after 1 second?
   (d) What is the terminal velocity of the object?

40. **(Autocatalytic Reaction)** In an autocatalytic reaction, chemical A is converted into chemical B at a rate proportional to the product of the amount of B produced and the amount of A that remains unconverted. Suppose that we initially had 200 grams of A and 50 grams of B. If the constant of proportionality is 0.3, use the logistic equation to find an expression giving the amount of B that has been produced t hours after the reaction has started.

41. **(Ecology)** A paper manufacturer starts a forest by planting 1280 trees. Ecological studies have determined that the forest can support a maximum of 6400 trees. After

80 years of growth there will be 3200 trees. The rate of growth of the number of trees at time t is proportional to the size of the population $P(t)$ and to $6400 - P(t)$. Using the logistic equation, find an expression for the size of the population after t years of growth.

*42. **(Gompertz Growth)** Suppose that the tree population of Exercise 41 follows the Gompertz growth model instead of the logistic model.
   (a) Write the differential equation that must be satisfied by the tree population at time t.
   (b) Find the size of the tree population after t years.

43. **(Trust Fund)** In his will, a wealthy businessman sets aside \$200,000 to establish a trust fund for his son, to help him get started in his own business. The trust fund earns 10% annual continuous compound interest, and is to pay the son continuously, t years later, at the rate of $50,000 - 2000t$ dollars per year. How much money remains in the account after t years? After 5 years? Will the account last through the sixth year at these rates? Explain.

44. **(Mixing)** A large tank initially contains 50 gallons of pure water. A valve is opened allowing contaminated water to flow into the tank at the rate of 6 gallons per minute. A second valve is also opened, which allows the thoroughly mixed contents of the tank to drain out at the rate of 5 gallons per minute. If the contaminated water flowing into the tank is 20% contaminant by volume, how much contaminant is in the tank after t minutes? After 10 minutes?

45. **(Supply–Demand Equilibrium)** The supply and demand functions for a certain commodity are $S(x) = 160 + 9x + 5x'$ and $D(x) = 400 + 80e^{-t} - 3x + x'$, where $x = x(t)$ represents the unit selling price after t months. The equilibrium price at time $t = 0$ is \$18 per unit. Find the equilibrium

price after t months, and the long-range equilibrium price if it exists.

*46. **(Supply, Demand, and Merchant's Selling Price)** The supply and demand functions for a commodity are $S = 1260 + 500x + 90x'$ and $D = 6460 + 370x + 50x'$, where $x = x(t)$ is the unit selling price. If the selling price is set by the merchant according to the model described in Section 10.6, with $k = 0.1$, and if $x(0) = 25$, $x'(0) = 3$, what is the merchant's selling price at time t? At time $t = 1$?

*47. Find the general solution of the system
$$\begin{cases} \dfrac{dx}{dt} = x - 2y \\ \dfrac{dy}{dt} = 4x + 5y. \end{cases}$$

*48. Find the general solution of the system
$$\begin{cases} \dfrac{dx}{dt} = 4x - y + 3 \\ \dfrac{dy}{dt} = 4x - 8. \end{cases}$$

*49. Solve the initial-value problem
$$\begin{cases} \dfrac{dx}{dt} = x - y + 2 \\ \dfrac{dy}{dt} = 2x - y + 5, \\ x(0) = 1, \ y(0) = 5. \end{cases}$$

*50. **(Symbiotic Species)** Two species with populations $x(t)$ and $y(t)$ are in a symbiotic relation if each benefits from the presence of the other. Find the population of each species at time t if they satisfy the model
$$\frac{dx}{dt} = -\frac{1}{3}x + y$$
$$\frac{dy}{dt} = \frac{2}{9}x - \frac{2}{3}y$$
given that $x(0) = 9000$ and $y(0) = 4000$. Find the limiting values of these populations as $t \to +\infty$.

(Richardson's Arms Race Models) In Exercises 51 and 52, a system of differential equations modeling an arms race between two nations is given. In each exercise, (a) solve the system to determine the level of expenditures on arms by each nation at time t, (b) graph the optimal lines for the two nations, (c) determine whether the arms race has a stable equilibrium, and (d) check to see if the arms race is runaway or stabilizing as $t \to +\infty$, and if total disarmament is possible.

*51. $\begin{cases} \dfrac{dx}{dt} = 3y - 5x + 4 \\ \dfrac{dy}{dt} = x - 3y + 2 \\ x(0) = 7, \ y(0) = 6 \end{cases}$

*52. $\begin{cases} \dfrac{dx}{dt} = 2y - 2x + 6 \\ \dfrac{dy}{dt} = 3x - y + 1 \\ x(0) = 5, \ y(0) = 3 \end{cases}$

CHAPTER TEST ||||||||||||||||||||||||||||||||||||||||||||||||||||||||||||||||||||||||||||||||||||

1. Verify that for every constant c, $y = \frac{1}{2}x^3 - \frac{3}{4}x^2 + \frac{3}{4}x - \frac{3}{8} + ce^{-2x}$ is a solution of the differential equation $y' + 2y = x^3$.

In Exercises 2–5, solve the given differential equation.

2. $3xy' = (x^2 - 5)\sqrt{y}$, $x > 0$, $y > 0$

3. $\dfrac{dy}{dx} + \dfrac{2y}{x} = 1 - x$, $x > 0$

4. (Optional) $y'' - 4y' + 5y = 12$

5. (Optional) $y'' + 8y' + 16y = 0$

6. Solve the initial-value problem $\dfrac{dy}{dx} = \dfrac{x+1}{y^2}$, $y(0) = 2$.

7. Suppose that a drug is administered intravenously to a patient at the constant rate of 0.6 gram per hour, and that the patient's system removes it from the bloodstream at the rate of 75% per hour.
   (a) How much of the drug is in the patient's bloodstream after t hours? After 1 hour, 2 hours, 5 hours?
   (b) Is there a limit to the amount of the drug in the patient's bloodstream as $t \to +\infty$?

8. A cold drink that is 36°F is set into a room that is 70°F. In 5 minutes its temperature is 48°.
   (a) Write the differential equation that must be statisfied by the temperature of the drink at time t.
   (b) Solve this differential equation.
   (c) Find the temperature after 10 minutes.

9. A flu epidemic breaks out in a college with 12,000 students. It spreads at a rate that is proportional to both the number of students at the college who have been infected and to the number who have not. By the time the outbreak is discovered, 200 students have been infected. Four days later, 500 students have been infected.
   (a) Let P denote the number of students infected by day t. Write the differential equation that is satisfied by $P = P(t)$.

    (b) What is the name of the type of growth model that satisfies this differential equation?

    (c) Find the equation that tells how many students will be infected t days after the outbreak is discovered.

10. A local service club has a savings account from which they make contributions to community welfare projects. Through a program of increased fund-raising they anticipate putting a steady flow of money into this fund at the rate of $5000 + 200t$ dollars per year, t years from now. The club plans to contribute steadily from this fund to welfare projects at the rate of $3000 + 500t$ dollars per year, t years from now. The fund pays 8% continuous compound interest, and the current balance is $20,000. Write a differential equation which must be satisfied by the amount $A(t)$ of money in the account t years from now. Solve this differential equation to determine $A(t)$.

11. (Optional) Solve the initial-value problem

$$\begin{cases} \dfrac{dx}{dt} = 3x - 2y + 1 \\[2mm] \dfrac{dy}{dt} = 2x - 2y - 1 \\[2mm] x(0) = 3,\ y(0) = -\dfrac{9}{2}. \end{cases}$$

CALCULUS IN PROBABILITY THEORY

A **random** phenomenon is one whose exact outcome cannot be predicted with certainty. When one spins the wheel of a carnival game, the exact stopping point is unpredictable. When one is dealt a poker hand, the exact five cards are likewise not predictable. Other examples of random phenomena include the next winner in the Pennsylvania State Lottery, the grade-point average of a student selected randomly, the telephone number of the next person to get on a bus, and the number of hours a light bulb will last before burning out. It is often necessary to make decisions about quantities that behave randomly. The theory of probability was developed to help make rational decisions in such situations. As we shall see, calculus is an important tool in probability theory.

11.1 |||||| BASIC NOTIONS OF PROBABILITY

The first ingredient of a probability model is an **experiment,** either real or imagined. For example, the experiment could be to spin the wheel of a carnival game, to deal a hand of cards, or to randomly select the winning number for a state lottery. The set of all possible **outcomes** of the experiment is called the **sample space.** It is common to omit the adjective "possible", and speak of the "outcomes" of an experiment rather than the "possible outcomes." Thus, if the experiment is to roll one die, we say that the outcomes are 1, 2, 3, 4, 5, and 6, and that the sample space is the set {1, 2, 3, 4, 5, 6}.

The outcomes of an experiment are not necessarily numbers. Nevertheless, it is usually possible to identify them with numbers in a natural way. For example, when a computer selects a person at random from a list, it does so only after each person has been represented by a number; then it selects a *number* at random from the list. **A random variable X** is a function that assigns a real number $X(t)$ to each outcome t of an experiment. Once a random variable has been specified, the outcomes and the sample space are expressed in terms of it, as shown in Example 1 below.

A **discrete random variable** is one with only a finite number of outcomes in the sample space.

EXAMPLE 1   Suppose a box contains 100 tickets, numbered 1 through 100, and we randomly select one ticket from the box. Specify the experiment, the random variable, the outcomes, and the sample space. Is this random variable discrete?

Solution   The experiment is to draw one ticket from a box of 100 tickets, numbered 1 through 100. The random variable is

$$X(t) = \text{the number on card } t.$$

The outcomes are the integers 1, 2, 3, . . . , 100 and the sample space is the set $\{1, 2, 3, \ldots, 100\}$. This is a discrete random variable because there are (only) 100 outcomes in the sample space.   ∎

The **probability** of a given outcome of an experiment is a mathematical measure of the likelihood that, when the experiment is performed once, the specified outcome will occur. In setting up a discrete probability model, each outcome of the experiment is first assigned its own probability in such a way that all probabilities are nonnegative and ≤ 1, and the sum of all the probabilities is 1. Thus, if the sample space is $\{t_1, t_2, \ldots, t_n\}$, then each outcome t_i is assigned a probability $P(t_i)$ such that

$$0 \leq P(t_i) \leq 1 \qquad \text{for each } i, \tag{1}$$

and

$$P(t_i) + P(t_2) + \cdots + P(t_n) = 1. \tag{2}$$

It is beyond the scope of this book to discuss the many ways in which this can be done or the methods used to combine these probabilities to obtain the probabilities of sets of outcomes. We stick to the fundamental concepts and the most basic procedures. Our motive is to provide a background for the use of calculus in continuous probability models.

The two extreme probabilities, 0 and 1, are of special significance. In a discrete probability model, when $P(t_i) = 0$ we can be certain that the outcome t_i *cannot* occur, and when $P(t_i) = 1$ we can be certain that the outcome t_i *must* occur.

EXAMPLE 2   Suppose a box contains 100 tickets, numbered 1 through 100, and we select 1 ticket at random. We use the random variable X described in Example 1.
  (a) The probability that the number on the ticket is 58 is $\frac{1}{100}$, because there are 100 tickets, each of which is equally likely to be drawn. In short,

$$P(X = 58) = \tfrac{1}{100}.$$

For the same reason, for each outcome t_i, $P(X = t_i) = \frac{1}{100}$.
  (b) The probability that the number on the ticket is an odd number is $\frac{1}{2}$, since exactly half the tickets are numbered odd. In short, $P(X \text{ is odd}) = \frac{1}{2}$.
  (c) Similarly, $P(X \text{ is even}) = \frac{1}{2}$.

(d) The probability that the number on the ticket is in the interval $[21, 30]$ is $\frac{10}{100}$ or $\frac{1}{10}$ because, of the 100 numbers in the sample space, exactly 10 of them are in this interval. In short, $P(21 \leq X \leq 30) = \frac{1}{10}$.

(e) $P(51 \leq X \leq 75) = \frac{1}{4}$ because, of the 100 numbers in the sample space, exactly 25 of them are in this interval.

(f) $P(X$ is divisible by 3$) = \frac{33}{100}$. (Why?)   ■

EXAMPLE 3   A box contains 16 balls: 4 yellow, 3 red, 6 blue, 1 green, and 2 white. The balls are identical except for color. If we reach into the box and randomly (without looking) select 1 ball, what is the probability that the ball is
(a) yellow?
(b) blue or green?
(c) not white?
(d) black?

Solution   The experiment is to randomly select 1 ball from 16, of which there are 4 yellow, 3 red, 6 blue, 1 green, and 2 white. If each ball is considered as a separate outcome, then the sample space may be listed as

$$\{Y, Y, Y, Y, R, R, R, B, B, B, B, B, B, G, W, W\},$$

and the probability of each outcome is $\frac{1}{16}$, since they are all equally likely.
(a) The probability that the ball selected is yellow is

$$P(Y) = \tfrac{4}{16} = \tfrac{1}{4},$$

because 4 of the 16 members of the sample space are yellow.
(b) The probability that the ball selected is blue or green is

$$P(B \text{ or } G) = \tfrac{7}{16},$$

because 7 of the 16 members of the sample space are blue or green.
(c) The probability that the ball selected is not white is

$$P(\text{not } W) = \tfrac{14}{16} = \tfrac{7}{8},$$

because 14 of the 16 members of the sample space are not white.
(d) The probability that the ball selected is black is

$$P(\text{black}) = 0,$$

because there are no black balls in the box. It is impossible that the ball selected is black.   ■

||||||||||||| **Alternate Methods in Setting up Example 3**

If we regard the 5 colors as the outcomes of the experiment in Example 3, then the outcomes and their probabilities are as listed in Table 1, which is called the **probability table** for the experiment.

TABLE 1

Outcome	Y	R	B	G	W
Probability	$\frac{1}{4}$	$\frac{3}{16}$	$\frac{3}{8}$	$\frac{1}{16}$	$\frac{1}{8}$

$(\frac{1}{4} + \frac{3}{16} + \frac{3}{8} + \frac{1}{16} + \frac{1}{8} = 1)$

If we introduce the random variable X, defined by $X(Y) = 1$, $X(R) = 2$, $X(B) = 3$, $X(G) = 4$ and $X(W) = 5$, then the sample space is $\{1, 2, 3, 4, 5\}$ and the outcomes and their probabilities are listed in Table 2.

TABLE 2

X	1	2	3	4	5
$P(X)$	$\frac{1}{4}$	$\frac{3}{16}$	$\frac{3}{8}$	$\frac{1}{16}$	$\frac{1}{8}$

From Table 2 we construct a **bar graph** for Example 3, which illustrates visually the random variable X and the associated probabilities. The height of each bar is the probability of the associated outcome. (See Figure 1.)

Figure 1

The bar graph of a discrete probability model allows us to see at a glance the relative sizes of the probabilities of the outcomes, and often displays significant characteristics of the probability model.

Histograms (Probabilities as Areas)

A **histogram** for a discrete probability model is a special kind of bar graph in which all rectangles (bars) have the same width, Δx, and adjacent rectangles are abutting. The height of the rectangle above the number i is the probability of the outcome $X = i$. Figure 2 shows a histogram for Example 3, using the data from Table 2.

As is customary, we have drawn the bars so that the base of the rectangle corresponding to the outcome i has the number i at its center. It is important to observe that in a histogram where $\Delta x = 1$, the probability of an outcome is the *area* of the associated bar.

Figure 2

More generally, consider an experiment whose outcomes are described by a discrete random variable X, with associated probabilities given as in Table 3.

TABLE 3

X	1	2	3	$\cdots$	n
$P(X)$	p_1	p_2	p_3	$\cdots$	p_n

An associated histogram may be drawn, as illustrated in Figure 3.

Figure 3

It is then possible to express probabilities as definite integrals. For example,

$$P(X = 1) = \int_{0.5}^{1.5} P(X)\,dX,$$

$$P(X = i) = \int_{i-0.5}^{i+0.5} P(X)\,dX, \tag{3}$$

$$P(X \le 4) = \int_{0.5}^{4.5} P(X)\,dX,$$

$$P(X \le i) = \int_{0.5}^{i+0.5} P(X)\,dX, \tag{4}$$

and

$$P(i \le X \le j) = \int_{i-0.5}^{j+0.5} P(X)\,dX. \qquad \text{(Assuming } i \text{ and } j \text{ are integers with } 1 \le i \le j \le n). \tag{5}$$

EXAMPLE 4   A wheel, divided into 4 numbered sectors as shown in Figure 4, is spun so that it stops randomly with the pointer pointing at one of the four sectors. Specify an appropriate random variable, tabulate the probabilities of all its outcomes, and draw a histogram. Use integrals to calculate $P(X = 2)$, $P(X \le 2)$, and $P(3 \le X \le 4)$.

Figure 4

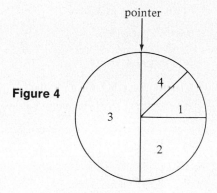

(Region 2 is half of region 3; regions 1 and 4 are half of region 2.)

Solution   An appropriate random variable and the probabilities for its outcomes are tabulated in Table 4.

TABLE 4

X	1	2	3	4
$P(X)$	$\frac{1}{8}$	$\frac{1}{4}$	$\frac{1}{2}$	$\frac{1}{8}$

The histogram is shown in Figure 5.

Figure 5

Using Figure 5, and Equations (3)–(5), we can calculate

$$P(X = 2) = \int_{1.5}^{2.5} P(X)\,dX = \int_{1.5}^{2.5} \tfrac{1}{4}\,dX = \tfrac{1}{4}X\Big]_{1.5}^{2.5} = \tfrac{1}{4}(2.5 - 1.5) = \tfrac{1}{4};$$

$$P(X \le 2) = \int_{0.5}^{2.5} P(X)\,dX = \int_{0.5}^{1.5} P(X)\,dX + \int_{1.5}^{2.5} P(X)\,dX$$

$$= \int_{0.5}^{1.5} \tfrac{1}{8}\,dX + \int_{1.5}^{2.5} \tfrac{1}{4}\,dX = \tfrac{1}{8}X\Big]_{0.5}^{1.5} + \tfrac{1}{4}X\Big]_{1.5}^{2.5}$$

$$= \tfrac{1}{8}(1.5 - 0.5) + \tfrac{1}{4}(2.5 - 1.5) = \tfrac{1}{8} + \tfrac{1}{4}$$

$$= \tfrac{3}{8};$$

$$P(3 \le X \le 4) = \int_{2.5}^{4.5} P(X)\,dX = \int_{2.5}^{3.5} P(X)\,dX + \int_{3.5}^{4.5} P(X)\,dX$$

$$= \int_{2.5}^{3.5} \tfrac{1}{2}\,dX + \int_{3.5}^{4.5} \tfrac{1}{8}\,dX = \tfrac{1}{2}X\Big]_{2.5}^{3.5} + \tfrac{1}{8}X\Big]_{3.5}^{4.5}$$

$$= \tfrac{1}{2}(3.5 - 2.5) + \tfrac{1}{8}(4.5 - 3.5) = \tfrac{1}{2} + \tfrac{1}{8}$$

$$= \tfrac{5}{8}. \qquad\blacksquare$$

11.1 EXERCISE SET ||||||||||||||||||||||||||||||||||||||||||||||||||||||||||||||||||||||||||||||||||||

1. A set of 15 billiard balls is numbered 1 through 15. If we randomly select one of them, what is the probability that we have selected
   (a) an even-numbered ball?
   (b) an odd-numbered ball?
   (c) a ball numbered less than 6?
   (d) the eight ball?

2. If we were to randomly select one letter from the word "CALCULUS," what is the probability that we would select
   (a) the letter "A"?
   (b) the letter "L"?
   (c) a vowel?
   (d) the letter "M"?

3. A carnival wheel is divided into 8 numbered sections of the same size, as shown in Figure 6. It is spun so that it stops randomly with the pointer pointing at one of the sectors. Find the probability that the pointer points to
   (a) a shaded sector.
   (b) an even-numbered sector.
   (c) an even-numbered shaded sector.
   (d) a sector that is adjacent to a shaded sector.
   (e) a sector that is adjacent to an unshaded sector.

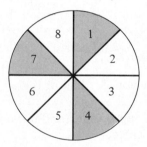

Figure 6

4. Don has 9 girlfriends: 3 freshmen, 4 sophomores, 2 juniors, and no seniors. If he selects at random one of these girls to ask for a date, what is the probability that she is
   (a) a freshman?
   (b) not a sophomore?
   (c) a sophomore or a junior?
   (d) not a senior?

5. Consider the experiment of randomly choosing one letter from the word "MISSISSIPPI." Let the sample space be the 4 distinct letters that appear in this word. Introduce an appropriate random variable for this experiment, tabulate the probabilities of all its outcomes, and draw a histogram.

6. For the experiment described in Exercise 4, specify an appropriate random variable, tabulate the probabilities of all its outcomes, and draw a histogram.

7. A box contains 15 balls, identical except for color. There are 2 green, 5 yellow, 4 red, 1 black, and 3 white balls. Let the sample space be the 5 colors: G, Y, R, B, W. For the experiment of randomly selecting one ball from the box, introduce an appropriate random variable, tabulate the probabilities of all its outcomes, and draw a histogram.

8. Rhonda has 30 books on her shelf: 4 mathematics books, 5 history books, 3 biographies, 1 dictionary, 12 novels, 3 psychology books, and 2 tennis manuals. Consider the experiment of pulling one book off the shelf at random. Introduce an appropriate random variable, tabulate the probabilities of all its outcomes, and draw a histogram.

9. When one coin is tossed, there are two equally likely possible outcomes: H (= head) and T (= tail). When two coins are tossed the sample space consists of four equally likely outcomes: HH, HT, TH, TT. Let the random variable X represent the number of heads that appear when two coins are tossed. Complete the following probability table for X:

X	0	1	2
$P(X)$			

Draw a histogram for the random variable X.

10. When three coins are tossed, there are eight equally likely outcomes in the sample space.
    (a) List them.
    (b) Let the random variable X represent the number of heads that appear

where 3 coins are tossed. Construct a probability table for X.
    (c) Draw a histogram for the random variable X.

11. Let the random variable X represent the number that appears on the top face of a die (a cube with its faces numbered 1, 2, 3, . . . , 6) when it is tossed. Compute
    (a) $P(X = 3)$.   (c) $P(X > 3)$.
    (b) $P(X < 3)$.   (d) $P(X < 6)$.

12. When a pair of dice is tossed the sample space consists of the 36 equally likely outcomes listed below:

1,1	1,2	1,3	1,4	1,5	1,6
2,1	2,2	2,3	2,4	2,5	2,6
3,1	3,2	3,3	3,4	3,5	3,6
4,1	4,2	4,3	4,4	4,5	4,6
5,1	5,2	5,3	5,4	5,5	5,6
6,1	6,2	6,3	6,4	6,5	6,6

Let the random variable X denote the sum of the numbers that appear on the top faces when a pair of dice is tossed. Find
    (a) $P(X \le 4)$.   (d) $P(X = 1)$.
    (b) $P(X = 5)$.   (e) $P(X \ge 9)$.
    (c) $P(X = 6)$.   (f) The most likely value of X.

13. (a) Construct a probability table for the random variable X of Exercise 12.
    (b) Draw a histogram for X.

14. Consider the experiment of tossing a pair of dice. This time, however, one die is normal while the other has a "0" instead of a "6".
    (a) List the 36 equally likely outcomes in the sample space, as in Exercise 12.
    Let the random variable X denote the sum of the numbers that appear on the top faces when these two dice are tossed. Find
    (b) $P(X \le 4)$.   (e) $P(X = 1)$.
    (c) $P(X = 5)$.   (f) $P(X \ge 9)$.
    (d) $P(X = 6)$.   (g) The most likely value of X.

15. (a) Construct a probability table for the random variable X of Exercise 14.
    (b) Draw a histogram for X.

16. From the histogram shown in Figure 7, use a definite integral to calculate the probabilities:
    (a) $P(X = 2)$.   (c) $P(2 \le X \le 4)$.
    (b) $P(X \le 3)$.   (d) $P(X \ge 4)$.

17. From the histogram below, calculate the following probabilities using a definite integral:
    (a) $P(X \le 4)$.   (c) $P(X \ge 6)$.
    (b) $P(3 \le X \le 6)$.   (d) $P(X \ge 0)$.

Figure 7

Figure 8

11.2 ‖‖‖‖ CONTINUOUS RANDOM VARIABLES

A **continuous random variable** is a random variable that can take on any value in an interval of real numbers. That interval can be finite, such as $[c, d]$, or infinite, such as $(-\infty, d)$, or $[c, +\infty)$, or $(-\infty, +\infty)$. Continuous random variables are used to model experiments in which the outcomes do not form a finite (discrete) set, but instead spread "continuously" over an interval of values. Examples are the weight of a human being selected at random, the temperature at a randomly selected point on the Earth, the life span of a TV set selected at random, and so on.

In contrast with discrete random variables, a continuous random variable can assume any one of an infinite set of values. For this reason we must expect the probability of any one of its outcomes to be 0. Thus, for continuous random variables we shall not be able to construct probability tables as in Section 11.1, or to make bar graphs or histograms.

We shall be interested in the probability that a continuous random variable X is between two values, say

$$P(a \le X \le b). \tag{1}$$

A basic principle of probability theory assures us that if $a < b$, then $P(X \le b) = P(X \le a) + P(a \le X \le b)$. Thus,

$$P(a \le X \le b) = P(X \le b) - P(X \le a). \tag{2}$$

We are now ready to indicate a mathematical procedure for obtaining (1).

Suppose X is a continuous random variable. We introduce the function $F(x)$ defined by

$$F(x) = P(X \leq x). \tag{3}$$

Equation (2) says that if $a \leq b$, then

$$P(a \leq X \leq b) = F(b) - F(a). \tag{4}$$

If the function F is differentiable, then Equation (4), along with our definition of definite integral, tells us that

$$P(a \leq X \leq b) = \int_a^b F'(x)\,dx$$
$$= \int_a^b f(x)\,dx, \tag{5}$$

where $f(x) = F'(x)$. The function F is called the **distribution function** for the random variable X, while its derivative f is called the **probability density function** for X. Therefore, we have the following principle:

> The probability that a continuous random variable X is between two numbers is equal to the definite integral of the *probability density* function of X between those two numbers.

Equations (4) and (5) tell us that to calculate $P(a \leq X \leq b)$, either the distribution function or the density function may be used.

EXAMPLE 1 A steel distributor finds that the amount x of steel sold daily varies randomly between 200 and 600 tons. Let the random variable X be measured in hundreds of tons so that $2 \leq X \leq 6$. Suppose that the probability density function for X is given by

$$f(x) = \tfrac{3}{160}(x^2 - x).$$

(See Figure 1.)

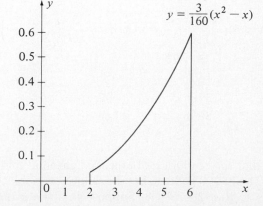

Figure 1

$$y = \frac{3}{160}(x^2 - x)$$

(a) What is the probability that at least 300 tons will be sold on a certain day?
(b) What is the probability that at most 400 tons will be sold on that day?
(c) What is the probability that at least 300 tons, but no more than 500 tons, will be sold on that day?

Solution (a) We want to know the probability that X lies in the interval $[3, 6]$ so that the day's sales will be at least 300 tons. The desired probability is given by Equation (5) as

$$P(3 \leq X \leq 6) = \frac{3}{160} \int_3^6 (x^2 - x)\, dx = \frac{3}{160} \left(\frac{x^3}{3} - \frac{x^2}{2} \right) \Big]_3^6$$

$$= \frac{3}{160} \left[\left(\frac{6^3}{3} - \frac{6^2}{2} \right) - \left(\frac{3^3}{3} - \frac{3^2}{2} \right) \right]$$

$$= \frac{3}{160} \left(\frac{189}{3} - \frac{27}{2} \right)$$

$$= \frac{297}{320}.$$

(b) We want to know the probability that X lies in the interval $[2, 4]$ so that the day's sales will be no more than 400 tons. The desired probability is given by Equation (5) as

$$P(2 \leq X \leq 4) = \frac{3}{160} \int_2^4 (x^2 - x)\, dx = \frac{3}{160} \left(\frac{x^3}{3} - \frac{x^2}{2} \right) \Big]_2^4$$

$$= \frac{3}{160} \left(\frac{56}{3} - 6 \right)$$

$$= \frac{19}{80}.$$

(c) We want to know the probability that X lies in the interval $[3, 5]$ so that the day's sales will be at least 300 tons, but no more than 500 tons. The desired probability is given by Equation (5) as

$$P(3 \leq X \leq 5) = \frac{3}{160} \int_3^5 (x^2 - x)\, dx = \frac{3}{160} \left(\frac{x^3}{3} - \frac{x^2}{2} \right) \Big]_3^5$$

$$= \frac{3}{160} \left(\frac{98}{3} - 8 \right)$$

$$= \frac{37}{80}. \qquad \blacksquare$$

EXAMPLE 2 A gasoline dealer finds that the amount x of unleaded gasoline sold weekly varies randomly between 2000 and 8000 gallons. Let the random variable X be measured in thousands of gallons so that $2 \leq X \leq 8$. Suppose that the probability density

function for X is given by

$$f(x) = \frac{20}{3} \frac{1}{(x + 2)^2}$$

(see Figure 2). If the dealer orders 5000 gallons of unleaded gasoline for a certain week, what is the probability of running out of gasoline before the end of the week?

Figure 2

Solution   We want the probability that X is at least 5. Thus, we need to know the probability that X lies in the interval [5, 8], which is given by Equation (5) as

$$\frac{20}{3} \int_5^8 \frac{1}{(x + 2)^2} \, dx.$$

Letting

$$u = x + 2,$$

so that

$$du = dx,$$

we have

$$\int \frac{1}{(x + 2)^2} \, dx = \int \frac{1}{u^2} \, du = \int u^{-2} \, du$$

$$= -\frac{1}{u} + C = -\frac{1}{x + 2} + C.$$

Then

$$\frac{20}{3} \int_5^8 \frac{1}{(x + 2)^2} \, dx = -\frac{20}{3} \frac{1}{x + 2} \Big]_5^8$$

$$= -\frac{20}{3} \left[\frac{1}{10} - \frac{1}{7} \right] = \frac{60}{210}$$

$$= \frac{2}{7}.$$

Thus, the probability of running out of gas before the end of the week is $\frac{2}{7}$.   ∎

|||||||||||| Probability Density Functions

In order that a continuous function $f(x)$ be suitable as a probability density function for some (unspecified) random variable X over an interval $[c, d]$ it must satisfy the following properties:

$$\text{I. } f(x) \geq 0, \text{ for all } x \text{ in } [c, d], \tag{6}$$

and

$$\text{II. } \int_c^d f(x)\,dx = 1. \tag{7}$$

Property I guarantees that when formula (5) is used to calculate $P(a \leq X \leq b)$, the result nonnegative for all intervals (a, b) which lie inside $[c, d]$, Property II says that the probability that X is in $[c, d]$ is 1; in other words, X *must* always lie in the interval $[c, d]$.

If f is any function with these two properties (I and II), then f is the probability density function of some random variable X defined over $[c, d]$. The corresponding probability distribution function is

$$F(x) = \int_c^x f(t)\,dt = P(X \leq x)$$

and the probability that X is in an interval $[a, b]$ is

$$P(a \leq X \leq b) = \int_a^b f(x)\,dx, \tag{8}$$

assuming that $c \leq a \leq b \leq d$.

Figure 3 shows a typical probability density function and a typical probability distribution function. Figure 4 illustrates Equation (8).

Figure 3

(a) Density function, $f(x)$

(b) Distribution function, $F(x)$

Figure 4

$$P(a \leq X \leq b) = \int_a^b f(x)dx$$

EXAMPLE 3 Show that the function f defined in Example 1 by

$$f(x) = \tfrac{3}{160}(x^2 - x), \qquad 2 \le x \le 6,$$

is a probability density function over the interval $[2, 6]$, and find the corresponding distribution function.

Solution A. To show that f is a probability density function on $[2, 6]$ we must show that it satisfies (6) and (7).

1. We are given

$$f(x) = \tfrac{3}{160} x(x - 1).$$

Since $x \ge 0$ and $(x - 1) \ge 0$ for x in the interval $[2, 6]$, we see that $f(x) \ge 0$ for all x in $[2, 6]$.

2. In this case, $[c, d] = [2, 6]$. We have

$$\int_2^6 \frac{3}{160} x(x - 1)\, dx = \frac{3}{160} \int_2^6 (x^2 - x)\, dx = \frac{3}{160} \left[\frac{x^3}{3} - \frac{x^2}{2} \right]_2^6$$

$$= \frac{3}{160} \left[(72 - 18) - \left(\frac{8}{3} - 2 \right) \right]$$

$$= \frac{3}{160} \left(54 - \frac{2}{3} \right) = \frac{3}{160} \left(\frac{160}{3} \right) = 1.$$

Therefore, f is a probability density function over $[2, 6]$.

B. By definition, the corresponding distribution function is

$$F(x) = \int_2^x \frac{3}{160} t(t - 1)\, dt$$

$$= \frac{3}{160} \left[\frac{t^3}{3} - \frac{t^2}{2} \right]_2^x = \frac{3}{160} \left[\left(\frac{x^3}{3} - \frac{x^2}{2} \right) - \left(\frac{8}{3} - 2 \right) \right]$$

$$= \frac{3}{160} \left[\frac{x^3}{3} - \frac{x^2}{2} - \frac{2}{3} \right]. \qquad \blacksquare$$

‖‖‖‖‖‖‖‖‖‖‖ **Probability Density Functions over Infinite Intervals**

To determine whether a function $f(x)$ is suitable as a probability density function for some (unspecified) random variable X over an *infinite* interval, we use Equations (6) and (7) applied to that interval rather than to $[c, d]$. Consequently, we shall need to use improper integrals, as presented in Section 7.6.

EXAMPLE 4   Prove that the function

$$f(t) = \frac{1}{(1 + t)^2}$$

is a probability density function for some random variable X defined over $[0, +\infty)$.

Solution   1. Clearly, $f(t) \geq 0$ for all t in $[0, +\infty)$.

2. $\displaystyle\int_0^{+\infty} \frac{1}{(1 + t)^2}\, dt = \lim_{b \to +\infty} \int_0^b \frac{1}{(1 + t)^2}\, dt$

$$= \lim_{b \to +\infty} \left(-\frac{1}{1 + t}\right)\Big]_0^b \qquad \text{(substitute } u = 1 + t)$$

$$= \lim_{b \to +\infty} \left[-\frac{1}{1 + b} - \left(-\frac{1}{1}\right)\right] = 1.$$

Therefore, f is a probability density function over $[0, +\infty)$. ∎

EXAMPLE 5
(Psychology)   A psychologist finds that the time t (in minutes) required by an average rat to go through a maze has the probability density function

$$f(t) = \frac{1}{(1 + t)^2} \qquad \text{for } t \geq 0.$$

Find the probability that a randomly chosen rat will take at least 3 minutes to go through the maze.

Solution   The desired probability is given by

$$P(3 \leq X < +\infty) = \int_3^{+\infty} \frac{1}{(1 + t)^2}\, dt = \lim_{b \to +\infty} \int_3^b \frac{1}{(1 + t)^2}\, dt$$

$$= \lim_{b \to +\infty} \left(-\frac{1}{1 + t}\right)\Big]_3^b$$

$$= \lim_{b \to +\infty} \left[-\frac{1}{1 + b} - \left(-\frac{1}{4}\right)\right] = \frac{1}{4}.$$

Roughly speaking, this answer means that one out of four randomly chosen rats will take 3 or more minutes to go through the maze. ∎

Probabilities over Intervals

We have been using integrals to calculate the probability that a continuous random variable lies within some interval. The careful reader has observed that we have always included the endpoints of the interval. However, since the probability that a continuous random variable equals any specific number is always 0, it makes no difference in the probability, whether we include the endpoints in the interval or not. Thus we have:

THEOREM 1

If X is a continuous random variable over an interval which includes the numbers a and b $(a < b)$, then

$$P(X < b) \quad = P(X \leq b),$$
$$P(X > a) \quad = P(X \geq a),$$
$$P(a < X < b) = P(a \leq X < b) = P(a < X \leq b) = P(a \leq X \leq b).$$

Another useful observation is that for a continuous random variable with probability density function $f(x)$ defined over an interval $[c, d]$, if a is any number in $[c, d]$, then by Property 5 of Section 7.2,

$$\int_c^a f(x)\, dx + \int_a^d f(x)\, dx = \int_c^d f(x)\, dx.$$

That is,

$$P(X \leq a) + P(X \geq a) = 1. \tag{9}$$

Moreover, Equation (9) remains true whether or not the interval contains its endpoints, c and d, and whether or not the interval is bounded. Thus, we have:

THEOREM 2

If X is a continuous random variable over an interval, then for any a in this interval,

$$P(X \geq a) = 1 - P(X \leq a). \tag{10}$$

EXAMPLE 6   In Example 5, the solution could have been found using Equation (10). Thus,

$$P(X \geq 3) = 1 - P(X \leq 3)$$
$$= 1 - \int_0^3 \frac{1}{(1 + t)^2}\, dt$$
$$= 1 - \left[-\frac{1}{1 + t} \right]_0^3$$
$$= 1 - \left(-\frac{1}{4} + 1 \right)$$
$$= \frac{1}{4}. \qquad \blacksquare$$

11.2 EXERCISE SET ||||||||||||||||||||||||||||||||||||||||||||||||||||||||||||||||||||||||||||

1. Let $f(x) = \frac{2}{21} x$ if $2 \leq x \leq 5$.
   (a) Show that $f(x)$ is a probability density function on the interval $[2, 5]$.
   (b) Find the probability that x lies in the interval $[2, 4]$.
   (c) Find the probability that x lies in the interval $[3, 5]$.

2. Let $f(x) = \dfrac{1 + 3x^2}{4}$.

(a) Show that $f(x)$ is a probability density function on the interval $[-1, 1]$.
(b) Find $P(X \le 0)$.
(c) Find $P(0 \le X \le \frac{1}{2})$.
(d) Find $P(X = \frac{1}{2})$.

3. Let $f(x) = e^{-x}$.

(a) Show that f is a probability density function on the interval $[0, +\infty)$.
(b) Find $P(X \le 1)$.
(c) Find $P(1 \le X \le 2)$.
(d) Find $P(X = 5)$.
(e) Find $P(X \ge 5)$.

4. Let $f(x) = \begin{cases} \dfrac{4x}{(x^2 + 2)^2} & \text{if } x \ge 0 \\ 0 & \text{otherwise.} \end{cases}$

(a) Show that f is a probability density function on the interval $(-\infty, +\infty)$.
(b) Find $P(X \le 0)$.
(c) Find $P(1 \le X \le 2)$.
(d) Find $P(X \ge 2)$.

In Exercises 5–8, determine the value of the constant k that will make the given function a probability density function over the given interval.

5. $f(x) = k\sqrt{x}, \ [0, 4]$

6. $f(x) = \dfrac{k}{x}, \ [1, 10]$

7. $f(x) = \dfrac{k}{x^3}, \ [2, +\infty)$

8. $f(x) = ke^{-x/2}, \ [0, +\infty)$

In Exercises 9 and 10, show that there is no value of k for which the given function is a probability density function on the given interval.

9. $f(x) = \dfrac{k}{\sqrt{x}}, \ [1, +\infty)$

10. $f(x) = e^{-kx}, \ (-\infty, +\infty)$

In Exercises 11–14, a probability density function $f(x)$ is given for a specified interval. Find the corresponding distribution function, $F(x)$.

11. $f(x) = \dfrac{2}{21} x, \ [2, 5]$

(See Exercise 1.)

12. $f(x) = \dfrac{1 + 3x^2}{4}, \ [-1, 1]$

(See Exercise 2.)

13. $f(x) = \dfrac{16}{7(1 + x)^2}, \ [1, 15]$

14. $f(x) = \dfrac{1}{18\sqrt{x}}, \ [1, 100]$

15. **(Reaction Time)** The time t (in seconds) that it takes a person to react to a changing traffic light has the probability density function

$$f(t) = \begin{cases} \dfrac{25}{12t^3} & \text{if } 1 \le t \le 5 \\ 0 & \text{otherwise.} \end{cases}$$

What is the probability that a person's reaction time will be at least 2 seconds, but no more than 4 seconds?

16. **(Waiting Time)** The waiting time t (in minutes) in a dentist's office has the probability density function

$$f(t) = \begin{cases} \frac{1}{30} & \text{if } 0 \le t \le 30 \\ 0 & \text{otherwise.} \end{cases}$$

(a) What is the probability of waiting at least 10 minutes?
(b) What is the probability of waiting at most 10 minutes?
(c) What is the probability of waiting at least 10 minutes, but no longer than 20 minutes?

17. **(Biology)** The probability that a plant of a certain species has a lifespan of t days has density function

$$f(t) = \dfrac{t}{1800}$$

over a 60-day period, $0 \leq t \leq 60$. Find
(a) the probability that such a plant lives at least 30 days.
(b) the probability that such a plant lives less than 10 days.
(c) the probability that such a plant dies sometime between the 40th day and the 50th day.

18. If X represents a number selected randomly from the interval $[0, 100]$, the probability density function is

$$f(x) = \frac{1}{100}.$$

(a) Find the corresponding distribution function.
(b) Find $P(40 \leq X \leq 60)$.
(c) Find $P(X_0 \leq X \leq X_0 + 20)$ and show that it is independent of X_0, as long as $0 \leq X_0 \leq 80$.
(d) Find $P(X_0 \leq X \leq X_0 + h)$, where $0 \leq X_0 \leq X_0 + h \leq 100$.

19. **(Length of Telephone Calls)** The length t (in minutes) of a certain business firm's telephone calls has the probability density function

$$f(t) = \begin{cases} \frac{1}{2}e^{-t/2} & \text{if } 0 \leq t < +\infty \\ 0 & \text{otherwise.} \end{cases}$$

What is the probability that a call will last 3 minutes or less?

20. **(Business)** A computer time-sharing service finds that the time t (in seconds) required by a job has the probability density function

$$f(t) = \begin{cases} \dfrac{2}{(2 + t)^2} & \text{if } t \geq 0 \\ 0 & \text{otherwise.} \end{cases}$$

What is the probability that a job will take at least 12 seconds?

21. **(Bacteria Culture)** The probability that a randomly chosen bacterium from a given culture lives exactly t days has density function

$$f(t) = 2^{-t} \ln 2$$

on the interval $[0, +\infty)$. (t is not necessarily a whole number.)
(a) Find the corresponding distribution function $F(t)$.
(b) Find the probability that a randomly chosen bacterium in the culture.
   (1) dies before the end of the 4th day.
   (2) dies sometime on the 4th day.
   (3) survives at least six days.

22. **(Education)** A study of the records of the mathematics department of a well-known university shows that the time t (in years) it takes the average candidate to complete the doctoral program has the probability density function

$$f(t) = \begin{cases} \frac{1}{5}e^{-t/5} & \text{if } t \geq 0 \\ 0 & \text{otherwise.} \end{cases}$$

Find the probability that a graduate student will take at least four years to receive her doctorate.

11.3 ‖‖‖‖‖ EXPECTED VALUE AND STANDARD DEVIATION

In this section we develop a generalization of the notion of the arithmetic mean of a set of numbers, and then introduce a mathematical measure of the total dispersion of a set of numbers away from its mean.

‖‖‖‖‖‖‖‖‖‖‖ **The Discrete Case**

EXAMPLE 1   Consider the set of ten numbers $x_1 = 2$, $x_2 = 6$, $x_3 = 8$, $x_4 = 2$, $x_5 = 2$, $x_6 = 1$, $x_7 = 1$, $x_8 = 2$, $x_9 = 5$, and $x_{10} = 5$. In elementary mathematics we learn that the arithmetic mean (or "average") of these ten numbers is

$$\mu = \frac{x_1 + x_2 + x_3 + \cdots + x_{10}}{10}$$

$$= \frac{2 + 6 + 8 + 2 + 2 + 1 + 1 + 2 + 5 + 5}{10}$$ (1)

$$= 3.4. \qquad\qquad \blacksquare$$

There is another way of viewing the arithmetic mean, which makes use of the concepts of probability and which leads to applications in probability theory.

EXAMPLE 2   Consider a box of ten Ping-Pong balls marked with the ten numbers given in Example 1, and consider the experiment of randomly selecting one ball from the box. Let the random variable X denote the number on the ball. The sample space is $\{1, 2, 5, 6, 8\}$* and the probability table for X is shown below.

X	1	2	5	6	8
$P(X)$	$\frac{1}{5}$	$\frac{2}{5}$	$\frac{1}{5}$	$\frac{1}{10}$	$\frac{1}{10}$

It is useful to know that the mean μ of X can be calculated from the above probability table. From (1), we have

$$\mu = \tfrac{2}{10} + \tfrac{6}{10} + \tfrac{8}{10} + \tfrac{2}{10} + \tfrac{2}{10} + \tfrac{1}{10} + \tfrac{1}{10} + \tfrac{2}{10} + \tfrac{5}{10} + \tfrac{5}{10}$$

$$= (\tfrac{1}{10} + \tfrac{1}{10}) + (\tfrac{2}{10} + \tfrac{2}{10} + \tfrac{2}{10} + \tfrac{2}{10}) + (\tfrac{5}{10} + \tfrac{5}{10}) + \tfrac{6}{10} + \tfrac{8}{10}$$

$$= \tfrac{1}{10} \cdot 2 + \tfrac{2}{10} \cdot 4 + \tfrac{5}{10} \cdot 2 + \tfrac{6}{10} \cdot 1 + \tfrac{8}{10} \cdot 1$$

$$= 1 \cdot \tfrac{2}{10} + 2 \cdot \tfrac{4}{10} + 5 \cdot \tfrac{2}{10} + 6 \cdot \tfrac{1}{10} + 8 \cdot \tfrac{1}{10}.$$

A glance at the probability table for X confirms that we have

$$\mu = 1 \cdot P(1) + 2 \cdot P(2) + 5 \cdot P(5) + 6 \cdot P(6) + 8 \cdot P(8). \qquad (2)$$
$$\blacksquare$$

Equation (2) is an instance of a more general principle, and leads us to the following definition.

Definition

> For a discrete random variable X, with sample space $\{x_1, x_2, \ldots, x_n\}$, the **mean** of X is
>
> $$\mu = x_1 P(x_1) + x_2 P(x_2) + \cdots + x_n P(x_n). \qquad (3)$$

---

* Strictly speaking, the sample space need not be a set of numbers. Nevertheless, the random variable *X represents* the sample space by a set of numbers. In the interest of simplicity, we also call the resulting set of numbers the sample space.

In the mathematical theory of probability, the "mathematical expectation" of a random variable is synonymous with its mean. Moreover, the term "mathematical expectation" is often preferred, because it more adequately conveys the meaning we intend to express by this concept. The following example will demonstrate the underlying concept of expectation and its equivalence with the concept of mean.

EXAMPLE 3   Consider the following game of chance. A player reaches into the box described in Example 2, randomly draws out a ball, and is paid the number of dollars shown on the ball. Thus, a player who draws the "5" ball will be paid $5. The ball is then put back into the box before the next player is allowed to draw. Since the arithmetic mean of the numbers on the balls is 3.4, we expect that in the long run the game pays out an average of $3.40 per player. Thus, we say that the "expected value" of the game to the player is $3.40. In order to break even in the long run, the operator of the game must charge at least $3.40 to play the game. To make a profit, he must charge more than this per game. ∎

Behind commercial games of chance there is a mathematical theory which enables the operators of the games to adjust charges and payoffs so that, in the long run, the game is profitable to the operator despite occasional high payoffs. In a less sensational vein, the same may be said of the insurance industry. Indeed, mathematicians called "actuaries" play a key role in determining insurance rates.

In general, **the expected value $E(X)$** of a random variable X is defined as its mean:

$$E(X) = \mu. \tag{4}$$

We often find it important to have a mathematical measure of the relative concentration of a given random variable X around its mean μ, or its dispersion away from μ. For a discrete random variable X with sample space $\{x_1, x_2, \ldots, x_n\}$ and mean μ, we define the **variance** of X to be

---

Var (X) $= E[(X - \mu)^2]$

$\qquad = (x_1 - \mu)^2 P(x_1) + (x_2 - \mu)^2 P(x_2) + \cdots + (x_n - \mu)^2 P(x_n). \tag{5}$

The expression $x_i - \mu$ measures the deviation of x_i from the expected value μ; it is squared in (5) so that the deviations to the left of μ do not cancel out deviations to the right of μ. Thus, the variance of X measures the average square of the deviation of X from its mean.

Since the terms of (5) involve squaring the deviations $x_i - \mu$, the variance involves a unit of measure different from that of the deviations themselves. For example, if the deviations are measured in inches then the variance is expressed in square inches. For this reason, the variance of X is usually replaced by its square root, which is called the **standard deviation** of X and is denoted by **σ**. Thus,

$$\sigma = \sqrt{\text{Var}\,(X)}. \qquad (6)$$

A small value of σ indicates that the values of X are concentrated near the expected value $\mu = E(X)$, while a larger value of σ indicates that the values of X are more dispersed, away from the expected value.

EXAMPLE 4   A professor teaching two sections of calculus, with 10 students in each section, gives the same examination to both sections. The exam scores are

Section A: 65, 66, 70, 71, 75, 75, 77, 78, 80, 83
Section B: 48, 55, 61, 64, 70, 75, 84, 90, 95, 98.

Calculate the mean, variance, and standard deviation for each section on this examination.

Solution   For Section A,

$$\mu = \frac{65 + 66 + 70 + 71 + 75 + 75 + 77 + 78 + 80 + 83}{10} = 74.$$

$$\begin{aligned}
\text{Var}\,(X) &= \tfrac{1}{10}[(65 - 74)^2 + (66 - 74)^2 + (70 - 74)^2 + (71 - 74)^2 \\
&\quad + (75 - 74)^2 + (75 - 74)^2 + (77 - 74)^2 + (78 - 74)^2 \\
&\quad + (80 - 74)^2 + (83 - 74)^2] \\
&= \tfrac{1}{10}[81 + 64 + 16 + 9 + 1 + 1 + 9 + 16 + 36 + 81] \\
&= 31.4.
\end{aligned}$$

$$\sigma = \sqrt{31.4} \approx 5.60.$$

For Section B,

$$\mu = \frac{48 + 55 + 61 + 64 + 70 + 75 + 84 + 90 + 95 + 98}{10} = 74.$$

$$\begin{aligned}
\text{Var}\,(X) &= \tfrac{1}{10}[(48 - 74)^2 + (55 - 74)^2 + (61 - 74)^2 + (64 - 74)^2 \\
&\quad + (70 - 74)^2 + (75 - 74)^2 + (84 - 74)^2 + (90 - 74)^2 \\
&\quad + (95 - 74)^2 + (98 - 74)^2] \\
&= \tfrac{1}{10}[676 + 361 + 169 + 100 + 16 + 1 + 100 + 256 + 441 + 576] \\
&= 269.6.
\end{aligned}$$

Thus,

$$\sigma = \sqrt{269.6} \approx 16.42. \qquad \blacksquare$$

Observe that in Example 4 both sections have the same average score, $\mu = 74$. However, the grades in Section A are concentrated closer to the mean whereas those in Section B are more dispersed away from the mean. The relative sizes of the standard deviations, $\sigma = 5.60$ for Section A and $\sigma = 16.42$ for Section B, reflect this difference.

The following theorem provides a useful formula, which will greatly simplify the calculation of the variance in many cases.

THEOREM 1

(Alternate Form of the Variance)
If X is a discrete random variable with sample space $\{x_1, x_2, \ldots, x_n\}$ and expected value $\mu = E(X)$, then

$$\text{Var}(X) = [x_1^2 P(x_1) + x_2^2 P(x_2) + \cdots + x_n^2 P(x_n)] - \mu^2. \qquad (7)$$

The proof of this theorem will be left as an exercise (Exercise 33).

||||||||||||| **The Continuous Case**

In our discussion of the expected value, variance, and standard deviation of a *discrete* random variable, calculus has played no role. The formulas defining $\mu = E(X)$ and Var (X) for a discrete random variable involve only summations. For a *continuous* random variable, however, these summations are replaced by integrals. We shall first define the relevant quantities, and then offer a justification for the use of integration.

Definition

If X is a continuous random variable with probability density function $f(x)$ on the interval $[c, d]$, then
(a) the **expected value** (or **mean**) of X is

$$\boldsymbol{E(X)} = \mu = \int_c^d x f(x)\, dx; \qquad (8)$$

(b) the **variance** of X is

$$\boldsymbol{\text{Var}(X)} = E[(X - \mu)^2]$$

$$= \int_c^d (x - \mu)^2 f(x)\, dx; \qquad (9)$$

(c) the **standard deviation** of X is

$$\sigma = \sqrt{\text{Var}(X)}. \qquad (10)$$

To understand why integrals are used to calculate these quantities, we recall the Riemann sum approach to definite integrals, as outlined in Section 7.4. We focus our attention on the integral used in Equation (8) to calculate $E(X)$. Using Riemann's approach, we subdivide the interval $[c, d]$ into n subintervals of equal width Δx, where

$$\Delta x = \frac{d - c}{n}. \tag{11}$$

Using the notation of Section 7.7 we label the endpoints of these subintervals

$$x_0, x_1, x_2, \ldots, x_n$$

where $x_0 = c$ and $x_n = d$. In each subinterval $[x_{i-1}, x_i]$ we choose a point $\tilde{x}_i$,

$$x_{i-1} \leq \tilde{x}_i \leq x_i.$$

Every Riemann sum over the interval $[c, d]$ for the function $xf(x)$ [the integrand in (8)] is of the form

$$R_n = \tilde{x}_1 f(\tilde{x}_1)\, \Delta x + \tilde{x}_2 f(\tilde{x}_2)\, \Delta x + \cdots + \tilde{x}_n f(\tilde{x})\, \Delta x. \tag{12}$$

We take a closer look at the terms in (12), with the help of the graph of the probability density function $y = f(x)$, shown in Figure 1.

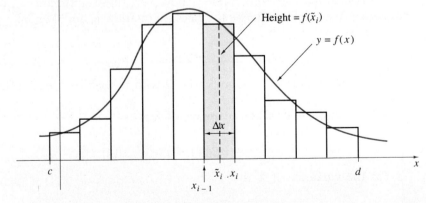

Figure 1

Each product $f(\tilde{x}_i)\, \Delta x$ in (12) is a rectangular approximation of the area under the curve $y = f(x)$ over the subinterval $[x_{i-1}, x_i]$. That is,

$$f(\tilde{x}_i)\, \Delta x \approx \int_{x_{i-1}}^{x_i} f(x)\, dx.$$

It can be shown using advanced techniques that, in fact, each $\tilde{x}_i$ can be chosen so that this approximation is an *equality:*

$$f(\tilde{x}_i)\,\Delta x = \int_{x_{i-1}}^{x_i} f(x)\,dx. \tag{13}$$

Now consider the discrete random variable X_n which has sample space $\{\tilde{x}_1, \tilde{x}_2, \ldots, \tilde{x}_n\}$ and probability table:

X_n	$\tilde{x}_1$	$\tilde{x}_2$	$\cdots$	$\tilde{x}_n$
$P(X_n)$	$f(\tilde{x}_1)\,\Delta x$	$f(\tilde{x}_2)\,\Delta x$	$\cdots$	$f(\tilde{x}_n)\,\Delta x$

(To confirm that this table legitimately defines a probability function we must show that $P(\tilde{x}_1) + P(\tilde{x}_2) + \cdots + P(\tilde{x}_n) = 1$. But this follows from the additive property of integrals:

$$
\begin{aligned}
f(\tilde{x}_1)\,\Delta x + f(\tilde{x}_2)\,\Delta x + \cdots + f(\tilde{x}_n)\,\Delta x &= \int_{x_0}^{x_1} f(x)\,dx + \int_{x_1}^{x_2} f(x)\,dx + \cdots + \int_{x_{n-1}}^{x_n} f(x)\,dx \\
&= \int_c^d f(x)\,dx \\
&= 1
\end{aligned}
$$

since f is a probability density function on $[c, d]$.)

By definition, the expected value of the random variable X_n is

$$
\begin{aligned}
E(X_n) &= \tilde{x}_1 f(\tilde{x}_1)\,\Delta x + \tilde{x}_2 f(\tilde{x}_2)\,\Delta x + \cdots + \tilde{x}_n f(\tilde{x}_n)\,\Delta x \\
&= R_n. \qquad \text{[from Equation (12)].}
\end{aligned}
$$

When we take the limit as $n \to +\infty$, it makes sense to say that the discrete random variable X_n "approaches" the continuous random variable X. Thus, we *define* the mean of X to be the limit

$$
\begin{aligned}
\mu = E(X) &= \lim_{n \to +\infty} E(X_n) = \lim_{n \to +\infty} R_n \\
&= \int_c^d x f(x)\,dx.
\end{aligned}
$$

EXAMPLE 5   Calculate the mean, variance, and standard deviation of the random variable X which has the probability density function

$$f(x) = \frac{25}{12x^3}$$

over the interval $[1, 5]$.

Solution   (a) By Equation (8), the mean of X is

$$\mu = \int_1^5 x \frac{25}{12x^3} \, dx = \frac{25}{12} \int_1^5 x^{-2} \, dx$$

$$= \frac{25}{12} [-x^{-1}]_1^5 = -\frac{25}{12} \left[\frac{1}{5} - 1 \right] = -\frac{25}{12} \left(-\frac{4}{5} \right)$$

$$= \frac{5}{3}.$$

(b) By Equation (9), the variance of X is

$$\text{Var}(X) = \int_1^5 \left(x - \frac{5}{3} \right)^2 \frac{25}{12x^3} \, dx$$

$$= \frac{25}{12} \int_1^5 \left(x^2 - \frac{10}{3} x + \frac{25}{9} \right) \cdot \frac{1}{x^3} \, dx$$

$$= \frac{25}{12} \int_1^5 \left(\frac{1}{x} - \frac{10}{3} x^{-2} + \frac{25}{9} x^{-3} \right) dx$$

$$= \frac{25}{12} \left[\ln x + \frac{10}{3} x^{-1} - \frac{25}{18} x^{-2} \right]_1^5$$

$$= \frac{25}{12} \left[\left(\ln 5 + \frac{10}{15} - \frac{25}{18 \cdot 25} \right) - \left(\ln 1 + \frac{10}{3} - \frac{25}{18} \right) \right]$$

$$= \frac{25}{12} \left[\ln 5 + \frac{2}{3} - \frac{1}{18} - \frac{10}{3} + \frac{25}{18} \right]$$

$$= \frac{25}{12} \left[\ln 5 - \frac{4}{3} \right]$$

$$\approx 0.575.$$

(c) By Equation (11), the standard deviation of X is

$$\sigma = \sqrt{0.575} \approx 0.758. \qquad \blacksquare$$

The following theorem establishes a formula for computing the variance of continuous random variables that is usually easier to use than (9).

THEOREM 2

(Alternate Form of the Variance)
If X is a continuous random variable with probability density function $f(x)$ over the interval $[c, d]$, then

$$\text{Var}(X) = \int_c^d x^2 f(x) \, dx - \mu^2 \qquad (14)$$

where $\mu = E(X)$.

Proof From Equation (9),

$$
\begin{aligned}
\text{Var}\,(X) &= \int_c^d (x - \mu)^2 f(x)\,dx \\
&= \int_c^d (x^2 - 2x\mu + \mu^2) f(x)\,dx \\
&= \int_c^d x^2 f(x)\,dx - 2\mu \int_c^d x f(x)\,dx + \mu^2 \int_c^d f(x)\,dx \\
&\quad \text{(properties of integrals, Section 7.2)} \\
&= \int_c^d x^2 f(x)\,dx - 2\mu \cdot \mu + \mu^2 \cdot 1 \\
&\left(\text{by definition, } \int_c^d x f(x)\,dx = \mu \text{ and } \int_c^d f(x)\,dx = 1 \right) \\
&= \int_c^d x^2 f(x)\,dx - \mu^2.
\end{aligned}
$$
∎

EXAMPLE 6 When Equation (14) is used to calculate the variance of the random variable X of Example 5, the computation shortens to

$$
\begin{aligned}
\text{Var}\,(X) &= \int_1^5 x^2 \frac{25}{12x^3}\,dx - \mu^2 \\
&= \frac{25}{12} \int_1^5 \frac{1}{x}\,dx - \left(\frac{5}{3}\right)^2 \\
&= \frac{25}{12} \ln 5 - \frac{25}{9} \\
&\approx 0.575.
\end{aligned}
$$
∎

While our definitions of expected value and variance, as written, apply to continuous random variables over *bounded* intervals, they may often be applied to random variables over *unbounded* intervals. Of course, these definitions extend to unbounded intervals only when the relevant improper integrals converge.

EXAMPLE 7 Calculate the mean and variance of the random variable X whose probability density function is $f(x) = 3/x^4$ on the interval $[1, +\infty)$.

Solution (a) Applying Equation (8) to the interval $[1, +\infty)$, we have

$$
\begin{aligned}
E(X) &= \int_1^{+\infty} x \frac{3}{x^4}\,dx = \lim_{b \to +\infty} \int_1^b \frac{3}{x^3}\,dx \\
&= 3 \lim_{b \to +\infty} \left[\frac{-1}{2x^2} \right]_1^b = \frac{3}{2} \lim_{b \to +\infty} \left[-\frac{1}{b^2} + 1 \right] \\
&= \frac{3}{2}.
\end{aligned}
$$

(b) Applying Equation (14) to the interval $[1, +\infty)$, we have

$$
\begin{aligned}
\text{Var }(X) &= \int_1^{+\infty} x^2 \frac{3}{x^4} \, dx - \mu^2 \\
&= \left\{ \lim_{b \to +\infty} \int_1^b \frac{3}{x^2} \, dx \right\} - \left(\frac{3}{2}\right)^2 \\
&= 3 \lim_{b \to +\infty} \left[-\frac{1}{x} \right]_1^b - \frac{9}{4} \\
&= 3 \lim_{b \to +\infty} \left[-\frac{1}{b} + 1 \right] - \frac{9}{4} \\
&= \frac{3}{4}.
\end{aligned}
$$

∎

The expected value and variance of a continuous random variable have the same intuitive meanings as the expected value and variance of a discrete random variable with a very large sample space. If the experiment represented by the random variable X is repeated many times, then $E(X)$ represents the "average" outcome and Var (X) represents the average squared deviation of the outcome from its expected value.

The expected value [mean, $\mu = E(X)$] of a continuous random variable X has an interesting geometric interpretation. It can be shown using more advanced techniques that the area between the probability density curve $y = f(x)$ and the x-axis will exactly balance across the vertical line $x = \mu$. (See Figure 2. Imagine cutting the shaded area out of a piece of cardboard and balancing it horizontally across the line.)

Figure 2

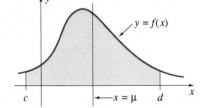

EXAMPLE 8   The amount of steel (in hundreds of tons) that a steel distributor sells daily is a random variable X with probability density function

$$
f(x) = \tfrac{3}{160} (x^2 - x)
$$

over the interval $[2, 6]$. (See Example 1, Section 11.2.)
(a) On the average, how much does the distributor expect to sell daily?
(b) Find the standard deviation of the random variable X.

Solution   (a) We want to calculate the expected value of X. By (8),

$$E(X) = \int_2^6 x \frac{3}{160} (x^2 - x)\, dx = \frac{3}{160} \int_2^6 (x^3 - x^2)\, dx$$

$$= \frac{3}{160} \left[\frac{x^4}{4} - \frac{x^3}{3} \right]_2^6$$

$$= \frac{3}{160} \left[\left(\frac{1296}{4} - \frac{216}{3} \right) - \left(\frac{16}{4} - \frac{8}{3} \right) \right]$$

$$= \frac{3}{160} \left[324 - 72 - 4 + \frac{8}{3} \right]$$

$$= 4.7.$$

Thus, the distributor's daily sales averages 470 tons of steel.
(b) By (14),

$$\text{Var}\,(X) = \int_2^6 x^2 \cdot \frac{3}{160} (x^2 - x)\, dx - \mu^2$$

$$= \frac{3}{160} \int_2^6 (x^4 - x^3)\, dx - (4.7)^2$$

$$= \frac{3}{160} \left[\frac{x^5}{5} - \frac{x^4}{4} \right]_2^6 - (4.7)^2$$

$$= \frac{3}{160} \left[\left(\frac{7776}{5} - \frac{1296}{4} \right) - \left(\frac{32}{5} - 4 \right) \right] - (4.7)^2$$

$$= 0.95.$$

Finally, $\sigma = \sqrt{0.95} \approx 0.975$.

Comments: Figure 3 shows the graph of the probability density function $f(x)$, and the vertical line $x = \mu$. Observe that the area would "balance" across

Figure 3

$$y = \frac{3}{160}(x^2 - x)$$

$x = \mu$

∎

the line $x = 4.7$. Observe also that $\sigma = 0.975$ is a kind of average deviation of X from 4.7.

|||||||||||| **The Median**

We define the **median** of a continuous random variable X to be the number m such that

$$P(X \le m) = \tfrac{1}{2}. \tag{15}$$

In other words, a randomly chosen member of the sample space has a 50% probability of being less than or equal to the median and a 50% probability of being greater than the median. The median is sometimes used as a substitute for the mean, especially when it is easier to calculate than the mean.

EXAMPLE 9   Find the median of the random variable whose probability density function is

$$f(x) = \frac{25}{12x^3}$$

over the interval $[1, 5]$. (See Example 5.)

Solution   By Equation (15), the median m must satisfy $P(X \le m) = \tfrac{1}{2}$. Now,

$$P(X \le m) = \int_1^m \frac{25}{12x^3}\, dx = \frac{25}{12}\left[-\frac{1}{2} x^{-2} \right]_1^m$$

$$= -\frac{25}{24}\left[\frac{1}{m^2} - 1 \right].$$

Thus the median m must satisfy the equation

$$\frac{25}{24}\left(1 - \frac{1}{m^2} \right) = \frac{1}{2}$$

$$1 - \frac{1}{m^2} = \frac{12}{25}$$

$$1 - \frac{12}{25} = \frac{1}{m^2}$$

$$\frac{13}{25} = \frac{1}{m^2}$$

$$m^2 = \frac{25}{13}.$$

Thus, the median is $m = 5/\sqrt{13} \approx 1.387$. [In Example 5 we found the mean $\mu = 5/3 \approx 1.667$.]   ∎

The median is especially useful for certain random variables X which have the probability density functions $f(x)$ defined over unbounded intervals. In such cases it may happen that the improper integral defining $E(X)$ diverges, leaving no possibility of finding the mean μ of X. When this happens it may be possible to calculate the median and use it instead of the mean.

EXAMPLE 10   In Examples 4 and 5 of Section 11.2 we used the random variable X whose probability density function is

$$f(t) = \frac{1}{(1 + t)^2}$$

on the interval $[0, +\infty)$.
(a) Show that the mean of X does not exist.
(b) Show that the median of X does exist, and find it.

Solution   (a) From Equation (8), the mean of X is

$$E(X) = \int_0^{+\infty} \frac{t}{(1 + t)^2}\, dt = \lim_{b \to +\infty} \int_0^b \frac{t}{(1 + t)^2}\, dt.$$

By Integral Formula 12,

$$\int \frac{t}{(1 + t)^2}\, dt = \frac{1}{1 + t} + \ln(1 + t) + C$$

and so

$$E(X) = \lim_{b \to +\infty} \left\{ \left[\frac{1}{1 + b} + \ln(1 + b) \right] - [1 + \ln 1] \right\}$$

$$= +\infty \qquad \left[\text{since } \lim_{b \to +\infty} \ln(1 + b) = +\infty \right].$$

In this case the improper integral defining $E(X)$ diverges. Therefore, X has no mean or expected value.
(b) We can calculate the median of X as follows:

$$P(X \le x) = \int_0^x \frac{1}{(1 + t)^2}\, dt = \left[\frac{-1}{1 + t} \right]_0^x = \frac{-1}{1 + x} - \frac{-1}{1 + 0}$$

$$= 1 - \frac{1}{1 + x}.$$

The median m is the number which satisfies

$$1 - \frac{1}{1 + m} = \frac{1}{2}$$

$$\frac{1}{2} = \frac{1}{1 + m}$$

$$2 = 1 + m$$

$$m = 1.$$

11.3 EXERCISE SET ||||||||||||||||||||||||||||||||||||||||||||||||||||||||||||||||||||||||||||||||||||

In Exercises 1 and 2 compute the mean, variance, and standard deviation of the discrete random variable X with the given probability table.

1.

X	-2	-1	0	1	2	3
$P(X)$	$\frac{1}{6}$	$\frac{1}{12}$	$\frac{1}{6}$	$\frac{1}{3}$	$\frac{1}{6}$	$\frac{1}{12}$

2.

X	1	2	3	4	5	6	7
$P(X)$	$\frac{1}{15}$	$\frac{2}{15}$	$\frac{2}{15}$	$\frac{1}{15}$	$\frac{1}{3}$	$\frac{1}{5}$	$\frac{1}{15}$

In Exercises 3 and 4 a discrete random variable X is given in tabular form. The values of X represent the possible outcomes of an experiment which is repeated many times. The entries in the second row of each table give the frequency (number of times it occurs) corresponding to each possible outcome of X. Construct a probability table for X and use it to find the expected value, variance, and standard deviation of X.

3.

X	-4	-3	-2	-1	0	1	2	3	4
Freq.	1	0	4	5	6	4	3	1	0

4.

X	0	1	2	3	4	5	6	7
Freq.	2	3	1	2	0	4	2	1

5. Let the random variable X denote the number of heads that appear when two fair coins are tossed. Find $E(X)$, Var (X), and σ.

6. Repeat Exercise 5 for three coins instead of two.

7. Let the random variable X denote the sum of the numbers that appear when two fair dice are tossed. Find $E(X)$, Var (X), and σ.

8. Repeat Exercise 7, assuming that one die is normal while the other has a "0" instead of a "6".

9. Find $E(X)$, Var (X), and σ for the random variable X of Exercise 16 of Section 11.1.

10. Find $E(X)$, Var (X), and σ for the random variable X of Exercise 17 of Section 11.1.

·In Exercises 11–16, find the mean, variance, and standard deviation of the random variable X which has the given probability density function over the indicated interval.

11. $f(x) = \frac{2}{21}x$, $[2, 5]$

12. $f(x) = \dfrac{1 + 3x^2}{4}$, $[-1, 1]$

13. $f(x) = \frac{1}{5}$, $[1, 6]$

14. $f(x) = \dfrac{1}{18\sqrt{x}}$, $[1, 100]$

15. $f(x) = \dfrac{4}{x^5}$, $[1, +\infty)$

16. $f(x) = \frac{7}{2}x^{-4.5}$, $[1, +\infty)$

In Exercises 17–24, find the median of the random variable X which has the given probability density function over the indicated interval.

17. $f(x) = \frac{2}{21}x$, $[2, 5]$

18. $f(x) = \frac{1}{5}$, $[1, 6]$

19. $f(x) = \dfrac{1}{18\sqrt{x}}$, $[1, 100]$

20. $f(x) = \dfrac{16}{7(1 + x)^2}$, $[1, 15]$

21. $f(x) = \dfrac{4}{x^5}$, $[1, +\infty)$

22. $f(x) = \frac{7}{2}x^{-4.5}$, $[1, +\infty)$

23. $f(x) = \dfrac{1}{x^2}$, $[1, +\infty)$

24. $f(x) = e^{-x}$, $[0, +\infty)$

25. **(Inventory Planning)** The number of gallons (in thousands) of unleaded gasoline

that a dealer sells weekly is a random variable with probability density function

$$f(x) = \frac{20}{3(x + 2)^2}$$

over the interval [2, 8]. (See Example 2, Section 11.2.)

(a) In an average week, how many gallons of unleaded gasoline can the dealer expect to sell?

(b) Find the standard deviation of this random variable.

26. **(Reaction Time)** The time t (in seconds) that it takes a randomly selected person to react to a changing traffic light is a random variable with probability density function

$$f(t) = \frac{25}{12t^3}$$

for $1 \leq t \leq 5$. (See Exercise 15, Section 11.2.)

(a) Find the expected reaction time for a randomly selected person.

(b) Find the standard deviation of this random variable.

27. **(Waiting Time)** The time t (in minutes) that a randomly selected patient must wait in a certain dentist's office is a random variable with probability density function

$$f(t) = \frac{1}{30}$$

for $0 \leq t \leq 30$. (See Exercise 16, Section 11.2.)

(a) Find the expected waiting time for a randomly selected patient.

(b) Find the standard deviation of this random variable.

28. **(Biology)** The lifespan (in days) of a randomly selected plant of a certain species is a random variable with probability density function

$$f(t) = \frac{t}{1800}$$

for $0 \leq t \leq 60$. (See Exercise 17, Section 11.2.)

(a) Find the average life expectancy of this species.

(b) Find the standard deviation of this random variable.

29. **(Length of Telephone Calls)** Find the median length of telephone calls for the business firm described in Exercise 19, Section 11.2.

30. **(Business)** Find the median length of time required by jobs for the computer time-sharing service described in Exercise 20, Section 11.2.

31. **(Bacteria Culture)** Find the median lifespan of bacteria in the culture described in Exercise 21, Section 11.2.

32. **(Life Expectancy of Equipment)** The lifetime (in years) of a certain type of electronic component is a random variable with probability density function

$$f(t) = \frac{1}{(1 + t^2)^{3/2}}$$

on the interval $[0, +\infty)$. Find the mean life expectancy for this type of component.

33. Prove Theorem 1. (*Hint:* follow the lines of the argument used in proving Theorem 2.)

11.4 ‖‖‖‖ UNIFORM, EXPONENTIAL, AND NORMAL PROBABILITY DENSITY FUNCTIONS

In this section we examine three special probability density functions that find frequent use in a large variety of applications. Each of these density functions furnishes a mathematical model of a special set of circumstances, and is used whenever these circumstances are met.

Uniform Density Functions

A uniform density function is a mathematical model of a random phenomenon in which the outcome is just as likely to be in one subinterval of its sample space as it is to be in any other subinterval of the same width. More precisely, a continuous random variable X over an interval $[c, d]$ is said to have a **uniform distribution** if its probability density function is constant:

$$f(x) = k \tag{1}$$

for all x in $[c, d]$. Such a function (1) is said to be a **uniform density function** because it distributes probability "uniformly" across the interval $[c, d]$.

Actually, there is only one permissible value of k in Equation (1); namely, $k = \dfrac{1}{d - c}$. To see this, we recall that in order for $f(x)$ to be a probability density function over $[c, d]$ we must have

$$\int_c^d f(x) = 1$$

or

$$\int_c^d k \, dx = 1$$
$$k[x]_c^d = 1$$
$$k(d - c) = 1$$
$$k = \frac{1}{d - c}.$$

Figure 1 shows the graph of the uniform density function over $[c, d]$.

Figure 1

We summarize the basic properties of uniform distributions in the following theorem.

THEOREM 1 | **(Uniform Probability Density Functions)**
The continuous random variable with uniform probability density function

$$f(x) = \frac{1}{d-c} \qquad (2)$$

on the interval $[c, d]$ has

(a) mean $\mu = \dfrac{c+d}{2}$, $\qquad (3)$

(b) median $m = \dfrac{c+d}{2}$, $\qquad (4)$

(c) standard deviation $\sigma = \dfrac{d-c}{\sqrt{12}}$. $\qquad (5)$

Proof   (a) By definition,

$$\mu = \int_c^d xf(x)\,dx = \int_c^d \frac{x}{d-c}\,dx$$

$$= \frac{1}{d-c}\left[\frac{x^2}{2}\right]_c^d = \frac{d^2 - c^2}{2(d-c)}$$

$$= \frac{(d+c)(d-c)}{2(d-c)} = \frac{d+c}{2}.$$

(b) To find the median, we must first find

$$P(X \le x) = \int_c^x f(t)\,dt = \int_c^x \frac{1}{d-c}\,dt$$

$$= \frac{1}{d-c}[t]_c^x = \frac{x-c}{d-c}.$$

The median m must satisfy $P(X \le m) = \frac{1}{2}$, or

$$\frac{m-c}{d-c} = \frac{1}{2}$$

$$2m - 2c = d - c$$

$$2m = d + c$$

$$m = \frac{d+c}{2}.$$

(c) By Equation (14) of Section 11.3,

$$\text{Var}(X) = \int_c^d x^2 f(x)\,dx - \mu^2$$

$$= \int_c^d \frac{x^2}{d-c}\,dx - \left(\frac{d+c}{2}\right)^2$$

$$= \frac{1}{d-c}\left[\frac{x^3}{3}\right]_c^d - \frac{d^2 + 2dc + c^2}{4}$$

$$= \frac{d^3 - c^3}{3(d-c)} - \frac{d^2 + 2cd + c^2}{4}$$

$$= \frac{d^2 + cd + c^2}{3} - \frac{d^2 + 2cd + c^2}{4}$$

$$= \frac{4d^2 + 4cd + 4c^2 - 3d^2 - 6cd - 3c^2}{12}$$

$$= \frac{d^2 - 2cd + c^2}{12}$$

$$= \frac{(d-c)^2}{12}.$$

Therefore, $\sigma = \sqrt{\text{Var}(X)} = \dfrac{d-c}{\sqrt{12}}.$ ■

EXAMPLE 1   A pointer is spun inside a circular dial, which is calibrated in real numbers, from 0 to 360 (degrees). If the pointer is "fair", then it is no more likely to stop at one region of the dial than at any other. If the random variable X represents the real number at which the pointer stops spinning, then the probability density function for X is the uniform density function

$$f(x) = \frac{1}{360 - 0} = \frac{1}{360}$$

over the interval [0, 360).

The mean, median, and standard deviation of the random variable X are found using Theorem 1:

$$\mu = \frac{0 + 360}{2} = 180,$$

$$m = \frac{0 + 360}{2} = 180,$$

$$\sigma = \frac{360 - 0}{\sqrt{12}} \approx 104.$$

■

IIIIIIIIIIIII   **Exponential Density Functions**

Let k be a positive real number. The function

$$f(x) = ke^{-kx} \tag{6}$$

is a probability density function on the interval $[0, +\infty)$, since $f(x) \geq 0$ and

$$\int_0^{+\infty} ke^{-kx}\,dx = \lim_{b \to +\infty} \int_0^b ke^{-kx}\,dx = \lim_{b \to +\infty} [-e^{-kx}]_0^b$$

$$= \lim_{b \to +\infty} \left[\frac{-1}{e^{kb}} - \left(\frac{-1}{e^0} \right) \right] = \lim_{b \to +\infty} \left[1 - \frac{1}{e^{kb}} \right]$$

$$= 1.$$

The function (6) is called the **exponential probability density function** with parameter k, and its graph is shown in Figure 2.

Figure 2

$$y = ke^{-kx}$$

In order to calculate the mean and variance of exponential density functions, and normal density functions to follow, we shall need the following special limit.

THEOREM 2

If n is any positive constant, then

$$\lim_{x \to +\infty} \frac{x^n}{e^x} = 0. \tag{7}$$

Equation (7) says that as $x \to +\infty$, e^x approaches $+\infty$ much faster than any constant power of x. The proof of this theorem is beyond the level of this book, but the following table provides some numerical confirmation.

TABLE 1 $\displaystyle \lim_{x \to +\infty} \frac{x^4}{e^x} = 0$

x	5	10	15	20	25
$\dfrac{x^4}{e^x}$	4.211	0.454	0.015	0.0003	0.000005

THEOREM 3

> **(Exponential Probability Density Functions)**
> The continuous random variable X with probability density function
>
> $$f(x) = ke^{-kx}$$
>
> on the interval $[0, +\infty)$ has
>
> (a) mean $\mu = \dfrac{1}{k}$, (8)
>
> (b) standard deviation $\sigma = \dfrac{1}{k}$, (9)
>
> (c) median $m = \dfrac{\ln 2}{k}$. (10)

Proof   (a) By definition, the mean of X is

$$
\begin{aligned}
\mu &= \int_0^{+\infty} xf(x)\,dx = \int_0^{+\infty} x \cdot ke^{-kx}\,dx \\
&= k \lim_{b \to +\infty} \int_0^b xe^{-kx}\,dx \\
&= k \lim_{b \to +\infty} \left[\frac{e^{-kx}}{k^2}(-kx-1) \right]_0^b \qquad \text{(Integral Formula 40)} \\
&= \frac{k}{k^2} \lim_{b \to +\infty} [e^{-kb}(-kb-1) - e^0(-1)] \\
&= \frac{1}{k} \lim_{b \to +\infty} \left[\frac{-kb}{e^{kb}} - \frac{1}{e^{kb}} + 1 \right] \\
&= \frac{1}{k}[0 - 0 + 1] \qquad \text{(using Theorem 2)} \\
&= \frac{1}{k}.
\end{aligned}
$$

(b) By Equation (14) of Section 11.3,

$$\text{Var}(X) = \int_0^{+\infty} x^2 f(x) - \mu^2. \tag{11}$$

Now,

$$
\begin{aligned}
\int_0^{+\infty} x^2 \cdot ke^{-kx}\,dx &= k \lim_{b \to +\infty} \int_0^b x^2 e^{-kx}\,dx \\
&= k \lim_{b \to +\infty} \left[\frac{x^2 e^{-kx}}{-k} - \frac{2}{-k} \int xe^{-kx}\,dx \right]_0^b
\end{aligned}
$$

$$\text{(Integral Formula 42)}$$

$$= k \lim_{b \to +\infty} \left(-\frac{1}{k} \right) \left[x^2 e^{-kx} - 2 \int x e^{-kx} \, dx \right]$$

$$= (-1) \lim_{b \to +\infty} \left[x^2 e^{-kx} - 2 \left\{ \frac{e^{-kx}}{k^2} (-kx - 1) \right\} \right]_0^b$$

(Integral Formula 40)

$$= - \lim_{b \to +\infty} \left[\frac{x^2}{e^{kx}} + \frac{2}{k} \cdot \frac{x}{e^{kx}} + \frac{2}{k^2} \cdot \frac{1}{e^{kx}} \right]_0^b$$

$$= - \lim_{b \to +\infty} \left[\left(\frac{b^2}{e^{kb}} + \frac{2}{k} \cdot \frac{b}{e^{kb}} + \frac{2}{k^2} \cdot \frac{1}{e^{kb}} \right) - \left(0 + 0 + \frac{2}{k^2} \right) \right]$$

$$= (-1) \left[0 + 0 + 0 - \frac{2}{k^2} \right]$$

$$= \frac{2}{k^2}.$$

Moreover, by part (a) above, $\mu = \dfrac{1}{k}$. Substituting these results into (11), we have

$$\text{Var}(X) = \frac{2}{k^2} - \frac{1}{k^2} = \frac{1}{k^2}.$$

Therefore,

$$\sigma = \sqrt{\text{Var}(X)} = \frac{1}{k}.$$

(c) By definition, the median m satisfies $P(X \le m) = \frac{1}{2}$. Thus,

$$\int_0^m k e^{-kx} \, dx = \frac{1}{2}$$

$$[-e^{-kx}]_0^m = \frac{1}{2}$$

$$\frac{-1}{e^{km}} - \frac{-1}{e^0} = \frac{1}{2}$$

$$1 - \frac{1}{e^{km}} = \frac{1}{2}$$

$$\frac{1}{2} = \frac{1}{e^{km}}$$

$$e^{km} = 2$$

$$km = \ln 2$$

$$m = \frac{\ln 2}{k}. \qquad \blacksquare$$

Exponential probability density functions are used to model random variables X defined over the interval $[0, +\infty)$ for which the probability of X lying in a small interval $[t, t + \Delta t]$ is greatest when $t = 0$ and decreases as t increases. Common examples of such phenomena include the length of time cars must wait at a stop sign, the length of time between two successive telephone calls at a busy switchboard, and the lifespan of an electronics component.

EXAMPLE 2 Suppose that the lifespan of a certain type of electrical fuse is a continuous random variable X with an exponential probability density function. If the expected value of a randomly selected fuse is 200 hours, find
(a) the probability density function for X.
(b) the probability that a randomly selected fuse has a lifespan of at least 300 hours.
(c) the probability that a randomly selected fuse lasts less than 100 hours.
(d) the probability that a randomly selected fuse burns out sometime during its 200th hour.
(e) the probability that a randomly selected fuse burns out during its first hour.

Solution (a) By definition, the exponential probability density function is of the form

$$f(t) = ke^{-kt}.$$

(We use the variable t to represent time, in hours.) We are given that the mean (expected value) is $\mu = 200$. From Equation (8), $\mu = 1/k$. Thus, $k = 1/200$. Therefore, the probability density function is

$$f(t) = \frac{1}{200} e^{-t/200}.$$

(b) $P(X \geq 300) = \int_{300}^{+\infty} \frac{1}{200} e^{-t/200} \, dt$

$$= \lim_{b \to +\infty} \int_{300}^{b} \frac{1}{200} e^{-t/200} \, dt$$

$$= \lim_{b \to +\infty} [-e^{-t/200}]_{300}^{b}$$

$$= \lim_{b \to +\infty} \left[\frac{-1}{e^{b/200}} - (-e^{-300/200}) \right]$$

$$= e^{-3/2}$$

$$\approx 0.223.$$

(c) $P(X \leq 100) = \int_{0}^{100} \frac{1}{200} e^{-t/200} \, dt$

$$= [-e^{-t/200}]_{0}^{100} = -e^{-100/200} - (-e^{0})$$

$$= 1 - e^{-1/2}$$

$$\approx 0.393.$$

(d) $P(199 \leq X \leq 200) = \int_{199}^{200} \frac{1}{200} e^{-t/200} \, dt$

$$= [-e^{-t/200}]_{199}^{200} = -e^{-1} - (-e^{-199/200})$$

$$\approx 0.00184.$$

(e) $P(0 \le X \le 1) = \int_0^1 \dfrac{1}{200} e^{-t/200} \, dt$

$\qquad = [-e^{-t/200}]_0^1 = -e^{-1/200} - (-e^0)$

$\qquad = 1 - e^{-1/200}$

$\qquad \approx 0.00499.$ ∎

EXAMPLE 3   The random variable X representing the length of time (in seconds) between successive telephone calls to a busy college information office has an exponential probability density function with mean $\mu = 90$ seconds. After receiving a call, what is the probability that
(a) another call will be received within the next 30 seconds?
(b) no call will be received in the next 5 minutes?

Solution   We are given that $\mu = 90$. From Equation (8), $\mu = 1/k$. Thus, $k = 1/90$, and the probability density function for X is

$$f(t) = \frac{1}{90} e^{-t/90}.$$

(a) $P(X \le 30) = \int_0^{30} \dfrac{1}{90} e^{-t/90} \, dt = [-e^{-t/90}]_0^{30} = -e^{-30/90} - (-e^0) = 1 - e^{-1/3}$

$\qquad \approx 0.283.$

(b) $P(X \ge 300) = \int_{300}^{+\infty} \dfrac{1}{90} e^{-t/90} \, dt$

$\qquad = \displaystyle\lim_{b \to +\infty} \int_{300}^b \dfrac{1}{90} e^{-t/90} \, dt = \lim_{b \to +\infty} [-e^{-t/90}]_{300}^b$

$\qquad = \displaystyle\lim_{b \to +\infty} \left[\dfrac{-1}{e^{b/90}} - (-e^{-300/90}) \right]$

$\qquad = \displaystyle\lim_{b \to +\infty} \left[e^{-10/3} - \dfrac{1}{e^{b/90}} \right]$

$\qquad = e^{-10/3}$

$\qquad \approx 0.0357.$ ∎

||||||||||||| **The Standard Normal Density Function**

The function

$$g(x) = e^{-x^2/2} \tag{12}$$

is continuous and differentiable over the entire real number line $(-\infty, +\infty)$. Using the techniques of graphing discussed in Sections 4.5 and 4.6 we can obtain its graph, as shown in Figure 3.

Figure 3

The graph has its highest point at $(0, 1)$ and inflection points at $x = \pm 1$. It is symmetric with respect to the y-axis, and has the x-axis as a horizontal asymptote. (See Exercises 31 and 32.)

Since g is continuous, the Second Fundamental Theorem of Calculus assures us that g has an antiderivative $\int g(x)\,dx$. Unfortunately, there is no elementary formula for this antiderivative; it cannot be expressed in terms of elementary functions. It can be shown, however, that the improper integral $\int_{-\infty}^{+\infty} g(x)\,dx$ converges, and that

$$\int_{-\infty}^{+\infty} e^{-x^2/2}\,dx = \sqrt{2\pi}. \tag{13}$$

We now define the **standard normal probability density function** by

$$f(x) = \frac{1}{\sqrt{2\pi}}\,e^{-x^2/2}. \tag{14}$$

This function is indeed a density function on $(-\infty, +\infty)$, since

1. $f(x) \geq 0$ for all x in $(-\infty, +\infty)$, and

2. $\displaystyle\int_{-\infty}^{+\infty} f(x)\,dx = \int_{-\infty}^{+\infty} \frac{1}{\sqrt{2\pi}}\,e^{-x^2/2}\,dx = \frac{1}{\sqrt{2\pi}} \int_{-\infty}^{+\infty} e^{-x^2/2}\,dx$

$$= \frac{1}{\sqrt{2\pi}} \cdot \sqrt{2\pi}$$

$$= 1.$$

The probability density function $f(x)$ defined by (14) is one of the most important in probability theory. It is so important that extensive tables with computed values of its distribution function

$$A(x) = P(X \leq x) = \int_{-\infty}^{x} \frac{1}{\sqrt{2\pi}}\,e^{-t^2/2}\,dt \tag{15}$$

are commonly available. Our (simplified) table is found as Table V in Appendix B.

Recall that the value of $A(x)$ is the area of the region shaded in Figure 4.

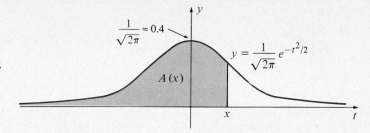

Figure 4

EXAMPLE 4   The following improper integrals are evaluated using Table V of Appendix B.

(a) $\displaystyle\int_{-\infty}^{0} \frac{1}{\sqrt{2\pi}}\, e^{-t^2/2}\, dt = 0.5000.$

(b) $\displaystyle\int_{-\infty}^{0.62} \frac{1}{\sqrt{2\pi}}\, e^{-t^2/2}\, dt = 0.7324.$

(c) $\displaystyle\int_{-\infty}^{2.87} \frac{1}{\sqrt{2\pi}}\, e^{-t^2/2}\, dt = 0.9979.$ ∎

Table V provides values for the integral (15) only when $x \geq 0$. To obtain the value of $A(x)$ when $x < 0$ we must perform some algebra. First we note that for any real number x,

$$\int_{-\infty}^{+\infty} e^{-t^2/2}\, dt = \int_{-\infty}^{x} e^{-t^2/2}\, dt + \int_{x}^{+\infty} e^{-t^2/2}\, dt. \tag{16}$$

Now when $x > 0$,

$$A(-x) = \frac{1}{\sqrt{2\pi}} \int_{-\infty}^{-x} e^{-t^2/2}\, dt$$

$$= \frac{1}{\sqrt{2\pi}} \int_{x}^{+\infty} e^{-t^2/2}\, dt \qquad \text{(since the curve is symmetric with respect to the y-axis),}$$

$$= \frac{1}{\sqrt{2\pi}} \int_{-\infty}^{+\infty} e^{-t^2/2}\, dt - \frac{1}{\sqrt{2\pi}} \int_{-\infty}^{x} e^{-t^2/2}\, dt \qquad \text{[from (16)]}$$

$$= 1 - A(x).$$

Therefore, for $A(x)$ defined by (15),

$$A(-x) = 1 - A(x). \tag{17}$$

EXAMPLE 5   Using Equations (15) and (17) along with Table V of Appendix B we obtain

(a) $\int_{-\infty}^{-1.0} \dfrac{1}{\sqrt{2\pi}} \, e^{-t^2/2} \, dt = A(-1.0) = 1 - A(1.0) = 1 - 0.8413 = 0.1587.$

(b) $\int_{-\infty}^{-2.54} \dfrac{1}{\sqrt{2\pi}} \, e^{-t^2/2} \, dt = A(-2.54) = 1 - A(2.54) = 1 - 0.9945 = 0.0055.$ ∎

Using Equations (2)–(5) of Section 11.2 with Equations (15) and (17) above, we can prove that the following rules hold for any random variable whose probability density function is the standard normal density function (See Exercises 33–36):

1. $P(X \le -x) = A(-x) = 1 - A(x)$ if $x > 0$ (18)
2. $P(a \le X \le b) = A(b) - A(a)$ (19)
3. $P(X \ge x) = 1 - A(x)$ (20)
4. $P(X \ge -x) = A(x)$ if $x > 0$ (21)
5. $P(-x \le X \le x) = 2A(x) - 1$ if $x > 0$ (22)

EXAMPLE 6   Suppose X is a continuous random variable with standard normal density function. Find each of the following:

(a) $P(0.25 \le X \le 1.78)$.   (e) $P(X \ge 2.51)$.
(b) $P(X \le -2.13)$.   (f) $P(X \ge -3.00)$.
(c) $P(-1.80 \le X \le 0)$.   (g) $P(-1.47 \le X \le 1.47)$.
(d) $P(-0.82 \le X \le -0.76)$.

Solution   (a) By (19), $P(0.25 \le X \le 1.78) = A(1.78) - A(0.25)$

$$= 0.9625 - 0.5987 = 0.3638.$$

(b) By (18), $P(X \le -2.13) = 1 - A(2.13) = 1 - 0.9834 = 0.0166.$

(c) By (19), $P(-1.80 \le X \le 0) = A(0) - A(-1.80)$

$$= 0.5000 - [1 - A(1.80)] \qquad [\text{by (16)}]$$
$$= 0.5000 - 1 + 0.9641 = 0.4641.$$

(d) By (19), $P(-0.82 \le X \le -0.76) = A(-0.76) - A(-0.82)$

$$= [1 - A(0.76)] - [1 - A(0.82)] \quad [\text{by (16)}]$$
$$= 1 - 0.7764 - 1 + 0.7939 = 0.0175.$$

(e) By (20), $P(X \ge 2.51) = 1 - A(2.51) = 1 - 0.9940 = 0.0060.$

(f) By (21), $P(X \ge -3.00) = A(3.00) = 0.9987.$

(g) By (22), $P(-1.47 \le X \le 1.47) = 2A(1.47) - 1 = 2(0.9292) - 1$

$$= 0.8584.$$ ∎

The following theorem summarizes some details regarding the standard normal density function.

THEOREM 4

(The Standard Normal Density Function)
The continuous random variable X with probability density function

$$f(x) = \frac{1}{\sqrt{2\pi}}\, e^{-x^2/2}$$

over the interval $(-\infty, +\infty)$ has

(a) mean $\mu = 0$,   (23)

(b) standard deviation $\sigma = 1$,   (24)

(c) mean $m = 0$.   (25)

Proof   (a) By definition,

$$\mu = \int_{-\infty}^{+\infty} xf(x)\,dx = \int_{-\infty}^{+\infty} x \cdot \frac{1}{\sqrt{2\pi}}\, e^{-x^2/2}\, dx$$

$$= \frac{1}{\sqrt{2\pi}} \int_{-\infty}^{0} xe^{-x^2/2}\, dx + \frac{1}{\sqrt{2\pi}} \int_{0}^{+\infty} xe^{-x^2/2}\, dx. \qquad \text{[Definition 3, Section 7.6]}$$

$$(26)$$

Evaluating the second integral on the right side of Equation (26),

$$\int_{0}^{+\infty} xe^{-x^2/2}\, dx = \lim_{b \to +\infty} \int_{0}^{b} xe^{-x^2/2}\, dx$$

$$= \lim_{b \to +\infty} [-e^{-x^2/2}]_{0}^{b}$$

$$= \lim_{b \to +\infty} \left[\frac{-1}{e^{b^2/2}} - \frac{-1}{e^0} \right]$$

$$= 1. \qquad (27)$$

We leave it to the reader to show (Exercise 37) that

$$\int_{-\infty}^{0} xe^{-x^2/2}\, dx = -1. \qquad (28)$$

Substituting (27) and (28) into (26), we have

$$\mu = \frac{1}{\sqrt{2\pi}}\,(-1) + \frac{1}{\sqrt{2\pi}}\,(1)$$

$$= 0.$$

(b) By definition, Var $(X) = \int_{-\infty}^{+\infty} (x - \mu)^2 f(x)\, dx$. Since $\mu = 0$, we have

$$\text{Var}\,(X) = \int_{-\infty}^{+\infty} x^2 f(x)\, dx = \int_{-\infty}^{+\infty} x^2 \cdot \frac{1}{\sqrt{2\pi}} e^{-x^2/2}\, dx$$

$$= \frac{1}{\sqrt{2\pi}} \int_{-\infty}^{0} x^2 e^{-x^2/2}\, dx + \frac{1}{\sqrt{2\pi}} \int_{0}^{+\infty} x^2 e^{-x^2/2}\, dx. \qquad (29)$$

Using integration by parts, with $u = x$ and $dv = xe^{-x^2/2}\, dx$, we have $du = dx$, $v = -e^{-x^2/2}$, and thus

$$\int x^2 e^{-x^2/2}\, dx = -xe^{-x^2/2} - \int -e^{-x^2/2}\, dx. \qquad (30)$$

Therefore, from (30),

$$\frac{1}{\sqrt{2\pi}} \int_{0}^{+\infty} x^2 e^{-x^2/2}\, dx = \frac{1}{\sqrt{2\pi}} \lim_{b \to +\infty} \left[-xe^{-x^2/2} + \int e^{-x^2/2}\, dx \right]_0^b$$

$$= \frac{1}{\sqrt{2\pi}} \left\{ \lim_{b \to +\infty} [-b/e^{b^2/2} - 0e^0] + \int_{0}^{+\infty} e^{-x^2/2}\, dx \right\}.$$

Now, $\displaystyle \lim_{b \to +\infty} \frac{-b}{e^{b^2/2}} = 0$ from Theorem 2, and

$$\frac{1}{\sqrt{2\pi}} \int_{0}^{+\infty} e^{-x^2/2}\, dx = \frac{1}{2}. \qquad \text{(Why?)} \qquad (31)$$

We leave the reader to show similarly (Exercise 38) that

$$\frac{1}{\sqrt{2\pi}} \int_{-\infty}^{0} e^{-x^2/2}\, dx = \frac{1}{2}. \qquad (32)$$

Substituting (31) and (32) into (29), we have

$$\text{Var}\,(X) = \tfrac{1}{2} + \tfrac{1}{2} = 1.$$

Therefore, the standard deviation is

$$\sigma = \sqrt{\text{Var}\,(X)} = 1.$$

(c) By definition, the median is the number m which satisfies $P(X \le m) = \tfrac{1}{2}$; that is,

$$A(m) = \tfrac{1}{2}.$$

By symmetry of the graph of $f(x)$ seen in Figure 3, or by Table V of Appendix B, this value is

$$m = 0. \qquad \blacksquare$$

The standard normal density function is often called the **Gaussian density function,** in honor of Carl Friedrich Gauss*, universally recognized as one of the greatest mathematicians of all times. It is perhaps the most important of all density functions, finding widespread use in applications of probability and statistics in both the natural sciences and the social sciences. The reader is likely well-acquainted with "grading on the curve", a procedure often used in scoring standardized examinations given to large numbers of students. The "curve" is understood to be a "bell-shaped" normal curve. The following example will show how the Gaussian density function arises in the theory of error.

**EXAMPLE 7
(Quality
Control)**

A precision machine produces parts for a NASA rocket that must weigh 550 grams. Due to normal variations in manufacturing, however, the actual weight varies somewhat. If the deviation from 550 grams is a random variable X with standard normal probability density function, what is the probability that a randomly chosen part produced by this machine will be within 2 grams of the specified 550 grams?

Solution

We want to find $P(548 \leq \text{weight} \leq 552)$, which is the same as $P(-2 \leq X \leq 2)$. By Equation (22),

$$
\begin{aligned}
P(-2 \leq X \leq 2) &= 2A(2) - 1 \\
&= 2(0.9772) - 1 \\
&= 0.9544.
\end{aligned}
$$

Thus, 95.44% of the parts produced by this machine will be within 2 grams of the specified 550 grams.   ∎

|||||||||||||| General Normal Density Functions

Given any real number μ and any positive real number σ, we define the function

$$
N(x) = \frac{1}{\sigma\sqrt{2\pi}} \, e^{-\frac{1}{2}\left(\frac{x-\mu}{\sigma}\right)^2}
\tag{33}
$$

to be the **normal probability density function with mean μ and standard deviation σ.**

To see that $N(x)$ is a probability density function, we first observe that $N(x) \geq 0$ and then set about proving that $\int_{-\infty}^{+\infty} N(x)\, dx = 1$. We do this by proving that

---

* Carl Friedrich Gauss (1777–1855) was born of poor, uneducated parents in Braunschwieg, Germany, where he showed genius in mathematics even as a very young child. At the age of 18 he invented the "method of least squares" and solved a problem in geometry that had been unsolved for more than 2000 years. He made significant contributions in physics and astronomy, but in mathematics his astonishing creativity produced brilliant discoveries in virtually every area. The impact of his genius is still being felt in various areas of contemporary mathematical research.

both

$$\int_{-\infty}^{\mu} N(x)\, dx = \tfrac{1}{2} \qquad \text{and} \qquad \int_{\mu}^{+\infty} N(x)\, dx = \tfrac{1}{2}. \tag{34}$$

We prove only the first equality in (34) and leave the second as an exercise (Exercise 39).

To evaluate the integral

$$\int_{-\infty}^{\mu} N(x)\, dx = \lim_{a \to -\infty} \int_{a}^{\mu} \frac{1}{\sigma \sqrt{2\pi}}\, e^{-\frac{1}{2}\left(\frac{x-\mu}{\sigma}\right)^2}\, dx$$

we make the following helpful substitution:

$$z = \frac{x - \mu}{\sigma}. \tag{35}$$

From (35), $dz = \dfrac{1}{\sigma}\, dx$ or $dx = \sigma\, dz$. Thus,

$$\int_{-\infty}^{\mu} N(x)\, dx = \lim_{a \to -\infty} \left[\int \frac{1}{\sigma \sqrt{2\pi}}\, e^{-z^2/2} \sigma\, dz \right]_{x=a}^{x=\mu} \tag{36}$$

We make the further substitution

$$c = \frac{a - \mu}{\sigma}. \tag{37}$$

From (35) and (37),

$$\text{when} \quad x = a, \qquad z = \frac{a-\mu}{\sigma} = c, \qquad \text{and}$$

$$\text{when} \quad x = \mu, \qquad z = 0.$$

Moreover, $c \to -\infty$ as $a \to -\infty$, and vice versa. Substituting these results into (36), we have

$$\int_{-\infty}^{\mu} N(x)\, dx = \lim_{c \to -\infty} \left[\int \frac{1}{\sqrt{2\pi}}\, e^{-z^2/2}\, dz \right]_{z=c}^{z=0}$$

$$= \int_{-\infty}^{0} \frac{1}{\sqrt{2\pi}}\, e^{-z^2/2}\, dz$$

$$= A(0) \qquad \text{(by definition)}$$

$$= \tfrac{1}{2} \qquad \text{by Table V, Appendix B.}$$

Therefore, the function $N(x)$ defined by (33) is a probability density function.

We leave the reader to prove (in Exercise 40) that $N(x)$ has mean μ and standard deviation σ. ∎

We now take a look at the geometry of the **normal curve** $y = N(x)$. Using the methods of Sections 4.5 and 4.6, we can show that the function $y = N(x)$ has only one relative extreme value, a relative maximum at $x = \mu$ (Exercise 41). We can also show that the curve $y = N(x)$ has two inflection points, at $x = \mu + \sigma$ and at $x = \mu - \sigma$ (Exercise 42). Figure 5 shows the graph of the normal curve $y = N(x)$. Notice that the curve has the vertical line $x = \mu$ as an axis of symmetry.

Figure 5

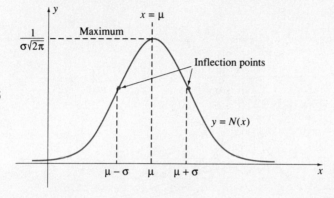

EXAMPLE 8   Graph three normal curves with mean $\mu = 3$: one with $\sigma = 1$, one with $\sigma = 2$, and one with $\sigma = \frac{1}{2}$.

Solution   We construct a table of values for each of the three probability density functions. The entries corresponding to inflection points are indicated with an asterisk.

TABLE 1 $\mu = 3$, $\qquad \sigma = 1$, $\qquad y = \dfrac{1}{\sqrt{2\pi}}\, e^{-\frac{1}{2}(x-3)^2}$

		*		
x	3	2 or 4	1 or 5	0 or 6
y	0.4	0.24	0.05	0.004

TABLE 2 $\mu = 3$, $\qquad \sigma = 2$, $\qquad y = \dfrac{1}{2\sqrt{2\pi}}\, e^{-\frac{1}{8}(x-3)^2}$

			*		
x	3	2 or 4	1 or 5	0 or 6	-1 or 7
y	0.2	0.18	0.12	0.06	0.03

TABLE 3 $\mu = 3$, $\quad \sigma = \dfrac{1}{2}$, $\quad y = \dfrac{2}{\sqrt{2\pi}}e^{-2(x-3)^2}$

*

x	3	2.5 or 3.5	2 or 4	1 or 5
y	0.8	0.48	0.11	0.0003

The resulting graphs are shown in Figure 6. The inflection points are indicated by heavy dots.

Figure 6

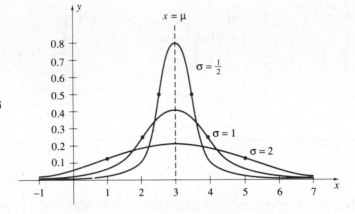

Observe that the standard deviation σ determines the *shape* of the normal curve. For values of σ less than 1, the bell curve is thinner, steeper, and reaches higher than the standard normal density function; for values of σ greater than 1, the bell curve is wider, flatter, and lower than the standard normal density function.

One interesting geometric feature is shared in common by all normal probability density functions $y = N(x)$. We can easily prove (Exercise 43) that

(a) 68.26% of the total area under a normal curve falls within one standard deviation of the mean. That is,

$$P(\mu - \sigma \le X \le \mu + \sigma) = 0.6826. \tag{38}$$

(b) 95.24% of the total area under a normal curve falls within two standard deviations of the mean. That is,

$$P(\mu - 2\sigma \le X \le \mu + 2\sigma) = 0.9524. \tag{39}$$

(c) 99.74% of the total area under a normal curve falls within three standard deviations of the mean. That is,

$$P(\mu - 3\sigma \le X \le \mu + 3\sigma) = 0.9974. \tag{40}$$

Figure 7 shows these areas.

Figure 7

68.26% of area

95.44% of area

99.74% of area

||||||||||||| Normally Distributed Random Variables

If a random variable X has a normal probability density function, then we shall say that X is a **normally distributed** random variable. As we have seen, it is important to be able to calculate probabilities such as $P(a \leq X \leq b)$. To facilitate these calculations we make a change of variables, converting to the *standard* normal random variable. This enables us to make use of the tabulated values of

$$A(x) = \int_{-\infty}^{x} \frac{1}{\sqrt{2\pi}} e^{-t^2/2} \, dt$$ given in Table V, Appendix B. The following theorem

and example show how this is done.

THEOREM 5

If X is a normally distributed random variable with mean μ and standard deviation σ, and if we define

$$Z = \frac{X - \mu}{\sigma}, \qquad \text{[Recall Equation (35).]}$$

then Z is the standard normal random variable, and

(a) $P(X \leq b) = P\left(Z \leq \frac{b - \mu}{\sigma}\right) = A\left(\frac{b - \mu}{\sigma}\right)$

(b) $P(X \geq a) = P\left(Z \geq \frac{a - \mu}{\sigma}\right) = 1 - A\left(\frac{a - \mu}{\sigma}\right)$

(c) $P(a \leq X \leq b) = A\left(\frac{b - \mu}{\sigma}\right) - A\left(\frac{a - \mu}{\sigma}\right),$

where $A(z) = \int_{-\infty}^{z} \frac{1}{\sqrt{2\pi}} e^{-t^2/2} \, dt$, whose values can be found using Table V, Appendix B.

We shall not write out a proof of this theorem. However, results (a)–(c) can be easily understood by observing, for example, that

$$P(X \le b) = P(X - \mu \le b - \mu)$$

$$= P\!\left(\frac{X - \mu}{\sigma} \le \frac{b - \mu}{\sigma}\right) \qquad \text{(since } \sigma > 0\text{)}$$

$$= P\!\left(Z \le \frac{b - \mu}{\sigma}\right)$$

$$= A\!\left(\frac{b - \mu}{\sigma}\right).$$

EXAMPLE 9   Suppose that X is a normally distributed random variable with mean $\mu = 15$ and standard deviation $\sigma = 4$. Find
(a) $P(X \le 25)$.
(b) $P(16 \le X \le 20)$.
(c) $P(X \ge 19)$.
(d) $P(X \le 12)$.
(e) $P(X \ge 13)$.
(f) $P(14 \le X \le 16)$.

Solution   Using Theorem 5, with $Z = \dfrac{X - 15}{4}$, and Equations (18)–(22), we have

(a) $P(X \le 25) = P\!\left(Z \le \dfrac{25 - 15}{4}\right) = P(Z \le 2.5)$

$\qquad = A(2.5) = 0.9938.$

(b) $P(16 \le X \le 20) = P\!\left(\dfrac{16 - 15}{4} \le Z \le \dfrac{20 - 15}{4}\right) = P\!\left(\dfrac{1}{4} \le Z \le \dfrac{5}{4}\right)$

$\qquad\qquad = A(1.25) - A(0.25) = 0.8944 - 0.5987$

$\qquad\qquad = 0.2957.$

(c) $P(X \ge 19) = P\!\left(Z \ge \dfrac{19 - 15}{4}\right) = P(Z \ge 1)$

$\qquad = 1 - A(1) \qquad \text{by Equation (20)}$

$\qquad = 1 - 0.8413 = 0.1587.$

(d) $P(X \le 12) = P\!\left(Z \le \dfrac{12 - 15}{4}\right) = P(Z \le -0.75)$

$\qquad = 1 - A(0.75) \qquad \text{by Equation (18)}$

$\qquad = 1 - 0.7734 = 0.2266.$

(e) $P(X \ge 13) = P\!\left(Z \ge \dfrac{13 - 15}{4}\right) = P(Z \ge -0.5)$

$\qquad = A(0.5) \qquad \text{by Equation (21)}$

$\qquad = 0.6915.$

(f) $P(14 \leq X \leq 16) = P\left(\dfrac{14 - 15}{4} \leq Z \leq \dfrac{16 - 15}{4}\right) = P\left(-\dfrac{1}{4} \leq Z \leq \dfrac{1}{4}\right)$

$\qquad\qquad\qquad = 2A(0.25) - 1 \qquad$ by Equation (22)

$\qquad\qquad\qquad = 2(0.5987) = -1$

$\qquad\qquad\qquad = 0.1974.$ ∎

EXAMPLE 10   In an automated soft drink bottling plant, the amount of soft drink that a machine dispenses into a bottle is a normally distributed random variable with mean $\mu = 12$ oz and standard deviation $\sigma = 0.2$ oz. Find the probability that a bottle selected at random contains
(a) less than 12 ounces
(b) less than 11.9 ounces
(c) more than 11.33 ounces
after passing through this machine. Assume, of course, that the bottle has a capacity somewhat greater than 12 ounces.

Solution   Let X denote the normally distributed random variable with mean $\mu = 12$ and standard deviation $\sigma = 0.2$. We make the substitution

$$Z = \frac{X - \mu}{\sigma} = \frac{X - 12}{0.2}.$$

Then Z is the standard normal random variable. By Theorem 5, we have

(a) $P(X < 12) = P\left(Z < \dfrac{12 - 12}{0.2}\right) = P(Z < 0)$

$\qquad\qquad = 0.5000. \qquad$ by Table V, Appendix B.

(b) $P(X < 11.9) = P\left(Z < \dfrac{11.9 - 12}{0.2}\right) = P(Z < -0.5)$

$\qquad\qquad = 1 - A(0.5) \qquad$ by Equation (18)

$\qquad\qquad = 1 - 0.6915 = 0.3085.$

(c) $P(X > 11.33) = P\left(Z > \dfrac{11.33 - 12}{0.2}\right) = P(Z > -3.35)$

$\qquad\qquad = A(3.35) \qquad$ by Equation (21)

$\qquad\qquad = 0.9996.$ ∎

11.4 EXERCISE SET ||||||||||||||||||||||||||||||||||||||||||||||||||||||||||||||||||||||||||||||||||||||||||||

In Exercises 1–4, find the formula for the uniform probability density function over the given interval, and find its mean, median, and standard deviation.

1. $[7, 15]$ 2. $[-3, 20]$

3. $[-10, 1]$ 4. $[-17, -3]$

5. A random variable has a uniform probability density function over the interval $[2, d]$ and has mean $\frac{18}{5}$. Find the value of d and the formula for the density function.

6. A random variable with uniform probability density function has mean $\frac{7}{2}$ and standard deviation $\frac{9}{\sqrt{12}}$. Find the formula for

the density function and the interval over which it is defined.

In Exercises 7 and 8, find the mean, median, and standard deviation of the random variable which has the given probability density function over the indicated interval.

7. $f(x) = 5e^{-5x}$, $[0, +\infty)$.

8. $f(x) = \frac{1}{4}e^{-x/4}$, $[0, +\infty)$.

9. For the random variable X whose probability density function is given in Exercise 7, find
   (a) $P(X \leq 1)$.
   (b) $P(X \geq 0.2)$.
   (c) $P(0.1 \leq X \leq 0.3)$.

10. For the random variable whose probability density function is given in Exercise 8, find
    (a) $P(X \leq 5)$.
    (b) $P(X \geq 6)$.
    (c) $P(2 \leq x \leq 7)$.

11. Use Table V in Appendix B to evaluate the following integrals:

    (a) $\int_{-\infty}^{1.58} \frac{1}{\sqrt{2\pi}} e^{-t^2/2} \, dt$.

    (b) $\int_{-\infty}^{-2.03} \frac{1}{\sqrt{2\pi}} e^{-t^2/2} \, dt$.

    (c) $\int_{1}^{+\infty} e^{-t^2/2} \, dt$.

12. Use Table V of Appendix B to evaluate the following integrals:

    (a) $\int_{-1}^{+1} \frac{1}{\sqrt{2\pi}} e^{-t^2/2} \, dt$.

    (b) $\int_{-\infty}^{1.50} e^{-t^2/2} \, dt$.

    (c) $\int_{2}^{+\infty} e^{-t^2/2} \, dt$.

13. Suppose that X is a continuous random variable with standard normal density function. Find each of the following probabilities:
    (a) $P(X \leq 2.17)$.
    (b) $P(0.34 \leq X \leq 1.55)$.

(c) $P(X \geq 2.26)$.
(d) $P(X \leq -0.93)$.

14. Suppose that X is a continuous random variable with standard normal density function. Find each of the following probabilities:
    (a) $P(X \geq 1.41)$.
    (b) $P(X \leq -1.08)$.
    (c) $P(1.02 \leq X \leq 1.69)$.
    (d) $P(-1.73 \leq X \leq 1.73)$.

15. Suppose X is a normally distributed random variable with mean $\mu = 10$ and standard deviation $\sigma = 2$. Find
    (a) $P(X \leq 13)$.
    (b) $P(X \geq 10)$.
    (c) $P(10 \leq X \leq 12)$.
    (d) $P(9 \leq X \leq 11)$.
    (e) $P(X \leq 9.5)$.
    (f) $P(X \geq 8)$.

16. Suppose X is a normally distributed random variable with mean $\mu = 45$ and standard deviation $\sigma = 4$. Find
    (a) $P(X \leq 48)$.
    (b) $P(X \leq 40)$.
    (c) $P(40 \leq X \leq 50)$.
    (d) $P(X \geq 46)$.
    (e) $P(49 \leq X \leq 51)$.
    (f) $P(X \geq 42)$.

In each of the following "word problems" (Exercise 17–30) specify an appropriate random variable and its probability density function, and express the solution as an integral before obtaining the numerical answer.

17. A real number from the interval $[20, 70]$ is chosen at random. What is the probability that it is less than 30? Between 30 and 50? Greater than 40?

18. A clock stops at some random time. What is the probability that the minute hand and hour hand form an angle between $45°$ and $90°$?

19. Subway trains arrive at a terminal every 10 minutes. What is the probability that a person arriving randomly will have to wait

less than 2 minutes for a train? Between 2 and 3 minutes? More than 7 minutes?

20. A point is chosen randomly on a line segment of length L, dividing it into two pieces. What is the probability that one of the pieces is at least twice as long as the other? (*Hint:* choose a point x in the interval $(0, L)$ and consider the ratio $X/(L - X)$.)

21. The lifespan of a well-known brand of light bulb has an expected value of 750 hours. Assuming that the lifespan is a random variable with exponential density function, find the probability that a randomly chosen light bulb of this brand will burn out
    (a) before 300 hours.
    (b) before 500 hours.
    (c) sometime after 700 hours but before 900 hours.
    (d) sometime after 1000 hours.

22. If 1000 of the light bulbs of Exercise 21 are used in an advertising display, after how many hours would you predict that half of them would be burned out?

23. In a certain factory, workers' accidents requiring medical attention occur, on the average, every 20 hours of factory operation time. The length of time between accidents is a random variable with exponential density function. If such an accident has just occurred, what is the probability that the next accident requiring medical attention will occur
    (a) in the next 4 hours of factory operation?
    (b) after 4 hours, but before 8 hours?
    (c) sometime after 10 hours?

24. A manufacturing machine produces an average of 4 defective items per hour. If the length of time (in minutes) between successive defective items is a random variable with exponential density function, and if a defective item has just been produced, what is the probability that

(a) a defective item will be produced in the next 5 minutes?
(b) no defective items will be produced in the next 10 minutes?
(c) no defective items will be produced in the next hour?

25. The temperature in a classroom at time t is a normally distributed random variable with mean $\mu = 70°F$ and standard deviation $\sigma = 2°F$. What is the probability that at a randomly selected time the temperature is at least 68°F? Between 68°F and 72°F? Above 76°F?

26. Suppose that the daily average temperature in New York City in January is a normally distributed random variable with mean 31°F and standard deviation 10°F. What is the probability that on a randomly chosen day in January the average temperature in New York City is less than 20°F? Between 30°F and 40°F? Above 40°F?

27. The electrical voltage at time t in a home is a normally distributed random variable with mean 115 volts and standard deviation 3 volts. At a random point in time, what is the probability that the voltage in the house is between 110 and 120 volts? Greater than 120 volts?

28. As part of a physical fitness program, an entire grade of elementary students in a city school district is weighed. The student's weights are found to be normally distributed, with mean 60 pounds and standard deviation 6 pounds. What percentage of these students weigh
    (a) under 50 pounds?
    (b) between 55 and 65 pounds?
    (c) between 65 and 75 pounds?
    (d) over 75 pounds?

29. The mathematics department at a university gives a common final examination to all 1200 students taking "Applied Calculus I." The resulting scores are normally distributed, with a mean score of 72% and a

standard deviation of 8%. How many of the students score
(a) under 60%?
(b) between 60% and 70%?
(c) between 70% and 80%?
(d) between 80% and 90%?
(e) over 90%?
(*Hint:* first find the fraction of the scores that fall in each category.)

30. On the exam described in Exercise 29, the university insists that 10% of the students taking the exam must get A's, 20% must get B's, 40% must get C's, 20% must get D's, and 10% must get F's. Determine the cut-off scores for each grade.

31. Let $f(x) = e^{-x^2/2}$. Use the methods of Sections 4.5 and 4.6 to show that
(a) the only critical number of f is $x = 0$.
(b) f has a relative maximum at $x = 0$.
(c) the graph of f is symmetric with respect to the y-axis.

32. Let $f(x) = e^{-x^2/2}$. Use the methods of Sections 4.5 and 4.6 to show that the graph of f has inflection points at $x = \pm 1$.

33. Prove Equation (18).

34. Prove Equation (20).

35. Prove Equation (21).

36. Prove Equation (22).

37. Prove Equation (28).

38. Prove Equation (32).

39. Prove the second equality in Equation (34).

40. Prove that the normal density function $N(x)$ defined in Equation (33) has mean μ and standard deviation σ.

41. Prove that the graph of the normal density function defined in Equation (33) has only one relative extreme value, a relative maximum at $x = \mu$.

42. Prove that the graph of the normal density function, Equation (33), has two inflection points, at $x = \mu + \sigma$ and at $x = \mu - \sigma$.

43. Let X be a normal random variable with mean μ and standard deviation σ. Prove Equations (38), (39), and (40).

KEY IDEAS FOR REVIEW ||||||||||||||||||||||||||||||||||||||||||||||||||||||||||||||||||||||||||

- A random variable assigns a real number to each outcome in a sample space of an experiment of chance.
- For a discrete random variable, the sample space consists of a finite set of outcomes t_1, $t_2, \ldots, t_n$. A probability is assigned to each one of the outcomes in such a way that
$$0 \le P(t_i) \le 1 \text{ for each } i,$$
and
$$P(t_1) + P(t_2) + \cdots + P(t_n) = 1.$$
- A histogram can be used to represent a discrete probability model. It shows probabilities as areas.
- A probability density function over an interval $[c, d]$ is a continuous function $f(x)$ such that $f(x) \ge 0$ for all x in $[c, d]$, and $\int_c^d f(x)\, dx = 1$. Moreover, the interval can be infinite, in which case an improper integral is used.

- For a continuous random variable with probability density function $f(x)$,
$$P(a \le X \le b) = \int_a^b f(x)\, dx.$$
- For a continuous random variable X,
$$P(X \ge a) = 1 - P(X \le a).$$
- For a discrete random variable X with sample space $\{x_1, x_2, \ldots, x_n\}$, the expected value, or mean, of X is
$$E(X) = x_1 P(x_1) + x_2 P(x_2) + \cdots + x_n P(x_n),$$
and the variance of X is
$$V(X) = (x_1 - \mu)^2 P(x_1) + (x_2 - \mu)^2 P(x_2)$$
$$+ \cdots + (x_n - \mu)^2 P(x_n)$$
$$= [x_1^2 P(x_1) + x_2^2 P(x_2)$$
$$+ \cdots + x_n^2 P(x_n)] - \mu^2.$$
- For a continuous random variable X with probability density function $f(x)$ over $[c, d]$,

the expected value, or mean, of X is

$$E(X) = \int_c^d xf(x)\,dx,$$

and the variance of X is

$$Var\,(X) = \int_c^d (x - \mu)^2 f(x)\,dx$$

$$= \left(\int_c^d x^2 f(x)\,dx \right) - \mu^2.$$

- For any random variable X, the standard deviation is $\sigma = \sqrt{Var\,(X)}$.

- The median of a continuous random variable X is the number m such that $P(X \le m) = \frac{1}{2}$.

- The uniform probability density function on $[c, d]$ is

$$f(x) = \frac{1}{d - c}.$$

It has mean and median $\mu = m = \dfrac{c + d}{2}$ and

standard deviation $\sigma = \dfrac{d - c}{\sqrt{12}}$.

- The exponential probability density function on $[0, +\infty)$ is

$$f(x) = ke^{-kx}$$

It has mean $\mu = \dfrac{1}{k}$, standard deviation $\sigma = \dfrac{1}{k}$, and median $m = \dfrac{\ln 2}{k}$.

- The standard normal probability density function on $(-\infty, +\infty)$ is

$$f(x) = \frac{1}{\sqrt{2\pi}}\,e^{-x^2/2}.$$

It has mean and median $\mu = m = 0$, and standard deviation $\sigma = 1$.

- The general normal probability density function on $(-\infty, +\infty)$ with mean (and median) μ and standard deviation σ is

$$N(x) = \frac{1}{\sigma\sqrt{2\pi}}\,e^{-\frac{1}{2}\left(\frac{x - \mu}{\sigma} \right)^2}$$

REVIEW EXERCISES ||||||||||||||||||||||||||||||||||||||||||||||||||||||||||||||||||||||||||||||||||||||

1. One letter is chosen at random from the word "ILLINOIS." What is the probability that it is
   (a) the letter "I"?
   (b) the letter "L"?
   (c) the letter "S"?
   (d) a consonant?

2. For the experiment described in Exercise 1, specify an appropriate random variable, tabulate the probabilities of its outcomes, and draw a histogram.

3. A box contains 14 balls: 5 red, 3 green, 4 black, and 2 white. Let the sample space be the 4 colors: R, G, B, W. For the experiment of randomly selecting 1 ball from the box, introduce an appropriate random variable, tabulate the probabilities of all its elementary outcomes, and draw a histogram.

4. Let the random variable X denote the sum of the numbers that appear on the top

faces when a pair of dice is tossed. Find
   (a) $P(X = 8)$.   (c) $P(X \le 5)$.
   (b) $P(X < 2)$.   (d) $P(X \le 13)$.

5. From the following histogram of a random variable X, calculate the following probabilities using definite integrals:
   (a) $P(X = 2)$.   (c) $P(2 \le X \le 4)$
   (b) $P(X \le 3)$.   (d) $P(X \ge 4)$.

6. From the histogram of Exercise 5,
   (a) construct the probability table for the random variable X.

(b) find $E(X)$.

(c) find Var (X) and σ.

7. Suppose that $f(x) = k(10x - 16 - x^2)$ is the probability density function for some random variable with sample space $[2, 8]$. Find each of the following:

(a) the value of k.   (c) $P(X = 4)$.

(b) $P(X \leq 4)$.   (d) $P(5 \leq X \leq 6)$.

8. For the random variable of Exercise 7, find

(a) the corresponding distribution function.

(b) $E(X)$.

(c) Var (X) and σ.

9. The lifetime t of a light bulb is a random variable with probability density function

$$f(t) = \begin{cases} \dfrac{2000}{9t^2} & \text{if } 200 \leq t \leq 2000 \\ 0 & \text{otherwise,} \end{cases}$$

where t is in hours.

(a) What is the probability that a bulb will last at least 600 hours?

(b) What is the probability that a bulb will last at most 1000 hours?

(c) What is the probability that a bulb will last at least 800 hours, but no more than 1600 hours?

10. For the random variable X described in Exercise 9,

(a) Show that indeed, $f(t)$ is a probability density function on the interval $[200, 2000]$.

(b) Find the corresponding distribution function.

(c) Find $E(X)$.

(d) Find σ.

11. Calculate the mean and standard deviation of the following set of exam scores: 55, 60, 67, 70, 70, 74, 75, 77, 85, 89, 92, 98.

In Exercises 12 and 13, find the value of k for which the given function $f(x)$ is a probability density function over the specified interval. Then find the expected value and

standard deviation of the associated random variable X.

12. $f(x) = k(16 - 4x)$, $[0, 4]$

13. $f(x) = \dfrac{k}{x^5}$, $[1, +\infty)$

14. Find the median of the random variable whose probability density function is the function $f(x)$ given in

(a) Exercise 12.   (b) Exercise 13.

15. The useful lifespan (in hundreds of hours of use) of a certain type of battery is a random variable with probability density function

$$f(x) = \frac{1}{(1 + x^2)^{3/2}}$$

on the interval $[0, +\infty)$. Find the expected lifespan of these batteries.

16. A damaged nuclear power plant is releasing iodine-131 into the atmosphere. The amount x (in milligrams) released daily has the probability density function

$$f(x) = \begin{cases} 2xe^{-x^2} & \text{if } x \geq 0 \\ 0 & \text{otherwise.} \end{cases}$$

Find the probability that on a given day, the plant will release at least 10 milligrams of iodine-131 into the atmosphere.

17. A radio station broadcasts the correct time every half-hour. What is the probability that a person who tunes in the station for the time will have to wait

(a) less than 12 minutes?

(b) more than 20 minutes?

(c) between 15 and 20 minutes?

18. A pointer is spun inside a circular dial which is calibrated with real numbers in the interval $[0, 100)$. Let the random variable X denote the number at which the pointer stops spinning. Find

(a) the probability density function for X.

(b) the associated distribution function.

(c) $P(30 \leq X \leq 45)$.

(d) $P(X \geq 80)$.

(e) $P(X = 50)$.

19. A department store finds that the time t (in months) that customers take to pay a bill is a random variable with probability density function

$$f(t) = \begin{cases} 3e^{-3t} & \text{if } t \geq 0 \\ 0 & \text{otherwise.} \end{cases}$$

(a) Show that $f(t)$ is a probability density function on $[0, +\infty)$.
(b) Find the probability that a customer chosen at random will pay his or her bill within 5 months.
(c) Find the probability that a customer will take at least 4 months to pay a bill.

20. A photocopy machine in a university library is in constant demand, 24 hours a day. The random variable X representing the length of time (in hours) between machine breakdowns has an exponential probability density function with an expected value of 8 hours. If the last breakdown was 2 hours ago, what is the probability that
(a) the machine will break down sometime in the next hour?
(b) the machine will work without a breakdown for at least the next three hours?

21. Use Table V in Appendix B to evaluate the following integrals:

(a) $\int_{-0.5}^{1.0} \frac{1}{\sqrt{2\pi}} e^{-t^2/2} \, dt.$

(b) $\int_{-\infty}^{2.24} e^{-t^2/2} \, dt.$

(c) $\int_{0.5}^{+\infty} e^{-t^2/2} \, dt.$

22. Assuming that the random variable X has the standard normal density function, find each of the following probabilities:
(a) $P(X \geq 1.32)$.
(b) $P(X \leq 1.09)$.
(c) $P(-0.83 \leq X \leq 0)$.

23. Assuming that the normally distributed random variable X has mean 320 and standard deviation 25, find
(a) $P(X \geq 350)$.
(b) $P(X \leq 340)$.
(c) $P(X \leq 300)$.
(d) $P(290 \leq X \leq 350)$.
(e) $P(330 \leq X \leq 340)$.
(f) $P(300 \leq X \leq 315)$.

24. Suppose that the height of female college students is a normally distributed random variable with mean 63 inches and standard deviation 4 inches. What percent of all female college students are
(a) less than 6 feet tall?
(b) less than 5 feet tall?
(c) between 5 feet and 5 feet 6 inches tall?

25. A machine produces ball bearings whose diameter is a normally distributed random variable X with mean $\mu = 25$ mm and standard deviation $\sigma = 0.03$ mm. A ball bearing is considered defective if $X < 24.95$ mm or $X > 25.05$ mm. Find the probability that a randomly selected ball bearing produced by this machine is defective. What percentage of the ball bearings produced by this machine are defective? (*Hint:* first, find the probability that a ball bearing is *not* defective.)

CHAPTER TEST ||||||||||||||||||||||||||||||||||||||||||||||||||||||||||||||||||||||||||||||||||||||

1. In the midst of the frustration of grading final exams for his class of 20 students, a professor decides to burn all his students' final exam papers and randomly assign 3 A's, 4 B's, 8 C's, 3 D's, and 2 F's as final grades. If you were a student in this class, what would be the probability that your final

grade is
   (a) an A?
   (b) no lower than a C?

2. For the final exam described in Exercise 1, let the random variable X assign numerical values to the letter grades in the traditional way: A = 4, B = 3, C = 2, D = 1, and F = 0.
   (a) Construct a probability table for the random variable X.
   (b) Draw the corresponding histogram.
   (c) Find $E(X)$.
   (d) Find σ.

3. Find the value of k for which the function

$$f(x) = \frac{k}{\sqrt{x}}$$

   is a probability density function on the interval $[1, 9]$, and find the corresponding distribution function.

4. Let $f(x) = \dfrac{2x}{(1 + x^2)^2}$.

   (a) Show that $f(x)$ is a probability density function on the interval $[0, +\infty)$.
   (b) Find $P(X \le 2)$.
   (c) Find $P(2 \le X \le 3)$.
   (d) Find $P(X > 5)$.

5. A random variable X has probability density function

$$f(x) = \tfrac{3}{4}(1 - x^2)$$

   on the interval $[-1, 1]$. Find its expected value and standard deviation.

6. Let the random variable X denote a randomly chosen number in the interval $[10, 50]$.
   (a) Write the probability density function for X.
   (b) Find $P(15 \le X \le 25)$.
   (c) Find $P(X \le 30)$.
   (d) Find $P(X \ge 45)$.

7. The distance between cars in a single lane of traffic is a random variable X with exponential density function having a mean of 100 feet. If you are in this lane of traffic, what is the probability that the distance between you and the car ahead of you is
   (a) less than 50 feet?
   (b) between 60 and 80 feet?
   (c) more than 100 feet?

8. If X is a normal random variable with mean 50 and standard deviation 8, find
   (a) $P(X \le 60)$.   (d) $P(X \ge 70)$.
   (b) $P(X \ge 40)$.   (e) $P(X \le 36)$.
   (c) $P(40 \le X \le 50)$.   (f) $P(40 \le X \le 60)$.

TAYLOR POLY-NOMIALS AND INFINITE SERIES

12.1 |||||| SEQUENCES OF REAL NUMBERS

The subject of calculus is distinguished by its reliance upon the concept of limit. In Chapter 2 we introduced the notion of

$$\lim_{x \to a} f(x) = L \tag{1}$$

and described some of its properties. We defined the derivative as a special limit:

$$f'(x_0) = \lim_{h \to 0} \frac{f(x_0 + h) - f(x_0)}{h} \tag{2}$$

$$= \lim_{x \to x_0} \frac{f(x) - f(x_0)}{x - x_0}. \tag{3}$$

In Chapter 7 we used Riemann's approach to show that the definite integral of a function $f(x)$ over an interval $[a, b]$ is another kind of limit:

$$\int_a^b f(x)\, dx = \lim_{n \to +\infty} R_n, \tag{4}$$

where R_n denotes the Riemann sum

$$R_n = \Delta x\,[f(x_1) + f(x_2) + \cdots + f(x_n)] \tag{5}$$

defined in Section 7.4.

Many limiting processes used in calculus involve the notion of infinity. In (2) we have h becoming *infinitely small* ($h \to 0$), while in (4) we have n becoming *infinitely large* ($n \to +\infty$). Both concepts, limits and infinity, were used together in our study of asymptotes to curves, in Sections 2.2 and 4.6.

The present chapter will require another kind of limit: the limit of a sequence. We have already seen examples of this type of limit, although we have not previously called attention to them.

By definition, a **sequence** (of real numbers) is an infinite succession* of real numbers:

$$a_1, a_2, a_3, a_4, a_5, \cdots, a_n, a_{n+1}, \cdots. \tag{6}$$

We call a_1 the **first term,** a_2 the **second term,** $\cdots$, a_n, the **nth term,** and so on. We shall denote the sequence (6) by the abbreviation $\{a_n\}$. Notice that this abbreviation merely encloses the formula for the nth term within braces (always braces, never parentheses or square brackets). The following example shows how this notation is used in practice.

EXAMPLE 1   Write out the first eight terms of the sequence $\{3(n - 2) + 7\}$.

Solution
$$\begin{aligned}
a_1 &= 3(1 - 2) + 7 = -3 + 7 = 4 \\
a_2 &= 3(2 - 2) + 7 = 0 + 7 = 7 \\
a_3 &= 3(3 - 2) + 7 = 3 + 7 = 10 \\
a_4 &= 3(4 - 2) + 7 = 6 + 7 = 13 \\
a_5 &= 3(5 - 2) + 7 = 9 + 7 = 16 \\
a_6 &= 3(6 - 2) + 7 = 12 + 7 = 19 \\
a_7 &= 3(7 - 2) + 7 = 15 + 7 = 22 \\
a_8 &= 3(8 - 2) + 7 = 18 + 7 = 25.
\end{aligned}$$

Thus, the first eight terms of this sequence are

$$4, 7, 10, 13, 16, 19, 22, 25. \qquad \blacksquare$$

EXAMPLE 2   Express each of the following sequences in the form $\{a_n\}$.
(a) $3, 5, 7, 9, 11, 13, 15, \cdots$
(b) $1, -\frac{1}{2}, \frac{1}{3}, -\frac{1}{4}, \frac{1}{5}, -\frac{1}{6}, \cdots$
(c) $\frac{1}{2}, \frac{3}{4}, \frac{7}{8}, \frac{15}{16}, \frac{31}{32}, \cdots$

Solution   (a) Each term of this sequence is 2 more than the previous term, and the first term is 2 more than 1. Thus, for each natural number n,

$$a_n = 1 + 2n.$$

Therefore, the sequence is $\{1 + 2n\}$.
(b) The nth term of this sequence is positive if n is odd and negative if n is even. We see that for each natural number n,

$$a_n = (-1)^{n+1} \frac{1}{n}.$$

---

* Formally speaking, it is more correct to say that a sequence is a function from the set of natural numbers $\{1, 2, 3, 4, \cdots, n, \cdots\}$ into the set of real numbers.

Therefore, the sequence is $\left\{\dfrac{(-1)^{n+1}}{n}\right\}$.

(c) The terms of this sequence are

$$a_1 = \frac{1}{2} = 1 - \frac{1}{2} = 1 - \frac{1}{2^1}$$

$$a_2 = \frac{3}{4} = 1 - \frac{1}{4} = 1 - \frac{1}{2^2}$$

$$a_3 = \frac{7}{8} = 1 - \frac{1}{8} = 1 - \frac{1}{2^3}$$

$$a_4 = \frac{15}{16} = 1 - \frac{1}{16} = 1 - \frac{1}{2^4}$$

and so on. In general, the nth term is

$$a_n = 1 - \frac{1}{2^n}.$$

Therefore, the sequence is $\left\{1 - \dfrac{1}{2^n}\right\}$. ∎

A sequence always has an infinite number of terms. However, these terms need not all be different. In fact, it is possible that a sequence has only finitely many *different* terms. The following is an example of a sequence whose terms take on only two different values.

EXAMPLE 3   Write out the terms of the sequence $\{1 + 2(-1)^n\}$.

Solution
$$a_1 = 1 + 2(-1)^1 = 1 - 2 = -1$$
$$a_2 = 1 + 2(-1)^2 = 1 + 2 = 3$$
$$a_3 = 1 + 2(-1)^3 = 1 - 2 = -1$$
$$a_4 = 1 + 2(-1)^4 = 1 + 2 = 3$$

and so on. It is clear that the successive terms of this sequence are alternately -1 and 3. Thus, the sequence is

$$-1, 3, -1, 3, -1, 3, -1, 3, \cdots.$$ ∎

Limits of Sequences; Convergence and Divergence

Thinking of the subscript n as an input variable with corresponding output a_n, we can "graph" a sequence $\{a_n\}$, using n on the horizontal axis and $y = a_n$ on

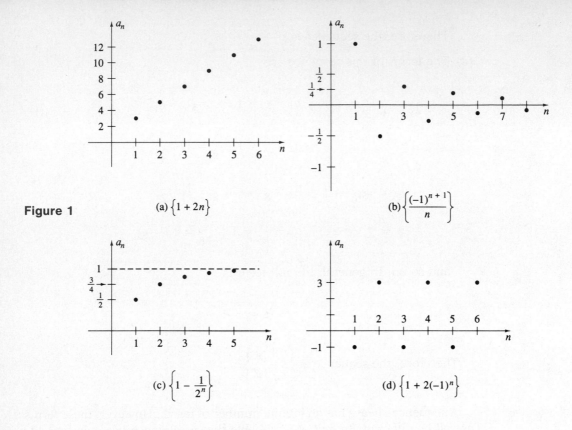

Figure 1

(a) $\{1 + 2n\}$

(b) $\left\{\dfrac{(-1)^{n+1}}{n}\right\}$

(c) $\left\{1 - \dfrac{1}{2^n}\right\}$

(d) $\{1 + 2(-1)^n\}$

the vertical axis. The graphs of the sequences given in Examples 2 and 3 are shown in Figure 1.

In each case, the graph of a sequence of real numbers is a collection of points in the plane, one point for each natural number subscript n. Of course, these graphs continue infinitely far to the right, beyond the points we have drawn. We are interested in whether or not these points approach a horizontal line as they go farther and farther to the right.

In (b) and (c) of Figure 1, the points get closer and closer to a horizontal line $y = L$ as $n \to +\infty$. When this happens for a sequence $\{a_n\}$ we say that the sequence has **limit** L, and we write

$$\lim_{n \to +\infty} a_n = L. \tag{7}$$

Equivalently, we say that the sequence $\{a_n\}$ **converges to L**, or in symbols,

$$a_n \to L. \tag{8}$$

Thus, corresponding to Figure 1(b) we say that

$$\lim_{n \to +\infty} \frac{(-1)^{n+1}}{n} = 0, \quad \text{or} \quad \frac{(-1)^{n+1}}{n} \to 0,$$

and corresponding to Figure 1(c) we say that

$$\lim_{n \to +\infty} \left(1 - \frac{1}{2^n}\right) = 1, \quad \text{or} \quad \left(1 - \frac{1}{2^n}\right) \to 1.$$

Observe that in Figure 1(a) and (d) the points do not get closer and closer to a horizontal line. When this happens for a sequence $\{a_n\}$, we say that the sequence **diverges,** or that it has **no limit**. Therefore, the sequences $\{1 + 2n\}$ and $\{1 + 2(-1)^n\}$ both diverge. For the sequence $\{1 + 2n\}$, however, we can be more specific about the behavior of a_n as $n \to +\infty$. As Figure 1 suggests,

$$\lim_{n \to +\infty} (1 + 2n) = +\infty,$$

or equivalently,

$$(1 + 2n) \to +\infty.$$

In general, when a sequence $\{a_n\}$ grows positively without bound, as $n \to +\infty$, we say that $\{a_n\}$ **diverges to $+\infty$** and we write

$$\lim_{n \to +\infty} a_n = +\infty \quad \text{or} \quad a_n \to +\infty. \tag{9}$$

Similarly, when a sequence $\{a_n\}$ grows negatively without bound, as $n \to +\infty$, we say that $\{a_n\}$ **diverges to $-\infty$,** and we write

$$\lim_{n \to +\infty} a_n = -\infty \quad \text{or} \quad a_n \to -\infty. \tag{10}$$

In both cases (9) and (10) we say that the sequence diverges. Even though limit notation is used in these cases, we are not claiming that the limit "exists." We must always remember that $+\infty$ and $-\infty$ are not numbers.

A simple way of discovering whether a sequence has a limit is by means of a table, as we have done for previous types of limits. The following example shows how this is done.

EXAMPLE 4 For each of the following sequences, construct a table and determine whether the sequence converges or diverges. Find the limit if possible.

(a) $\left\{\dfrac{4n - 5}{2n}\right\}.$ (b) $\left\{\dfrac{n + 7}{n^2}\right\}.$

(c) $\left\{(-1)^n + \dfrac{1}{n}\right\}.$ (d) $\{4n - 11\}.$

Solution   (a) For the sequence $\left\{\dfrac{4n-5}{2n}\right\}$ we have

n	1	2	5	10	20	30	50	100	1000
$a_n = \dfrac{4n-5}{2n}$	-0.5	0.75	1.5	1.75	1.875	1.917	1.95	1.975	1.9975

Hence, we conclude that $\displaystyle\lim_{n\to+\infty} \dfrac{4n-5}{2n} = 2.$

(b) For the sequence $\left\{\dfrac{n+7}{n^2}\right\}$ we have

n	1	2	5	10	50	100	200	500	1000
$a_n = \dfrac{n+7}{n^2}$	8	2.25	0.48	0.17	0.0228	0.0107	0.0052	0.0020	0.0010

Hence, we conclude that $\displaystyle\lim_{n\to+\infty} \dfrac{n+7}{n^2} = 0.$

(c) For the sequence $\left\{(-1)^n + \dfrac{1}{n}\right\}$ we have

n	1	2	3	4	5	6	7	8	9	10	11
$a_n = (-1)^n + \dfrac{1}{n}$	0	1.5	-0.67	1.25	-0.8	1.167	-0.857	1.125	-0.889	1.10	-0.909

Observe that the terms are alternately close to 1 and close to -1. There is no single number to which the terms get closer and closer. Thus,

$$\lim_{n\to+\infty}\left[(-1)^n + \dfrac{1}{n}\right] \text{ does not exist.}$$

(d) For the sequence $\{4n - 11\}$ we have

n	1	2	3	4	5	6	10	20	50	100
$a_n = 4n - 11$	-7	-3	1	5	9	13	29	69	189	389

Hence, $\displaystyle\lim_{n\to+\infty}(4n-11) = +\infty.$ ∎

Caution Convergence or divergence of a sequence $\{a_n\}$ is not determined by its first few terms. It is not determined by the first hundred terms, the first thousand terms, or even the first million terms. Indeed, the sequence $\{a_n\}$ given by

$$a_n = \begin{cases} \dfrac{1}{n} & \text{if } n \leq 1,000,000 \\[2mm] n & \text{if } n > 1,000,000 \end{cases}$$

appears to be converging to 0, through its first million terms. But beginning with the 1,000,001st term, the terms of this sequence grow without bound. Thus, this sequence diverges to $+\infty$. For this reason, neither graphs and tables nor calculators and computers can be used to *prove* that a sequence converges to a certain limit. They are used only to investigate the trend of the terms of a sequence and to *suggest* a limit.

Algebra of Limits

A rigorous treatment of limits of sequences would require that we develop some theorems about their behavior. We present several important theorems here. Their proofs, however, are beyond the scope of this book.

THEOREM 1

> Every convergent sequence is **bounded**. That is, if $\{a_n\}$ converges then there exist real numbers A and B such that for all natural numbers n,
>
> $$A \leq a_n \leq B.$$
>
> Equivalently, every unbounded sequence diverges.

THEOREM 2

> **(Some Special Limits)**
>
> (a) $\displaystyle\lim_{n \to +\infty} c = c$
> (limit of a constant sequence)
>
> (b) $\displaystyle\lim_{n \to +\infty} \frac{1}{n} = 0$
>
> (c) $\displaystyle\lim_{n \to +\infty} n = +\infty$
>
> (d) $\displaystyle\lim_{n \to +\infty} \left(1 + \frac{1}{n}\right)^n = e$
> [Equation (3) of Section 5.1]
>
> (e) $\displaystyle\lim_{n \to +\infty} c^n \begin{cases} = 0 & \text{if } |c| < 1 \\ = 1 & \text{if } c = 1 \\ = +\infty & \text{if } c > 1 \\ \text{does not exist if } c \leq -1 \end{cases}$
>
> (f) $\displaystyle\lim_{n \to +\infty} \frac{n^k}{e^n} = 0$ for any constant k
>
> (g) $\displaystyle\lim_{n \to +\infty} \frac{n^k}{a^n} = 0$ $\quad$ if $|a| > 1$ and k is any constant

THEOREM 3

(Algebra of Limits)

If $\{a_n\}$ and $\{b_n\}$ are convergent sequences and c is any constant real number, then

(a) $\displaystyle\lim_{n \to +\infty} (a_n \pm b_n) = \left(\lim_{n \to +\infty} a_n \right) \pm \left(\lim_{n \to +\infty} b_n \right)$

(b) $\displaystyle\lim_{n \to +\infty} (ca_n) = c \lim_{n \to +\infty} a_n$

(c) $\displaystyle\lim_{n \to +\infty} (a_n b_n) = \left(\lim_{n \to +\infty} a_n \right) \left(\lim_{n \to +\infty} b_n \right)$

(d) $\displaystyle\lim_{n \to +\infty} \left(\frac{1}{a_n} \right) = \frac{1}{\displaystyle\lim_{n \to +\infty} a_n}$ if $\displaystyle\lim_{n \to +\infty} a_n \neq 0$

(e) $\displaystyle\lim_{n \to +\infty} \left(\frac{a_n}{b_n} \right) = \frac{\displaystyle\lim_{n \to +\infty} a_n}{\displaystyle\lim_{n \to +\infty} b_n}$ if $\displaystyle\lim_{n \to +\infty} b_n \neq 0$

(f) $\displaystyle\lim_{n \to +\infty} a_n^c = \left(\lim_{n \to +\infty} a_n \right)^c$ if c is a positive constant such that a_n^c is a real number for every n.

EXAMPLE 5   Use Theorems 1–3 to determine the limits of the following sequences, where they exist:

(a) $\{3\}$ (e) $\left\{ \dfrac{5 + 2n}{2^n} \right\}$

(b) $\{\ln n\}$ (f) $\left\{ \dfrac{n - 3}{2^n} \right\}$

(c) $\left\{ -\dfrac{4}{n} \right\}$ (g) $\left\{ \sqrt{9 - \dfrac{1}{n}} \right\}$

(d) $\left\{ \dfrac{7}{n^2} \right\}$

Solution   (a) By Theorem 2(a), the limit of the constant sequence $\{3\}$ is $\displaystyle\lim_{n \to +\infty} 3 = 3$.

(b) Recall from Chapter 5 that the function $f(x) = \ln x$ is unbounded as $x \to +\infty$. Thus, by Theorem 1,

$$\lim_{n \to +\infty} \ln n = +\infty.$$

(c) $\displaystyle\lim_{n \to +\infty} -\frac{4}{n} = -4 \lim_{n \to +\infty} \frac{1}{n}$ [Theorem 3(b)]

$\qquad = -4 \cdot 0$ [Theorem 2(b)]

$\qquad = 0$.

(d) $\lim\limits_{n \to +\infty} \dfrac{7}{n^2} = 7 \lim\limits_{n \to +\infty} \dfrac{1}{n^2}$ [Theorem 3(b)]

$\qquad = 7 \left(\lim\limits_{n \to +\infty} \dfrac{1}{n} \right)\left(\lim\limits_{n \to +\infty} \dfrac{1}{n} \right)$ [Theorem 3(c)]

$\qquad = 7 \cdot 0 \cdot 0$ [Theorem 2(b)]

$\qquad = 0.$

(e) $\lim\limits_{n \to +\infty} \dfrac{5 + 2n}{n} = \lim\limits_{n \to +\infty} \left(\dfrac{5}{n} + 2 \right)$

$\qquad = \lim\limits_{n \to +\infty} \dfrac{5}{n} + \lim\limits_{n \to +\infty} 2$ [Theorem 3(a)]

$\qquad = 5 \lim\limits_{n \to +\infty} \dfrac{1}{n} + \lim\limits_{n \to +\infty} 2$ [Theorem 3(b)]

$\qquad = 5 \cdot 0 + 2$ [Theorem 2(b) and (a)]

$\qquad = 2.$

(f) $\lim\limits_{n \to +\infty} \dfrac{n - 3}{2^n} = \lim\limits_{n \to +\infty} \left(\dfrac{n}{2^n} - \dfrac{3}{2^n} \right)$

$\qquad = \left(\lim\limits_{n \to +\infty} \dfrac{n}{2^n} \right) - 3\left(\lim\limits_{n \to +\infty} \dfrac{1}{2^n} \right)$ [Theorem 3(a) and (b)]

$\qquad = 0 - 3 \cdot \lim\limits_{n \to +\infty} \left(\dfrac{1}{2} \right)^n$ [Theorem 2(g)]

$\qquad = -3 \cdot 0$ [Theorem 2(e)]

$\qquad = 0.$

(g) $\lim\limits_{n \to +\infty} \sqrt{9 - \dfrac{1}{n}} = \lim\limits_{n \to +\infty} \left(9 - \dfrac{1}{n} \right)^{1/2}$

$\qquad = \left[\lim\limits_{n \to +\infty} \left(9 - \dfrac{1}{n} \right) \right]^{1/2}$ [Theorem 3(f)]

$\qquad = \left[\lim\limits_{n \to +\infty} 9 - \lim\limits_{n \to +\infty} \dfrac{1}{n} \right]^{1/2}$ [Theorem 3(a)]

$\qquad = [9 - 0]^{1/2}$ [Theorem 2(a) and (b)] ·

$\qquad = 3.$ ∎

As an aid in calculating the limit of a quotient of two polynomials in n, as $n \to +\infty$, we often divide the numerator and denominator of the fraction by the *highest power of n that occurs in the denominator.* The following example illustrates this procedure.

EXAMPLE 6   Investigate the limits of the following sequences:

(a) $\left\{ \dfrac{n^2 + 3n - 1}{9 - 2n^2} \right\}$ (b) $\left\{ \dfrac{n^4 + 7n - 1}{5n^3 - 3} \right\}$

Solution (a) $\displaystyle \lim_{n \to +\infty} \frac{n^2 + 3n - 1}{9 - 2n^2} = \lim_{n \to +\infty} \frac{(n^2 + 3n - 1) \div n^2}{(9 - 2n^2) \div n^2}$

$$= \lim_{n \to +\infty} \frac{1 + \dfrac{3}{n} - \dfrac{1}{n^2}}{\dfrac{9}{n^2} - 2}$$

$$= \frac{\displaystyle \lim_{n \to +\infty} \left(1 + \frac{3}{n} - \frac{1}{n^2}\right)}{\displaystyle \lim_{n \to +\infty} \left(\frac{9}{n^2} - 2\right)} \qquad \text{[Theorem 3(e)]}$$

$$= \frac{1 + 3\left(\displaystyle \lim_{n \to +\infty} \frac{1}{n}\right) - \left(\displaystyle \lim_{n \to +\infty} \frac{1}{n}\right)^2}{9 \left(\displaystyle \lim_{n \to +\infty} \frac{1}{n}\right)^2 - 2} \qquad \text{[Theorem 3]}$$

$$= \frac{1 + 3 \cdot 0 - 0^2}{9 \cdot 0 - 2} \qquad \text{[Theorem 2]}$$

$$= -\frac{1}{2}.$$

(b) $\displaystyle \lim_{n \to +\infty} \frac{n^4 + 7n - 1}{5n^3 - 3} = \lim_{n \to +\infty} \frac{(n^4 + 7n - 1) \div n^3}{(5n^3 - 3) \div n^3}$

$$= \lim_{n \to +\infty} \frac{n + \dfrac{7}{n^2} - \dfrac{1}{n^3}}{5 - \dfrac{3}{n^3}}.$$

At this point we can see that the numerator diverges to $+\infty$, since

$$\lim_{n \to +\infty} \left(n + \frac{7}{n^2} - \frac{1}{n^3}\right) = +\infty.$$

However, the limit of the denominator is $\displaystyle \lim_{n \to +\infty} \left(5 - \frac{3}{n^2}\right) = 5$. In this example we cannot apply Theorem 3(e), since it requires that both the numerator and denominator sequences *converge*. Instead, we see directly from the above analysis that

$$\lim_{n \to +\infty} \frac{n^4 + \dfrac{7}{n^2} - \dfrac{1}{n^3}}{5 - \dfrac{3}{n^2}} = +\infty,$$

and hence our given sequence (b) diverges to $+\infty$. ∎

|||||||||||||| **Monotone Sequences**

We often encounter sequences $\{a_n\}$ for which one of the following conditions holds:

(1) $a_1 \leq a_2 \leq \cdots \leq a_n \leq a_{n+1} \leq \cdots$; in this case we say that $\{a_n\}$ is a **monotone increasing** sequence.

(2) $a_1 \geq a_2 \geq \cdots \geq a_n \geq a_{n+1} \geq \cdots$; in this case we say that $\{a_n\}$ is a **monotone decreasing** sequence.

If a sequence is one of these two types, then we say that it is a **monotone** sequence.

One way to determine whether a sequence is monotone is by examining the sequence of differences $\{a_{n+1} - a_n\}$.

THEOREM 4

(a) If $a_{n+1} - a_n \geq 0$ for all natural numbers n, then the sequence $\{a_n\}$ is monotone increasing.

(b) If $a_{n+1} - a_n \leq 0$ for all natural numbers n, then the sequence $\{a_n\}$ is monotone decreasing.

EXAMPLE 7 Determine whether each of the following sequences is monotone increasing, monotone decreasing, or neither.

(a) $\left\{ \dfrac{3n - 2}{n} \right\}$ (b) $\left\{ \dfrac{3}{2^n} \right\}$ (c) $\{5 + (-1)^n\}$.

Solution (a) For this sequence, $a_n = \dfrac{3n - 2}{n} = 3 - \dfrac{2}{n}$, so that

$$a_{n+1} - a_n = \left(3 - \frac{2}{n+1}\right) - \left(3 - \frac{2}{n}\right)$$

$$= \frac{2}{n} - \frac{2}{n+1} = 2\left(\frac{1}{n} - \frac{1}{n+1}\right)$$

$$= 2\frac{(n+1) - n}{n(n+1)}$$

$$= 2 \cdot \frac{1}{n(n+1)}$$

$$> 0.$$

Since $a_{n+1} - a_n > 0$ for all natural numbers n, the sequence $\{a_n\}$ is monotone increasing.

(b) For this sequence, $a_n = \dfrac{3}{2^n}$, so that

$$a_{n+1} - a_n = \frac{3}{2^{n+1}} - \frac{3}{2^n}$$

$$= \frac{3}{2^n}\left(\frac{1}{2} - 1\right) = \frac{3}{2^n}\left(-\frac{1}{2}\right)$$

$$< 0.$$

Since $a_{n+1} - a_n < 0$ for all natural numbers n, the sequence $\{a_n\}$ is monotone decreasing.

(c) For this sequence,

$$
\begin{aligned}
a_{n+1} - a_n &= [5 + (-1)^{n+1}] - [5 + (-1)^n] \\
&= (-1)^{n+1} - (-1)^n \\
&= (-1)^n[(-1) - 1] = (-1)^n(-2) \\
&= (-1)^{n+1} \cdot 2.
\end{aligned}
$$

This expression alternates sign between positive and negative as $n \to +\infty$. Therefore, $\{a_n\}$ is neither monotone increasing nor monotone decreasing. It is not a monotone sequence. ■

Sometimes the following theorem is useful in determining monotonicity of a sequence $\{a_n\}$.

THEOREM 5

Suppose $\{a_n\}$ is a sequence and f is a function defined and differentiable on $[1, +\infty)$ such that for all natural numbers n,

$$f(n) = a_n.$$

Then
(a) if $f'(x) \geq 0$ throughout $[1, +\infty)$, then $\{a_n\}$ is a monotone increasing sequence.
(b) if $f'(x) \leq 0$ throughout $[1, +\infty)$, then $\{a_n\}$ is a monotone decreasing sequence.

EXAMPLE 8   Use Theorem 5 to determine whether the sequences (a) and (b) in Example 7 are monotone.

Solution   (a) Let $f(x) = \dfrac{3x - 2}{x} = 3 - \dfrac{2}{x} = 3 - 2x^{-1}$. Then f is differentiable on $[1, +\infty)$ and

$$f'(x) = 0 - 2(-1)x^{-2} = \frac{2}{x^2}$$
$$> 0,$$

for all x in $[1, +\infty)$. Therefore, by Theorem 5, the sequence $\left\{\dfrac{3n - 2}{n}\right\}$ is monotone increasing.

(b) Let $f(x) = \dfrac{3}{2^x} = 3 \cdot 2^{-x}$. By equation (9) of Section 5.3, this function is differentiable on $[1, +\infty)$ and

$$
\begin{aligned}
f(x) &= 3 \cdot 2^{-x} \cdot \ln 2 \cdot (-1) \\
&= -\frac{3 \ln 2}{2^x} \\
&< 0,
\end{aligned}
$$

for all x in $[1, +\infty)$. Therefore, by Theorem 5, the sequence $\{a_n\}$ is monotone decreasing. ∎

An important property of monotone sequences is given in the following remarkable theorem.

THEOREM 6	A monotone sequence converges if and only if it is bounded.

We already know from Theorem 1 that if a sequence converges it must be bounded. We also know that there are bounded sequences that do not converge (as in Example 3). Thus, in general, from knowing that a sequence is bounded we cannot conclude that it converges. Theorem 6 tells us, however, that if we know a *monotone* sequence is bounded, then we can conclude that it converges. This theorem reflects a profound property of the real number system, called the "completeness property." We shall need to use this theorem later in the chapter. Its proof, however, is well beyond the scope of this book.

12.1 EXERCISE SET ||||||||||||||||||||||||||||||||||||||||||||||||||||||||||||||||||||||||||||||

In Exercises 1–8, write out the first six terms of the given sequence.

1. $\{7n - 5\}$
2. $\{19 - 6n\}$
3. $\left\{\dfrac{3n}{2^n}\right\}$
4. $\left\{\dfrac{4n - 1}{2^n + 1}\right\}$
5. $\left\{\dfrac{(-1)^n}{5n}\right\}$
6. $\left\{3 + \dfrac{(-1)^n}{n}\right\}$
7. $\{4 + (-1)^n\}$
8. $\{3(-1)^n - 1\}$

In Exercises 9–14, find a formula for a_n, the nth term of the given sequences.

9. $\{3, 6, 9, 12, 15, \ldots\}$
10. $\{1, 3, 5, 7, 9, 11, \ldots\}$

11. $\left\{\dfrac{1}{2}, \dfrac{2}{3}, \dfrac{3}{4}, \dfrac{4}{5}, \ldots\right\}$
12. $\left\{\dfrac{1}{3}, \dfrac{1}{9}, \dfrac{1}{27}, \dfrac{1}{81}, \ldots\right\}$
13. $\{-2, 4, -8, 16, -32, \ldots\}$
14. $\{2, 0, 2, 0, 2, 0, \ldots\}$

In Exercises 15–18, use a calculator to complete the table and decide whether the given sequence converges or diverges. Find the limit where possible.

15.

n	1	5	10	20	50	100	200
$\dfrac{6n}{11 - 3n}$							

16.

n	1	2	5	10	20	50
$\dfrac{n+15}{2n^2-1}$						

17.

n	1	5	10	25	50	101
$\dfrac{3n+(-1)^n}{n+2}$						

18.

n	1	2	5	10	20	50	100
$\dfrac{3n^2+5}{n+2}$							

In Exercises 19–36, use Theorems 2 and 3 and the methods of Examples 5 and 6, as needed, to determine whether the given sequence converges or diverges. Find the limit where possible.

19. $\left\{\dfrac{3+8n}{n}\right\}$

20. $\left\{\dfrac{n-1}{2n+5}\right\}$

21. $\left\{-\dfrac{3}{n}\right\}$

22. $\{n-3\}$

23. $\left\{\left(\dfrac{7}{8}\right)^n\right\}$

24. $\{7\}$

25. $\{(1.27)^n\}$

26. $\left\{\left(-\dfrac{2}{3}\right)^n\right\}$

27. $\{(-0.99)^n\}$

28. $\left\{\left(1+\dfrac{1}{n}\right)^{2n}\right\}$

29. $\left\{\dfrac{2^n+5^n}{7^n}\right\}$

30. $\left\{\dfrac{1}{\sqrt{n}}\right\}$

31. $\{\sqrt{n+1}\}$

32. $\left\{\sqrt{3-\dfrac{1}{n}}\right\}$

33. $\left\{\dfrac{4n-12}{n^2+6}\right\}$

34. $\left\{\dfrac{n-5}{10-n}\right\}$

35. $\left\{\dfrac{n^3+3n^2-7}{5n^2+8}\right\}$

36. $\left\{\dfrac{n^2-2}{1-n}\right\}$

In Exercises 37–44, determine whether the given sequence is monotone increasing, monotone decreasing, or neither.

37. $\left\{\dfrac{1}{n+2}\right\}$

38. $\{n^2+2n\}$

39. $\{(-1)^n-3\}$

40. $\left\{\dfrac{n+5}{n+1}\right\}$

41. $\left\{\dfrac{8-7n}{n}\right\}$

42. $\left\{\dfrac{(-1)^n}{n+3}\right\}$

43. $\{n^2-10n\}$

44. $\left\{\dfrac{1}{2n^2-1}\right\}$

12.2 ‖‖‖‖ SUMMATION NOTATION AND POLYNOMIALS

We are often interested in sums of the form

$$a_1 + a_2 + \cdots + a_n. \tag{1}$$

In fact, a special notation has been devised to represent such a sum. By definition,

$$\sum_{i=1}^{n} a_i = a_1 + a_2 + \cdots + a_n. \tag{2}$$

The Greek capital letter "Σ" (sigma) is used to represent "sum"; the variable i representing the subscript begins with $i = 1$ in the first term, advances by an increment of 1 with each successive term of the sum, and terminates with $i = n$ when the last term has been reached.

The subscripts in (2) need not begin with $i = 1$; they may begin with any integer. For example,

$$\sum_{i=4}^{9} a_i = a_4 + a_5 + a_6 + a_7 + a_8 + a_9.$$

Moreover, the letter i is strictly a dummy variable; it does not appear on the right side of Equation (2). Thus, any letter may be used instead of i. For example,

$$a_1 + a_2 + \cdots + a_n = \sum_{i=1}^{n} a_i = \sum_{j=1}^{n} a_j = \sum_{k=1}^{n} a_k. \qquad (3)$$

EXAMPLE 1 Calculate each of the following sums.

(a) $\sum_{i=1}^{6} (2i - 1)$ (b) $\sum_{i=1}^{5} i^2$ (c) $\sum_{j=1}^{4} \frac{j+1}{j}$ (d) $\sum_{k=3}^{7} \frac{3k+2}{k^2-1}$

Solution (a) $\sum_{i=1}^{6} (2i - 1) = (2 - 1) + (4 - 1) + (6 - 1) + (8 - 1) + (10 - 1) + (12 - 1)$

$$= 1 + 3 + 5 + 7 + 9 + 11$$

$$= 36.$$

(b) $\sum_{i=1}^{5} i^2 = 1^2 + 2^2 + 3^2 + 4^2 + 5^2 = 1 + 4 + 9 + 16 + 25 = 55.$

(c) $\sum_{j=1}^{4} \frac{j+1}{j} = \frac{2}{1} + \frac{3}{2} + \frac{4}{3} + \frac{5}{4} = \frac{73}{12}.$

(d) $\sum_{k=3}^{7} \frac{3k+2}{k^2-1} = \frac{11}{8} + \frac{14}{15} + \frac{17}{24} + \frac{20}{35} + \frac{23}{48} \approx 4.067.$ ∎

There are some simple algebraic properties of summations that will be very helpful to know. The following theorem summarizes them:

THEOREM 1

If $a_1, a_2, \ldots, a_n, b_1, b_2, \ldots, b_n$ and c are any real numbers, then

(a) $\sum_{i=1}^{n} (a_i + b_i) = \sum_{i=1}^{n} a_i + \sum_{i=1}^{n} b_i$

(b) $\sum_{i=1}^{n} (a_i - b_i) = \sum_{i=1}^{n} a_i - \sum_{i=1}^{n} b_i$

(c) $\sum_{i=1}^{n} ca_i = c \sum_{i=1}^{n} a_i$

(d) $\sum_{i=1}^{n} c = nc.$

Proof (a) $\displaystyle\sum_{i=1}^{n} (a_i + b_i) = (a_1 + b_1) + (a_2 + b_2) + \cdots + (a_n + b_n)$

$$= (a_1 + a_2 + \cdots + a_n) + (b_1 + b_2 + \cdots + b_n)$$

$$= \sum_{i=1}^{n} a_i + \sum_{i=1}^{n} b_i.$$

(b) The proof of (b) is left as an exercise (Exercise 35).

(c) $\displaystyle\sum_{i=1}^{n} (ca_i) = (ca_1) + (ca_2) + \cdots + (ca_n)$

$$= c(a_1 + a_2 + \cdots + a_n)$$

$$= c\sum_{i=1}^{n} a_i.$$

(d) $c = \underbrace{c + c + \cdots + c}_{n\ c\text{'s}} = nc.$

$\blacksquare$

|||||||||||||| **Geometric Progressions**

By definition, a **geometric progression** is a sequence of the form

$$a, \ ar, \ ar^2, \ ar^3, \ldots, \ ar^n, \ ar^{n+1}, \ldots \ . \tag{4}$$

We shall need to have a formula for the *sum* of the first $n + 1$ terms of the sequence (4),

$$a + ar + ar^2 + \cdots + ar^n = \sum_{i=0}^{n} ar^i.$$

The following theorem gives us the formula we need.

THEOREM 2

Let a denote any real number, and $r \neq 1$. Then

$$\sum_{i=0}^{n} ar^i = a\left(\frac{1 - r^{n+1}}{1 - r}\right). \tag{5}$$

Proof Let $r \neq 1$, and define

$$S_n = a + ar + ar^2 + \cdots + ar^n. \tag{6}$$

Multiplying both sides by r, we have

$$S_n r = ar + ar^2 + ar^3 + \cdots + ar^{n+1}. \tag{7}$$

Subtracting Equation (7) from Equation (6), we obtain

$$S_n - S_n r = a + ar + ar^2 + \cdots + ar^n - ar - ar^2 - \cdots - ar^{n+1}$$
$$= a - ar^{n+1}. \qquad \text{[All terms cancel out except the first and last.]}$$

Thus,

$$S_n(1 - r) = a(1 - r^{n+1})$$

and so

$$S_n = \frac{a(1 - r^{n+1})}{1 - r}. \qquad \blacksquare$$

EXAMPLE 2 For each of the following geometric progressions, find a and r, and use Theorem 2 to find the sum.

(a) $3 + 6 + 12 + 24 + 48 + 96$.

(b) $5 + \frac{5}{3} + \frac{5}{9} + \frac{5}{27} + \frac{5}{81}$.

(c) $1 - \frac{1}{2} + \frac{1}{4} - \frac{1}{8} + \frac{1}{16} - \frac{1}{32} + \frac{1}{64}$.

Solution (a) In this geometric progression, $a = 3$ and $r = 2$. The last term is $96 = 3 \cdot 32 = 3 \cdot 2^5$, so $n = 5$. Thus, the sum is

$$\sum_{i=0}^{5} 3 \cdot 2^i = 3\left(\frac{1 - 2^6}{1 - 2}\right) = 3\frac{1 - 64}{-1}$$
$$= 3(63)$$
$$= 189.$$

(b) Here $a = 5$, $r = \frac{1}{3}$, and since the last term is $\frac{5}{81} = 5 \cdot (\frac{1}{3})^4$, we have $n = 4$. Thus, the sum is

$$\sum_{i=0}^{4} 5 \cdot (\tfrac{1}{3})^i = 5\left[\frac{1 - (\frac{1}{3})^5}{1 - \frac{1}{3}}\right]$$
$$= 5\frac{1 - \frac{1}{243}}{\frac{2}{3}} = 5 \cdot \frac{242}{162}$$
$$= \frac{605}{81} \approx 7.469.$$

(c) Here $a = 1$, $r = -\frac{1}{2}$ and since the last term is $\frac{1}{64} = (-\frac{1}{2})^6$, we have $n = 6$. Thus, the sum is

$$\sum_{i=0}^{5} 1(-\tfrac{1}{2})^i = \frac{1 - (-\frac{1}{2})^7}{1 - (-\frac{1}{2})} = \frac{1 + \frac{1}{128}}{\frac{3}{2}}$$
$$= \frac{43}{64} \approx 0.672. \qquad \blacksquare$$

IIIIIIIIIIIII **Applications in Business: Annuities**

An **ordinary annuity** is a sequence of equal payments made at equal time intervals into an interest-bearing account, where the interest is compounded with the same frequency as that with which the payments are made.

Suppose that an investor makes regular deposits of **P** dollars each, over n equal time periods into an account that pays interest at the rate i per time period, compounded at the end of each time period. For convenience, it is customary to assume that the first payment is made at the *end* of the first time period. We shall derive a formula for A_n, the value of the account at the end of the nth time period (immediately after the nth payment).

At the end of the first time period we have

$$A_1 = P.$$

At the end of the second time period we have

$$A_2 = \text{new deposit} + A_1 + \text{interest earned by } A_1$$
$$= P + P + Pi$$
$$= P[1 + (1 + i)].$$

At the end of the third time period we have

$$A_3 = \text{new deposit} + A_2 + \text{interest earned by } A_2$$
$$= P + P[1 + (1 + i)] + P[1 + (1 + i)]i$$
$$= P\{1 + [1 + (1 + i)] \cdot 1 + [1 + (1 + i)] \cdot i\}$$
$$= P\{1 + [1 + (1 + i)](1 + i)\} \quad \text{(factoring by grouping)}$$
$$= P\{1 + (1 + i) + (1 + i)^2\}.$$

In general, at the end of the nth time period we have

$$A_n = P\{1 + (1 + i) + (1 + i)^2 + \cdots + (1 + i)^{n-1}\}.$$

Inside the braces we recognize the sum of a geometric progression. Thus, we may apply Theorem 2 to obtain

$$A_n = P \sum_{k=0}^{n-1} (1 + i)^k$$
$$= P\left[\frac{1 - (1 + i)^{(n-1)+1}}{1 - (1 + i)}\right] \quad \text{(by Theorem 2)}$$
$$= P\left[\frac{1 - (1 + i)^n}{-i}\right]$$

We have thus derived the following result:

THEOREM 3

If regular payments of P dollars each are made over n equal time periods into an account paying interest at the rate i per time period, compounded at the end of each time period, then the value of the annuity at the end of the nth time period (immediately after the nth payment has been made) is

$$A_n = P\left[\frac{(1 + i)^n - 1}{i}\right]. \tag{8}$$

The quantity inside the brackets in Equation (8) is used so frequently that it is given a special symbol,

$$S_{\overline{n}|i} = \frac{(1 + i)^n - 1}{i} \tag{9}$$

(read "S angle n at i"). Typical books on elementary mathematics of finance have tables from which $S_{\overline{n}|i}$ can be conveniently found. Since this topic is somewhat tangential for our purposes, we do not include such a table in this book. Instead, we shall depend upon a calculator to evaluate (9).

EXAMPLE 3 Suppose \$500 is deposited every six months over a period of 5 years (10 payments) into an account bearing interest at the rate of 8% per year, compounded every six months after successive payments are made. Find the value of the annuity just after the last payment has been made.

Solution In this annuity, $P = 500$, $i = .08/2 = .04$, and $n = 10$. Thus,

$$A_n = 500\left[\frac{(1.04)^{10} - 1}{.04}\right].$$

Using a calculator, we find

$$A_n \approx 500[12.006] \approx \$6003. \qquad \blacksquare$$

||||||||||||| **Polynomials in Taylor Form**

In Section 1.5 we defined a polynomial as a function of the form

$$p(x) = a_0 + a_1 x + a_2 x^2 + \cdots + a_n x^n. \tag{10}$$

In summation notation,

$$p(x) = \sum_{k=0}^{n} a_k x^k.$$

Recall from Section 10.1 that the kth derivative of $p(x)$ is represented by

$$p^{(k)}(x) = \frac{d^k}{dx^k}[p(x)]. \tag{11}$$

For completeness, we define the 0th derivative of $p(x)$ to be $p(x)$ itself:

$$p^{(0)}(x) = \frac{d^0}{dx^0}[p(x)] = p(x). \tag{12}$$

We now develop a relationship between the coefficients a_k in (10) and the kth derivatives evaluated at 0, $p^{(k)}(0)$. First, observe that

$$p^{(0)}(0) = p(0)$$
$$= a_0.$$

Differentiating both sides of (10), we find

$$p^{(1)}(x) = a_1 + 2a_2x^1 + 3a_3x^2 + 4a_4x^3 + \cdots + na_nx^{n-1}. \tag{13}$$

Thus,

$$p^{(1)}(0) = a_1.$$

Differentiating both sides of (13), we have

$$p^{(2)}(x) = 2a_2 + 2 \cdot 3a_3x + 3 \cdot 4a_4x^2 + 4 \cdot 5a_5x^3 + \cdots + (n-1)na_nx^{n-2}. \tag{14}$$

Thus,

$$p^{(2)}(0) = 2a_2.$$

Continuing, we differentiate both sides of (14):

$$p^{(3)}(x) = 2 \cdot 3a_3 + 2 \cdot 3 \cdot 4a_4x + 3 \cdot 4 \cdot 5a_5x^2 + \cdots + (n-2)(n-1)na_nx^{n-3}, \tag{15}$$

from which we obtain

$$p^{(3)}(0) = 2 \cdot 3a_3.$$

Looking ahead, from (15) we can also see that

$$p^{(4)}(0) = 2 \cdot 3 \cdot 4a_4.$$

In general, for $k = 1, 2, 3, \ldots, n$, we have

$$p^{(k)}(0) = 1 \cdot 2 \cdot 3 \cdot \cdots \cdot k\, a_k.$$

This gives us a formula for the kth coefficient of $p(x)$:

$$a_k = \frac{p^{(k)}(0)}{1 \cdot 2 \cdot 3 \cdot \cdots \cdot k}. \tag{16}$$

The denominator in (16) occurs often enough in mathematics to be given a special name and symbol $k!$ (read "**k factorial**") which is defined for every positive integer k as

$$k! = 1 \cdot 2 \cdot 3 \cdot \cdots \cdot k. \tag{17}$$

Thus,

$$1! = 1$$
$$2! = 1 \cdot 2 = 2$$
$$3! = 1 \cdot 2 \cdot 3 = 6$$
$$4! = 1 \cdot 2 \cdot 3 \cdot 4 = 24$$
$$5! = 1 \cdot 2 \cdot 3 \cdot 4 \cdot 5 = 120$$
$$6! = 1 \cdot 2 \cdot 3 \cdot 4 \cdot 5 \cdot 6 = 720$$

and so on.

For completeness we wish to define $0!$ as well. To make formulas work out conveniently we define

$$0! = 1. \tag{18}$$

With these definitions, we can now summarize our results in the following theorem:

THEOREM 4

If $p(x) = a_0 + a_1 x + a_2 x^2 + \cdots + a_n x^n$ is any polynomial, then for each $k = 0, 1, 2, \ldots, n$,

$$a_k = \frac{p^{(k)}(0)}{k!}. \tag{19}$$

Thus, we can rewrite $p(x)$ in the form

$$p(x) = \sum_{k=0}^{n} \frac{p^{(k)}(0)}{k!} x^k. \tag{20}$$

We call the right side of Equation (20) the **Maclaurin* form** of the polynomial $p(x)$.

A polynomial is a sum of constant multiples of power of x. In some applications it is helpful to express a polynomial in powers of $(x - c)$ for some constant real number c. That is, for a given polynomial

$$p(x) = a_0 + a_1 x + a_2 x^2 + \cdots + a_n x^n$$

we may wish to find constants $b_0, b_1, b_2, \ldots, b_n$ such that

$$p(x) = b_0 + b_1(x - c) + b_2(x - c)^2 + b_3(x - c)^3 + \cdots + b_n(x - c)^n. \quad (21)$$

We now derive a formula for the coefficients $b_0, b_1, b_2, \ldots, b_n$.

First, observe that

$$\begin{aligned} p(c) &= b_0 + b_1 \cdot 0 + \cdots + b_n \cdot 0 \\ &= b_0. \end{aligned}$$

Differentiating both sides of Equation (21), we have

$$p^{(1)}(x) = b_1 + 2b_2(x - c) + 3b_3(x - c)^2 + \cdots + nb_n(x - c)^{n-1}, \quad (22)$$

and so

$$p^{(1)}(c) = b_1.$$

Differentiating both sides of Equation (22), we have

$$p^{(2)}(x) = 2b_2 + 2 \cdot 3b_3(x - c) + 3 \cdot 4b_4(x - c)^2 + \cdots + n(n-1)b_n(x - c)^{n-2},$$

and so

$$p^{(2)}(c) = 2b_2.$$

In general we find that for $k = 0, 1, 2, \ldots, n$,

$$p^{(k)}(c) = k! b_k.$$

We have therefore derived a formula for the coefficients b_k:

$$b_k = \frac{p^{(k)}(c)}{k!}. \quad (23)$$

The reader should note the similarity between this formula for b_k and Equation (16) for a_k.

---

* See footnote, page 718.

We summarize these results in the following theorem:

THEOREM 5

Given any polynomial $p(x) = \sum_{k=0}^{n} a_k x^k$ and any real number c, the polynomial $p(x)$ can be written in the form

$$p(x) = \sum_{k=0}^{n} \frac{p^{(k)}(c)}{k!} (x - c)^k. \qquad (24)$$

We call the right hand side of Equation (24) the **Taylor* form of $p(x)$** about c.

EXAMPLE 4   Express the polynomial $p(x) = 7 - 8x + 4x^2 + 5x^3 - 3x^4$ in Taylor form about 1.

Solution   We shall use Theorem 5 with $c = 1$. First,

$$p^{(0)}(1) = p(1) = 7 - 8 + 4 + 5 - 3 = 5.$$

The first derivative of $p(x)$ is

$$p'(x) = -8 + 8x + 15x^2 - 12x^3$$

and so

$$p^{(1)}(1) = -8 + 8 + 15 - 12 = 3.$$

The second derivative of $p(x)$ is

$$p''(x) = 8 + 30x - 36x^2$$

and so

$$p^{(2)}(1) = 8 + 30 - 36 = 2.$$

The third derivative of $p(x)$ is

$$p'''(x) = 30 - 72x,$$

and the fourth derivative is $p^{(4)}(x) = -72$. Thus, $p^{(3)}(1) = 30 - 72 = -42$ and $p^{(4)}(1) = -72$. From Theorem 5, therefore, we have

$$p(x) = \frac{p^{(0)}(1)}{0!} + \frac{p^{(1)}(1)}{1!}(x-1) + \frac{p^{(2)}(1)}{2!}(x-1)^2 + \frac{p^{(3)}(1)}{3!}(x-1)^3 + \frac{p^{(4)}(1)}{4!}(x-1)^4$$

$$= \frac{5}{1} + \frac{3}{1}(x-1) + \frac{2}{2}(x-1)^2 + \frac{-42}{6}(x-1)^3 + \frac{-72}{24}(x-1)^4$$

$$= 5 + 3(x-1) + (x-1)^2 - 7(x-1)^3 - 3(x-1)^4. \qquad \blacksquare$$

---

* See footnote, page 718.

12.2 EXERCISE SET ||||||||||||||||||||||||||||||||||||||||||||||||||||||||||||||||||||||||||||||||||

In Exercises 1–10, calculate the indicated sum.

1. $\displaystyle\sum_{i=1}^{6} (3i - 7)$

2. $\displaystyle\sum_{i=1}^{5} [2(i - 3) + 1]$

3. $\displaystyle\sum_{i=1}^{5} (i^2 - 3i)$ 4. $\displaystyle\sum_{i=1}^{4} i(i + 2)$

5. $\displaystyle\sum_{i=1}^{4} \frac{i}{i + 1}$ 6. $\displaystyle\sum_{i=1}^{6} (-2)^n$

7. $\displaystyle\sum_{k=2}^{7} \frac{(-1)^k}{k}$ 8. $\displaystyle\sum_{k=3}^{9} \frac{4k - 5}{2k + 3}$

9. $\displaystyle\sum_{k=0}^{6} \frac{(-1)^k}{2^k}$ 10. $\displaystyle\sum_{k=0}^{5} \frac{k - 1}{k^2 + 2}$

In Exercises 11–18, use Theorems 1 and 2 to calculate the indicated sum.

11. $\displaystyle\sum_{k=0}^{7} \left(\frac{2}{3}\right)^k$ 12. $\displaystyle\sum_{k=0}^{10} 16$

13. $\displaystyle\sum_{k=0}^{7} \left[10 - \left(\frac{2}{3}\right)^k\right]$ 14. $\displaystyle\sum_{k=0}^{5} 3^k$

15. $\displaystyle\sum_{k=0}^{5} \frac{3^k}{2^{k+1}}$ 16. $\displaystyle\sum_{k=0}^{6} \frac{2^k + 3^k}{4^k}$

17. $1 + 0.03 + (0.03)^2 + (0.03)^3 + \cdots + (0.03)^8$

18. $3 - 3(0.2) + 3(0.2)^2 - 3(0.2)^3 + \cdots + 3(0.2)^8$

19. Suppose $200 is deposited quarterly into an account paying 8% interest per year, compounded after each successive payment is made. Find the value of the account after the 20th payment has been made.

20. Suppose $25 is deposited each month into an account that pays 6% interest per year, compounded monthly after each payment is made. Find the value of the account after 10 years (i.e., after the 120th payment has been made).

21. Mr. and Mrs. Smith want to make monthly deposits into an account that pays 6% annual interest, compounded monthly after each payment is made, in order to have $20,000 after 17 years to put toward their son's college education. If they make $50 monthly payments, will they meet this objective? What if they make $60 payments? What monthly payment would ensure that after 17 years the amount would be worth $30,000?

22. A college professor deposits $150 quarterly into a retirement account paying 8% annual interest, compounded quarterly after each payment is made. What amount will be in the account when she retires, 20 years after she starts the account?

In Exercises 23–26 a polynomial $p(x)$ is given. Find the indicated derivatives.

23. $p(x) = 3 - x^2 + 5x^3 - 7x^5 + x^6$; find $p^{(5)}(0)$ and $p^{(7)}(0)$.

24. $p(x) = 18 + 5x - 6x^2 + 9x^4 + 7x^6$; find $p^{(2)}(0)$ and $p^{(4)}(0)$.

25. $p(x) = 15 - 3(x - 2) + 8(x - 2)^3 + 5(x - 2)^4 - (x - 2)^6$; find $p^{(4)}(2)$ and $p^{(5)}(2)$.

26. $p(x) = -7 + 8(x + 3) + 9(x + 3)^2 - 10(x + 3)^3 + 11(x + 3)^4$; find $p^{(0)}(-3)$, $p^{(3)}(-3)$, and $p^{(4)}(-3)$.

In Exercises 27–34, express the given polynomial $p(x)$ in Taylor form about the indicated number c.

27. $p(x) = 3 + 2x - x^2$; $c = 3$

28. $p(x) = 2x^2 - 5x + 1$; $c = -1$

29. $p(x) = 1 + x + x^2 + x^3$; $c = 1$

30. $p(x) = 3x - 5x^2 + x^4$; $c = 1$

31. $p(x) = 1 + x + x^2 + x^3$; $c = 2$

32. $p(x) = 3x - 5x^2 + x^4$; $c = 2$

33. $p(x) = x^5$; $c = -2$

34. $p(x) = 2x^5 - 4x^3 + x$; $c = -1$

35. Prove part (b) of Theorem 1.

12.3 ⅠⅠⅠⅠⅠ   TAYLOR POLYNOMIALS; TAYLOR'S THEOREM

Polynomials are among the simplest of functions. To calculate the values of a polynomial

$$p(x) = a_0 + a_1 x + a_2 x^2 + \cdots + a_n x^n,$$

one only needs to be able to multiply and add (or subtract). Many mathematical functions, however, are much more difficult to evaluate. With the availability of tables and calculators, it is perhaps hard for today's students to see what is so difficult about finding the values of $\sin x$, $\ln x$, or e^x. Imagine that we did not have calculators or tables for these functions. How then would we find $\sin 34°$, $\ln 29$, or $e^{1.84}$? One of the classic problems in mathematics has been that of calculating values of functions such as these. In this section we show how this problem is solved by approximating these functions with carefully constructed polynomials.

Once we understand the use of Taylor polynomials, as presented in this section, we can begin to understand how calculators calculate $\sin x$, $\ln x$, e^x, and so on. It would be a gross misunderstanding to think that a calculator "knows" $\sin 34°$, $\ln 29$, or $e^{1.84}$. Indeed, it doesn't know these numbers any more than you do. In fact, a calculator *calculates* them by procedures similar to the one you are about to learn.

ⅠⅠⅠⅠⅠⅠⅠⅠⅠⅠⅠⅠⅠ **Taylor Polynomials for a Function**

Suppose that we have a function f, and that we wish to calculate values of $f(x)$, for x belonging to some interval. We select some convenient number c in this interval, and pick a natural number n such that f has derivatives of order $1, 2, \ldots, n$ in this interval. We shall derive the formula for a polynomial $p(x)$ of degree $\leq n$ that approximates $f(x)$ for values of x near c.

In order for a polynomial $p(x)$ to be a good approximation of $f(x)$ for values of x near c, we require that the graph of $p(x)$ be tangent to the graph of $f(x)$ at the point $(c, f(c))$. This means that we require

$$p(c) = f(c) \qquad \text{and} \qquad p'(c) = f'(c).$$

Similarly, we want the graphs of $p(x)$ and $f(x)$ to have the same concavity at c, which means

$$p''(c) = f''(c).$$

In fact, to assure that the curve $y = p(x)$ conforms as closely as possible to the curve $y = f(x)$ near c, we would require

$$p^{(k)}(c) = f^{(k)}(c) \qquad \text{for } k = 0, 1, 2, \ldots, n. \tag{1}$$

In Theorem 5 of Section 12.2, we showed that $p(x)$ can always be written in Taylor form:

$$p(x) = \sum_{k=0}^{n} \frac{p^{(k)}(c)}{k!} (x - c)^k. \tag{2}$$

Substituting (1) into Equation (2) we have a formula for a polynomial $p(x)$ which approximates $f(x)$, for x near c:

$$p(x) = \sum_{k=0}^{n} \frac{f^{(k)}(c)}{k!} (x - c)^k.$$

We formalize the above ideas in the following definition.

Definition

If f has derivatives of order $1, 2, \ldots, n$ in an interval containing c, then the **nth Taylor*** **polynomial for f about c** is

$$T_n(x) = \sum_{k=0}^{n} \frac{f^{(k)}(c)}{k!} (x - c)^k. \tag{3}$$

If $c = 0$, the resulting polynomial

$$M_n(x) = \sum_{k=0}^{n} \frac{f^{(k)}(0)}{k!} x^k \tag{4}$$

is called the **nth Maclaurin**** **polynomial for f.**

EXAMPLE 1   Find the sixth Taylor polynomial for $f(x) = \sin x$

(a) about $\dfrac{\pi}{6}$ (b) about $\dfrac{\pi}{2}$.

Solution   To use Equation (3) we shall need the first six derivatives of $\sin x$ at $\pi/6$ and at $\pi/2$.

---

* Brook Taylor (1685–1731) was born into a well-to-do family in Middlesex, England, and earned a doctorate from Cambridge in 1714. Elected as a Fellow of the Royal Society in 1712, he was its secretary, 1714–1718. He was estranged from his father when he married a woman of lower economic status. After she died in childbirth he married a second wife, one who met with his father's approval. She also died in childbirth, although their daughter survived.

   During Taylor's career as a productive mathematician, he was closely associated with such leading scientific figures as Isaac Newton and Edmund Halley. His writings were often not well received because their terse style made them hard to follow. In his major book of 1715, Taylor developed and used the series that bears his name, although he was not the first to consider such series. His published works also show his interest in art, music, and philosophy.

** Colin Maclaurin (1698–1746) was born in Kilmodan, Scotland, and educated at the University of Glasgow. The son of a well-educated minister, Maclaurin lost both parents while still a child, and was raised by an uncle who was also a minister. He entered the university to study divinity, but was drawn into mathematics after a year. Before the age of 20, he was appointed professor of mathematics at Marischal College, Aberdeen. And upon Newton's recommendation, he was appointed professor at Edinburgh. Maclaurin's rigorous development of geometric methods and his systematic defense of Newton's calculus earned him the status of the most influential mathematican in the British Isles in the generation that followed Newton.

$$f^{(0)}(x) = \sin x \qquad f^{(0)}\left(\frac{\pi}{6}\right) = \frac{1}{2} \qquad f^{(0)}\left(\frac{\pi}{2}\right) = 1$$

$$f^{(1)}(x) = \cos x \qquad f^{(1)}\left(\frac{\pi}{6}\right) = \frac{\sqrt{3}}{2} \qquad f^{(1)}\left(\frac{\pi}{2}\right) = 0$$

$$f^{(2)}(x) = -\sin x \qquad f^{(2)}\left(\frac{\pi}{6}\right) = -\frac{1}{2} \qquad f^{(2)}\left(\frac{\pi}{2}\right) = -1$$

$$f^{(3)}(x) = -\cos x \qquad f^{(3)}\left(\frac{\pi}{6}\right) = -\frac{\sqrt{3}}{2} \qquad f^{(3)}\left(\frac{\pi}{2}\right) = 0$$

$$f^{(4)}(x) = \sin x \qquad f^{(4)}\left(\frac{\pi}{6}\right) = \frac{1}{2} \qquad f^{(4)}\left(\frac{\pi}{2}\right) = 1$$

$$f^{(5)}(x) = \cos x \qquad f^{(5)}\left(\frac{\pi}{6}\right) = \frac{\sqrt{3}}{2} \qquad f^{(5)}\left(\frac{\pi}{2}\right) = 0$$

$$f^{(6)}(x) = -\sin x \qquad f^{(6)}\left(\frac{\pi}{6}\right) = -\frac{1}{2} \qquad f^{(6)}\left(\frac{\pi}{2}\right) = -1$$

Then we have

(a) about $\dfrac{\pi}{6}$,

$$T_6(x) = \sum_{k=0}^{6} \frac{f^{(k)}\left(\frac{\pi}{6}\right)}{k!}\left(x - \frac{\pi}{6}\right)^k$$

$$= \frac{f^{(0)}\left(\frac{\pi}{6}\right)}{0!} + \frac{f^{(1)}\left(\frac{\pi}{6}\right)}{1!}\left(x - \frac{\pi}{6}\right) + \frac{f^{(2)}\left(\frac{\pi}{6}\right)}{2!}\left(x - \frac{\pi}{6}\right)^2$$

$$+ \frac{f^{(3)}\left(\frac{\pi}{6}\right)}{3!}\left(x - \frac{\pi}{6}\right)^3 + \frac{f^{(4)}\left(\frac{\pi}{6}\right)}{4!}\left(x - \frac{\pi}{6}\right)^4$$

$$+ \frac{f^{(5)}\left(\frac{\pi}{6}\right)}{5!}\left(x - \frac{\pi}{6}\right)^5 + \frac{f^{(6)}\left(\frac{\pi}{6}\right)}{6!}\left(x - \frac{\pi}{6}\right)^6$$

$$= \frac{\frac{1}{2}}{1} + \frac{\frac{\sqrt{3}}{2}}{1}\left(x - \frac{\pi}{6}\right) + \frac{-\frac{1}{2}}{2}\left(x - \frac{\pi}{6}\right)^2 + \frac{-\frac{\sqrt{3}}{2}}{6}\left(x - \frac{\pi}{6}\right)^3$$

$$+ \frac{\frac{1}{2}}{24}\left(x - \frac{\pi}{6}\right)^4 + \frac{\frac{\sqrt{3}}{2}}{120}\left(x - \frac{\pi}{6}\right)^5 + \frac{-\frac{1}{2}}{720}\left(x - \frac{\pi}{6}\right)^6$$

$$= \frac{1}{2} + \frac{\sqrt{3}}{2}\left(x - \frac{\pi}{6}\right) - \frac{1}{4}\left(x - \frac{\pi}{6}\right)^2 - \frac{\sqrt{3}}{12}\left(x - \frac{\pi}{6}\right)^3 + \frac{1}{48}\left(x - \frac{\pi}{6}\right)^4$$

$$+ \frac{\sqrt{3}}{240}\left(x - \frac{\pi}{6}\right)^5 - \frac{1}{1440}\left(x - \frac{\pi}{6}\right)^6.$$

(b) about $\dfrac{\pi}{2}$,

$$T_6(x) = \sum_{k=0}^{6} \frac{f^{(k)}\left(\dfrac{\pi}{2}\right)}{k!}\left(x - \frac{\pi}{2}\right)^k$$

$$= \frac{f^{(0)}\left(\dfrac{\pi}{2}\right)}{0!} + \frac{f^{(1)}\left(\dfrac{\pi}{2}\right)}{1!}\left(x - \frac{\pi}{2}\right) + \frac{f^{(2)}\left(\dfrac{\pi}{2}\right)}{2!}\left(x - \frac{\pi}{2}\right)^2 + \frac{f^{(3)}\left(\dfrac{\pi}{2}\right)}{3!}\left(x - \frac{\pi}{2}\right)^3$$

$$+ \frac{f^{(4)}\left(\dfrac{\pi}{2}\right)}{4!}\left(x - \frac{\pi}{2}\right)^4 + \frac{f^{(5)}\left(\dfrac{\pi}{2}\right)}{5!}\left(x - \frac{\pi}{2}\right)^5 + \frac{f^{(6)}\left(\dfrac{\pi}{2}\right)}{6!}\left(x - \frac{\pi}{2}\right)^6$$

$$= 1 + 0\left(x - \frac{\pi}{2}\right) + \left(\frac{-1}{2!}\right)\left(x - \frac{\pi}{2}\right)^2 + 0\left(x - \frac{\pi}{2}\right)^3$$

$$+ \frac{1}{4!}\left(x - \frac{\pi}{2}\right)^4 + 0\left(x - \frac{\pi}{2}\right)^5 + \left(\frac{-1}{6!}\right)\left(x - \frac{\pi}{2}\right)^6$$

$$= 1 - \frac{\left(x - \dfrac{\pi}{2}\right)^2}{4} + \frac{\left(x - \dfrac{\pi}{2}\right)^4}{24} - \frac{\left(x - \dfrac{\pi}{2}\right)^6}{720}. \qquad \blacksquare$$

Example 1 shows that Taylor polynomials $T_n(x)$ for the same function $f(x)$ about different points c can look quite different. In fact, the Taylor polynomial $T_6(x)$ for $\sin x$ about $\pi/2$ is much simpler than the Taylor polynomial $T_6(x)$ for $\sin x$ about $\pi/3$.

EXAMPLE 2 Find the eighth Maclaurin polynomial for $f(x) = e^x$.

Solution To use Equation (4) we need the first eight derivatives of e^x at 0. But this is trivial, since for all $k = 0, 1, 2, \ldots, 8$,

$$f^{(k)}(x) = \frac{d^k}{dx^k}\left[e^x\right] = e^x$$

and thus

$$f^{(k)}(0) = 1.$$

By equation (4), therefore

$$M_8(x) = \sum_{k=0}^{8} \frac{f^{(k)}(0)}{k!} x^k$$

$$= \sum_{k=0}^{8} \frac{x^k}{k!}$$

$$= \frac{x^0}{0!} + \frac{x^1}{1!} + \frac{x^2}{2!} + \frac{x^3}{3!} + \cdots + \frac{x^8}{8!}$$

$$= 1 + x + \frac{x^2}{2} + \frac{x^3}{6} + \frac{x^4}{24} + \frac{x^5}{120} + \frac{x^6}{720} + \frac{x^7}{5040} + \frac{x^8}{40,320}. \qquad \blacksquare$$

unction f with derivatives of order 0, 1,

$c)^k$ (5)

some interval containing c,
essed:

$\approx f(x)$?

n the approximation is "good"?

und by understanding Taylor's celebrated
efully here, but its proof is beyond the level

---

has derivative of order $n + 1$ in some open interval
mber c. Then for each x in this interval, there is a number
c such that

$$f(x) = \left\{ \sum_{k=0}^{n} \frac{f^{(k)}(c)}{k!} (x - c)^k \right\} + \frac{f^{(n+1)}(t)}{(n+1)!} (x - c)^{n+1}.$$ (6)

Because Taylor's Theorem may seem quite difficult to comprehend on first
reading, we take a closer look. When we ask questions about how well $T_n(x)$ approximates $f(x)$, we are really asking about the "remainder" function,

$$R_n(x) = f(x) - T_n(x).$$ (7)

Taylor's theorem tells us that when f has an $(n + 1)$st derivative throughout some
interval containing c, then this remainder is expressible in the form

$$R_n(x) = \frac{f^{(n+1)}(t)}{(n+1)!} (x - c)^{n+1},$$ (8)

for some number t in the interval between x and c. Thus, *Taylor's Theorem is
really a statement about the accuracy of the approximation $T_n(x) \approx f(x)$.*

Accuracy of the Approximation $T_n(x) \approx f(x)$ (Optional)

The trouble we encounter when we try to use Equation (8) is that Taylor's theorem does not give us any idea about the value of t, except that it is in the interval between x and c. Since we do not know the va

value of $R_n(x)$. Fortunately, we can often resort to

Consider some interval I containing c. Suppose

number M such that

$$\left|f^{(n+1)}(t)\right| \le M$$

for all t in the interval I. Then, substituting (9) into Equa

$$\left|R_n(x)\right| \le \frac{M\left|x - c\right|^{n+1}}{(n+1)!}$$

for all x in this interval. We can see by the following example how

(10) truly is.

EXAMPLE 3 How accurate is the sixth Taylor polynomial for $\sin x$ about $\pi/2$,
(optional)

$$T_6(x) = 1 - \frac{(x - \frac{\pi}{2})^2}{4} + \frac{(x - \frac{\pi}{2})^4}{24} - \frac{(x - \frac{\pi}{2})^6}{720}$$

[see Example 1] when used in calculating $\sin x$ for values of x between
$\pi/2$ (that is, between 45° and 90°)?

Solution The accuracy of the approximation is measured by the remainder, $R_6(x)$. Acc
ing to Equation (8),

$$R_6(x) = \frac{f^{(7)}(t)}{7!}\left(x - \frac{\pi}{2}\right)^7$$

for some t between x and $\pi/2$. Now, when $f(x) = \sin x$,

$$f^{(7)}(x) = -\cos x.$$

Regardless of the value of x, $\left|-\cos x\right| \le 1$. Thus, we can take $M = 1$ in (9) and
(10). Therefore, from (10),

$$\left|R_6(x)\right| \le \frac{1}{7!}\left|x - \frac{\pi}{2}\right|^7. \tag{12}$$

As stated in the problem, we are considering the approximation $T_6(x) \approx f(x)$ for
values of x between $\pi/4$ and $\pi/2$. Thus

$$\frac{\pi}{4} \le x \le \frac{\pi}{2}$$

so that

$$\frac{\pi}{4} - \frac{\pi}{2} \leq x - \frac{\pi}{2} \leq \frac{\pi}{2} - \frac{\pi}{2}$$

$$-\frac{\pi}{4} \leq x - \frac{\pi}{2} \leq 0.$$

Thus

$$\left| x - \frac{\pi}{2} \right| \leq \frac{\pi}{4} \approx 0.785 < 0.8,$$

or

$$\left| x - \frac{\pi}{2} \right| < 0.8. \tag{13}$$

Substituting this result into (12), we have

$$|R_6(x)| \leq \frac{(0.8)^7}{7!} = 0.0000416.$$

Therefore, in using the Taylor polynomial (11) to calculate $\sin x$, when x is between $\pi/4$ and $\pi/2$, the value of $T_6(x)$ will deviate from $\sin x$ by no more than 0.0000416. Put another way, $T_6(x)$ will provide at least 4 decimal place accuracy when used to approximate $\sin x$, for $\pi/4 \leq x \leq \pi/2$. ∎

There are at least two ways to improve the accuracy of Taylor approximations:

(1) by increasing the value of n; that is, by using a Taylor polynomial of higher degree, or
(2) by decreasing the size of the interval.

In the following two examples we examine these two alternatives.

**EXAMPLE 4
(optional)** How is the accuracy of the approximation

$$T_n(x) \approx \sin x$$

about $\pi/2$ affected when n is changed from 6 to 8, but the interval $\pi/4 \leq x \leq \pi/2$ is maintained?

Solution We have,

$$T_8(x) = 1 - \frac{\left(x - \frac{\pi}{2} \right)^2}{4} + \frac{\left(x - \frac{\pi}{2} \right)^4}{24} - \frac{\left(x - \frac{\pi}{2} \right)^6}{720} + \frac{\left(x - \frac{\pi}{2} \right)^8}{40,320}. \tag{14}$$

According to Equation (8), the remainder is

$$R_8(x) = \frac{f^{(9)}(t)}{9!} \left(x - \frac{\pi}{2} \right)^9.$$

Since $f^{(9)}(x) = \sin x$, we have $|f^{(9)}(t)| \leq 1$ regardless of the value of t. Thus

$$|R_8(x)| \leq \frac{1}{9!} \left|x - \frac{\pi}{2}\right|^9$$

$$\leq \frac{(0.8)^9}{9!} \quad \text{[from (13)]}$$

$$= 0.00000037.$$

Therefore, the value of $T_8(x)$ will deviate from $\sin x$ by no more than 0.00000037, when x is between $\pi/4$ and $\pi/2$. The accuracy has been improved from 4 decimal places to 6 decimal places. ∎

EXAMPLE 5 (optional)   How is the reliability of the approximation

$$T_n(x) \approx \sin x$$

about $\pi/2$ affected when n is kept at 6 but when the interval is halved to $3\pi/8 \leq x \leq \pi/2$?

Solution   Going back to Example 3, we recall that

$$|R_6(x)| \leq \frac{1}{7!} \left(x - \frac{\pi}{2}\right)^7. \tag{12}$$

From

$$\frac{3\pi}{8} \leq x \leq \frac{\pi}{2}$$

we find

$$\frac{3\pi}{8} - \frac{\pi}{2} \leq x - \frac{\pi}{2} \leq \frac{\pi}{2} - \frac{\pi}{2}$$

$$-\frac{\pi}{8} \leq x - \frac{\pi}{2} \leq 0$$

so that

$$\left|x - \frac{\pi}{2}\right| \leq \frac{\pi}{8} \approx 0.393$$

or

$$\left|x - \frac{\pi}{2}\right| < 0.4.$$

Substituting this result into (12) we have

$$|R_6(x)| \leq \frac{(0.4)^7}{7!} = 0.000000325$$

$$< 0.00000033.$$

Therefore, the value of $T_6(x)$ will deviate from sin x by no more than 0.00000033, when x is between $3\pi/8$ and $\pi/2$. The accuracy has been improved from 4 decimal places to 6 decimals places. ∎

Graphs of Taylor Polynomials

When we view the graph of a Taylor polynomial of a function f about c, we can get a good idea of how well the Taylor polynomial $T_n(x)$ approximates the function $f(x)$ for values of x near c.

In Figure 1, we graph $y = e^x$ along with its Maclaurin polynomial $M_n(x)$, for $n = 1, 2, 3,$ and 4.

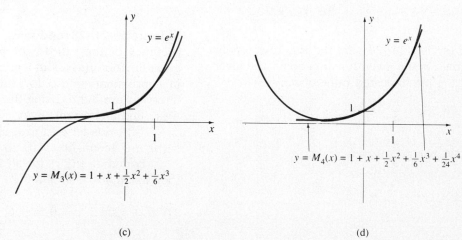

Figure 1
Maclaurin Polynomials for $f(x) = e^x$

In Figure 2 we graph $y = \sin x$ along with its Maclaurin polynomial $M_n(x)$, for $n = 1, 2, 3,$ and 4.

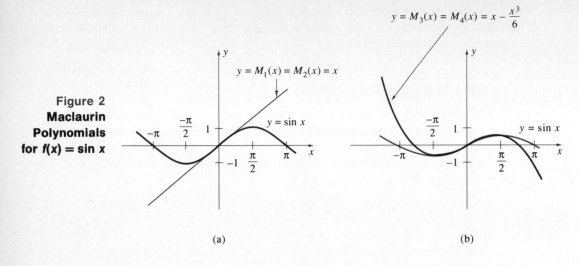

Figure 2
Maclaurin
Polynomials
for $f(x) = \sin x$

(a)

(b)

12.3 EXERCISE SET ||||||||||||||||||||||||||||||||||||||||||||||||||||||||||||||||||||||||||||||||||||||

In Exercises 1–10, find the nth Maclaurin polynomial for the given function about the given cated value of n.

1. $\sin x$, $n = 8$

2. $\cos x$, $n = 8$

3. $\ln(1 + x)$, $n = 8$

4. e^{-2x}, $n = 6$

5. $\dfrac{1}{1 - x}$, $n = 6$

6. $\dfrac{1}{1 + 2_x}$, $n = 6$

7. $\sqrt{1 + x}$, $n = 5$

8. $\sqrt[3]{1 + x}$, $n = 6$

9. $\tan x$, $n = 4$

10. $\sec x$, $n = 4$

In Exercises 11–20, find the nth Taylor polynomial for the given function about the given c, using the indicated value of n.

11. $\sin x$, $c = \dfrac{\pi}{4}$, $n = 6$

12. $\sin x$, $c = \dfrac{\pi}{3}$, $n = 6$

13. $\cos x$, $c = \dfrac{\pi}{2}$, $n = 6$

14. $\cos x$, $c = \dfrac{\pi}{4}$, $n = 6$

15. $\dfrac{1}{x}$, $c = 1$, $n = 8$

16. $\sqrt{x}$, $c = 4$, $n = 1$

17. e^x, $c = 1$, $n = 6$

18. $\ln x$, $c = 1$, $n = 8$

19. $\tan x$, $c = \dfrac{\pi}{4}$, $n = 4$

20. $\sec x$, $c = \dfrac{\pi}{4}$, $n = 4$

In Exercises 21–28 (**optional**), we examine the accuracy of the nth Taylor polynomial of a given function $f(x)$ about a given c, when used in approximating $f(x)$ for values of x in an interval I around c. Using the given n and I,
  (a) find a constant M for which Inequality (9) holds for all t in I, and
  (b) use Inequality (10) to find an upper bound on $|R_n(x)|$ for all x in I.
[See Examples 3, 4, and 5.]

21. $f(x) = \sin x$, $c = 0$, $n = 6$, $I = \left(0, \dfrac{\pi}{4}\right)$

22. $f(x) = \cos x$, $c = 0$, $n = 8$, $I = \left(-\dfrac{\pi}{4}, \dfrac{\pi}{4}\right)$

23. $f(x) = e^x$, $c = 0$, $n = 6$, $I = (0, 1)$

24. $f(x) = e^x$, $c = 0$, $n = 8$, $I = (0, 1)$

25. $f(x) = \ln x$, $c = 1$, $n = 6$, $I = \left(\dfrac{1}{2}, \dfrac{3}{2}\right)$

26. $f(x) = \ln x$, $c = 4$, $n = 6$, $I = (3, 5)$

27. $f(x) = \sqrt{x}$, $c = 1$, $n = 4$, $I = (0, 2)$

28. $f(x) = \sqrt{x}$, $c = 5$, $n = 4$, $I = (4, 6)$

12.4 ‖‖‖‖   INFINITE SERIES

If $\{a_n\}$ is a sequence of real numbers, then the infinite sum

$$\sum_{k=1}^{\infty} a_k = a_1 + a_2 + a_3 + \cdots + a_n + \cdots \tag{1}$$

is called an **infinite series.** The number a_k is called the **kth term** of the series $\sum_{k=1}^{\infty} a_k$, as well as the kth term of the sequence $\{a_k\}$.

One might argue that such a sum does not make sense; one cannot possibly add *infinitely many* numbers. Indeed, what we have in mind by (1) is something a bit different from ordinary addition.

We may begin by thinking of an infinite series $\sum_{k=1}^{\infty} a_k$ as a purely formal symbolism, constructed from a sequence $\{a_k\}$ of real numbers. We then construct a sequence $\{S_n\}$ of sums

$$S_1 = a_1$$
$$S_2 = a_1 + a_2$$
$$S_3 = a_1 + a_2 + a_3$$
$$\vdots$$
$$S_n = a_1 + a_2 + \cdots + a_n = \sum_{k=1}^{n} a_k$$
$$\vdots$$

We call the sum S_n the **nth partial sum** of the series $\sum_{k=1}^{\infty} a_k$, and we call the sequence $\{S_n\}$ the **sequence of partial sums** of the series.

Definition 1

When we say

$$\sum_{k=1}^{\infty} a_k = S,$$

we mean that the sequence $\{S_n\}$ of partial sums

$$S_n = \sum_{k=1}^{n} a_k \tag{2}$$

converges, and has limit S. That is,

$$S_n \to S, \text{ or } \lim_{n \to +\infty} \sum_{k=1}^{n} a_k = S. \tag{3}$$

Definiton 2   When the sequence of partial sums $\{S_n\}$ converges to a real number S, we say that the series $\sum_{k=1}^{\infty} a_k$ **converges** to S, and that $\sum_{k=1}^{\infty} a_k$ is a **convergent series.** Moreover, we say that S is the **sum** of the series. If the sequence of partial sums $\{S_n\}$ is a divergent sequence, then we say that $\sum_{k=1}^{\infty} a_k$ **diverges** or is a **divergent series.** If $\lim_{n \to +\infty} S_n = +\infty$, then we say that $\sum_{k=1}^{\infty} a_k$ diverges to $+\infty$ and we write $\sum_{k=1}^{\infty} a_k = +\infty$.

EXAMPLE 1   Determine whether the series $\sum_{k=1}^{\infty} \dfrac{1}{k(k+1)}$ converges or diverges. Find the sum, if it converges.

Solution   Our analysis of this series involves a "trick." The kth term of the series is

$$a_k = \frac{1}{k(k+1)}$$

which can be rewritten as

$$a_k = \frac{1}{k} - \frac{1}{k+1} \qquad \text{(verify this).}$$

Thus, the nth partial sum is

$$S_n = \sum_{k=1}^{n} \left(\frac{1}{k} - \frac{1}{k+1} \right)$$

$$= \left(1 - \frac{1}{2} \right) + \left(\frac{1}{2} - \frac{1}{3} \right) + \left(\frac{1}{3} - \frac{1}{4} \right) + \cdots + \left(\frac{1}{n} - \frac{1}{n+1} \right).$$

Notice that all terms in this sum cancel out, except the first and the last terms. Thus,

$$S_n = 1 - \frac{1}{n+1}.$$

Then,

$$\lim_{n \to +\infty} S_n = \lim_{n \to +\infty} \left(1 - \frac{1}{n+1} \right)$$

$$= 1.$$

Therefore, by Definition 1,

$$\sum_{k=1}^{\infty} \frac{1}{k(k+1)} = 1.$$

∎

EXAMPLE 2 Show that each of the following series diverges.

(a) $\sum_{k=1}^{\infty} (-1)^k,$ (b) $\sum_{k=1}^{\infty} 2.$

Solution (a) The nth partial sum of the series $\sum_{k=1}^{\infty} (-1)^k$ is

$$S_n = \sum_{k=1}^{n} (-1)^k = -1 + 1 - 1 + 1 - 1 + \cdots + (-1)^n$$

$$= \begin{cases} -1 & \text{if } n \text{ is odd} \\ 0 & \text{if } n \text{ is even.} \end{cases}$$

Since the sequence $\{S_n\}$ diverges, the given series diverges.

(b) The nth partial sum is

$$S_n = \sum_{k=1}^{n} 2 = \underbrace{2 + 2 + 2 + \cdots + 2}_{n \ 2\text{'s}}$$

$$= 2n.$$

Thus, $\lim\limits_{n \to +\infty} S_n = +\infty$, so we say that the series $\sum_{k=1}^{\infty} 2$ diverges to $+\infty$. ∎

As in Section 12.2, the variable k used in the subscript is strictly a dummy variable. Any letter may be used instead of k. Thus,

$$\sum_{k=1}^{\infty} a_k, \qquad \sum_{i=1}^{\infty} a_i, \qquad \text{and} \qquad \sum_{j=1}^{\infty} a_j$$

all represent the same series

$$a_1 + a_2 + a_3 + \cdots + a_n + a_{n+1} + \cdots.$$

Moreover, it is not necessary that the first term of a series have 1 as its subscript. For example, the following are acceptable infinite series:

$$a_0 + a_1 + a_2 + \cdots + a_n + \cdots = \sum_{k=0}^{\infty} a_k$$

$$a_4 + a_5 + a_6 + \cdots + a_n + \cdots = \sum_{k=4}^{\infty} a_k.$$

However, we must realize that, for example,

$$\sum_{k=0}^{\infty} a_k = a_0 + \sum_{k=1}^{\infty} a_k, \qquad \text{and} \qquad \sum_{k=4}^{\infty} a_k = \sum_{k=1}^{\infty} a_k - (a_1 + a_2 + a_3).$$

Caution.   Many students are "sloppy" about the distinction between series and sequences. A sequence $\{a_n\}$ must never be called a series, and vice versa. Remember that in a series, the terms are being *added*.

|||||||||||| **Geometric Series**

An infinite series of the form

$$\sum_{k=0}^{\infty} ar^k = a + ar + ar^2 + ar^3 + \cdots + ar^n + \cdots \tag{4}$$

is called a **geometric series.** The nth partial sum of this series can easily be obtained, using Theorem 2 of Section 12.2. When $r \neq 1$ we have

$$S_n = \sum_{k=0}^{n} ar^k$$

$$= a\left(\frac{1 - r^{n+1}}{1 - r}\right)$$

$$= \frac{a}{1 - r} - \left(\frac{a}{1 - r}\right)r^{n+1}.$$

Therefore, when $r \neq 1$ the limit of the sequence of nth partial sums is

$$\lim_{n \to +\infty} S_n = \frac{a}{1 - r} - \frac{a}{1 - r}\left(\lim_{n \to +\infty} r^{n+1}\right). \tag{5}$$

By Theorem 2 of Section 12.1,

$$\lim_{n \to +\infty} r^{n+1} = r \lim_{n \to +\infty} r^n$$

$$= r \begin{cases} 0 & \text{if } |r| < 1 \\ 1 & \text{if } r = 1 \\ +\infty & \text{if } r > 1 \\ \text{(does not exist if } r \leq -1). \end{cases}$$

When $|r| < 1$, this result substituted into Equation (5) yields

$$\lim_{n \to +\infty} S_n = \frac{a}{1 - r} - \frac{a}{1 - r} \cdot 0$$

$$= \frac{a}{1 - r}.$$

It is not hard to see that when $|r| \geq 1$, $\{S_n\}$ diverges.

THEOREM 1

> **(Geometric Series)**
> The geometric series
>
> $$\sum_{k=0}^{\infty} ar^k = a + ar + ar^2 + \cdots + ar^n + \cdots$$
>
> (a) converges to $\dfrac{a}{1-r}$ if $|r| < 1$,
>
> (b) diverges if $|r| \geq 1$.

EXAMPLE 3   Investigate each of the following geometric series for convergence or divergence. Find the sum where possible.

(a) $3 + \dfrac{3}{2} + \dfrac{3}{4} + \dfrac{3}{8} + \dfrac{3}{16} + \cdots + \dfrac{3}{2^n} + \cdots$

(b) $5 - \dfrac{5}{3} + \dfrac{5}{9} - \dfrac{5}{27} + \cdots + \dfrac{5}{(-3)^n} + \cdots$

(c) $\displaystyle\sum_{k=0}^{\infty} \left(\dfrac{3}{2}\right)^k$

(d) $\displaystyle\sum_{k=3}^{\infty} 7\left(-\dfrac{1}{4}\right)^k$

Solution   (a) In this geometric series, $a = 3$ and $r = \frac{1}{2}$. Since $|r| < 1$, the series converges and we have

$$\sum_{k=0}^{\infty} 3\left(\dfrac{1}{2}\right)^k = \sum_{k=0}^{\infty} ar^k$$

$$= \dfrac{a}{1-r} \qquad \text{(by Theorem 1)}$$

$$= \dfrac{3}{1-\frac{1}{2}} = \dfrac{3}{\left(\frac{1}{2}\right)}$$

$$= 6.$$

(b) In this geometric series, $a = 2$ and $r = -\frac{1}{3}$. Since $|r| < 1$, the series converges and we have

$$\sum_{k=0}^{\infty} 5\left(-\dfrac{1}{3}\right)^k = \sum_{k=0}^{\infty} ar^k$$

$$= \dfrac{a}{1-r} \qquad \text{(by Theorem 1)}$$

$$= \dfrac{5}{1-\left(-\frac{1}{3}\right)} = \dfrac{5}{\left(\frac{4}{3}\right)} = \dfrac{15}{4}$$

$$= 3.75.$$

(c) In this geometric series we have $a = 1$ and $r = \frac{3}{2}$. Since $|r| > 1$, the series diverges, by Theorem 1. Further, since all terms are positive, we have

$$\sum_{k=0}^{\infty} \left(\frac{3}{2}\right)^k = +\infty.$$

(d) First, we notice that the index k begins with 3, rather than 0. To use Theorem 1 the index must begin with $k = 0$. Thus, we write

$$\sum_{k=3}^{\infty} 7\left(-\frac{1}{4}\right)^k = \sum_{k=0}^{\infty} 7\left(-\frac{1}{4}\right)^k - \left[7\left(-\frac{1}{4}\right)^0 + 7\left(-\frac{1}{4}\right)^1 + 7\left(-\frac{1}{4}\right)^2\right]. \quad (6)$$

Now, for the geometric series $\sum_{k=0}^{\infty} 7(-\frac{1}{4})^k$, we have $a = 7$ and $r = -\frac{1}{4}$. Since $|r| < 1$, this series converges and we have

$$\sum_{k=0}^{\infty} 7\left(-\frac{1}{4}\right)^k = \frac{7}{1 - \left(-\frac{1}{4}\right)} = \frac{7}{1 + \frac{1}{4}} = \frac{7}{\left(\frac{5}{4}\right)}$$

$$= \frac{28}{5}.$$

Substituting this result into Equation (6), we have

$$\sum_{k=3}^{\infty} 7\left(-\frac{1}{4}\right)^k = \frac{28}{5} - \left[7 - \frac{7}{4} + \frac{7}{16}\right]$$

$$= \frac{28}{5} - \frac{91}{16} = -\frac{7}{80}$$

$$= -0.0875. \qquad \blacksquare$$

EXAMPLE 4 (Drug Therapy) In the treatment of certain diseases, the patient must take a single daily dose of a drug. The body does not completely eliminate this drug in one day, but only a certain fixed percentage of it. Thus, the amount of the drug that is in the body after each dose increases day by day. We shall see that this amount approaches a "plateau."

Suppose that the daily dosage is 25 milligrams and that in 24 hours the body eliminates 30% of the amount of the drug that is in the body after the dose is taken. Let S_n denote the amount of the drug that is in the body on the nth day, just after the nth dose has been taken. Then

$$S_1 = 25 \quad \text{(milligrams)}$$
$$S_2 = 25 + 25(0.70)$$
$$S_3 = 25 + [25 + 25(0.70)](0.70) = 25 + 25(0.70) + 25(0.70)^2$$
$$\vdots$$
$$S_n = 25 + 25(0.70) + 25(0.70)^2 + \cdots + 25(0.70)^{n-1} = \sum_{k=0}^{n-1} 25(0.70)^k.$$

Then,

$$\lim_{n \to \infty} S_n = \sum_{k=0}^{\infty} 25(0.70)^k$$

$$= \frac{25}{(1 - 0.70)} = \frac{25}{0.30}$$

$$= 83\tfrac{1}{3} \text{ milligrams.}$$

That is, in the long run, the amount of the drug in the body each day, after the daily dose has been taken, approaches a plateau of $83\tfrac{1}{3}$ milligrams. ∎

|||||||||||||| **The Multiplier Effect in Economics**

Suppose that, on the average, each component of a national economy spends* 80 cents of every dollar it earns. In the language of economics, the nation's **marginal propensity to consume (MPC)** is 0.80. To stimulate the economy, the government injects R dollars into the economy, through some means such as a tax cut or a deficit-spending program. As this money circulates through the economy, each component receiving a portion of it spends 80% of what it receives.

To determine the total increase in spending throughout the nation, caused by these R dollars, we reason as follows:

$$R = \text{original government spending}$$
$$0.8R = \text{amount spent by the recipients of these } R \text{ dollars}$$
$$(0.8)^2R = \text{amounts spent by the recipients of the previous } 0.8R \text{ dollars}$$
$$(0.8)^3R = \text{amount spent by the recipients of the previous } (0.8)^2R \text{ dollars}$$

and so on. Thus, the total spending impact of the original R dollars on the national economy is

$$S = R + 0.8R + (0.8)^2R + (0.8)^3R + \cdots$$

$$= \sum_{k=0}^{\infty} R(0.8)^k$$

$$= \frac{R}{1 - 0.8}$$

$$= 5R.$$

That is, when the MPC is 0.8, an infusion of R dollars into the economy has the effect of inducing an increase of $5R$ dollars in the total national spending. The number 5, in this case, is called the **multiplier.**

---

* For simplicity, all expenditures cited here take place within the nation, and not in foreign markets.

EXAMPLE 5 Suppose that the government of a nation whose marginal propensity to consume is 0.95 initiates an 8-billion-dollar tax cut. Determine the total effect on the nation's spending. What is the multiplier?

Solution Using the above reasoning, the total increase in the nation's spending (in billions of dollars) is

$$S = 8 + 8(0.95) + 8(0.95)^2 + 8(0.95)^3 + \cdots$$

$$= \sum_{k=0}^{\infty} 8(0.95)^k$$

$$= \frac{8}{1 - 0.95} = \frac{8}{0.05}$$

$$= 160.$$

Therefore, the 8-billion-dollar tax cut will increase spending throughout the nation by 160 billion dollars. The multiplier is $1/(1 - 0.95) = 20$. ∎

|||||||||||| Perpetuities

A **perpetuity** is a sequence of equal payments paid at regular time intervals, beginning on some initial date and continuing without end. In Section 7.6 we calculated the present value of a perpetuity by treating the perpetuity as a continuous stream of income. That method, however, gives only an *approximate* answer, since the perpetuity is not actually a *continuous* stream of income. To calculate the actual present value, we must use infinite series. More precisely, the answer is found as the sum of a geometric series.

Suppose that we want to set up an interest-bearing account that will make equal payments of R dollars (to a charity, for example) at the end of regular time intervals. Suppose that the account pays interest at the rate i per time period, compounded at the end of each time period, just before the next payment is made. Let P denote the initial principal that must be invested in the account in order to set up the perpetuity. Assume that the first payment is to be made immediately.

The present value of a payment of R dollars made k time periods from now is

$$R(1 + i)^{-k}$$

by Equation (8) of Section 5.1. Thus, the total present value of all the payments made by the perpetuity is

$$P = R + R(1 + i)^{-1} + R(1 + i)^{-2} + R(1 + i)^{-3} + \cdots$$

or

$$P = \sum_{k=0}^{\infty} R(1 + i)^{-k} = \sum_{k=0}^{\infty} R\left(\frac{1}{1 + i}\right)^k. \tag{7}$$

We observe that the series in Equation (7) is a geometric series with $|r| = \left|\dfrac{1}{1 + i}\right| < 1$. Thus, the sum may be found using Theorem 1. The present value P

is sometimes called the **capitalized value** of the perpetuity. It may be thought of as the principal that must be invested now to set up the perpetuity.

EXAMPLE 6
(Perpetuities)   A loyal university alumnus wishes to establish a perpetuity that will contribute $1000 each year to his Alma Mater. If these contributions are made from an interest-bearing account that pays 8% per year, compounded annually after each contribution has been made, what is the capitalized value of this perpetuity? Assume that the first $1000 contribution is made immediately.

Solution   From Equation (7), the present value of this perpetuity is

$$P = \sum_{k=0}^{\infty} 1000(1 + 0.08)^{-k}.$$

By Theorem 1,

$$P = \frac{1000}{1 - (1.08)^{-1}}$$
$$= \$13,500.$$

Therefore, the capitalized value of the perpetuity is $13,500.   ∎

NOTE   In Example 5 of Section 7.6 the answer to a nearly identical question was $12,500. The discrepancy arises from two sources. First, in the present example the alumnus must pay $1000 immediately, rather than spread that amount continuously over one year. Second, the interest is added to the account only once each year, rather than continuously.

EXAMPLE 7
(Perpetuities)   Suppose that in Example 6 the interest is compounded semiannually instead of annually, but all other specifications are kept unchanged. How much money must the alumnus invest today to set up the fund? Compare this amount with the answer to Example 6.

Solution   This time the interest rate is $i = 0.04$ per time period, but payments are made only at the end of the 0th time period, the 2nd time period, the 4th time period, the 6th time period, and so on. Thus, the total present value of all the payments is

$$P = 1000 + 1000(1.04)^{-2} + 1000(1.04)^{-4} + 1000(1.04)^{-6} + \cdots$$
$$= \sum_{k=0}^{\infty} 1000(1.04)^{-2k}.$$

This is a geometric series with $a = 1000$ and $r = (1.04)^{-2}$. By Theorem 1,

$$P = \frac{1000}{1 - (1.04)^{-2}}$$
$$\approx \$13,254.90.$$

Therefore, when the account pays interest semiannually, the fund can be started with $13,254.90, a savings of $245.10 over the amount required in Example 4. ∎

|||||||||||| **Basic Algebraic Properties of Series**

The following theorem is used frequently when series are combined in linear combinations.

THEOREM 2

(Linear Combinations of Series)

If $\sum_{k=1}^{\infty} a_k$ and $\sum_{k=1}^{\infty} b_k$ are convergent series, and c is any real number, then the series $\sum_{k=1}^{\infty} (a_k + b_k)$, $\sum_{k=1}^{\infty} (a_k - b_k)$, and $\sum_{k=1}^{\infty} ca_k$ are also convergent, and

(a) $\displaystyle\sum_{k=1}^{\infty} (a_k + b_k) = \sum_{k=1}^{\infty} a_k + \sum_{k=1}^{\infty} b_k$

(b) $\displaystyle\sum_{k=1}^{\infty} (a_k - b_k) = \sum_{k=1}^{\infty} a_k - \sum_{k=1}^{\infty} b_k$

(c) $\displaystyle\sum_{k=1}^{\infty} ca_k = c \sum_{k=1}^{\infty} a_k$

Proof Suppose $\sum_{k=1}^{\infty} a_k = S$ and $\sum_{k=1}^{\infty} b_k = T$. Denote the sequences of partial sums by

$$S_n = \sum_{k=1}^{n} a_k \quad \text{and} \quad T_n = \sum_{k=1}^{n} b_k.$$

Then, by definition, $S_n \to S$ and $T_n \to T$. We shall prove (a) and (c) leaving the proof of (b) as an exercise (Exercise 35).

(a) Denote the nth partial sum of $\sum_{k=1}^{n} (a_k + b_k)$ by U_n; that is,

$$U_n = \sum_{k=1}^{n} (a_k + b_k).$$

By Theorem 1 of Section 12.2,

$$U_n = \sum_{k=1}^{n} a_k + \sum_{k=1}^{n} b_k.$$

Then

$$\lim_{n \to +\infty} U_n = \lim_{n \to +\infty} \left(\sum_{k=1}^{n} a_k + \sum_{k=1}^{n} b_k \right)$$

$$= \left(\lim_{n \to +\infty} \sum_{k=1}^{n} a_k \right) + \left(\lim_{n \to +\infty} \sum_{k=1}^{n} b_k \right) \qquad \text{[Theorem 3, Section 12.1]}$$

$$= \lim_{n \to +\infty} S_n + \lim_{n \to +\infty} T_n$$

$$= S + T.$$

By Definition 1, this means that the series $\sum_{k=1}^{\infty} (a_k + b_k)$ converges, and

$$\sum_{k=1}^{\infty} (a_k + b_k) = S + T$$

$$= \sum_{k=1}^{\infty} a_k + \sum_{k=1}^{\infty} b_k.$$

(c) Denote the nth partial sum of $\sum_{k=1}^{\infty} ca_k$ by V_n; that is,

$$V_n = \sum_{k=1}^{n} ca_k.$$

By Theorem 1 of Section 12.2,

$$V_n = c \sum_{k=1}^{n} a_k.$$

Then

$$\lim_{n \to +\infty} V_n = \lim_{n \to +\infty} \left(c \sum_{k=1}^{n} a_k \right)$$

$$= c \lim_{n \to +\infty} \sum_{k=1}^{n} a_k \qquad \text{[Theorem 3, Section 12.1]}$$

$$= c \lim_{n \to +\infty} S_n$$

$$= cS.$$

By Definition 1, this means that the series $\sum_{k=1}^{\infty} ca_k$ converges, and

$$\sum_{k=1}^{\infty} ca_k = cS$$

$$= c \sum_{k=1}^{\infty} a_k.$$ ■

EXAMPLE 8   Find the sum of each of the following series:

(a) $\displaystyle\sum_{k=0}^{\infty} \frac{2^k + 3^k}{5^k}$ (b) $\displaystyle\sum_{k=0}^{\infty} \frac{4}{k^2 + k}$

Solution   (a) $\displaystyle\sum_{k=0}^{\infty} \frac{2^k + 3^k}{5^k} = \sum_{k=0}^{\infty} \left[\frac{2^k}{5^k} + \frac{3^k}{5^k} \right] = \sum_{k=0}^{\infty} \left[\left(\frac{2}{5}\right)^k + \left(\frac{3}{5}\right)^k \right]$

$$= \sum_{k=0}^{\infty} \left(\frac{2}{5}\right)^k + \sum_{k=0}^{\infty} \left(\frac{3}{5}\right)^k \qquad \text{by Theorem 2(a)}$$

$$= \frac{1}{1 - \frac{2}{5}} + \frac{3}{1 - \frac{3}{5}} \qquad \text{by Theorem 1}$$

$$= \frac{5}{3} + \frac{5}{2}$$

$$= \frac{25}{6}.$$

(b) $\displaystyle\sum_{k=0}^{\infty} \frac{4}{k^2 + k} = 4 \sum_{k=0}^{\infty} \frac{1}{k^2 + k}$ by Theorem 2(c)

$$= 4 \sum_{k=0}^{\infty} \frac{1}{k(k + 1)}$$

$$= 4 \cdot 1 \quad \text{by Example 1}$$

$$= 4.$$ ∎

12.4 EXERCISE SET ||||||||||||||||||||||||||||||||||||||||||||||||||||||||||||||||||||||||||||

In Exercises 1–6, write out the sum of the first six terms. Do not calculate the sum.

1. $\displaystyle\sum_{k=1}^{\infty} \frac{3k}{k^2 + 1}$

2. $\displaystyle\sum_{k=1}^{\infty} \frac{k + 4}{2k - 1}$

3. $\displaystyle\sum_{k=0}^{\infty} \frac{(-1)^k}{3^k}$

4. $\displaystyle\sum_{k=0}^{\infty} \frac{2^k}{\sqrt{k + 1}}$

5. $\displaystyle\sum_{k=5}^{\infty} \frac{k - 3}{k + 1}$

6. $\displaystyle\sum_{k=7}^{\infty} \frac{(-1)^k}{2k}$

In Exercises 7 and 8, use the method of Example 1 to demonstrate that the series converges, and find its sum.

7. $\displaystyle\sum_{k=50}^{\infty} \left(\frac{1}{k} - \frac{1}{k + 1} \right)$

8. $\displaystyle\sum_{k=1}^{\infty} \left(\frac{1}{k} - \frac{1}{k + 2} \right)$

In Exercises 9–16, determine whether the given geometric series converges or diverges, and find its sum where possible.

9. $\dfrac{4}{3} + \dfrac{4}{9} + \dfrac{4}{27} + \dfrac{4}{81} + \cdots$

10. $\dfrac{2}{5} - \dfrac{6}{25} + \dfrac{18}{125} - \dfrac{54}{625} + \cdots$

11. $0.5 + 0.05 + 0.005 + 0.0005 + \cdots$

12. $0.21 + 0.0021 + 0.000021 + \cdots$

13. $1 - \dfrac{4}{3} + \dfrac{16}{9} - \dfrac{64}{27} + \cdots$

14. $0.1 - 0.2 + 0.4 - 0.8 + \cdots$

15. $0.1 - 0.1 + 0.1 - 0.1 + 0.1 + \cdots$

16. $\dfrac{1}{\sqrt{2}} + \dfrac{1}{2} + \dfrac{1}{2^{3/2}} + \dfrac{1}{4} + \dfrac{1}{2^{5/2}} + \cdots$

17. **(Drug Therapy)** In the treatment of a certain disease the patient must take 12 milligrams of a drug daily. If the body eliminates 20% of the amount of the drug in the body each day, find S_n, the amount of the drug that is in the body on the nth day, just after the nth dose has been taken. In the long run, how much of the drug will be in the body each day after the daily dose is taken? Express the answer as the sum of a geometric series.

18. **(Drug Therapy)** Redo Exercise 17, given that the daily dosage is 5 milligrams and that the body eliminates 10% of the drug daily.

19. **(Economics)** Use an infinite series to express the total increase in spending that will be generated by a 9-billion-dollar tax cut in a nation whose MPC is 85%. Find the sum of this series. What is the "multiplier" in this case?

20. **(Economics)** In a nation whose MPC is 92%, the government decides to stimulate the economy by spending an extra 20 billion dollars on public renewal projects. Use an infinite series to express the total impact on spending in this nation, and find the sum of this series. What is the multiplier in this case?

21. **(Bouncing Ball)** A ball is dropped from a height of 9 feet. It rebounds to height of 6 feet, then to a height of 4 feet, and so on. Each time it hits the ground it rebounds to a height that is $\frac{2}{3}$ of its previous rebound. Use an infinite series to express the total distance traveled by the ball in coming (bouncing) to rest.

22. **(Fine Arts Grant)** In order to help establish a new community theatre, a local fine arts benefactor awards a 1-year grant of $50,000, with the understanding that the grant will be renewed each succeeding year for an amount equal to $\frac{3}{4}$ of the previous year's grant. What is the total amount of these annual grants, in the long run?

23. **(Perpetuity)** A corporation decides to create a perpetuity scholarship fund that will pay $1000 to each of four college students chosen annually by its scholarship committee. These payments will be made from an interest-bearing account that earns 6% annual interest, compounded annually after each payment of $4000.

    (a) Express the capitalized value of this perpetuity as an infinite series.
    (b) Find the sum of this series.

24. **(Perpetuity)** Redo Exercise 23, assuming that the 6% annual interest is compounded semiannually.

25. **(Perpetuity)** Redo Exercise 23, assuming that the 6% annual interest is compounded quarterly.

26. **(Perpetuity)** A businessman wishes to set up a perpetuity that will pay $500 each year to the winner of an annual Peace Essay contest. Express the capitalized value of this perpetuity as an infinite series, if the awards are being funded from an account earning 8% annual interest, compounded quarterly. Assume that the first quarter begins as the first award is made. Find the sum of this series.

27. **(Perpetuity)** Redo Exercise 26, assuming that the 8% annual interest is compounded continuously.

28. **(Perpetuity)** A local service club wishes to set up a perpetuity that will pay $100 each month to the winner of their monthly "good citizen" award. Express the capitalized value of this perpetuity, given that the money for these awards is paid from an account earning 7% annual continuous compound interest.

In Exercises 29–34, find the sum of the given series.

29. $\displaystyle\sum_{k=0}^{\infty} \frac{3^k - 4^k}{5^k}$

30. $\displaystyle\sum_{k=0}^{\infty} \frac{1 + 2^k}{3^k}$

31. $\displaystyle\sum_{k=0}^{\infty} \frac{(-1)^k}{2^k}$

32. $\displaystyle\sum_{k=0}^{\infty} \frac{7^{k+2}}{8^k}$

33. $\displaystyle\sum_{k=0}^{\infty} \frac{4^k - 3}{6^k}$

34. $\displaystyle\sum_{k=0}^{\infty} (-1)^{k+1}(-0.68)^k$

35. Prove part (b) of Theorem 2.

12.5 ‖‖‖‖‖ TESTS FOR CONVERGENCE AND DIVERGENCE OF SERIES

This section presents the technical tools necessary for the correct handling of infinite series. A working knowledge of these convergence and divergence tests will enable the reader to understand the development of power series and Taylor series presented in the next section.

There is a simple criterion, called the general term test, that is sometimes useful in showing that a series diverges.

THEOREM 1

(General Term Test—a test for divergence)

(a) If a series $\sum_{k=1}^{\infty} a_k$ converges, then

$$\lim_{k \to +\infty} a_k = 0. \tag{1}$$

(b) If the sequence $\{a_k\}$ does not converge to 0, then the series $\sum_{k=1}^{\infty} a_k$ diverges.

Proof   (a) Consider a convergent series $\sum_{k=1}^{\infty} a_k$, with $\sum_{k=1}^{\infty} a_k = S$. Let $\{S_n\}$ denote its sequence of partial series. Then $\lim_{n \to +\infty} S_n = S$.

For every natural number $n > 1$, we have

$$S_n = a_1 + a_2 + \cdots + a_n$$
$$= (a_1 + a_2 + \cdots + a_{n-1}) + a_n$$
$$= S_{n-1} + a_n.$$

Thus,

$$a_n = S_n - S_{n-1}.$$

We now take the limit as $n \to +\infty$. By Theorem 3 of Section 12.1,

$$\lim_{n \to +\infty} a_n = \lim_{n \to +\infty} S_n - \lim_{n \to +\infty} S_{n-1}$$
$$= S - \lim_{(n-1) \to +\infty} S_{n-1}$$
$$= S - S$$
$$= 0.$$

(b) Since (b) is equivalent to (a), it has already been proved.   ∎

Caution   The general term test can never be used to prove that a series converges; it can only be used to prove that a series *diverges*. Equation (1) does not guarantee that a series $\sum_{k=1}^{\infty} a_k$ converges. Rather, *failure* of Equation (1) to hold guarantees that a series $\sum_{k=1}^{\infty} a_k$ diverges.

EXAMPLE 1   Consider the series $\sum_{k=1}^{\infty} \dfrac{k}{k+1}$. The general term is

$$a_k = \frac{k}{k+1}.$$

Taking the limit as $k \to +\infty$, we have

$$\lim_{k \to +\infty} a_k = \lim_{k \to +\infty} \frac{k}{k+1} = \lim_{k \to +\infty} \frac{1}{1 + 1/k} = 1.$$

Therefore, by the general term test, $\sum_{k=1}^{\infty} \dfrac{k}{k+1}$ diverges.   ∎

||||||||||||| Series of Nonnegative Terms

An infinite series $\sum_{k=1}^{\infty} a_k$ such that $a_k \geq 0$ for all k is said to be a **series of nonnegative terms.** There are special tests that apply to such series. As we shall see, these tests form the basis for more general tests, which come later.

The first test, Theorem 2, is basic. It may seem unimportant at first glance, but it is fundamental to the tests which follow.

THEOREM 2

A series $\sum_{k=1}^{\infty} a_k$ of nonnegative terms converges if and only if its sequence of partial sums $\{S_n\}$ is bounded.*

Proof Consider a series $\sum_{k=1}^{\infty} a_k$ of nonnegative terms; that is, $a_k \geq 0$ for all k. Then, for each natural number n,

$$S_{n+1} = a_1 + a_2 + \cdots + a_n + a_{n+1}$$
$$= S_n + a_{n+1}$$
$$\geq S_n \text{ since } a_{n+1} \geq 0.$$

Therefore, $\{S_n\}$ is a monotone increasing sequence. By Theorem 6 of Section 12.1, $\{S_n\}$ converges if and only if it is bounded.   ∎

A convergent series $\sum_{k=1}^{\infty} a_k$ is an example of an infinite process that yields a finite result: we add infinitely many numbers and yet obtain a finite sum! In Section 7.6 we studied a similar mathematical construction, improper integrals. Associated with each convergent improper integral is a plane region that is infinite in extent and yet finite in area. In fact, the two ideas are related. We can use our knowledge of improper integrals to develop a test for convergence of infinite series.

Suppose that $\sum_{k=1}^{\infty} a_k$ is a series of nonnegative terms, and that f is a function that is continuous, nonnegative and decreasing on the interval $[1, +\infty)$, with

$$f(n) = a_n \tag{2}$$

for all natural numbers n. Figure 1 suggests a relationship between the nth partial sums $S_n = \sum_{k=1}^{n} a_k$ of such a series and $\int_1^{\infty} f(x)\, dx$.

---

* See the definition of bounded sequence, page 699.

Figure 1

In Figure 1, each rectangle has base $\Delta x = 1$. Thus, the area of the first rectangle is $f(2) \Delta x = f(2) = x_2$. The area of the second rectangle is $f(3) \Delta x = f(3) = x_3$, and so on. By comparing the sum of the rectangular areas and the area under the curve from 1 to n, we see from Figure 1 that for all natural numbers n,

$$\sum_{k=2}^{n} a_k \leq \int_1^n f(x)\, dx.$$

Thus,

$$S_n = a_1 + \sum_{k=2}^{n} a_k$$

$$\leq a_1 + \int_1^n f(x)\, dx. \tag{3}$$

Suppose $\int_1^{+\infty} f(x)\, dx$ converges. It follows from Equation (3) that for all natural numbers n,

$$0 \leq S_n \leq a_1 + \int_1^{+\infty} f(x)\, dx. \tag{4}$$

But then $\{S_n\}$ is a bounded sequence, so by Theorem 2, $\sum_{k=1}^{\infty} a_k$ converges. Therefore, if $\int_1^{+\infty} f(x)\, dx$ converges, so does $\sum_{k=1}^{\infty} a_k$.

Figure 2 shows another relationship between S_n and $\int_1^{+\infty} f(x)\, dx$. Comparing the sum of the rectangular areas and the area under the curve from 1 to n, we

Figure 2

see from Figure 2 that for all natural numbers n,

$$S_n = \sum_{k=1}^{n} a_k \geq \sum_{k=1}^{n-1} a_k = \sum_{k=1}^{n-1} f(k)\,\Delta x, \qquad \text{where } \Delta x = 1$$

and since this represents the sum of the areas of the shaded rectangles, we have

$$S_n \geq \int_1^n f(x)\,dx. \tag{5}$$

Hence,

$$\lim_{n \to +\infty} S_n \geq \lim_{n \to +\infty} \int_1^n f(x)\,dx.$$

That is,

$$\sum_{k=1}^{\infty} a_n \geq \int_1^{+\infty} f(x)\,dx \tag{6}$$

Now, suppose $\int_1^{+\infty} f(x)\,dx$ diverges. Then, since f is nonnegative and decreasing on $[1, +\infty)$ we must have $\int_1^{+\infty} f(x)\,dx = +\infty$. But then, by Equation (6) we must also have $\sum_{k=1}^{\infty} a_k = +\infty$. Thus, if $\int_1^{+\infty} f(x)\,dx$ diverges, so does $\sum_{k=1}^{\infty} a_k$.

We summarize the above discussion in the following theorem.

THEOREM 3

> **(The Integral Test)**
> Suppose $\sum_{k=1}^{\infty} a_k$ is a series of nonnegative terms. If $f(x)$ is a function which is continuous and decreasing on $[1, +\infty)$ and
>
> $$f(n) = a_n$$
>
> for all natural numbers n, then
>
> $$\sum_{k=1}^{\infty} a_k \text{ converges if and only if } \int_1^{+\infty} f(x)\,dx \text{ converges.}$$
>
> That is, if one converges so does the other, and if one diverges so does the other.

From the above discussion we can extract a little more information that is occasionally of some value in obtaining a crude estimate of the sum $\sum_{k=1}^{\infty} a_k$. The following result can be derived by piecing together Inequalities (4) and (6).

Corollary

> Under the conditions of Theorem 3, if $\int_1^{+\infty} f(x)\,dx$ converges, then
>
> $$\int_1^{+\infty} f(x)\,dx \leq \sum_{k=1}^{\infty} a_k \leq \int_1^{+\infty} f(x)\,dx + a_1. \tag{7}$$

EXAMPLE 2 Investigate each of the following series for convergence or divergence:

(a) $\displaystyle\sum_{k=1}^{\infty} \frac{1}{k^2}$ (b) $\displaystyle\sum_{k=1}^{\infty} \frac{1}{k}$ (the **harmonic series**)

Solution (a) The function $f(x) = 1/x^2$ satisfies the requirements of Theorem 3. Moreover,

$$\int_1^{+\infty} \frac{1}{x^2} \, dx = \lim_{b \to +\infty} \int_1^b \frac{1}{x^2} \, dx = \lim_{b \to +\infty} \left[-\frac{1}{x} \right]_1^b$$

$$= \lim_{b \to +\infty} \left[1 - \frac{1}{b} \right]$$

$$= 1.$$

Therefore, $\displaystyle\sum_{k=1}^{\infty} \frac{1}{k^2}$ converges. In fact, using the corollary, we have

$$\int_1^{+\infty} \frac{1}{x^2} \, dx \le \sum_{k=1}^{\infty} \frac{1}{k^2} \le \int_1^{+\infty} \frac{1}{x^2} \, dx + \frac{1}{1^2}$$

or

$$1 \le \sum_{k=1}^{\infty} \frac{1}{k^2} \le 1 + 1.$$

Therefore, $\displaystyle\sum_{k=1}^{\infty} \frac{1}{k^2}$ converges to a number between 1 and 2.

(b) The function $f(x) = 1/x$ satisfies the requirement of Theorem 3. Moreover

$$\int_1^{+\infty} \frac{1}{x} \, dx = \lim_{b \to +\infty} \int_1^b \frac{1}{x} \, dx = \lim_{b \to +\infty} \left[\ln b \right]$$

$$= +\infty.$$

Therefore, by Theorem 3, the harmonic series $\displaystyle\sum_{k=1}^{\infty} \frac{1}{k}$ diverges. ∎

EXAMPLE 3 For each positive real number p, the series

$$\sum_{k=1}^{\infty} \frac{1}{k^p}$$

is called a **p-series**. For what values of p does the p-series converge?

Solution Since we have already considered the case $p = 1$ in Example 2, we asssume $p \ne 1$, and let

$$f(x) = x^{-p}.$$

The derivative of f is

$$f'(x) = -px^{-p-1} = \frac{-p}{x^{(p+1)}}.$$

Since $f'(x) < 0$ on $[1, +\infty)$, f is decreasing on $[1, +\infty)$. Thus, f satisfies the requirements of Theorem 3. Now,

$$\int_1^{+\infty} x^{-p}\,dx = \lim_{b \to +\infty} \left[\frac{x^{-p+1}}{-p+1}\right]_1^b$$

$$= \frac{1}{-p+1} \lim_{b \to +\infty} [x^{-p+1}]_1^b$$

$$= \frac{1}{1-p} \lim_{b \to +\infty} [b^{1-p} - 1]$$

$$= \frac{1}{1-p} \left\{\left(\lim_{b \to +\infty} b^{1-p}\right) - 1\right\}. \tag{8}$$

To evaluate this improper integral we need to know the following limit.

If m is a constant real number, then

$$\lim_{b \to +\infty} b^m = \begin{cases} +\infty & \text{if } m > 0 \\ 1 & \text{if } m = 0 \\ 0 & \text{if } m < 0 \end{cases} \tag{9}$$

Thus, if $p > 1$, we have $1 - p < 0$, and from (8) and (9) we have

$$\int_1^\infty x^{-p}\,dx = \frac{1}{1-p}(0 - 1) = \frac{1}{p-1}. \tag{10}$$

Moreover, if $0 < p < 1$ we have $1 - p > 0$, and from (8) and (9) we have

$$\int_1^\infty x^{-p}\,dx = \frac{1}{1-p}(+\infty) = +\infty.$$

Therefore, by the integral test, $\sum_{k=1}^\infty \frac{1}{k^p}$ converges if $p > 1$ and diverges if $0 < p < 1$.

Equation (10), together with (7), tells us that when $p > 1$,

$$\frac{1}{p-1} < \sum_{k=1}^\infty k^{-p} < \frac{1}{p-1} + \frac{1}{1^p} = \frac{1 + (p-1)}{p+1} = \frac{p}{p-1}.$$

Putting these results together with Example 2, we have the following:

THEOREM 4

For each positive real number p, the p-series

$$\sum_{k=1}^\infty \frac{1}{k^p}$$

(a) converges to a number between $\dfrac{1}{p-1}$ and $\dfrac{p}{p-1}$, if $p > 1$

(b) diverges to $+\infty$ if $0 < p \le 1$.

EXAMPLE 4   Investigate the convergence or divergence of the series

$$\sum_{k=1}^{\infty} \frac{1}{\sqrt{3k + 5}}$$

SOLUTION   The function $f(x) = \dfrac{1}{\sqrt{3x + 5}} = (3x + 5)^{-1/2}$ has derivative

$$f'(x) = -\tfrac{1}{2}(3x + 5)^{-3/2}(3) = \frac{-3}{2(3x + 5)^{3/2}},$$

which is negative on the interval $[1, +\infty)$. Thus, f fulfills the requirements of Theorem 3. Then

$$\int_1^{\infty} \frac{1}{\sqrt{3x + 5}} \, dx = \lim_{b \to +\infty} \int_1^b (3x + 5)^{-1/2} \, dx$$
$$= \lim_{b \to +\infty} \left[\tfrac{2}{3}(3x + 5)^{1/2}\right]_1^b$$
$$= \tfrac{2}{3} \lim_{b \to +\infty} \left[\sqrt{3b + 5} - \sqrt{8}\right] = +\infty.$$

Therefore, the series $\displaystyle\sum_{k=1}^{\infty} \frac{1}{\sqrt{3k + 5}}$ diverges, by Theorem 3.   ∎

The following theorem allows us to add a wide variety of series to our collection of known convergent series.

THEOREM 5

(Comparison Test)
Suppose $\sum_{k=1}^{\infty} a_k$ and $\sum_{k=1}^{\infty} b_k$ are series of nonnegative terms and that there is a constant k such that, for all $k \geq k_0$,

$$a_k \leq b_k.$$

Then
  (a) if $\sum_{k=1}^{\infty} b_k$ converges, so does $\sum_{k=1}^{\infty} a_k$.
  (b) if $\sum_{k=1}^{\infty} a_k$ diverges (to $+\infty$), so does $\sum_{k=1}^{\infty} b_k$.

Proof   Statements (a) and (b) are equivalent, so we need only prove (a).
Let $S_n = \sum_{k=1}^{n} a_k$ and $T_n = \sum_{k=1}^{n} b_k$ denote the nth partial sums of their respective series. Then $\{S_n\}$ and $\{T_n\}$ are monotone increasing sequences such that for all n,

$$0 \leq S_n \leq T_n. \tag{11}$$

If $\sum_{k=1}^{\infty} b_k$ converges, then by Theorem 2, $\{T_n\}$ is a bounded sequence. But then by (11), $\{S_n\}$ is also a bounded sequence. Therefore, by Theorem 2, the series $\sum_{k=1}^{\infty} a_k$ converges. ■

EXAMPLE 5   Determine whether each of the following series converges or diverges.

(a) $\displaystyle\sum_{k=1}^{\infty} \frac{1}{k^2 + 3}$ (b) $\displaystyle\sum_{k=1}^{\infty} \frac{k}{k^2 - 5}$ (c) $\displaystyle\sum_{k=1}^{\infty} \frac{\sqrt{k+2}}{k+3}$ (d) $\displaystyle\sum_{k=1}^{\infty} \frac{1}{2^k + k}$

Solution   (a) For all positive integers k, $\dfrac{1}{k^2 + 3} < \dfrac{1}{k^2}$. By Theorem 4, $\displaystyle\sum_{k=1}^{\infty} \frac{1}{k^2}$ converges.

Therefore, by the comparison test, $\displaystyle\sum_{k=1}^{\infty} \frac{1}{k^2 + 3}$ converges.

(b) The first two terms of the given series are negative, so we ignore them. When $k \geq 3$, we have $\dfrac{k}{k^2 - 5} > \dfrac{k}{k^2} = \dfrac{1}{k}$. By Theorem 4, $\displaystyle\sum_{k=1}^{\infty} \frac{1}{k}$ diverges. Therefore, by the comparison test, $\displaystyle\sum_{k=1}^{\infty} \frac{k}{k^2 - 5}$ diverges.

(c) We have $\dfrac{\sqrt{k+2}}{k+3} > \dfrac{\sqrt{k}}{k+3} > \dfrac{\sqrt{k}}{k+k}$ when $k > 3$

$$= \frac{\sqrt{k}}{2k}$$

$$= \frac{1}{2\sqrt{k}}.$$

Thus, after the first three terms, $\dfrac{\sqrt{k+2}}{k+3} > \dfrac{1}{2\sqrt{k}}$. The series $\displaystyle\sum_{k=1}^{\infty} \frac{1}{2\sqrt{k}} = \dfrac{1}{2} \displaystyle\sum_{k=1}^{\infty} \frac{1}{k^{1/2}}$ diverges, by Theorem 4. Therefore, by the comparison test, $\displaystyle\sum_{k=1}^{\infty} \frac{\sqrt{k+2}}{k+3}$ diverges.

(d) For all positive integers k, $\dfrac{1}{2^k + k} < \dfrac{1}{2^k}$. The series $\displaystyle\sum_{k=1}^{\infty} \frac{1}{2^k}$ is a convergent geometric series. Therefore, by comparison test, $\displaystyle\sum_{k=1}^{\infty} \frac{1}{2^k + k}$ converges. ■

Note   Successful application of the comparison test requires that we first make an educated guess as to whether the given series converges or diverges. Based on a preliminary inspection of the given series, we must make a decision: whether to show that it is term-by-term less than or equal to some convergent series, or greater than or equal to some divergent series. This decision affects every step of the solution process, and if we find that we have made the wrong preliminary decision, we go back and start over.

The next test is one of the most powerful and popular of all the convergence tests. Its proof, which is based on the comparison test together with geometric series, involves sophisticated methods that are beyond the scope of this book.

THEOREM 6

(Ratio Test)
Suppose $\sum_{k=1}^{\infty} a_k$ is a series of positive terms, and

$$\lim_{k \to +\infty} \frac{a_{k+1}}{a_k} = \rho \tag{12}$$

(a) If $\rho < 1$, the series converges.
(b) If $\rho > 1$, the series diverges.
(c) If $\rho = 1$, this test provides no help; another test must be used.

EXAMPLE 6   Use the ratio test to decide whether the following series converge or diverge.

(a) $\displaystyle\sum_{k=1}^{\infty} \frac{k}{3^k}$ (b) $\displaystyle\sum_{k=1}^{\infty} \frac{1}{k!}$ (c) $\displaystyle\sum_{k=1}^{\infty} \frac{4^k}{k^2}$ (d) $\displaystyle\sum_{k=1}^{\infty} \frac{k!}{2^k}$

Solution   (a) For this series, $a_k = \dfrac{k}{3^k}$ and $a_{k+1} = \dfrac{k+1}{3^{k+1}}$. Thus,

$$\lim_{k \to +\infty} \frac{a_{k+1}}{a_k} = \lim_{k \to +\infty} \frac{\dfrac{k+1}{3^{k+1}}}{\dfrac{k}{3^k}} = \lim_{k \to +\infty} \left(\frac{k+1}{3^{k+1}} \cdot \frac{3^k}{k} \right)$$

$$= \lim_{k \to +\infty} \left[\frac{3^k}{3^{k+1}} \cdot \frac{k+1}{k} \right]$$

$$= \lim_{k \to +\infty} \left[\frac{1}{3} \cdot \frac{1+1/k}{1} \right] = \frac{1}{3} \cdot \frac{1+0}{1}$$

$$= \frac{1}{3}.$$

Thus $\rho = \frac{1}{3}$. Since $\rho < 1$, the series converges.

(b) For this series, $a_k = \dfrac{1}{k!}$ and $a_{k+1} = \dfrac{1}{(k+1)!}$. Thus,

$$\lim_{k \to +\infty} \frac{a_{k+1}}{a_k} = \lim_{k \to +\infty} \frac{1}{(k+1)!} \cdot \frac{k!}{1} = \lim_{k \to +\infty} \frac{1}{(k+1) \cdot k!} \cdot \frac{k!}{1}$$

$$= \lim_{k \to +\infty} \frac{1}{k+1}$$

$$= 0.$$

Thus $\rho = 0$. Since $\rho < 1$, the series converges.

(c) For this series, $a_k = \dfrac{4^k}{k^2}$ and $a_{k+1} = \dfrac{4^{k+1}}{(k+1)^2}$. Thus,

$$\frac{a_{k+1}}{a_k} = \frac{4^{k+1}}{(k+1)^2} \cdot \frac{k^2}{4^k} = \frac{4 \cdot 4^k \cdot k^2}{(k^2 + 2k + 1) \cdot 4^k}$$

$$= \frac{4k^2}{k^2 + 2k + 1} = \frac{4}{1 + \dfrac{2}{k} + \dfrac{1}{k^2}}.$$

Therefore,

$$\rho = \lim_{k \to +\infty} \frac{a_{k+1}}{a_k} = \lim_{k \to +\infty} \frac{4}{1 + \dfrac{2}{k} + \dfrac{1}{k^2}} = 4.$$

Since $\rho > 1$, the series diverges.

(d) For this series, $a_k = \dfrac{k!}{2^k}$ and $a_{k+1} = \dfrac{(k+1)!}{2^{k+1}}$. Thus,

$$\frac{a_{k+1}}{a_k} = \frac{(k+1)!}{2^{k+1}} \cdot \frac{2^k}{k!} = \frac{(k+1)k!}{2k!} = \frac{k+1}{2}.$$

Therefore,

$$\rho = \lim_{k \to +\infty} \frac{a_{k+1}}{a_k} = \lim_{k \to +\infty} \frac{k+1}{2} = +\infty.$$

Since $\rho > 1$, the series diverges. ∎

Series with Positive and Negative Terms

In general, series may have both positive and negative terms. Series with only nonnegative terms are special cases. To handle the more general series we need a few convergence tests. The first test we give is for an **alternating series**—one whose terms are alternately positive and negative.

THEOREM 7

(Alternating Series Test)
A series of the form

$$\sum_{k=1}^{\infty} (-1)^k a_k \quad \text{or} \quad \sum_{k=1}^{\infty} (-1)^k a_k$$

where

$$(1) \quad a_1 \geq a_2 \geq a_3 \geq \cdots \geq a_k \geq a_{k+1} \geq \cdots \geq 0$$

and

$$(2) \quad \lim_{k \to +\infty} a_k = 0$$

always converges.

Thus, if the terms of a series alternate in sign, decrease in absolute value, and satisfy the general term test, then the series converges. The proof of this theorem is beyond the scope of this book.

EXAMPLE 7   The alternating harmonic series

$$\sum_{k=1}^{\infty} \frac{(-1)^{k+1}}{k} = 1 - \frac{1}{2} + \frac{1}{3} - \frac{1}{4} + \frac{1}{5} - \frac{1}{6} + \cdots$$

satisfies the conditions of the alternating series test, and is therefore a convergent series.   ∎

A series $\sum_{k=1}^{\infty} a_k$ is said to be **absolutely convergent** if the series $\sum_{k=1}^{\infty} |a_k|$ converges. It turns out that if we can show that a series is absolutely convergent, then it will follow that the series converges, but not vice versa.

THEOREM 8   Every absolutely convergent series converges, but *not* every convergent series is absolutely convergent.

Proof   (a) Suppose $\sum_{k=1}^{\infty} a_k$ is an absolutely convergent series. Then the series $\sum_{k=1}^{\infty} |a_k|$ converges. Define the series $\sum_{k=1}^{\infty} b_k$ by

$$b_k = a_k + |a_k|. \tag{13}$$

Then, for all k,

$$b_k = \begin{cases} 2|a_k| & \text{if } a_k \geq 0 \\ 0 & \text{if } a_k < 0. \end{cases}$$

Thus,

$$0 \leq b_k \leq 2|a_k|. \tag{14}$$

But the series $\sum_{k=1}^{\infty} 2|a_k| = 2 \sum_{k=1}^{\infty} |a_k|$ converges. Therefore, by (14) and the comparison test, $\sum_{k=1}^{\infty} b_k$ converges. Now, from Equation (13),

$$a_k = b_k - |a_k|.$$

Since both series $\sum_{k=1}^{\infty} b_k$ and $\sum_{k=1}^{\infty} |a_k|$ converge, Theorem 2 of Section 12.4 guarantees that $\sum_{k=1}^{\infty} a_k$ converges. Therefore, every absolutely convergent series converges.

(b) Consider the alternating harmonic series $\sum_{k=1}^{\infty} \frac{(-1)^{k+1}}{k}$. In Example 7 this series was shown to be convergent. The series $\sum_{k=1}^{\infty} \left| \frac{(-1)^{k+1}}{k} \right| = \sum_{k=1}^{\infty} \frac{1}{k}$ is the harmonic series, which was shown to be divergent in Example 2. Thus, the alternating harmonic series is not absolutely convergent. Therefore, not every convergent series is absolutely convergent.   ■

12.5 EXERCISE SET ||||||||||||||||||||||||||||||||||||||||||||||||||||||||||||||||||||||||||||||||||||||||||

In Exercises 1–24, use an appropriate test to determine whether the given series of non-negative terms converges or diverges.

1. $\displaystyle\sum_{k=1}^{\infty} \frac{1}{k\sqrt{k}}$

2. $\displaystyle\sum_{k=1}^{\infty} \frac{1}{k^3}$

3. $\displaystyle\sum_{k=1}^{\infty} \frac{1}{3\sqrt{k}}$

4. $\displaystyle\sum_{k=1}^{\infty} \frac{1}{\sqrt[3]{k}}$

5. $\displaystyle\sum_{k=1}^{\infty} \frac{k}{k^3 + 7}$

6. $\displaystyle\sum_{k=0}^{\infty} \frac{k-1}{10k + 9}$

7. $\displaystyle\sum_{k=1}^{\infty} \frac{1}{k^2 + 5k}$

8. $\displaystyle\sum_{k=1}^{\infty} \frac{1}{2^k + 3^k}$

9. $\displaystyle\sum_{k=0}^{\infty} \frac{k}{99k - 1}$

10. $\displaystyle\sum_{k=0}^{\infty} \frac{3}{5 + \sqrt{k}}$

11. $\displaystyle\sum_{k=0}^{\infty} \frac{\sqrt{k}}{1 + k^2}$

12. $\displaystyle\sum_{k=0}^{\infty} ke^{-k^2}$

13. $\displaystyle\sum_{k=2}^{\infty} \frac{\ln k}{k}$

14. $\displaystyle\sum_{k=2}^{\infty} \frac{1}{k \ln k}$

15. $\displaystyle\sum_{k=1}^{\infty} ke^{-k}$

16. $\displaystyle\sum_{k=1}^{\infty} \frac{(k+1)^2}{e^k}$

17. $\displaystyle\sum_{k=1}^{\infty} \frac{2^k}{100k}$

18. $\displaystyle\sum_{k=0}^{\infty} \frac{k-3}{3^k}$

19. $\displaystyle\sum_{k=0}^{\infty} \frac{k+5}{k!}$

20. $\displaystyle\sum_{k=1}^{\infty} k^2 e^{-k}$

21. $\displaystyle\sum_{k=0}^{\infty} \frac{\sqrt{k}}{2^k}$

22. $\displaystyle\sum_{k=1}^{\infty} \frac{\sqrt{k}}{k!}$

23. $\displaystyle\sum_{k=1}^{\infty} \frac{k!}{10^k}$

24. $\displaystyle\sum_{k=1}^{\infty} \frac{e^{\sqrt{k}}}{\sqrt{k}}$

In Exercises 25–28, use the corollary to Theorem 3 to estimate the sum of the given series.

25. $\displaystyle\sum_{k=1}^{\infty} \frac{1}{k^5}$

26. $\displaystyle\sum_{k=4}^{\infty} \frac{1}{k^3}$

27. $\displaystyle\sum_{k=1}^{\infty} ke^{-k}$

28. $\displaystyle\sum_{k=1}^{\infty} ke^{-k^2}$

In Exercises 29–38, determine whether the given series converges. If so, does it converge absolutely?

29. $\displaystyle\sum_{k=1}^{\infty} \frac{(-1)^k}{k^2}$

30. $\displaystyle\sum_{k=1}^{\infty} \frac{(-1)^k}{\sqrt{k}}$

31. $\displaystyle\sum_{k=2}^{\infty} \frac{(-1)^k k}{k-1}$

32. $\displaystyle\sum_{k=0}^{\infty} \frac{(-1)^k}{3^k}$

33. $\displaystyle\sum_{k=1}^{\infty} \frac{(-1)^{k+1}}{2k^2 + 10}$

34. $\displaystyle\sum_{k=0}^{\infty} \frac{(-1)^k}{k!}$

35. $\displaystyle\sum_{k=2}^{\infty} \frac{(-1)^k}{k \ln k}$

36. $\displaystyle\sum_{k=1}^{\infty} \frac{(-1)^{k+1} e^{\sqrt{k}}}{\sqrt{k}}$

37. $\displaystyle\sum_{k=0}^{\infty} \sin\left(\frac{k\pi}{2}\right)$

38. $\displaystyle\sum_{k=1}^{\infty} \frac{\cos(k\pi)}{k}$

12.6 |||||| POWER SERIES: TAYLOR SERIES AND MACLAURIN SERIES

This section is the culmination of the entire chapter. All the various strands of theory developed in this chapter come together in the study of Taylor series. We begin by considering the general concept of "power" series.

Definition

A series of the form

$$\sum_{k=0}^{\infty} a_k x^k = a_0 + a_1 x + a_2 x^2 + a_3 x^3 + \cdots \qquad (1)$$

where $a_0, a_1, a_2, \ldots$ are constants, is called a **power series in x.** A series of the form

$$\sum_{k=0}^{\infty} a_k (x - c)^k = a_0 + a_1 (x - c) + a_2 (x - c)^2 + a_3 (x - c)^3 + \cdots \qquad (2)$$

where $c, a_0, a_1, a_2, \ldots$ are constants, is called a **power series in $(x - c)$.**

Notice that series (1) is a special case of series (2), with $c = 0$. When we speak of a power series, in general, we shall mean series (2), understanding that series (1) is included among the possibilities.

A power series is a function of x. We shall be interested in knowing whether such a function can be differentiated, integrated, and so on. But we first, need to know the *domain* of such a function. The domain is the set of all real numbers x for which the power series converges. To determine this set we need to use the convergence tests developed in Section 12.5.

||||||||||||| **Interval of Convergence of a Power Series**

A power series $\sum_{k=0}^{\infty} a_k (x - c)^k$ always converges for at least one number x; namely, $x = c$. But what about other values of x? We shall discover that the set of all numbers x for which the power series converges is always an interval centered at c. The following theorem provides the basis for establishing this fact. Its proof is somewhat more complicated than is appropriate for this text, so we omit it.

THEOREM 1

(a) If a power series $\sum_{k=0}^{\infty} a_k x^k$ converges at $x = x_1$, then it converges absolutely for all $|x| < |x_1|$. [See Figure 1(a).]

(a)   (b)

Figure 1

(b) If a power series $\sum_{k=0}^{\infty} a_k (x - c)^k$ converges at $x = c + r$, then it converges absolutely for all x in the interval $|x - c| < |r|$. [See Figure 1(b).]

Thus, to determine the set of all real numbers for which a power series $\sum_{k=0}^{\infty} a_k(x - c)^k$ converges, we seek the largest value of $r > 0$ such that the series converges absolutely for all values of x in the interval $(c - r, c + r)$. The largest such value of r is called the **radius of convergence** of the series, and is denoted R. If the series converges only at $x = c$, we say that $R = 0$. If the series converges for all real numbers x, then we say that $R = +\infty$.

We shall describe a method for finding R, the radius of convergence. Once we have found R, if $0 < R < +\infty$, we then test the endpoints of the interval $(c - R, c + R)$ or $(-R, R)$ to determine whether the power series converges at these points. Finally, the set of all numbers for which a power series converges is seen to be an interval, which is called its **interval of convergence.**

The following examples show how to use the ratio test, along with other appropriate tests from Section 12.5, to find the interval of convergence.

EXAMPLE 1   Find the interval of convergence of the following power series.

$$\sum_{k=0}^{\infty} \frac{x^k}{k + 1} = 1 + \frac{x}{2} + \frac{x^2}{3} + \frac{x^3}{4} + \cdots \tag{3}$$

Solution   We use the ratio test to determine where the given series is absolutely convergent; that is, where

$$\sum_{k=0}^{\infty} \left| \frac{x^k}{k + 1} \right| \tag{4}$$

converges. To use the ratio test we must calculate the limit

$$
\begin{aligned}
\rho &= \lim_{k \to +\infty} \frac{|x|^{k+1}}{k + 2} \div \frac{|x|^k}{k + 1} \\
&= \lim_{k \to +\infty} \frac{|x|^{k+1}}{k + 2} \cdot \frac{k + 1}{|x|^k} = \lim_{k \to +\infty} \frac{|x|(k + 1)}{k + 2} \\
&= |x| \lim_{k \to \infty} \frac{1 + 1/k}{1 + 2/k} = |x| \cdot \frac{1 + 0}{1 + 0} \\
&= |x|.
\end{aligned}
$$

By the ratio test, the series (4) converges if $\rho < 1$; that is, if

$$|x| < 1.$$

Therefore, the radius of convergence is

$$R = 1.$$

The interval of convergence is thus $(-1, 1)$, possibly including one or both endpoints as well.

We test the left endpoint. Substituting $x = -1$ into Equation (3), we have the series

$$\sum_{k=0}^{\infty} \frac{(-1)^k}{k+1} = 1 - \frac{1}{2} + \frac{1}{3} - \frac{1}{4} + \cdots.$$

This is the alternating harmonic series, which converges, as shown in Example 7 of Section 12.5.

Finally, we test the right endpoint. Substituting $x = 1$ into Equation (3) yields the series

$$\sum_{k=0}^{\infty} \frac{(1)^k}{k+1} = 1 + \frac{1}{2} + \frac{1}{3} + \frac{1}{4} + \cdots.$$

This is the harmonic series, which diverges, as shown in Example 2 of Section 12.5.

Putting together the above results we see that the interval of convergence for the series (3) is the half-open interval $[-1, 1)$. ∎

EXAMPLE 2 Find the interval of convergence of the following power series.

$$\sum_{k=0}^{\infty} \frac{(x-3)^k}{2^k} = 1 + \frac{x-3}{2} + \frac{(x-3)^2}{4} + \frac{(x-3)^3}{8} + \cdots \tag{5}$$

Solution We use the ratio test to determine where the given series is absolutely convergent; that is, where

$$\sum_{k=0}^{\infty} \left| \frac{(x-3)^k}{2^k} \right| \tag{6}$$

converges. To use the ratio test we calculate

$$\begin{aligned}
\rho &= \lim_{k \to +\infty} \frac{|x-3|^{k+1}}{2^{k+1}} \div \frac{|x-3|^k}{2^k} \\
&= \lim_{k \to +\infty} \frac{|x-3|^{k+1}}{2^{k+1}} \cdot \frac{2^k}{|x-3|^k} \\
&= \lim_{k \to +\infty} \frac{|x-3|}{2} \\
&= \frac{|x-3|}{2}.
\end{aligned}$$

By the ratio test, the series (6) converges when $\rho < 1$; that is, when

$$\frac{|x-3|}{2} < 1,$$

which is equivalent to

$$|x - 3| < 2 \tag{7}$$

or

$$-2 < x - 3 < 2, \quad \text{or} \quad 1 < x < 5.$$

Therefore, the given series converges absolutely in the interval $(1, 5)$. We now test the endpoints of the interval.

Substituting $x = 1$ into Equation (5) we have the series

$$\sum_{k=0}^{\infty} \frac{(1 - 3)^k}{2^k} = \sum_{k=0}^{\infty} \frac{(-2)^k}{2^k} = \sum_{k=0}^{\infty} (-1)^k.$$

This series diverges, by the general term test.

Substituting $x = 5$ into Equation (5) we have the series

$$\sum_{k=0}^{\infty} \frac{(5 - 3)^k}{2^k} = \sum_{k=0}^{\infty} 1,$$

which also diverges by the general term test.

Therefore, the interval of convergence for this series is the open interval $(-1, 1)$.

Finally, note that Inequality (7) tells us that the radius of convergence of the series (5) is

$$R = 2. \qquad \blacksquare$$

In Examples 1 and 2 we have found the radius of convergence indirectly, after first calculating ρ and applying the ratio test. We can streamline the procedure somewhat by finding R directly, from the following theorem.

THEOREM 2

(Radius of Convergence)
For a power series $\sum_{k=0}^{\infty} a_k x^k$ or $\sum_{k=0}^{\infty} a_k(x - c)^k$, the radius of convergence is

$$R = \lim_{k \to +\infty} \left| \frac{a_k}{a_{k+1}} \right|, \qquad (8)$$

if this limit exists.

Caution   In Equation (8) for R the subscripts in the numerator and denominator are k and $k + 1$, respectively, whereas in the equation for ρ (Section 12.5) they are $k + 1$ and k, respectively. It is easy to confuse the two situations.

EXAMPLE 3   Find the interval of convergence of the following series.

$$\sum_{k=1}^{\infty} \frac{3^k(x + 1)^k}{k^2} = 3(x + 1) + \frac{9}{4}(x + 1)^2 + \frac{27}{9}(x + 1)^3 + \frac{81}{16}(x + 1)^4 + \cdots \quad (9)$$

Solution   By Theorem 2 the radius of convergence is

$$
\begin{aligned}
R &= \lim_{k \to +\infty} \left| \frac{a_k}{a_{k+1}} \right| = \lim_{k \to +\infty} \frac{3^k}{k^2} \div \frac{3^{k+1}}{(k+1)^2} \\
&= \lim_{k \to +\infty} \frac{3^k}{k^2} \cdot \frac{(k+1)^2}{3^{k+1}} = \lim_{k \to +\infty} \frac{1}{3} \cdot \frac{k^2 + 2k + 1}{k^2} \\
&= \frac{1}{3} \lim_{k \to +\infty} \frac{1 + 2/k + 1/k^2}{1} \\
&= \frac{1}{3}.
\end{aligned}
$$

Therefore, the given series (9) converges absolutely in the interval

$$|x + 1| < \tfrac{1}{3}$$

or

$$-\tfrac{1}{3} < x + 1 < \tfrac{1}{3}, \qquad \text{or} \qquad -\tfrac{4}{3} < x < -\tfrac{2}{3}.$$

We now test the endpoints of this interval.

When $x = -\tfrac{4}{3}$ the series (9) becomes

$$\sum_{k=1}^{\infty} \frac{3^k(-\tfrac{1}{3})^k}{k^2} = \sum_{k=1}^{\infty} \frac{(-1)^k}{k^2}.$$

This series converges by the alternating series test.

When $x = -\tfrac{2}{3}$ the series (9) becomes

$$\sum_{k=1}^{\infty} \frac{3^k(\tfrac{1}{3})^k}{k^2} = \sum_{k=1}^{\infty} \frac{1}{k^2}.$$

This is a p-series with $p = 2$. By Theorem 4 of Section 12.5, it converges.

Therefore, the interval of convergence for the given series (9) is the closed interval $[-\tfrac{4}{3}, -\tfrac{2}{3}]$.   ∎

Remark   Examples 1, 2, and 3 show that the interval of convergence of a power series can be an open interval, a closed interval, or a half-open interval. The following example will show that it can also be the entire real number line or a single point.

EXAMPLE 4   Find the interval of convergence of each of the following power series:

$$\text{(a)} \sum_{k=0}^{\infty} \frac{x^k}{k!} \qquad \text{(b)} \sum_{k=0}^{\infty} k!(x - 4)^k$$

Solution   (a) The radius of convergence for this series is

$$
\begin{aligned}
R &= \lim_{k \to +\infty} \left| \frac{a_k}{a_{k+1}} \right| = \lim_{k \to +\infty} \frac{1}{k!} \div \frac{1}{(k+1)!} = \lim_{k \to +\infty} \frac{(k+1)!}{k!} \\
&= \lim_{k \to +\infty} (k + 1) \\
&= +\infty.
\end{aligned}
$$

Since $R = +\infty$, the interval of convergence is the entire real number line $(-\infty, +\infty)$.

(b) The radius of convergence of this series is

$$R = \lim_{k \to +\infty} \left| \frac{a^k}{a_{k+1}} \right| = \lim_{k \to +\infty} \frac{k!}{(k+1)!} = \lim_{k \to +\infty} \frac{1}{k+1}$$
$$= 0.$$

Since $R = 0$, the interval of convergence is $[4, 4]$, the single point $x = 4$. ∎

|||||||||||| Functions Defined by Power Series

The expression

$$f(x) = \sum_{k=0}^{\infty} a_k x^k \tag{10}$$

or

$$f(x) = \sum_{k=0}^{\infty} a_k (x - c)^k \tag{11}$$

(where $c, a_0, a_1, a_2, \ldots$ are constant real numbers) defines a function of x. The domain of the function is the interval of convergence of the associated power series. Since form (11) includes form (10) as a special case, we shall often state our results only for power series of the form (11), leaving the reader to conclude that the results also apply to series of the form (10).

Of fundamental importance here is that functions defined by power series have derivatives of all orders, and can be differentiated and integrated term by term.

THEOREM 3

(Differentiation of Power Series)
If a function f is defined by a power series

$$f(x) = \sum_{k=0}^{\infty} a_k (x - c)^k, \tag{11}$$

then at every point x inside the interval of convergence of (11), f is differentiable and

$$f'(x) = \sum_{k=0}^{\infty} k a_k (x - c)^{k-1}. \tag{12}$$

Moreover, the two series (11) and (12) have the same radius of convergence. (However, the two series may behave differently at the endpoints of the interval of convergence.)

EXAMPLE 5 Find the derivative of the function

$$f(x) = \sum_{k=0}^{\infty} \frac{x^k}{k+1} = 1 + \frac{x}{2} + \frac{x^2}{3} + \frac{x^3}{4} + \cdots \tag{13}$$

and compare the intervals of convergence for the two series.

Solution By Theorem 3, the derivative of f is

$$f'(x) = \sum_{k=0}^{\infty} \frac{kx^k}{k+1} = \frac{1}{2} + \frac{2}{3}x + \frac{3}{4}x^2 + \frac{4}{5}x^3 + \cdots. \tag{14}$$

In Example 1 we found that the interval of convergence of the series (13) is the half-open interval $[-1, 1)$.

According to Theorem 3, the radius of convergence of both series (13) and (14) is the same, $R = 1$. Thus, to find the interval of convergence of (14) we need only test the endpoints of the interval.

Substituting $x = -1$ into (14), we have the series

$$\sum_{k=0}^{\infty} \frac{k(-1)^k}{k+1}.$$

But this series diverges, by the general term test. When we substitute $x = 1$ into (14) we have the series

$$\sum_{k=0}^{\infty} \frac{k(1)^k}{k+1},$$

which also diverges by the general term test.

Therefore the interval of convergence of the series (14) is the open interval $(-1, 1)$. ∎

The following result is an immediate consequence of Theorem 3.

Corollary

> A function $f(x)$ which is representable as a power series has derivatives of all orders at every x in the interior of the interval of convergence.

The power of Theorem 3 is greater than one might at first suspect. The following examples show some of its power.

EXAMPLE 6 Find the sum of the power series

$$f(x) = \sum_{k=0}^{\infty} \frac{x^k}{k!} \tag{15}$$

and the interval over which this sum is valid.

Solution In Example 4 we found that the interval of convergence of this power series is $(-\infty, +\infty)$. Theorem 3 tells us that at every point of this interval, the derivative of f is

$$f'(x) = \frac{d}{dx} \sum_{k=0}^{\infty} \frac{x^k}{k!}$$

$$= \frac{d}{dx}\left[1 + x + \frac{x^2}{2!} + \frac{x^3}{3!} + \frac{x^4}{4!} + \frac{x^5}{5!} + \cdots\right]$$

$$= \left[0 + 1 + \frac{2}{2!}x + \frac{3}{3!}x^2 + \frac{4}{4!}x^3 + \frac{5}{5!}x^4 + \cdots\right]$$

$$= \left[1 + x + \frac{x^2}{2!} + \frac{x^3}{3!} + \frac{x^4}{4!} + \cdots\right]$$

$$= f(x).$$

Thus, the function $y = f(x)$ satisfies the differential equation

$$\frac{dy}{dx} = y.$$

In Chapter 10 we saw that the general solution of this differential equation is

$$y = y_0 e^x, \tag{16}$$

where $y_0 = f(0)$. Now, for our function f,

$$f(0) = 1 + 0 + \frac{0^2}{2!} + \frac{0^3}{3!} + \cdots = 1,$$

so that

$$y_0 = 1.$$

Substituting this result into Equation (16), we have

$$y = e^x.$$

Therefore, we have found that the function $f(x)$ defined by the power series (15) is

$$f(x) = e^x. \tag{17}$$

Since the power series representing $f(x)$ and $f'(x)$ have $(-\infty, +\infty)$ as interval of convergence, Equation (17) is valid for all real numbers ∎

In Example 6 we have derived the following important result.

The power series expansion

$$e^x = 1 + x + \frac{x^2}{2!} + \frac{x^3}{3!} + \frac{x^4}{4!} + \cdots$$

$$= \sum_{k=0}^{\infty} \frac{x^k}{k!} \tag{18}$$

is valid for every real number x.

EXAMPLE 7   Find the sum of the power series

$$f(x) = \sum_{k=0}^{\infty} \frac{(-1)^k x^{2k+1}}{(2k+1)!} = x - \frac{x^3}{3!} + \frac{x^5}{5!} - \frac{x^7}{7!} + \cdots, \tag{19}$$

and the interval over which this sum is valid.

Solution   Let $f(x) = \sum_{k=0}^{\infty} \frac{(-1)^k x^{2k+1}}{(2k+1)!}$. The radius of convergence of this series is

$$R = \lim_{k \to +\infty} \left| \frac{a_k}{a_{k+1}} \right| = \lim_{k \to +\infty} \left| \frac{(-1)^k}{(2k+1)!} \div \frac{(-1)^{k+1}}{[2(k+1)+1]!} \right|$$

$$= \lim_{k \to +\infty} \frac{1}{(2k+1)!} \cdot \frac{(2k+3)!}{1}$$

$$= \lim_{k \to +\infty} \frac{(2k+3)(2k+2)(2k+1)!}{(2k+1)!}$$

$$= \lim_{k \to +\infty} (2k+3)(2k+2)$$

$$= +\infty.$$

Since $R = +\infty$, the interval of convergence is $(-\infty, +\infty)$.
By Theorem 3, the derivative of $f(x)$ is

$$f'(x) = 1 - \frac{3x^2}{3!} + \frac{5x^4}{5!} - \frac{7x^6}{7!} + \frac{9x^8}{9!} - \cdots$$

$$= 1 - \frac{x^2}{2!} + \frac{x^4}{4!} - \frac{x^6}{6!} + \frac{x^8}{8!} - \cdots, \tag{20}$$

and the second derivative of f is

$$f''(x) = -\frac{2x}{2!} + \frac{4x^3}{4!} - \frac{6x^5}{6!} + \frac{8x^7}{8!} - \frac{10x^8}{10!} + \cdots$$

$$= -x + \frac{x^3}{3!} - \frac{x^5}{5!} + \frac{x^7}{7!} - \frac{x^8}{8!} + \cdots$$

$$= -\left[x - \frac{x^3}{3!} + \frac{x^5}{5!} - \frac{x^7}{7!} + \frac{x^8}{8!} - \cdots \right]$$

$$= -f(x).$$

Thus, the function $y = f(x)$ satisfies the differential equation

$$\frac{d^2y}{dx^2} = -y.$$

That is, the function $y = f(x)$ satisfies the second-order linear differential equation

$$y'' + y = 0.$$

Using the method of Section 10.5, we find that the general solution of this differential equation is

$$f(x) = A \sin(x + \theta). \tag{21}$$

We now determine the values of the constants A and θ. From Equation (19),

$$f(0) = 0, \tag{22}$$

while, from Equation (21),

$$f(0) = A \sin \theta. \tag{23}$$

Putting Equations (22) and (23) together, we have

$$0 = A \sin \theta. \tag{24}$$

From Equation (21) we know that $A \neq 0$, since $f(x)$ is not identically 0. Thus, Equation (24) tells us that

$$\sin \theta = 0.$$

So we may take $\theta = 0$. Therefore, Equation (21) becomes

$$f(x) = A \sin x. \tag{25}$$

From Equation (20),

$$f'(0) = 1 \tag{26}$$

while, from Equation (25),

$$f'(x) = A \cos x.$$

Thus,

$$f'(0) = A \cos 0 = A. \tag{27}$$

Putting Equations (26) and (27) together, we have

$$1 = A.$$

Finally, substituting this result into Equation (25), we see that our function $f(x)$ is

$$f(x) = \sin x.$$

Since the power series representing $f(x)$, $f'(x)$, and $f''(x)$ all have $(-\infty, +\infty)$ as the interval of convergence, the above calculations are valid for all values of x. ∎

In Example 7 we have derived the following important result.

The power series expansion

$$\sin x = x - \frac{x^3}{3!} + \frac{x^5}{5!} - \frac{x^7}{7!} + \cdots$$

$$= \sum_{k=0}^{\infty} \frac{(-1)^k x^{2k+1}}{(2k+1)!} \qquad (28)$$

is valid for every real number x.

EXAMPLE 8   Derive a power series expansion for the cosine function which is valid for all real numbers x.

Solution   We know that $\cos x = \dfrac{d}{dx} [\sin x]$. Applying Theorem 3 to the power series (28) representing $\sin x$, we have

$$\cos x = \frac{d}{dx} \left[x - \frac{x^3}{3!} + \frac{x^5}{5!} - \frac{x^7}{7!} + \frac{x^9}{9!} - \cdots \right]$$

$$= 1 - \frac{3x^2}{3!} + \frac{5x^4}{5!} - \frac{7x^7}{7!} + \frac{9x^8}{9!} - \cdots$$

$$= 1 - \frac{x^2}{2!} + \frac{x^4}{4!} - \frac{x^6}{6!} + \frac{x^8}{8!} - \cdots .$$

Since a power series and its derivative have the same radius of convergence, this power series representation of $\cos x$ is valid for all real numbers. ∎

In Example 8 we have derived the following important result.

The power series expansion

$$\cos x = 1 - \frac{x^2}{2!} + \frac{x^4}{4!} - \frac{x^6}{6!} + \cdots$$

$$= \sum_{k=0}^{\infty} \frac{(-1)^k x^{2k}}{(2k)!} \qquad (29)$$

is valid for every real number x.

|||||||||||| **Algebraic Manipulation of Power Series**

Once we have obtained several valid power series expansions of functions, we may wish to transform them algebraically to obtain valid power series expansions of other functions. The following theorems provide rules for doing this.

THEOREM 4

(**Substitution in Power Series**)

Suppose that

$$f(x) = \sum_{k=0}^{\infty} a_k x^k \qquad \left[\text{or } f(x) = \sum_{k=0}^{\infty} a_k (x - c)^k \right]$$

is a valid power series expansion in some interval of convergence I. If u is any algebraic expression, then the substitution

$$f(u) = \sum_{k=0}^{\infty} a_k u^k \qquad \left[\text{or } f(u) = \sum_{k=0}^{\infty} a_k (u - c)^k \right]$$

is valid as long as the value of u falls within the interval I.

When we apply Theorem 2 of Section 12.4 to power series, we have the following results.

THEOREM 5

For all x in the interval of convergence of both $\sum_{k=0}^{\infty} a_k x^k$ and $\sum_{k=0}^{\infty} b_k x^k$, we have

(a) $\displaystyle \sum_{k=0}^{\infty} a_k x^k + \sum_{k=0}^{\infty} b_k x^k = \sum_{k=0}^{\infty} (a_k + b_k) x^k$

(b) $\displaystyle \sum_{k=0}^{\infty} a_k x^k - \sum_{k=0}^{\infty} b_k x^k = \sum_{k=0}^{\infty} (a_k - b_k) x^k$

(c) $\displaystyle u(x) \sum_{k=0}^{\infty} a_k x^k = \sum_{k=0}^{\infty} a_k u(x) x^k$

as long as $u(x)$ is defined for each x in the interval of convergence.

For simplicity we have stated Theorem 5 for power series in x, but it is equally true in general for power series in $(x - c)$.

EXAMPLE 9 For each of the following functions, find a power series representation that is valid for all real numbers x.

(a) $f(x) = e^{3x}$

(b) $g(x) = x \sin(x^2)$

(c) $h(x) = \sin x - \cos x$

Solution   (a) Replacing x by $3x$ in Equation (18), we have

$$e^{3x} = 1 + 3x + \frac{9x^2}{2!} + \frac{27x^3}{3!} + \frac{81x^4}{4!} + \cdots$$

$$= \sum_{k=0}^{\infty} \frac{(3x)^k}{k!}.$$

(b) Replacing x by x^2 in Equation (28), we have

$$\sin(x^2) = x^2 - \frac{x^6}{3!} + \frac{x^{10}}{5!} - \frac{x^{14}}{7!} + \cdots$$

$$= \sum_{k=0}^{\infty} \frac{(-1)^k x^{4k+2}}{(2k+1)!}.$$

Then, using Theorem 5(c), we multiply both sides by x to obtain

$$x\sin(x^2) = x^3 - \frac{x^7}{3!} + \frac{x^{11}}{5!} - \frac{x^{15}}{7!} + \cdots$$

$$= \sum_{k=0}^{\infty} \frac{(-1)^k x^{4k+3}}{(2k+1)!}.$$

(c) By Equations (28) and (29),

$$\sin x - \cos x = \left[x - \frac{x^3}{3!} + \frac{x^5}{5!} - \frac{x^7}{7!} + \cdots \right] - \left[1 - \frac{x^2}{2!} + \frac{x^4}{4!} - \frac{x^6}{6!} + \cdots \right]$$

By Theorem 5, we may combine these two series to obtain

$$\sin x - \cos x = -1 + x + \frac{x^2}{2!} - \frac{x^3}{3!} - \frac{x^4}{4!} + \frac{x^5}{5!} + \frac{x^6}{6!} - \frac{x^7}{7!} - \frac{x^8}{8!} + \cdots \quad \blacksquare$$

|||||||||||| Taylor Series and Maclaurin Series

We know that any function representable as a power series,

$$f(x) = \sum_{k=0}^{\infty} a_k (x - c)^k, \tag{30}$$

has derivatives of all orders in the interior of its interval of convergence (Theorem 3). We now make some observations about these derivatives, at c_0. The function and its first few derivatives are

$$f^{(0)}(x) = f(x) = a_0 + a_1(x - c) + a_2(x - c)^2 + a_3(x - c)^3 + \cdots$$
$$f^{(1)}(x) = a_1 + 2a_2(x - c)^1 + 3a_3(x - c)^2 + 4a_4(x - c)^3 + \cdots$$
$$f^{(2)}(x) = 2a_2 + 2 \cdot 3a_3(x - c)^1 + 3 \cdot 4a_4(x - c)^2 + 4 \cdot 5a_5(x - c)^3 + \cdots$$
$$f^{(3)}(x) = 2 \cdot 3a_3 + 2 \cdot 3 \cdot 4a_4(x - c)^1 + 3 \cdot 4 \cdot 5a_5(x - c)^2$$
$$+ 4 \cdot 5 \cdot 6a_6(x - c)^3 + \cdots$$

$$\vdots$$

We can see that at $x = x_0$, these derivatives are

$$f^{(0)}(x_0) = a_0 = 0!a_0$$
$$f^{(1)}(x_0) = a_1 = 1!a_1$$
$$f^{(2)}(x_0) = 2a_2 = 2!a_2$$
$$f^{(3)}(x_0) = 2 \cdot 3a_3 = 3!a_3$$
$$\vdots$$
$$f^{(k)}(x_0) = 2 \cdot 3 \cdots ka_k = k!a_k$$
$$\vdots$$

Hence, the coefficients a_k in the power series (30) are

$$a_k = \frac{f^{(k)}(c)}{k!} \tag{31}$$

for $k = 0, 1, 2, 3, \ldots$. Therefore, we have the following theorem.

THEOREM 6

If a function $f(x)$ is representable as a power series (30), then for every x in the interior of the interval of convergence, the power series is, in fact,

$$f(x) = \sum_{k=0}^{\infty} \frac{f^{(k)}(c_0)}{k!} (x - c)^k \tag{32}$$

or, if $c = 0$,

$$f(x) = \sum_{k=0}^{\infty} \frac{f^{(k)}(0)}{k!} x^k. \tag{33}$$

Any power series of the form (32) is called the **Taylor series of f about c,** and any power series of the form (33) is called the **Maclaurin series of f.** Theorem 6 tells us that *every power series is the Taylor series (or Maclaurin series) of its sum function.* Thus, for example, the Maclaurin series of e^x is given by Equation (18), while the Maclaurin series of $\sin x$ and $\cos x$ are given by Equations (28) and (29), respectively.

EXAMPLE 10 Find the Maclaurin series for the function

$$. f(x) = \frac{1}{1 - x}$$

and determine its interval of convergence.

Solution In our study of geometric series (Section 12.4) we saw that if $|x| < 1$, then

$$1 + x + x^2 + x^3 + \cdots = \frac{1}{1 - x}$$

By Theorem 6, this series must be the Maclaurin expansion of $1/(1 - x)$. When we test the endpoints of the inverval $(-1, 1)$, we see that when $x = \pm 1$, the series diverges, by the general term test. Therefore, the interval of convergence is $(-1, 1)$. ∎

Example 10 establishes the following important fact.

The Maclaurin expansion

$$\frac{1}{1 - x} = 1 + x + x^2 + x^3 + \cdots$$

$$= \sum_{k=0}^{\infty} x^k \tag{34}$$

is valid for all real numbers x in the interval $(-1, 1)$.

It is occasionally useful to be able to integrate a function representable by a power series. The following theorem justifies "integrating term by term."

THEOREM 7

(Integration of Power Series)
If $f(x) = \sum_{k=0}^{\infty} a_k(x - c)^k$, and if a and b are any two points in the interior of the interval of convergence, then

$$\int_a^b f(x)\,dx = \sum_{k=0}^{\infty} a_k \int_a^b (x - c)^k\,dx$$

$$= \sum_{k=0}^{\infty} \frac{a_k}{k + 1} [(b - c)^{k+1} - (a - c)^{k+1}] \tag{35}$$

EXAMPLE 11   Find a Maclaurin series converging to the function $\ln(1 + x)$, for $x > -1$, and determine its interval of convergence.

Solution   To obtain this series we use a clever trick. We know that for $x > -1$,

$$\ln(1 + x) = [\ln(1 + t)]_0^x$$

$$= \int_0^x \frac{dt}{1 + t}. \tag{36}$$

Substituting $x = -t$ into Equation (34) we obtain

$$\frac{1}{1 + t} = \frac{1}{1 - (-t)}$$

$$= 1 + (-t) + (-t)^2 + (-t)^3 + \cdots$$

$$= 1 - t + t^2 - t^3 + t^4 - t^5 + \cdots. \tag{37}$$

In Example 10 we showed that the interval of convergence of this series is $-1 < (-t) < 1$, which is equivalent to $-1 < t < 1$. Thus, for any x in this interval, Equations (36) and (37) tell us that

$$\ln (1 + x) = \int_0^x (1 - t + t^2 - t^3 + t^4 - t^5 + \cdots) \, dt$$

$$= \left[t - \frac{t^2}{2} + \frac{t^3}{3} - \frac{t^4}{4} + \frac{t^5}{5} - \frac{t^6}{6} + \cdots \right]_0^x \qquad \text{(Theorem 7)}$$

$$= x - \frac{x^2}{2} + \frac{x^3}{3} - \frac{x^4}{4} + \frac{x^5}{5} - \frac{x^6}{6} + \cdots. \qquad \blacksquare$$

In Example 11 we have derived the following important fact.

The power series expansion

$$\ln (1 + x) = x - \frac{x^2}{2} + \frac{x^3}{3} - \frac{x^4}{4} + \frac{x^5}{5} - \frac{x^6}{6} + \cdots$$

$$= \sum_{k=1}^{\infty} \frac{(-1)^{k+1} x^k}{k} \qquad (38)$$

is valid for all x in the interval $(-1, 1)$.

EXAMPLE 12 Find the Taylor series for the function $f(x) = \sin x$ about $c = \pi/6$.

Solution In Example 1 of Section 12.3 we calculated the sixth Taylor polynomial of $\sin x$ about $c = \pi/6$. In this example we extend that result to the infinite series

$$\sum_{k=0}^{\infty} \frac{f^{(k)}\left(\dfrac{\pi}{6}\right)}{k!} \left(x - \frac{\pi}{6} \right)^k. \qquad (39)$$

We observe that

$$f^{(0)}\left(\frac{\pi}{6}\right) = \sin \frac{\pi}{6} = \frac{1}{2}$$

$$f^{(1)}\left(\frac{\pi}{6}\right) = \cos \frac{\pi}{6} = \frac{\sqrt{3}}{2}$$

$$f^{(2)}\left(\frac{\pi}{6}\right) = -\sin \frac{\pi}{6} = -\frac{1}{2}$$

$$f^{(3)}\left(\frac{\pi}{6}\right) = -\cos \frac{\pi}{6} = -\frac{\sqrt{3}}{2}$$

$$f^{(4)}\left(\frac{\pi}{6}\right) = \sin \frac{\pi}{6} = \frac{1}{2}$$

and so on. Observe that in this sequence $\{f^{(k)}(\pi/6)\}$, the four numbers $1/2$, $\sqrt{3}/2$, $-1/2$, $-\sqrt{3}/2$ repeat in cyclic fashion. Thus, the Taylor series (39) is

$$\frac{1}{2} + \frac{\sqrt{3}}{2}\left(x - \frac{\pi}{6}\right) - \frac{1}{2}\frac{\left(x - \dfrac{\pi}{6}\right)^2}{2!} - \frac{\sqrt{3}}{2}\frac{\left(x - \dfrac{\pi}{6}\right)^3}{3!} + \frac{1}{2}\frac{\left(x - \dfrac{\pi}{6}\right)^4}{4!} + \frac{\sqrt{3}}{2}\frac{\left(x - \dfrac{\pi}{6}\right)^5}{5!}$$
$$- \frac{1}{2}\frac{\left(x - \dfrac{\pi}{6}\right)^6}{6!} - \frac{\sqrt{3}}{2}\frac{\left(x - \dfrac{\pi}{6}\right)^7}{7!} + \cdots.$$

∎

‖‖‖‖‖‖‖‖‖‖‖‖ Some Finishing Touches (Optional)

The topic of this chapter is a profound one, and few students ever feel comfortable with the level of understanding they are able to achieve at first exposure. A few remarks made at this point may provide some welcome perspective.

A power series can be thought of as a polynomial of infinite degree. Both polynomials and power series are continuous functions that have derivatives of all orders, and that can be differentiated term by term. However, while polynomials are defined over the entire real number line $(-\infty, +\infty)$, power series are restricted to their interval of convergence.

The nth Taylor polynomial of a function f about c is identical with the sum of the first $n + 1$ terms of the Taylor series of f about c.

In applications, Taylor series are frequently used to calculate the numerical value of a quantity. Calculators and computers do not "know" the numerical values of such quantities as $\sin 34°$, $\ln 29$, or $e^{1.84}$. Instead, they calculate such values using techniques related to Taylor series.

A calculator or computer cannot use *all* of the terms of a Taylor series, since it would take an infinite amount of time to perform an infinite number of additions. Thus, it will use only the first n terms, for some n. In this situation, Taylor's Theorem is an essential tool in determining the degree of accuracy to be expected and how many terms of the series must be used to guarantee a specified accuracy.

Not every function has a Taylor series about c. In order for $f(x)$ to have a Taylor series about c it must have derivatives of all orders at c.

Even if the Taylor series of $f(x)$ about c converges at x, it does not necessarily converge to $f(x)$. In the notation of Section 12.3, we can write

$$f(x) = T_n(x) + R_n(x),$$

where $T_n(x)$ is the nth Taylor polynomial of f about c, and $R_n(x)$ is the "nth remainder." In order for $f(x)$ to *equal* its Taylor series we must have

$$f(x) = \lim_{n \to +\infty} T_n(x),$$

or equivalently,

$$\lim_{n \to +\infty} R_n(x) = 0. \tag{40}$$

The problem of verifying condition (40) can be quite complicated. We have avoided this problem by using methods that circumvent it. Actually, all functions (with Taylor series) that fall within the scope of this book do satisfy Equation (40). In fact, for all functions $f(x)$ that you are likely to encounter in practice, you can expect that if $f(x)$ has a Taylor series about c, then that Taylor series converges to $f(x)$ everywhere in its interval of convergence.

We have seen the alternating harmonic series

$$1 - \frac{1}{2} + \frac{1}{3} - \frac{1}{4} + \frac{1}{5} - \frac{1}{6} + \cdots$$

several times in this chapter. We know that it converges, by the alternating series test. For the curious, it is time to reveal its sum. In Example 11 we saw that the power series expansion

$$\ln(1 + x) = x - \frac{x^2}{2} + \frac{x^3}{3} - \frac{x^4}{4} + \frac{x^5}{5} - \frac{x^6}{6} + \cdots \tag{41}$$

is valid for all x in the interval $(-1, 1)$. It is very tempting to substitute $x = 1$ into Equation (41) and thereby obtain

$$\ln 2 = 1 - \frac{1}{2} + \frac{1}{3} - \frac{1}{4} + \frac{1}{5} - \frac{1}{6} + \cdots. \tag{42}$$

We would like to believe this, because it would provide an easy answer for the sum of the alternating harmonic series. Unfortunately, none of the theorems we have presented so far justifies substituting $x = 1$ into Equation (41). In the interest of completeness, we state the theorem that allows us to make this substitution with confidence.

THEOREM 8

Suppose that $f(x)$ is continuous on a domain which includes a half-open interval $(a, b]$ or $([a, b))$ containing c, and that

$$f(x) = \sum_{k=0}^{\infty} a_k(x - c)^k$$

for all x in the open interval (a, b). If this power series converges at $x = b$ (or $x = a$), then it must converge to $f(b)$ (or $f(a)$).

Because of this theorem, we see that Equation (42) is indeed true. The theorem seems simple enough, but it rests upon a deeper theoretical result, known as Abel's* Theorem.

One area of mathematics that makes heavy use of power series is differential equations, where power series are used to obtain solutions. Unfortunately, limitations of time and space prohibit us from presenting illustrative examples here.

Another area in which infinite series find frequency application is that of probability and mathematical statistics. We conclude by showing how the Maclaurin series representation of e^x allows us to calculate areas under the normal probability density curve.

In Section 11.4 we saw that the probability that a standard normal random variable X is in an interval (a, b) is found using the area function

$$A(x) = \int_{-\infty}^{x} \frac{1}{\sqrt{2\pi}} e^{-t^2/2}\, dt,$$

whose values for $x > 0$ are tabulated in Table V, Appendix B. It was noted that there is no combination of elementary functions that can be used to calculate the values of $A(x)$. How, then, can a table for $A(x)$ be obtained? One way is to use infinite series.

First, observe that for $x > 0$,

$$A(x) = \int_{-\infty}^{0} \frac{1}{\sqrt{2\pi}} e^{-t^2/2}\, dt + \int_{0}^{x} \frac{1}{\sqrt{2\pi}} e^{-t^2/2}\, dt. \tag{43}$$

Since the normal curve is symmetric relative to the y-axis,

$$\int_{-\infty}^{0} \frac{1}{\sqrt{2\pi}} e^{-t^2/2}\, dt = \frac{1}{2} \int_{-\infty}^{+\infty} \frac{1}{\sqrt{2\pi}} e^{-t^2/2}\, dt$$

$$= \frac{1}{2} \cdot 1. \qquad \text{[See Section 11.4]}$$

Thus, Equation (43) becomes

$$A(x) = \frac{1}{2} + \int_{0}^{x} \frac{1}{\sqrt{2\pi}} e^{-t^2/2}\, dt, \tag{44}$$

and we may focus our attention on evaluating the integral in (44).

---

* Niels Henrik Abel (1802–1829), was born in the village of Findö, Norway, where his father was pastor in a small church. There were seven children in the family, and life was poor but happy. Abel's mathematical genius was discovered at age 15. When Abel was 18 years old his father died, leaving him to provide for his mother and six siblings. In his brief lifetime, beset with ill health, poverty, and lack of recognition, Abel affected the course of mathematical history with several pivotal contributions in abstract algebra and analysis.

Substituting $x = -t^2/2$ into the Maclaurin series (18) for e^x, we have

$$e^{-t^2/2} = 1 + \left(-\frac{t^2}{2}\right) + \frac{1}{2!}\left(-\frac{t^2}{2}\right)^2 + \frac{1}{3!}\left(-\frac{t^2}{2}\right)^3 + \frac{1}{4!}\left(-\frac{t^2}{2}\right)^4 + \cdots$$

$$= \frac{1}{2^0 \cdot 0!} - \frac{1}{2^1 \cdot 1!}\, t^2 + \frac{1}{2^2 \cdot 2!}\, t^4 - \frac{1}{2^3 \cdot 3!}\, t^6 + \frac{1}{2^4 \cdot 4!}\, t^8 - \cdots$$

$$= \sum_{k=0}^{\infty} \frac{(-1)^k\, t^{2k}}{2^k \cdot k!}.$$

Integrating term by term we have

$$\int_0^x e^{-t^2/2}\, dt = \sum_{k=0}^{\infty} \int_0^x \frac{(-1)^k t^{2k}}{2^k \cdot k!}\, dt$$

$$= \sum_{k=0}^{\infty} \frac{(-1)^k x^{2k+1}}{2^k \cdot k!(2k+1)}.$$

Thus,

$$\int_0^x \frac{1}{\sqrt{2\pi}}\, e^{-t^2/2}\, dt = \frac{1}{\sqrt{2\pi}}\left[x - \frac{x^3}{3 \cdot 2^1 \cdot 1!} + \frac{x^5}{5 \cdot 2^2 \cdot 2!}\right.$$

$$\left. - \frac{x^7}{7 \cdot 2^3 \cdot 3!} + \frac{x^9}{9 \cdot 2^4 \cdot 4!} - \cdots \right]. \tag{45}$$

As an example, we show how to calculate the value of $A(1)$ found in Table V of Appendix B. Using the first five terms of the series (45) for $x = 1$, we calculate

$$\int_0^1 \frac{1}{\sqrt{2\pi}}\, e^{-t^2/2}\, dt \approx \frac{1}{\sqrt{2\pi}}\left[1 - \frac{1}{6} + \frac{1}{40} - \frac{1}{336} + \frac{1}{3456}\right]$$

$$\approx 0.34135.$$

Substituting this result into Equation (44) we obtain

$$A(1) \approx 0.50000 + 0.34135$$

$$\approx 0.84135.$$

This result would round off to 0.8414, which is slightly larger than the value $A(1) = 0.8413$ found in Table V of Appendix B. Notice that the next term in the series (45) is negative, so that by using six terms of the series (45), the sum will be somewhat less. In fact, using six terms of the series (45), our approximation would be

$$A(1) \approx 0.84134.$$

When rounded off to four decimal places, this figure agrees with the value found in Table V.

The problem of using Taylor's Theorem to determine the number of terms of the power series (45) that must be used to guarantee a specified degree of accuracy must be left to more advanced courses.

12.6 EXERCISE SET ||||||||||||||||||||||||||||||||||||||||||||||||||||||||||||||||||||||||||

In Exercises 1–20, find the radius of convergence and the interval of convergence. (Be sure to test for convergence at the endpoints of the interval.)

1. $\displaystyle\sum_{k=1}^{\infty} \frac{x^k}{k}$

2. $\displaystyle\sum_{k=1}^{\infty} \frac{x^k}{k^2}$

3. $\displaystyle\sum_{k=1}^{\infty} \frac{(x-1)^k}{3^k}$

4. $\displaystyle\sum_{k=1}^{\infty} \frac{(x+2)^k}{\sqrt{k}}$

5. $\displaystyle\sum_{k=1}^{\infty} \frac{(-x)^k}{k^3}$

6. $\displaystyle\sum_{k=1}^{\infty} (-2x)^k$

7. $\displaystyle\sum_{k=1}^{\infty} 4k(x+3)^k$

8. $\displaystyle\sum_{k=1}^{\infty} k^2(x-1)^k$

9. $\displaystyle\sum_{k=0}^{\infty} \frac{(2x)^k}{k!}$

10. $\displaystyle\sum_{k=0}^{\infty} \frac{(-x)^k}{k!}$

11. $\displaystyle\sum_{k=0}^{\infty} k!\left(\frac{x}{3}\right)^k$

12. $\displaystyle\sum_{k=0}^{\infty} \frac{k!x^k}{k+1}$

13. $\displaystyle\sum_{k=0}^{\infty} (-e)^k(x-2)^k$

14. $\displaystyle\sum_{k=0}^{\infty} \frac{(-1)^k kx^k}{4^k}$

15. $\displaystyle\sum_{k=1}^{\infty} \frac{x^k}{k(k+1)}$

16. $\displaystyle\sum_{k=1}^{\infty} \frac{(x-5)^k}{k^2}$

17. $\displaystyle\sum_{k=1}^{\infty} \frac{(x-4)^k}{\sqrt{k}}$

18. $\displaystyle\sum_{k=1}^{\infty} 2^k(x+1)^k$

19. $\displaystyle\sum_{k=1}^{\infty} \frac{(-1)^k 3^k}{k} (x+2)^k$

20. $\displaystyle\sum_{k=1}^{\infty} \frac{(-1)^k k^3}{5^k} (x-3)^k$

In Exercises 21–28, find a power series expansion of the given function that is valid for all real numbers x, except as indicated. As in Example 9, apply the techniques of Theorems 4 and 5 to the known power series expansions (18), (28), and (29).

21. xe^x

22. $(e^x)^3$

23. $\cos(2x)$

24. $\sin\left(x - \dfrac{\pi}{3}\right)$

25. $x^2 \sin(-x)$

26. $\dfrac{\sin x}{x}$ $(x \neq 0)$

27. $\cos x + \sin x$

28. $\dfrac{x^2}{e^x}$

In Exercises 29–36, by differentiating or integrating a known power series term-by-term or by using methods used in Exercises 21–28, find the Maclaurin series of the indicated function. In each case find the interval of convergence.

29. $\dfrac{1}{(1-x)^2}$ [*Hint:* differentiate $\dfrac{1}{1-x}$.]

30. $\dfrac{x}{1+x^2}$

31. $\dfrac{1}{(1-x)^3}$

32. $\dfrac{x^3}{1+5x}$

33. $x^2 \ln(1+x)$

34. $\dfrac{\ln(1+x)}{x}$ $(x \neq 0)$

35. $\displaystyle\int_0^x \frac{1}{1+t^2}\,dt$

36. $\displaystyle\int_0^x \ln(1+t)\,dt$

In Exercises 37–44, use Theorem 6 to write out the first n terms of the Taylor series for the given function $f(x)$ about the specified number c.

37. $f(x) = \sin x$, $c = \pi/3$, $n = 6$

38. $f(x) = \cos x$, $c = \pi/3$, $n = 6$

39. $f(x) = \ln x$, $c = 1$, $n = 7$

40. $f(x) = \ln x$, $c = 2$, $n = 7$

41. $f(x) = \sqrt{x}$, $c = 1$, $n = 6$

42. $f(x) = \sqrt[3]{x}$, $c = 1$, $n = 5$

43. $f(x) = \tan x$, $c = 0$, $n = 5$

44. $f(x) = 2^x$, $c = 0$, $n = 5$

KEY IDEAS FOR REVIEW ‖‖‖

■ A sequence $\{a_n\}$ is an infinite succession of numbers $a_1, a_2, a_3, \ldots, a_n, a_{n+1}, \ldots$.

■ The convergence or divergence of a sequence is not affected by its first few terms, or even by its first million terms.

■ $\lim\limits_{n \to +\infty} c = c$

$$\lim_{n \to +\infty} \frac{1}{n} = 0 \qquad \lim_{n \to +\infty} c^n = \begin{cases} = 0 & \text{if } |c| < 1 \\ = 1 & \text{if } c = 1 \\ = +\infty & \text{if } c > 1 \\ \text{does not exist if } c \le 1.. \end{cases}$$

$$\lim_{n \to +\infty} n = +\infty$$

$$\lim_{n \to +\infty} \left(1 + \frac{1}{n}\right)^n = e \qquad \lim_{n \to +\infty} \frac{n}{a^n} = 0 \quad \text{if } |a| > 1$$

and k is any constant

■ If $\lim\limits_{n \to +\infty} a_n$ and $\lim\limits_{n \to +\infty} b_n$ exist and c is any real number, then

$$\lim_{n \to +\infty} (a_n \pm b_n) = \lim_{n \to +\infty} a_n \pm \lim_{n \to +\infty} b_n$$

$$\lim_{n \to +\infty} c a_n = c \lim_{n \to +\infty} a_n$$

$$\lim_{n \to +\infty} \frac{a_n}{b_n} = \frac{\lim\limits_{n \to +\infty} a_n}{\lim\limits_{n \to +\infty} b_n} \qquad \text{if } \lim_{n \to +\infty} b_n \ne 0$$

$\lim\limits_{n \to +\infty} a_n^c = (\lim\limits_{n \to +\infty} a_n)^c$ if c is a positive constant such that a_n^c is always a real number.

■ A monotone sequence converges if and only if it is bounded.

■ $\displaystyle\sum_{k=1}^{n} a_k = a_1 + a_2 + \cdots + a_n$.

■ A geometric progression is a sequence of the form

$$a, ar, ar^2, ar^3, \ldots, ar^n, \ldots.$$

The sum of its first $n + 1$ terms is

$$\sum_{k=0}^{n} ar^k = a\left(\frac{1 - r^{n+1}}{1 - r}\right).$$

■ $k! = 1 \cdot 2 \cdot 3 \cdots k$, $0! = 1$.

■ If $p(x) = a_0 + a_1 x + \cdots + a_n x^n$, then for each k,

$$a_k = \frac{p^{(k)}(0)}{k!}.$$

■ Every polynomial $p(x) = \sum_{k=0}^{n} a_k x^k$ can be rewritten in Taylor form

$$p(x) = \sum_{k=0}^{n} \frac{p^{(k)}(c)}{k!}(x - c)^k$$

for any constant c.

■ If f is any function that is differentiable n times in an interval containing c, then the nth Taylor polynomial for f about c is

$$T_n(x) = \sum_{k=0}^{n} \frac{f^{(k)}(c)}{k!}(x - c)^k.$$

If $c_0 = 0$, the resulting polynomial $M_n(x)$ is called the nth Maclaurin polynomial for f.

■ Taylor's Theorem: If $f(x)$ has derivative of order $n + 1$ in an open interval containing c, then for each x in this interval, there is a number t between x and x_0 such that

$$f(x) = \left\{\sum_{k=0}^{n} \frac{f^k(c)}{k!}(x - c)^k\right\}$$

$$+ \frac{f^{k(n+1)}(t)}{(n+1)!}(x - c)^{n+1}$$

$$= T_n(x) + R_n(x).$$

■ An infinite series $\sum_{k=1}^{\infty} a_k$ converges if and only if its sequence of partial sums $\{S_n\}$ converges, where

$$S_n = \sum_{k=1}^{n} a_k.$$

■ The geometric series $\sum_{k=0}^{\infty} ar^k = a + ar + ar^2 + \cdots + ar^n + \cdots$

(a) converges to $\dfrac{a}{1 - r}$ if $|r| < 1$,

(b) diverges if $|r| > 1$.

■ If $\sum_{k=1}^{\infty} a_k$ and $\sum_{k=1}^{\infty} b_k$ are convergent series, and c is any real number, then

$$\sum_{k=1}^{\infty} (a_k \pm b_k) = \sum_{k=1}^{\infty} a_k \pm \sum_{k=1}^{\infty} b_k$$

$$\sum_{k=1}^{\infty} ca_k = c \sum_{k=1}^{\infty} a_k.$$

■ General Term Test: If $\sum_{k=1}^{\infty} a_k$ converges, then $\lim_{k \to +\infty} a_k = 0$. *Caution:* the converse is not true.

■ A series of nonnegative terms converges if and only if its sequence of partial sums is bounded.

■ Integral Test: For a series $\sum_{k=1}^{\infty} a_k$ of nonnegative terms, if $f(x)$ is a continuous, decreasing function on $[1, +\infty)$ such that $f(n) = a_n$ for all natural numbers n, then $\sum_{k=1}^{\infty} a_k$ converges if and only if $\int_1^{+\infty} f(x)\, dx$ converges.

■ The p-series $\sum_{k=1}^{\infty} \dfrac{1}{k^p}$ (p is a positive constant) converges if $p > 1$, and diverges to $+\infty$ if $0 < p \le 1$.

■ Comparison Test: Suppose $\sum_{k=1}^{\infty} a_k$ and $\sum_{k=1}^{\infty} b_k$ are series of nonnegative terms, and k_0 is a constant such that $a_k \le b_k$, for all natural numbers $k \ge k_0$. If $\sum_{k=1}^{\infty} b_k$ converges, so does $\sum_{k=1}^{\infty} a_k$. If $\sum_{k=1}^{\infty} a_k$ diverges, so does $\sum_{k=1}^{\infty} b_k$.

■ Ratio Test: Suppose $\sum_{k=1}^{\infty} a_k$ is a series of positive terms, and $\rho = \lim_{k \to +\infty} \dfrac{a_{k+1}}{a_k}$. The series converges if $\rho < 1$ and diverges if $\rho > 1$.

■ Alternating Series Test: An alternating series converges if its general term has limit 0.

■ Every absolutely convergent series converges, but not every convergent series is absolutely convergent.

■ A power series is any series of the form $\sum_{k=0}^{\infty} a_k x^k$ or $\sum_{k=0}^{\infty} a_k (x - c)^k$.

■ A power series $\sum_{k=0}^{\infty} a_k (x - x_0)^k$ always converges in an interval centered at c. The radius of convergence is

$$R = \lim_{k \to +\infty} \left| \frac{a_k}{a_{k+1}} \right|.$$

■ A power series can be differentiated and integrated term-by-term. The radius of convergence is unchanged under these operations.

■ The Maclaurin series for $f(x)$ is

$$\sum_{k=0}^{\infty} \frac{f^{(k)}(0)}{k!} x^k.$$

■ The Taylor series for $f(x)$ about c is

$$\sum_{k=0}^{\infty} \frac{f^{(k)}(c)}{k!} (x - c)^k.$$

■ Some special Maclaurin series:

$$e^x = 1 + x + \frac{x^2}{2!} + \frac{x^3}{3!} + \frac{x^4}{4!} + \cdots$$
$$\text{on } (-\infty, +\infty)$$

$$\sin x = x - \frac{x^3}{3!} + \frac{x^5}{5!} - \frac{x^7}{7!} + \cdots$$
$$\text{on } (-\infty, +\infty)$$

$$\cos x = 1 - \frac{x^2}{2!} + \frac{x^4}{4!} - \frac{x^6}{6!} + \cdots$$
$$\text{on } (-\infty, +\infty)$$

$$\frac{1}{1 - x} = 1 + x + x^2 + x^3 + \cdots$$
$$\text{on } (-1, +1)$$

$$\ln (1 + x) = x - \frac{x^2}{2} + \frac{x^3}{3} - \frac{x^4}{4} + \cdots$$
$$\text{on } (-1, +1]$$

REVIEW EXERCISES ||||||||||||||||||||||||||| |||||||||||||||||||||||||||||||||||||||||||||||||

In Exercises 1–4, write out the first six terms of the given sequence.

1. $\{11 - 3n\}$

2. $\left\{ \dfrac{(-2)^n}{5n^2} \right\}$

3. $\left\{ \dfrac{1 + (-1)^n}{2} \right\}$

4. $\left\{ \dfrac{n^2 - 10}{n + 1} \right\}$

In Exercises 5–12, determine whether the given sequence converges or diverges. Find the limit, where possible.

5. $\left\{\dfrac{(-1)^n}{n}\right\}$

6. $\left\{\dfrac{n+2}{100-3n}\right\}$

7. $\left\{\left(\dfrac{9}{13}\right)^n\right\}$

8. $\left\{\dfrac{n^2+1}{5n+3}\right\}$

9. $\{5\}$

10. $\left\{\sqrt{\dfrac{n+1}{4n+3}}\right\}$

11. $\left\{\dfrac{3n^2-2n+11}{12n^2+5}\right\}$

12. $\left\{\left(1+\dfrac{1}{n}\right)^{-n}\right\}$

In Exercises 13 and 14 calculate the indicated sum.

13. $\displaystyle\sum_{k=2}^{6}(k^2-5k)$

14. $\displaystyle\sum_{k=0}^{5}5\left(\dfrac{2}{3}\right)^k$

15. Suppose $50 is deposited quarterly into an account paying 6% annual interest, compounded after each successive payment is made. Find the value of the account after 8 years (after the 32nd payment has been made).

16. Express the polynomial $p(x)=10-x+5x^3$ in Taylor form about $c=2$.

17. Express the polynomial $p(x)=3-x+x^2-3x^3+5x^4$ in Taylor form about $c=-1$.

18. Find the fifth Maclaurin polynomial for the function $f(x)=\sin 2x$.

19. Find the third Maclaurin polynomial for the function $f(x)=\sqrt{2x+9}$.

20. Find the fourth Taylor polynomial for the function $f(x)=1/(x-1)$ about $c=2$.

21. Find the fourth Taylor polynomial of the function $f(x)=\sqrt[3]{x}$ about $c=1$.

In Exercises 22–25, write out the sum of the first six terms of the given series. Do not evaluate the sum.

22. $\displaystyle\sum_{k=1}^{\infty}\dfrac{(-1)^k}{3k+1}$

23. $\displaystyle\sum_{k=1}^{\infty}\dfrac{2^k}{k^2-2}$

24. $\displaystyle\sum_{k=0}^{\infty}\dfrac{x^{3k+2}}{3k+2}$

25. $\displaystyle\sum_{k=0}^{\infty}\dfrac{(-1)^k x^{2k+1}}{k!}$

In Exercises 26–29, determine whether the given geometric series converges or diverges. Find the sum, where possible.

26. $\frac{1}{2}+\frac{3}{4}+\frac{9}{8}+\frac{27}{16}+\cdots$

27. $.3+.03+.003+.0003+\cdots$

28. $\frac{1}{3}-\frac{1}{9}+\frac{1}{27}-\frac{1}{81}+\cdots$

29. $.001-.003+.009-.027+\cdots$

30. A ball is dropped from a height of 10 feet. Each time it hits the ground it rebounds to a height that is 60% of its previous rebound. Use an infinite series to express the total distance traveled by the ball in coming to rest and find its sum.

31. Use an infinite series to express the total increase in spending that will be generated by a 6-billion-dollar tax cut in a nation whose MPC is 70%.

32. The alumni association of a college wishes to set up a perpetuity that will pay $500 annually to the winner of their special honors award. The award money will come from a fund that pays 9% interest, compounded annually after each award is made. Express the capitalized value of this perpetuity as an infinite series and find its sum.

In Exercises 33–40, use an appropriate test to determine convergence or divergence.

33. $\displaystyle\sum_{k=0}^{\infty}\dfrac{k+2}{3k-1}$

34. $\displaystyle\sum_{k=1}^{\infty}\dfrac{1}{\sqrt[3]{k}}$

35. $\displaystyle\sum_{k=1}^{\infty}\dfrac{(-1)^k}{\sqrt{k}}$

36. $\displaystyle\sum_{k=0}^{\infty}\dfrac{5}{k^2+3}$

37. $\displaystyle\sum_{k=0}^{\infty}ke^{-2k}$

38. $\displaystyle\sum_{k=0}^{\infty}\dfrac{2k}{k!}$

39. $\displaystyle\sum_{k=0}^{\infty}\dfrac{1}{3k-1}$

40. $\displaystyle\sum_{k=1}^{\infty}\dfrac{(-1)^k k}{k^2+4k-2}$

In Exercises 41 and 42, determine whether the series converges, converges absolutely, or diverges.

41. $\displaystyle\sum_{k=1}^{\infty} \frac{(-1)^k}{k+3}$

42. $\displaystyle\sum_{k=1}^{\infty} \frac{(-1)^k}{k!}$

In Exercises 43–48, find the radius of convergence and the interval of convergence. (Be sure to test the endpoints of the interval.)

43. $\displaystyle\sum_{k=0}^{\infty} (3x)^k$

44. $\displaystyle\sum_{k=1}^{\infty} \frac{x^k}{k^2}$

45. $\displaystyle\sum_{k=1}^{\infty} \frac{(-x)^k}{\sqrt{k}}$

46. $\displaystyle\sum_{k=0}^{\infty} \frac{(x+3)^k}{k!}$

47. $\displaystyle\sum_{k=0}^{\infty} \frac{k(x+2)^k}{2^k}$

48. $\displaystyle\sum_{k=0}^{\infty} \frac{(-2)^k(x-1)^k}{k+1}$

In Exercises 49–54, find a power series expansion of the given function by transforming one or more known power series. In each case, find the interval of convergence.

49. $\sin(3x)$

50. e^{3x-1}

51. $\dfrac{x^2}{1-x}$

52. $2 + x^2 - xe^x$

53. $e^x + \sin x$

54. $\dfrac{1}{(1+3x)^2}$

55. $x \ln(1+x)$

56. $\cos\sqrt{x+2}$

In Exercises 57–59, find the first n terms of the Taylor series for the given function $f(x)$ about the given c_0, for the specified values of n.

57. $f(x) = \sin 2x$, $c = \pi/4$, $n = 7$

58. $f(x) = \ln(1+x)$, $c = 1$, $n = 6$

59. $f(x) = \sqrt{1+x}$, $c = 8$, $n = 4$

CHAPTER TEST ||||||||||||||||||||||||||||||||||||||||||||||||||||||||||||||||||||||||||||||||||||||

1. Determine whether the following sequences converge or diverge. Find the limit, where possible.

   (a) $\left\{ \dfrac{(-1)^n}{n+1} \right\}$ (b) $\left\{ \left(\dfrac{6}{5}\right)^n \right\}$ (c) $\left\{ \dfrac{1-n^2}{5n+3} \right\}$

2. Find the fourth Taylor polynomial for the function $f(x) = \cos x$ about $c = \pi/3$.

3. Determine whether the following series converge or diverge. Find the sum, if possible.

   (a) $\displaystyle\sum_{k=1}^{\infty} \frac{1}{9}\left(\frac{8}{7}\right)^k$ (b) $\displaystyle\sum_{k=0}^{\infty} 5\left(\frac{3}{4}\right)^k$

4. In the treatment of a certain disease a patient must take 50 milligrams of a drug daily, over a period of many months. The body eliminates 20% of the drug from the body each day. In the long run, how much of the drug will be in the body each day after the daily dose is taken? Express your answer as the sum of an infinite series.

5. For each of the following series, use an appropriate test to determine convergence or divergence.

   (a) $\displaystyle\sum_{k=1}^{\infty} \frac{k+3}{k^2}$ (b) $\displaystyle\sum_{k=0}^{\infty} \frac{8k}{5^k}$

6. Find the radius of convergence and the interval of convergence for the series

$$\sum_{k=0}^{\infty} \frac{(x-1)^k}{2^k}.$$

7. Use the known Maclaurin series expansion of $\cos x$ to obtain the Maclaurin series for the function

$$f(x) = x \cos (2x).$$

8. Use the known Maclaurin series expansion of $1/(1 + x)$ to obtain the Maclaurin series for the function

$$f(x) = \frac{1}{(1 + x)^2}.$$

9. Find the first four terms of the Maclaurin series expansion of the function

$$f(x) = \sqrt{x + 4}.$$

APPENDIX A: REVIEW OF BASIC ALGEBRA

Almost every student taking this course will have studied the basic high school algebra used in this book. Appendix A provides a quick review of the necessary algebraic techniques. It starts with a diagnostic test so that students can determine which material, if any, needs reviewing. This review may be done quickly and students can consult the chapter while studying the calculus, as the need arises. We have keyed each question in the test to the section in this chapter that reviews the material covered by the question. The answers to the diagnostic test appear in the answer section at the back of the book.

A.1 |||||||| DIAGNOSTIC TEST
(for class use or as a self-test)

Time alloted: fifty minutes.

|||||||||||||| Instructions to the Student
(if taken as a self-test)

Treat this diagnostic test as you would any other test. Thus,

1. Take the test in a quiet place.
2. Do not consult any books or material.
3. Do not ask anyone for help.
4. Spend no more than *fifty minutes* on this test.
5. Follow directions carefully.
6. Do not use a calculator.

|||||||||||||| Test Questions

1. (Section A.2) Which of the following numbers, $\frac{3}{7}$, -3, $\sqrt{3}$, 5, $\sqrt{\frac{4}{9}}$, 0 is (are)
   (a) rational?
   (b) a natural number?
   (c) an integer?
   (d) irrational?

2. (Section A.2) What type of decimal represents an irrational number?

3. (Section A.2) For $x = 3$ and $y = -4$, compute $2x - y$.

4. (Section A.2) Compute
$9 - [-3(6 - 8) + 3]$.

5. (Section A.2) Compute $(\frac{3}{8}) \div (\frac{7}{12} + \frac{2}{3})$ and simplify your answer.

6. (Section A.2) Compute
$$\frac{|4 - 6|}{2 - |-5|}.$$

7. (Sections A.2 and A.4) Which of the following is (are) undefined? $\frac{0}{3}$, $5 \div 0$, 0×4, $\frac{0}{0}$, 0×0, 4^0.

8. (Section A.3) Sketch the interval $[-1, 4)$ on the real number line.

9. (Section A.3) Solve the equation $5x + 4 = 3x$ for x.

10. (Section A.3) Find all real numbers x for which $3 - 2x \le -7$ holds, and sketch the solution set on a number line.

11. (Section A.3) Solve for x and y in the simultaneous system
$$x + 2y = 8$$
$$3x - y = 17.$$

12. (Section A.3) If four nurds and 5 blurbs cost a total of $71, and five nurds and 3 blurbs cost a total of $66, how much do each nurd and each blurb cost?

13. (Section A.4) Compute $16^{3/2}$.

14. (Section A.4) For $x = \frac{2}{3}$, compute and simplify x^{-2}.

15. (Section A.4) Using only positive exponents, rewrite
$$\left(\frac{a^{-2}b^3}{c^{-1}} \right)^3.$$

16. (Section A.5) Multiply and simplify:
$$(x - 2a)(x^2 + 3ax - 2a).$$

17. (Section A.5) Factor $2x^3 - 7x^2 - 15x$ completely.

18. (Section A.5) Compute and simplify
$$\frac{2x}{x^2 - 4} + \frac{1}{x(x - 2)} - \frac{1}{x - 2}.$$

19. (Section A.6) By factoring, solve
$$6x^2 - x - 2 = 0.$$

20. (Section A.6) Using the quadratic formula, solve
$$2x^2 - 3x - 2 = 0.$$

21. (Section A.6) Find all real numbers x for which $(x - 3)(x + 1) < 0$ holds.

22. (Section A.6) Solve for x:
$$\frac{x^3 - x}{x^2 - 36} = 0.$$

23. (Section A.8) Draw a Cartesian coordinate system, plot the indicated point, and name the quadrant in which the point lies.
(a) $(-4, 2)$
(b) $(3, -5)$
(c) $(0, 3)$

24. (Section A.8) Draw a Cartesian coordinate system and sketch the graph of the equation $2x - 3y = 10$.

A.2 ||||||| THE REAL NUMBER SYSTEM

The numbers used in this book are the **real numbers.** We begin by reviewing the composition of this system of numbers.

The numbers 1, 2, 3, . . . , which we use to count objects, are called **natural** or **counting numbers.** The system of natural numbers is adequate for recording the assets or profits of a person or business, but is inadequate when the need arises to record a net loss or no profit.

To indicate no profit we need zero. To indicate losses we need negative numbers. The natural number system contains neither negative numbers nor zero.

The system of **integers** fills these needs. It consists of (a) the natural numbers, (b) the negatives of natural numbers, and (c) zero:

$$\ldots, -4, -3, -2, -1, 0, 1, 2, 3, 4, \ldots$$

We assume that you remember the rules for adding signed integers. For example,

$$3 + 4 = 7 \qquad 6 + (-2) = 4 \qquad 5 + (-8) = -3 \qquad -4 + (-5) = -9.$$

The system of integers proves inadequate for division.

When we try to divide two apples equally among four people, we find that no integer will express how many apples each person should get. We need **rational numbers** to solve this problem. A **rational number** is a number that can be written as a ratio $\frac{p}{q}$ of two integers, p and q, where q does not equal zero. Examples of rational numbers are $0, \frac{2}{3}, -4, \frac{7}{5}, -\frac{3}{4}$. Thus, when we divide two apples equally among four people, each person gets $\frac{1}{2}$ apple. Since every integer n can be written as $\frac{n}{1}$, we see that every integer is a rational number. Thus, the set of rational numbers includes the set of integers.

The **Real Number System** includes the rational numbers (and hence the integers and natural numbers), as well as other numbers, as will be seen below. To obtain a simple and useful geometric description of the set of real numbers we draw a horizontal straight line, which we will call the **real number line;** pick a point, label it with the number 0, and call it the **origin;** and choose the **positive direction** to the right of the origin and the **negative direction** to the left of the origin. An arrow indicates the positive direction.

Next we select a unit of length, by locating a point on the positive side of the origin, and placing the number 1 at this point.

Having determined the location of 0 and 1 we continue, locating the remaining integers by counting off the positive integers in the positive direction and the negative integers in the negative direction. We space them so that the successive integers are one unit of length apart.

The location of any rational number p/q on the number line is determined geometrically as follows: divide the unit length between 0 and 1 into q parts (assuming $q > 0$) and lay off p of those parts starting from 0, in the direction consistent with the sign of p.

The Real Number System is identified with the set of all points on a straight line. Every point on the line corresponds to a real number, called its **coordinate,** and for every real number there is a point on the line. We often say that the set of real numbers and the set of points on the real number line are in **one-to-one correspondence.** The numbers to the right of the origin are called **positive.** The numbers to the left of the origin are called **negative.** The positive numbers and zero together are called the **nonnegative** numbers, whereas the negative numbers and zero together are called the **nonpositive** numbers.

The rational numbers are so densely scattered along the real number line that they appear to completely fill the line. However, there are points on the number line not occupied by rational numbers. Indeed there are many such points. One can show that $\sqrt{2}$, $\sqrt{3}$, $\sqrt[3]{7}$, and so on, are such numbers. Another such number is π, the ratio of the circumference of a circle to its diameter. Real numbers that are not rational numbers are called **irrational numbers.** Thus, the Real Number System consists of the set of all rational numbers together with the set of all irrational numbers (Figure 1).

Figure 1

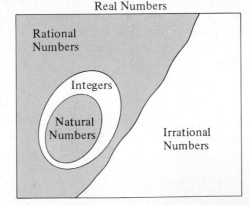

As electronic calculators prepare us to believe, real numbers may be represented as decimals. **Finite decimals,** such as 31.87, always represent rational numbers. The number $31.87 = \frac{3187}{100}$ is rational because it is a quotient of integers. It is called a finite decimal (or a **terminating decimal**) because the number of digits following to the right of the decimal point is finite. Many rational numbers have decimal representations that are nonterminating; for example,

$$\tfrac{1}{3} = .333333\ldots,$$
$$\tfrac{5}{11} = .45454545\ldots.$$

Notice that these are **repeating decimals;** they each have a block of digits that repeats over and over, ad infinitum. It is a fact that the decimal representation of a rational number must be either terminating or repeating. By contrast, the decimal representation of an irrational number must be both nonterminating and nonrepeating. This provides us with an important and useful distinction between rational and irrational numbers. The real number system may thus be viewed as the set of all possible decimals, terminating or nonterminating. The nonterminating, nonrepeating decimals represent irrational numbers; all others represent rational numbers (Figure 2).

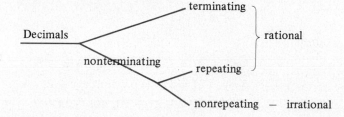

Figure 2

IIIIIIIIIIIII ## Parentheses and the Order of Arithmetic Operations

To avoid ambiguity, when several of the four fundamental arithmetic operations (addition, subtraction, multiplication, and division) occur in the same problem, an **order of priority** is observed. In the absence of parentheses the following rules apply.

(a) When additions and/or subtractions (only) occur, they are performed in the order in which they appear, left to right, Thus,

$$13 - 4 + 7 + 2 - 1 = 9 + 7 + 2 - 1 = 18 - 1 = 17.$$

(b) When multiplications and/or divisions (only) occur, they are performed in the order in which they appear, left to right. Thus,

$$18 \div 2 \times 5 \div 3 = 9 \times 5 \div 3 = 45 \div 3 = 15.$$

(c) When operations from (a) are mixed with operations from (b), all multiplications and divisions are performed before additions and subtractions. Thus

$$24 \div 6 + 2 \times 2 = 4 + 4 = 8,$$
$$17 - 3 \times 4 = 17 - 12 = 5,$$
$$12 \times 3 \div 9 - 1 = 36 \div 9 - 1 = 4 - 1 = 3.$$

Parentheses are used to override the above rules. Observe how parentheses affect the following expressions.

With Parentheses	Without Parentheses
$10-(7+1)=10-8=2$	$10-7+1=3+1=4$
$(12-2)\times 5=10\times 5=50$	$12-2\times 5=12-10=2$
$3\times(2+5)-4=3\times 7-4=21-4=17$	$3\times 2+5-4=6+5-4=7$
$24\div(4+2)+1=24\div 6+1=4+1=5$	$24\div 4+2+1=6+2+1=9$

The operations of multiplication and addition (or subtraction) obey the **distributive laws:**

$$a(b + c) = ab + ac,$$

and

$$a(b - c) = ab - ac.$$

Thus, for example,

$$5a(7 + b) = 35a + 5ab, \text{ not } 35a + b;$$
$$3x(y - 2) = 3xy - 6x, \text{ not } 3xy - 2;$$
$$-2(x - 5) = -2x + 10, \text{ not } -2x - 5;$$
$$a(2b + 3c - 4d) = 2ab + 3ac - 4ad.$$

The following computations illustrate some familiar basic properties of the real numbers.

Example 1

$$2 + [-3 + 4] = [2 + (-3)] + 4 = 3$$
$$[3 \cdot (-4)] \cdot 2 = 3 \cdot [(-4) \cdot 2] = -24$$
$$2 \cdot (3 + 5) = (2 \cdot 3) + (2 \cdot 5) = 16$$
$$(3 + 4) \cdot (-2) = 3 \cdot (-2) + 4 \cdot (-2) = -14.$$

The **negative** $-a$ of a is the number located on the opposite side of 0, but at the same distance from 0, as a (Figure 3).

Figure 3   or

Observe that $-a$ need not be a negative number. For example, if $a = -3$ then $-a = 3$, and in this case $-a$ is positive. The negative $-a$ of a satisfies the property

$$a + (-a) = 0. \qquad \left[\text{Thus, } \frac{2}{3} + \left(-\frac{2}{3}\right) = 0.\right]$$

If a and b are real numbers, the **difference** between a and b, denoted by $a - b$ is defined by

$$a - b = a + (-b), \tag{1}$$

and the operation is called **subtraction.** Thus,

$$6 - 2 = 4 \qquad 2 - 2 = 0 \qquad 0 - 8 = -8.$$

The usual rules of signs in multiplication may be summarized as follows. Let a and b denote real numbers.

Property	Example
$-(-a) = a$	$-(-\frac{1}{3}) = \frac{1}{3}$
$(-a)(b) = -(ab) = a(-b)$	$(-2)(3) = -6 = (2)(-3)$
$(-1)(a) = -a$	$(-1)(\frac{1}{4}) = -\frac{1}{4}$
$(-a)(-b) = ab$	$(-2)(-4) = 8 = 2 \cdot 4$
$-(a + b) = (-a) + (-b)$	$-(2 + 5) = (-2) + (-5) = -7$

Terms versus Factors: The real numbers a and b are said to be **terms** of the sum $a + b$, whereas they are said to be **factors** of the product ab. For example,

$$3x, \ 4xy, \ \text{and} \ -11y^2 \ \text{are terms in the sum} \ 3x + 4xy - 11y^2,$$

and

$$3, \ c, \ \text{and} \ d \ \text{are factors in the product} \ 3cd.$$

The **multiplicative property of zero:** if a is any real number, then

$$0 \cdot a = 0. \tag{2}$$

The **reciprocal** $1/a$ of a nonzero number a satisfies the property

$$a \cdot \frac{1}{a} = 1. \quad \left[\text{e.g., } 4 \cdot \frac{1}{4} = 1. \right]$$

The reciprocal of a nonzero fraction a/b is the fraction b/a, because $\dfrac{a}{b} \cdot \dfrac{b}{a} = 1$. That is,

$$\frac{1}{\dfrac{a}{b}} = \frac{b}{a}. \tag{3}$$

Using Equation (2), we see that the number 0 has no reciprocal, since there is no number that when multiplied by 0 yields 1.

Division: To divide a real number a by a nonzero real number b means to multiply a by the reciprocal of b. The resulting **quotient of a by b** is denoted $\dfrac{a}{b}$, or $a \div b$, or a/b. Thus,

$$a \div b = a \cdot \frac{1}{b} = \frac{a}{b}. \tag{4}$$

For example,

$$x \div 2 = \frac{x}{2} = x \cdot \frac{1}{2}.$$

In formula (4) the number a is called the **numerator** and the number b is called the **denominator.**

To divide one fraction by another, we use rules (3) and (4) together:

$$\frac{a}{b} \div \frac{c}{d} = \frac{a}{b} \cdot \frac{1}{\dfrac{c}{d}} = \frac{a}{b} \cdot \frac{d}{c}.$$

That is, we multiply the first fraction by the reciprocal of the second fraction. For example,

$$\frac{3}{4} \div \frac{1}{2} = \frac{3}{4} \cdot \frac{2}{1} = \frac{3}{2}; \qquad \frac{5}{7} \div \frac{2}{5} = \frac{5}{7} \cdot \frac{5}{2} = \frac{25}{14}.$$

Division as the inverse of multiplication: When we divide 48 by 16 we seek a number that, when multiplied by 16, yields 48. In this case,

$$48 \div 16 = 3 \qquad \text{because} \qquad 48 = 16 \times 3.$$

The general rule is

$$a \div b = c \qquad \text{if and only if*} \qquad a = b \times c.$$

This provides another way of seeing that division by 0 is impossible. For,

$$a \div 0 = c \qquad \text{if and only if*} \qquad a = 0 \times c.$$

But if $a \neq 0$, then there is **no** such number c, whereas if $a = 0$, then **any** number c would qualify. Therefore, we repeat,

$$\textit{Division by zero is not permitted.}$$

We summarize the familiar properties of fractions.

---

* The connector "if and only if" means that either both statements are true or both statements are false.

Let a, b, c, and d denote real numbers with $b \neq 0$, $d \neq 0$.

Property	Example
$\dfrac{a}{b} = \dfrac{c}{d}$ if and only if $ad = bc$	$\dfrac{2}{3} = \dfrac{4}{6}$ since $2 \cdot 6 = 3 \cdot 4$
$\dfrac{a}{b} = \dfrac{ad}{bd}$	$\dfrac{6}{12} = \dfrac{6 \cdot 3}{12 \cdot 3} = \dfrac{18}{36}$
$\dfrac{a}{b} + \dfrac{c}{b} = \dfrac{a+c}{b}$	$\dfrac{2}{3} + \dfrac{5}{3} = \dfrac{7}{3}$
$\dfrac{a}{b} + \dfrac{c}{d} = \dfrac{ad + bc}{bd}$	$\dfrac{2}{5} + \dfrac{3}{4} = \dfrac{2 \cdot 4 + 5 \cdot 3}{5 \cdot 4} = \dfrac{23}{20}$
$\dfrac{a}{b} \cdot \dfrac{c}{d} = \dfrac{ac}{bd}$	$\dfrac{2}{3} \cdot \dfrac{4}{5} = \dfrac{2 \cdot 4}{3 \cdot 5} = \dfrac{8}{15}$
$\dfrac{a}{b} \div \dfrac{c}{d} = \dfrac{a/b}{c/d} = \dfrac{a}{b} \cdot \dfrac{d}{c}$	$\dfrac{2/3}{4/5} = \dfrac{2}{3} \cdot \dfrac{5}{4} = \dfrac{2 \cdot 5}{3 \cdot 4} = \dfrac{5}{6}$
(where $b \neq 0$, $c \neq 0$, $d \neq 0$)	
$\dfrac{a}{0}$ is undefined. (does not exist)	$\dfrac{2}{0}$ is undefined; $\dfrac{0}{0}$ is undefined

If a and b are real numbers, we can compare their relative positions on the real number line by defining the relations of **less than, greater than, less than or equal to,** and **greater than or equal to.** These relations are denoted by the symbols $<$, $>$, $\leq$, and $\geq$, respectively.

Let a and b denote real numbers.

Algebraic Statement	Equivalent Statement	Geometric Statement
$a > 0$	a is positive	a lies to the right of the origin
$a < 0$	a is negative	a lies to the left of the origin
$a > b$	a is greater than b	a lies to the right of b
$a < b$	a is less than b	a lies to the left of b
$a \geq b$	a is greater than or equal to b	a is not to the left of b
$a \leq b$	a is less than or equal to b	a is not to the right of b

The symbols $<$, $>$, $\leq$, and $\geq$ are called **inequality symbols** and expressions such as $a < b$, $a \geq b$ are called **inequalities.** We often combine these symbols. For example, $a \leq b < c$ means that $a \leq b$ and $b < c$, simultaneously.

EXAMPLE 2   We have

$$-1 < 3, \quad 2 \leq 2, \quad -2.7 < -1.2, \quad -4 < -2 < 0, \quad \text{and} \quad -\tfrac{7}{2} < \tfrac{7}{2} < 7.$$

Observe the following properties of inequalities.

Let a, b, and c denote real numbers.

Property	Example
If $a \geq b$ and $b \geq c$, then $a \geq c$.	Since $8 \geq 4$ and $4 \geq 3$, then $8 \geq 3$.
If $a \leq b$ and $b \leq c$, then $a \leq c$.	Since $-5 \leq -3$ and $-3 \leq -1$, then $-5 \leq -1$.
If $a > 0$, then $\dfrac{1}{a} > 0$.	Since $2 > 0$, then $\dfrac{1}{2} > 0$.
If $a < 0$, then $\dfrac{1}{a} < 0$.	Since $-3 < 0$, then $-\dfrac{1}{3} < 0$.

Absolute Value

If a is any real number, the **absolute value** of a, denoted by $|a|$, is defined as the distance between a and 0 on the real number line. Thus,

$$|a| = \begin{cases} a & \text{if } a \geq 0, \\ -a & \text{if } a < 0. \end{cases}$$

(See Figure 4.)

Figure 4

EXAMPLE 3   We have

$$|4| = 4, \quad |-4| = 4, \quad \text{and} \quad |0| = 0.$$

Also, the distance between the point labeled 3.4 and the origin on the real number line is $|3.4| = 3.4$. Similarly, the distance between the point labeled -2.3 and the origin on the real number line is $|-2.3| = 2.3$.

In working with the notation of absolute value, it is important to perform the operations within the bars first, as in Example 4.

EXAMPLE 4   (a) $|5 - 2| = |3| = 3$ (b) $|3 - 5| - |8 - 6| = |-2| - |2| = 2 - 2 = 0$

The basic properties of absolute value may be summarized as follows. Let a and b denote real numbers.

Property	Example
$|a| \geq 0$	$|-1.4| = 1.4 \geq 0$
$|a| = |-a|$	$|2.5| = |-2.5| = 2.5$
$|a - b| = |b - a|$	$|3 - 5| = |-2| = |5 - 3| = |2| = 2$
$|a||b| = |ab|$	$|2||-3| = |-6| = 6$
$|a|^2 = a^2$	$|3|^2 = 9; |-4|^2 = 16$

A.2 EXERCISE SET ||||||||||||||||||||||||||||||||||||||||||||||||||||||||||||||||||||||||||||||

1. Which of the numbers $\sqrt{4}, \frac{8}{3}, 1.7, 28, -6,$ $\sqrt{10}, \pi, \frac{20}{5}$ are
   (a) integers?
   (b) rational numbers?
   (c) natural numbers?
   (d) irrational numbers?

2. Describe the decimal representation of rational numbers.

3. Find
   (a) the negative of -8.
   (b) the reciprocal of -5.
   (c) the reciprocal of $\frac{2}{9}$.
   (d) the negative of 0.

4. Find all natural number factors of 24.

5. Draw a real number line and plot the points whose coordinates are
   (a) 4
   (b) -5
   (c) $\frac{3}{2}$
   (d) -2.5
   (e) 0

6. Draw a real number line and plot the points whose coordinates are
   (a) -3
   (b) 2
   (c) 3.5
   (d) -4.5
   (e) $-\sqrt{2}$

7. On the following real number line, give the coordinates of the points A, B, C, and D.

In Exercises 8–31, perform the indicated operations and simplify.

8. $3 - (-5)$ 9. $4 \cdot (-2) + 5$

10. $3 - (2 + 5)$ 11. $4 - (-4 + 10)$

12. $8 - (-3) \cdot (-2)$

13. $-(3 - 9) - (-2)$

14. $\frac{1}{2} + \frac{2}{3}$ 15. $\frac{2}{5} + \frac{3}{10}$

16. $\frac{1}{2} - \frac{4}{5}$ 17. $\frac{5}{8} - \frac{7}{12}$

18. $\left(\frac{1}{2}\right) \cdot \left(\frac{8}{9}\right)$ 19. $\left(\frac{4}{3}\right) \cdot \left(\frac{27}{20}\right)$

20. $\left(4 + \frac{9}{2}\right)\left(\frac{4}{3}\right)$ 21. $4 + \left(\frac{9}{2}\right)\left(\frac{4}{3}\right)$

22. $(-\frac{2}{3}) \cdot (\frac{1}{2}) + \frac{4}{5}$

23. $(-\frac{2}{3})(\frac{1}{2} + \frac{4}{5})$

24. $\frac{2}{3} \div \frac{5}{4} + 1$

25. $\frac{5}{4} \div \frac{15}{18} - 1$

26. $\dfrac{2}{\frac{4}{3}}$

27. $\dfrac{\frac{2}{7}}{\frac{5}{4}}$

28. $4 + \dfrac{\frac{8}{3}}{2}$

29. $10 - \dfrac{\frac{1}{2}}{\frac{3}{4}}$

30. $\dfrac{1 + \frac{2}{7}}{3 - \frac{1}{2}}$

31. $\dfrac{\frac{4}{9} + \frac{1}{3}}{2 - \frac{5}{6}}$

In Exercises 32–43, rewrite each statement using the symbols $<$, $>$, $\leq$, and $\geq$.

32. 4 is greater than 2.

33. -3 is less than -2.

34. -9 is less than 6.

35. 2 is greater than -7.

36. x is less than or equal to 7.

37. a is greater than or equal to 5.

38. a is not greater than 4.

39. x is not less than 2.

40. c is nonnegative.

41. y is nonpositive.

42. a is greater than $\frac{1}{4}$ and less than $\frac{1}{2}$.

43. c is less than or equal to -5.

In Exercises 44–53, replace the square with the symbol $<$, $>$, $\leq$, or $\geq$ to make the statement true.

44. $2 \,\square\, 4$

45. $-3 \,\square\, 4$

46. $\frac{2}{3} \,\square\, \frac{1}{3}$

47. $-2.8 \,\square\, -2.2$

48. $2 \,\square\, -2$

49. $-4 \,\square\, 3$

50. $0 \,\square\, 1.3$

51. $-1.8 \,\square\, 2.4$

52. $\frac{1}{5} \,\square\, \frac{2}{5}$

53. $-4 \,\square\, 4$

Compute:

54. $|2|$

55. $\left|-\frac{2}{3}\right|$

56. $|1.5|$

57. $|-0.8|$

58. $-|2|$

59. $-\left|-\frac{2}{5}\right|$

60. $|2 - 3|$

61. $|2 - 2|$

62. $|2 - (-2)|$

63. $|2| + |-3|$

64. $\dfrac{|14 - 8|}{|-3|}$

65. $\dfrac{|2 - 12|}{|1 - 6|}$

66. $\dfrac{|3| - |2|}{|3| + |2|}$

67. $\dfrac{|4| - 2|4||3|}{|4 - 3|}$

A.3 ||||||| LINEAR EQUATIONS AND INEQUALITIES

Many applied problems in a variety of areas call for the solution of linear equations. Recent work in areas such as operations research has increased the need for solving inequalities as well. The solutions to these problems enable many manufacturing concerns, such as oil refineries and steel plants, to optimize their "product mix" and thereby maximize their profits.

||||||||||||| **Linear Equations in One Unknown**

A **linear equation** in the unknown x is an equation that can be written in the form

$$ax = b, \tag{1}$$

where a and b are given real numbers, and $a \neq 0$. A **solution** of this linear equation is a number x that satisfies (1). Thus, $x = \frac{4}{3}$ is a solution of the equation $3x = 4$, but $x = 2$ is not a solution.

Examples of linear equations are

$$2x + 3 = 5, \qquad \tfrac{3}{2} + 4(x - 1) = 6, \qquad 2 + \tfrac{1}{2}(3x - 1) = 3 + \tfrac{1}{2}x.$$

We can rewrite each of these equations in the form given in (1) by using the following properties of the real numbers.

Let a, b, and c denote real numbers.

Property   **Example(s)**

I. The same number may be added to, or subtracted from, both sides of an equation.

If $b = c$, then $a + b = a + c$ If $b = c$, then $2 + b = 2 + c$
    and $b - a = c - a$.   and $b - 3 = c - 3$.

II. Both sides of an equation may be multiplied by, or divided by, the same nonzero number.

If $b = c$, then $a \cdot b = a \cdot c$ for   If $b = c$, then $5b = 5c$.
any a.

If $b = c$ and $a \neq 0$, then $\dfrac{b}{a} = \dfrac{c}{a}$.   If $b = c$, then $\dfrac{b}{4} = \dfrac{c}{4}$.

III. Cancellation properties:

If $a + b = a + c$, then $b = c$ for   If $-5 + b = -5 + c$, then $b = c$.
any real number a.

If $a \neq 0$ and $ab = ac$, then $b = c$.   If $4b = 4c$, then $b = c$.

If $a \neq 0$ and $\dfrac{b}{a} = \dfrac{c}{a}$, then $b = c$.   If $\dfrac{b}{3} = \dfrac{c}{3}$, then $b = c$.

EXAMPLE 1   Consider the equation

$$\tfrac{3}{2} + 4(x - 1) = 5. \tag{2}$$

We can use the above properties to rewrite Equation (2) in the form of Equation (1) as follows:

$$\tfrac{3}{2} + 4(x - 1) = 5$$
$$\tfrac{3}{2} + 4x - 4 = 5 \qquad \text{(we multiplied 4 by } x - 1\text{)}$$
$$4x - \tfrac{5}{2} = 5 \qquad \text{(we added } \tfrac{3}{2} \text{ to } -4\text{)}$$
$$4x = 5 + \tfrac{5}{2} \qquad \text{(we added } \tfrac{5}{2} \text{ to both sides)}$$
$$4x = \tfrac{15}{2}. \qquad \text{(we combined 5 and } \tfrac{5}{2}\text{)}$$

This equation is in the form of Equation (1).

We can always obtain a solution of (1). Since $a \neq 0$, we multiply both sides of (1) by $1/a$.

$$\frac{1}{a} \cdot (ax) = \frac{1}{a} \cdot b$$

$$\left(\frac{1}{a} \cdot a\right) x = \frac{b}{a}$$

$$1 \cdot x = \frac{b}{a}$$

$$x = \frac{b}{a}.$$

The only solution of Equation (1) is $x = b/a$.

EXAMPLE 2   Solve $2x + 7 = 4x - 15$.

Solution
$$2x + 7 = 4x - 15$$
$$2x = 4x - 22 \qquad \text{(we added } -7 \text{ to both sides)}$$
$$-2x = -22 \qquad \text{(we added } -4x \text{ to both sides)}$$
$$x = \frac{-22}{-2} = 11. \qquad \text{(we multiplied both sides by } -\tfrac{1}{2})$$ ■

||||||||||||| Linear Inequalities in One Unknown

A **solution** of a linear inequality in the unknown x is a number x that satisfies the given inequality. Thus, $x = 2$ is a solution of the inequality

$$2x + 3 > 5$$

since

$$2(2) + 3 = 7 > 5.$$

However, $x = 1$ is not a solution, since

$$2(1) + 3 = 5.$$

The following properties of inequalities will enable us to follow procedures for solving inequalities that are similar to those used in solving equations.

Let a, b, and c denote real numbers

Property	Example
I. The same number may be added to, or subtracted from, both sides of an inequality.	
If $a < b$, then $a + c < b + c$.	Since $2 < 5$, then $2 + 3 < 5 + 3$.
If $a > b$, then $a + c > b + c$.	Since $-2 > -4$, then $-2 + 3 > -4 + 3$.
If $a < b$, then $a - c < b - c$.	Since $3 < 6$, then $3 - 8 < 6 - 8$.
If $a > b$, then $a - c > b - c$.	Since $2 > -3$, then $2 - 4 > -3 - 4$.
II. Both sides of an inequality may be multiplied by, or divided by, the same *positive* number.	
Suppose $c > 0$.	Since $4 < 6$, then $4 \cdot 3 < 6 \cdot 3$.
If $a < b$, then $ac < bc$.	Since $-2 > -5$, then $-2 \cdot 3 > -5 \cdot 3$.
If $a > b$, then $ac > bc$. (The sense of an inequality *remains unchanged* when it is multiplied by a *positive* number.)	
III. When both sides of an inequality are multiplied or divided by the same *negative* number, the inequality sign must be reversed.	
Suppose $c < 0$.	Since $3 < 5$, then $3(-2) > 5(-2)$.
If $a < b$, then $ac > bc$.	Since $-3 > -5$, then $(-3)(-2) < (-5)(-2)$.
If $a > b$, then $ac < bc$. (The sense of an inequality *is reversed* when it is multiplied by a *negative* number.)	

EXAMPLE 3   Solve the linear inequality

$$2x + 11 \geq 5x - 1.$$

Solution   We will perform addition and subtraction to collect terms in x, just as we did for linear equations.

$$2x + 11 \geq 5x - 1$$
$$2x \geq 5x - 12 \qquad \text{(we added } -11 \text{ to both sides)}$$
$$-3x \geq -12. \qquad \text{(we added } -5x \text{ to both sides)}$$

We now divide both sides of the inequality by -3, a negative number, and therefore reverse the sense of the inequality.

$$\frac{-3x}{-3} \leq \frac{-12}{-3}$$
$$x \leq 4.$$

∎

|||||||||||| **Intervals**

Certain sets of numbers on the real number line occur quite often, so it is useful to have a convenient notation for them. In particular, we can describe the sets under consideration using inequalities.

An **interval** is a set of numbers on the real number line that forms a line segment, a half-line, or the entire real number line itself. Let a and b be real numbers with $a < b$. The **closed interval** from a to b, denoted by $[a, b]$ is the set of numbers x such that

$$a \leq x \leq b,$$

and the **open interval** from a to b, denoted by (a, b), is the set of numbers x such that

$$a < x < b.$$

These sets are shown in Figure 1. The numbers a and b are called the **endpoints** of the interval.

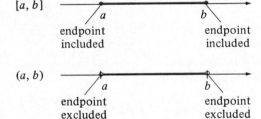

Figure 1

A closed interval includes its endpoints; this condition is indicated graphically by darkening the circles at a and b. An open interval *does not* include its endpoints; this condition is indicated graphically with an open circle at each endpoint. A bracket, [or], indicates notationally that an endpoint is part of the interval and a parenthesis, (or), indicates notationally that the endpoint is not part of the interval.

There are two **half-open intervals** from a to b, $[a, b)$ and $(a, b]$. The half-open interval $[a, b)$ is the set of numbers x such that $a \leq x < b$ and the half-open interval $(a, b]$ is the set of numbers x such that $a < x \leq b$. These sets are shown in Figure 2.

Finally, we also need intervals that extend indefinitely in one or both directions. Several intervals of this type and their notations are shown in Figure 3. In this notation, the symbol $+\infty$ means that the interval extends indefinitely to the right (in the positive direction), and the symbol $-\infty$ means that the interval extends indefinitely to the left (in the negative direction). The symbols $+\infty$ and $-\infty$ do not denote numbers.

Figure 2

Figure 3

$(-\infty, +\infty)$ ————————————————————————→
the entire real number line

EXAMPLE 4   Sketch each of the following intervals on the real number line:

$$(-1, 2) \qquad (-\infty, 3].$$

Solution   The intervals are sketched in Figure 4.

$(-1, 2)$ ←—+——+——+——+——+——+——+——+——+—→
                   -3 -2 -1 $\;\;0\;\;$ 1 $\;\;2\;\;$ 3 $\;\;4$

Figure 4

$(-\infty, 3]$ ←—+——+——+——+——+——+——+——+—→
                   -3 -2 -1 $\;\;0\;\;$ 1 $\;\;2\;\;$ 3 ■

By the **solution set** of an inequality in x, we mean the set of all solutions to the inequality. Thus, the solution set of the inequality given in Example 3 is the set of all real numbers less than or equal to 4; that is, the interval $(-\infty, 4]$.

EXAMPLE 5   Solve the inequality $5x - 3 > 2x + 1$, and sketch the solution set.

Solution   Operating on the inequality, we have

$$5x - 3 > 2x + 1$$
$$5x > 2x + 4 \qquad \text{(we added 3 to both sides)}$$
$$3x > 4 \qquad \text{(we subtracted } 2x \text{ from both sides)}$$
$$x > 4/3. \qquad \text{(we divided both sides by 3)}$$

Thus, the solution set consists of all real numbers greater than 4/3.

$$-2 \quad -1 \quad 0 \quad 1\tfrac{4}{3} \; 2 \quad 3 \quad 4$$

■

EXAMPLE 6   Solve the inequality $1 < 3x - 2 \le 7$, and sketch the solution set.

Solution   Operating on the inequality, we have

$$1 < 3x - 2 \le 7$$
$$3 < 3x \le 9 \qquad \text{(we added 2 to each member)}$$
$$1 < x \le 3. \qquad \text{(we divided each member by 3)}$$

The solution set consists of all points in the half-open interval $(1, 3]$.

$$-2 \quad -1 \quad 0 \quad 1 \quad 2 \quad 3 \quad 4$$

■

Linear Equations in Two Unknowns

A **linear equation** in the unknowns x and y is an equation of the form

$$ax + by = c,$$

where a, b, and c are constants, and a and b are not both zero. In many applications we need to solve a **system** of simultaneous linear equations

$$
\begin{aligned}
ax + by &= c \\
dx + ey &= f.
\end{aligned}
\tag{3}
$$

A **solution** of system (3) is a pair of numbers x, y that satisfy *both* equations in (3). A geometric approach to solving (3) will be presented in Section 1.3. Here we shall discuss an algebraic approach, which we shall illustrate with the following example.

EXAMPLE 7   Solve the linear system

$$
\begin{aligned}
x + 2y &= 8 \\
2x - 3y &= -5.
\end{aligned}
$$

Solution   Eliminate one of the unknowns by adding a multiple of the first equation to the second equation. Thus, we eliminate x by adding -2 times the first equation to the second equation, obtaining

$$-2x - 4y = -16$$
$$\underline{2x - 3y = -5}$$
$$\text{(add)}$$
$$2x + (-2x) - 3y + (-4y) = -5 + (-16)$$

or

$$-7y = -21$$
$$y = 3.$$

Substituting $y = 3$ in the first equation, we obtain

$$x + 2(3) = 8$$

or

$$x = 2.$$

Then, the solution of the given system is

$$x = 2, \; y = 3.$$ ∎

EXAMPLE 8   Solve the linear system

$$x + 2y = 8$$
$$2x + 4y = 10.$$

Solution   Eliminating x, we add -2 times the first equation to the second equation, obtaining

$$0x + 0y = -6,$$

which obviously makes no sense. This result means that the given system of equations has *no* solution. ∎

A system of equations having no simultaneous solution is called an **inconsistent** system.

EXAMPLE 9   Solve the linear system

$$x + 2y = 8$$
$$2x + 4y = 16.$$

Solution   Eliminating x, we add -2 times the first equation to the second equation, obtaining

$$0x + 0y = 0.$$

Thus, the equations are multiples of each other (the second equation is twice the first one). The given system has infinitely many solutions. If we choose any value for y, we obtain a corresponding value for x by solving

$$x = 8 - 2y.$$

Thus, if $y = 3$, we obtain $x = 2$; if $y = 2$, we obtain $x = 4$. ∎

Applications of Equations

Equations provide a powerful tool for solving problems in the everyday world. Seldom do these problems present themselves in equation form, however. They are stated in words, and before attempting to solve a problem, we must translate the words into mathematics. Following are some suggestions for solving word problems.

Suggested Procedure for Solving Word Problems

Step 1. Read the problem carefully.

Step 2. Identify the unknown quantities in the problem. Denote the unknown(s) by a letter(s), and specify the units (feet, pounds, dollars, and so on).

Step 3. Set up equations or inequalities to describe the relationships stated in the problem.

Step 4. Solve the equations or the inequalities.

Step 5. Check to see if quantities other than the one(s) described by the letter(s) are required. Be sure to answer all parts of the problem.

Step 6. Check your solutions, to verify that they meet all the conditions specified in the problem.

In setting up Step 2, avoid vague expressions like "let x = Mary." To be precise, let x = Mary's age in years, or x = Mary's height in inches, or whatever the problem suggests.

EXAMPLE 10   A rectangular field whose length is 10 meters longer than its width is to be enclosed with 100 meters of fencing material. What are the dimensions of the field?

Solution   Let

$$x = \text{the width of the field (in meters).}$$

Then

$$x + 10 = \text{the length of the field (in meters).}$$

The total amount of fencing material used is

$$2(\text{length} + \text{width}).$$

Thus,

$$2(x + 10 + x) = 100$$
$$2(2x + 10) = 100$$
$$4x + 20 = 100$$
$$4x = 80$$
$$x = 20. \tag{4}$$

The solution $x = 20$ satisfies Equation (4). Thus, the field is 20 meters wide and 30 meters long.   ■

EXAMPLE 11
(Business)

The Jones Growth Fund has current assets of \$120 million and wants to have assets of more than \$255 million so that it will make the list of "The Top 50 Mutual Funds." If its assets have been growing at the rate of \$15 million per year, how long will it be before the fund is listed, assuming that growth continues at the same rate?

Solution

Let x be the number of years that it will take the fund to make the list. During this period of time its assets will have increased by $15x$. Thus, its total assets at the end of x years will be

$$15x + 120.$$

We now want to find x so that

$$15x + 120 > 255,$$

since we want total assets of *more than* \$255 million. Then,

$$15x > 255 - 120 = 135$$

or

$$x > \tfrac{135}{15} = 9.$$

Thus, the fund will be listed after 9 years. ■

EXAMPLE 12 If 3 sulfa pills and 4 penicillin pills cost 69 cents and 5 sulfa pills and 2 penicillin pills cost 73 cents, what is the cost of each type of pill?

Solution Let

$$x = \text{the cost (in cents) of each sulfa pill, and}$$
$$y = \text{the cost (in cents) of each penicillin pill.}$$

Then

$$3x + 4y = 69$$
$$5x + 2y = 73.$$

Multiplying the second equation by -2 and adding to eliminate y, we obtain

$$
\begin{aligned}
3x + 4y &=\ \ \ \ 69 \\
-10x - 4y &= -146 \\
\hline
-7x\quad\ \ \ &= -\ \ 77 \\
x &= 11.
\end{aligned}
$$

Substituting $x = 11$ in the first equation, we obtain (verify)

$$y = 9.$$

The solution $x = 11$, $y = 9$ satisfies both equations. Thus, each sulfa pill costs 11 cents and each penicillin pill costs 9 cents. ■

A.3 EXERCISE SET ||||||||||||||||||||||||||||||||||||||||||||||||||||||||||||||||||||||||||||

In Exercises 1–12, solve the linear equations for x.

1. $3x + 5 = -1$

2. $5x + 10 = 0$

3. $\frac{1}{2}x + 2 = 4$

4. $\frac{3}{2}x - 2 = 7$

5. $-5x + 8 = 3x - 4$

6. $2x - 1 = 3x + 2$

7. $-3x + 8 = x - 16$

8. $6x + 4 = -3x - 5$

9. $2(3x + 1) = 3x - 4$

10. $-3(2x + 1) = -8x + 1$

11. $kx + 6 = 4x$

12. $rx - 8 = 8x$

In Exercises 13–18, sketch the intervals.

13. $(-2, 3)$ 14. $[1, 5]$

15. $(-3, -1]$

16. $(-1, +\infty)$

17. $(-\infty, 2]$

18. $[2, 6)$

In Exercises 19–22, write the sets in interval notation.

19.
$$-3 \ -2 \ -1 \ \ 0 \ \ 1 \ \ 2 \ \ 3$$

20.
$$-3 \ -2 \ -1 \ \ 0 \ \ 1 \ \ 2 \ \ 3$$

21.
$$-3 \ -2 \ -1 \ \ 0 \ \ 1 \ \ 2 \ \ 3$$

22.
$$-3 \ -2 \ -1 \ \ 0 \ \ 1 \ \ 2 \ \ 3$$

In Exercises 23–28, express the sets in interval notation.

23. $-5 \le x < 1$

24. $-4 \le x \le 1$

25. $x > 9$

26. $x \le -2$

27. $-12 \le x \le -3$

28. $-6 \le x < 2$

In Exercises 29–34, express the intervals as inequalities.

29. $(-4, 3]$

30. $[5, 8]$

31. $(-\infty, -2]$

32. $(-2, -1)$

33. $(3, +\infty)$

34. $[0, 5)$

In Exercises 35–43, solve the inequalities, and sketch the solution set.

35. $3x + 4 < 1$

36. $2x - 5 \ge 15$

37. $4x + 3 \le 11$

38. $-2x + 3 \ge 12$

39. $2x + 4 \le 3x - 2$

40. $3x - 5 \ge 5x + 2$

41. $-4 \le 2x + 2 \le 3$

42. $5 < 3x - 1 \le 11$

43. $3 \le 1 - 2x \le 7$

In Exercises 44–51, solve the given linear systems.

44. $x + y = 2$
$2x + 3y = 6$

45. $x - 2y = 5$
$3x + 4y = 5$

46. $x - 4y = 4$
$2x - 8y = 2$

47. $x + 2y = 3$
$3x + 6y = 7$

48. $x - 2y = 5$
$-5x + 10y = -25$

49. $x + 3y = 10$
$2x + y = 10$

50. $x - 3y = -13$
$2x - y = -1$

51. $2x - 2y = 12$
$3x - 3y = 18$

52. The width of a rectangle is 4 centimeters less than twice its length. If the perimeter is 12 centimeters, find the dimensions of the rectangle.

53. If you pay $63 for a car radio after receiving a 30% discount, what was the price of the radio before the discount?

54. A purse contains $3.20 in quarters and dimes. If there are three more quarters than dimes, how many coins of each type are there?

55. A machine is designed to package 100 ounces of liquid vitamin per bottle. Due to poor design, the machine can make an error of at most 2 ounces per bottle. If x is the number of ounces in a selected bottle, write an inequality, using absolute value, that indicates that the maximum error is 2 ounces per bottle. Solve the inequality.

56. Firm A rents a compact car for $180 a week, mileage unlimited, and firm B rents the same for $100 per week plus $.20 for each mile driven. Under what conditions does it cost less to rent from firm A?

57. A vacuum cleaner salesperson is paid $20 per day plus $30 for each vacuum cleaner sold. At what point does the salesperson's income exceed $110 per day?

58. A photography store sells sampler A, consisting of six rolls of color film and four rolls of black and white film, for $21. It sells sampler B, consisting of four rolls of color film and six rolls of black and white film, for $19. What is the cost per roll of each type of film?

59. A landscaping firm prepares two plans for a homeowner. Plan A includes eight hemlocks and twelve junipers at a cost of $520; plan B includes six hemlocks and fifteen junipers at a cost of $510. What is the cost of each type of tree?

60. A supermarket mixes coffee that sells for $1.20 per pound with coffee that sells for $1.80 per pound to obtain 24 pounds of coffee that sells for $1.60 per pound. How much of each type of coffee is used in the mix?

A.4 ||||||| EXPONENTS AND RADICALS

We begin by summarizing the facts about **integral exponents.** Suppose that a and b are real numbers and n is a positive integer.

Definition	Examples
$a^1 = a$	$2^1 = 2$
$a^2 = a \cdot a$	$2^2 = 2 \cdot 2 = 4$
$a^n = \underbrace{a \cdot a \cdot \ldots \cdot a}_{n \text{ factors}}$	$(-3)^3 = (-3)(-3)(-3) = -27$
	$(1.5)^3 = (1.5)(1.5)(1.5) = 3.375$
n is called the **exponent** and a is called the **base.**	$(\frac{1}{2})^3 = \frac{1}{2} \cdot \frac{1}{2} \cdot \frac{1}{2} = \frac{1}{8}$
$a^0 = 1 \ (a \neq 0)$	$3^0 = 1$
	$(-2)^0 = 1$
	$(\frac{1}{3})^0 = 1$
$a^{-n} = \dfrac{1}{a^n} \ (a \neq 0)$	$3^{-4} = \dfrac{1}{3^4} = \dfrac{1}{3 \cdot 3 \cdot 3 \cdot 3} = \dfrac{1}{81}$
	$(\frac{1}{2})^{-3} = \dfrac{1}{(\frac{1}{2})^3} = \dfrac{1}{\frac{1}{2} \cdot \frac{1}{2} \cdot \frac{1}{2}} = \dfrac{1}{\frac{1}{8}} = 8$
	$(-\frac{1}{4})^{-3} = \dfrac{1}{(-\frac{1}{4})^3} = \dfrac{1}{(-\frac{1}{4}) \cdot (-\frac{1}{4}) \cdot (-\frac{1}{4})} = -\dfrac{1}{-\frac{1}{64}}$
	$= -64$
$\left(\dfrac{a}{b}\right)^{-n} = \left(\dfrac{b}{a}\right)^n$	$\left(\dfrac{2}{3}\right)^{-4} = \left(\dfrac{3}{2}\right)^4 = \dfrac{81}{16}.$

The number a is called an **nth root of b** if

$$a^n = b.$$

When $n = 2$, we say that a is a **square root of b**; when $n = 3$, we say that a is a **cube root of b.** Thus, 5 and -5 are square roots of 25 because $5^2 = (-5)^2 = 25$. Also, -2 is a cube root of -8 because $(-2)^3 = -8$. In the real number system, every positive number has two square roots (one positive, one negative) but only one cube root (positive), whereas every negative number has no square root but one cube root (negative). The general situation is summarized in the following table.

b	$n(\geq 2)$	Number of nth roots of b	Form of nth roots
> 0	Even	2	$a, -a$
< 0	Even	None	None
> 0	Odd	1	$a > 0$
< 0	Odd	1	$a < 0$
0	All	1	0

For every integer $n \geq 2$, we let

$$b^{1/n} \quad \text{or} \quad \sqrt[n]{b}$$

denote the nth root of b if n is odd and the positive nth root of b if n is even. Thus,

$$16^{1/2} = \sqrt[2]{16} = 4$$
$$8^{1/3} = \sqrt[3]{8} = 2$$
$$(-8)^{1/3} = \sqrt[3]{-8} = -2.$$

In the real number system, the symbol $\sqrt[n]{b}$ has no meaning when n is even and b is negative. For $n = 2$, we write

$$b^{1/2} = \sqrt[2]{b} = \sqrt{b}.$$

The symbols $\sqrt{}$ and $\sqrt[n]{}$ are commonly called **radicals.**

Warning   Beware of the common mistake.

$$\sqrt{b^2} = b.$$

This equation is true for $b \geq 0$, but false for $b < 0$. For example, if $b = -6$, then

$$\sqrt{b^2} = \sqrt{(-6)^2} = \sqrt{36} = 6 \neq b$$

In general, for any b,

$$\sqrt{b^2} = |b|.$$

Thus,

$$\sqrt{(-6)^2} = |-6| = 6. \qquad \blacksquare$$

The algebraic properties of exponents are summarized in the following table. Let a and b denote real numbers and let m and n denote integers.

Property	Examples	
$a^m \cdot a^n = a^{m+n}$	$a^3 \cdot a^2 = a^{3+2} = a^5$	$2^3 \cdot 2^2 = 2^{3+2} = 2^5 = 32$
	$a^4 \cdot a^{-2} = a^{4+(-2)} = a^2$	$3^4 \cdot 3^{-2} = 3^{4+(-2)} = 3^2 = 9$
	$a^3 \cdot a^{-5} = a^{3+(-5)} = a^{-2}$	$3^3 \cdot 3^{-5} = 3^{3+(-5)} = 3^{-2}$
	$= \dfrac{1}{a^2}$	$= \dfrac{1}{3^2} = \dfrac{1}{9}$
$\dfrac{a^m}{a^n} = a^{m-n}, \; a \neq 0$	$\dfrac{a^5}{a^3} = a^{5-3} = a^2$	$\dfrac{2^5}{2^3} = 2^{5-3} = 2^2 = 4$
	$\dfrac{a^2}{a^4} = a^{2-4} = a^{-2} = \dfrac{1}{a^2}$	$\dfrac{5^2}{5^4} = 5^{2-4} = 5^{-2} = \dfrac{1}{5^2} = \dfrac{1}{25}$
	$\dfrac{a^{-2}}{a^3} = a^{-2-3} = a^{-5}$	$\dfrac{3^{-2}}{3^3} = 3^{-2-3} = 3^{-5} = \dfrac{1}{3^5}$
	$= \dfrac{1}{a^5}$	$= \dfrac{1}{243}$
	$\dfrac{a^3}{a^3} = a^{3-3} = a^0 = 1$	$\dfrac{4^3}{4^3} = 4^{3-3} = 4^0 = 1$
$(a^m)^n = a^{mn}$	$(a^2)^3 = a^{2 \cdot 3} = a^6$	$(2^2)^3 = 2^{2 \cdot 3} = 2^6 = 64$
	$(a^{-2})^3 = a^{-2 \cdot 3} = a^{-6}$	$(2^{-2})^3 = 2^{-2 \cdot 3} = 2^{-6} = \dfrac{1}{2^6}$
	$= \dfrac{1}{a^6}$	$= \dfrac{1}{64}$
	$(a^3)^{-4} = a^{3 \cdot (-4)}$	$(2^3)^{-4} = 2^{3 \cdot (-4)} = 2^{-12}$
	$= a^{-12} = \dfrac{1}{a^{12}}$	$= \dfrac{1}{2^{12}} = \dfrac{1}{4096}$
$(a \cdot b)^m = a^m b^m$	$(a \cdot b)^3 = a^3 \cdot b^3$	$(2 \cdot 3)^3 = 2^3 \cdot 3^3 = 8 \cdot 27 = 216$
$\left(\dfrac{a}{b}\right)^m = \dfrac{a^m}{b^m}, \; b \neq 0$	$\left(\dfrac{a}{b}\right)^{-3} = \dfrac{a^{-3}}{b^{-3}} = \dfrac{b^3}{a^3}$	$\left(\dfrac{4}{3}\right)^{-3} = \dfrac{4^{-3}}{3^{-3}} = \dfrac{3^3}{4^3} = \dfrac{27}{64}$

All the properties listed in the table above for integer exponents also hold for rational exponents. Thus, if $a > 0$ and $b > 0$,

$$a^{-1/2} = \frac{1}{a^{1/2}} \qquad\qquad 9^{-1/2} = \frac{1}{9^{1/2}} = \frac{1}{3}$$

$$a^{1/2} \cdot a^{1/3} = a^{1/2 + 1/3} = a^{5/6} \qquad\qquad 64^{1/2} \cdot 64^{1/3} = 64^{5/6} = (64^{1/6})^5 = 2^5 = 32$$

$$\frac{a^{1/2}}{a^{3/2}} = a^{1/2 - 3/2} = a^{-1} = \frac{1}{a} \qquad\qquad \frac{9^{1/2}}{9^{3/2}} = 9^{1/2 - 3/2} = 9^{-1} = \frac{1}{9}$$

$$(a^{1/2})^{1/2} = a^{(1/2) \cdot (1/2)} = a^{1/4} = \sqrt[4]{a} \qquad\qquad (81^{1/2})^{1/2} = 81^{(1/2) \cdot (1/2)} = 81^{1/4} = \sqrt[4]{81} = 3$$

$$(a \cdot b)^{3/2} = a^{3/2} \cdot b^{3/2} \qquad\qquad (4 \cdot 25)^{3/2} = 4^{3/2} \cdot 25^{3/2} = (4^{1/2})^3 \cdot (25^{1/2})^3$$
$$= 2^3 \cdot 5^3 = 8 \cdot 125 = 1000$$

$$\left(\frac{a}{b}\right)^{3/2} = \frac{a^{3/2}}{b^{3/2}} \qquad\qquad \left(\frac{9}{25}\right)^{3/2} = \frac{9^{3/2}}{25^{3/2}} = \frac{(9^{1/2})^3}{(25^{1/2})^3} = \frac{3^3}{5^3} = \frac{27}{125}$$

Fractional exponents can be converted to radicals, and vice versa, as can be seen as follows:

$$a^{m/n} = (a^m)^{1/n} = \sqrt[n]{a^m},$$

and

$$a^{m/n} = (a^{1/n})^m = (\sqrt[n]{a})^m.$$

Thus, for example, $8^{2/3} = (8^2)^{1/3} = 64^{1/3} = \sqrt[3]{64} = 4$, and also, $8^{2/3} = (8^{1/3})^2 = 2^2 = 4$.

The algebraic properties of radicals are summarized in the following table. Let a and b denote real numbers, positive whenever necessary, and let m and n denote positive integers.

Property	Examples
$\sqrt[n]{a^n} = (\sqrt[n]{a})^n = a$	$\sqrt[3]{5^3} = (\sqrt[3]{5})^3 = 5$
$\sqrt[n]{a^m} = (\sqrt[n]{a})^m = a^{m/n}$	$\sqrt[3]{16} = \sqrt[3]{2^4} = (\sqrt[3]{2})^4 = 2^{3/4}$
$\sqrt[n]{ab} = \sqrt[n]{a}\,\sqrt[n]{b}$	$\sqrt{144} = \sqrt{9 \cdot 16} = \sqrt{9}\,\sqrt{16} = 3 \cdot 4 = 12$
$\sqrt[n]{a/b} = \sqrt[n]{a}/\sqrt[n]{b}$	$\sqrt[4]{5/8} = \sqrt[4]{5}/\sqrt[4]{8}$
$\sqrt[n]{\sqrt[m]{a}} = \sqrt[m]{\sqrt[n]{a}} = \sqrt[mn]{a}$	$\sqrt[3]{\sqrt{64}} = \sqrt[6]{64} = \sqrt{\sqrt[3]{64}}$
	$(\sqrt[3]{8} = 2 = \sqrt{4})$

Warning $\sqrt[n]{a + b}$ does *not* equal $\sqrt[n]{a} + \sqrt[n]{b}$.

$\sqrt[n]{a}\,\sqrt[m]{b}$ does *not* equal $\sqrt[nm]{ab}$ or $\sqrt[n+m]{ab}$.

A.4 EXERCISE SET ||||||||||||||||||||||||||||||||||||||||||||||||||||||||||||||||||||||||||

In Exercises 1–36, compute the value of the expressions, in the simplest form.

1. 3^4

2. 2^5

3. 5^{-3}

4. 4^{-2}

5. $(2/3)^5$

6. $(3/5)^3$

7. $(1/2)^{-3}$

8. $(2/7)^{-2}$

9. 8^0

10. $(-3/7)^0$

11. $32^{1/5}$

12. $27^{2/3}$

13. $64^{-2/3}$

14. $8^{-4/3}$

15. $3^4 \cdot 3^5$

16. $(-2)^3(-2)^5$

17. $(-8)^{-1/3}(-8)^{-1/3}$

18. $7^{3/4} \cdot 7^{5/4}$

19. $2^{1/2} \cdot 2^{3/2}$

20. $81^{3/4} \cdot 81^{1/2}$

21. $4^{10}/4^8$

22. $5^6/5^9$

23. $7^3/7^{1/2}$

24. $2^{7/2}/2^{3/2}$

25. $3^{1/4}/3^{-3/4}$

26. $4^{-1/3}/4^{5/3}$

27. $(2^3)^2$

28. $(-2^2)^4$

29. $(3^{1/2})^6$

30. $(2^{1/3})9$

31. $[(-64)^4]^{1/3}$

32. $(16^3)^{1/4}$

33. $(\sqrt{27})^{1/3}$

34. $\sqrt{(-3)^2}$

35. $\sqrt[3]{(-27)^2}$

36. $\sqrt[4]{(81)^3}$

In Exercises 37–54, rewrite the expressions using only positive exponents.

37. 5^{-4}

38. 3^{-3}

39. $(1/3)^{-4}$

40. $(1/2)^{-3}$

41. $(3/2)^{-2}$

42. $(2/5)^{-2}$

43. a^{-3}

44. b^{-4}

45. $a^2 \cdot a^{-5}$

46. $b^4 b^{-9}$

47. a^{-6}/a^{-2}

48. b^{-5}/b^2

49. $a/a^{3/2}$

50. $b^3/b^{2/3}$

51. $a^{-4}b^{-4}$

52. $a^{-2}b^{-2}$

53. a^{-5}/b^{-5}

54. a^{-8}/b^8

In Exercises 55–66, simplify the expressions. Assume that all symbols represent positive real numbers.

55. $a^4 a^2$

56. $\dfrac{a^{-5}}{a^{-2}}$

57. $\dfrac{a^3 b}{a^4 b^2}$

58. $\dfrac{a^m}{a^{m/2}}$

59. $\left(\dfrac{a^{-2}b}{c^2}\right)^3$

60. $\left(\dfrac{a^{-3}b^{-2}c^{-2}}{a^2 b^3 c^2}\right)^2$

61. $(a^{1/4}b^{1/2})^8$

62. $\left(\dfrac{a^{1/3}}{b^{1/6}}\right)^{12}$

63. $\left(\dfrac{a^8}{b^4}\right)^{1/2}$

64. $\left(\dfrac{a^2 b^4}{25}\right)^{1/2}$

65. $\left(\dfrac{a^2 b^3 c^{-2}}{a^3 b^2 c}\right)^{-1}$

66. $a^2(a^{-3} + a^{-2})$

A.5 ||||||| POLYNOMIALS AND RATIONAL EXPRESSIONS

A **variable** is a symbol to which we can assign various real numbers. An **algebraic expression** is obtained by combining variables and constants by means of the operations of addition, subtraction, multiplication, division, and taking roots. Examples of algebraic expressions are

$$2x^2 + 3x - 2 \qquad \frac{3xy + 2x}{x - 1} \qquad \frac{2x - y + \sqrt{z}}{3x + 2y}.$$

Since variables in algebraic expressions represent real numbers, we can use the familiar properties of the real numbers discussed earlier in operations with algebraic expressions.

||||||||||||| **Polynomials**

A **polynomial in one variable x** is an algebraic expression of the form

$$c_0 x^n + c_1 x^{n-1} + \cdots + c_{n-1}x + c_n, \tag{1}$$

where n is a nonnegative integer, the **coefficients** $c_0, c_1, \ldots, c_n$ are real numbers, and $c_0 \neq 0$. We say that the **degree** of the polynomial (1) is n, the highest power on x occuring in the polynomial. The following are polynomials in x:

$$3x^4 + 2x + 5 \text{ (degree is 4)} \quad 2x^3 + x - 1 \text{ (degree is 3)}$$
$$\tfrac{2}{3}x \text{ (degree is 1)} \quad 2 \text{ (degree is 0)}.$$

The following algebraic expressions are *not* polynomials in x:

$$2x^{1/2} + 3, \qquad 3 - \frac{4}{x}, \qquad \frac{1}{2x^2 + 3x}, \qquad \frac{2x - 1}{x - 2}.$$

A **polynomial in two variables x and y** is a sum of expressions of the form $cx^m y^n$, where the coefficient c is a real number and m and n are nonnegative integers. Examples of polynomials in x and y are

$$4x^2 + 2xy - 6xy^2 \quad \text{and} \quad 2x^2 y + 12xy - x + 2y.$$

||||||||||||| **Operations with Polynomials**

Addition and *subtraction* of polynomials can be viewed as adding or subtracting *like terms* (terms in which the powers of x and y are the same). Thus,

$$(3x^2 - 2x + 5) + (x^2 + 4x - 9) = (3x^2 + x^2) + (-2x + 4x) + (5 - 9)$$
$$= 4x^2 + 2x - 4,$$

and

$$(3x^2 - 4xy + y^2 + 2x) - (5x^2 - 6xy + 3y^2 - y)$$
$$= (3x^2 - 5x^2) + (-4xy + 6xy) + (y^2 - 3y^2) + 2x + y$$
$$= -2x^2 + 2xy - 2y^2 + 2x + y.$$

Notice that the degree of the sum or difference of two polynomials is less than or equal to the higher of the degrees of the two polynomials. For example,

$$(x^4 + 11x^3 - 7x) - (x^4 - 2x^2 + 5) = 11x^3 + 2x^2 - 7x - 5.$$

To *multiply a polynomial by a constant*, the distributive law tells us that we merely multiply each coefficient of the polynomial by the constant. Thus,

$$-3(2x^2 - 4x + 3) = -6x^2 + 12x - 9.$$

Notice that multiplying a polynomial by a constant does not affect the degree of the polynomial, unless the constant is 0.

To *multiply* two polynomials, we use the distributive law several times. Thus,

$$\begin{aligned}
(3x + 2)(2x^2 - 4x + 3) &= 3x(2x^2 - 4x + 3) + 2(2x^2 - 4x + 3) \\
&= 6x^3 - 12x^2 + 9x + 4x^2 - 8x + 6 \\
&= 6x^3 - 8x^2 + x + 6,
\end{aligned}$$

and

$$\begin{aligned}
(2x - y)(x^2 - xy + y) &= 2x(x^2 - xy + y) - y(x^2 - xy + y) \\
&= 2x^3 - 2x^2y + 2xy - x^2y + xy^2 - y^2 \\
&= 2x^3 - 3x^2y + 2xy + xy^2 - y^2.
\end{aligned}$$

Observe that the degree of the product of two nonzero polynomials is the sum of the degrees of the two polynomials.

Factoring

The process of factoring a polynomial is the reverse of the process of multiplication, described above. It is generally more difficult to factor than to multiply; skill in factoring develops through practice. The following examples review some basic techniques.

EXAMPLE 1   (Factoring a common factor from each term)

(a) $x^2 - 2x = x(x - 2)$

(b) $4x^2 + 2xy - 6xy^2 = 2x(2x + y - 3y^2)$

(c) $4x^2y + 12xy - 8xy^2 = 4xy(x + 3 - 2y)$

(d) $2x(x + 2y) + 3y(x + 2y) = (2x + 3y)(x + 2y)$.

EXAMPLE 2   (Factoring certain second degree polynomials in x)

Factor $x^2 + x - 12$.

Solution   We write

$$\begin{aligned}
x^2 + x - 12 &= (ax + b)(cx + d) \\
&= acx^2 + (ad + bc)x + bd,
\end{aligned}$$

and look for suitable values of a, b, c, and d. Thus, we must have

$$ac = 1, \quad ad + bc = 1, \quad \text{and} \quad bd = -12.$$

Because it is most convenient to work with integers, we try to find integers a, b, c, and d satisfying these equations. Of course, a, b, c, and d, in general, need not be integers. By intelligent trial and error, we find that $a = 1$, $b = 4$, $c = 1$, and $d = -3$ are correct choices. Thus,

$$x^2 + x - 12 = (x + 4)(x - 3).$$ ∎

EXAMPLE 3 Factor $6x^2 + 5x - 4$.

Solution We write

$$6x^2 + 5x - 4 = (ax + b)(cx + d)$$
$$= acx^2 + (ad + bc)x + bd,$$

and look for suitable values of a, b, c, and d. We must have

$$ac = 6, \quad ad + bc = 5, \quad \text{and} \quad bd = -4.$$

By intelligent trial and error we find that $a = 2$, $b = -1$, $c = 3$, and $d = 4$ are correct choices. Thus,

$$6x^2 + 5x - 4 = (2x - 1)(3x + 4).$$ ∎

EXAMPLE 4 Factor $6x^2 - 7xy - 3y^2$.

Solution We write

$$6x^2 - 7xy - 3y^2 = (ax + by)(cx + dy)$$
$$= acx^2 + (ad + bc)xy + bdy^2.$$

We must have

$$ac = 6, \quad ad + bc = -7, \quad \text{and} \quad bd = -3.$$

A solution is (verify) $a = 3$, $b = 1$, $c = 2$, $d = -3$. Hence,

$$6x^2 - 7xy - 3y^2 = (3x + y)(2x - 3y).$$ ∎

EXAMPLE 5 Factor $x^3 + x^2 - 6x$.

Solution We first factor the common factor x from each term and then factor the resulting polynomial, obtaining

$$x^3 + x^2 - 6x = x(x^2 + x - 6)$$
$$= x(x + 3)(x - 2).$$ ∎

||||||||||||| **Factoring Special Forms**

Some special-case polynomials that occur frequently are easily factored. It is important to memorize these forms so that you will recognize them, and be able to factor them, whenever you encounter them.

General Form   **Example**

I. Second degree

Perfect square

$a^2 + 2ab + b^2 = (a + b)^2$ $x^2 + 10 + 25 = (x + 5)^2$

$a^2 - 2ab + b^2 = (a - b)^2$ $4x^2 - 12x + 9 = (2x - 3)^2$

Difference of two squares

$a^2 - b^2 = (a - b)(a + b)$ $4x^2 - 9 = (2x - 3)(2x + 3)$

II. Higher degrees

Sum or difference of two cubes

$a^3 - b^3 = (a - b)(a^2 + ab + b^2)$ $x^3 - 8 = (x - 2)(x^2 + 2x + 4)$

$a^3 + b^3 = (a + b)(a^2 - ab + b^2)$ $8x^3 + 27 = (2x + 3)(4x^2 - 6x + 9)$

Difference of two nth powers

$a^n - b^n = (a - b)(a^{n-1} + a^{n-2}b + a^{n-3}b^2$ $x^5 - y^5 = (x - y)(x^4 + x^3y$

$\qquad + \cdots + ab^{n-2} + b^{n-1})$ $\qquad\qquad + x^2y^2 + xy^3 + y^4)$

Warning   A second degree sum of two perfect squares,

$$a^2 + b^2$$

cannot be factored (within the system of real numbers).   ∎

EXAMPLE 6   Factor
(Second-   (a) $x^2 - 16$ (e) $(3x + y)^2 - (4x - 8y)^2$
degree   (b) $25 - 4x^2$ (f) $x^2 - 6x + 9$
Special Forms)   (c) $9x^2 - 16y^2$ (g) $16x^2 - 8x + 1$
(d) $(3x + y)^2 + (4x - 8y)^2$ (h) $4x^2 + 20x + 25.$

Solution   (a) $x^2 - 16 = x^2 - 4^2 = (x + 4)(x - 4).$
(b) $25 - 4x^2 = 5^2 - (2x)^2 = (5 + 2x)(5 - 2x).$
(c) $9x^2 - 16y^2 = (3x)^2 - (4y)^2 = (3x + 4y)(3x - 4y).$
(d) Cannot be factored because this expression is a second degree sum of two perfect squares.
(e) $(3x + y)^2 - (4x - 8y)^2 = [(3x + y) - (4x - 8y)][(3x + y) + (4x - 8y)]$
$\qquad\qquad\qquad = [-x + 9y][7x - 7y] = 7(-x + 9y)(x - y)$
(f) $x^2 - 6x + 9 = (x)^2 - 2(x)(3) + (3)^2 = (x - 3)^2$
(g) $16x^2 - 8x + 1 = (4x)^2 - 2(4x)(1) + 1^2 = (4x - 1)^2$
(h) $4x^2 + 20x + 25 = (2x)^2 + 2(2x)(5) + 5^2 = (2x + 5)^2.$ ∎

EXAMPLE 7
(Higher-
degree
Special Forms)

Factor
(a) $x^3 + 125$ (c) $32x^5 - 243$
(b) $27a^3 - 8b^3$ (d) $x^4 - 16$.

Solution
(a) $x^3 + 125 = x^3 + 5^3 = (x + 5)(x^2 - 5x + 25)$
(b) $27a^3 - 8b^3 = (3a)^3 - (2b)^3 = (3a - 2b)[(3a)^2 + (3a)(2b) + (2b)^2]$
$\qquad = (3a - 2b)(9a^2 + 6ab + 4b^2)$.
(c) $32x^5 - 243 = (2x)^5 - 3^5$
$\qquad = (2x - 3)[(2x)^4 + (2x)^3(3) + (2x)^2 3^2 + (2x)3^3 + 3^4]$
$\qquad = (2x + 3)(16x^4 + 24x^3 + 36x^2 + 54x + 81)$.
(d) $x^4 - 16$ can be factored into two ways. First, as a difference of 4th powers,

$$x^4 - 16 = x^4 - 2^4 = (x - 2)(x^3 + 2x^2 + 4x + 8).$$

Second, as a difference of squares,

$$x^4 - 16 = (x^2)^2 - 4^2 = (x^2 - 4)(x^2 + 4) = (x^2 - 2^2)(x^2 + 4)$$
$$= (x - 2)(x + 2)(x^2 + 4).$$

To see that these two factorizations are not contradictory, multiply: $(x + 2)(x^2 + 4) = (x^3 + 2x^2 + 4x + 8)$. Thus, the second approach to factoring $x^4 - 16$ is better because it leads to a more complete factorization. ■

EXAMPLE 8
(Factoring by
Grouping)

Factor
(a) $x^3 + x^2 + 3x + 3$
(b) $x^3 - 5x^2 + 4x - 20$
(c) $2x^3 + 3x^2 - 8x - 12$

Solution
(a) We group together the first two terms, and then the second two:

$$(x^3 + x^2) + (3x + 3),$$

and then factor each grouping separately

$$x^2(x + 1) + 3(x + 1).$$

We notice that $(x + 1)$ is a factor of each term, so by the distributive law we have

$$(x^2 + 3)(x + 1).$$

(b) Again, grouping the first two terms and then the second two terms, we have

$$(x^3 - 5x^2) + (4x - 20),$$
or $\qquad x^2(x - 5) + 4(x - 5),$
or $\qquad (x^2 + 4)(x - 5).$

(c) By grouping, we have

$$(2x^3 + 3x^2) + (-8x - 12),$$

or

$$x^2(2x + 3) + (-4)(2x + 3)$$

or

$$(x^2 - 4)(2x + 3).$$

Since $(x^2 - 4) = (x - 2)(x + 2)$, our complete factorization is

$$(x - 2)(x + 2)(2x + 3). \qquad ■$$

|||||||||||| **Rational Expressions**

A **rational** expression is a quotient of two polynomials. Examples of rational expressions are

$$\frac{5x + 4}{x - 2}, \qquad \frac{2y^3 + y + 2}{y^2 - 2y - 3}, \qquad \text{and} \qquad \frac{2x^2 + 3xy - y^2}{3x^2 + 2xy}.$$

Since division by zero is not defined, it is always understood that the denominator of a rational expression is never zero for the value of the variable used. This condition sometimes imposes restrictions on the values that the variable may assume. For example, in the rational expression

$$\frac{x^2 + 2x + 5}{x - 1}$$

we must always have $x \neq 1$ since otherwise $x - 1 = 0$.

When performing the basic arithmetic operations of addition, subtraction, multiplication, and division with rational expressions, we treat them exactly as if they were rational numbers. The following examples illustrate the basic arithmetic operations on rational expressions and their simplification.

|||||||||||| **Addition and Subtraction**

To add or subtract two fractions,
(a) if the denominators are the same, retain that denominator and add or subtract the numerators.
(b) if the denominators are not the same, convert the fractions to equivalent fractions with the same denominator and then apply rule (a).

EXAMPLE 9

(a) $\dfrac{x + 3}{x - 2} + \dfrac{x}{x - 2} = \dfrac{x + 3 + x}{x - 2} = \dfrac{2x + 3}{x - 2}$

(b) $\dfrac{x + 3}{x - 1} + \dfrac{x}{x - 2} = \dfrac{x + 3}{x - 1} \cdot \dfrac{x - 2}{x - 2} + \dfrac{x}{x - 2} \cdot \dfrac{x - 1}{x - 1} = \dfrac{(x + 3)(x - 2) + x(x - 1)}{(x - 1)(x - 2)}$

$\qquad = \dfrac{x^2 + x - 6 + x^2 - x}{(x - 1)(x - 2)} = \dfrac{2x^2 - 6}{(x - 1)(x - 2)} = \dfrac{2(x^2 - 3)}{(x - 1)(x - 2)}$

EXAMPLE 10 (a) $\dfrac{3x}{x-4} - \dfrac{2x}{x-4} = \dfrac{3x-2x}{x-4} = \dfrac{x}{x-4}$

(b) $\dfrac{x}{x+3} - \dfrac{x-2}{x+2} = \dfrac{x}{x+3} \cdot \dfrac{x+2}{x+2} - \dfrac{x-2}{x+2} \cdot \dfrac{x+3}{x+3} = \dfrac{x(x+2)-(x+3)(x-2)}{(x+3)(x+2)}$

$$= \dfrac{x^2+2x-(x^2+x-6)}{(x+3)(x+2)} = \dfrac{x+6}{(x+3)(x+2)}$$

|||||||||||| **Simplification**

To simplify (or "reduce") a rational expression, factor the numerator and denominator completely; then cancel any factors common to both the numerator and the denominator.

EXAMPLE 11 Simplify

$$\frac{x^3 - 25x}{x^4 - 3x^3 - 10x^2}.$$

Solution

$$\frac{x^3 - 25x}{x^4 - 3x^3 - 10x^2} = \frac{x(x^2 - 25)}{x^2(x^2 - 3x - 10)} = \frac{\cancel{x}(x-5)(x+5)}{x\cancel{x}(x-5)(x+2)}$$

$$= \frac{x+5}{x(x+2)}.$$ ∎

|||||||||||| **Multiplication**

To multiply two fractions, merely multiply the numerators together and the denominators together, and then simplify. It is advisable to do the simplification *before* the multiplications are actually performed.

EXAMPLE 12 $\dfrac{x^2-4}{x^2-2x} \cdot \dfrac{x^3+3x^2}{x^2-x-6} = \dfrac{(x-2)(x+2)}{x(x-2)} \cdot \dfrac{x^2(x+3)}{(x-3)(x+2)}$

$$= \frac{x(x+3)}{x-3}.$$

|||||||||||| **Division**

To divide one fraction by another, invert the second fraction and multiply.

EXAMPLE 13 (a) $\dfrac{10x}{9y} \div \dfrac{25x}{4yz} = \dfrac{10x}{9y} \cdot \dfrac{4yz}{25x} = \dfrac{2 \cdot 5 \cdot 4 \cdot xyz}{9 \cdot 5 \cdot 5xy} = \dfrac{8z}{45}$

(b) $\dfrac{x+2}{x+7} \div \dfrac{x^2-4}{x^2-49} = \dfrac{x+2}{x+7} \cdot \dfrac{x^2-49}{x^2-4} = \dfrac{(x+2)(x-7)(x+7)}{(x+7)(x-2)(x+2)} = \dfrac{x-7}{x-2}$

A **complex fraction** is a fraction whose numerator or denominator contains one or more fractions. It may be simplified in either of two ways:

(a) by treating the complex fraction as a division of one rational expression by another, or

(b) by multiplying both the numerator and the denominator by the lowest common denominator of all the fractions contained in the numerator or the denominator.

EXAMPLE 14 (a) $\dfrac{y + \dfrac{2}{y}}{y - \dfrac{1}{y}} = \dfrac{\dfrac{y^2 + 2}{y}}{\dfrac{y^2 - 1}{y}} = \dfrac{y^2 + 2}{y} \cdot \dfrac{y}{y^2 - 1} = \dfrac{y^2 + 2}{y^2 - 1} = \dfrac{y^2 + 2}{(y + 1)(y - 1)}.$

(b) $\dfrac{y + \dfrac{2}{y}}{y - \dfrac{1}{y}} = \dfrac{\left(y + \dfrac{2}{y}\right)y}{\left(y - \dfrac{1}{y}\right)y} = \dfrac{y^2 + 2}{y^2 - 1}.$

(c) $\dfrac{1 - \dfrac{1}{1 + x}}{2 + \dfrac{1}{x}} = \dfrac{\dfrac{(1 + x) - 1}{1 + x}}{\dfrac{2x + 1}{x}} = \dfrac{x}{1 + x} \cdot \dfrac{x}{2x + 1} = \dfrac{x^2}{2x^2 + 3x + 1}.$

(d) $\dfrac{1 - \dfrac{1}{1 + x}}{2 + \dfrac{1}{x}} = \dfrac{\left(1 - \dfrac{1}{1 + x}\right)(x)(1 + x)}{\left(2 + \dfrac{1}{x}\right)(x)(1 + x)} = \dfrac{x(1 + x) - x}{2x(1 + x) + (1 + x)}$

$$= \dfrac{x + x^2 - x}{2x + 2x^2 + 1 + x} = \dfrac{x^2}{2x^2 + 3x + 1}.$$

||||||||||| Rational Equations

We may have occasion to solve a **rational** equation of the type given in the following two examples.

EXAMPLE 15 Solve the rational equation

$$\frac{3}{2(x + 2)} + 5 = \frac{4}{x + 2}.$$

Solution To eliminate fractions we multiply both sides of the equation by the lowest common denominator, $2(x + 2)$, obtaining

$$3 + 5(2)(x + 2) = 4(2)$$
$$10x + 20 = 8 - 3$$
$$10x = -15$$
$$x = -\tfrac{3}{2}.$$

Substituting in the algebraic equation we find (verify) that $x = -\frac{3}{2}$ satisfies the equation.

■

Warning Whenever we multiply both sides of an equation by an expression involving the unknown, we risk introducing **extraneous solutions** (solutions of the new equation that do not satisfy the original equation). For example, the equation

$$x = 5$$

has only one solution, but when we multiply both sides of this equation by $x - 3$ we obtain the equation

$$x(x - 3) = 5(x - 3),$$

which has *two* solutions, $x = 5$ and $x = 3$. The "solution" $x = 3$ is said to be extraneous because it does not satisfy the original equation $x = 5$.

Therefore, the following rule is followed: ***whenever both sides of an equation are multiplied by an expression involving the unknown, all resulting solutions must be checked.***

■

The following example illustrates this situation.

EXAMPLE 16 Solve

$$\frac{x}{x - 3} + 8 = \frac{3}{x - 3}.$$

Solution To clear fractions we multiply by $x - 3$:

$$x + 8(x - 3) = 3$$
$$x + 8x - 24 = 3$$
$$9x = 27$$
$$x = 3.$$

Substituting in the given equation, we have

$$\frac{3}{3 - 3} + 8 = \frac{3}{3 - 3}.$$

■

Since division by zero is undefined, we conclude that $x = 3$ is an extraneous solution. Our analysis shows that the above equation has *no solution*, because if the equation had a solution it would have to be 3. However, 3 is *not* a solution.

A.5 EXERCISE SET ||||||||||||||||||||||||||||||||||||||||||||||||||||||||||||||||||||||||||||

In Exercises 1—16, perform the indicated operations and simplify.

1. $(4x^2 + 3x + 2) + (3x^2 - 2x - 5)$

2. $(2x^2 + 3x - 8) - (5 - 2x + 2x^2)$

3. $(4xy^2 + 2xy + 2x + 3) -$
$$(-2xy^2 + xy - y + 2)$$

4. $(2x^2y^2 + xy^2 - 2xy + x - y) -$
$$(3xy^2 + 2x^2y - 3x + y)$$

5. $(x + 2)(x - 6)$

6. $(x - 8)(x + 3)$

7. $(2x + 3)(3x - 5)$

8. $(4x + 5)(3x - 7)$

9. $(2x + 3)(2x^2 - x + 1)$

10. $(3x - 4)(x^3 + 2x^2 - x + 3)$

11. $(3x + 5)^2$

12. $(5 - 4x)^2$

13. $(3x - y)(2x + y)$

14. $(2x^2 + y)(3x - y^2)$

15. $(2x - y)(x^2 + 2xy - y)$

16. $(x^2 + y^2)(2x^2 - 3xy + y^2)$

In Exercises 17–36, factor the expressions.

17. $x^2 - 2x$

18. $x - x^3$

19. $x^2 - 49$

20. $81 - 64x^2$

21. $25y^2 - 9x^2$

22. $100x^2 - 4y^2$

23. $3x(2x + 5y) - 4y(2x + 5y)$

24. $4x(5x - y) + 3y(5x - y)$

25. $x^2 - 6x + 8$

26. $x^2 - 2x - 15$

27. $6x^2 - 11x - 10$

28. $3x^2 + 10x + 8$

29. $2x^3 - x^2 + 3x$

30. $6x^3 + 18x^2 + 12x$

31. $x^2 - xy - 6y^2$

32. $2x^3y + 5x^2y^2 - 3xy^3$

33. $x^3 + 2x^2 - x - 2$

34. $3x^3 - 4x^2 + 6x - 8$

35. $x^4 - x^2 + x^3 - x$

36. $2x^4 - 6x^3 + x^2 - 3x$

In Exercises 37–54, perform the indicated operations and simplify.

37. $\dfrac{5x}{x + 3} + \dfrac{3x}{x + 3}$

38. $\dfrac{3a}{a + 2} + \dfrac{6}{a + 2}$

39. $\dfrac{3x}{x - 1} + \dfrac{2x}{x - 2}$

40. $\dfrac{y}{y^2 - 4} + \dfrac{2}{y + 2}$

41. $\dfrac{2a + 3}{a^2 + a - 6} - \dfrac{a + 2}{a - 2}$

42. $\dfrac{2a}{a^2 - 4} + \dfrac{3a + 1}{a - 2}$

43. $\dfrac{a^2 - 9}{3a^2 - 9a} \cdot \dfrac{a^2}{a + 3}$

44. $\dfrac{b^2 + 2b - 8}{b^4 + 4b^3} \cdot \dfrac{b}{b^2 - 4b + 4}$

45. $\dfrac{a^2 + a - 2}{a^2 + 2a - 15} \div \dfrac{a^2 + 2a - 3}{a - 3}$

46. $\dfrac{x^2 + 4x - 5}{x^2 - 1} \div \dfrac{x}{x^2 + 2x + 1}$

47. $\dfrac{y}{y-1} + \dfrac{y}{y+1}$

48. $\dfrac{x}{x+y} + \dfrac{y}{x-y}$

49. $\dfrac{1 + (1/x)}{1 - (1/x)}$

50. $\dfrac{\dfrac{1}{x} + \dfrac{1}{y}}{x+y}$

51. $\dfrac{\dfrac{2}{x} + \dfrac{3}{x-1}}{\dfrac{1}{x}}$

52. $\dfrac{\dfrac{1}{x-2} + \dfrac{2}{x+3}}{\dfrac{1}{x-2}}$

53. $\dfrac{5 - \dfrac{1}{x+4}}{\dfrac{3}{x+4}}$

54. $\dfrac{\dfrac{1}{x+h} - \dfrac{1}{x}}{h}$

In Exercises 55–62, solve the equations. Check for extraneous solutions.

55. $\dfrac{2}{x-1} + \dfrac{1}{3} = \dfrac{1}{x-1}$

56. $\dfrac{8x+1}{x-2} + 4 = \dfrac{7x+3}{x-2}$

57. $-\dfrac{2x}{x+1} = 1 + \dfrac{2}{x+1}$

58. $\dfrac{3}{x} - 1 = \dfrac{1}{2} - \dfrac{6}{x}$

59. $\dfrac{1}{x-1} + \dfrac{1}{x+2} = \dfrac{7}{x^2+x-2}$

60. $\dfrac{1}{x} + \dfrac{x}{x-1} = 1$

61. $\dfrac{2}{x-2} + \dfrac{2}{x^2-4} = \dfrac{3}{x+2}$

62. $\dfrac{2}{x-2} + 1 = \dfrac{x+2}{x-2}$

A.6 ||||||| QUADRATIC EQUATIONS AND INEQUALITIES

Quadratic Equations

A **quadratic equation in x** is an equation that can be written in the form

$$ax^2 + bx + c = 0, \tag{1}$$

where a, b, and c are real numbers and $a \neq 0$.

If $b = 0$, the equation (1) reduces to

$$ax^2 + c = 0,$$

which we can easily solve for x when a and c have opposite signs, as in the next example.

EXAMPLE 1   Solve the quadratic equation $3x^2 - 8 = 0$.

Solution   We have

$$x^2 = \frac{8}{3}.$$

Consequently, solutions of the given quadratic equation are

$$x = \sqrt{\tfrac{8}{3}} \quad \text{and} \quad x = -\sqrt{\tfrac{8}{3}}.$$

Thus, $x = \pm\sqrt{\tfrac{8}{3}}$. ■

Solution by Factoring

Fundamental Principle: If a product of real numbers is zero, then at least one of the factors must be zero. That is, if $AB = 0$ then either $A = 0$ or $B = 0$ (or both).

This principle is extremely useful in solving the quadratic equation (1) in cases where $ax^2 + bx + c$ can be factored.

EXAMPLE 2   Solve the quadratic equation $2x^2 - x - 3 = 0$.

Solution   Factoring the given equation, we obtain

$$2x^2 - x - 3 = (2x - 3)(x + 1) = 0.$$

Thus, by the fundamental principle,

$$2x - 3 = 0 \quad \text{or} \quad x + 1 = 0$$

so

$$x = \frac{3}{2} \quad \text{or} \quad x = -1.$$

Therefore, the solutions of the given equation are $x = \tfrac{3}{2}$ and $x = -1$. ■

The Quadratic Formula

The quadratic equation (1) can always be solved, even when $ax^2 + bx + c$ cannot be factored into first degree factors with rational coefficients. An algebraic technique called "completing the square" can be applied to (1), producing the solutions

$$x = \frac{-b \pm \sqrt{b^2 - 4ac}}{2a}. \tag{2}$$

Equation (2) is called the **quadratic formula.**

EXAMPLE 3   Solve each of the following quadratic equations using the quadratic formula.
(a) $2x^2 + x - 15 = 0$ (c) $2x^2 - x - 3 = 0$ (Example 2).
(b) $x^2 + 2x + 3 = 0$

Solution Using the quadratic formula (2), we find
(a) $a = 2$, $b = 1$, $c = -15$,

$$x = \frac{-1 + \sqrt{1 - 4 \cdot 2 \cdot (-15)}}{2 \cdot 2} = \frac{-1 + \sqrt{121}}{4} = \frac{-1 + 11}{4} = \frac{10}{4} = \frac{5}{2}$$

and

$$x = \frac{-1 - \sqrt{1 - 4 \cdot 2 \cdot (-15)}}{2 \cdot 2} = \frac{-1 - \sqrt{121}}{4} = \frac{-1 - 11}{4} = \frac{-12}{4} = -3.$$

Therefore, the solutions of the given equation are $x = \frac{5}{2}$ and $x = -3$. The reader will find it instructive to solve this equation by factoring; the same answers should be obtained.
(b) $a = 1$, $b = 2$, $c = 3$. So

$$x = \frac{-2 \pm \sqrt{4 - 4 \cdot 1 \cdot 3}}{2 \cdot 1} = \frac{-2 \pm \sqrt{-8}}{2}.$$

There are no real number solutions, because there is no real number whose square is -8 ($\sqrt{-8}$ is not a real number).
(c) $a = 2$, $b = -1$, $c = -3$. So

$$x = \frac{1 \pm \sqrt{1 - 4 \cdot 2(-3)}}{2 \cdot 2} = \frac{1 \pm \sqrt{25}}{4}.$$

That is, the solutions of the given quadratic equation are

$$x = \frac{1 + 5}{4} = \frac{3}{2} \quad \text{and} \quad x = \frac{1 - 5}{4} = -1.$$

These solutions are identical with those found in Example 2, where the same equation was solved by factoring. ■

The Discriminant Test

A quadratic equation $ax^2 + bx + c = 0$ may have two, one, or no real number solutions. From the quadratic formula we can see that

If $b^2 - 4ac > 0$, then Equation (1) has two different real number solutions.
If $b^2 - 4ac = 0$, then Equation (1) has only one real number solution.
If $b^2 - 4ac < 0$, then Equation (1) has no real number solution.

The number $b^2 - 4ac$ is called the **discriminant** of Equation (1).

||||||||||||| **Solving Nonlinear Inequalities by Factoring**

We cannot use the quadratic formula to solve a second-degree inequality such as

$$(x - 2)(x + 1) > 0.$$

To obtain the solution we may use the following method. The product of the two factors is positive only when both factors $(x - 2)$ and $(x + 1)$ have the same sign. Thus, we examine the sign of each factor separately. First, $(x - 2) > 0$ when $x > 2$ and $(x + 1) > 0$ when $x > -1$. Second, $(x - 2) < 0$ when $x < 2$ and $(x + 1) < 0$ when $x < -1$. The dependence of the signs of $(x - 2)$ and $(x + 1)$ upon x is shown graphically in Figure 1.

(a) Sign of $(x - 2)$ $- - - - - - - - - - - - 0 + + + + + +$

Figure 1

(b) Sign of $(x + 1)$ $- - - - - 0 + + + + + + + + + + + + +$

In Figure 1, we see that both factors are positive when $x > 2$ and both are negative when $x < -1$. Thus, x is a solution of the quadratic inequality if

$$x > 2 \quad \text{or} \quad x < -1.$$

The set of points satisfying the quadratic inequality is shown in color in Figure 2.

Figure 2

We can also say that the solution of this quadratic inequality consists of all points in the interval $(-\infty, -1)$ together with all points in the interval $(2, \infty)$.

EXAMPLE 4 Solve the inequality

$$\frac{x + 1}{2 - x} \leq 0.$$

Solution Figure 3 gives an analysis of the signs of $x + 1$ and $2 - x$.

Sign of $(x + 1)$ $- - - -$ 0 $+ + + + + + +$ $+ + +$

Figure 3 Sign of $(2 - x)$ $+ + + +$ $+$ $+ + + + + +$ 0 $- - -$

Since $(x + 1)/(2 - x)$ can be negative only if the factors $x + 1$ and $2 - x$ have opposite signs, we conclude that x is a solution of the given inequality if

$$x \le -1 \quad \text{or} \quad x > 2.$$

Note that we must have $x \ne 2$ since $2 - x$ appears in the denominator. The solution set is thus $(-\infty, -1]$ together with $(2, +\infty)$.   ∎

|||||||||||||| **Applications**

EXAMPLE 5   The length of a rectangular pool is 3 times its width and the pool is surrounded by 4 feet of walk (Figure 4). If the total area enclosed by the walk and the pool is 684 square feet, find the dimensions of the pool.

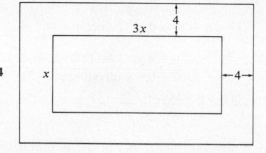

Figure 4

Solution   If we let $x =$ the width (in feet) of the pool, then $3x$ is the length (in feet) of the pool. The region enclosed by the walk has length $3x + 8$ and width $x + 8$. Thus, the area of this region is $(3x + 8)(x + 8)$. Consequently,

$$(3x + 8)(x + 8) = 684$$
$$3x^2 + 32x + 64 = 684$$
$$3x^2 + 32x - 620 = 0$$
$$(3x + 62)(x - 10) = 0$$
$$x = 10.$$

(Reject $x = -62/3$ because x represents width, which cannot be negative.) The dimensions of the pool are thus 10 feet by 30 feet.   ∎

EXAMPLE 6   An investor purchases a number of shares of stock for $600. If the investor had paid $2 less per share, the number of shares would number 10 more than the amount purchased. How many shares did the investor buy? What was the price per share?

Solution Let n = number of shares purchased and p = price paid (in dollars per share). Because

$$\left(\begin{array}{c}\text{number of}\\\text{shares}\end{array}\right) \times \left(\begin{array}{c}\text{price per}\\\text{share}\end{array}\right) = \begin{array}{c}\text{total dollars}\\\text{invested}\end{array}$$

we must have

$$n \cdot p = 600$$

and

$$(n + 10)(p - 2) = 600.$$

Then

$$np + 10p - 2n - 20 = 600$$
$$600 + 10p - 2n - 620 = 0 \qquad \text{(since } np = 600\text{)}$$
$$10p - 2n - 20 = 0$$
$$10\left(\frac{600}{n}\right) - 2n - 20 = 0 \qquad \left(\text{since } p = \frac{600}{n}\right)$$
$$6000 - 2n^2 - 20n = 0 \qquad \text{(multiplying by } n\text{)}$$
$$n^2 + 10n - 3000 = 0 \qquad \text{(dividing by } -2\text{)}$$
$$(n - 50)(n + 60) = 0$$
$$n = 50.$$

(Reject $n = -60$ since n represents the number of shares purchased, which cannot be negative.) The investor purchased 50 shares of stock. Then

$$p = \frac{600}{n} = \frac{600}{50} = 12,$$

so that the price per share was \$12.   ∎

A.6 EXERCISE SET ||||||||||||||||||||||||||||||||||||||||||||||||||||||||||||||||||||||||

In Exercises 1–20, solve the given quadratic equation (that is, find all real number solutions).

1. $x^2 - 5 = 0$

2. $x^2 + 2 = 0$

3. $3x^2 + 4 = 0$

4. $2x^2 - 3 = 0$

5. $x^2 - 3x = 0$

6. $2x^2 + 5x = 0$

7. $y^2 - y - 6 = 0$

8. $y^2 + 3y - 10 = 0$

9. $3u^2 - 18u + 15 = 0$

10. $4x^2 - 4x - 3 = 0$

11. $2x^2 + 2x - 5 = 0$

12. $u^2 - 4u + 2 = 0$

13. $y^2 + 2y + 4 = 0$

14. $2x^2 + 5x + 4 = 0$

15. $3u^2 + u - 4 = 0$

16. $2y^2 - 7y + 3 = 0$

17. $x^2 + 3x + 3 = 0$

18. $x^2 + 3x + 1 = 0$

19. $2x^2 - 2x - 1 = 0$

20. $2x^2 - x + 1 = 0$

In Exercises 21–26, use the method of factoring to solve the equation.

21. $x^3 + 2x^2 - 8x = 0$

22. $2x^3 + x^2 - 3x = 0$

23. $x^3 = 4x$

24. $2x^3 - 3x^2 = 9x$

25. $\dfrac{3x^3 - 4x^2}{x^2 - 5} = 0$

26. $\dfrac{16x^2 - 4x^4}{x^2 - 64} = 0$

In Exercises 27–34, solve the inequalities.

27. $x^2 + x - 6 > 0$

28. $x^2 - 3x - 10 \geq 0$

29. $2x^2 - 3x - 5 < 0$

30. $3x^2 - 4x - 4 \leq 0$

31. $\dfrac{2x + 3}{2x - 1} < 0$

32. $\dfrac{3x + 2}{2x - 3} \geq 0$

33. $\dfrac{x - 1}{x + 1} \geq 0$

34. $\dfrac{2x - 1}{x + 2} \leq 0$

In Exercises 35–38, determine all values of x for which the expressions are real numbers.

35. $\sqrt{x^2 - 9}$

36. $\sqrt{250 - 10x^2}$

37. $\dfrac{1}{\sqrt{64 - 4x^2}}$

38. $\dfrac{1}{\sqrt{4x^2 - 144}}$

39. A 16×20-inch rectangular mounting board is used to mount a rectangular photograph. How wide is the uniform border if the photograph occupies $\frac{3}{5}$ of the mounting board?

40. The length of a rectangle exceeds twice its width by 4 feet. If the area of the rectangle is 48 square feet, find the dimensions.

41. The base of a triangle is 2 feet greater than twice its height. If the area is 12 square feet, find the dimensions.

42. A business machine dealer purchased a number of used printing calculators at an auction for a total expenditure of $240. After giving one of the calculators to his daughter, he sold the remaining calculators at a price of $15 each for a profit of $30 on the entire transaction. How many printing calculators did he buy?

43. A number of students rented a car for a one-week camping trip for $160. If another student had joined the original group, each person's share of expenses would have been reduced by $8. How many students were in the original group?

44. A freelance photographer worked a number of days for a newspaper to earn $480. If she had been paid $8 less per day, she would have had to work two more days to earn the same amount. What was her daily rate of pay?

A.7 |||||||| SOME COMMON ALGEBRAIC MISTAKES

For the reader's convenience, we provide here a list of common mistakes often made in the course of algebraic manipulations. All of them can be avoided by exercising reasonable *care* in one's work.

Incorrect	Correct
$2(a + 3) = 2a + 3$	$2(a + 3) = 2a + 6$
$x - (a + 2) = x - a + 2$	$x - (a + 2) = x - a - 2$
$-x^2 = (-x)^2$	$-x^2 = -(-x)^2 = -(x^2)$
$\dfrac{1}{0} = 0; \dfrac{0}{0} = 1$	Both $\dfrac{1}{0}$ and $\dfrac{0}{0}$ are undefined
$(a + b)^2 = a^2 + b^2$	$(a + b)^2 = a^2 + 2ab + b^2$
$(3a)^2 = 3a^2$	$(3a)^2 = 9a^2$
$\dfrac{1}{2a^3} = 2a^{-3}$	$\dfrac{1}{2a^3} = \dfrac{1}{2} a^{-3}$
$a^2 \cdot a^3 = a^6$	$a^2 \cdot a^3 = a^5$
$(a^2)^3 = a^5$	$(a^2)^3 = a^6$
$\sqrt{9} = \pm 3$	$\sqrt{9} = 3$
$\sqrt{4a} = 2a$	$\sqrt{4a} = \sqrt{4}\sqrt{a} = 2\sqrt{a}$
$\sqrt{2a} = 2\sqrt{a}$ or $\sqrt{2}(a)$	$\sqrt{2a} = \sqrt{2}\sqrt{a}$
$\sqrt{3 - 2x} = -\sqrt{2x - 3}$	$\sqrt{3 - 2x}$ cannot be simplified
$\sqrt{a + b} = \sqrt{a} + \sqrt{b}$	$\sqrt{a + b}$ cannot be simplified
$\sqrt{a^2 + b^2} = a + b$	$\sqrt{a^2 + b^2}$ cannot be simplified
$\dfrac{x}{a} + \dfrac{y}{b} = \dfrac{x + y}{a + b}$	$\dfrac{x}{a} + \dfrac{y}{b} = \dfrac{bx + ay}{ab}$
$\dfrac{a + c}{b + c} = \dfrac{a}{b}$	$\dfrac{a + c}{b + c}$ cannot be simplified
$\dfrac{a}{b + c} = \dfrac{a}{b} + \dfrac{a}{c}$	$\dfrac{a}{b + c}$ cannot be simplified
$\dfrac{2a}{c} - \dfrac{a - b}{c} = \dfrac{2a - a - b}{c} = \dfrac{a - b}{c}$	$\dfrac{2a}{c} - \dfrac{a - b}{c} = \dfrac{2a - (a - b)}{c}$ $= \dfrac{2a - a + b}{c}$ $= \dfrac{a + b}{c}$
$\dfrac{ax}{bx + cy} = \dfrac{a}{b + cy}$	$\dfrac{ax}{bx + cy}$ cannot be simplified

A.8 ||||||| THE RECTANGULAR COORDINATE SYSTEM

In Section A.2 we identified the system of real numbers with the points on a straight line; that is, the system of real numbers is in one-to-one correspondence with the points on the real line. In this section we shall show that the points in a plane are in one-to-one correspondence with the set of all ordered pairs of real numbers. An ordered pair (a, b) of the real numbers a and b is a listing in the indicated order. Thus, the ordered pairs $(2, 3)$ and $(3, 2)$ are different.

Draw a pair of perpendicular lines, intersecting at a point that we shall denote by the letter O, and call the **origin.** One of the lines, the x-axis, usually has a horizontal orientation. The other line, the y-axis, then has a vertical orientation. Suppose that the x-axis is a real line, with its number 0 located at the origin. We now choose a unit of length on the y-axis (often the same as the unit of length on the x-axis) and label the y-axis so that it also is a real line, on which the segment above the origin represents positive numbers and the segment below the origin represents negative numbers. The x-axis and y-axis are then called the **coordinate axes,** and together they constitute a **rectangular coordinate system,** or (after René Descartes*) a **Cartesian coordinate system** (Figure 1).

Figure 1

Consider a point P in the plane of the paper. From P draw a perpendicular to the x-axis and suppose that this perpendicular intersects the x-axis at the point $x = a$. Now draw a perpendicular from P to the y-axis and suppose that this perpendicular intersects the y-axis at the point $y = b$ (Figure 2). We associate the ordered pair (a, b) with the point P, and we denote the point P by (a, b) or by $P(a, b)$.

Figure 2

---

* René Descartes (1596–1650) was born in France to an old noble family. He was a great philosopher, scientist, and mathematician, and is considered one of the fathers of modern physiology. His invention of analytic geometry greatly influenced the future development of science. When he was a young man of nineteen or so, he became a gambler and profligate in Paris for a brief time. Fascinated by warfare, he enlisted in the army and fought in several major battles.

The number a is called the **x-coordinate,** or **abscissa,** of P and the number b is called the **y-coordinate,** or **ordinate,** of P. We also say that a and b are the **coordinates** of P. It is clear from Figure 2 that each point in the plane determines a unique, ordered pair (a, b) of real numbers. Conversely, for each ordered pair (a, b) of real numbers, there is a unique point P in the plane whose coordinates are (a, b) (see Figure 2). Thus, we have established a one-to-one correspondence between the set of all points in the plane and the set of all ordered pairs of real numbers.

EXAMPLE 1   Figure 3 shows a number of points and their coordinates. Observe that the coordinates of the origin are $(0, 0)$; if r is a real number, the ordered pair $(r, 0)$ represents a point on the x-axis. Similarly, an ordered pair of the form $(0, r)$ represents a point on the y-axis.

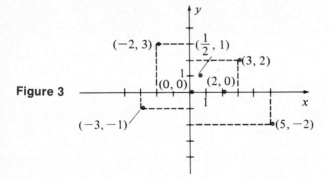

Figure 3

The coordinate axes divide the plane into four numbered parts, called **quadrants** (see Figure 4). Thus, a point with a negative x-coordinate and a positive y-coordinate lies in the second quadrant and a point having two negative coordinates lies in the third quadrant.

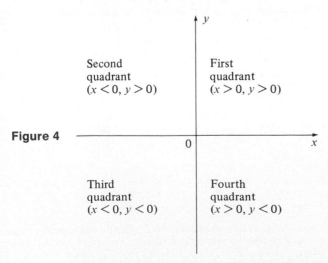

Figure 4

The points lying *on* the axes do not belong to any quadrant.

|||||||||||| **Graphs of Equations**

We shall use a Cartesian coordinate system to provide a "picture" of an equation in two variables. Examples of such equations are

$$2x + y = 4, \qquad 2x^2 + y = 7, \qquad 4x^2 + 4y^2 = 16, \quad \text{and} \quad y = \frac{2}{x-1}.$$

An ordered pair of real numbers (a, b) is said to be a **solution** of an equation in two variables x and y if the equation is satisfied when $x = a$ and $y = b$ are substituted.

EXAMPLE 2   Consider the equation

$$2x + y = 4.$$

The following ordered pairs of real numbers are solutions:

$$(0, 4), \quad (-1, 6), \quad (2, 0) \quad \text{and} \quad (3, -2).$$

For if $x = -1$ and $y = 6$ are substituted in the equation we obtain

$$2(-1) + 6 = 4.$$

On the other hand, the following ordered pairs of real numbers are not solutions:

$$(0, 5), \qquad (-2, 4), \qquad (3, 0), \quad \text{and} \quad (5, -1).$$

For if $x = 5$ and $y = -1$ are substituted in the equation, we obtain

$$2(5) + (-1) = 10 - 1 = 9 \neq 4.$$

The set of all solutions of an equation is called its **solution set**. The **graph** of an equation in x and y is the geometric or graphical representation of the solution set obtained by plotting the points of the solution set in the Cartesian coordinate system.

EXAMPLE 3   Sketch the graph of the equation

$$2x + y = 4. \tag{1}$$

Solution   The solution set of Equation (1) consists of infinitely many points, so we can obviously not find and plot them all. We sketch the graph by finding and plotting enough solutions to obtain a good approximation of the graph.

We now rewrite (1) as

$$y = 4 - 2x, \tag{2}$$

and obtain Table 1 by substituting the indicated value of x in Equation (2) to calculate the corresponding value of y.

TABLE 1

x	-2	-1	0	1	2	3
$y = 4 - 2x$	8	6	4	2	0	-2

The graph is sketched in Figure 5. It appears to be a straight line. The fact that it is a straight line is established in Section 1.3.

Figure 5

$y = 4 - 2x$

EXAMPLE 4   Sketch the graph of

$$y = x^2 - 4. \tag{3}$$

Solution   Proceeding as in Example 3, we choose values of x and calculate the corresponding values of y from (3). The results are shown in Table 2.

TABLE 2

x	-3	-2	$-\frac{3}{2}$	-1	0	$\frac{1}{2}$	1	2	3
$y = x^2 - 4$	5	0	$-\frac{7}{4}$	-3	-4	$-\frac{15}{4}$	-3	0	5

When we plot these points, we obtain the graph shown in Figure 6. This curve is called a **parabola.**

Figure 6

$$y = x^2 - 4$$

EXAMPLE 5   Sketch the graph of

$$y = 2x - x^2. \tag{4}$$

Solution   We choose values of x arbitrarily and calculate the corresponding values of y from (4). The results are shown in Table 3, and when the points in this table are plotted we obtain the graph shown in Figure 7.

TABLE 3

x	-2	-1	$-\frac{1}{2}$	0	$\frac{1}{2}$	1	$\frac{3}{2}$	2	$\frac{5}{2}$	3	4
$y = 2x - x^2$	-8	-3	$-\frac{5}{4}$	0	$\frac{3}{4}$	1	$\frac{3}{4}$	0	$-\frac{5}{4}$	-3	-8

Figure 7

$$y = 2x - x^2$$

A.8 EXERCISE SET ||||||||||||||||||||||||||||||||||||||||||||||||||||||||||||||||||||||||||||

1. Draw a Cartesian coordinate system and plot the points whose coordinates are
   (a) $(3, 1)$ (c) $(-3, -2)$
   (b) $(-2, 3)$ (d) $(4, -5)$

2. Draw a Cartesian coordinate system and plot the points whose coordinates are
   (a) $(2, 0)$ (c) $(\frac{1}{2}, -\frac{5}{3})$
   (b) $(0, -2)$ (d) $(-1.5, -2.5)$

For Exercises 3 and 4, consider the following Cartesian coordinate system.

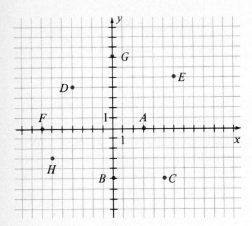

3. Give the coordinates of the points
   (a) A (c) C
   (b) B (d) D

4. Give the coordinates of the points
   (a) E (c) G
   (b) F (d) H

5. Name the quadrant in which each of the following points is located.
   (a) $(4, -5)$ (c) $(2, 3)$
   (b) $(-2, 4)$ (d) $(-\frac{2}{3}, -\frac{4}{5})$

6. Which of the following are solutions of
   $$2x - 3y = 12$$
   (a) $(0, -4)$ (c) $(3, 1)$
   (b) $(1, 3)$ (d) $(3, -2)$

7. Which of the following are solutions of
   $$2x + 3y^2 = 18$$
   (a) $(3, -2)$ (c) $(9, 0)$
   (b) $(2, 1)$ (d) $(15, 4)$

8. Consider the equation
   $$4x + 3y = 12$$
   Complete the following table so that the ordered pair (x, y) is a solution of the given equation.

x	1		0		-3	
y		-2	0			2

In Exercises 9–16, sketch the graph of the equations.

9. $y = 2x + 4$ 10. $3x + 5y = 15$

11. $y = x^2 + 3$ 12. $2x^2 + y = 4$

13. $y = x^3$ 14. $y = x - x^2 + 1$

15. $3x^2 - 2y + 5 = 0$ 16. $y = \frac{1}{2}x^3 - 1$

17. Sketch the set of all points in the plane whose y coordinate is 3.

18. Sketch the set of all points in the plane whose x-coordinate is 5.

19. What is the equation of the graph shown below?

20. What is the equation of the graph shown below?

21. The points $A(2, 7)$, $B(4, 3)$ and $C(x, y)$ determine a right triangle, with two sides parallel to the coordinate axes, whose hypotenuse is AB. Find x and y. (*Hint:* There is more than one answer.)

22. The points $A(2, 6)$, $B(4, 6)$, $C(4, 8)$, and $D(x, y)$ form a rectangle. Find x and y.

APPENDIX B: TABLES

Table of Exponentials and Their Reciprocals

x	e^x	e^{-x}	x	e^x	e^{-x}
0.00	1.0000	1.0000	1.3	3.6693	0.2725
0.01	1.0101	0.9900	1.4	4.0552	0.2466
0.02	1.0202	0.9802	1.5	4.4817	0.2231
0.03	1.0305	0.9704	1.6	4.9530	0.2019
0.04	1.0408	0.9608	1.7	5.4739	0.1827
0.05	1.0513	0.9512	1.8	6.0496	0.1653
0.06	1.0618	0.9418	1.9	6.6859	0.1496
0.07	1.0725	0.9324	2.0	7.3891	0.1353
0.08	1.0833	0.9231	2.1	8.1662	0.1225
0.09	1.0942	0.9139	2.2	9.0250	0.1108
0.10	1.1052	0.9048	2.3	9.9742	0.1003
0.11	1.1163	0.8958	2.4	11.023	0.0907
0.12	1.1275	0.8869	2.5	12.182	0.0821
0.13	1.1388	0.8781	2.6	13.464	0.0743
0.14	1.1503	0.8694	2.7	14.880	0.0672
0.15	1.1618	0.8607	2.8	16.445	0.0608
0.16	1.1735	0.8521	2.9	18.174	0.0550
0.17	1.1853	0.8437	3.0	20.086	0.0498
0.18	1.1972	0.8353	3.1	22.198	0.0450
0.19	1.2092	0.8270	3.2	24.533	0.0408
0.20	1.2214	0.8187	3.3	27.113	0.0369
0.21	1.2337	0.8106	3.4	29.964	0.0334
0.22	1.2461	0.8025	3.5	33.115	0.0302
0.23	1.2586	0.7945	3.6	36.598	0.0273
0.24	1.2712	0.7866	3.7	40.447	0.0247
0.25	1.2840	0.7788	3.8	44.701	0.0224
0.26	1.2969	0.7711	3.9	49.402	0.0202
0.27	1.3100	0.7634	4.0	54.598	0.0183
0.28	1.3231	0.7558	4.1	60.340	0.0166
0.29	1.3364	0.7483	4.2	66.686	0.0150
0.30	1.3499	0.7408	4.3	73.700	0.0136
0.35	1.4191	0.7047	4.4	81.451	0.0123
0.40	1.4918	0.6703	4.5	90.017	0.0111
0.45	1.5683	0.6376	4.6	99.484	0.0101
0.50	1.6487	0.6065	4.7	109.95	0.0091
0.55	1.7333	0.5769	4.8	121.51	0.0082
0.60	1.8221	0.5488	4.9	134.29	0.0074
0.65	1.9155	0.5220	5	148.41	0.0067
0.70	2.0138	0.4966	6	403.43	0.0025
0.75	2.1170	0.4724	7	1096.6	0.0009
0.80	2.2255	0.4493	8	2981.0	0.0003
0.85	2.3396	0.4274	9	8103.1	0.0001
0.90	2.4596	0.4066	10	22026	0.00005
0.95	2.5857	0.3867	11	59874	0.00002
1.0	2.7183	0.3679	12	162,754	0.000006
1.1	3.0042	0.3329	13	442,413	0.000002
1.2	3.3201	0.3012	14	1,202,604	0.0000008
			15	3,269,017	0.0000003

APPENDIX B
TABLE II

Table of Natural Logarithms

n	$\log_e n$	n	$\log_e n$	n	$\log_e n$
		4.5	1.5041	9.0	2.1972
0.1	−2.3026	4.6	1.5261	9.1	2.2083
0.2	−1.6094	4.7	1.5476	9.2	2.2192
0.3	−1.2040	4.8	1.5686	9.3	2.2300
0.4	−0.9163	4.9	1.5892	9.4	2.2407
0.5	−0.6931	5.0	1.6094	9.5	2.2513
0.6	−0.5108	5.1	1.6292	9.6	2.2618
0.7	−0.3567	5.2	1.6487	9.7	2.2721
0.8	−0.2231	5.3	1.6677	9.8	2.2824
0.9	−0.1054	5.4	1.6864	9.9	2.2925
1.0	0.0000	5.5	1.7047	10	2.3026
1.1	0.0953	5.6	1.7228	11	2.3979
1.2	0.1823	5.7	1.7405	12	2.4849
1.3	0.2624	5.8	1.7579	13	2.5649
1.4	0.3365	5.9	1.7750	14	2.6391
1.5	0.4055	6.0	1.7918	15	2.7081
1.6	0.4700	6.1	1.8083	16	2.7726
1.7	0.5306	6.2	1.8245	17	2.8332
1.8	0.5878	6.3	1.8405	18	2.8904
1.9	0.6419	6.4	1.8563	19	2.9444
2.0	0.6931	6.5	1.8718	20	2.9957
2.1	0.7419	6.6	1.8871	25	3.2189
2.2	0.7885	6.7	1.9021	30	3.4012
2.3	0.8329	6.8	1.9169	35	3.5553
2.4	0.8755	6.9	1.9315	40	3.6889
2.5	0.9163	7.0	1.9459	45	3.8067
2.6	0.9555	7.1	1.9601	50	3.9120
2.7	0.9933	7.2	1.9741	55	4.0073
2.8	1.0296	7.3	1.9879	60	4.0943
2.9	1.0647	7.4	2.0015	65	4.1744
3.0	1.0986	7.5	2.0149	70	4.2485
3.1	1.1314	7.6	2.0281	75	4.3175
3.2	1.1632	7.7	2.0142	80	4.3820
3.3	1.1939	7.8	2.0541	85	4.4427
3.4	1.2238	7.9	2.0669	90	4.4998
3.5	1.2528	8.0	2.0794	95	4.5539
3.6	1.2809	8.1	2.0919	100	4.6052
3.7	1.3083	8.2	2.1041		
3.8	1.3350	8.3	2.1163		
3.9	1.3610	8.4	2.1282		
4.0	1.3863	8.5	2.1401		
4.1	1.4110	8.6	2.1518		
4.2	1.4351	8.7	2.1633		
4.3	1.4586	8.8	2.1748		
4.4	1.4816	8.9	2.1861		

Table of Compound Interest

i = rate of interest per period n = number of periods

$i = .005$ $i = .01$

n	$(1 + i)^n$	$(1 + i)^{-n}$	n	$(1 + i)^n$	$(1 + i)^{-n}$
1	1.0050 0000	0.9950 2488	1	1.0100 0000	0.9900 9901
2	1.0100 2500	0.9900 7450	2	1.0201 0000	0.9802 9605
3	1.0150 7513	0.9851 4876	3	1.0303 0100	0.9705 9015
4	1.0201 5050	0.9802 4752	4	1.0406 0401	0.9609 8034
5	1.0252 5125	0.9753 7067	5	1.0510 1005	0.9514 6569
6	1.0303 7751	0.9705 1808	6	1.0615 2015	0.9420 4524
7	1.0355 2940	0.9656 8963	7	1.0721 3535	0.9327 1805
8	1.0407 0704	0.9608 8520	8	1.0828 5671	0.9234 8322
9	1.0459 1058	0.9561 0468	9	1.0936 8527	0.9143 3982
10	1.0511 4013	0.9513 4794	10	1.1046 2213	0.9052 8695
11	1.0563 9583	0.9466 1487	11	1.1156 6835	0.8963 2372
12	1.0616 7781	0.9419 0534	12	1.1268 2503	0.8874 4923
13	1.0669 8620	0.9372 1924	13	1.1380 9328	0.8786 6260
14	1.0723 2113	0.9325 5646	14	1.1494 7421	0.8699 6297
15	1.0776 8274	0.9279 1688	15	1.1609 6896	0.8613 4947
16	1.0830 7115	0.9233 0037	16	1.1725 7864	0.8528 2126
17	1.0884 8651	0.9187 0684	17	1.1843 0443	0.8443 7749
18	1.0939 2894	0.9141 3616	18	1.1961 4748	0.8360 1731
19	1.0993 9858	0.9095 8822	19	1.2081 0895	0.8277 3992
20	1.1048 9558	0.9050 6290	20	1.2201 9004	0.8195 4447
21	1.1104 2006	0.9005 6010	21	1.2323 9194	0.8114 3017
22	1.1159 7216	0.8960 7971	22	1.2447 1586	0.8033 9621
23	1.1215 5202	0.8916 2160	23	1.2571 6302	0.7954 4179
24	1.1271 5978	0.8871 8567	24	1.2697 3465	0.7875 6613
25	1.1327 9558	0.8827 7181	25	1.2824 3200	0.7797 6844
26	1.1384 5955	0.8783 7991	26	1.2952 5631	0.7720 4796
27	1.1441 5185	0.8740 0986	27	1.3082 0888	0.7644 0392
28	1.1498 7261	0.8696 6155	28	1.3212 9097	0.7568 3557
29	1.1556 2197	0.8653 3488	29	1.3345 0388	0.7493 4215
30	1.1614 0008	0.8610 2973	30	1.3478 4892	0.7419 2292
31	1.1672 0708	0.8567 4600	31	1.3613 2740	0.7345 7715
32	1.1730 4312	0.8524 8358	32	1.3749 4068	0.7273 0411
33	1.1789 0833	0.8482 4237	33	1.3886 9009	0.7201 0307
34	1.1848 0288	0.8440 2226	34	1.4025 7699	0.7129 7334
35	1.1907 2689	0.8398 2314	35	1.4166 0276	0.7059 1420
36	1.1966 8052	0.8356 4492	36	1.4307 6878	0.6989 2495
37	1.2026 6393	0.8314 8748	37	1.4450 7647	0.6920 0490
38	1.2086 7725	0.8273 5073	38	1.4595 2724	0.6851 5337
39	1.2147 2063	0.8232 3455	39	1.4741 2251	0.6783 6967
40	1.2207 9424	0.8191 3886	40	1.4888 6373	0.6716 5314
41	1.2268 9821	0.8150 6354	41	1.5037 5237	0.6650 0311
42	1.2330 3270	0.8110 0850	42	1.5187 8989	0.6584 1892
43	1.2391 9786	0.8069 7363	43	1.5339 7779	0.6518 9992
44	1.2453 9385	0.8029 5884	44	1.5493 1757	0.6454 4546
45	1.2516 2082	0.7989 6402	45	1.5648 1075	0.6390 5492
46	1.2578 7892	0.7949 8907	46	1.5804 5885	0.6327 2764
47	1.2641 6832	0.7910 3390	47	1.5962 6344	0.6264 6301
48	1.2704 8916	0.7870 9841	48	1.6122 2608	0.6202 6041
49	1.2768 4161	0.7831 8250	49	1.6283 4834	0.6141 1921
50	1.2832 2581	0.7792 8607	50	1.6446 3182	0.6080 3882

TABLE III

	$i = .015$			$i = .02$	
n	$(1 + i)^n$	$(1 + i)^{-n}$	n	$(1 + i)^n$	$(1 + i)^{-n}$
1	1.0150 0000	0.9852 2167	1	1.0200 0000	0.9803 9216
2	1.0302 2500	0.9706 6175	2	1.0404 0000	0.9611 6878
3	1.0456 7838	0.9563 1699	3	1.0612 0800	0.9423 2233
4	1.0613 6355	0.9421 8423	4	1.0824 3216	0.9238 4543
5	1.0772 8400	0.9282 6033	5	1.1040 8080	0.9057 3081
6	1.0934 4326	0.9145 4219	6	1.1261 6242	0.8879 7138
7	1.1098 4491	0.9010 2679	7	1.1486 8567	0.8705 6018
8	1.1264 9259	0.8877 1112	8	1.1716 5938	0.8534 9037
9	1.1433 8998	0.8745 9224	9	1.1950 9257	0.8367 5527
10	1.1605 4083	0.8616 6723	10	1.2189 9442	0.8203 4830
11	1.1779 4894	0.8489 3323	11	1.2433 7431	0.8042 6304
12	1.1956 1817	0.8363 8742	12	1.2682 4179	0.7884 9318
13	1.2135 5244	0.8240 2702	13	1.2936 0663	0.7730 3253
14	1.2317 5573	0.8118 4928	14	1.3194 7876	0.7578 7502
15	1.2502 3207	0.7998 5150	15	1.3458 6834	0.7430 1473
16	1.2689 8555	0.7880 3104	16	1.3727 8571	0.7284 4581
17	1.2880 2033	0.7763 8526	17	1.4002 4142	0.7141 6256
18	1.3073 4064	0.7649 1159	18	1.4282 4625	0.7001 5937
19	1.3269 5075	0.7536 0747	19	1.4568 1117	0.6864 3076
20	1.3468 5501	0.7424 7042	20	1.4859 4740	0.6729 7133
21	1.3670 5783	0.7314 9795	21	1.5156 6634	0.6597 7582
22	1.3875 6370	0.7206 8763	22	1.5459 7967	0.6468 3904
23	1.4083 7715	0.7100 3708	23	1.5768 9926	0.6341 5592
24	1.4295 0281	0.6995 4392	24	1.6084 3725	0.6217 2149
25	1.4509 4535	0.6892 0583	25	1.6406 0599	0.6095 3087
26	1.4727 0953	0.6790 2052	26	1.6734 1811	0.5975 7928
27	1.4948 0018	0.6689 8574	27	1.7068 8648	0.5858 6204
28	1.5172 2218	0.6590 9925	28	1.7410 2421	0.5743 7455
29	1.5399 8051	0.6493 5887	29	1.7758 4469	0.5631 1231
30	1.5630 8022	0.6397 6243	30	1.8113 6158	0.5520 7089
31	1.5865 2642	0.6303 0781	31	1.8475 8882	0.5412 4597
32	1.6103 2432	0.6209 9292	32	1.8845 4059	0.5306 3330
33	1.6344 7918	0.6118 1568	33	1.9222 3140	0.5202 2873
34	1.6589 9637	0.6027 7407	34	1.9606 7603	0.5100 2817
35	1.6838 8132	0.5938 6608	35	1.9998 8955	0.5000 2761
36	1.7091 3954	0.5850 8974	36	2.0398 8734	0.4902 2315
37	1.7347 7663	0.5764 4309	37	2.0806 8509	0.4806 1093
38	1.7607 9828	0.5679 2423	38	2.1222 9879	0.4711 8719
39	1.7872 1025	0.5595 3126	39	2.1647 4477	0.4619 4822
40	1.8140 1841	0.5512 6232	40	2.2080 3966	0.4528 9042
41	1.8412 2868	0.5431 1559	41	2.2522 0046	0.4440 1021
42	1.8688 4712	0.5350 8925	42	2.2972 4447	0.4353 0413
43	1.8968 7982	0.5271 8153	43	2.3431 8936	0.4267 6875
44	1.9253 3302	0.5193 9067	44	2.3900 5314	0.4184 0074
45	1.9542 1301	0.5117 1494	45	2.4378 5421	0.4101 9680
46	1.9835 2621	0.5041 5265	46	2.4866 1129	0.4021 5373
47	2.0132 7910	0.4967 0212	47	2.5363 4352	0.3942 6836
48	2.0434 7829	0.4893 6170	48	2.5870 7039	0.3865 3761
49	2.0741 3046	0.4821 2975	49	2.6388 1179	0.3789 5844
50	2.1052 4242	0.4750 0468	50	2.6915 8803	0.3715 2788

TABLE III

	$i = .025$			$i = .03$	
n	$(1 + i)^n$	$(1 + i)^{-n}$	n	$(1 + i)^n$	$(1 + i)^{-n}$
1	1.0250 0000	0.9756 0976	1	1.0300 0000	0.9708 7379
2	1.0506 2500	0.9518 1440	2	1.0609 0000	0.9425 9591
3	1.0768 9063	0.9285 9941	3	1.0927 2700	0.9151 4166
4	1.1038 1289	0.9059 5064	4	1.1255 0881	0.8884 8705
5	1.1314 0821	0.8838 5429	5	1.1592 7407	0.8626 0878
6	1.1596 9342	0.8622 9687	6	1.1940 5230	0.8374 8426
7	1.1886 8575	0.8412 6524	7	1.2298 7387	0.8130 9151
8	1.2184 0290	0.8207 4657	8	1.2667 7008	0.7894 0923
9	1.2488 6297	0.8007 2836	9	1.3047 7318	0.7664 1673
10	1.2800 8454	0.7811 9840	10	1.3439 1638	0.7440 9391
11	1.3120 8666	0.7621 4478	11	1.3842 3387	0.7224 2128
12	1.3448 8882	0.7435 5589	12	1.4257 6089	0.7013 7988
13	1.3785 1104	0.7254 2038	13	1.4685 3371	0.6809 5134
14	1.4129 7382	0.7077 2720	14	1.5125 8972	0.6611 1781
15	1.4482 9817	0.6904 6556	15	1.5579 6742	0.6418 6195
16	1.4845 0562	0.6736 2493	16	1.6047 0644	0.6231 6694
17	1.5216 1826	0.6571 9506	17	1.6528 4763	0.6050 1645
18	1.5596 5872	0.6411 6591	18	1.7024 3306	0.5873 9461
19	1.5986 5019	0.6255 2772	19	1.7535 0605	0.5702 8603
20	1.6386 1644	0.6102 7094	20	1.8061 1123	0.5536 7575
21	1.6795 8185	0.5953 8629	21	1.8602 9457	0.5375 4928
22	1.7215 7140	0.5808 6467	22	1.9161 0341	0.5218 9250
23	1.7646 1068	0.5666 9724	23	1.9735 8651	0.5066 9175
24	1.8087 2595	0.5528 7535	24	2.0327 9411	0.4919 3374
25	1.8539 4410	0.5393 9059	25	2.0937 7793	0.4776 0557
26	1.9002 9270	0.5262 3472	26	2.1565 9127	0.4636 9473
27	1.9478 0002	0.5133 9973	27	2.2212 8901	0.4501 8906
28	1.9964 9502	0.5008 7778	28	2.2879 2768	0.4370 7675
29	2.0464 0739	0.4886 6125	29	2.3565 6651	0.4243 4636
30	2.0975 6758	0.4767 4269	30	2.4272 6247	0.4119 8676
31	2.1500 0677	0.4651 1481	31	2.5000 8035	0.3999 8715
32	2.2037 5694	0.4537 7055	32	2.5750 8276	0.3883 3703
33	2.2588 5086	0.4427 0298	33	2.6523 3524	0.3770 2625
34	2.3153 2213	0.4319 0534	34	2.7319 0530	0.3660 4490
35	2.3732 0519	0.4213 7107	35	2.8138 6245	0.3553 8340
36	2.4325 3532	0.4110 9372	36	2.8982 7833	0.3450 3243
37	2.4933 4870	0.4010 6705	37	2.9852 2668	0.3349 8294
38	2.5556 8242	0.3912 8492	38	3.0747 8348	0.3252 2615
39	2.6195 7448	0.3817 4139	39	3.1670 2698	0.3157 5355
40	2.6850 6384	0.3724 3062	40	3.2620 3779	0.3065 5684
41	2.7521 9043	0.3633 4695	41	3.3598 9893	0.2976 2800
42	2.8209 9520	0.3544 8483	42	3.4606 9589	0.2889 5922
43	2.8915 2008	0.3458 3886	43	3.5645 1677	0.2805 4294
44	2.9638 0808	0.3374 0376	44	3.6714 5227	0.2723 7178
45	3.0379 0328	0.3291 7440	45	3.7815 9584	0.2644 3862
46	3.1138 5086	0.3211 4576	46	3.8950 4372	0.2567 3653
47	3.1916 9713	0.3133 1294	47	4.0118 9503	0.2492 5876
48	3.2714 8956	0.3056 7116	48	4.1322 5188	0.2419 9880
49	3.3532 7680	0.2982 1576	49	4.2562 1944	0.2349 5029
50	3.4371 0872	0.2909 4221	50	4.3839 0602	0.2281 0708

TABLE III

	$i = .035$			$i = .04$	
n	$(1 + i)^n$	$(1 + i)^{-n}$	n	$(1 + i)^n$	$(1 + i)^{-n}$
1	1.0350 0000	0.9661 8357	1	1.0400 0000	0.9615 3846
2	1.0712 2500	0.9335 1070	2	1.0816 0000	0.9245 5621
3	1.1087 1788	0.9019 4271	3	1.1248 6400	0.8889 9636
4	1.1475 2300	0.8714 4223	4	1.1698 5856	0.8548 0419
5	1.1876 8631	0.8419 7317	5	1.2166 5290	0.8219 2711
6	1.2292 5533	0.8135 0064	6	1.2653 1902	0.7903 1453
7	1.2722 7926	0.7859 9096	7	1.3159 3178	0.7599 1781
8	1.3168 0904	0.7594 1156	8	1.3685 6905	0.7306 9021
9	1.3628 9735	0.7337 3097	9	1.4233 1181	0.7025 8674
10	1.4105 9876	0.7089 1881	10	1.4802 4428	0.6755 6417
11	1.4599 6972	0.6849 4571	11	1.5394 5406	0.6495 8093
12	1.5110 6866	0.6617 8330	12	1.6010 3222	0.6245 9705
13	1.5639 5606	0.6394 0415	13	1.6650 7351	0.6005 7409
14	1.6186 9452	0.6177 8179	14	1.7316 7645	0.5774 7508
15	1.6753 4883	0.5968 9062	15	1.8009 4351	0.5552 6450
16	1.7339 8604	0.5767 0591	16	1.8729 8125	0.5339 0818
17	1.7946 7555	0.5572 0378	17	1.9479 0050	0.5133 7325
18	1.8574 8920	0.5383 6114	18	2.0258 1652	0.4936 2812
19	1.9225 0132	0.5201 5569	19	2.1068 4918	0.4746 4242
20	1.9897 8886	0.5025 6588	20	2.1911 2314	0.4563 8695
21	2.0594 3147	0.4855 7090	21	2.2787 6807	0.4388 3360
22	2.1315 1158	0.4691 5063	22	2.3699 1879	0.4219 5539
23	2.2061 1448	0.4532 8563	23	2.4647 1554	0.4057 2633
24	2.2833 2849	0.4379 5713	24	2.5633 0416	0.3901 2147
25	2.3632 4498	0.4231 4699	25	2.6658 3633	0.3751 1680
26	2.4459 5856	0.4088 3767	26	2.7724 6978	0.3606 8923
27	2.5315 6711	0.3950 1224	27	2.8833 6858	0.3468 1657
28	2.6201 7196	0.3816 5434	28	2.9987 0332	0.3334 7747
29	2.7118 7798	0.3687 4815	29	3.1186 5145	0.3206 5141
30	2.8067 9370	0.3562 7841	30	3.2433 9751	0.3083 1867
31	2.9050 3148	0.3442 3035	31	3.3731 3341	0.2964 6026
32	3.0067 0759	0.3325 8971	32	3.5080 5875	0.2850 5794
33	3.1119 4235	0.3213 4271	33	3.6483 8110	0.2740 9417
34	3.2208 6033	0.3104 7605	34	3.7943 1634	0.2635 5209
35	3.3335 9045	0.2999 7686	35	3.9460 8899	0.2534 1547
36	3.4502 6611	0.2898 3272	36	4.1039 3255	0.2436 6872
37	3.5710 2543	0.2800 3161	37	4.2680 8986	0.2342 9685
38	3.6960 1132	0.2705 6194	38	4.4388 1345	0.2252 8543
39	3.8253 7171	0.2614 1250	39	4.6163 6599	0.2166 2061
40	3.9592 5972	0.2525 7247	40	4.8010 2063	0.2082 8904
41	4.0978 3381	0.2440 3137	41	4.9930 6145	0.2002 7793
42	4.2412 5799	0.2357 7910	42	5.1927 8391	0.1925 7493
43	4.3897 0202	0.2278 0590	43	5.4004 9527	0.1851 6820
44	4.5433 4160	0.2201 0231	44	5.6165 1508	0.1780 4635
45	4.7023 5855	0.2126 5924	45	5.8411 7568	0.1711 9841
46	4.8669 4110	0.2054 6787	46	6.0748 2271	0.1646 1386
47	5.0372 8404	0.1985 1968	47	6.3178 1562	0.1582 8256
48	5.2135 8898	0.1918 0645	48	6.5705 2824	0.1521 9476
49	5.3960 6459	0.1853 2024	49	6.8333 4937	0.1463 4112
50	5.5849 2686	0.1790 5337	50	7.1066 8335	0.1407 1262

TABLE III

	$i = .045$			$i = .05$	
n	$(1 + i)^n$	$(1 + i)^{-n}$	n	$(1 + i)^n$	$(1 + i)^{-n}$
1	1.0450 0000	0.9569 3780	1	1.0500 0000	0.9523 8095
2	1.0920 2500	0.9157 2995	2	1.1025 0000	0.9070 2948
3	1.1411 6613	0.8762 9660	3	1.1576 2500	0.8638 3760
4	1.1925 1860	0.8385 6134	4	1.2155 0625	0.8227 0247
5	1.2461 8194	0.8024 5105	5	1.2762 8156	0.7835 2617
6	1.3022 6012	0.7678 9574	6	1.3400 9564	0.7462 1540
7	1.3608 6183	0.7348 2846	7	1.4071 0042	0.7106 8133
8	1.4221 0061	0.7031 8513	8	1.4774 5544	0.6768 3936
9	1.4860 9514	0.6729 0443	9	1.5513 2822	0.6446 0892
10	1.5529 6942	0.6439 2768	10	1.6288 9463	0.6139 1325
11	1.6228 5305	0.6161 9874	11	1.7103 3936	0.5846 7929
12	1.6958 8143	0.5896 6386	12	1.7958 5633	0.5568 3742
13	1.7721 9610	0.5642 7164	13	1.8856 4914	0.5303 2135
14	1.8519 4492	0.5399 7286	14	1.9799 3160	0.5050 6795
15	1.9352 8244	0.5167 2044	15	2.0789 2818	0.4810 1710
16	2.0223 7015	0.4944 6932	16	2.1828 7459	0.4581 1152
17	2.1133 7681	0.4731 7639	17	2.2920 1832	0.4362 9669
18	2.2084 7877	0.4528 0037	18	2.4066 1923	0.4155 2065
19	2.3078 6031	0.4333 0179	19	2.5269 5020	0.3957 3396
20	2.4117 1402	0.4146 4286	20	2.6532 9771	0.3768 8948
21	2.5202 4116	0.3967 8743	21	2.7859 6259	0.3589 4236
22	2.6336 5201	0.3797 0089	22	2.9252 6072	0.3418 4987
23	2.7521 6635	0.3633 5013	23	3.0715 2376	0.3255 7131
24	2.8760 1383	0.3477 0347	24	3.2250 9994	0.3100 6791
25	3.0054 3446	0.3327 3060	25	3.3863 5494	0.2953 0277
26	3.1406 7901	0.3184 0248	26	3.5556 7269	0.2812 4073
27	3.2820 0956	0.3046 9137	27	3.7334 5632	0.2678 4832
28	3.4296 9999	0.2915 7069	28	3.9201 2914	0.2550 9364
29	3.5840 3649	0.2790 1502	29	4.1161 3560	0.2429 4632
30	3.7453 1813	0.2670 0002	30	4.3219 4238	0.2313 7745
31	3.9138 5745	0.2555 0241	31	4.5380 3949	0.2203 5947
32	4.0899 8104	0.2444 9991	32	4.7649 4147	0.2098 6617
33	4.2740 3018	0.2339 7121	33	5.0031 8854	0.1998 7254
34	4.4663 6154	0.2238 9589	34	5.2533 4797	0.1903 5480
35	4.6673 4781	0.2142 5444	35	5.5160 1537	0.1812 9029
36	4.8773 7846	0.2050 2817	36	5.7918 1614	0.1726 5741
37	5.0968 6049	0.1961 9921	37	6.0814 0694	0.1644 3563
38	5.3262 1921	0.1877 5044	38	6.3854 7729	0.1566 0536
39	5.5658 9908	0.1796 6549	39	6.7047 5115	0.1491 4797
40	5.8163 6454	0.1719 2870	40	7.0399 8871	0.1420 4568
41	6.0781 0094	0.1645 2507	41	7.3919 8815	0.1352 8160
42	6.3516 1548	0.1574 4026	42	7.7615 8756	0.1288 3962
43	6.6374 3818	0.1506 6054	43	8.1496 6693	0.1227 0440
44	6.9361 2290	0.1441 7276	44	8.5571 5028	0.1168 6133
45	7.2482 4843	0.1379 6437	45	8.9850 0779	0.1112 9651
46	7.5744 1961	0.1320 2332	46	9.4342 5818	0.1059 9668
47	7.9152 6849	0.1263 3810	47	9.9059 7109	0.1009 4921
48	8.2714 5557	0.1208 9771	48	10.4012 6965	0.0961 4211
49	8.6436 7107	0.1156 9158	49	10.9213 3313	0.0915 6391
50	9.0326 3627	0.1107 0965	50	11.4673 9979	0.0872 0373

TABLE III

	$i = .06$			$i = .08$	
n	$(1 + i)^n$	$(1 + i)^{-n}$	n	$(1 + i)^n$	$(1 + i)^{-n}$
1	1.0600 0000	0.9433 9623	1	1.0800 0000	0.9259 2593
2	1.1236 0000	0.8899 9644	2	1.1664 0000	0.8573 3882
3	1.1910 1600	0.8396 1928	3	1.2597 1200	0.7938 3224
4	1.2624 7696	0.7920 9366	4	1.3604 8896	0.7350 2985
5	1.3382 2558	0.7472 5817	5	1.4693 2808	0.6805 8320
6	1.4185 1911	0.7049 6054	6	1.5868 7432	0.6301 6963
7	1.5036 3026	0.6650 5711	7	1.7138 2427	0.5834 9040
8	1.5938 4807	0.6274 1237	8	1.8509 3021	0.5402 6888
9	1.6894 7896	0.5918 9846	9	1.9990 0463	0.5002 4897
10	1.7908 4770	0.5583 9478	10	2.1589 2500	0.4631 9349
11	1.8982 9856	0.5267 8753	11	2.3316 3900	0.4288 8286
12	2.0121 9647	0.4969 6936	12	2.5181 7012	0.3971 1376
13	2.1329 2826	0.4688 3902	13	2.7196 2373	0.3676 9792
14	2.2609 0396	0.4423 0096	14	2.9371 9362	0.3404 6104
15	2.3965 5819	0.4172 6506	15	3.1721 6911	0.3152 4170
16	2.5403 5168	0.3936 4628	16	3.4259 4264	0.2918 9047
17	2.6927 7279	0.3713 6442	17	3.7000 1805	0.2702 6895
18	2.8543 3915	0.3503 4379	18	3.9960 1950	0.2502 4903
19	3.0255 9950	0.3305 1301	19	4.3157 0106	0.2317 1206
20	3.2071 3547	0.3118 0473	20	4.6609 5714	0.2145 4821
21	3.3995 6360	0.2941 5540	21	5.0338 3372	0.1986 5575
22	3.6035 3742	0.2775 0510	22	5.4365 4041	0.1839 4051
23	3.8197 4966	0.2617 9726	23	5.8714 6365	0.1703 1528
24	4.0489 3464	0.2469 7855	24	6.3411 8074	0.1576 9934
25	4.2918 7072	0.2329 9863	25	6.8484 7520	0.1460 1790
26	4.5493 8296	0.2198 1003	26	7.3963 5321	0.1352 0176
27	4.8223 4594	0.2073 6795	27	7.9880 6147	0.1251 8682
28	5.1116 8670	0.1956 3014	28	8.6271 0639	0.1159 1372
29	5.4183 8790	0.1845 5674	29	9.3172 7490	0.1073 2752
30	5.7434 9117	0.1741 1013	30	10.0626 5689	0.0993 7733
31	6.0881 0064	0.1642 5484	31	10.8676 6944	0.0920 1605
32	6.4533 8668	0.1549 5740	32	11.7370 8300	0.0852 0005
33	6.8405 8988	0.1461 8622	33	12.6760 4964	0.0788 8893
34	7.2510 2528	0.1379 1153	34	13.6901 3361	0.0730 4531
35	7.6860 8679	0.1301 0522	35	14.7853 4429	0.0676 3454
36	8.1472 5200	0.1227 4077	36	15.9681 7184	0.0626 2458
37	8.6360 8712	0.1157 9318	37	17.2456 2558	0.0579 8572
38	9.1542 5235	0.1092 3885	38	18.6252 7563	0.0536 9048
39	9.7035 0749	0.1030 5552	39	20.1152 9768	0.0497 1341
40	10.2857 1794	0.0972 2219	40	21.7245 2150	0.0460 3093
41	10.9028 6101	0.0917 1905	41	23.4624 8322	0.0426 2123
42	11.5570 3267	0.0865 2740	42	25.3394 8187	0.0394 6411
43	12.2504 5463	0.0816 2962	43	27.3666 4042	0.0365 4084
44	12.9854 8191	0.0770 0908	44	29.5559 7166	0.0338 3411
45	13.7646 1083	0.0726 5007	45	31.9204 4939	0.0313 2788
46	14.5904 8748	0.0685 3781	46	34.4740 8534	0.0290 0730
47	15.4659 1673	0.0646 5831	47	37.2320 1217	0.0268 5861
48	16.3938 7173	0.0609 9840	48	40.2105 7314	0.0248 6908
49	17.3775 0403	0.0575 4566	49	43.4274 1899	0.0230 2693
50	18.4201 5427	0.0542 8836	50	46.9016 1251	0.0213 2123

APPENDIX B   Table of Integrals
TABLE IV

Basic Forms

1. $\int du = u + C$

2. $\int af(u)\, du = a \int f(u)\, du$

3. $\int [f(u) \pm g(u)]\, du = \int f(u)\, du \pm \int g(u)\, du$

4. $\int u^n\, du = \dfrac{u^{n+1}}{n+1} + C$ if $n \neq -1$

5. $\int \dfrac{1}{u}\, du = \ln |u| + C$

6. $\int e^u\, du = e^u + C$

7. $\int f(x)g'(x)\, dx = f(x)g(x) - \int g(x)f'(x)\, dx$ $\left(\int u\, dv = uv - \int v\, du \right)$

8. $\int (au + b)^n\, du = \dfrac{1}{a} \dfrac{(au + b)^{n+1}}{n+1} + C$ $(n \neq -1)$

Integrals Involving $au + b$

9. $\int \dfrac{1}{au + b}\, du = \dfrac{1}{a} \ln |au + b| + C$

10. $\int \dfrac{u}{au + b}\, du = \dfrac{u}{a} - \dfrac{b}{a^2} \ln |au + b| + C$

11. $\int \dfrac{u^2}{au + b}\, du = \dfrac{1}{a^3} \left[\dfrac{1}{2}(au + b)^2 - 2b(au + b) + b^2 \ln |au + b| \right] + C$

12. $\int \dfrac{u\, du}{(au + b)^2} = \dfrac{1}{a^2} \left[\dfrac{b}{au + b} + \ln |au + b| \right] + C$

13. $\int \dfrac{u^2\, du}{(au + b)^2} = \dfrac{1}{a^3} \left[(au + b) - \dfrac{b^2}{au + b} - 2b \ln |au + b| \right] + C$

14. $\int \dfrac{du}{u(au + b)} = \dfrac{1}{b} \ln \left| \dfrac{u}{au + b} \right| + C$

15. $\int \dfrac{du}{u^2(au + b)} = \dfrac{-1}{bu} + \dfrac{a}{b^2} \ln \left| \dfrac{au + b}{u} \right| + C$

16. $\int \dfrac{du}{u(au + b)^2} = \dfrac{1}{b(au + b)} + \dfrac{1}{b^2} \ln \left| \dfrac{u}{au + b} \right| + C$

17. $\int \dfrac{du}{(au + b)(cu + d)} = \dfrac{1}{bc - ad} \ln \left| \dfrac{cu + d}{au + b} \right| + C$ if $bc - ad \neq 0$

18. $\int \dfrac{u\, du}{(au + b)(cu + d)} = \dfrac{1}{bc - ad} \left\{ \dfrac{b}{a} \ln |au + b| - \dfrac{d}{c} \ln |cu + d| \right\} + C$ (if $bc - ad \neq 0$)

Integrals Involving $\sqrt{au + b}$

19. $\displaystyle\int \sqrt{au + b}\, du = \frac{2}{3a}(au + b)^{3/2} + C$

20. $\displaystyle\int u\sqrt{au + b}\, du = \frac{2(3au - 2b)(au + b)^{3/2}}{15a^2} + C$

21. $\displaystyle\int u^2\sqrt{au + b}\, du = \frac{2}{105a^3}(15a^2u^2 - 12abu + 8b^2)(au + b)^{3/2} + C$

22. $\displaystyle\int u^n\sqrt{au + b}\, du = \frac{2u^n(au + b)^{3/2}}{a(2n + 3)} - \frac{2bn}{a(2n + 3)}\int u^{n-1}\sqrt{au + b}\, du + C$

23. $\displaystyle\int \frac{u\, du}{\sqrt{au + b}} = \frac{2}{3a^2}(au - 2b)\sqrt{au + b} + C$

24. $\displaystyle\int \frac{u^2\, du}{\sqrt{au + b}} = \frac{2}{15a^3}(3a^2u^2 - 4abu + 8b^2)\sqrt{au + b} + C$

25. $\displaystyle\int \frac{u^n\, du}{\sqrt{au + b}} = \frac{2u^n\sqrt{au + b}}{a(2n + 1)} - \frac{2bn}{a(2n + 1)}\int \frac{u^{n-1}\, du}{\sqrt{au + b}}$

26. $\displaystyle\int \frac{du}{u\sqrt{au + b}} = \frac{1}{\sqrt{b}}\ln\left|\frac{\sqrt{au + b} - \sqrt{b}}{\sqrt{au + b} + \sqrt{b}}\right| + C \qquad (\text{if } b > 0)$

27. $\displaystyle\int \frac{\sqrt{au + b}\, du}{u} = 2\sqrt{au + b} + b\int \frac{du}{u\sqrt{au + b}}$

Integrals Involving $u^2 \pm a^2$, $a > 0$

Note: When the symbol "$\pm$" is used several times in the same formula, it is understood to be "$+$" *throughout* the formula or "$-$" throughout the formula.

28. $\displaystyle\int \frac{du}{u^2 - a^2} = \frac{1}{2a}\ln\left|\frac{u - a}{u + a}\right| + C$

29. $\displaystyle\int \frac{du}{a^2 - u^2} = \frac{1}{2a}\ln\left|\frac{u + a}{u - a}\right| + C$

30. $\displaystyle\int \frac{u^2\, du}{\sqrt{u^2 \pm a^2}} = \frac{u}{2}\sqrt{u^2 \pm a^2} - \frac{\pm a^2}{2}\ln\left|u + \sqrt{u^2 \pm a^2}\right| + C$

31. $\displaystyle\int \frac{du}{\sqrt{u^2 \pm a^2}} = \ln\left|u + \sqrt{u^2 \pm a^2}\right| + C$

32. $\displaystyle\int \sqrt{u^2 \pm a^2}\, du = \frac{u}{2}\sqrt{u^2 \pm a^2} \pm \frac{a^2}{2}\ln\left|u + \sqrt{u^2 \pm a^2}\right| + C$

33. $\displaystyle\int u^2\sqrt{u^2 \pm a^2}\, du = \frac{u}{8}(2u^2 \pm a^2)\sqrt{u^2 \pm a^2} - \frac{a^4}{8}\ln\left|u + \sqrt{u^2 \pm a^2}\right| + C$

34. $\displaystyle\int \frac{\sqrt{a^2 \pm u^2}\, du}{u} = \sqrt{a^2 \pm u^2} - a\ln\left|\frac{a + \sqrt{a^2 \pm u^2}}{u}\right| + C$

35. $\int \dfrac{\sqrt{u^2 \pm a^2}\, du}{u^2} = \dfrac{-\sqrt{u^2 \pm a^2}}{u} + \ln\left|u + \sqrt{u^2 \pm a^2}\right| + C$

36. $\int \dfrac{du}{u\sqrt{a^2 \pm u^2}} = \dfrac{1}{a}\ln\left|\dfrac{u}{a + \sqrt{a^2 \pm u^2}}\right| + C$

37. $\int \dfrac{du}{u^2\sqrt{u^2 \pm a^2}} = \dfrac{-\sqrt{u^2 \pm a^2}}{\pm a^2 u} + C$

38. $\int \dfrac{du}{u^2\sqrt{a^2 - u^2}} = \dfrac{-\sqrt{a^2 - u^2}}{a^2 u} + C$

Integrals Involving Exponential and Logarithmic Functions

39. $\int e^{au}\, du = \dfrac{1}{a}\, e^{au} + C$

40. $\int u e^{au}\, du = \dfrac{e^{au}}{a^2}(au - 1) + C$

41. $\int a^u\, du = \dfrac{a^u}{\ln a} + C,$ for $a > 0,\ a \neq 1$

42. $\int u^n e^{ku}\, du = \dfrac{u^n e^{ku}}{k} - \dfrac{n}{k}\int u^{n-1} e^{ku}\, du$ $(k \neq 0)$

43. $\int \dfrac{e^{ku}\, du}{u^n} = -\dfrac{e^{ku}}{(n-1)u^{n-1}} + \dfrac{k}{n-1}\int \dfrac{e^{ku}\, du}{u^{n-1}}$ $(n \neq 1)$

44. $\int \ln u\, du = u \ln u - u + C$

45. $\int u^n \ln u\, du = \dfrac{u^{n+1}}{(n+1)^2}\left[(n+1)\ln u - 1\right] + C$

Integrals Involving Trigonometric Functions

46. $\int \sin u\, du = -\cos u + C$

47. $\int \cos u\, du = \sin u + C$

48. $\int \tan u\, du = -\ln|\cos u| + C$

49. $\int \cot u\, du = \ln|\sin u| + C$

50. $\int \sec u\, du = \ln|\sec u + \tan u| + C$

51. $\int \csc u\, du = -\ln|\csc u + \cot u| + C$

52. $\int \sec^2 u\, du = \tan u + C$

53. $\int \csc^2 u\, du = -\cot u + C$

54. $\int \sec u \tan u\, du = \sec u + C$

55. $\int \csc u \cot u \, du = -\csc u + C$

56. $\int \sin^2 u \, du = \dfrac{u}{2} - \dfrac{\sin 2u}{4} + C$

57. $\int \cos^2 u \, du = \dfrac{u}{2} + \dfrac{\sin 2u}{4} + C$

58. $\int \sin^n u \, du = \dfrac{-1}{n} \sin^{n-1} u \cos u + \dfrac{n-1}{n} \int \sin^{n-2} u \, du$

59. $\int \cos^n u \, du = \dfrac{1}{n} \cos^{n-1} u \sin u + \dfrac{n-1}{n} \int \cos^{n-2} u \, du$

60. $\int \tan^n u \, du = \dfrac{1}{n-1} \tan^{n-1} u - \int \tan^{n-2} u \, du$

61. $\int \cot^n u \, du = \dfrac{-1}{n-1} \cot^{n-1} u - \int \cot^{n-2} u \, du$

Areas under the Standard Normal Curve

$$y = \frac{1}{\sqrt{2\pi}} e^{-t^2/2}$$

The table gives the area $A(x)$ of the shaded region, for $x > 0$. To obtain $A(-x)$ use the formula $A(-x) = 1 - A(x)$. See Section 11.4 for explanation.

$$A(x) = \int_{-\infty}^{x} \frac{1}{\sqrt{2\pi}} e^{-t^2/2} \, dt.$$

x	0	1	2	3	4	5	6	7	8	9
0.0	.5000	.5040	.5080	.5120	.5160	.5199	.5239	.5279	.5319	.5359
0.1	.5398	.5438	.5478	.5517	.5557	.5596	.5636	.5675	.5714	.5753
0.2	.5793	.5832	.5871	.5910	.5948	.5987	.6026	.6064	.6103	.6141
0.3	.6179	.6217	.6255	.6293	.6331	.6368	.6406	.6443	.6480	.6517
0.4	.6554	.6591	.6628	.6664	.6700	.6736	.6772	.6808	.6844	.6879
0.5	.6915	.6950	.6985	.7019	.7054	.7088	.7123	.7157	.7190	.7224
0.6	.7257	.7291	.7324	.7357	.7389	.7422	.7454	.7486	.7517	.7549
0.7	.7580	.7611	.7642	.7673	.7704	.7734	.7764	.7794	.7823	.7852
0.8	.7881	.7910	.7939	.7967	.7995	.8023	.8051	.8078	.8106	.8133
0.9	.8159	.8186	.8212	.8238	.8264	.8289	.8315	.8340	.8365	.8389
1.0	.8413	.8438	.8461	.8485	.8508	.8531	.8554	.8577	.8599	.8621
1.1	.8643	.8665	.8686	.8708	.8729	.8749	.8770	.8790	.8810	.8830
1.2	.8849	.8869	.8888	.8907	.8925	.8944	.8962	.8980	.8997	.9015
1.3	.9032	.9049	.9066	.9082	.9099	.9115	.9131	.9147	.9162	.9177
1.4	.9192	.9207	.9222	.9236	.9251	.9265	.9279	.9292	.9306	.9319
1.5	.9332	.9345	.9357	.9370	.9382	.9394	.9406	.9418	.9429	.9441
1.6	.9452	.9463	.9474	.9484	.9495	.9505	.9515	.9525	.9535	.9545
1.7	.9554	.9564	.9573	.9582	.9591	.9599	.9608	.9616	.9625	.9633
1.8	.9641	.9649	.9656	.9664	.9671	.9678	.9686	.9693	.9699	.9706
1.9	.9713	.9719	.9726	.9732	.9738	.9744	.9750	.9756	.9761	.9767
2.0	.9772	.9778	.9783	.9788	.9793	.9798	.9803	.9808	.9812	.9817
2.1	.9821	.9826	.9830	.9834	.9838	.9842	.9846	.9850	.9854	.9857
2.2	.9861	.9864	.9868	.9871	.9875	.9878	.9881	.9884	.9887	.9890
2.3	.9893	.9896	.9898	.9901	.9904	.9906	.9909	.9911	.9913	.9916
2.4	.9918	.9920	.9922	.9925	.9927	.9929	.9931	.9932	.9934	.9936
2.5	.9938	.9940	.9941	.9943	.9945	.9946	.9948	.9949	.9951	.9952
2.6	.9953	.9955	.9956	.9957	.9959	.9960	.9961	.9962	.9963	.9964
2.7	.9965	.9966	.9967	.9968	.9969	.9970	.9971	.9972	.9973	.9974
2.8	.9974	.9975	.9976	.9977	.9977	.9978	.9979	.9979	.9980	.9981
2.9	.9981	.9982	.9982	.9983	.9984	.9984	.9985	.9985	.9986	.9986
3.0	.9987	.9987	.9987	.9988	.9988	.9989	.9989	.9989	.9990	.9990
3.1	.9990	.9991	.9991	.9991	.9992	.9992	.9992	.9992	.9993	.9993
3.2	.9993	.9993	.9994	.9994	.9994	.9994	.9994	.9995	.9995	.9995
3.3	.9995	.9995	.9995	.9996	.9996	.9996	.9996	.9996	.9996	.9997
3.4	.9997	.9997	.9997	.9997	.9997	.9997	.9997	.9997	.9997	.9998

ANSWERS TO ODD-NUMBERED EXERCISES AND CHAPTER TESTS

|||||||||||| **CHAPTER 1**

1.1 Exercise Set, p. 8

1. (a) 7 (b) -3 (c) 22 (d) $-5a + 7$ (e) $-5a + 2$ (f) $-5a - 5h + 7$ (g) $5x + 7$ (h) $5x - 7$
(i) $-4x + 7$

3. (a) 1 (b) $-\frac{5}{7}$ (c) $\frac{17}{11}$ (d) -1 (e) $\dfrac{a^2 + 1}{3a - 1}$ (f) $\dfrac{a^2 - 4a + 5}{3a - 7}$ (g) $\dfrac{x^2 + 1}{-3x - 1}$ (h) $\dfrac{x^4 + 1}{3x^2 - 1}$ (i) $\dfrac{3x - 1}{x^2 + 1}$

5. (a) $C = \frac{5}{9}(F - 32)$ (b) (i) $-\frac{140}{9}°$ (ii) $-\frac{160}{9}°$ (iii) $-\frac{70}{3}°$ (iv) $0°$ (v) $37°$ (vi) $100°$ 7. all real numbers
9. all real numbers greater than or equal to 1   11. all real numbers except -1 13. all real numbers
15. all real numbers x for which $|x| \geq 2$; i.e. $x \geq 2$ or $x \leq -2$.

17.

19.

21.

23.

25. is a function of x 27. is a function of x 29. not a function of x

1.2 Exercise Set, p. 19

1.

3.

5.

7.

9.

11. (a) 53   (b) 125   (c) $2x^3 - 1$ (d) $(2x - 1)^3$ (e) $4x - 3$ (f) x^9

13. (a) $\frac{3}{5}$ (b) 11   (c) $\dfrac{3}{x^2 + 1}$ (d) $\dfrac{2x^2 - 4x + 11}{x^2 - 2x + 1}$ (e) does not exist   (f) 38

15. $f(x) = x^8$, $g(x) = 5x - 3$, $h(x) = (f \circ g)(x)$ 17. $f(x) = 3x + 11$, $g(x) = \sqrt{x + 2}$, $h(x) = (f \circ g)(x)$

19. $f(x) = x^{1/3}$, $g(x) = \dfrac{3x - 5}{x + 4}$, $h(x) = (f \circ g)(x)$

21. (a) [0, 8]   (b) \$600, \$550   (c) falling, rising, falling   (d) waiting till later
(e) at the end of day 8, or the beginning of day 1   (f) at the end of day 3

23. (a) $C(w) = \begin{cases} 0.25, & 0 < w \le 1 \\ 0.45, & 1 < w \le 2 \\ 0.65, & 2 < w \le 3 \end{cases}$ (b)   (c) \$0.65

25. (a) $C(x) = 4x^2 - 2x + 2438$ (b) \$12,338 27. $R(x) = \begin{cases} 300x + 300{,}000 & \text{, if } x \le 100 \\ -2x^2 + 500x + 300{,}000 & \text{, if } 100 < x < 150 \\ 200x + 300{,}000 & \text{, if } x \ge 150 \end{cases}$

29. $A(x) = 1.07x$

1.3 Exercise Set, p. 32

1. (a) $\frac{5}{6}$
   (b) $-\frac{5}{9}$
   (c) no slope
   (d) 0

3. (a) $y = -2x - 2$

(b) $y = \frac{3}{2}x + 2$

5. (a) 3 (b) 0 (c) $\frac{3}{2}$ 7. (a) rises (b) falls (c) rises (d) falls
9. (a) increases by 2 units   (b) increases by 6 units

11. (a)

not parallel

(b)

coincident
(parallel)

(c)

not parallel

13. (a)   (b)   (c)

15. (a) $y - 5 = \frac{3}{4}(x - 3)$ or $y - 2 = \frac{3}{4}(x + 1)$ (b) $y = \frac{4}{3}x$ 17. $x + 2y = -3$ or $y = -\frac{1}{2}x - \frac{3}{2}$

19. (a) $y = -\frac{3}{4}x + \frac{5}{2}$ (b) $y = 3$ (c) $y = \frac{2}{3}x$ (d) $x = -2$

21. (a) $2x - y = 3$; $A = 2$, $B = -1$, $C = 3$ (b) $3x - 2y = 8$; $A = 3$, $B = -2$, $C = 8$
    (c) $y = 3$; $A = 0$, $B = 1$, $C = 3$ (d) $4x - 3y = -4$; $A = 4$, $B = -3$, $C = -4$

23. (a) intersect at only one point (b) parallel (c) identical (d) intersect at only one point

1.4 Exercise Set, p. 46

1. (a) $S = 8000 + (0.12)(8000)t = 8000 + 960t$ (b)   (c) \$13,760
   (d) \$8,720

3. $r = 0.06$ (6%)   5. (a) \$47.50 (b) \$43.96 (c) \$73.13

7. $S = 1.04x$, 4% profit

9. (a) $V = 400,000 - 40,000t$ (b)   (c) \$200,000 (d) 10 years

11. $r = 0.2$ (20%)   13. (a)

(b) $x = 50$ million tons
(c) $R(50) = \$40$ million

15. $P(x) = 40x - 24,000$; $P(560) = -1600$. Thus, the manufacturer lost $1600.

17. (a) $N = 0.08t + 3.4$ (b) 4.6 million   19. (a) $A = 0.1w + 0.5$ (b) 6.5 kg.   21. $y = \frac{9}{5}x - \frac{19}{5}$
23. $y = \frac{9}{5}x - \frac{17}{5}$ 25. (a) $y = \frac{1}{1000}(\frac{3036}{8700}x + \frac{23108}{8700})$ (b) 0.03895 ppm

1.5 Exercise Set, p. 58

1.

Vertex $= (\frac{1}{2}, -\frac{1}{4})$
Axis of symmetry:
$x = \frac{1}{2}$

3.

Vertex: $(\frac{1}{2}, 6\frac{1}{4})$
Axis of symmetry:
$x = \frac{1}{2}$

5.

Vertex: $(2, 7)$
Axis of symmetry:
$x = 2$

7.

Vertex: $(2, 1)$
Axis of symmetry:
$x = 2$

9. $D = 0$, 1 x-intercept   11. $D = -3$, 0 x-intercepts   13. $D = 33$, 2 x-intercepts

15. $D = 0$, 1 x-intercept   17. (a) $A = (12 - 2x)(16 - 2x) = 192 - 56x + 4x^2$ (b) $A = 56x - 4x^2$

19. (a) $A = 50l - l^2$ (b) $l = w = 25$ ft.

21. (a) $P(x) = x(100 - x) = 100x - x^2$

   (b) The graph of $y = 100x - x^2$ is a parabola with the highest point at $x = 50$. The two numbers are 50 and 50. The largest product is $P(50) = 2500$.

23. (a) $P(x) = -5x^2 + 850x - 30,000$ (b) $P(70) = \$50$ (c) 85¢

25. (a) \$20,000   (b) $R(x) = -100x^2 + 1700x + 20,000$ (c) $P(x) = -100x^2 + 1700x$

   (d) $x = \$8.50$, so price per ticket = \$16.50; Attendance = 1650; profit = \$7225

1.6 Exercise Set, p. 65

1. $(0, 0)$, $(2, 4)$ 3. $(3, 12)$, $(1, 4)$ 5. $(-3, 6)$, $(1, 2)$ 7. $(2, 4)$, $(-2, 4)$ 9. $(1, 1)$

11. $(4, 3)$, $(-4, -3)$ 13. $(2, 1)$, $(-1, -2)$ 15. $(3, 1)$, $(-3, 1)$, $(6\sqrt{2}, 8)$, $(-6\sqrt{2}, 8)$

17. $(1, 3)$, $(1, -3)$, $(-1, 3)$, $(-1, -3)$

19. The equation $x^2 - x = x - 2$ is equivalent to $x^2 - 2x + 2 = 0$. The discriminant of this equation is negative, so it has no real number solution.

21. (a) $C(x) = 1 + \sqrt{x + 5}$ (thousands), $R(x) = x$ (b) 4 units   23. 3 million pens at 15 cents each.

Review Exercises, Chapter 1, p. 67

1. (a) 1   (b) -2 (c) 10   (d) $12x^2 - 2$ 3. all real numbers greater than or equal to $\frac{1}{2}$.

5. (b) only   7. $\frac{3}{2}$ 9. identical   11.   13. $y - 2 = x - 3$ 15. $y = -\frac{5}{6}x + \frac{4}{3}$

17. (b) only   19. $S(t) = 5000 + 350t$, $S(4) = \$6400$, $S(10) = \$8500$, 14.28 years or 14 years, 4 months

21. (a) $C(x) = \begin{cases} 7.98x & \text{if } 0 \le x \le 3 \\ 3.99x + 11.97 & \text{if } 3 < x \le 10 \end{cases}$ (b) \$43.89

23. (a) \$700 or 9.6%   (d)

   (b) $700t$

   (c) $V(t) = 7300 - 700t$

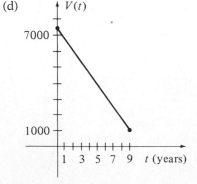

25. $(2, 1)$ and $(4, 11)$ 27. $D = 0$; 1 x-intercept

29. (a) $P(x) = -450 + 68x - 2x^2$ (b) $P(10) = \$30$, $P(13) = \$96$, $P(20) = \$110$ (c) \$17

31. $y = \frac{11}{19}x + \frac{349}{19}$

Chapter Test, Chapter 1, p. 69

1. $f(0) = -5$, $f(3) = -\frac{1}{5}$, $f(2x) = \dfrac{2x - 5}{6x + 1}$

2. $[-\frac{5}{3}, +\infty)$ or $x \geq -\frac{5}{3}$

3.

4. (a) $C(x) = 2.30x + 160$, $R(x) = 3.50x$, $P(x) = 1.20x - 160$
   (b) $C(400) = \$1080$, $R(400) = \$1400$, $P(400) = \$320$

5. $3x - 2y = 8$

6. $2x + 3y = 7$

7. (a) fall
   (b) decrease
   (c) decrease by 6 units

(d)

8. (a)

Units (weekly)

(b) break-even at $x = 50$
(c) $45,000

9. (a) $400
   (b) $v(t) = 50,000 - 4000t$
   (c)

(d) 7.5 years

10. $D = -8$, no x-intercepts

11.

Vertex $= (\frac{5}{4}, -\frac{49}{8})$

12. (a) $C(x) = 6000 - 30x$, $R(x) = 100x - \dfrac{x^2}{2}$, $P(x) = \dfrac{-x^2}{2} + 130x - 6000$ (b) \$1.30; $P(130) = \$24.50$

13. $(3, 6)$ and $(-1, 2)$

|||||||||||| CHAPTER 2

2.1 Exercise Set, p. 84

1. (a)

x	4.4	4.2	4.1	4.01	4.001
$2x - 3$	5.8	5.4	5.2	5.02	5.002

$\displaystyle\lim_{x \to 4} (2x - 3) = 5$

x	3.6	3.8	3.9	3.99	3.999
$2x - 3$	4.2	4.6	4.8	4.98	4.998

(b)

3. (a)

x	-1	-1.5	-1.9	-1.99	-1.999
$\dfrac{2x + 4}{\lvert x + 2 \rvert}$	2	2	2	2	2

x	-3	-2.5	-2.1	-2.01	-2.001
$\dfrac{2x + 4}{\lvert x + 2 \rvert}$	-2	-2	-2	-2	-2

$\displaystyle\lim_{x \to -2^-} f(x) = -2$; $\displaystyle\lim_{x \to -2^+} f(x) = 2$; $\displaystyle\lim_{x \to -2} f(x)$ does not exist

(b)

5. (a)

x	1	.5	.1	.01	.001
f(x)	1	.25	.01	.0001	.000001

x	−1	−.5	−.1	−.01	−.001
f(x)	−1	−.5	−.1	−.01	−.001

$\lim_{x \to 0^-} f(x) = 0$; $\lim_{x \to 0^+} f(x) = 0$; $\lim_{x \to 0} f(x) = 0$

(b)

7. 6 9. −2 11. 11 13. $\frac{13}{25}$ 15. 6 17. −2 19. does not exist 21. 0 23. 0
25. −1 27. does not exist 29. −1 31. does not exist 33. 3 35. $2x$

37. (a) 5 (b)

39. $\lim_{x \to -5^-} f(x) = -10$

$\lim_{x \to -5^+} f(x) = -10$

$\lim_{x \to -5} f(x) = -10$

41. $\lim_{x \to -1^-} f(x) = -\infty$

$\lim_{x \to -1^+} f(x) = +\infty$

$\lim_{x \to -1} f(x)$ does not exist

43. $\lim_{x \to a^-} f(x) = \lim_{x \to a^+} f(x) = 3$, $\lim_{x \to a} f(x) = 3$ 45. $\lim_{x \to a^-} f(x) = 3$, $\lim_{x \to a^+} f(x) = 6$, $\lim_{x \to a} f(x)$ does not exist

47. $\lim_{x \to a^-} f(x) = +\infty$, $\lim_{x \to a^+} f(x) = +\infty$, $\lim_{x \to a} f(x) = +\infty$

2.2 Exercise Set, p. 97

1. $-\infty$ 3. $-\infty$ 5. $+\infty$ 7. $-\infty$ 9. $+\infty$ 11. $+\infty$ 13. $\frac{1}{3}$ 15. $-\infty$
17. does not exist 19. 0 21. 5 23. $+\infty$ 25. 5 27. 2 29. $-\infty$ 31. $\frac{2}{5}$ 33. 0
35. $+\infty$ 37. vertical: $x = -2$; horizontal: $y = 1$ 39. vertical: $x = -1$, $x = 2$; horizontal: $y = 0$
41. vertical: $x = -4$; horizontal: $y = 0$ 43. 0

45. (a) 20; before instruction the average student types 20 words per minute
    (b) 100; with unlimited instruction the average student types 100 words per minute.

2.3 Exercise Set, p. 103

1. (a) continuous everywhere except at $x = a$ (b) continuous everywhere (c) continuous everywhere
   (d) continuous everywhere except at $x = a$
3. (a)

    (b)

    (c)

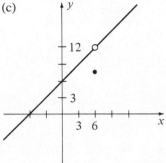

   not continuous, $f(6)$ is not defined   continuous   not continuous, $f(6) \neq \lim_{x \to 6} f(x)$

5. continuous   7. not continuous   9. not continuous   11. continuous   13. $x \neq 3$ 15. all x
17. all x 19. $x > -2$ 21. $x \neq 2, x \neq -3$
23. $w = 0, 1, 2, 3, \ldots$ The function has a "jump" at each integer.
25. (a) No (b)

    (c) $P(5.5) = 12.5$, $P(6.1) = 12.3$ The overtime initially lowers the profit even though production is increased.
27. 1

2.4 Exercise Set, p. 116

1. (a) 64 ft/sec (d) 0 ft/sec
   (b) 48 ft/sec   At the beginning of the third second (on the way up) the projectile has the same height
   (c) 32 ft/sec   as at the end of the third second (on the way down). Thus, the net change in height
                    during the third second is 0. That is, $h(3) - h(2) = 0$.
3. 3   5. 6   7. 0   9. (a) 1 (b)   (c) h (d) 0 (e) 0

11. $2x_0 - 2$ 13. (a) $\frac{20}{3}$ (b) 4   (c) 0   15. 3   17. 2
19. (a) $0.2°F$ per hour   (b) $-\frac{1}{6}°F$ per hour
    (c) The values of the temperature necessary to compute the relevant limit are unknown.
21. (a) $-\$1.50$ per gallon   (b) $\$4$ per gallon   (c) $-\$2$ per gallon   (d) $\$0$ per gallon   (e) $\$2$ per gallon
23. (a) 26 mph   (b) 44 mph   (c) 38 mph

Review Exercises, Chapter 2, p. 120

1. 7   3. $+\infty$ 5. -16 7. $-\frac{2}{5}$ 9. $\lim\limits_{x\to 1^-} f(x) = 1$, $\lim\limits_{x\to 1^+} f(x) = 1$, $\lim\limits_{x\to 1} f(x) = 1$ 11. 2

13. $f(-2) = -7$ 15. continuous   17. not continuous

19. not continuous   21. all x 23. $x \neq 0$ 25. 2   27. 12
29. (a) $\$1212$ per liter   (b) $\$612$ per liter   31. (a) 42 grams per week   (b) 120 grams per week

Chapter Test, Chapter 2, p. 122

1. (a) $\frac{5}{2}$ (b) 2   2. (a) does not exist   (b) $-\infty$ 3. (a) $-\frac{1}{2}$ (b) 0
4. vertical: $x = 2$, $x = -2$; horizontal: $y = 1$ 5. (a) continuous everywhere   (b) discontinuous at $x = a$
   (c) discontinuous at $x = c$ (d) discontinuous at $x = a$ 6. (a) no   (b) yes   (c) no
7. continuous everywhere except at $x = 4$, $x = -\frac{1}{2}$ 8. (a) $-\frac{1}{2}$ (b) -1
9. (a) 2   (b)   (c) $h - 1$ (d) -1 10. 4

|||||||||||| CHAPTER 3

3.1 Exercise Set, p. 134

1. (a) $\frac{3}{4}$ (b) 0   3. 3   5. $6x + 3h$ 7. 2   9. 0   11. (a) $2x - 2$ (e)
                                                                        (b) 2
                                                                        (c) 2
                                                                        (d) $y = 2x$

13. $f'(-2) = -11$ 15. (a) $(2, 4)$ 17. (a) $32t$ 19. (a) 20 cases per day
    $f'(0) = -3$ (b) $(\frac{3}{4}, \frac{7}{8})$ (b) 160 ft/sec   (b) 0 cases per day
    $f'(3) = 9$ (c) $(0, 2)$ (c) 4 sec   (c) -20 cases per day

21. not differentiable at $x = 1$ 23. not differentiable at $x = 1$

25. (a) At a, d, and g, where the graph has sharp corners.
    (b) At f and h. Since the function is not continuous at these points it cannot be differentiable there.

3.2 Exercise Set, p. 143

1. $\Delta x = 4, \Delta y = 20$ 3. $\Delta x = 7, \Delta y = 1$ 5. $\Delta x = 3, \Delta y = -15$ 7. 0   9. $7x^6$ 11. $18t^2$
13. $12x^3 - 2$ 15. 0   17. 2   19. $55u^4 - 28u^3$ 21. $3x^2 + 2x$ 23. $6x^2 + 6x - 2$
25. (a) 36 (b) 0 (c) 36   27. (a) 0, $-\frac{2}{3}$ (b) $-2, \frac{4}{3}$ (c) 2, $-\frac{8}{3}$
29. (a) $50 - 2x$ (b) $N'(5) = 40, N'(30) = -10$
    (c) When \$5000 is spent, the number of units sold is increasing at the rate of 40 units per thousand dollars
        spent. When \$30,000 is spent, the number of units sold is decreasing at the rate of 10 units per
        thousand dollars spent.
31. $2\pi r$

3.3 Exercise Set, p. 152

1. $40x + 11$ 3. $18x^5 + 28x^3$ 5. $3x^2 - 4x - 3$ 7. $5s^4 + 4s^3 + 9s^2 - 10s - 5$ 9. $\dfrac{-5}{(x-1)^2}$

11. $\dfrac{12u}{(u^2 + 6)^2}$ 13. $\dfrac{6x}{(x^2 - 1)^2}$ 15. $-8x^{-3}$ 17. $\dfrac{-4}{x^3} + \dfrac{12}{x^5} + 20x^3$ 19. $\frac{1}{3}x^{-2/3}$ 21. $-12x^{-5/2}$

23. $44t^{8/3} + 240t^3 - \frac{20}{3}t^{2/3} - 40t$ 25. $\dfrac{4x^{3/2} + 6x^2 + 3}{\sqrt{x}(2 + 4\sqrt{x})^2}$

27. (a) $6x + \dfrac{1}{\sqrt{x}}$ (b) $3t^2 - 4t - 3$ (c) $\dfrac{1 - 2u - u^2}{(u^2 - u)^2}$ 29. (a) $5x^4 - 8x^3 - 9x^2$ (b) $\dfrac{8x^3 - 18x^2}{(2x - 3)^2}$
31. (a) $h'(-3) = \frac{16}{49}$ (b) $h'(0) = 4$ (c) $h'(2) = \frac{1}{4}$ 33. (a) -1 (b) $-\frac{1}{4}$ (c) $-\frac{1}{4}$ 35. 64 years

3.4 Exercise Set, p. 159

1. (a) $MC = 40 - \dfrac{x}{10}$, $MR = 80 + \dfrac{x}{5}$, $MP = 40 + \dfrac{3x}{10}$ (b) \$37 per cycle (c) \$86 per cycle (d) \$37 (e) \$86

3. (a) $P(x) = 40x + \dfrac{3x^2}{20} - 2000$, $MP = 40 + \dfrac{3x}{10}$ (b) \$52 per unit (c) \$52

5. (a) $R(x) = \frac{1}{5}x^2 + 80x$, $MR = \frac{2}{5}x + 80$ (b) cost $\approx$ \$10, revenue $\approx$ \$120, profit $\approx$ \$110

7. 16 cameras per month   9. (a) \$1.20 per racket (b) 100 rackets

11. (a) $100x - \dfrac{x^2}{20}$ (b) \$20 per calculator (c) $80 - \dfrac{x}{10}$

(d) \$60 per calculator, \$0 per calculator, $-\$920$ per calculator

(e) The maximum profit occurs when 800 calculators are sold per week.

3.5 Exercise Set, p. 167

1. $y = u^6$, $u = 3x + 2$ 3. $y = u^{1/3}$, $u = x^3 - 2x^2$ 5. $120x(3x^2 + 1)^{19}$ 7. $\dfrac{-1}{2\sqrt{4 - x}}$

9. $100x(2 - 5x^2)^{-11}$ 11. $2x(2x + 3)^4(7x + 3)$ 13. $\dfrac{2x^2(5 - 3x)}{\sqrt{4x - 2x^2}}$

15. $5(4x + 9)^9(5x + 2)^6(68x - 47)$ 17. $\frac{1}{5}(15x^2 + 8x - 2)(5x^3 + 4x^2 - 2x + 4)^{-4/5}$

19. $-\dfrac{2x + 4}{3(x^2 + 4x - 3)^{4/3}}$ 21. $20x^2(2x^5 + 6)^9(3x^3 - 5)^{19}(33x^5 - 25x^2 + 54)$ 23. $\dfrac{7}{(x + 5)^2}$

25. $y = -\frac{4}{9}x + \frac{20}{9}$ 27. $\frac{12}{5}$

3.6 Exercise Set, p. 173

1. \$1 per week increase   3. (a) π ft/sec^2 (b) 2π ft/sec^2 (c) 3π ft/sec^2

5. (a) $\dfrac{32}{9\pi}$ ft/min (b) $\dfrac{8}{9\pi}$ ft/min (c) $\dfrac{2}{9\pi}$ ft/min   7. (a) $\dfrac{1}{3\pi}$ in/sec (b) $\dfrac{4}{3\pi}$ in/sec

9. $1600.8 \approx 1601$ eagles per year   11. 5 units per month increase

3.7 Exercise Set, p. 178

1. $-\dfrac{2x + 3}{2y}$ 3. $-\dfrac{6xy + 2y^3}{3x^2 + 6xy^2}$ 5. $\dfrac{4x - 2xy^2}{2x^2y + 9y^2}$ 7. $-\dfrac{x}{y}$ 9. $\dfrac{y + 3}{2y - x}$ 11. $\dfrac{y^2 + 2x}{3y^2 - 2xy}$ 13. $\dfrac{2}{7}$

15. $-\frac{3}{5}$ 17. $y = -\frac{3}{4}x + \frac{25}{4}$ 19. -1.92 lbs/wk

21. (a) $\frac{3}{2}$ ft/sec (b) $\sqrt{96} \approx 9.80$ ft/sec (c) $2\sqrt{99} \approx 19.90$ ft/sec   23. $\frac{48}{13}$ ft/sec

25. 3 units per month   27. 3.4 ft/sec

Review Exercises, Chapter 3, p. 181

1. $\frac{2}{3}$ 3. -3 5. $-\dfrac{1}{x^2}$ 7. $9x^2 - 4x + 5$ 9. $\dfrac{1}{\sqrt{x}} + \dfrac{1}{x^2}$ 11. $\dfrac{2 - 4s - 3s^2}{(s^2 - s)^2}$ 13. $\dfrac{1 + 2x - x^2}{(1 + x^2)^2}$

15. (a) $4x + 1$ (d)   17. $(-2, 12)$ and $(\frac{1}{3}, -\frac{19}{27})$ 19. $-16(2x + 3)^{-9}$

(b) 13

(c) $y = 9x - 9$

21. $\dfrac{2x + 3}{2\sqrt{x^2 + 3x}}$ 23. $\frac{3}{5}(12x^3 - 6x^2 + 2x)(3x^4 - 2x^3 + x^2)^{-2/5}$

25. 17 million dollars profit per million dollars spent on research   27. (a) $8t^3 - 240t^2$ (b) 30 seconds

29. $.00048\pi$ cm^2 per month   31. (a) \$3 per pen (b) \$3   33. $\dfrac{1}{9\pi}$ in./min.   35. $\dfrac{x+y}{y-x}$

37. $y = \frac{1}{5}(7 - x)$

Chapter Test, Chapter 3, p. 184

1. -3 2. $6x_0$
3. f fails to be differentiable at $x = -3$, because the graph of f has a "corner" there; f is continuous everywhere.
4. $35x^4 - 16x - 3$ 5. 2   6. $\dfrac{-12}{(3x-11)^5}$ 7. $\dfrac{-(3x^2 + 16x + 3)}{(x^2 - 1)^2}$ 8. $\frac{8}{3}x^{1/3} + \frac{10}{3}x^{-5/3}$
9. $(3x^4 + 7x + 2)^7[(3x^4 + 7x + 2)(3x^2 - 8x) + 8(12x^3 + 7)(x^3 - 4x^2)]$
10. (a) $V = \frac{1}{4} - t^{-1/2}$ (b) 16 seconds   11. $y = \frac{14}{3}x - \frac{25}{3}$ 12. 300π in^3/sec

|||||||||||| CHAPTER 4

4.1 Exercise Set, p. 191

1. (a) $(b, c), (d, e)$ (b) $(a, b), (c, d), (e, f)$ 3. increasing on $(-\infty, +\infty)$, no horizontal tangents
5. increasing on $(2, +\infty)$, decreasing on $(-\infty, 2)$, horizontal tangent at $x = 2$
7. increasing on $(0, +\infty)$, decreasing on $(-\infty, 0)$, horizontal tangent at $x = 0$
9. increasing on $(-\infty, -1)$ and $(1, +\infty)$, decreasing on $(-1, 1)$, horizontal tangents at $x = -1$ and $x = 1$
11. increasing on $(-\infty, +\infty)$, horizontal tangent at $x = 1$
13. increasing on $(-1, 0)$ and $(1, +\infty)$, decreasing on $(-\infty, -1)$ and $(0, 1)$, horizontal tangents at $x = -1$, $x = 0$, and $x = 1$ 15. never increasing, decreasing on $(-\infty, 0)$ and $(0, +\infty)$, no horizontal tangents
17. increasing on $(0, +\infty)$, decreasing on $(-\infty, 0)$, horizontal tangent at $x = 0$
19.
21.
23.

25. (a) increasing for $x < 25{,}000$, decreasing for $x > 25{,}000$ (b) $x = 25{,}000$
27. increasing for $t < 40$, decreasing for $t > 40$

4.2 Exercise Set, p. 202

1. relative maxima at x_1, x_3, and x_6 3. relative minimum at $x = 4$
5. relative maximum at $x = 3$ 7. there are no critical numbers   9. relative minimum at $x = 0$
11. relative maximum at $x = -1$, relative minimum at $x = 2$ 13. there are no critical numbers
15. relative maximum at $x = 1$, relative minimum at $x = 0$
17. $x = 1$ is a critical number but no relative extremum occurs there
19. $x = 0$ is a critical number but no relative extremum occurs there
21. relative maximum at $x = 0$ 23. relative maximum at $x = 1$, relative minimum at $x = -1$
25. relative minimum at $x = \frac{3}{2}$ and $x = 0$ is a critical number but no relative extremum occurs there

4.3 Exercise Set, p. 207

1. (a) x_1 (b) x_3 (c) x_3 (d) x_6 (e) none (f) none
3. The absolute maximum value is 7 occurring at $x = 2$ and at $x = -6$. The absolute minimum value is -9 occurring at $x = -2$.
5. (a) The absolute maximum value is 4 occurring at $x = 2$. The absolute minimum value is 0 occurring at $x = 1$.
   (b) The absolute maximum value is 4 occurring at $x = -1$ and at $x = 2$. The absolute minimum value is 0 occurring at $x = -2$ and at $x = 1$.
7. (a) The absolute maximum value is 10 occurring at $x = 0$. The absolute minimum value is $\frac{10}{3}$ occurring at $x = -2$.
   (b) The absolute maximum value is $15\frac{1}{3}$ occurring at $x = 4$. The absolute minimum value is $\frac{26}{3}$ occurring at $x = 2$.
9. The absolute maximum value is 1 occurring at $x = 1$. The absolute minimum value is $\frac{1}{3}$ occurring at $x = 3$.
11. The absolute maximum value is 3 occurring at $x = 4$. The absolute minimum value is 0 occurring at $x = 1$.
13. The absolute maximum value is 6 occurring at $x = -8$. The absolute minimum value is 2 occurring at $x = 0$.   15. 20 cars   17. 3 days, 620 bacteria per cubic centimeter   19. 150 books per week

4.4 Exercise Set, p. 216

1. (a) (x_1, x_3) and (x_5, x_7) (b) (x_3, x_5) and (x_7, x_8)
   (c) inflection points occur at $(x_3, f(x_3))$, $(x_5, f(x_5))$, and $(x_7, 0)$
3. concave upward on $(-\infty, \infty)$
5. inflection point at $(-2, 54)$, concave upward for $x > -2$, concave downward for $x < -2$
7. inflection point at $(1, -\frac{2}{3})$, concave upward for $x > 1$, concave downward for $x < 1$
9. inflection points at $(0, 6)$ and $(2, -10)$, concave upward for $x < 0$ and for $x > 2$, concave downward for $0 < x < 2$ 11. relative minimum is $f(3) = -4$
13. relative maximum is $f(-1) = 12$, relative minimum is $f(2) = -15$
15. relative maximum is $f(-1) = 3$, relative minimum is $f(-\frac{1}{3}) = \frac{73}{27}$
17. there are no relative extrema   19. relative minimum is $f(-3) = -12$
21. relative maximum is $f(1) = 8$, relative minimum is $f(2) = 7$
23. relative maximum is $f(2) = 32$, relative minimum is $f(6) = 0$ 25. relative minimum is $f(3) = 0$
27. relative maximum is $f(0) = 1$
29. relative maximum is $f(0) = 2$, relative minima are $f(1) = f(-1) = 1.75$ 31. 4 weeks
33. (a) the unemployment rate begins to decrease   (b) unemployment begins to decrease

4.5 Exercise Set, p. 223

1.

3.

5.

7.

9.

11.

13.

15.

4.6 Exercise Set, p. 235

1.

3.

5.

7.

9.

11.

13.

15.

4.7 Exercise Set, p. 246

1. $60,000   3. 80 feet by 80 feet   5.

7. An increase of $5 per room will maximize the daily revenue. 90 rooms will be occupied daily, and the maximum daily revenue will be $4050.
9. Eighty trees will maximize the total yield. The average yield per tree will be 60 bushels, and the maximum total yield will be 4800 bushels.
11. 2 hours   13. 500 mph; $4000   15. 80 orders of lot size 24 boats
17. cut 10/3 inches to obtain maximum volume of 16,000/27 cubic inches
19. cut the wire so that $\dfrac{40\pi}{4+\pi}$ cm (≈ 17.6 cm) is used for the circle, and the remainder (≈ 22.4 cm) is used for the square
21. $144   23. $r = \dfrac{8}{\sqrt[3]{\pi}}$ cm, $h = \dfrac{16}{\sqrt[3]{\pi}}$ cm   25. $x = 8''$ and $y = 16''$ (the page is $10''$ by $20''$)
27. $2\frac{2}{3}$ miles from C

4.8 Exercise Set, p. 256

1. $\frac{73}{12} \approx 6.08$ 3. $\frac{83}{27} \approx 3.07$ 5. 1.99   7. $\frac{129}{256} \approx 0.504$ 9. 0.05   11. $\frac{4}{15} \approx 0.267$ 13. $2\,dx$
15. $(6x^2 - 6x + 5)\,dx$ 17. $\dfrac{2}{(x+1)^2}\,dx$ 19. $10x(x^2+1)^4\,dx$ 21. 1.08 cubic cm.
23. approximately 10.05 cubic cm.   25. (a) change in revenue $\approx$ $200, change in profit $\approx -$200
                                        (b) change in revenue $\approx$ $400, change in profit $\approx$ $1200
27. 0.253 square cm   29. $0.625

4.9 Exercise Set, p. 264

1. 2.24   3. 2.71   5. 3.16   7. 1.90   9. 0.62   11. 2.52   13. -1.32 15. 3.5 seconds
17. 1.46 milligrams

Review Exercises, Chapter 4, p. 265

1. (x_1, x_2), (x_4, x_6), and (x_7, x_8) 3. x_2, x_6, and x_8 5. (x_3, x_5) 7. x_3, x_5
9. decreasing on $(-\infty, +\infty)$, no horizontal tangents
11. increasing on $(-\infty, +\frac{3}{2})$, decreasing on $(\frac{3}{2}, +\infty)$, horizontal tangents at $x = \frac{3}{2}$ and at $x = 0$
13. relative maximum at $x = 1$ 15. $x = 2$ is a critical number but no relative extremum occurs there
17. The absolute maximum value is 1 occurring at $x = 1$. The absolute minimum value is -15 occurring at $x = -3$.
19. The absolute maximum value is 1 occurring at $x = \frac{1}{8}$. The absolute minimum value is -26 occurring at $x = 8$.
21. concave upward on $(-\infty, +\infty)$

23. inflection point at $(-0.5, 9.5)$, concave upward for $x > -0.5$, concave downward for $x < -0.5$

25. relative maximum is $f(-2) = 24$, relative minimum is $f(2) = -8$ 27. no relative extrema occur

29.

31.

$(-2, -8)$ $(2, -8)$

33.

35.

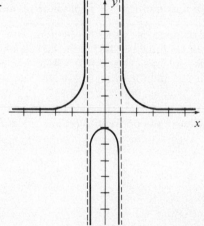

37. (a) 6.95 (b) $\frac{2027}{225} \approx 9.01$ 39. $(3x^2 - 1)\,dx$

41. (a) increasing for $x < 7500$, decreasing for $x > 7500$ (b) 7500 typewriters

43. increasing for $x < 3$ and $x > 4$ 45. 25 orders per year of lot size 100 stoves

47. $r = \sqrt[3]{\dfrac{16}{\pi}}$ feet and $h = 2\sqrt[3]{\dfrac{16}{\pi}}$ feet   49. 2π cubic feet   51. 2.65

Chapter Test, Chapter 4, p. 268

1. (a) increasing for $x < -3$ and $x > 1$ (b) decreasing for $-3 < x < 1$
   (c) at $(-3, 20)$ and $(1, -12)$
   (d) relative maximum $f(-3) = 20$, relative minimum $f(1) = -12$
   (e) $(-1, +\infty)$ (f) yes, at $(-1, 4)$

2.

$(-3, 20)$

$(-1, 4)$

2

-2

2

x

$(1, -12)$

3. relative maximum $f(-1) = -2$, relative minimum $f(1) = 2$

4. absolute maximum $f(8) = \frac{2}{3}$, absolute minimum $f(1) = -\frac{2}{3}$ 5. $dy = \dfrac{5x}{\sqrt{5x^2 + 6}}\, dx$

6. Annual demand decreases by 300 tons.   7. 1.5 hours

8. Maximum volume is 74.07 cubic inches, obtained by cutting out $\frac{5}{3}$ inch by $\frac{5}{3}$ inch corners.

9. (a) Since $f(2) = -14$ and $f(3) = 4$, and since $-14 < 0 < 4$, the intermediate value property assures us that $f(c) = 0$ for some c between 2 and 3.

   (b) $x_1 = 3$, $x_2 = 2.8462$, $x_3 = 2.8372$

10. (a) x-intercept $(0, 0)$, y-intercept $(0, 0)$

    (b) $f(x)$ has 2 discontinuities (at $x = -1$ and $x = 2$), so the graph of f is in 3 continuous pieces

    (d) vertical asymptotes $x = -1$ and $x = 2$, horizontal asymptote $y = 0$

(c)

$x = 2$

$x = -1$

$x = -1$

3

-3

3

x

-3

$x = 2$

(e) decreasing on $(-\infty, -1)$, $(-1, 2)$ and $(2, +\infty)$

|||||||||||| **CHAPTER 5**

5.1 Exercise Set, p. 282

1.

t	1	10	20	120
P	208.16	298.36	445.11	24,302

3.

5.

7.

9.

11.

13.

15.

17. (a) $1574.86 (b) $1622.94 (c) $1613.91 (d) $1600.62 (e) $1627.49 19. (a) $24,268.44 (b) $24,653.20 21. $6634.50 23. $4165.78 25. (a) 8.16% (b) 8.24% (c) 8.30% (d) 8.33% 27. compounded semiannually 29. (a) 2984 bacteria (b) 49,065 bacteria
31. 0.00026976 lumens

5.2 Exercise Set, p. 290

1. (a) $3^2 = 9$ (b) $4^3 = 64$ (c) $2^{-2} = \frac{1}{4}$ 3. (a) 4 (b) -1 (c) 7 (d) $\frac{5}{3}$ 5. (a) 4 (b) -2
(c) -5 (d) 6 (e) 10 (f) 45 (g) $\frac{2}{3}$ (h) $-\frac{1}{2}$ 7. (a) 1.1761 (b) 0.2219 (c) 3.4771
(d) -2.6990 (e) 1.8751 (f) 0.8266   9. (a) 64 (b) $\frac{1}{5}$ (c) 3 (d) $\frac{1}{2}$

11.

13.

15.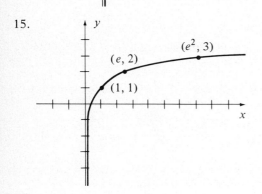

17.

19. (a) 1.3863 (b) 0.8959 (c) -3.4657 (d) -0.0912 (e) 2.5541 (f) 3.4903   21. (a) $3b - 2c$
(b) $\dfrac{b - c - d}{2}$ (c) $b + \dfrac{1}{3}c$ 23. (a) 3.8671 (b) 3.3219 (c) -4.3219
25. 11.45 years $\approx$ 11 years, 165 days   27. 11.57 years $\approx$ 11 years, 207 days   29. 27.47 days
31. \$52,598   33. (a) $10^{8.25} \approx 177,827,941$ (b) 7.7

5.3 Exercise Set, p. 301

1. $2e^{2x}$ 3. $-1.08e^{-1.08x}$ 5. $3x^2 e^{x^3}$ 7. $\dfrac{3}{2\sqrt{3x+2}} e^{\sqrt{3x+2}}$ 9. $6x - 3x^2 e^{-x^3} + 8xe^{x^2}$

11. $5(e^x + e^{-x})^4(e^x - e^{-x})$ 13. $(3x^2 + 4x^3)e^{4x}$ 15. $\dfrac{1-x}{e^x}$ 17. $\dfrac{4}{(e^x + e^{-x})^2}$ 19. $\dfrac{1}{x-2}$

21. $\dfrac{4x+3}{2x^2 + 3x - 1}$ 23. $\dfrac{5(\ln x)^4}{x}$ 25. $-\dfrac{1}{x}$ 27. $\dfrac{1}{2x}$ 29. $\dfrac{1}{2x\sqrt{\ln x}}$ 31. $\dfrac{6}{x} + 8x$

33. $x(1 + 2\ln x)$ 35. $\dfrac{2\ln x(1 - \ln x)}{x^3}$ 37. $\dfrac{e^x + 1}{e^x + x}$ 39. $\dfrac{(x\ln x - 1)e^x}{x(\ln x)^2}$ 41. $3^{2x} \cdot 2\ln 3$

43. $4^{-3x}(-3 \ln 4)$ 45. $x2^x(2 + x \ln 2)$ 47. $\dfrac{500}{3}(1.02)^{x/3} \ln (1.02)$ 49. $\dfrac{1}{x \ln 3}$

51. $\dfrac{2x - 5}{(x^2 - 5x) \ln 2}$ 53. $\dfrac{1}{2x \ln 10}$ 55. $x^2(3 \log x + \dfrac{1}{\ln 10})$

57. $f(x) = \ln x$, $x > 0$; $f'(x) = 1/x > 0$. Thus f is increasing on $(0, +\infty)$. Moreover, $f'(x) = -1/x^2 < 0$. Thus, the graph of f is concave downward on $(0, +\infty)$.

59. 2   61. no relative extrema, no inflection points   63. relative minimum $f(0) = 2$, no inflection points

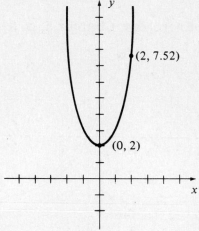

65. relative minimum $f(-1) = -\dfrac{1}{e}$, inflection point $(-2, \dfrac{-2}{e^2})$

67. relative minimum $f\left(\dfrac{1}{e}\right) = -\dfrac{1}{e}$, no inflection points   69. $\dfrac{2x + 5}{x^2 + 5x}$

5.4 Exercise Set, p. 311

1. (a) 5000   (b) 0.4   (c) $\dfrac{dQ}{dt} = 0.04Q$ (d) 6000 per day   (e) $5000e^2 \approx 36{,}945$

3. (a) $\dfrac{dQ}{dt} = -0.05Q$ (b) $Q = 8000e^{-0.05t}$ (c) 6215 lb   5. 183.9   7. (a) $\dfrac{dQ}{dt} = 0.034Q$

(b) 235,720,000   (c) 2020   9. 20.39 years   11. 8.7 % per hour   13. (a) 6.44 hours
(b) 0.0294 mg/ml   15. 16.26 years   17. 1860 years   19. (a) 0.0354   (b) 5.87 billion barrels
21. 93 years

Review Exercises, Chapter 5, p. 314

1. 268.13   3.

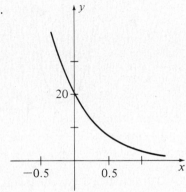

5. $5718.01   7. $3653.45   9. 9.42%
11. (a) -4 (b) 2   (c) $\frac{1}{2}$ (d) -3 (e) -2
(f) 4   (g) $-\frac{2}{3}$ (h) $0.02t$ (i) -5 (j) 7
(k) 1.02   (l) $\frac{1}{2}$

13. (a) $\dfrac{v}{u}$ (b) $\dfrac{u+v}{w}$ 15. 51.598 million   17. $1.03e^{1.03x}$ 19. $\dfrac{3x+1}{\sqrt{2x+1}}\,e^{x\sqrt{2x+1}}$

21. $\dfrac{(x-2)e^x}{x^3}$ 23. $\dfrac{1}{x+5}$ 25. $\dfrac{2}{3x}$ 27. $\dfrac{3x^2+5}{x^3+5x+3}$ 29. $2 \cdot 3^{2x}\ln 3$ 31. $\dfrac{1}{x\ln 2}$ 33. 5

35. $\ln(x+2) + \dfrac{x}{x+2}$ 37. $\dfrac{dQ}{dt} = -0.023Q$; $Q = Q_0 e^{-0.023t}$; half-life = 30.1 years

39. (a) $\dfrac{dQ}{dt} = -0.133Q$, $Q = Q_0 e^{-0.133t}$ (b) 5.21 years   (c) 10.42 years   41. 13.15% per day

Chapter Test, Chapter 5, p. 315

1.

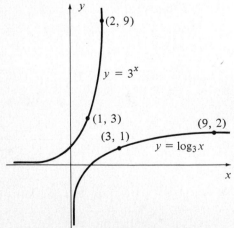

2. (a) $6341.20   3. 6.2 years $\approx$ 6 years, 79 days
(b) $6356.25
4. (a) 4   (d) -2
(b) -2 (e) 5
(c) 49   (f) $\frac{1}{5}$

5. $\frac{1}{2}r + \frac{3}{2}s - 2t$ 6. $3e$ 7. (a) $5e^{0.1x}$ (b) $\dfrac{12x^2 + 1}{4x^3 + x}$ (c) $\dfrac{4x^3 - 3x^4}{e^{3x}}$ (d) $\dfrac{e^{\sqrt{x}}}{2\sqrt{x}}$

8. (a) $2x(4^{x^2}) \ln 4$ (b) $2x \log_5 x + \dfrac{x}{\ln 5}$. Alternate: (a) $\dfrac{10x[\ln(x^2 + 1)]^4}{x^2 + 1}$ (b) $3x^2 \ln(2x) + x^2$

9. (a) $\dfrac{dQ}{dt} = 0.03Q$ (b) 5399   (c) 23.1 hours   10. (a) 3.0% per year   (b) 23 years   (c) 76.75 years

|||||||||||| **CHAPTER 6**

6.1 Exercise Set, p. 327

1. $3x + C$ 3. $\dfrac{s^7}{7} + C$ 5. $\dfrac{5x^{6/5}}{6} + C$ 7. $\dfrac{3u^8}{8} + C$ 9. $\dfrac{6}{x} + C$ 11. $3e^u + C$

13. $10 \ln|x| + C$ 15. $\frac{3}{4}x^{4/3} + C$ 17. $\dfrac{3^t}{\ln 3} + C$ 19. $\dfrac{x^2}{2} + \ln|x| + C$

21. $t^4 - \frac{12}{5}t^{5/4} + 3t + C$ 23. $\frac{6}{5}t^{5/3} - 3e^t + 2\ln|t| + C$ 25. $x^5 - \dfrac{6}{7}x^{7/2} - \dfrac{2}{15x^3} + C$ 27. no

29. no   31. yes   33. incorrect   35. incorrect   37. $-\dfrac{1}{3x} - \dfrac{1}{3x^2} + C$

39. $\frac{4}{3}x^3 - 14x^2 + 49x + C$ 41. $\frac{2}{5}x^{5/2} + 2x^{3/2} + C$ 43. $2x^{3/2} + x^2 + 5x + 2000$
45. $y = \frac{2}{3}x^3 - \frac{3}{2}x^2 + x + \frac{43}{3}$ 47. 660,000 people   49. $197\frac{1}{3}$ million cubic feet

6.2 Exercise Set, p. 338

1. $\frac{1}{20}(5x - 3)^4 + C$ 3. $\frac{1}{10}(s^2 + 1)^5 + C$ 5. $-\frac{1}{24}(1 - w^3)^8 + C$ 7. $-\dfrac{1}{4(x^2 + 9)^2} + C$

9. $\ln|x + 5| + C$ 11. $\dfrac{-1}{9(x^3 + 2)^9} + C$ 13. $\frac{2}{3}(x^3 + 2)^{3/2} + C$ 15. $\frac{1}{6}(t^4 + 5)^{3/2} + C$

17. $\dfrac{e^{2x}}{2} + C$ 19. $\frac{1}{2}e^{4s-1} + C$ 21. $e^{x^2} + C$ 23. $e^{x^3 + x^2} + C$ 25. $2e^{\sqrt{t}} + C$
27. $\frac{1}{2}(1 + e^y)^2 + C$ 29. $\frac{1}{2}\ln|x^2 + 2x| + C$ 31. $-\ln|2 - x| + C$ 33. $\frac{1}{6}\ln(3s^2 + 2) + C$
35. $\frac{2}{3}\sqrt{t^3 + 4} + C$ 37. $\frac{1}{3}(\ln x)^3 + C$ 39. $\ln|\ln x| + C$ 41. 1 million dollars   43. 9.16 grams

6.3 Exercise Set, p. 345

1. $xe^{3x} - \frac{1}{3}e^{3x} + C$ 3. $t^2 e^t - 2te^t + 2e^t + C$ 5. $\frac{2}{3}t^{3/2} \ln t - \frac{4}{9}t^{3/2} + C$

7. $\dfrac{x^2}{2}(\ln x)^2 - \dfrac{x^2}{2}\ln x + \dfrac{x^2}{4} + C$ 9. $-\dfrac{x}{5}e^{-5x} - \dfrac{e^{-5x}}{25} + C$ 11. $\frac{2}{3}x(x + 3)^{3/2} - \frac{4}{15}(x + 3)^{5/2} + C$

13. $\dfrac{4x}{3}(3x + 1)^{1/2} - \dfrac{8}{27}(3x + 1)^{3/2} + C$ 15. $(x + 1)\ln(x + 1) - x + C$

17. $\left(\dfrac{x^2}{2} + 2x\right)\ln x - \dfrac{x^2}{4} - 2x + C$ 19. $\dfrac{x^4 \ln x}{4} - \dfrac{x^4}{16} + C$ 21. $(x^3 - 3x^2 + 6x - 6)e^x + C$

23. $x \log x - \dfrac{x}{\ln 10} + C$ 25. $\left(\dfrac{x}{\ln 3} - \dfrac{1}{(\ln 3)^2}\right) \cdot 3^x + C$ 27. $(25 + 50e^{-1})$ million dollars $\approx \$43,393,972$

6.4 Exercise Set, p. 353

1. Formula 10, with $u = x$: $\dfrac{x}{5} - \dfrac{2}{5}\ln|5x + 10| + C$

3. Formula 31, with $u = 3x$: $\frac{1}{3}\ln|3x + \sqrt{9x^2 - 25}| + C$ 5. Formula 28, with $u = 7x$: $\dfrac{1}{28}\ln\left|\dfrac{7x - 2}{7x + 2}\right| + C$

7. Formula 36, with $u = 4t$: $\dfrac{1}{3}\ln\left|\dfrac{4t}{3 + \sqrt{9 + 16t^2}}\right| + C$ 9. Formula 29, with $u = 3x$: $\dfrac{1}{6}\ln\left|\dfrac{3x + 1}{3x - 1}\right| + C$

11. Formula 39, with $u = x$: $\frac{2}{5}e^{5x/2} + C$ 13. Formula 4, with $u = 4t + 3$: $\dfrac{-1}{8(4t + 3)^2} + C$

15. Formula 36, with $u = 2x$: $\dfrac{1}{3}\ln\left|\dfrac{2x}{3 + \sqrt{9 - 4x^2}}\right| + C$ 17. Formula 5, with $u = x - 4$: $\frac{2}{3}\ln|x - 4| + C$

19. Formula 11, with $u = x$: $\frac{1}{27}[\frac{1}{2}(3x - 5)^2 + 10(3x - 5) + 25\ln|3x - 5|] + C$

21. Formula 17, with $u = t$: $-\dfrac{1}{14}\ln\left|\dfrac{t + 4}{3t - 2}\right| + C$

23. Formula 21, with $u = x$: $\dfrac{2}{13125}(375x^2 - 480x + 512)(5x + 8)^{3/2} + C$

25. Formula 32, with $u = 2x$: $x\sqrt{4x^2 - 9} - \dfrac{9}{2}\ln|2x + \sqrt{4x^2 - 9}| + C$

27. Formula 18, with $u = s$: $\frac{1}{5}\{\frac{3}{2}\ln|2s + 3| + \ln|s - 1|\} + C$

29. Formula 26, with $u = t$: $\dfrac{1}{2}\ln\left|\dfrac{\sqrt{t + 4} - 2}{\sqrt{t + 4} + 2}\right| + C$

31. Formula 34, with $u = 3x$: $\sqrt{2 - 9x^2} - \sqrt{2}\ln\left|\dfrac{\sqrt{2} + \sqrt{2 - 9x^2}}{3x}\right| + C$

33. Formulas 42 and 40, with $n = 2$ and $u = t$: $\frac{1}{5}t^2 e^{5t} - \frac{2}{25}t e^{5t} + \frac{2}{125}e^{5t} + C$

35. Formula 31, with $u = x + 1$: $\ln|(x + 1) + \sqrt{x^2 + 2x}| + C$

Review Exercises, Chapter 6, p. 355

1. $\dfrac{x^6}{6} + C$ 3. $3e^t + C$ 5. $\frac{1}{18}(3x + 4)^6 + C$ 7. $3x^{1/3} + C$ 9. $\frac{3}{7}x^{7/3} - 4\ln|x| + e^x + C$

11. $\ln|t + 3| + C$ 13. $\dfrac{(x^3 + 2)^9}{27} + C$ 15. $\ln|x^3 + x| + C$ 17. $\frac{5}{2}\ln(e^{2x} + 1) + C$

19. $\frac{1}{3}(\ln x)^3 + C$ 21. $\dfrac{2t}{3}(t - 3)^{3/2} - \dfrac{4}{15}(t - 3)^{5/2} + C$ 23. $\frac{1}{2}e^{2x}[x^3 - \frac{3}{2}x^2 + \frac{3}{2}x - \frac{3}{4}] + C$

25. $(x + 5)\ln(x + 5) - x + C$ 27. $\frac{1}{5}\ln|5x + \sqrt{16 + 25x^2}| + C$

29. $\dfrac{x}{18}\sqrt{1 + 9x^2} - \dfrac{1}{54}\ln|3x + \sqrt{1 + 9x^2}| + C$ 31. $-\frac{1}{64}[\frac{1}{2}(5 - 4t)^2 - 10(5 - 4t) + 25\ln|5 - 4t|] + C$

33. $\dfrac{5}{7}\left[\dfrac{1}{2}\ln|2x - 1| + 3\ln|x + 3|\right] + C$ 35. $\sqrt{t^2 + 4} - 2\ln\left|\dfrac{2 + \sqrt{t^2 + 4}}{t}\right| + C$

37. $y = (x^4 + 9)^{3/2} + 125$ 39. $40 - 200e^{-4}$ tons ≈ 36.337 tons

Chapter Test, Chapter 6, p. 356

1. $\frac{1}{42}(7x + 2)^6 + C$ 2. $3x^{4/3} - 10\sqrt{x} + C$ 3. $\frac{1}{2}\ln(x^2 + 5) + C$ 4. $\dfrac{-1}{6(x^3 - 4)^2} + C$

5. $\dfrac{2}{15}(3t-4)(t+2)^{3/2} + C$ 6. $\dfrac{e^{3t}}{9}(3t-1) + C$ 7. $\frac{3}{2}e^{x^2} + C$ 8. $\frac{1}{2}(\ln x)^2 + C$

9. $\dfrac{3}{25}\left(\dfrac{-4}{5x-4} + \ln|5x-4|\right) + C$ 10. $\dfrac{x}{2}\sqrt{x^2+4} + 2\ln\left|x+\sqrt{x^2+4}\right| + C$ 11. $\dfrac{1}{4}\ln\left|\dfrac{x-1}{x+3}\right| + C$

12. $P(x) = 80x - 0.1x^2 - 2750$

|||||||||||| **CHAPTER 7**

7.1 Exercise Set, p. 364

1. (a)

(b) $A = (4)(3) = 12$
(c) 12

3. (a)

(b) $A = \frac{1}{2}bh = \frac{1}{2}(3)(6) = 9$
(c) 9

5. (a)

(b) $A = \frac{1}{2}(3)(6) + (3)(4) = 21$
(c) 21

7. (a)

(b) $A = \frac{1}{2}(3)(3) + (3)(3) = \frac{27}{2}$
(c) $\frac{27}{2}$

9.

$A = \frac{28}{3}$

11.

$A = \frac{16}{3}$

13.

$$A = \frac{9}{2}$$

15.

$$A = e - 1$$

17.

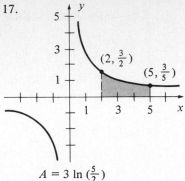

$$A = 3 \ln \left(\tfrac{5}{2}\right)$$

19. 3   21. $-\frac{19}{5}$ 23. $e^2 - 1$ 25. $\frac{2}{3}$ 27. $\frac{14}{3}$ 29. $\frac{9}{2}$ 31. $e - \dfrac{1}{e^2}$

7.2 Exercise Set, p. 377

1. $\int_2^2 (3x^2 + x)\,dx = [x^3 + \frac{1}{2}x^2]_2^2 = (8 + 2) - (8 + 2) = 0$ 3. $\int_2^3 (2x + x^2)\,dx = \frac{34}{3}$,
while $\int_3^2 (2x + x^2)\,dx = -\frac{34}{3}$ 5. (a) $\frac{81}{4}$ (b) $\int_0^3 x^3\,dx = \frac{81}{4}$ 7. $\frac{32}{3}$ 9. $\frac{821}{12}$ 11. $\frac{27}{4}$ 13. 0
15. -18 17. $-\frac{1}{2}(15 + e^{-4})$ 19. 2052   21. $-\frac{7}{4} - 9\sqrt[3]{4}$ 23. $\frac{73}{3}$ 25. $\frac{32}{3}$ 27. $\frac{1}{2} + \ln 3$
29. 13   31. $\frac{31}{6}$

33.

$$A = \frac{1}{6}$$

35.

$$A = \frac{32}{3}$$

37.

$$A = \frac{64}{3}$$

39.

$$A = \frac{32}{3}$$

41. $\frac{26}{3}$ 43. 3   45. $\frac{1}{2}$

7.3 Exercise Set, p. 388

1. (a) 8 (b) $e^{-2} - 1$ 3. $3000(e^{0.8} - e^{0.4}) \approx 2201$ flies   5. (a) 140 million barrels
   (b) 59 million barrels   7. 1282.5 gallons   9. (a) $8060 (b) $20,125 (c) $19,950 (d) $175
11. 11.07 million barrels   13. 12.3 million troy ounces   15. (a) 96 ft/sec, 96 ft/sec
   (b) 144 ft, 432 ft   17. $346,000; $150,000   19. (a) $972,000; $468,000
   (b) $584,000; $114,000 (c) 10 years (d) $746,700

7.4 Exercise Set, p. 402

1. 9   3. $\frac{21}{2}$ 5. $\frac{16}{3}$ 7. $\frac{14}{3}$ 9. 4   11. $\frac{5}{2}$ 13. 1   15. $\frac{22}{3}$
17. $\dfrac{2}{\sqrt{3}}$ 19. $\frac{1}{2} \ln 3$ 21. $\frac{665}{108} \approx 6.16°$ 23. $36.60 per barrel

7.5 Exercise Set, p. 414

1. (a) $194,600 (b) $79,100   3. $11,800   5. $85,000   7. 72   9. 144
11. consumers' surplus = producers' surplus = 6.125
13. consumers' surplus = 4, producers' surplus = $\frac{16}{3}$

7.6 Exercise Set, p. 422

1. $\frac{1}{2}$ 3. $\frac{1}{2}$ 5. diverges   7. diverges   9. $\frac{1}{3}$ 11. diverges   13. diverges
15. diverges   17. diverges   19. 1   21. $\frac{1}{4}$ million trees   23. $30,000   25. $7,778,000
27. $2069   29. $\frac{1}{3}$ 31. diverges   33. diverges   35. 0   37. 0

7.7 Exercise Set, p. 431

1. $L_n = 5$, $R_n = 7$, $M_n = 6$, $T_n = 6$ 3. $L_n = 1.25$, $R_n = 1.25$, $M_n = 1.375$, $T_n = 1.25$
5. $L_n = 91$, $R_n = 139$, $M_n = 113.5$, $T_n = 115$ 7. $L_n = -27$, $R_n = 27$, $M_n = 0$, $T_n = 0$
9. $L_n = 1.83$, $R_n = 1.43$, $M_n = 1.60$, $T_n = 1.63$ 11. $L_n = 41.3$, $R_n = 68.1$, $M_n = 53.0$, $T_n = 54.7$
13. $L_n = 2.78$, $R_n = 3.78$, $M_n = 3.22$, $T_n = 3.28$ 15. $L_n = 3.43$, $R_n = 3.88$, $M_n = 3.66$, $T_n = 3.65$

Review Exercises, Chapter 7, p. 433

1.
$A = \frac{15}{4}$

3.
$A = \frac{21}{2}$

5.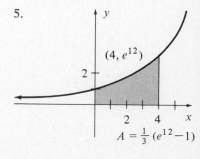
$A = \frac{1}{3}(e^{12} - 1)$

7. 3   9. -70 11. $2(e^{1/2} - 1)$ 13. 2   15. $\frac{110}{3} - \ln 2$ 17. 53   19. $4(e^{-2} - e^{-9/2}) - 663$

21. $\frac{8}{3}$ 23. area of shaded region is $\frac{1}{3}$ 25. $\frac{23}{3}$ 27. $\frac{1}{2}$ 29. $\frac{32}{3}$ 31. 9.152 milliseconds

$A = \frac{1}{3}$

33. $V(t) = 2(t + 1)^{1/2} + 5$, $x(t) = \frac{4}{3}(t + 1)^{3/2} + 5t$ 35. (a) 12 years   (b) $3168   37. $e^2 - e$
39. $\frac{3}{2}$ 41. 1   43. diverges   45. $\frac{1}{2}$ 47. $\frac{1}{2}$ 49. $92,229
51. consumers' surplus = producers' surplus = 8   53. $2963
55. $L_n = 1.28$, $R_n = .95$, $M_n = 1.09$, $T_n = 1.12$

Chapter Test, Chapter 7, p. 436

1. $\frac{13}{3}$ 2. $\frac{9}{2}$ 3. (a) -4 (b) 6   (c) 0

4. $\Delta x = \dfrac{2 - 0}{n} = \dfrac{2}{n}$; $x_0 = 0$, $x_1 = \dfrac{2}{n}$, $x_2 = \dfrac{4}{n}$, $x_3 = \dfrac{6}{n}$, $x_4 = \dfrac{8}{n}$, ..., $x_n = \dfrac{2n}{n} = 2$.

$$R_n = f\left(\frac{2}{n}\right)\Delta x + f\left(\frac{4}{n}\right)\Delta x + f\left(\frac{6}{n}\right)\Delta x + f\left(\frac{8}{n}\right)\Delta x + \cdots + f\left(\frac{2n}{n}\right)\Delta x$$

$$= \left(\frac{2}{n}\right)^2 \cdot \frac{2}{n} + \left(\frac{4}{n}\right)^2 \cdot \frac{2}{n} + \left(\frac{6}{n}\right)^2 \cdot \frac{2}{n} + \left(\frac{8}{n}\right)^2 \cdot \frac{2}{n} + \cdots + \left(\frac{2n}{n}\right)^2 \cdot \frac{2}{n}$$

$$= \frac{2}{n}\left[\frac{2^2}{n^2} + \frac{4^2}{n^2} + \frac{6^2}{n^2} + \frac{8^2}{n^2} + \cdots + \frac{(2n)^2}{n^2}\right]$$

$$= \frac{2}{n}\left[\frac{2^2 \cdot 1^2}{n^2} + \frac{2^2 \cdot 2^2}{n^2} + \frac{2^2 \cdot 3^2}{n^2} + \frac{2^2 \cdot 4^2}{n^2} + \cdots + \frac{2^2 \cdot n^2}{n^2}\right] = \frac{2}{n} \cdot \frac{2^2}{n^2}[1^2 + 2^2 + 3^2 + 4^2 + \cdots + n^2]$$

$$= \frac{8}{n^3} \cdot \frac{n(n + 1)(2n + 1)}{6} \qquad \text{by Equation (7) of Section 7.4}$$

$$= \frac{4}{3} \cdot \frac{n}{n} \cdot \frac{n + 1}{n} \cdot \frac{2n + 1}{n}$$

$$= \frac{4}{3} \cdot \frac{n + \dfrac{1}{n}}{1} \cdot \frac{2 + \dfrac{1}{n}}{1}$$

$$\lim_{n \to \infty} R_n = \frac{4}{3} \cdot \frac{1 + 0}{1} \cdot \frac{2 + 0}{1} = \frac{8}{3}. \qquad \int_0^2 x^2 \, dx = \frac{8}{3}.$$

5. 16.667 million ounces   6. $\frac{1}{6}$ 7. (a) $\frac{1}{4}$ (b) diverges
8. present value is $82,600; future value is $183,831
9. consumers' surplus is 24; producers' surplus is 48
10. $L_n = 26$, $R_n = 42$, $M_n = 33$, $T_n = 34$

||||||||||||| **CHAPTER 8**

8.1 Exercise Set, p. 446

1. (a) 4 (b) -7 (c) -3 (d) 8 (e) all points (x, y) 3. (a) 1 (b) $2e^2 + 4 + \ln 2$
   (c) all (x, y) for which $y > 0$ 5. (a) 2 (b) 16 (c) 34 (d) 13 (e) all points (x, y, z)
7. (a) 6 (b) 70   9. (a) e^{11} (b) $3e$ 11. (a) $13 + 2h$ (b) $2x + 2h + 3y$ (c) $2h$ (d) $2h$ (e) 2

13.

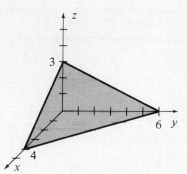

15. (a) 207
    (b) 77

17. (a) \$64,000
    (b) \$484,000

19. The surface is a plane
    intersecting the coordinate axes
    at $(4, 0, 0)$, $(0, 6, 0)$, and $(0, 0, 3)$.

21. The surface is a plane
    intersecting the coordinate axes
    at $(8, 0, 0)$, $(0, -8, 0)$, and $(0, 0, -4)$.

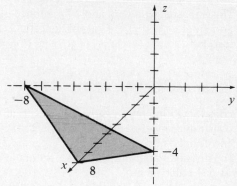

23. The surface is a plane
    intersecting the coordinate axes
    at $(-2, 0, 0)$, $(0, 2, 0)$, and $(0, 0, 2)$.

8.2 Exercise Set, p. 456

1. $f_x = 3x^2 + y$, $f_y = x + 6y$, $f_{xy} = 1$ 3. $f_x = 8x(x^2 + 5y)^3$, $fy = 20(x^2 + 5y)^3$, $f_{xy} = 120x(x^2 + 5y)^2$
5. $f_x = \dfrac{3x^2}{y^2}$, $fy = \dfrac{-2x^3}{y^3}$, $f_{xy} = \dfrac{-6x^2}{y^3}$ 7. $f_x = y^2 e^{xy^2}$, $fy = 2xy e^{xy^2}$, $f_{xy} = (2xy^3 + 2y)e^{xy^2}$

9. $f_x = e^y + y^2 e^x$, $f_y = xe^y + 2ye^x - 4y^3$, $f_{xy} = e^y + 2ye^x$ 11. $f_x = 1/x$, $f_y = 1/y$, $f_{xy} = 0$
13. $f_x(-1, 2) = 8$, $f_y(-1, 2) = -18$, $f_{xy}(-1, 2) = 6$ 15. $f_x(0, 1) = e + 1$, $f_y(0, 1) = 2$, $f_{xy}(0, 1) = e + 2$
17. $f_{xx} = 4$, $f_{yy} = -6y$, $f_{xy} = f_{yx} = 1$ 19. $f_{xx} = f_{yy} = f_{xy} = f_{yx} = 0$
21. $f_{xx} = (2 + x)e^{x+y^2}$, $f_{yy} = 2x(1 + 2y^2)e^{x+y^2}$, $f_{xy} = f_{yx} = 2y(1 + x)e^{x+y^2}$
25. $f_{xx}(-3, 2) = 20$, $f_{yy}(-3, 2) = -12$, $f_{xy}(-3, 2) = f_{yx}(-3, 2) = -30$
27. $f_{xx}(3, 5) = 0$, $f_{yy}(3, 5) = -\frac{3}{25}$, $f_{xy}(3, 5) = f_{yx}(3, 5) = \frac{1}{5}$ 29. $f_x = -2$, $f_y = 1$, $f_z = 7$
31. $f_u = \frac{2}{3}$, $f_v = -\frac{4}{3}$, $f_x = \frac{7}{9}$, $f_y = -\frac{14}{9}$ 33. $f_u = -3$, $f_v = -3$, $f_x = -8$, $f_y = 4$, $f_z = 8$

35. (a) $\dfrac{\partial Q}{\partial x} = 40x^{-1/3}y^{1/3}$, $\dfrac{\partial Q}{\partial y} = 20x^{2/3}y^{-2/3}$ (b) $\dfrac{\partial Q}{\partial x}\Big|_{(8,27)} = 60$, $\dfrac{\partial Q}{\partial y}\Big|_{(8,27)} = \dfrac{80}{9}$

   (c) When 8 units of labor and 27 units of capital are used, an increase of 1 unit of labor produces approximately 60 additional cameras and an increase of 1 unit of capital produces approximately $\frac{80}{9}$ additional cameras.
37. (a) $4x + 3y$ (b) $3x$ 39. (a) 56 roofs (b) 84 roofs

8.3 Exercise Set, p. 468

1. $(3, -\frac{3}{2})$ 3. $(\frac{9}{10}, -\frac{7}{10})$ 5. $(-\frac{1}{19}, \frac{46}{57})$ 7. $(0, 0), (1, 0)$ 9. relative minimum is $f(1, 2) = -7$
11. relative maximum is $f(3, 1) = 22$ 13. relative minimum is $f(\frac{1}{3}, -\frac{1}{3}) = -\frac{1}{27}$
15. relative minimum is $f(2, 5) = -47$ 17. relative minimum at any point on the line $y = x$
19. Fifty million dollars on research and \$4 million for development creates the maximum profit of \$$4292\frac{2}{3}$ million.   21. Sixty salespeople in 10 stores creates the maximum daily profit of \$120,000.
23. Three master plumbers and 2 apprentice plumbers yields a minimum cost of \$11,959.
25. Four meters wide by 4 meters long by 2 meters high yields the minimum surface area of 48 square meters.

8.4 Exercise Set, p. 477

1. constrained minimum value of $\frac{9}{5}$ occurs at $(\frac{3}{5}, \frac{6}{5})$
3. constrained maximum value of 12 occurs at $(6, 6)$, constrained minimum value of -12 occurs at $(-6, -6)$
5. constrained maximum value of $\frac{25}{3}$ occurs at $(\frac{10}{3}, -\frac{5}{3})$
7. constrained minimum value of $\frac{1}{6}$ occurs at $(\frac{1}{6}, \frac{1}{6}, \frac{1}{3})$ 9. 8, 8, 8   11. 324,000 square feet
13. 11,664 cubic inches   15. 10, 10, 10
17. Forty model A air conditioners and 30 model B air conditioners produces the constrained minimum weekly cost of \$8,450.
19. 398 compact cars and 1602 standard cars.
21. 10 meters long by 10 meters wide by 20 meters high yields the maximum volume of 2000 cubic meters.
23. $x = 200$, $y = 120$, $z = 300$

8.5 Exercise Set, p. 484

1. $df = (15x^2y + 7)\,dx + (5x^3 - 16y)\,dy$ 3. $df = \dfrac{x}{\sqrt{x^2 + y^3}}\,dx + \dfrac{3y^2}{2\sqrt{x^2 + y^3}}\,dy$

5. $df = 2xe^{3y}\,dx + 3x^2e^{3y}\,dy$ 7. $df = \dfrac{-2y}{(x - y)^2}\,dx + \dfrac{2x}{(x - y)^2}\,dy$ 9. $df = \dfrac{2}{x}\,dx + \dfrac{5}{y}\,dy$ 11. 0.37

13. -0.10 15. -0.12 17. 2.88   19. 8.96   21. $71\frac{2}{3}$ units of cameras
23. 0.46 square meters   25. $df = (4y^3 + 16xz^3)\,dx + (12xy^2 - 10z)\,dy + (-10y + 24x^2z^2)\,dz$

27. $df = \dfrac{2xz^4}{1 + y^3}\,dx - \dfrac{3x^2y^2z^4}{(1 + y^3)^2}\,dy + \dfrac{4x^2z^3}{1 + y^3}\,dz$ 29. $df = \dfrac{10x}{\sqrt{z}}\,dx - \dfrac{3}{\sqrt{z}}\,dy + \left(\dfrac{-5x^2}{2z^{3/2}} + \dfrac{3y}{2z^{3/2}}\right)dz$

31. $df = 2uve^{u^2v}(x^2 + y^3)\, du + u^2 e^{u^2v}(x^2 + y^3)\, dv + 2xe^{u^2v}\, dx + 3y^2 e^{u^2v}\, dy$

33. $df = \dfrac{2}{x-y}\, du + \dfrac{1}{x-y}\, dv - \dfrac{3}{x-y}\, dw + \dfrac{-2u - v + 3w}{(x-y)^2}\, dx + \dfrac{2u + v - 3w}{(x-y)^2}\, dy$ 35. -0.0053

37. 6.50   39. (a) 0.54 square feet   (b) 0.28 square feet

8.6 Exercise Set, p. 497

1. 15   13. $e + \dfrac{1}{e} - 2$ 5. $\dfrac{\ln 2}{2}(e^2 - 1)$ 7. $\frac{1}{3}$ 9. $9 \ln 3 - 4$ 11. $\frac{2}{3}$ 13. 20   15. 27

17. 1   19. $\frac{5}{28}$ 21. 64   23. 42   25. $\frac{43}{24}$ 27. 32   29. $\frac{16}{3}$ 31. $2e - 4$ 33. $-\frac{21}{5}$

35. $\dfrac{3}{5}\dfrac{80^{4/3} - 30^{4/3}}{2000}(50^{5/3} - 20^{5/3}) \approx 40.08$ thousand gallons   37. 15.5 feet

Review Exercises, Chapter 8, p. 500

1. (a) 36
   (b) 0
   (c) -18
   (d) -36

3. (a) 4960
   (b) 7500

5. The surface is a plane
   intersecting the coordinate
   axes at $(6, 0, 0)$, $(0, 2, 0)$,
   and $(0, 0, -3)$.

7. $f_x = (2x + x^2)e^{x-y^2}$, $f_y = -2x^2 y e^{x-y^2}$, $f_{xy} = -2y(2x + x^2)e^{x-y^2}$ 9. $f_{xx} = 24x + (4x^3 + 6x)e^{x^2-y^2}$,
   $f_{yy} = -2x + (4xy^2 - 2x)e^{x^2-y^2}$, $f_{xy} = -2y - 2y(2x^2 + 1)e^{x^2-y^2}$, $f_{yx} = -2y - 2y(2x^2 + 1)e^{x^2-y^2}$

11. $f_x = \dfrac{15x^4 - y^4}{z^3}$, $f_y = \dfrac{-4xy^3}{z^3}$, $f_z = \dfrac{-3(3x^5 - xy^4)}{z^4}$ 13. $(-\frac{7}{5}, \frac{3}{5})$ 15. no relative extrema

17. relative maximum is $f(2, 2) = 8$ 19. 200, at (10, 10)

21. constrained minimum value of $\frac{64}{3}$ occurs at $(\frac{8}{3}, \frac{8}{3}, \frac{8}{3})$ 23. $df = 3x^2 e^{y^2z}\, dx + 2x^3 yze^{y^2z}\, dy + x^3 y^2 e^{y^2z}\, dz$

25. -4.72 27. (a) $6x + 2y$ (b) $2x$ 29. 2 meters on each side

31. 9 type A engines and 5 type B engines for $793 profit   33. $20,000 each   35. 12.04

37. $11.25\pi \approx 35.34$ square inches   39. -6 41. $4(e - 1)$ 43. $\frac{3}{4}e^{4/3} - e + \frac{1}{4}$ 45. $\dfrac{5e^3 - e}{2}$

47. $\frac{32}{35}$ 49. $\frac{7}{10}$ 51. $\frac{19}{6}$ mg/liter

Chapter Test, Chapter 8, p. 503

1. (a) -1 (b) -2 (c) $-\frac{7}{8}$ (d) all points (x, y) where $x \neq y$,   2. (a) 1920 (b) 1900

3. $f_x = -2ye^{-2x} + 8x$, $f_y = e^{-2x}$, $f_{xx} = 4ye^{-2x} + 8$, $f_{yy} = 0$, $f_{xy} = -2e^{-2x}$, $f_{yx} = -2e^{-2x}$

4. $f_x = 4z$, $f_y = -2ye^{y^2}z$, $f_z = 4x - e^{y^2}$ 5. -0.56

6. At the critical point $(1, 0)$, f has relative minimum -2. At the critical point $(-1, 0)$, f has no relative
   extremum.   7. Maximum value 2 occurs at $(\frac{3}{2}, -\frac{1}{2})$; minimum value -2 occurs at $(-\frac{3}{2}, \frac{1}{2})$.

8. (a) 0.9 minutes per day weightless (b) 3.2 minutes per hour of exercise

9. plant A producing 1800 coats and plant B producing 2200 coats weekly

10. 12 units of A, 16 units of B, 20 units of C   11. 88 cubic feet   12. $\frac{1}{6}$ 13. 78

||||||||||||| **CHAPTER 9**

9.1 Exercise Set, p. 520

1. IV   3. I   5. II   7. $\frac{4}{9}\pi$ 9. $-\frac{5}{6}\pi$ 11. $\frac{5}{12}\pi$ 13. $240°$ 15. $-105°$ 17. $450°$

19. $\frac{5}{2}\pi$ 21. 0   23. -1 25. $-\frac{1}{2}$ 27. $5, 5\sqrt{3}$ 29. $4\sqrt{2}$ 31. $\sin\theta = \frac{3}{5}, \cos\theta = \frac{4}{5}, \tan\theta = \frac{3}{4}$

33. $\sin\theta = \dfrac{x}{\sqrt{x^2+1}}, \cos\theta = \dfrac{1}{\sqrt{x^2+1}}, \tan\theta = x$ 35. $\sin\theta = \frac{12}{13}, \cos\theta = -\frac{5}{13}, \tan\theta = -\frac{12}{5}$

37. $\sin\theta = -\frac{5}{13}, \cos\theta = \frac{12}{13}, \tan\theta = -\frac{5}{12}$ 39. $\sin\theta = \dfrac{-2\sqrt{5}}{5}, \cos\theta = -\dfrac{\sqrt{5}}{5}, \tan\theta = 2$ 41. $4\cot\theta$

43. $7\sin\theta$ 45. $150\sin35° \approx 86$ meters   47. $\dfrac{80}{\tan 8°} \approx 569.2$ ft

9.2 Exercise Set, p. 529

1. $-3\sin 3x$ 3. $2x\cos(x^2+1)$ 5. $3\sin^2 x\cos x$ 7. $2x\cos x - x^2\sin x$ 9. $\cos^2 x - \sin^2 x$

11. $-2x\sin x^2 - 3x^2\sin x^3$ 13. $2\tan x\sec^2 x$ 15. $3x^2\tan x + x^3\sec^2 x$ 17. $\dfrac{\cos x}{2\sqrt{\sin x}}$

19. $-2e^{2x}\sin e^{2x}$ 21. $e^x(\sin x + \cos x)$ 23. $\dfrac{\cos\sqrt{x-1}}{2\sqrt{x-1}}$ 25. $2\sin x\cos^2 x - \sin^3 x$ 27. $\cot x$

29. $\dfrac{\sin x - x\cos x}{\sin^2 x}$ 35. $z_x = -\sin x\sin y, z_y = \cos x\cos y$ 37. $z_x = \dfrac{1}{y}\cos\left(\dfrac{x}{y}\right), z_y = \dfrac{-x}{y^2}\cos\left(\dfrac{x}{y}\right)$

39. $z_x = 3x^2 y^4\sec^2(x^3 y^4), z_y = 4x^3 y^3\sec^2(x^3 y^4)$ 41. $-\sqrt{3}$ 43. $y = x - \pi$

45. -2000π gallons per day per month   47. 20.9 feet per hour

9.3 Exercise Set, p. 536

1. $-\dfrac{\cos 5x}{5} + C$ 3. $\sin(x+4) + C$ 5. $\sin x - \cos x + C$ 7. $\dfrac{\sin^2 x}{2} + C$ 9. $-\frac{1}{4}\cos^4 x + C$

11. $\dfrac{-\cos x^3}{3} + C$ 13. $-\csc x + C$ 15. $2\sqrt{\sin x} + C$ 17. $\sin x - x\cos x + C$

19. $-\frac{1}{3}\ln|\cos 3x| + C$ 21. $\frac{1}{2}\tan^2 x + C$ 23. $\dfrac{\sqrt{3}-1}{2}$ 25. 0   27. $\dfrac{\sqrt{2}}{2}$ 29. $\dfrac{1}{4} - \dfrac{\cos\pi^4}{4}$

31. $\dfrac{\pi - 2}{8}$ 33. $2 - \sqrt{2}$ 35. $\sqrt{2} - 1$ 37. $y = x - \sin 2x$ 39. (a) 10 ft. (b) $7 + \dfrac{2}{\pi}$ ft.

Review Exercises, Chapter 9, p. 538

1. II   3. $\dfrac{7\pi}{4}$ 5. $330°$ 7. $\dfrac{7\pi}{2}$ 9. $\sqrt{3}$ 11. $\sin\theta = \frac{4}{5}, \cos\theta = \frac{3}{5}, \tan\theta = \frac{4}{3}$

13. $\sin\theta = -\frac{4}{5}, \cos\theta = -\frac{3}{5}, \tan\theta = \frac{4}{3}$ 15. $\cos x + 2\sin 2x$ 17. $2\cos^2 x\cos 2x - 2\cos x\sin x\sin 2x$

19. $\dfrac{3}{\sin x} - \dfrac{3x\cos x}{\sin^2 x}$ 21. $\dfrac{\sec^2 x}{2\sqrt{\tan x}}$ 23. $z_x = 6\cos 2x, z_y = 6\sin 3y$ 25. $y = 0$ 27. $-e^{\cos x} + C$

29. $-2x\cos\frac{1}{2}x + 4\sin\frac{1}{2}x + C$ 31. $\dfrac{\pi}{2}$ 33. 4   35. $\sin\theta = \frac{4}{5}, \theta \approx 53.13°$

37. (a) $5\pi\sqrt{2}$ people per day   (b) -5π people per day

Chapter Test, Chapter 9, p. 540

1. (a) $\dfrac{5\pi}{6}$ (b) $420°$ 2. $\sin \theta = \dfrac{\sqrt{10}}{10}$, $\cos \theta = \dfrac{-3\sqrt{10}}{10}$, $\tan \theta = -\dfrac{1}{3}$ 3. $-3 \sin\left(3x + \dfrac{\pi}{2}\right)$

4. $-2 \sin x + 3 \sec^2 x$ 5. $3x^2 \sin x^2 + 2x^4 \cos x^2$ 6. $3 \tan^2 x \sec^2 x$ 7. $\frac{2}{3}(\sin x)^{3/2} + C$

8. $\frac{1}{2} \sin x^2 + C$ 9. $-\frac{1}{3} \ln |\cos (3x - 4)| + C$ 10. $N(t) = 12 + 10 \sin\left(\dfrac{\pi t}{6}\right)$ million plants

|||||||||||| **CHAPTER 10**

10.1 Exercise Set, p. 547

1. 30   3. $81e^{-3x}$ 5. $48(1 - x)^{-5}$ 7. $\frac{105}{32}x^{-9/2}$ 9. 3   11. 2   13. 4

23. $\frac{3}{2}x^2 - \ln |x| + C$ 25. $\frac{1}{6}x^3 + 2x^2 + Cx + D$ 27. $3 \ln |x| - 5x^{-1} + C$ 29. $-\frac{3}{2} \sin (2x) + Cx + D$

31. $y = \frac{1}{3}e^{2x} + \frac{2}{3}$ 33. $y = \frac{1}{2} \sin (2x) + \frac{1}{3} \cos (3x) - \frac{1}{3}$ 35. $y = 3x - \sin x + 4$

10.2 Exercise Set, p. 560

1. $y = Ce^{2x}$ 3. $y = Ce^{x^2/2}$ 5. $y = Cx^2$ 7. $y = Ce^{x^3}$ 9. $y = Ce^x - 5$ 11. $y = \dfrac{1 - Ce^{-x^2}}{2}$

13. $y = \sqrt[3]{\frac{3}{2}x^2 + C}$ 15. $y = \dfrac{x^2}{4} + \dfrac{1}{2} \ln |x| + C$ 17. $y = Ce^{x + (5/2)x^2}$ 19. $y = \frac{1}{2} \ln (4 + x^2) + C$

21. $N = 50 + Ce^{-5t}$ 23. $y = \sqrt{10t + 9}$ 25. $y = 5e^{-x^2}$ 27. $y = \sqrt{3x^2 + 36}$

29. $y = \frac{1}{2}[\ln (x^2 + 2) + 6 - \ln 6]$ 31. (a) $Q = 5000e^{0.08t}$ (b) $5000e^{0.8} \approx \$11,127.70$

33. (a) $A = 80e^{\ln (13/16)t} = 80(\frac{13}{16})^t$ (b) 57 milligrams   35. (a) $(5 - t)^2$ feet (b) 5 hours

37. $r = k \ln\left(\dfrac{S}{S_0}\right)$ 39. 1.60 units, 1.78 units, 1.88 units, 2.00 units   41. 4494   43. 50,759,000

45. (a) 1326 grams (b) 25.2 minutes

47. 2.78 cubic feet, 1.59 cubic feet, 1.15 cubic feet, 0.99 cubic feet; if the room were filled with "fresh" air it would contain 0.9 cubic feet of carbon dioxide.

49. (a) $\dfrac{dx}{dt} = -0.5x + 0.2xy$, $\dfrac{dy}{dx} = y - 0.3xy$ (b) $-5 \ln y + 2y = 10 \ln x - 3x + 2300 - 35 \ln 10$

51. 852   57. (a) $z = 120 \dfrac{1 - e^{0.021t}}{2 - 3e^{0.021t}}$ (b) $z(3) = 6.53$ grams, $z(10) = 16.48$ grams

59. $z = \dfrac{180t}{2t + 13}$; when $t = 5$, $z \approx 39$ grams

10.3 Exercise Set, p. 581

1. $P = 80 - Ce^{-4t}$ 3. $P = \dfrac{100}{1 + Ce^{-1000t}}$ 5. $P = 30e^{-Ce^{-5t}}$ 7. $P = 50 - 38e^{-5t}$

9. $P = \dfrac{80}{1 + 7e^{-240t}}$ 11. $P = 16(8)^{-e^{-3t}}$

13. (a) $Q(t) = 200 - 120e^{-0.036t} = 200 - 120(\frac{5}{6})^{t/5}$; $Q(15) \approx 130$, $Q(30) \approx 159$ (b) on the 20th day

15. (a) $\dfrac{dS}{dx} = k(20 - S)$ (b) $S(x) = 20 - 19e^{-kt}$ (c) $S(x) = 20 - 19e^{-0.0859t}$ (d) $S(5) \approx 7630$ units

17. (a) $T = 400 - 320(\frac{1}{2})^{t/3}$ (b) $T(5) = 299.2°$

19. (a) $V = 213\frac{1}{3}(1 - e^{-0.15t})$ (b) $V(10) \approx 165.7$ ft/sec (c) $213\frac{1}{3}$ ft/sec ≈ 145 miles/hour

23. (a) $P = \dfrac{21{,}000}{1 + 20e^{-0.301t}}$, $P(5) \approx 3860$ (b) 9.64 hours $\approx$ 9 hours, 38 minutes

25. $P = \dfrac{48{,}000}{1 + 15e^{-0.0766t}}$ 27. 19%, 33% 29. (a) $dP/dt = 0.317P(1 - P)$ (b) 9.29 months

31. (a) $P = 21{,}000(21)^{-e^{-0.1119t}}$; $P(5) \approx 3686$
   (b) From part (a), $P(4) = 2999.9 \approx 3000$. From Exercise 23, the logistic model predicts
   $P(4) = 3000.07 \approx 3000$.
   (c) 12.6 hours.

33. $P = 48{,}000(16)^{-e^{-0.0330t}}$, $P(5) \approx 44{,}000$ 35. $L/2$

37. (a) $P'' = kP'(\ln L - \ln P - 1)$ (b) $\left(\dfrac{1}{k} \ln \left[\ln \left(\dfrac{L}{P_0} \right) \right], \dfrac{L}{e} \right)$

10.4 Exercise Set, p. 594

1. $y = \frac{7}{4} + Ce^{-4x}$ 3. $y = -\frac{2}{3}x + \frac{2}{9} + Ce^{3x}$ 5. $y = x + \dfrac{C}{x}$ 7. $y = \frac{1}{3}e^{5x} + Ce^{2x}$

9. $y = \dfrac{\ln x}{x} + \dfrac{x}{2} + \dfrac{C}{x}$ 11. $y = (x + 1)e^x$ 13. $y = 2 + 8e^{-x^2}$ 15. $y = \dfrac{\ln x + 2}{x}$

17. $y = -\frac{b}{a} + Ce^{ax}$ 19. 601,000 21. \$112,712; \$99,495

23. (a) $y = 40(50 + t) - \dfrac{187{,}500{,}000}{(50 + t)^3}$ (b) $y(10) \approx 1532$ grams

25. $y = 12{,}000 - 6000e^{-0.005t}$; $y(12) = 6349$; $y(24) = 6678$; $y(60) = 7555$
27. $x = 10 + 13e^{-t}$, $\bar{x} = 10$ 29. $x = 4 + 5e^{-0.2t} - 6e^{-3t}$, $\bar{x} = 4$

10.5 Exercise Set, p. 614

1. $y = c_1 x + c_2/x$ 3. $y = c_1/\sqrt{x} + c_2 x^3$ 5. $y = c_1 e^{5x} + c_2 e^{-2x}$ 7. $y = c_1 e^{2x} + c_2 e^{-2x}$

9. $y = c_1 \cos 5x + c_2 \sin 5x$ 11. $y = c_1 e^{-x/2} + c_2 x e^{-x/2}$ 13. $y = e^{-x/2}\left(c_1 \cos \dfrac{\sqrt{3}}{2} x + c_2 \sin \dfrac{\sqrt{3}}{2} x\right)$

15. $y = c_1 + c_2 e^{x/3}$ 17. $y = c_1 e^{2x/5} + c_2 e^{-3x/2} - 5$ 19. $y = c_1 e^{0.4x} + c_2 x e^{0.4x} + 6.25$

21. $y = e^{-x/2}(c_1 \cos 2x + c_2 \sin 2x) + \dfrac{5}{17}$ 23. $y = 6e^{x/3} + 2e^{-3x}$ 25. $y = 2e^{-x} + 3xe^{-x}$

27. $y = e^{-x}(-3 \cos \sqrt{2}\, x - \sqrt{2} \sin \sqrt{2}\, x) + 4$
29. $y = 5e^{2x/5} - 2x + 5$
31. $\sqrt{10} \sin (4x + 1.249)$; amplitude $= \sqrt{10}$, period $= \pi/2$, frequency $= 2/\pi$, phase shift $= 0.312$ (leftward).
33. $\sqrt{29} \sin (\frac{1}{3}x + 5.903)$; amplitude $= \sqrt{29}$, period $= 6\pi$, frequency $= 1/6\pi$, phase shift $= 17.708$ (leftward).

35. $y = A \sin (5x + \theta)$ 37. $y = Ae^{-x/2} \sin \left(\dfrac{\sqrt{3}}{2} x + \theta\right)$ 39. $y = Ae^{-x/2} \sin (2x + \theta) + \dfrac{5}{17}$

10.6 Exercise Set, p. 624

1. $x(t) = 4e^{-t} - 3e^{-5t/3} + 2$, $x(5) = \$2.03$, $\bar{x} = \$2.00$

3. $x(t) = -\dfrac{5}{2} e^{t/10} - \dfrac{5}{2} e^{-t/2} + 8$, $x(5) = \$3.67$, $\bar{x}$ does not exist

5. $x(t) = e^{-t}(-5 \cos 3t - 2 \sin 3t) + 25$, $x(1) = \$26.72$ 7. $x = c_1 e^{7t} + c_2 e^{2t}$, $y = -\frac{3}{2}c_1 e^{7t} + c_2 e^{2t}$
9. $x = c_1 \cos 2t + c_2 \sin 2t$, $y = 2(c_1 - c_2) \cos 2t + 2(c_1 + c_2) \sin 2t$
11. $x = c_1 e^{3t} + c_2 e^{-2t} - \frac{11}{3}$, $y = c_1 e^{3t} + 6c_2 e^{-2t} - \frac{23}{3}$
13. $x = 500 \cos (0.4t) + 400 \sin (0.4t) = 640.3 \sin (0.4t + 0.896)$
   $y = 800 \cos (0.4t) - 1000 \sin (0.4t) = 1280.6 \sin (0.4t + 2.467)$

15. (a) $x = 18e^{-t} - 3e^{-7t} + 3$, $y = 9e^{-t} + 3e^{-7t} + 2$

(b)

(c) Stable equilibrium at $(3, 2)$

(d) Stabilizing

17. (a) $x = -\frac{3}{2}e^{-4t} + 7e^{-t} - \frac{3}{2}$, $y = \frac{3}{4}e^{-4t} + 7e^{-t} - \frac{7}{4}$

(b)

(c) no stable equilibrium.

(d) total disarmament is possible.

23. The given system yields $J'' + abJ = 0$; by Theorem 9 the solution is $J = A \sin(\sqrt{ab}\,t + \theta)$. Similarly, $R'' + abR = bc$, so that $R = B \sin(\sqrt{ab}\,t + \phi) + c/a$. The frequency is $2\pi/\sqrt{ab}$.

Review Exercises, Chapter 10, p. 628

1. $360x^2 - \frac{15}{16}x^{-7/2}$ 5. $y = \dfrac{-1}{2x^2 + C}$ 7. $y = -e^{-3x} + Ce^{-2x}$ 9. $y = c_1 e^{x/3} + c_2 e^{-7x}$

11. $y = Ce^{3x^2 + 2x}$ 13. $y = Ce^{5x} - \dfrac{2}{5}$ 15. $y = \dfrac{x^4 + 2x^2 + C}{4(x^2 + 1)}$ 17. $y = (e^x - x + C)^{-1}$

19. $y = e^{x/4}\left[c_1 \cos\left(\dfrac{\sqrt{7}}{4}x\right) + c_2 \sin\left(\dfrac{\sqrt{7}}{4}x\right)\right] + 3$ 21. $P = \dfrac{30}{1 + 0.5e^{-150t}}$ 23. $y = 2e^{x^2/2 - 3x} - 1$

25. $y = 10e^{2x/5} - 2e^{-x}$ 27. $y = \dfrac{x^2}{4} + \dfrac{x}{3} - \dfrac{1}{2x^2}$ 29. $\sqrt{29}\sin(3x + 1.9513)$

31. $D(x) = \dfrac{-10 \ln x}{\ln 5} + 12$; $D(2) \approx 7.69$, $D(3) \approx 5.17$ 33. $A = 10e^{-0.01t}$; $A(60) \approx 5.49$ lb; 4.99 hours

35. $-2 \ln y + 5y = -3 \ln x + 4x + C$

37. (a) $\dfrac{dz}{dt} = \dfrac{12k}{49}(98 - z)^2$ (b) $z = \dfrac{98t}{t + 12}$ (c) $z(10) \approx 44.55$ grams

39. (a) $\dfrac{dv}{dt} = 2(16 - v)$ (b) $v = 16 - 7e^{-2t}$ (c) 15.05 ft/sec   (d) 16 ft/sec   41: $P = \dfrac{6400}{1 + 4e^{-0.0173t}}$

43. $A = 300 - 20t - 100e^{0.1t}$ thousands; $A(5) \approx \$35{,}128$; no, since $A(6) \approx -\$2212$.

45. $x(t) = 10e^{-t} + 20 - 12e^{-3t}$; $\bar{x} = \$20$

47. $x = e^{3t}[c_1 \cos 2t + c_2 \sin 2t]$; $y = e^{3t}[(-c_1 - c_2) \cos 2t + (c_1 - c_2) \sin 2t]$

49. $x = 4 \cos t - 2 \sin t - 3$, $y = 6 \cos t + 2 \sin t - 1$

51. (a) $x = \tfrac{1}{2}e^{-6t} + 5e^{-2t} + \tfrac{3}{2}$, $y = -\tfrac{1}{6}e^{-6t} + 5e^{-2t} + \tfrac{7}{6}$

(b)

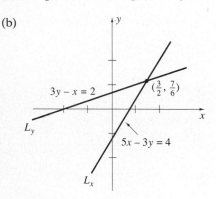

(c) Stable equilibrium
(d) Stabilizing

Chapter Test, Chapter 10, p. 632

2. $y = (\tfrac{1}{12}x^2 - \tfrac{5}{6} \ln |x| + C)^2$ 3. $y = \dfrac{x}{3} - \dfrac{x^2}{4} + \dfrac{C}{x^2}$ 4. $y = e^{2x}(c_1 \cos x + c_2 \sin x) + \dfrac{12}{5}$

5. $y = C_1 e^{-4x} + C_2 x e^{-4x}$ 6. $y = (\tfrac{3}{2}x^2 + 3x + 8)^{1/3}$

7. (a) $A = 0.8(1 - e^{-0.75t})$; $A(1) = 0.422$ grams, $A(2) = 0.621$ grams, $A(5) = 0.781$ grams
   (b) Yes; $\lim\limits_{t \to +\infty} A(t) = 0.8$ grams

8. (a) $dT/dt = k(70 - T)$ (b) $T = 70 - 34e^{-0.087t}$ (c) $T(10) \approx 55.76°\text{F}$

9. (a) $dP/dt = kP(12{,}000 - P)$ (b) logistic   (c) $P = \dfrac{12{,}000}{1 + 59e^{-0.2355t}}$

10. $dA/dt = 0.08A + 2000 - 300t$; $A = 21{,}875 + 3750t - 1875e^{0.08t}$

11. $x = -3e^{-t} + 8e^{2t} - 2$, $y = -6e^{-t} + 4e^{2t} - \tfrac{5}{2}$

|||||||||||| **CHAPTER 11**

11.1 Exercise Set, p. 640

1. (a) $\tfrac{7}{15}$ (b) $\tfrac{8}{15}$ (c) $\tfrac{1}{3}$ (d) $\tfrac{1}{15}$ 3. (a) $\tfrac{3}{8}$ (b) $\tfrac{1}{2}$ (c) $\tfrac{1}{8}$ (d) $\tfrac{5}{8}$ (e) $\tfrac{7}{8}$

5. $X(M) = 1$, $X(I) = 2$, $X(S) = 3$, $X(P) = 4$

X	1	2	3	4
$P(X)$	$\tfrac{1}{11}$	$\tfrac{4}{11}$	$\tfrac{4}{11}$	$\tfrac{2}{11}$

7. $X(G) = 1$, $X(Y) = 2$, $X(R) = 3$, $X(B) = 4$, $X(W) = 5$

X	1	2	3	4	5
$P(X)$	$\frac{2}{15}$	$\frac{1}{3}$	$\frac{4}{15}$	$\frac{1}{15}$	$\frac{3}{15}$

9. (b)

11. (a) $\frac{1}{6}$
   (b) $\frac{1}{3}$
   (c) $\frac{1}{2}$
   (d) $\frac{5}{6}$

13. (a)

X	2	3	4	5	6	7	8	9	10	11	12
$P(X)$	$\frac{1}{36}$	$\frac{1}{18}$	$\frac{1}{12}$	$\frac{1}{9}$	$\frac{5}{36}$	$\frac{1}{6}$	$\frac{5}{36}$	$\frac{1}{9}$	$\frac{1}{12}$	$\frac{1}{18}$	$\frac{1}{36}$

(b)

15. (a)

X	1	2	3	4	5	6	7	8	9	10	11
$P(X)$	$\frac{1}{36}$	$\frac{1}{18}$	$\frac{1}{12}$	$\frac{1}{9}$	$\frac{5}{36}$	$\frac{1}{6}$	$\frac{5}{36}$	$\frac{1}{9}$	$\frac{1}{12}$	$\frac{1}{18}$	$\frac{1}{36}$

(b)

17. (a) $\frac{5}{8}$ (b) $\frac{5}{8}$ (c) $\frac{1}{4}$ (d) 1

11.2 Exercise Set, p. 649

1. (a) $f(x) \geq 0$ for all x in $[2, 5]$, and $\int_2^5 \frac{2}{21} x \, dx = 1$.   (b) $\frac{4}{7}$ (c) $\frac{16}{21}$

3. (a) $f(x) \geq 0$ for all x in $[0, +\infty)$, and $\int_0^{+\infty} e^{-x} \, dx = 1$.   (b) 0.632   (c) 0.233   (d) 0   (e) 0.00674

5. $\frac{3}{16}$ 7. 8   9. $\int_1^{+\infty} \frac{k}{\sqrt{x}} \, dx$ diverges if $k \neq 0$.   11. $F(x) = \frac{x^2 - 4}{21}$ 13. $\frac{8}{7} - \frac{16}{7(x+1)}$

15. $\frac{25}{128} \approx 0.195$ 17. (a) $\frac{3}{4}$ (b) $\frac{1}{36}$ (c) $\frac{1}{4}$ 19. $1 - e^{-1.5} \approx 0.7769$

21. (a) $F(t) = 1 - 2^{-t}$ (b) $\frac{15}{16}$; $\frac{1}{16}$; $\frac{1}{64}$

11.3 Exercise Set, p. 664

1. $\mu = \frac{1}{2}$, Var $(\bar{X}) = 2.25$, $\sigma = 1.5$

3.

X	-4	-3	-2	-1	0	1	2	3	4
$P(X)$	$\frac{1}{24}$	0	$\frac{1}{6}$	$\frac{5}{24}$	$\frac{1}{4}$	$\frac{1}{6}$	$\frac{1}{8}$	$\frac{1}{24}$	0

$\mu = -\frac{1}{6}$, Var $(X) = 2.56$, $\sigma = 1.60$

5. $\mu = 1$, Var $(X) = \frac{1}{2}$, $\sigma = 0.707$ 7. $\mu = 7$, Var $(X) = 5.833$, $\sigma = 2.415$

9. $\mu = 2.56$, Var $(X) = 1.87$, $\sigma = 1.37$ 11. $\mu = \frac{26}{7}$, Var $(X) = 0.704$, $\sigma = 0.839$

13. $\mu = \frac{7}{2}$, Var $(X) = 2.08$, $\sigma = 1.44$ 15. $\mu = \frac{4}{3}$, Var $(X) = \frac{2}{9}$, $\sigma = \frac{\sqrt{2}}{3}$ 17. $m = \sqrt{\frac{29}{2}} \approx 3.81$

19. $m = 30.25$ 21. $m = \sqrt[4]{2} \approx 1.189$ 23. $m = 2$ 25. (a) 4.11 gal.   (b) 1.64 gal.

27. (a) 15 min.   (b) 8.66 min   29. $m = 1.386$ min   31. $m = 1$ day

11.4 Exercise Set, p. 685

1. $f(x) = \frac{1}{8}$, $\mu = 11$, $m = 11$, $\sigma = \frac{4\sqrt{3}}{3}$ 3. $f(x) = \frac{1}{11}$, $\mu = -\frac{9}{2}$, $m = -\frac{9}{2}$, $\sigma = \frac{11\sqrt{3}}{6}$

5. $d = \frac{26}{5}$, $f(x) = \frac{5}{16}$ 7. $\mu = \frac{1}{5}$, $m = \frac{\ln 2}{5}$, $\sigma = \frac{1}{5}$ 9. (a) 0.9933   (b) 0.3679   (c) 0.3834

11. (a) 0.9429   (b) 0.0212   (c) 0.1587   13. (a) 0.9850   (b) 0.3063   (c) 0.0119   (d) 0.1762

15. (a) 0.9332   (b) 0.5000   (c) 0.3413   (d) 0.3830   (e) 0.4013   (f) 0.8413

17. $X =$ the real number chosen; $f(x) = \frac{1}{50}$; $\frac{1}{5}$, $\frac{2}{5}$, $\frac{3}{5}$

19. $X =$ the waiting time in minutes; $f(x) = \frac{1}{10}$ on $[0, 10]$; $\frac{1}{5}$, $\frac{1}{10}$, $\frac{3}{10}$

21. $X =$ lifespan of a lightbulb; $f(x) = \frac{1}{750} e^{-x/750}$ on $[0, +\infty)$;   (a) 0.3297   (b) 0.4866   (c) 0.0920
   (d) 0.2636

23. $X =$ length of time until the next accident; $f(x) = \frac{1}{20} e^{-x/20}$ on $[0, +\infty)$;   (a) 0.1813
   (b) 0.1484   (c) 0.6065

25. $X =$ temperature of the classroom; $f(x) = A(z)$, where $Z = \dfrac{X - 70}{2}$;   (a) 0.8413   (b) 0.6826   (c) 0.0013

27. $X =$ voltage in the house; $f(x) = A(z)$, where $Z = \dfrac{X - 115}{3}$; 0.9050; 0.0475

29. $X =$ the student's score; $f(x) = A(z)$, where $Z = \dfrac{X - 72}{8}$;   (a) 80   (b) 401   (c) 528   (d) 176   (e) 15

Review Exercises, Chapter 11, p. 689

1. (a) $\frac{3}{8}$ (b) $\frac{1}{4}$ (c) $\frac{1}{8}$ (d) $\frac{1}{2}$ 3. $X(R) = 1$, $X(G) = 2$, $X(B) = 3$, $X(w) = 4$

X	1	2	3	4
$P(X)$	$\frac{5}{14}$	$\frac{3}{14}$	$\frac{2}{7}$	$\frac{1}{7}$

5. (a) $\int_{1.5}^{2.5} P(X)\,dX = 0.3$ (b) $\int_{0.5}^{3.5} P(X)\,dX = 0.6$ (c) $\int_{1.5}^{4.5} P(X)\,dX = 0.8$ (d) $\int_{3.5}^{5.5} P(X)\,dX = 0.4$

7. (a) $\frac{1}{36}$ (b) $\frac{7}{27}$ (c) 0 (d) $\frac{13}{54}$ 9. (a) $\frac{7}{27}$ (b) $\frac{8}{9}$ (c) $\frac{5}{36}$ 11. $\mu = 76$, $\sigma = 12.40$

13. $k = 4$, $\mu = \frac{4}{3}$, $\sigma = \dfrac{\sqrt{2}}{3}$ 15. 100 hours   17. (a) $\frac{2}{5}$ (b) $\frac{1}{3}$ (c) $\frac{1}{6}$

19. (a) $f(t) \geq 0$ and $\int_{0}^{+\infty} 3e^{-3t}\,dt = 1$ (b) $1 - e^{-15}$ (c) e^{-12} 21. (a) 0.5328 (b) 0.9875 (c) 0.3085

23. (a) 0.1151 (b) 0.7881 (c) 0.2119 (d) 0.7698 (e) 0.1327 (f) 0.2088   25. 0.0950; 9.50%

Chapter Test, Chapter 11, p. 691

1. (a) $\frac{3}{20}$ (b) $\frac{3}{4}$

2. (a)

X	1	2	3	4	5
$P(X)$	$\frac{3}{20}$	$\frac{1}{5}$	$\frac{2}{5}$	$\frac{3}{20}$	$\frac{1}{10}$

(b) $\frac{8}{20}$

(c) 2.85
(d) 1.152

3. $k = \frac{1}{4}$; $F(x) = \frac{1}{2}\sqrt{x} - \frac{1}{2}$ 4. (b) $\frac{4}{5}$ (c) $\frac{1}{10}$ (d) $\frac{1}{26}$ 5. $E(X) = 0$; $\sigma = \sqrt{\frac{1}{5}} \approx 0.447$
6. (a) $f(x) = \frac{1}{40}$ (b) $\frac{1}{4}$ (c) $\frac{1}{2}$ (d) $\frac{1}{8}$ 7. (a) 0.3935 (b) 0.0995 (c) 0.3679
8. (a) 0.8944 (b) 0.7734 (c) 0.4649 (d) 0.0062 (e) 0.0401 (f) 0.7888

||||||||||||| **CHAPTER 12**

12.1 Exercise Set, p. 705

1. 2, 9, 16, 23, 30, 37   3. $\frac{3}{2}, \frac{3}{2}, \frac{9}{8}, \frac{3}{4}, \frac{15}{32}, \frac{9}{32}$ 5. $-\frac{1}{5}, \frac{1}{10}, -\frac{1}{15}, \frac{1}{20}, -\frac{1}{25}, \frac{1}{30}$ 7. 3, 5, 3, 5, 3, 5
9. $\{3n\}$ 11. $\left\{\dfrac{n}{n+1}\right\}$ 13. $\{(-2)^n\}$

15.

n	1	5	10	20	50	100	200
$\dfrac{6n}{11-3n}$	0.75	-7.50	-3.158	-2.449	-2.158	-2.076	-2.037

The sequence converges to -2.

17.

n	1	5	10	25	50	101
$\dfrac{3n+(-1)^n}{n+2}$	0.667	2.000	2.583	2.741	2.904	2.932

The sequence converges to 3.

19. converges to 8   21. converges to 0   23. converges to 0   25. diverges to $+\infty$
27. converges to 0   29. converges to 0   31. diverges to $+\infty$ 33. converges to 0
35. diverges to $+\infty$ 37. monotone decreasing   39. neither   41. monotone decreasing
43. neither

12.2 Exercise Set, p. 716

1. 21   3. 10   5. $\frac{163}{60}$ 7. $\frac{101}{420}$ 9. $\frac{43}{64}$ 11. $\frac{6305}{2187} \approx 2.88$ 13. 77.12   15. $\frac{665}{64}$
17. $\dfrac{1-(0.03)^9}{0.97} \approx 1.031$ 19. \$4859.47   21. No; Yes; \$84.94   23. $-840, 0$ 25. 120, 0
27. $p(x) = -4(x-3) - (x-3)^2$ 29. $p(x) = 4 + 6(x-1) + 4(x-1)^2 + (x-1)^3$
31. $p(x) = 15 + 17(x-2) + 7(x-2)^2 + (x-2)^3$
33. $p(x) = -32 + 80(x+2) - 80(x+2)^2 + 40(x+2)^3 - 10(x+2)^4 + (x+2)^5$

12.3 Exercise Set, p. 726

1. $x - \dfrac{x^3}{3!} + \dfrac{x^5}{5!} - \dfrac{x^7}{7!}$ 3. $x - \frac{1}{2}x^2 + \frac{1}{3}x^3 - \frac{1}{4}x^4 + \frac{1}{5}x^5 - \frac{1}{6}x^6 + \frac{1}{7}x^7 - \frac{1}{8}x^8$
5. $1 + x + x^2 + x^3 + x^4 + x^5 + x^6$ 7. $1 + \frac{1}{2}x - \frac{1}{8}x^2 + \frac{1}{16}x^3 - \frac{5}{128}x^4 + \frac{7}{256}x^5$ 9. $x + \frac{1}{3}x^3$
11. $\dfrac{\sqrt{2}}{2}\left[1 + \left(x - \dfrac{\pi}{4}\right) - \dfrac{1}{2!}\left(x - \dfrac{\pi}{4}\right)^2 - \dfrac{1}{3!}\left(x - \dfrac{\pi}{4}\right)^3 + \dfrac{1}{4!}\left(x - \dfrac{\pi}{4}\right)^4 + \dfrac{1}{5!}\left(x - \dfrac{\pi}{4}\right)^5 - \dfrac{1}{6!}\left(x - \dfrac{\pi}{4}\right)^6\right]$
13. $-\left(x - \dfrac{\pi}{2}\right) + \dfrac{1}{3!}\left(x - \dfrac{\pi}{2}\right)^3 - \dfrac{1}{5!}\left(x - \dfrac{\pi}{2}\right)^5$
15. $1 - (x-1) + (x-1)^2 - (x-1)^3 + (x-1)^4 - (x-1)^5 + (x-1)^6 - (x-1)^7 + (x-1)^8$
17. $e\left[1 + (x-1) + \dfrac{(x-1)^2}{2!} + \dfrac{(x-1)^3}{3!} + \dfrac{(x-1)^4}{4!} + \dfrac{(x-1)^5}{5!} + \dfrac{(x-1)^6}{6!}\right]$
19. $1 + 2\left(x - \dfrac{\pi}{4}\right) + 2\left(x - \dfrac{\pi}{4}\right)^2 + \dfrac{8}{3}\left(x - \dfrac{\pi}{4}\right)^3 + \dfrac{10}{3}\left(x - \dfrac{\pi}{4}\right)^4$ 21. $M = 1; |R_6(x)| \leq \dfrac{1}{7!}\left(\dfrac{\pi}{4}\right)^7 < 0.00004$
23. $M = e; |R_6(x)| \leq \dfrac{e}{7!} < 0.00054$ 25. $M = 6!2^7; |R_6(x)| \leq \dfrac{1}{7} < 0.143$
27. $M = \dfrac{105}{2^9\sqrt{2}}; |R_4(x)| \leq \dfrac{105}{2^9\sqrt{2}\,5!} < 0.00121$

12.4 Exercise Set, p. 738

1. $\frac{3}{2} + \frac{6}{5} + \frac{9}{10} + \frac{12}{17} + \frac{15}{26} + \frac{18}{37}$ 3. $1 - \frac{1}{3} + \frac{1}{9} - \frac{1}{27} + \frac{1}{81} - \frac{1}{243}$ 5. $\frac{1}{3} + \frac{3}{7} + \frac{1}{2} + \frac{5}{9} + \frac{3}{5} + \frac{7}{11}$ 7. $\frac{1}{50}$
9. converges to 2   11. converges to $\frac{5}{9}$ 13. diverges   15. diverges   17. $\sum_{k=0}^{\infty} 12(0.8)^k = 60$ mg.

19. $\sum_{k=0}^{\infty} 9(0.85)^k = 60$ billion dollars. The multiplier is $6\frac{2}{3}$.   21. $2[\sum_{k=0}^{\infty} 9(\frac{2}{3})^k] - 9 = 45$ feet

23. (a) $\sum_{k=0}^{\infty} 4000(1 + 0.06)^{-k}$ (b) \$70,666.67   25. (a) $\sum_{k=0}^{\infty} 4000\left(1 + \dfrac{0.06}{4}\right)^{-4k}$ (b) \$69,185.28

27. $\sum_{k=0}^{\infty} 500e^{-0.08k} = 6503.33$ 29. $-\frac{5}{2}$ 31. $\frac{2}{3}$ 33. $-\frac{3}{5}$

12.5 Exercise Set, p. 751

1. converges   3. diverges   5. converges   7. converges   9. diverges   11. converges
13. diverges   15. converges   17. diverges   19. converges   21. converges   23. diverges
25. between 1/4 and 5/4   27. between $2/e$ and $3/e$ 29. converges absolutely   31. diverges
33. converges absolutely   35. converges, but not absolutely   37. diverges

12.6 Exercise Set, p. 772

1. $R = 1$; $[-1, 1)$ 3. $R = 3$; $(-2, 4)$ 5. $R = 1$; $[-1, 1]$ 7. $R = 1$; $(-4, -2)$

9. $R = +\infty$; $(-\infty, +\infty)$ 11. $R = 0$; $[0, 0]$ 13. $R = \dfrac{1}{e}$; $\left(2 - \dfrac{1}{e}, 2 + \dfrac{1}{e}\right)$ 15. $R = 1$; $[-1, 1]$

17. $R = 1$; $[3, 5)$ 19. $R = \dfrac{1}{3}$; $(-\frac{7}{3}, -\frac{5}{3}]$ 21. $\sum_{k=0}^{\infty} \dfrac{x^{k+1}}{k!}$ 23. $\sum_{k=0}^{\infty} \dfrac{(-1)^k 2^{2k} x^{2k}}{(2k)!}$

25. $\sum_{k=0}^{\infty} \dfrac{(-1)^{k+1} x^{2k+3}}{(2k+1)!}$ 27. $1 + x - \dfrac{x^2}{2!} - \dfrac{x^3}{3!} + \dfrac{x^4}{4!} + \dfrac{x^5}{5!} - \dfrac{x^6}{6!} - \dfrac{x^7}{7!} + \cdots$ 29. $\sum_{k=0}^{\infty} (k+1)x^k$; $(-1, 1)$

31. $\sum_{k=0}^{\infty} \dfrac{(k+1)(k+2)}{2} x^k$; $(-1, 1)$ 33. $\sum_{k=0}^{\infty} \dfrac{(-1)^k x^{k+3}}{k+1}$; $(-1, 1]$ 35. $\sum_{k=0}^{\infty} \dfrac{(-1)^k x^{2k+1}}{2k+1}$; $[-1, 1]$

37. $\dfrac{\sqrt{3}}{2} + \dfrac{1}{2}\left(x - \dfrac{\pi}{3}\right) - \dfrac{\sqrt{3}}{2} \dfrac{\left(x - \dfrac{\pi}{3}\right)^2}{2!} - \dfrac{1}{2} \dfrac{\left(x - \dfrac{\pi}{3}\right)^3}{3!} + \dfrac{\sqrt{3}}{2} \dfrac{\left(x - \dfrac{\pi}{3}\right)^4}{4!} + \dfrac{1}{2} \dfrac{\left(x - \dfrac{\pi}{3}\right)^5}{5!}$

39. $0 + (x - 1) - \frac{1}{2}(x - 1)^2 + \frac{1}{3}(x - 1)^3 - \frac{1}{4}(x - 1)^4 + \frac{1}{5}(x - 1)^5 - \frac{1}{6}(x - 1)^6$

41. $1 + \dfrac{1}{2}(x - 1) - \dfrac{1}{2^2} \dfrac{(x - 1)^2}{2!} + \dfrac{1 \cdot 3}{2^3} \dfrac{(x - 1)^3}{3!} - \dfrac{1 \cdot 3 \cdot 5}{2^4} \dfrac{(x - 1)^4}{4!} + \dfrac{1 \cdot 3 \cdot 5 \cdot 7}{2^5} \dfrac{(x - 1)^5}{5!}$

43. $0 + x + 0 + \dfrac{x^3}{3} + 0 + \dfrac{2x^5}{15}$

Review Exercises, Chapter 12, p. 774

1. $8, 5, 2, -1, -4, -7$ 3. $0, 1, 0, 1, 0, 1$ 5. converges to 0   7. converges to 0
9. converges to 5   11. converges to $\frac{1}{4}$ 13. -10 15. \$2034.41
17. $13 - 32(x + 1) + 40(x + 1)^2 - 23(x + 1)^3 + 5(x + 1)^4$ 19. $3 + \frac{1}{3}x - \frac{1}{54}x^2 + \frac{1}{486}x^3$
21. $1 + \frac{1}{3}(x - 1) - \frac{1}{9}(x - 1)^2 + \frac{5}{81}(x - 1)^3 - \frac{10}{243}(x - 1)^4$ 23. $-2 + 2 + \frac{8}{7} + \frac{8}{7} + \frac{32}{23} + \frac{32}{17}$
25. $x - x^3 + \frac{1}{2}x^5 - \frac{1}{6}x^7 + \frac{1}{24}x^9 - \frac{1}{120}x^{11}$ 27. converges to $\frac{1}{3}$ 29. diverges
31. $\sum_{k=0}^{\infty} 6(0.7)^k = 20$ (billion)   33. diverges   35. converges   37. converges   39. diverges
41. converges, but not absolutely   43. $R = \frac{1}{3}$; $(-\frac{1}{3}, \frac{1}{3})$ 45. $R = 1$; $(-1, 1]$ 47. $R = 2$; $(-4, 0)$
49. $\sum_{k=0}^{\infty} \dfrac{(-1)^k (3x)^{2k+1}}{(2k+1)!}$; $(-\infty, +\infty)$ 51. $\sum_{k=0}^{\infty} x^{k+2}$; $(-1, 1)$
53. $1 + 2x + \dfrac{x^2}{2!} + \dfrac{x^4}{4!} + \dfrac{2x^5}{5!} + \dfrac{x^6}{6!} + \dfrac{x^8}{8!} + \dfrac{2x^9}{9!} + \cdots$; $(-\infty, +\infty)$ 55. $\sum_{k=0}^{\infty} \dfrac{(-1)^k x^{k+2}}{k+1}$; $(-1, 1]$
57. $1 + 0 - 2\left(x - \dfrac{\pi}{4}\right)^2 + 0 + \dfrac{2}{3}\left(x - \dfrac{\pi}{4}\right)^4 + 0 - \dfrac{4}{45}\left(x - \dfrac{\pi}{4}\right)^6$
59. $3 + \frac{1}{6}(x - 8) - \frac{1}{216}(x - 8)^2 + \frac{1}{3888}(x - 8)^3$

Chapter Test, Chapter 12, p. 776

1. (a) converges to 0   (b) diverges to $+\infty$ (c) diverges to $-\infty$.

2. $\dfrac{1}{2} - \dfrac{\sqrt{3}}{2}\left(x - \dfrac{\pi}{3}\right) - \dfrac{1}{4}\left(x - \dfrac{\pi}{3}\right)^2 + \dfrac{\sqrt{3}}{12}\left(x - \dfrac{\pi}{3}\right)^3 + \dfrac{1}{48}\left(x - \dfrac{\pi}{3}\right)^4$ 3. (a) diverges to $+\infty$

   (b) converges to 20   4. $\sum_{k=0}^{\infty} 50(0.8)^k = 250$ mg   5. (a) diverges   (b) converges

6. $R = 2$; $(-1, 3)$ 7. $\sum_{k=0}^{\infty} \dfrac{(-1)^k 2^{2k} x^{2k+1}}{(2k)!}$ 8. $\sum_{k=0}^{\infty} (-1)^{k+1}(k+1)x^k$

9. $2 + \dfrac{1}{4}x - \dfrac{1}{64}x^2 + \dfrac{1}{512}x^3$

||||||||||||| **APPENDIX A**

A.1 Diagnostic Test, p. A1

1. (a) $\frac{3}{7}, -3, 5, \sqrt{\frac{4}{9}}, 0$ (b) 5 (c) $-3, 5, 0$ (d) $\sqrt{3}$ 2. a nonrepeating decimal   3. 10
4. 0   5. $\frac{3}{10}$ 6. $-\frac{2}{3}$.   7. $5 \div 0$, %   8. ━┼━◆━┼━┼━┼━┼━◆━┼━   9. $x = -2$
$\qquad\qquad\qquad\qquad\qquad\qquad\qquad\qquad\qquad$ $-2 -1\ 0\ 1\ 2\ 3\ 4\ 5$

10. ━┼━┼━┼━┼━┼━◆━┼━┼━   11. $x = 1, y = 6$ 12. each nurd costs \$9; each blurb costs \$7
$\qquad$ $0\ 1\ 2\ 3\ 4\ 5\ 6\ 7$

13. 64   14. $\dfrac{9}{14}$ 15. $\dfrac{c^3 b^9}{a^6}$ 16. $x^3 + ax^2 - 2ax - 6a^2x + 4a^2$ 17. $x(2x+3)(x-5)$

18. $\dfrac{x+1}{x(x+2)}$ 19. $x = \frac{2}{3}$ or $-\frac{1}{2}$ 20. $x = 2$ or $-\frac{1}{2}$ 21. $-1 < x < 3$ 22. $x = 0$ or 1 or -1

23. (a)

second quadrant

(b)

fourth quadrant

(c)

no quadrant

24.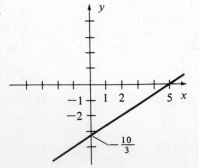

A.2 Exercise Set, p. A11

1. (a) $\sqrt{4}$, 28, -6, $\frac{20}{5}$ (b) $\sqrt{4}$, $\frac{8}{3}$, 1.7, 28, -6, $\frac{20}{5}$ (c) $\sqrt{4}$, 28, $\frac{20}{5}$ (d) $\sqrt{10}$, π
3. (a) 8 (b) $-\frac{1}{5}$ (c) $\frac{9}{2}$ (d) 0   5.

7. $A = \frac{3}{2}$, $B = -3$, $C = -\frac{9}{2}$, $D = 3$ 9. -3 11. -2 13. 8   15. $\frac{7}{10}$ 17. $\frac{1}{24}$ 19. $\frac{9}{5}$
21. 10   23. $-\frac{13}{15}$ 25. $\frac{1}{2}$ 27. $\frac{8}{35}$ 29. $\frac{28}{3}$ 31. $\frac{2}{3}$ 33. $-3 < -2$ 35. $2 > -7$
37. $a \ge 5$ 39. $x \ge 2$ 41. $y \le 0$ 43. $c \le -5$ 45. $<$ 47. $<$ 49. $<$ 51. $<$
53. $<$ 55. $\frac{2}{3}$ 57. 0.8   59. $-\frac{2}{5}$ 61. 0   63. 5   65. 2   67. -20

A.3 Exercise Set, p. A22

1. $x = -2$ 3. $x = 4$ 5. $x = \frac{3}{2}$ 7. $x = 6$ 9. $x = -2$ 11. $x = \dfrac{6}{4-k}$

13. ![number line from -2 to 3]   15. ![number line -3 to -1]   17. ![number line at 2]

19. $(-1, 2]$ 21. $(-3, +\infty)$ 23. $[-5, 1)$ 25. $(9, +\infty)$ 27. $[-12, -3]$ 29. $-4 < x \le 3$
31. $x \le -2$ 33. $x > 3$ 35. $x < -1$

![number line at -1]

37. $x \le 2$

![number line at 2]

39. $x \ge 6$

![number line at 6]

41. $-3 \le x \le \frac{1}{2}$

![number line -3 to 0]

43. $-3 \le x \le -1$

![number line -3 to -1]

45. $x = 3$, $y = -1$ 47. no solution   49. $x = 4$, $y = 2$
51. any pair (x, y) satisfying $x - y = 6$ is a solution   53. $90   55. $|100 - x| \le 2$, $98 \le x \le 102$
57. when more than 3 vacuum cleaners are sold on one day   59. $35 for hemlock, $20 for juniper

A.4 Exercise Set, p. A28

1. 81   3. $\frac{1}{125}$ 5. $\frac{32}{243}$ 7. 8   9. 1   11. 2   13. $\frac{1}{16}$ 15. 3^9 17. $\frac{1}{4}$ 19. 4   21. 16
23. $7^{5/2}$ 25. 3   27. 64   29. 27   31. 256   33. $\sqrt{3}$ 35. 9   37. $\dfrac{1}{5^4}$ 39. 81   41. $\frac{4}{9}$
43. $\dfrac{1}{a^3}$ 45. $\dfrac{1}{a^3}$ 47. $\dfrac{1}{a^4}$ 49. $\dfrac{1}{a^{1/2}}$ 51. $\dfrac{1}{a^4 b^4}$ 53. $\dfrac{b^5}{a^5}$ 55. a^6 57. $\dfrac{1}{ab}$ 59. $\dfrac{b^3}{a^6 c^6}$
61. $a^2 b^4$ 63. $\dfrac{a^4}{b^2}$ 65. $\dfrac{ac^3}{b}$

A.5 Exercise Set, p. A38

1. $7x^2 + x - 3$ 3. $6xy^2 + xy + 2x + y + 1$ 5. $x^2 - 4x - 12$ 7. $6x^2 - x - 15$
9. $4x^3 + 4x^2 - x + 3$ 11. $9x^2 + 30x + 25$ 13. $6x^2 + xy - y^2$ 15. $2x^3 + 3x^2 y + y^2 - 2xy^2 - 2xy$

17. $x(x - 2)$ 19. $(x + 7)(x - 7)$ 21. $(5y + 3x)(5y - 3x)$ 23. $(3x - 4y)(2x + 5y)$
25. $(x - 4)(x - 2)$ 27. $(3x + 2)(2x - 5)$ 29. $x(2x^2 - x + 3)$ 31. $(x - 3y)(x + 2y)$
33. $(x + 2)(x - 1)(x + 1)$ 35. $x(x - 1)(x + 1)(x + 1)$ 37. $\dfrac{8x}{x + 3}$ 39. $\dfrac{x(5x - 8)}{(x - 1)(x - 2)}$
41. $\dfrac{-(a^2 + 3a + 3)}{(a + 3)(a - 2)}$ 43. $\dfrac{a}{3}$ 45. $\dfrac{a + 2}{(a + 5)(a + 3)}$ 47. $\dfrac{2y^2}{(y - 1)(y + 1)}$ 49. $\dfrac{x + 1}{x - 1}$ 51. $\dfrac{5x - 2}{x - 1}$
53. $\dfrac{5x + 19}{3}$ 55. $x = -2$ 57. no solution   59. $x = 3$ 61. $x = 12$

A.6 Exercise Set, p. A44

1. $\sqrt{5}, -\sqrt{5}$ 3. no solution   5. $0, 3$ 7. $3, -2$ 9. $5, 1$ 11. $\dfrac{-1 \pm \sqrt{11}}{2}$ 13. no solution
15. $-\frac{4}{3}, 1$ 17. no solution   19. $\dfrac{1 \pm \sqrt{3}}{2}$ 21. $0, -4, 2$ 23. $0, -2, 2$ 25. $0, \frac{4}{3}$
27. $x < -3$ or $x > 2$ 29. $-1 < x < \frac{5}{2}$ 31. $-\frac{3}{2} < x < \frac{1}{2}$ 33. $x < -1$ or $x \geq 1$
35. $x \leq -3$ or $x \geq 3$ 37. $-4 < x < 4$ 39. 2 inches   41. base 8 feet, height 3 feet
43. 4 students

A.8 Exercise Set, p. A52

1.

3. (a) $(3, 0)$ 5. (a) IV   7. (a) and (c)
   (b) $(0, -5)$ (b) II
   (c) $(5, -5)$ (c) I
   (d) $(-4, 4)$ (d) III

9.

11.

13.

15.

17.

19. $x = -2$

21. Either $C(2, 3)$ or $C(4, 7)$

KEY IDEAS AND FORMULAS OF ELEMENTARY CALCULUS

The Definite Integral

- If $f(x)$ is a continuous function on the interval $[a, b]$, we define the definite integral of f from a to b by

$$\int_a^b f(x)\,dx = F(b) - F(a)$$

where $F(x)$ is any antiderivative of $f(x)$.

- If $f(x)$ is continuous and nonnegative on $[a, b]$, then $\int_a^b f(x)\,dx$ represents the area of the region under the curve $y = f(x)$ over the interval $[a, b]$.

- **Riemann Sums** Divide the interval $[a, b]$ into n subintervals of equal width $\Delta x = (b - a)/n$. Let x_k be an arbitrary point in the kth subinterval. The expression

$$f(x_1)\,\Delta x + f(x_2)\,\Delta x + \cdots + f(x_n)\,\Delta x$$

is called a Riemann sum.

If $f(x)$ is continuous and nonnegative on $[a, b]$, then the limit of the Riemann sums as $n \to \infty$ is $\int_a^b f(x)\,dx$.

Basic Integral Forms

- $\int du = u + C$

- $\int af(u)\,du = a \int f(u)\,du$

- $\int [f(u) \pm g(u)]\,du = \int f(u)\,du \pm \int g(u)\,du$

- $\int u^n\,du = \dfrac{u^{n+1}}{n+1} + C$ (if $n \neq -1$)

- $\int \dfrac{1}{u}\,du = \ln |u| + C$

- $\int e^u\,du = e^u + C$

- Integration by parts:

$$\int f(x)g'(x)\,dx = f(x)g(x) - \int g(x)f'(x)\,dx$$

$$\left(\int u\,dv = uv - \int v\,du \right)$$

- $\int (au + b)^n\,dx = \dfrac{1}{a}\dfrac{(au + b)^{n+1}}{n+1} + C$, if $n \neq -1$

- $\int \dfrac{1}{au + b}\,du = \dfrac{1}{a}\ln |au + b| + C$

Trigonometric Functions

- $\dfrac{d}{dx}[\sin u] = \cos u \cdot \dfrac{du}{dx}$

- $\dfrac{d}{dx}[\cos u] = -\sin u \cdot \dfrac{du}{dx}$

- $\dfrac{d}{dx}[\tan u] = \sec^2 u \cdot \dfrac{du}{dx}$

- $\int \sin u\,du = -\cos u + C$

- $\int \cos u\,du = \sin u + C$

- $\int \tan u\,du = -\ln |\cos u| + C$

Partial Derivatives

- If $z = f(x, y)$ then f_x (or $\partial f/\partial x$ or $\partial z/\partial x$) is the ordinary derivative of f with respect to x, holding y constant. Similarly, f_y (or $\partial f/\partial y$ or $\partial z/\partial y$) is the ordinary derivative of f with respect to y, holding x constant.

- The notion of partial derivative extends to functions of three or more variables.